AF698119

Selected Titles in This Series

223 **Bernard Chazelle, Jacob E. Goodman, and Richard Pollack, Editors,** Advances in discrete and computational geometry, 1999

222 **Kang-Tae Kim and Steven G. Krantz, Editors,** Complex geometric analysis in Pohang, 1999

221 **J. Robert Dorroh, Giséle Ruiz Goldstein, Jerome A. Goldstein, and Michael Mudi Tom, Editors,** Applied analysis, 1999

220 **Mark Mahowald and Stewart Priddy, Editors,** Homotopy theory via algebraic geometry and group representations, 1998

219 **Marc Henneaux, Joseph Krasil′shchik, and Alexandre Vinogradov, Editors,** Secondary calculus and cohomological physics, 1998

218 **Jan Mandel, Charbel Farhat, and Xiao-Chuan Cai, Editors,** Domain decomposition methods 10, 1998

217 **Eric Carlen, Evans M. Harrell, and Michael Loss, Editors,** Advances in differential equations and mathematical physics, 1998

216 **Akram Aldroubi and EnBing Lin, Editors,** Wavelets, multiwavelets, and their applications, 1998

215 **M. G. Nerurkar, D. P. Dokken, and D. B. Ellis, Editors,** Topological dynamics and applications, 1998

214 **Lewis A. Coburn and Marc A. Rieffel, Editors,** Perspectives on quantization, 1998

213 **Farhad Jafari, Barbara D. MacCluer, Carl C. Cowen, and A. Duane Porter, Editors,** Studies on composition operators, 1998

212 **E. Ramírez de Arellano, N. Salinas, M. V. Shapiro, and N. L. Vasilevski, Editors,** Operator theory for complex and hypercomplex analysis, 1998

211 **Józef Dodziuk and Linda Keen, Editors,** Lipa's legacy: Proceedings from the Bers Colloquium, 1997

210 **V. Kumar Murty and Michel Waldschmidt, Editors,** Number theory, 1998

209 **Steven Cox and Irena Lasiecka, Editors,** Optimization methods in partial differential equations, 1997

208 **Michel L. Lapidus, Lawrence H. Harper, and Adolfo J. Rumbos, Editors,** Harmonic analysis and nonlinear differential equations: A volume in honor of Victor L. Shapiro, 1997

207 **Yujiro Kawamata and Vyacheslav V. Shokurov, Editors,** Birational algebraic geometry: A conference on algebraic geometry in memory of Wei-Liang Chow (1911–1995), 1997

206 **Adam Korányi, Editor,** Harmonic functions on trees and buildings, 1997

205 **Paulo D. Cordaro and Howard Jacobowitz, Editors,** Multidimensional complex analysis and partial differential equations: A collection of papers in honor of François Treves, 1997

204 **Yair Censor and Simeon Reich, Editors,** Recent developments in optimization theory and nonlinear analysis, 1997

203 **Hanna Nencka and Jean-Pierre Bourguignon, Editors,** Geometry and nature: In memory of W. K. Clifford, 1997

202 **Jean-Louis Loday, James D. Stasheff, and Alexander A. Voronov, Editors,** Operads: Proceedings of Renaissance Conferences, 1997

201 **J. R. Quine and Peter Sarnak, Editors,** Extremal Riemann surfaces, 1997

200 **F. Dias, J.-M. Ghidaglia, and J.-C. Saut, Editors,** Mathematical problems in the theory of water waves, 1996

199 **G. Banaszak, W. Gajda, and P. Krasoń, Editors,** Algebraic K-theory, 1996

198 **Donald G. Saari and Zhihong Xia, Editors,** Hamiltonian dynamics and celestial mechanics, 1996

(*Continued in the back of this publication*)

Advances in Discrete and Computational Geometry

CONTEMPORARY MATHEMATICS

223

Advances in Discrete and Computational Geometry

Proceedings of the 1996 AMS-IMS-SIAM
Joint Summer Research Conference
Discrete and Computational Geometry: Ten Years Later
July 14–18, 1996
Mount Holyoke College

Bernard Chazelle
Jacob E. Goodman
Richard Pollack
Editors

American Mathematical Society
Providence, Rhode Island

QA
167
.J65
1996

Editorial Board

Dennis DeTurck, managing editor

Andreas Blass Andy R. Magid Michael Vogelius

The 1996 AMS-IMS-SIAM Joint Summer Research Conference in the Mathematical Sciences, "Discrete and Computational Geometry: Ten Years Later", was held at Mount Holyoke College, South Hadley, MA, July 14–18, 1996, with support from the National Science Foundation, Grant DMS-9632966, and the National Security Agency, Grant MDA 904-96-1-0094.

1991 *Mathematics Subject Classification.* Primary 52–XX, 68–XX.

Any opinions, findings, and conclusions or recommendations expressed in this material are those of the authors and do not necessarily reflect the views of the National Science Foundation.

Library of Congress Cataloging-in-Publication Data

Joint Summer Research Conference "Discrete and Computational Geometry: Ten Years Later" (1996 : Mount Holyoke College)

Advances in discrete and computational geometry / Bernard Chazelle, Jacob E. Goodman, Richard Pollack, editors.

p. cm. — (Contemporary mathematics, ISSN 0271-4132 ; 223)

"1996 Joint Summer Research Conference "Discrete and Computational Geometry: Ten Years Later", July 14–18, 1996, Mount Holyoke College, South Hadley, MA."

Includes bibliographical references.

ISBN 0-8218-0674-2 (alk. paper)

1. Combinatorial geometry—Congresses. 2. Geometry—Data processing—Congresses. I. Chazelle, B. (Bernard) II. Goodman, Jacob E. III. Pollack, Richard. IV. Title. V. Series: Contemporary mathematics (American Mathematical Society) ; v. 223.

QA167.J65 1996
516′.13–dc21 98-34538
CIP

Copying and reprinting. Material in this book may be reproduced by any means for educational and scientific purposes without fee or permission with the exception of reproduction by services that collect fees for delivery of documents and provided that the customary acknowledgment of the source is given. This consent does not extend to other kinds of copying for general distribution, for advertising or promotional purposes, or for resale. Requests for permission for commercial use of material should be addressed to the Assistant to the Publisher, American Mathematical Society, P. O. Box 6248, Providence, Rhode Island 02940-6248. Requests can also be made by e-mail to `reprint-permission@ams.org`.

Excluded from these provisions is material in articles for which the author holds copyright. In such cases, requests for permission to use or reprint should be addressed directly to the author(s). (Copyright ownership is indicated in the notice in the lower right-hand corner of the first page of each article.)

© 1999 by the American Mathematical Society. All rights reserved.
The American Mathematical Society retains all rights
except those granted to the United States Government.
Printed in the United States of America.

∞ The paper used in this book is acid-free and falls within the guidelines
established to ensure permanence and durability.
Visit the AMS home page at URL: `http://www.ams.org/`

10 9 8 7 6 5 4 3 2 1 04 03 02 01 00 99

Contents

Preface

On July 14–18, 1996, Mt. Holyoke College hosted an AMS-IMS-SIAM Summer Research Conference on the theme "Discrete and Computational Geometry: Ten Years Later." The meeting was organized by B. Chazelle, J. E. Goodman, J. O'Rourke, J. Pach, and R. Pollack. The title harked back to a 1986 conference in Santa Cruz, organized by J. E. Goodman and R. Pollack. Its contagious success heralded years of fruitful interaction between the established field of discrete geometry and the then youthful offshoot from computer science, computational geometry.

This volume presents a collection of research and survey articles, most of which were presented at the Mt. Holyoke meeting. The mix of articles in this volume touches on a wide variety of current topics in discrete and computational geometry. It includes surveys on range searching, computational topology, packing and covering, real algebraic geometry, visibility, universality theorems, tilings, arrangements, geometric discrepancy theory, transversal theory, and polytopes, as well as research contributions on monotone paths, flag complexes, discrepancy theory, geometric graph theory, polygonal covers, and Helly-type theorems.

In "Geometric range searching and its relatives," Agarwal and Erickson provide a comprehensive survey of the vast recent literature on the computational complexity of searching in higher-dimensional space. Geometric sampling and space partitioning techniques (e.g., cuttings) are discussed in great detail.

The next article, "Deformed products and maximal shadows of polytopes," by Amenta and Ziegler, builds on the classical result of Klee and Minty on the exponential length of pivoting sequences for the simplex algorithm. It presents sharp bounds on the maximal lengths of increasing paths and planar projections.

In "Flag complexes, labelled rooted trees, and star shellings," Billera, Chan, and Liu establish and investigate an intriguing correspondence between flag complexes and labelled rooted trees, and they derive enumerative results concerning complexes of cubes called star-shellable.

In the following article, "Discrepancy bounds for geometric set systems with square incidence matrices," Chazelle proves the existence of n triangles and n points such that for any two-coloring of the points there is at least one triangle which contains at least $\Omega(n^{1/4})$ more points of one color than those of the other color. The result was previously known only if the number of triangles is asymptotically greater than the number of points.

"Computational topology," by Dey, Edelsbrunner, and Guha, provides a thorough, accessible introduction to the nascent field of computational topology: applications can be found in robotics, molecular modeling, image processing, cartography, mesh generation, etc.

In "Recent progress on packing and covering," G. Fejes Tóth surveys the last ten years of research in the theory of packing and covering, one of the oldest and most developed areas in discrete geometry.

"Acoptic polyhedra" are polyhedra with no self-intersections. Grünbaum surveys their basic properties in 3-space. The motivation is to study how crucial a requirement convexity is in deriving standard properties of polytopes.

The article by Holt and Klee, "A proof of the strict monotone 4-step conjecture," establishes a variant of Hirsch's celebrated conjecture for the case of a 4-polytope with 8 facets, along with many related facts.

In "Interactions between real algebraic geometry and discrete and computational geometry," Itenberg and Roy introduce and survey the fast-evolving field of real algebraic geometry, highlighting its discrete-geometric, combinatorial, topological, and algorithmic aspects. Many of the most recent results on combinatorial patchworking, roadmaps, and quantifier elimination can be found in this article.

The next contribution, "Open problems in the combinatorics of visibility and illumination," is a large collection of open combinatorial and computational problems concerning art gallery theorems and related visibility questions.

"Halving lines and perfect cross-matchings," by Pach and Solymosi, characterizes the sets of $2n$ points in the plane (in general position) having a perfect matching with pairwise crossings as being precisely those admitting n halving lines. This leads to an efficient algorithm for checking the existence of such matchings.

In "Three-dimensional grid drawings of graphs," Pach, Thiele, and Tóth prove that any n-vertex graph can be embedded on a three-dimensional grid of volume $O(n^2)$, provided that its chromatic number is bounded.

The next article, "On polygonal covers," by Pocchiola and Vegter, gives a constructive proof of a result by L. Fejes Tóth asserting the possibility of covering n disjoint convex bodies in the plane by polygons totaling at most $6n - 9$ edges.

In "The universality theorems for oriented matroids and polytopes," Richter-Gebert surveys the main developments in the polyhedral theory of universality: these concern the complexity (e.g., computational, topological, algebraic) of the realization spaces for polytopes and oriented matroids.

Senechal's contribution, "Periodic and aperiodic tilings of E^n," is an accessible survey of the beautiful developments on planar tilings and, in particular, the relations between local properties and global ones (such as periodicity).

If computational geometry has a father, Shamos's claim to the title is as strong as anyone's. He is the one who massaged an eclectic collection of problems and curiosities into a new field of science. In "The early years of computational geometry—a personal memoir," Shamos reminisces. His account is mostly personal, often anecdotal, always enjoyable.

The next article, "Arrangements of surfaces in higher dimensions," by Sharir, reviews recent progress in the study of arrangements for surfaces embedded in d-space. Numerous applications can be found in motion planning, geometric optimization, and transversal theory. The results—many by the author himself—represent the culmination of over a decade of active research on this topic.

In "Geometric discrepancy theory," Spencer gives an engaging account of a subject at the crossroads of geometry, number theory and combinatorics, with an emphasis on some of the most exciting recent developments.

The following contribution, "Proof of Reay's conjecture on certain positive-dimensional intersections," by Tverberg, settles a conjecture of Reay on an extension of Radon's theorem to one-dimensional intersections.

The last ten years have witnessed an explosion of results in geometric transversal theory. In "Progress in geometric transversal theory," Wenger walks us through its main developments, beginning at the starting point of the subject, i.e., Hadwiger's seminal extension of Helly's theorem to line transversals, and touching on such contemporary concepts as combinatorial complexity, oriented matroids, and piercing numbers.

The next article, "Recent progress on polytopes," by Ziegler, offers seven open problems in the combinatorial theory of convex polytopes. The discussion centers around five main subjects: universality theorems, triangulations, 0/1 polytopes, neighborly polytopes, and monotone paths. This contribution ties in beautifully with the articles above by Richter-Gebert, Holt and Klee, and Amenta and Ziegler.

The volume ends with an appendix devoted to the Computational Geometry Impact Task Force Report. This document was prepared by a panel, chaired by Chazelle, consisting of 30-odd computer scientists and geometers. It discusses the challenges and opportunities ahead for the field of computational geometry in view of its recent accomplishments.

The editors extend their warm gratitude to the authors for their willing participation, and to the many anonymous referees whose help was indispensable in bringing this volume to fruition.

Happy reading!

Bernard Chazelle
Jacob E. Goodman
Richard Pollack

Contemporary Mathematics
Volume **223**, 1999

Geometric Range Searching and Its Relatives

Pankaj K. Agarwal and Jeff Erickson

ABSTRACT. A typical range-searching problem has the following form: Preprocess a set S of points in $\mathbb{R}^d$ so that the points of S lying inside a query region can be reported or counted quickly. We survey the known techniques and data structures for range searching and describe their application to other related searching problems.

1. Introduction

About ten years ago, the field of range searching, especially simplex range searching, was wide open. At that time, neither efficient algorithms nor nontrivial lower bounds were known for most range-searching problems. A series of papers by Haussler and Welzl [**HW**], Clarkson [**Cl, Cl2**], and Clarkson and Shor [**CS**] not only marked the beginning of a new chapter in geometric searching, but also revitalized computational geometry as a whole. Led by these and a number of subsequent papers, tremendous progress has been made in geometric range searching, both in terms of developing efficient data structures and proving nontrivial lower bounds. From a theoretical point of view, range searching is now almost completely solved. The impact of general techniques developed for geometric range searching—ε-nets, $1/r$-cuttings, partition trees, multi-level data structures, to name a few—is evident throughout computational geometry. This volume provides an excellent opportunity to recapitulate the current status of geometric range searching and to summarize the recent progress in this area.

1991 *Mathematics Subject Classification.* Primary 68-02, 68P05, 68P10, 68P20, 68Q05, 68Q25.

Key words and phrases. Data structure, multidimensional searching, range searching, arrangement, partition tree, random sampling.

Pankaj Agarwal's work on this paper was supported by National Science Foundation Grant CCR-93-01259, by Army Research Office MURI grant DAAH04-96-1-0013, by a Sloan fellowship, by an NYI award and matching funds from Xerox Corporation, and by a grant from the U.S.-Israeli Binational Science Foundation.

Jeff Erickson's work was supported by National Science Foundation grant DMS-9627683 and by Army Research Office MURI grant DAAH04-96-1-0013.

Tables 1, 3, 6, 7, 8, and 9 reprinted with permission from *Handbook of Discrete and Computational Geometry*, Jacob E. Goodman and Joseph O'Rourke, eds., CRC Press, 1997. Copyright CRC Press, Boca Raton, Florida, ©1997.

© 1999 American Mathematical Society

Range searching arises in a wide range of applications, including geographic information systems, computer graphics, spatial databases, and time-series databases. Furthermore, a variety of geometric problems can be formulated as a range-searching problem. A typical range-searching problem has the following form. Let S be a set of n points in $\mathbb{R}^d$, and let $\mathcal{R}$ be a family of subsets of $\mathbb{R}^d$; elements of $\mathcal{R}$ are called *ranges*. We wish to preprocess S into a data structure so that for a query range $\gamma \in \mathcal{R}$, the points in $S \cap \gamma$ can be reported or counted efficiently. Typical examples of ranges include rectangles, halfspaces, simplices, and balls. If we are only interested in answering a single query, it can be done in linear time, using linear space, by simply checking for each point $p \in S$ whether p lies in the query range. Most applications, however, call for querying the same point set S several times (and sometimes we also insert or delete a point periodically), in which case we would like to answer a query faster by preprocessing S into a data structure.

Range counting and range reporting are just two instances of range-searching queries. Other examples include *emptiness queries*, where one wants to determine whether $S \cap \gamma = \emptyset$, and *optimization queries*, where one wants to choose a point with certain property (*e.g.*, a point in γ with the largest x_1-coordinate). In order to encompass all different types of range-searching queries, a general range-searching problem can be defined as follows.

Let $(\mathbf{S}, +)$ be a commutative semigroup.[1] For each point $p \in S$, we assign a weight $w(p) \in \mathbf{S}$. For any subset $S' \subseteq S$, let $w(S') = \sum_{p \in S'} w(S)$, where addition is taken over the semigroup.[2] For a query range $\gamma \in \mathcal{R}$, we wish to compute $w(S \cap \gamma)$. For example, counting queries can be answered by choosing the semigroup to be $(\mathbb{Z}, +)$, where $+$ denotes standard integer addition, and setting $w(p) = 1$ for every $p \in S$; emptiness queries by choosing the semigroup to be $(\{0,1\}, \vee)$ and setting $w(p) = 1$; reporting queries by choosing the semigroup to be $(2^S, \cup)$ and setting $w(p) = \{p\}$; and optimization queries by choosing the semigroup to be $(\mathbb{R}, \max)$ and choosing $w(p)$ to be, for example, the x_1-coordinate of p.

We can, in fact, define a more general (decomposable) *geometric searching* problem. Let S be a set of *objects* in $\mathbb{R}^d$ (*e.g.*, points, hyperplanes, balls, or simplices), $(\mathbf{S}, +)$ a commutative semigroup, $w : S \to \mathbf{S}$ a weight function, $\mathcal{R}$ a set of ranges, and $\Diamond \subseteq S \times \mathcal{R}$ a "spatial" relation between objects and ranges. Then for a range $\gamma \in \mathcal{R}$, we want to compute $\sum_{p \Diamond \gamma} w(p)$. Range searching is a special case of this general searching problem, in which S is a set of points in $\mathbb{R}^d$ and $\Diamond = \in$. Another widely studied searching problem is *intersection searching*, where $p \Diamond \gamma$ if p intersects γ. As we will see below, range-searching data structures are useful for many other geometric searching problems.

The performance of a data structure is measured by the time spent in answering a query, called the *query time*, by the *size* of the data structure, and by the time constructed in the data structure, called the *preprocessing time*. Since the data structure is constructed only once, its query time and size are generally more important than its preprocessing time. If a data structure supports insertion and deletion operations, its *update time* is also relevant. We should remark that the query time of a range-reporting query on any reasonable machine depends on the

[1] A *semigroup* $(\mathbf{S}, +)$ is a set $\mathbf{S}$ equipped with an associative addition operator $+ : \mathbf{S} \times \mathbf{S} \to \mathbf{S}$. A semigroup is *commutative* if $x + y = y + x$ for all $x, y \in \mathbf{S}$.

[2] Since $\mathbf{S}$ need not have an additive identity, we may need to assign a special value *nil* to the empty sum.

output size, so the query time for a range-reporting query consists of two parts—*search time*, which depends only on n and d, and *reporting time*, which depends on n, d, and the output size. Throughout this survey paper we will use k to denote the output size.

We assume that d is a small fixed constant, and that big-Oh and big-Omega notation hides constants depending on d. The dependence on d of the performance of almost all the data structures mentioned in this survey is exponential, which makes them unsuitable in practice for large values of d.

The size of any range-searching data structure is at least linear, since it has to store each point (or its weight) at least once, and the query time in any reasonable model of computation such as pointer machines, RAMs, or algebraic decision trees is $\Omega(\log n)$ even when $d = 1$. Therefore, we would like to develop a linear-size data structure with logarithmic query time. Although near-linear-size data structures are known for orthogonal range searching in any fixed dimension that can answer a query in polylogarithmic time, no similar bounds are known for range searching with more complex ranges such as simplices or disks. In such cases, we seek a tradeoff between the query time and the size of the data structure—How fast can a query be answered using $O(n \operatorname{polylog} n)$ space, how much space is required to answer a query in $O(\operatorname{polylog} n)$ time, and what kind of tradeoff between the size and the query time can be achieved?

In this paper we survey the known techniques and data structures for range-searching problems and describe their applications to other related searching problems. As mentioned in the beginning, the quest for efficient range-searching data structure has led to many general, powerful techniques that have had a significant impact on several other geometric problems. The emphasis of this survey is on describing known results and general techniques developed for range searching, rather than on open problems. The paper is organized as follows. We describe, in Section 2, different models of computation that have been used to prove upper and lower bounds on the performance of data structures. Next, in Section 3, we review data structures for orthogonal range searching and its variants. Section 4 surveys known techniques and data structures for simplex range searching, and Section 5 discusses some variants and extensions of simplex range searching. Finally, we review data structures for intersection searching and optimization queries in Sections 6 and 7, respectively.

2. Models of computation

Most algorithms and data structures in computational geometry are implicitly described in the familiar *random access machine* (RAM) model, described in [**AHU**], or the *real RAM* model described by Preparata and Shamos [**PrS**]. In the traditional RAM model, memory cells can contain arbitrary $(\log n)$-bit integers, which can be added, multiplied, subtracted, divided (computing $\lfloor x/y \rfloor$), compared, and used as pointers to other memory cells in constant time. A few algorithms rely on a variant of the RAM model, proposed by Fredman and Willard [**FW**], that allows memory cells to contain w-bit integers, for some parameter $w \geq \log n$, and permits both arithmetic and bitwise logical operations in constant time. In a real RAM, we also allow memory cells to store arbitrary real numbers (such as coordinates of points). We allow constant-time arithmetic on and comparisons between real numbers, but we do not allow conversion between integers and reals. In the case of range searching over a semigroup other than the integers, we also allow memory cells to contain arbitrary values from the semigroup, but these values can only be added (using the semigroup's addition operator, of course).

Almost all known range-searching data structures can be described in the more restrictive *pointer machine* model, originally developed by Tarjan [**T**].[3] The main difference between the two models is that on a pointer machine, a memory cell can be accessed only through a series of pointers, while in the RAM model, any memory cell can be accessed in constant time. Tarjan's basic pointer machine model is most suitable for studying range-reporting problems. In this model, a data structure is a directed graph with outdegree 2. To each node v in this graph, we associate a label $\ell(v)$, which is an integer between 0 and n. Nonzero labels are indices of the points in S. The query algorithm, given a range γ, begins at a special starting node and performs a sequence of the following operations: (1) visit a new node by traversing an edge from a previously visited node, (2) create a new node v with $\ell(v) = 0$, whose outgoing edges point to previously visited nodes, and (3) redirect an edge leaving a previously visited node, so that it points to another previously visited node. When the query algorithm terminates, the set of visited nodes $W(\gamma)$, called the *working set*, is required to contain the indices of all points in the query range; that is, if $p_i \in \gamma$, then there must be a node $v \in W(\gamma)$ such that $\ell(v) = i$. The working set $W(\gamma)$ may contain labels of points that are not in the query range. The size of the data structure is the number of nodes in the graph, and the query time for a range γ is the size of the smallest possible working set $W(\gamma)$. The query time ignores the cost of other operations, including the cost of deciding which edges to traverse. There is no notion of preprocessing or update time in this model. Note that the model accommodates both static and self-adjusting data structures.

Chazelle [**Ch4**] defines several generalizations of the pointer-machine model that are more appropriate for answering counting and semigroup queries. In Chazelle's generalized pointer-machine models, nodes are labeled with arbitrary $O(\log n)$-bit integers. In addition to traversing edges in the graph, the query algorithm is also allowed to perform various arithmetic operations on these integers. An *elementary* pointer machine can add and compare integers; in an *arithmetic* pointer machine, subtraction, multiplication, integer division, and shifting ($x \mapsto 2^x$) are also allowed. When the query algorithm terminates in these models, some node in the working set is required to contain the answer. If the points have weights from an additive semigroup other than the integers, nodes in the data structure can also be labeled with semigroup values, but these values can only be added.

Most lower bounds, and a few upper bounds, are described in the so-called *semigroup arithmetic model*, which was originally introduced by Fredman [**Fr4**] and refined by Yao [**Y2**]. In the semigroup arithmetic model, a data structure can be *informally* regarded as a set of precomputed partial sums in the underlying semigroup. The size of the data structure is the number of sums stored, and the query time is the minimum number of semigroup operations required (on the precomputed sums) to compute the answer to a query. The query time ignores the cost of various auxiliary operations, including the cost of determining which of the precomputed sums should be added to answer a query. Unlike the pointer

[3] Several very different models of computation with the name "pointer machine" have been proposed; these are surveyed by Ben-Amram [**BA2**], who suggests the less ambiguous term *pointer algorithm* for the model we describe.

machine model, the semigroup model allows immediate access, at no cost, to any precomputed sum.

The informal model we have just described is much too powerful. For example, in this informal model, the optimal data structure for counting queries consists of the $n + 1$ integers $0, 1, \ldots, n$. To answer a counting query, we simply return the correct answer; since no additions are required, we can answer queries in zero "time," using a "data structure" of only linear size!

Here is a more formal definition that avoids this problem. Let $(\mathbf{S}, +)$ be a commutative semigroup. A *linear form* is a sum of variables over the semigroup, where each variable can occur multiple times, or equivalently, a homogeneous linear polynomial with positive integer coefficients. The semigroup is *faithful* if any two identically equal linear forms have the same set of variables, although not necessarily with the same set of coefficients.[4]

For example, the semigroups $(\mathbb{Z}, +)$, $(\mathbb{R}, \min)$, $(\mathbb{N}, \gcd)$, and $(\{0,1\}, \vee)$ are faithful, but the semigroup $(\{0, 1\}, + \bmod 2)$ is not faithful.

Let $S = \{p_1, p_2, \ldots, p_n\}$ be a set of objects, $\mathbf{S}$ a faithful semigroup, $\mathcal{R}$ a set of ranges, and $\Diamond$ a relation between objects and ranges. (Recall that in the standard range-searching problem, the objects in S are points, and $\Diamond$ is containment.) Let $x_1, x_2, \ldots, x_n$ be a set of n variables over $\mathbf{S}$, each corresponding to a point in S. A *generator* $g(x_1, \ldots, x_n)$ is a linear form $\sum_{i=1}^{n} \alpha_i x_i$, where the α_i's are non-negative integers, not all zero. (In practice, the coefficients α_i are either 0 or 1.) A *storage scheme* for $(S, \mathbf{S}, \mathcal{R}, \Diamond)$ is a collection of generators $\{g_1, g_2, \ldots, g_s\}$ with the following property: For any query range $\gamma \in \mathcal{R}$, there is an set of indices $I_\gamma \subseteq \{1, 2, \ldots, s\}$ and a set of labeled nonnegative integers $\{\beta_i \mid i \in I_\gamma\}$ such that the linear forms

$$\sum_{p_i \Diamond \gamma} x_i \quad \text{and} \quad \sum_{i \in I_\gamma} \beta_i g_i$$

are identically equal. In other words, the equation

$$\sum_{p_i \Diamond \gamma} w(p_i) = \sum_{i \in I_\gamma} \beta_i g_i(w(p_1), w(p_2), \ldots, w(p_n))$$

holds for *any* weight function $w : S \to \mathbf{S}$. (Again, in practice, $\beta_i = 1$ for all $i \in I_\gamma$.) The size of the smallest such set I_γ is the query time for γ; the time to actually choose the indices I_γ is ignored. The space used by the storage scheme is measured by the number of generators. There is no notion of preprocessing time in this model.

We emphasize that although a storage scheme can take advantage of special properties of the set S or the semigroup $\mathbf{S}$, it must work for *any* assignment of weights to S. In particular, this implies that lower bounds in the semigroup model do not apply to the problem of counting the number of points in the query range, even though $(\mathbb{N}, +)$ is a faithful semigroup, since a storage scheme for the counting problem only needs to work for the particular weight function $w(p) = 1$ for all $p \in S$. Similar arguments apply to emptiness, reporting, and optimization queries, even though the semigroups $(\{0,1\}, \vee)$, $(2^S, \cup)$, and $(\mathbb{R}, \min)$ are all faithful.

The requirement that the storage scheme must work for any weight assignment even allows us to model problems where the weights depend on the query. For example, suppose for some set S of objects with real weights, we have a storage

[4]More formally, $(\mathbf{S}, +)$ is faithful if for each $n > 0$, for any sets of indices $I, J \subseteq \{1, \ldots, n\}$ so that $I \neq J$, and for every sequence of positive integers α_i, β_j $(i \in I, j \in J)$, there are semigroup values $s_1, s_2, \ldots, s_n \in \mathbf{S}$ such that $\sum_{i \in I} \alpha_i s_i \neq \sum_{j \in J} \beta_j s_j$.

scheme that lets us quickly determine the minimum weight of any object hit by a query ray. In other words, we have a storage scheme for S under the semigroup $(\mathbb{R}, \min)$ that supports intersection searching, where the query ranges are rays. We can use such a storage scheme to answer ray-shooting queries, by letting the weight of each object be its distance along the query ray from the basepoint. If we want the first object hit by the query ray instead of just its distance, we can use the faithful semigroup $(S \times \mathbb{R}, \diamond)$, where

$$(p_1, \delta_1) \diamond (p_2, \delta_2) = \begin{cases} (p_1, \delta_1) & \text{if } \delta_1 \leq \delta_2, \\ (p_2, \delta_2) & \text{otherwise,} \end{cases}$$

and letting the weight of an object $p \in S$ be (p, δ), where δ is the distance along the query ray between the basepoint and p. We reiterate, however, that lower bounds in the semigroup model do not imply lower bounds on the complexity of ray shooting.

Although in principle, storage schemes can exploit of special properties of the semigroup $\mathbf{S}$, in practice, they never do. All known upper and lower bounds in the semigroup arithmetic model hold for all faithful semigroups. In other models of computation where semigroup values can be manipulated, such as RAMs and elementary pointer machines, slightly better upper bounds are known for some problems when the semigroup is $(\mathbb{N}, +)$ **[Ch4]**.

The semigroup model is formulated slightly differently for offline range-searching problems. Here we are given a set of weighted points S and a finite set of query ranges $\mathcal{R}$, and we want to compute the total weight of the points in each query range. This is equivalent to computing the product Aw, where A is the incidence matrix of the points and ranges, and w is the vector of weights. In the offline semigroup model, introduced by Chazelle **[Ch11]**, an algorithm can be described as a circuit (or straight-line program) with one input for every point and one output for every query range, where every gate (respectively, statement) performs a binary semigroup addition. The running time of the algorithm is the total number of gates (respectively, statements). For any weight function $w : S \to \mathbf{S}$, the output associated with a query range γ is $w(S \cap \gamma)$. Just as in the online case, the circuit is required to work for any assignment of weights to the points; in effect, the outputs of the circuit are the linear forms $\sum_{p_i \in \gamma} x_i$. See Figure 1 for an example.

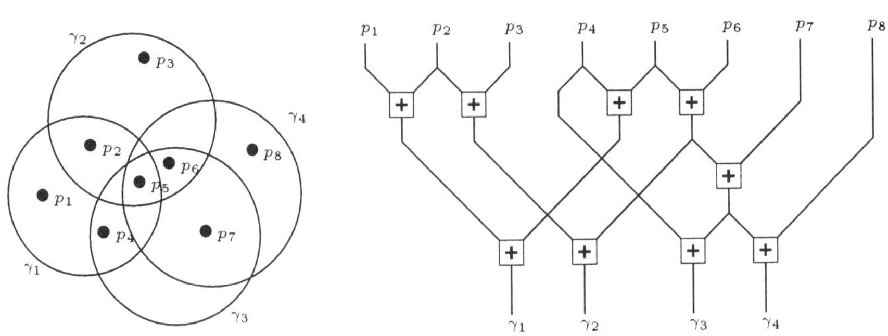

FIGURE 1. A set of eight points and four disks, and an offline semigroup arithmetic algorithm to compute the total weight of the points in each disk.

A serious weakness of the semigroup model is that it does not allow subtractions even if the weights of the points belong to a group. Therefore, we will also consider the *group model*, in which both additions and subtractions are allowed [**Wi2, Ch10, Ch11**]. Chazelle [**Ch11**] considers an extension of the offline group model in which circuits are allowed a limited number of *help gates*, which can compute arbitrary binary functions.

Of course it is natural to consider arithmetic circuits that also allow multiplication ("the ring model"), division ("the field model"), or even more general functions such as square roots or exponentiation. There is a substantial body of literature on the complexity of various types of arithmetic circuits [**vzG, Str, BCS**], but almost nothing is known about the complexity of geometric range searching in these models. Perhaps the only relevant result is that any circuit with operations $+, -, \times, \div, \sqrt{}$ requires $\Omega(\log n)$ time to answer any reasonable range query, or $\Omega(n \log n)$ time to solve any reasonable offline range searching problem, since such a circuit can be modeled as an algebraic computation tree with no branches [**BO**] or as a straight-line program on a real RAM [**BA**]. (Computation trees with more general functions are considered in [**GV**].)

Almost all geometric range-searching data structures are constructed by subdividing space into several regions with nice properties and recursively constructing a data structure for each region. Range queries are answered with such a data structure by performing a depth-first search through the resulting recursive space partition. The *partition graph* model, recently introduced by Erickson [**Er, Er2, Er3**], formalizes this divide-and-conquer approach, at least for hyperplane and halfspace range searching data structures. The partition graph model can be used to study the complexity of emptiness queries, unlike the semigroup arithmetic and pointer machine models, in which such queries are trivial.

Formally, a partition graph is a directed acyclic graph with constant outdegree, with a single source, called the *root*, and several sinks, called *leaves*. Associated with each internal node is a cover of $\mathbb{R}^d$ by a constant number of connected subsets called *query regions*, each associated with an outgoing edge. Each internal node is labeled either *primal* or *dual*, indicating whether the query regions should be considered a decomposition of "primal" or "dual" space. (Point-hyperplane duality is discussed in Section 4.2.) Any partition graph defines a natural search structure, which is used both to preprocess a set of points and to perform a query for a hyperplane or halfspace. The points are preprocessed one at a time. To preprocess a single point, we perform a depth-first search of the graph, starting at the root. At each primal node, we traverse the outgoing edges corresponding to the query regions that contain the point; at each dual node, we traverse the edges whose query regions intersect the point's dual hyperplane. For each leaf ℓ of the partition graph, we maintain a set P_ℓ containing the points that reach ℓ during the preprocessing phase. The query algorithm for hyperplanes is an exactly symmetric depth-first search—at primal nodes, we look for query regions that intersect the hyperplane, and at dual nodes, we look for query regions that contain its dual point. The answer to a query is determined by the sets P_ℓ associated with the leaves ℓ of the partition graph that the query algorithm reaches. For example, the output of an emptiness query is "yes" (*i.e.*, the query hyperplane contains none of the points) if and only if $P_\ell = \emptyset$ for every leaf ℓ reached by the query algorithm. The size of the partition graph is the number of edges in the graph; the complexity of the query regions and the sizes of the sets P_ℓ are not considered. The preprocessing time for a single point and the query time for a hyperplane are given by the number of edges traversed during the search; the time required to actually construct the partition graph and to test the query regions is ignored.

We conclude this section by noting that most of the range-searching data structures discussed in this paper (halfspace range-reporting data structures being a notable exception) are based on the following general scheme. Given a point set S, they precompute a family $\mathcal{F} = \mathcal{F}(S)$ of *canonical subsets* of S and store the weight $w(C) = \sum_{p \in C} w(p)$ of each canonical subset $C \in \mathcal{F}$. For a query range γ, they determine a partition $\mathcal{C}_\gamma = \mathcal{C}(S, \gamma) \subseteq \mathcal{F}$ of $S \cap \gamma$ and add the weights of the subsets in $\mathcal{C}_\gamma$ to compute $w(S \cap \gamma)$. Borrowing terminology from **[M6]**, we will refer to such a data structure as a *decomposition scheme.*

There is a close connection between decomposition schemes and storage schemes in the semigroup arithmetic model described earlier. Each canonical subset $C = \{p_i \mid i \in I\} \in \mathcal{F}$, where $I \subseteq \{1, 2, \ldots, n\}$, corresponds to the generator $\sum_{i \in I} x_i$. In fact, because the points in any query range are always computed as the *disjoint* union of canonical subsets, any decomposition scheme corresponds to a storage scheme that is valid for *any* semigroup. Conversely, lower bounds in the semigroup model imply lower bounds on the complexity of any decomposition scheme.

How exactly the weights of canonical subsets are stored and how $\mathcal{C}_\gamma$ is computed depends on the model of computation and on the specific range-searching problem. In the semigroup (or group) arithmetic model, the query time depends only on the number of canonical subsets in $\mathcal{C}_\gamma$, regardless of how they are computed, so the weights of canonical subsets can be stored in an arbitrary manner. In more realistic models of computation, however, some additional structure must be imposed on the decomposition scheme in order to efficiently compute $\mathcal{C}_\gamma$. In a *hierarchical* decomposition scheme, the weights are stored in a tree T. Each node v of T is associated with a canonical subset $C_v \in \mathcal{F}$, and the children of v are associated with subsets of C_v. Besides the weight of C_v, some auxiliary information is also stored at v, which is used to determine whether $C_v \in \mathcal{C}_\gamma$ for a query range γ. Typically, this auxiliary information consists of some geometric object, which plays the same role as a query region in the partition graph model.

If the weight of each canonical subset can be stored in $O(1)$ memory cells, then the total size of the data structure is just $O(|\mathcal{F}|)$. If the underlying searching problem is a range-reporting problem, however, then the "weight" of a canonical subset is the set itself, and thus it is not realistic to assume that each "weight" requires only constant space. In this case, the size of the data structure is $O(\sum_{C \in \mathcal{F}} |C|)$ if each subset is stored explicitly at each node of the tree. As we will see below, the size can be reduced to $O(|\mathcal{F}|)$ by storing the subsets implicitly (*e.g.*, storing points only at leaves).

To determine the points in a query range γ, a query procedure performs a depth-first search of the tree T, starting from the root. At each node v, using the auxiliary information stored at v, the procedure determines whether the query range γ contains C_v, intersects C_v, or is disjoint from C_v. If γ contains C_v, then C_v is added to $\mathcal{C}_\gamma$ (rather, the weight of C_v is added to a running counter). Otherwise, if γ intersects C_v, the query procedure identifies a subset of children of v, say $\{w_1, \ldots, w_a\}$, so that the canonical subsets $C_{w_i} \cap \gamma$, for $1 \le i \le a$, form a partition of $C_v \cap \gamma$. Then the procedure searches each w_i recursively. The total query time is $O(\log n + |\mathcal{C}_\gamma|)$, provided constant time is spent at each node visited.

3. Orthogonal range searching

In d-dimensional orthogonal range searching, the ranges are d-rectangles, each of the form $\prod_{i=1}^{d}[a_i, b_i]$, where $a_i, b_i \in \mathbb{R}$. This is an abstraction of multi-key searching [**BF, Wi4**], which is a central problem in statistical and commercial databases. For example, the points of S may correspond to employees of a company, each coordinate corresponding to a key such as age, salary, or experience. Queries such as "Report all employees between the ages of 30 and 40 who earn more than \$30,000 and who have worked for more than 5 years" can be formulated as orthogonal range-reporting queries. Because of its numerous applications, orthogonal range searching has been studied extensively for the last 25 years. A survey of earlier results can be found in the books by Mehlhorn [**Meh**] and Preparata and Shamos [**PrS**]. In this section we review more recent data structures and lower bounds.

3.1. Upper bounds. Most of the recent orthogonal range-searching data structures are based on *range trees*, introduced by Bentley [**Be2**]. For $d = 1$, the range tree of S is either a minimum-height binary search tree on S or an array storing S in sorted order. For $d > 1$, the range tree of S is a minimum-height binary tree T with n leaves, whose ith leftmost leaf stores the point of S with the ith smallest x_1-coordinate. To each interior node v of T, we associate a canonical subset $C_v \subseteq S$ containing the points stored at leaves in the subtree rooted at v. For each v, let a_v (resp. b_v) be the smallest (resp. largest) x_1-coordinate of any point in C_v, and let C_v^* denote the projection of C_v onto the hyperplane $x_1 = 0$. The interior node v stores a_v, b_v, and a $(d-1)$-dimensional range tree constructed on C_v^*. For any fixed dimension d, the size of the overall data structure is $O(n \log^{d-1} n)$, and it can be constructed in time $O(n \log^{d-1} n)$. The range-reporting query for a rectangle $\gamma = \prod_{i=1}^{d}[a_i, b_i]$ can be answered as follows. If $d = 1$, the query can be answered by a binary search. For $d > 1$, we traverse the range tree as follows. Suppose we are at a node v. If v is a leaf, then we report its corresponding point if it lies inside γ. If v is an interior node and the interval $[a_v, b_v]$ does not intersect $[a_1, b_1]$, there is nothing to do. If $[a_v, b_v] \subseteq [a_1, b_1]$, we recursively search in the $(d-1)$-dimensional range tree stored at v, with the $(d-1)$-rectangle $\prod_{i=2}^{d}[a_i, b_i]$. Otherwise, we recursively visit both children of v. The query time of this procedure is $O(\log^d n + k)$, which can be improved to $O(\log^{d-1} n + k)$ using the *fractional-cascading* technique [**CG, Lu**]. A range tree can also answer a range-counting query in time $O(\log^{d-1} n)$. Range trees are an example of a multi-level data structure, which we will discuss in more detail in Section 5.1.

The best data structures known for orthogonal range searching are by Chazelle [**Ch, Ch4**], who used compressed range trees and other techniques to improve the storage and query time. His results in the plane, under various models of computation, are summarized in Table 1; the preprocessing time of each data structure is $O(n \log n)$. If the query rectangles are "three-sided rectangles" of the form $[a_1, b_1] \times [a_2, \infty]$, then one can use a *priority search tree* of size $O(n)$ to answer a planar range-reporting query in time $O(\log n + k)$ [**Mc**].

Each of the two-dimensional results in Table 1 can be extended to queries in $\mathbb{R}^d$ at a cost of an additional $\log^{d-2} n$ factor in the preprocessing time, storage, and query-search time. For $d \geq 3$, Subramanian and Ramaswamy [**SR**] have proposed a data structure that can answer a range-reporting query in time $O(\log^{d-2} n \log^* n + k)$ using $O(n \log^{d-1} n)$ space, and Bozanis *et al.* [**BKMT2**]

Problem	Model	Size	Query time
Counting	RAM	n	$\log n$
	APM	n	$\log n$
	EPM	n	$\log n + k$
Reporting	RAM	n	$\log n + k \log^{\varepsilon}(2n/k)$
		$n \log\log n$	$\log n + k \log\log(4n/k)$
		$n \log^{\varepsilon} n$	$\log n + k$
	APM	n	$k \log(2n/k)$
	EPM	n	$k \log^2(2n/k)$
		$\frac{n \log n}{\log\log n}$	$\log n + k$
Semigroup	Semigroup	m	$\frac{\log n}{\log(2m/n)}$
	RAM	n	$\log^{2+\varepsilon} n$
		$n \log\log n$	$\log^2 n \log\log n$
		$n \log^{\varepsilon} n$	$\log^2 n$
	APM	n	$\log^3 n$
	EPM	n	$\log^4 n$

TABLE 1. Asymptotic upper bounds for planar orthogonal range searching, due to Chazelle [**Ch, Ch4**], in the random access machine (RAM), arithmetic pointer machine (APM), elementary pointer machine (EPM), and semigroup arithmetic models.

have proposed a data structure with $O(n \log^d n)$ size and $O(\log^{d-2} n + k)$ query time. The query time (or the query-search time in the range-reporting case) can be reduced to $O((\log n/\log\log n)^{d-1})$ in the RAM model by increasing the space to $O(n \log^{d-1+\varepsilon} n)$. In the semigroup arithmetic model, a query can be answered in time $O((\log n/\log(m/n))^{d-1})$ using a data structure of size m, for any $m = \Omega(n \log^{d-1+\varepsilon} n)$ [**Ch7**]. Willard [**Wi3**] proposed a data structure of size $O(n \log^{d-1} n/\log\log n)$, based on fusion trees, that can answer an orthogonal range-reporting query in time $O(\log^{d-1} n/\log\log n + k)$. Fusion trees were introduced by Fredman and Willard [**FW**] for an $O(n\sqrt{\log n})$ sorting algorithm in a RAM model that allows bitwise logical operations.

Overmars [**Ov2**] showed that if S is a subset of a $u \times u$ grid U in the plane and the vertices of query rectangles are also a subset of U, then a range-reporting query can be answered in time $O(\sqrt{\log u} + k)$, using $O(n \log n)$ storage and preprocessing, or in $O(\log\log u + k)$ time, using $O(n \log n)$ storage and $O(u^3 \log u)$ preprocessing. See [**KV**] for some other results on range-searching for points on integer grids.

Orthogonal range-searching data structures based on range trees can be extended to handle c-oriented ranges in a straightforward manner. The performance of such a data structure is the same as that of a c-dimensional orthogonal range-searching structure. If the ranges are homothets of a given triangle, or translates of a convex polygon with constant number of edges, a two-dimensional range-reporting query can be answered in $O(\log n + k)$ time using linear space [**CE, CE2**]. If the ranges are octants in $\mathbb{R}^3$, a range-reporting query can be answered in either $O((k+1)\log n)$ or $O(\log^2 n + k)$ time using linear space [**CE2**].

3.2. Lower bounds. Fredman [**Fr2, Fr3, Fr4, FV**] was the first to prove nontrivial lower bounds on orthogonal range searching, in a version of semigroup arithmetic model in which the points can be inserted and deleted dynamically. He showed that a mixed sequence of n insertions, deletions, and queries requires $\Omega(n \log^d n)$ time. These bounds were extended by Willard [**Wi2**] to the group model, under some fairly restrictive assumptions.

Yao [**Y2**] proved a lower bound for two-dimensional static data structures in the semigroup arithmetic model. He showed that if only m units of storage is available, a query takes $\Omega(\log n/\log((m/n)\log n))$ in the worst case. Vaidya [**V**] proved lower bounds for orthogonal range searching in higher dimensions, which were later improved by Chazelle [**Ch7**]. In particular, Chazelle proved the following strong result about the *average*-case complexity of orthogonal range searching:

THEOREM 1 (Chazelle [**Ch7**]). *Let d, n, m be positive integers with $m \geq n$. If only m units of storage are available, then the expected query time for a random orthogonal range query in a random set of n points in the unit hypercube $[0,1]^d$ is $\Omega((\log n/\log(2m/n))^{d-1})$ in the semigroup arithmetic model.*

A rather surprising result of Chazelle [**Ch6**] shows that any data structure on a basic pointer machine that answers a d-dimensional range-reporting query in $O(\text{polylog}\, n + k)$ time must have size $\Omega(n(\log n/\log\log n)^{d-1})$; see also [**AnS**]. Notice that this lower bound is greater than the $O(n \log^{d-2+\varepsilon} n)$ upper bound in the RAM model (see Table 1).

These lower bounds do not hold for offline orthogonal range searching, where given a set of n weighted points in $\mathbb{R}^d$ and a set of n rectangles, one wants to compute the weight of the points in each rectangle. Recently, Chazelle [**Ch11**] proved that the offline version takes $\Omega(n(\log n/\log\log n)^{d-1})$ time in the semigroup model, and $\Omega(n \log\log n)$ time in the group model. An $\Omega(n \log n)$ lower bound also holds in the algebraic decision tree and algebraic computation tree models [**SY, BO**].

3.3. Secondary memory structures. If the input point set is rather large and does not fit into main memory, then the data structure must be stored in secondary memory—on disk, for example—and portions of it must moved into main memory when needed to answer a query. In this case the bottleneck is the time spent in transferring data between main and secondary memory. We assume that data is stored in secondary memory in blocks of size B, where B is a parameter. Each access to the secondary memory transfers one block (*i.e.*, B words), and we count this as one input/output (I/O) operation. The size of a data structure is the number of blocks required to store it, and the query (resp. preprocessing) time is defined as the number of I/O operations required to answer a query (resp. to construct the structure). To simplify our notation, let $N = n/B$, the number of blocks required to hold the input, and let $\text{Log}\, n = \log_B n$. Under this model, the size of any data structure is at least N, and the query time is at least $\text{Log}\, n$. I/O-efficient orthogonal range-searching structures have received much attention recently, but most of the results are known only for the planar case. The main idea underlying these structures is to construct high-degree trees instead of binary trees. For example, variants of B-trees are used to answer 1-dimensional range-searching queries [**BM, Com**]. A number of additional tricks are developed to optimize the size and the query time. See [**Ar, AV, OSdBvK, VV**] for I/O efficient data structures that have been used for answering range searching and related queries.

Table 2 summarizes the known results on secondary-memory structures for orthogonal range searching. The data structure by Subramanian and Ramaswamy [**SR**] for 3-sided queries supports insertion/deletion of a point in time

$$O(\operatorname{Log} n + (\operatorname{Log}^2 n)/B).$$

Using the argument by Chazelle [**Ch6**], they proved that any secondary-memory data structure that answers a range-reporting query using $O(\operatorname{polyLog} n + k/B)$ I/O operations requires $\Omega(N \log N / \log \operatorname{Log} n)$ storage. Hellerstein *et al.* [**HKP**] have shown that if a data structure for two-dimensional range-reporting query uses at most $O(N)$ disk blocks, then a query requires at least

$$\Omega((k/B)\sqrt{\log B / \log\log B})$$

disk accesses; this extends an earlier lower bound by Kanellakis *et al.* [**KRVV**].

d	Range	Size	Query Time	Source
$d = 1$	Interval	N	$\operatorname{Log} n + k/B$	[**BM, Com**]
$d = 2$	Quadrant	$N \log\log B$	$\operatorname{Log} n + k/B$	[**RS**]
	3-sided rectangle	N	$\operatorname{Log} n + k/B + \log^* B$	[**SR**]
	3-sided rectangle	$N \log B \log\log B$	$\operatorname{Log} n + k/B$	[**RS**]
	Rectangle	$N \log N / \log \operatorname{Log} n$	$\operatorname{Log} n + k/B + \log^* B$	[**SR**]
	Rectangle	cN	$k/B^{1-1/2c}$	[**SR**]
$d = 3$	Octant	$N \log N$	$\beta(n) \operatorname{Log} n + k/B$	[**VV**]
	Rectangle	$N \log^4 N$	$\beta(n) \operatorname{Log} n + k/B$	[**VV**]

TABLE 2. Asymptotic upper bounds for secondary memory structures; here $N = n/B$, $\operatorname{Log} n = \log_B n$, and $\beta(n) = \log\log \operatorname{Log} n$.

3.4. Practical data structures. None of the data structures described in Section 3.1 are used in practice, even in two dimensions, because of the polylogarithmic overhead in the size and the query time. In many real applications, the input is too large to be stored in the main memory, so the number of disk accesses is a major concern. On the other hand, the range-searching data structures described in Section 3.3 are not simple enough to be of practical use for $d \geq 2$. For a data structure to be used in real applications, its size should be at most cn, where c is a very small constant, the time to answer a typical query should be small—the lower bounds proved in Section 3.2 imply that we cannot hope for small worst-case bounds—and it should support insertions and deletions of points. Keeping these goals in mind, a plethora of data structures have been proposed. We will sketch the general ideas and mention some of the data structures in a little detail. For the sake of simplicity, we will present most of the data structures in two dimensions. The book by Samet [**Sam2**] is an excellent survey of data structures developed in 1970s and 80s; more recent results are described in the survey papers [**Gr, Güt, HoS, NW2, O**].

The most widely used data structures for answering 1-dimensional range queries are B-trees and their variants [**BM, Com**]. Since a B-tree requires a linear order on the input elements, one needs to define such an ordering on points in higher dimensions in order to store them into a B-tree. An obvious choice is lexicographical ordering, also known as the *bit concatenation* method, but this ordering performs rather poorly for higher dimensional range searching because a separate disk access may be required to report each point. A better scheme for ordering the points is the *bit-interleaving* method, proposed by Morton [**Mo**]. A point $p = (x, y)$, where the binary representations of x and y are $x = x_{m-1}x_{m-2}\dots x_0$ and $y = y_{m-1}y_{m-2}\dots y_0$, is regarded as the integer whose binary representation is $x_{m-1}y_{m-1}x_{m-2}\dots y_0$. A B-tree storing points based on the bit-interleaving ordering is referred to as an *N-tree* [**Wh**] or a *zkd-tree* [**OM**] in the literature. See [**Sam2**] for a more detailed discussion on the applications of bit interleaving in spatial data structures. Faloutsos [**F**] suggested using Gray codes to define a linear order on points. In general, space-filling curves[5] can be used to define a linear ordering on input points; Hilbert and Morton curves, shown in Figure 2, are the some of the space-filling curves commonly used for this purpose. See [**AM, ARWW, FRo, J**] for a comparison of the performance of various space-filling curves in the context of range searching. Since B-trees require extra space to store pointers, several hashing schemes, including *linear hashing* [**KrS**], *dynamic z-hashing* [**HSW**] and *spiral hashing* schemes [**Mu**] are proposed to minimize the size of the data structure. The performance of any method that maps higher-dimensional points to a set of points in one dimension deteriorates rapidly with the dimension because such a mapping does not preserve neighborhoods, though there has been some recent work on locality preserving hashing schemes [**IMRV**].

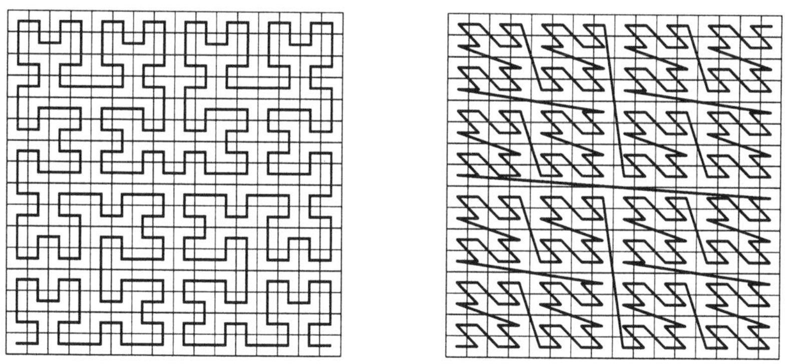

(i) (ii)

FIGURE 2. Examples of space-filling curves used for range searching. (i) Hilbert curve. (ii) Morton curve.

A more efficient approach to answer range queries is to construct a recursive partition of space, typically into rectangles, and to construct a tree induced by this partition. The simplest example of this type of data structure is the *quad tree* in the plane. A quad tree is a 4-way tree, each of whose nodes is associated with a square

[5]Formally speaking, a curve $\mathbb{R} \to [0,1]^d$ is called a *space-filling curve* if it visits each point of the unit hypercube exactly once. However, the same term often refers to approximations of space-filling curves that visit every point in a cubical lattice, such as the curves drawn in Figure 2. See the book by Sagan [**S**] for a detailed discussion on space-filling curves and [**Bi**] for some other applications of these curves.

B_v. B_v is partitioned into four equal-size squares, each of which is associated with one of the children of v. The squares are partitioned until at most one point is left inside a square. Quad trees can be extended to higher dimensions in an obvious way (they are called *oct-trees* in 3-space). In d-dimensions, a node has 2^d children. A range-search query can be answered by traversing the quad tree in a top-down fashion. Because of their simplicity, quad trees are one of the most widely used data structures for a variety of problems. For example, they were used as early as in 1920s, by Weyl [**We**] for computing the complex roots of a univariate polynomial approximately; Greengard used them for the so-called n-body problem [**Gre**]. See the books by Samet [**Sam, Sam2**] for a detailed discussion on quad trees and their applications.

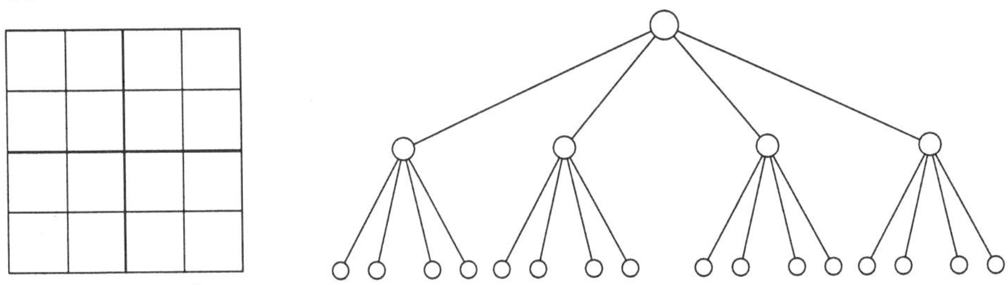

FIGURE 3. A quad tree.

One disadvantage to quad trees is that arbitrarily many levels of partitioning may be required to separate tightly clustered points. Finkel and Bentley [**FB**] described a variant of the quad tree for range searching, called a *point quad-tree*, in which each node is associated with a rectangle and the rectangle is partitioned into four rectangles by choosing a point in the interior and drawing horizontal and vertical lines through that point. Typically the point is chosen so that the height of the tree is $O(\log n)$. A recent paper by Faloutsos and Gaede [**FG**] analyzes the performance of quad trees using Hausdorff fractal dimension. See also [**FBF, Ho**] for other data structures based on quad trees.

In order to minimize the number of disk accesses, one can partition the square into many squares (instead of four) by a drawing either a uniform or a nonuniform grid. The *grid file*, introduced by Nievergelt *et al.* [**NHS**] is based on this idea. Since grid files are used frequently in geographic information systems, we describe them briefly. A grid file partitions the plane into a nonuniform grid by drawing horizontal and vertical lines. The grid lines are chosen so that the points in each cell can be stored in a single block of the disk. The grid is then partitioned into rectangles, each rectangle being the union of a subset of grid cells, so that the points in each rectangle can still be stored in a single block of the disk. The data within each block can be organized in an arbitrary way. The grid file maintains two pieces of information: a *grid directory*, which stores the index of the block that stores the points lying in each grid cell, and two arrays, called *linear scales*, which store the x-coordinates (resp. y-coordinates) of the vertical (resp. horizontal) lines. It is assumed that the linear scales are small enough to be stored in main memory. A point can be accessed by two disk accesses as follows. By searching with the x- and y-coordinates of the points in the linear scales, we determine the grid cell that contains the point. We then access that cell of the grid directory (using one disk access) to determine the index of the block that stores p, and finally we access that block and retrieve the point p (second disk access). A range query is answered by locating the cells that contain the corners of the query rectangle and thus determining all the grid cells that intersect the query rectangle. We then access each of these cells to report all the points lying in the query rectangle. Several heuristics are used to minimize the number of disk accesses required to answer a query and to update the structures as points are inserted or deleted. Note that a range query reduces to another range query on the grid directory, so one can store the grid directory itself as a grid file. This notion of a hierarchical grid file was proposed by Hinrichs [**Hi**] and Krishnamurthy and Wang [**KW**]. A related data structure, known as the *BANG file*, was proposed by Freestone [**Fre**]; other variants of grid files are proposed in [**Hi, HSW2, Ou**].

Quad trees, grid files, and their variants construct a grid on a rectangle containing all the input points. One can instead partition the enclosing rectangle into two rectangles by drawing a horizontal or a vertical line and partitioning each of the two rectangles independently. This is the idea behind the so called *kd-tree* due to Bentley [**Be**]. In particular, a *kd*-tree is a binary tree, each of whose nodes v is associated with a rectangle B_v. If B_v does not contain any point in its interior, v is a leaf. Otherwise, B_v is partitioned into two rectangles by drawing a horizontal or vertical line so that each rectangle contains at most half of the points; splitting lines are alternately horizontal and vertical. A *kd*-tree can be extended to higher dimensions in an obvious manner.

In order to minimize the number of disk accesses, Robinson [**R**] suggested the following generalization of a *kd*-tree, which is known as a *kd-B-tree*. One can construct a B-tree instead of a binary tree on the recursive partition of the enclosing rectangle, so all leaves of the tree are at the same level and each node has between $B/2$ and B children. The rectangles associated with the children are obtained by splitting B_v recursively, as in a *kd*-tree approach; see Figure 4(i). Let $w_1, \ldots w_s$ be the children of v. Then $B_{w_1}, \ldots, B_{w_s}$ can be stored implicitly at v by storing them as a *kd*-tree, or the coordinates of their corners can be stored explicitly. If points are dynamically inserted into a *kd*-B-tree, then some of the nodes may have to be split, which is an expensive operation, because splitting a node may require reconstructing the entire subtree rooted at that node; see Figure 4(ii). Several variants of *kd*-B-trees have been proposed to minimize the number of splits, to optimize the space, and to improve the query time [**BSW, ELS, Free, H, LS, SeK, SK, SRF, SSA**]. We mention only two of the variants here: *Buddy trees* [**SRF**] and *hB-trees* [**LS, ELS**]. A buddy tree is a combination of a quad tree and *kd*-B-tree in the sense that rectangles are split into sub-rectangles only at some specific locations, which simplifies the split procedure; see Seeger and Kriegel [**SK**] for details. If points are in degenerate position, then it may not be possible to split them into two halves by a line. Lomen and Salzberg [**LS**] circumvent this problem by introducing a new data structure, called *hB-tree*, in which the region associated with a node is allowed to be $R_1 \setminus R_2$ where R_1 and R_2 are rectangles. A more refined version of this data structure, known as hB^{Π}*-tree*, is presented in [**ELS**].

In a *kd*-tree, a rectangle is partitioned into two rectangles by drawing a horizontal or vertical line. One can instead associate a convex polygon B_v with each node v of the tree, use an arbitrary line to partition B_v into two convex polygons, and associate the two polygons with the children of v. This idea is the same as in binary space partition trees [**FKN, PY**]. Again, one can construct a B-tree on this recursive partitioning scheme to reduce the number of disk accesses. The resulting structure called *cell trees* is studied in [**Gü, GB**].

All the data structures described in this section construct a recursive partition of the space. There are other data structures (of which the R-tree is perhaps the most famous example) that construct a hierarchical cover of the space. We will discuss some of these data structures in the next subsection.

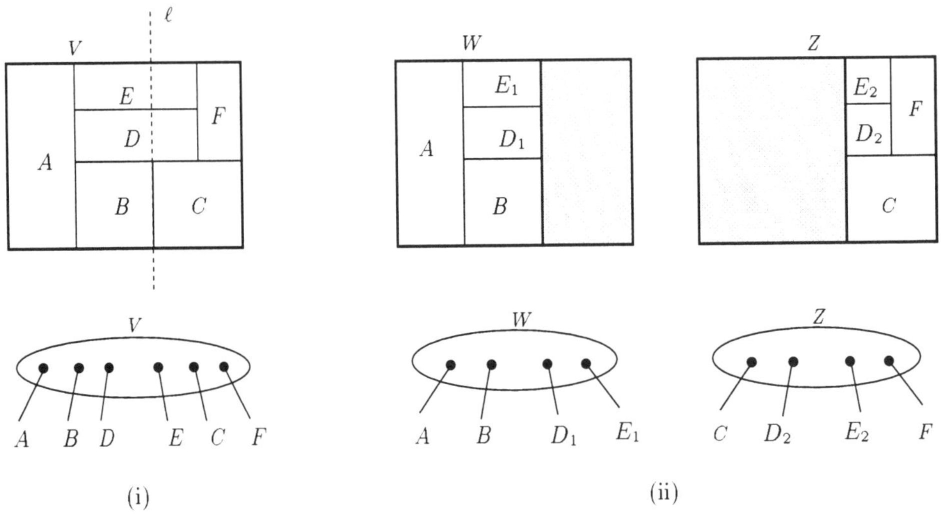

FIGURE 4. Splitting a node V of a kd-B-tree by a line ℓ. (i) V before the split. (ii) V is split into two nodes W and Z; the subtrees rooted at D and E are split recursively.

3.5. The partial sum problem. Preprocess a d-dimensional array A with n entries, in an additive semigroup, into a data structure, so that for a d-dimensional rectangle $\gamma = [a_1, b_1] \times \cdots \times [a_d, b_d]$, the sum

$$\sigma(A, \gamma) = \sum_{(k_1, k_2, \ldots, k_d) \in \gamma} A[k_1, k_2, \ldots, k_d]$$

can be computed efficiently. In the offline version, given A and m rectangles $\gamma_1, \gamma_2, \ldots, \gamma_m$, we wish to compute $\sigma(A, \gamma_i)$ for each i. Note that this is just a special case of orthogonal range searching, where the points lie on a regular d-dimensional lattice.

Partial-sum queries are widely used for on-line analytical processing (OLAP) of commercial databases. OLAP allows companies to analyze aggregate databases built from their data warehouses. A popular data model for OLAP applications is the multidimensional database, known as *data cube* [**GBLP**], which represents the data as d-dimensional array. Thus, an aggregate query can be formulated as a partial-sum query. Driven by this application, several heuristics have been proposed to answer partial-sum queries on data cubes [**GHRU, HAMS, HBA, HRU, RKR, Sh**].

Yao [**Y**] showed that, for $d = 1$, a partial-sum query can be answered in $O(\alpha(n))$ time using $O(n)$ space.[6] If the additive operator is *max* or *min*, then a partial-sum

[6]Here, $\alpha(n)$ and $\alpha(m, n)$ denote functional inverses of Ackermann's function. These functions grow extremely slowly; for example, $\alpha(n) \le 4$ for all $n \le 2 \uparrow (2 \uparrow (2 \uparrow 2)) = 2 \uparrow 2^{16}$, where for any

query can be answered in $O(1)$ time under the RAM model using a Cartesian tree, developed by Vuillemin [**Vu**], and the nearest-common-ancestor algorithm of Harel and Tarjan [**HT**].

For $d > 1$, Chazelle and Rosenberg [**CR**] gave a data structure of size

$$O(n \log^{d-1} n)$$

that can answer a partial-sum query in time $O(\alpha(n) \log^{d-2} n)$. They also showed that the offline version takes $\Omega(n+m\alpha(m,n))$ time for any fixed $d \geq 1$. If points are allowed to insert into S, the query time is $\Omega(\log n/ \log\log n)$ for the one-dimensional case [**Fr**, **Y2**]; the bounds were extended by Chazelle [**Ch7**] to

$$\Omega((\log n/ \log\log n)^d),$$

for any fixed dimension d. All of these lower bounds hold in the semigroup arithmetic model. Chazelle [**Ch2**] extended the data structure by Yao to the following variant of the partial-sum problem: Let T be a rooted tree with n nodes, each of whose node is associated with an element of a commutative semigroup. Preprocess T so that for a query node v, the sum of the weights in the subtree rooted at v can be computed efficiently. Chazelle showed that such a query can be answered in $O(\alpha(n))$ time, using $O(n)$ space.

3.6. Rectangle-rectangle searching. Preprocess a set S of n rectangles in $\mathbb{R}^d$ so that for a query rectangle q, the rectangles of S that intersect q can be reported (or counted) efficiently. Rectangle-rectangle searching is central to many applications because, in practice, polygonal objects are approximated by rectangles. Chazelle [**Ch4**] has shown that the bounds mentioned in Table 1 also hold for this problem.

In practice, two general approaches are used to answer a query. A rectangle $\prod_{i=1}^{d}[a_i, b_i]$ in $\mathbb{R}^d$ can be mapped to the point $(a_1, a_2, \ldots, a_d, b_1, b_2, \ldots, b_d)$ in $\mathbb{R}^{2d}$, and a rectangle-intersection query can be reduced to orthogonal range searching. Many heuristic data structures based on this scheme have been proposed; see [**FR**, **PST**, **SRF**] for a sample of such results. The second approach is to construct a data structure on S directly in $\mathbb{R}^d$. The most popular data structure based on this approach is the *R-tree*, originally introduced by Guttman [**Gu**].

An R-tree is a multiway tree (like a B-tree), each of whose nodes stores a set of rectangles. Each leaf stores a subset of input rectangles, and each input rectangle is stored at exactly one leaf. For each node v, let R_v be the smallest rectangle containing all the rectangles stored at v; R_v is stored at the parent of v (along with the pointer to v). R_v induces the subspace corresponding to the subtree rooted at v, in the sense that for any query rectangle intersecting R_v, the subtree rooted at v is searched. Rectangles stored at a node are allowed to overlap. Therefore, unlike all the data structures discussed in Section 3.4, a R-tree forms a recursive cover of the data space, instead of a recursive partition. Although allowing rectangles to overlap helps reduce the size of the data structure, answering a query becomes more expensive. Guttman suggests some heuristics to construct a R-tree so that the overlap is minimized. Better heuristics for minimizing the overlap were developed by Beckmann *et al.* [**BKSS**], Green [**Gr**], and Kamal and Faloutsos [**KF**, **KF2**, **KF3**]. There are many variants of R-tree, depending on the

positive integer k, $2 \uparrow k = 2^{2\uparrow(k-1)}$ denotes an exponential tower of k twos. For formal definitions, see [**SA**].

application: an *R^*-tree* [**BKSS**] uses more sophisticated techniques to minimize the overlap; a *Hilbert-R-tree* [**KF3**] defines a linear ordering on the rectangles, by sorting their centers along the Hilbert space-filling curve, and constructs a B-tree based on this ordering of rectangles; and an *R^+-tree* avoids overlapping directory cells by clipping rectangles [**SRF**]. Additional variants are suggested to avoid overlap in higher dimensions. Berchtold *et al.* [**BKK**] define the *X-tree*, in which the interior nodes are allowed to be arbitrarily large; Lin *et al.* [**LJF**] project rectangles onto a lower dimensional space and construct an R-tree (or some variant thereof) on these projections. Leutenegger *et al.* [**LLE**] compare different variants of R-trees and discuss advantages of different heuristics used to minimize the overlap of rectangles.

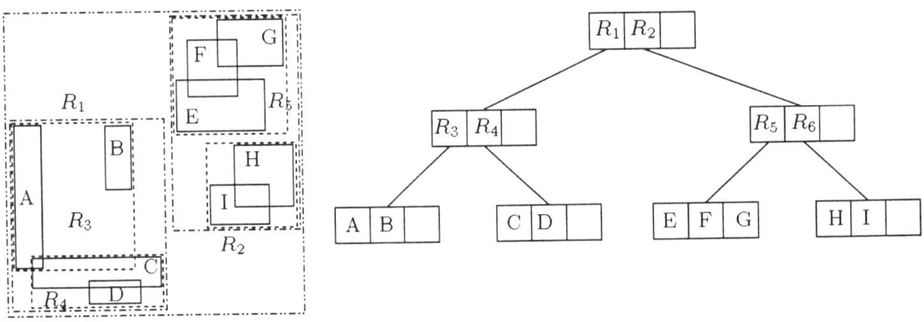

FIGURE 5. An R-tree.

We discuss some more general rectangle-intersection searching problems in Section 6.3.

4. Simplex range searching

As mentioned in the introduction, simplex range searching has received considerable attention during the last few years. Besides its direct applications, simplex range-searching data structures have provided fast algorithms for numerous other geometric problems. See the survey paper by Matoušek [**M7**] for an excellent review of techniques developed for simplex range searching.

Unlike orthogonal range searching, no simplex range-searching data structure is known that can answer a query in polylogarithmic time using near-linear storage. In fact, the lower bounds stated below indicate that there is very little hope of obtaining such a data structure, since the query time of a linear-size data structure, under the semigroup model, is roughly at least $n^{1-1/d}$ (thus saving only a factor of $n^{1/d}$ over the naïve approach). Since the size and query time of any data structure have to be at least linear and logarithmic, respectively, we consider these two ends of the spectrum: (i) How fast can a simplex range query be answered using a linear-size data structure, and (ii) how large should the size of a data structure be in order to answer a query in logarithmic time. By combining these two extreme cases, as we describe below, we obtain a tradeoff between space and query time.

Unless stated otherwise, each of the data structures we describe in this section can be constructed in time that is only a polylogarithmic or n^ε factor larger than its size.

4.1. Linear-size data structures. Most of the linear-size data structures for simplex range searching are based on so-called *partition trees*, originally introduced by Willard [**Wi**]. Roughly speaking, partition trees are based on the following idea: Given a set S of points in $\mathbb{R}^d$, partition the space into a few, say, a constant number of, regions, each containing roughly equal number of points, so that for any hyperplane h, the number of points lying in the regions that intersect h is much less than the total number of points. Then recursively construct a similar partition for the subset of points lying in each region.

Willard's original partition tree for a set S of n points in the plane is a 4-way tree, constructed as follows. Let us assume that n is of the form 4^k for some integer k, and that the points of S are in general position. If $k = 0$, the tree consists of a single node that stores the coordinates of the only point in S. Otherwise, using the ham-sandwich theorem [**E**], find two lines ℓ_1, ℓ_2 so that each quadrant Q_i, for $1 \leq i \leq 4$, induced by ℓ_1, ℓ_2 contains exactly $n/4$ points. The root stores the equations of ℓ_1, ℓ_2 and the value of n. For each quadrant, recursively construct a partition tree for $S \cap Q_i$ and attach it as the ith subtree of the root. The total size of the data structure is linear, and it can be constructed in $O(n \log n)$ time. A halfplane range-counting query can be answered as follows. Let h be a query halfplane. Traverse the tree, starting from the root, and maintain a global count. At each node v storing n_v nodes in its subtree, perform the following step: If the line ∂h intersects the quadrant Q_v associated with v, recursively visit the children of v. If $Q_v \cap h = \emptyset$, do nothing. Otherwise, since $Q_v \subseteq h$, add n_v to the global count. The quadrants associated with the four children of any interior node are induced by two lines, so ∂h intersects at most three of them, which implies that the query procedure does not explore the subtree of one of the children. Hence, the query time of this procedure is $O(n^\alpha)$, where $\alpha = \log_3 4 \leq 0.7925$. A similar procedure can answer a simplex range-counting query within the same time bound, and a simplex range-reporting query in time $O(n^\alpha + k)$. Edelsbrunner and Welzl [**EW**] described a simple variant of Willard's partition tree that improves the exponent in the query-search time to $\log_2(1 + \sqrt{5}) - 1 \approx 0.695$.

A partition tree for points in $\mathbb{R}^3$ was first proposed by Yao [**Ya**], which can answer a query in time $O(n^{0.98})$. This bound was improved slightly in subsequent papers [**DE, EH, YDEP**]. Using the Borsuk-Ulam theorem, Yao *et al.* [**YDEP**] showed that, given a set S of n points in $\mathbb{R}^3$, one can find three planes so that each of the eight (open) octants determined by them contains at most $\lfloor n/8 \rfloor$ points of S. Avis [**Av**] proved that such a partition of $\mathbb{R}^d$ by d hyperplanes is not always possible for $d \geq 5$; the problem is still open for $d = 4$. Weaker partitioning schemes were proposed in [**Co, YY**].

After the initial improvements and extensions on Willard's partition tree, a major breakthrough was made by Haussler and Welzl [**HW**]. They formulated range searching in an abstract setting and, using elegant probabilistic methods, gave a randomized algorithm to construct a linear-size partition tree with $O(n^\alpha)$ query time, where $\alpha = 1 - \frac{1}{d(d-1)+1} + \varepsilon$ for any $\varepsilon > 0$. The major contribution of their paper is the abstract framework and the notion of ε-nets. A somewhat different abstract framework for randomized algorithms was proposed by Clarkson [**Cl, CS**] around the same time; see also [**Mul**]. These abstract frameworks and the general results attained under these frameworks popularized randomized algorithms in computational geometry [**Mul4**]. We briefly describe the framework and the main result by Haussler and Welzl because they are most pertinent to range searching.

A *range space* is a set system $\Sigma = (X, \mathcal{R})$ where X is a set of objects and $\mathcal{R}$ is a family of subsets of X. The elements of $\mathcal{R}$ are called the *ranges* of Σ. Σ is called a *finite range space* if the ground set X is finite. Here are a few examples of geometric range spaces:

(i) $\Sigma_1 = (\mathbb{R}^d, \{h \mid h \text{ is a halfspace in } \mathbb{R}^d\})$,

(ii) $\Sigma_2 = (\mathbb{R}^d, \{B \mid B \text{ is a ball in } \mathbb{R}^d\})$,

(iii) Let H be the set of all hyperplanes in $\mathbb{R}^d$. For a segment s, let $H_s \subseteq H$ be the set of all hyperplanes intersecting s. Define the range space $\Sigma_3 = (H, \{H_s \mid s \text{ is a segment in } \mathbb{R}^d\})$.

For a finite range space $\Sigma = (X, \mathcal{R})$, a subset $N \subseteq X$ is called an *ε-net* if $N \cap \gamma \neq \emptyset$ for every range $\gamma \in \mathcal{R}$ with $|\gamma| \geq \varepsilon|X|$. That is, N intersects every "large" range of Σ. (The notion of ε-nets can be extended to infinite range spaces as well.) A subset $A \subseteq X$ can be *shattered* if every subset of A has the form $A \cap \gamma$ for some $\gamma \in \mathcal{R}$. The *Vapnik-Chervonenkis dimension*, or *VC-dimension*, of a range space Σ is the size of the largest subset A that can be shattered. For example, the VC-dimensions of Σ_1, Σ_2, and Σ_3 are $d+1$, $d+2$, and $2d$, respectively. The main result of Haussler and Welzl is that, given a finite range space $\Sigma = (X, \mathcal{R})$ and parameters $0 < \varepsilon, \delta \leq 1$, if we choose a random subset $N \subset X$ of size

$$\max\left\{\frac{8d}{\varepsilon}\log\frac{8d}{\varepsilon}, \frac{4}{\varepsilon}\log\frac{2}{\delta}\right\},$$

then N is an ε-net of Σ with probability at least $1 - \delta$. The bound on the size of ε-nets was improved by Blumer *et al.* [**BEHW**] and Komlós *et al.* [**KPW**].

THEOREM 2 (Komlós *et al.* [**KPW**]). *For any finite range space $(X, \mathcal{R})$ of VC-dimension d and for any $0 < \varepsilon < 1$, if N is a subset of X obtained by*

$$\frac{d}{\varepsilon}\left(\log\frac{1}{\varepsilon} + 2\log\log\frac{1}{\varepsilon} + 3\right)$$

random independent draws, then N is an ε-net of $(X, \mathcal{R})$ with probability at least $1 - e^{-d}$.

Theorem 2 and some other similar results [**Cl, CS**] have been used extensively in computational geometry and learning theory; see the books by Motwani and Raghavan [**MR**], Mulmuley [**Mul4**], and Anthony and Biggs [**AB**] and the survey papers [**Cl3, M3, Se**].

The first linear-size data structure with near-optimal query time for simplex range queries in the plane was developed by Welzl [**W**]. His algorithm is based on the following idea. A *spanning path* of a set S of points is a polygonal chain whose vertices are the points of S. The *crossing number* of a polygonal path is the maximum number of its edges that can be crossed by a hyperplane. Using Theorem 2, Welzl constructs a spanning path $\Pi = \Pi(S)$ of any set S of n points in $\mathbb{R}^d$ whose crossing number is $O(n^{1-1/d}\log n)$. The bound on the crossing number was improved by Chazelle and Welzl [**CW**] to $O(n^{1-1/d})$, which is tight in the worst case. Let $p_1, p_2, \ldots, p_n$ be the vertices of Π. If we know the edges of Π that cross h, then the weight of points lying in one of the halfspaces bounded by h can be computed by answering $O(n^{1-1/d})$ partial-sum queries on the sequence

$W = \langle w(p_1), \ldots, w(p_n) \rangle$. Hence, by processing W for partial-sum queries, we obtain a linear-size data structure for simplex range searching, with $O(n^{1-1/d}\alpha(n))$ query time, in the semigroup arithmetic model. (Recall that the time spent in finding the edges of Π crossed by h is not counted in the semigroup model.) In any realistic model of computation such as pointer machines or RAMs, however, we also need an efficient linear-size data structure for computing the edges of Π crossed by a hyperplane. Chazelle and Welzl [**CW**] produced such a data structure for $d \leq 3$, but no such structure is known for higher dimensions. Although spanning paths were originally introduced for simplex range searching, they have been successfully applied to solve a number of other algorithmic as well as combinatorial problems; see, for example, [**A2, CJ2, EGHS+, MWW, P, W2**].

Matoušek and Welzl [**MW**] gave an entirely different algorithm for the halfspace range-counting problem in the plane, using a combinatorial result of Erdős and Szekeres [**ES**]. The query time of their data structure is $O(\sqrt{n}\log n)$, and it uses $O(n)$ space and $O(n^{3/2})$ preprocessing time. If subtractions are allowed, their algorithm can be extended to the triangle range-counting problem. An interesting open question is whether the preprocessing time can be improved to near linear. In order to make this improvement, we need a near-linear time algorithm for the following problem, which is interesting its own right: Given a sequence X of n integers, partition X into $O(\sqrt{n})$ subsequences, each of which is either monotonically increasing or decreasing. The existence of such a partition of X follows from the result by Erdős and Szekeres, but the best known algorithm for computing such a partition runs in time $O(n^{3/2})$ [**BYF**]. However, a longest monotonically increasing subsequence of X can be computed in $O(n\log n)$ time. The technique by Matoušek and Welzl has also been applied to solve some other geometric-searching problems, including ray shooting and intersection searching [**BYF2**].

The first data structure with roughly $n^{1-1/d}$ query time and near-linear space, for $d > 3$, was obtained by Chazelle *et al.* [**CSW**]. Given a set S of n points in $\mathbb{R}^d$, they construct a family $\mathcal{F} = \{\Xi_1, \ldots, \Xi_k\}$ of triangulations of $\mathbb{R}^d$, each of size $O(r^d)$. For any hyperplane h, there is at least one Ξ_i so that only $O(n/r)$ points lie in the simplices of Ξ_i that intersect h. Applying this construction recursively, they obtain a tree structure of size $O(n^{1+\varepsilon})$ that can answer a halfspace range-counting query in time $O(n^{1-1/d})$. The extra n^ε factor in the size is due to the fact that they maintain a family of partitions instead of a single partition. Another consequence of maintaining a family of partitions is that, unlike partition trees, this data structure cannot be used directly to answering simplex range queries. Instead, Chazelle *et al.* [**CSW**] construct a multi-level data structure (which we describe in Section 5.1) to answer simplex range queries.

Matoušek [**M6**] developed a simpler, slightly faster data structure for simplex range queries, by returning to the theme of constructing a single partition, as in the earlier partition-tree papers. His algorithm is based on the following partition theorem, which can be regarded as an extension of the result by Chazelle and Welzl.

THEOREM 3 (Matoušek [**M**]). *Let S be a set of n points in $\mathbb{R}^d$, and let $1 < r \leq n/2$ be a given parameter. Then there exists a family of pairs*

$$\Pi = \{(S_1, \Delta_1), \ \ldots, (S_m, \Delta_m)\}$$

such that each $S_i \subseteq S$ lies inside the simplex Δ_i, $n/r \leq |S_i| \leq 2n/r$, $S_i \cap S_j = \emptyset$ for all $i \neq j$, and every hyperplane crosses at most $cr^{1-1/d}$ simplices of Π; here

c is a constant. If $r \le n^{\alpha}$ for some suitable constant $0 < \alpha < 1$, then Π can be constructed in $O(n \log r)$ time.

Note that although S is being partitioned into a family of subsets, unlike the earlier results on partition trees, it does not partition $\mathbb{R}^d$ because Δ_i's may intersect. In fact, it is an open problem whether $\mathbb{R}^d$ can be partitioned into $O(r)$ disjoint simplices that satisfy the above theorem.

Using Theorem 3, a partition tree T can be constructed as follows. Each interior node v of T is associated with a canonical subset $C_v \subseteq S$ and a simplex Δ_v containing C_v; if v is the root of T, then $C_v = S$ and $\Delta_v = \mathbb{R}^d$. Choose r to be a sufficiently large constant. If $|S| \le 4r$, T consists of a single node, and it stores all points of S. Otherwise, we construct a family of pairs $\Pi = \{(S_1, \Delta_1), \dots, (S_m, \Delta_m)\}$ using Theorem 3. The root u stores the value of n. We recursively construct a partition tree T_i for each S_i and attach T_i as the ith subtree of u. The root of T_i also stores Δ_i. The total size of the data structure is linear, and it can be constructed in time $O(n \log n)$. A simplex range-counting query can be answered in the same way as with Willard's partition tree. Since any hyperplane intersects at most $cr^{1-1/d}$ simplices of Π, the query time is $O(n^{1-1/d+\log_r c})$; the $\log_r c$ term in the exponent can be reduced to any arbitrarily small positive constant ε by choosing r sufficiently large. The query time can be improved to $O(n^{1-1/d} \operatorname{polylog} n)$ by choosing $r = n^{\varepsilon}$.

In a subsequent paper Matoušek [**M6**] proved a stronger version of Theorem 3, using some additional sophisticated techniques (including Theorem 5 described below), that gives a linear-size partition tree with $O(n^{1-1/d})$ query time.

If the points in S lie on a k-dimensional algebraic surface of constant degree, the *crossing number* in Theorem 3 can be improved to $O(r^{1-1/\gamma})$, where $\gamma = 1/\lfloor (d+k)/2 \rfloor$ [**AgM2**], which implies that in this case a simplex range query can be answered in time $O(n^{1-1/\gamma+\varepsilon})$ using linear space.

Finally, we note that better bounds can be obtained for the halfspace range-reporting problem, using the so-called *filtering search* technique introduced by Chazelle [**Ch**]. All the data structured mentioned above answer a range-reporting query in two stages. The first stage "identifies" the k points of a query output, in time $f(n)$ that is independent of the output size, and the second stage explicitly reports these points in $O(k)$ time. Chazelle observes that since $\Omega(k)$ time will be spent in reporting k points, the first stage can compute in $f(n)$ time a superset of the query output of size $O(k)$, and the second stage can "filter" the actual k points that lie in the query range. This observation not only simplifies the data structure but also gives better bounds in many cases, including halfspace range reporting. See [**AvKO, Ch, CCPY, CP**] for some applications of filtering search.

An optimal halfspace reporting data structure in the plane was proposed by Chazelle *et al.* [**CGL**]. They compute *convex layers* $L_1, \dots, L_m$ of S, where L_i is the set of points lying on the boundary of the convex hull of $S \setminus \bigcup_{j<i} L_j$, and store them in a linear-size data structure, so that a query can be answered in $O(\log n + k)$ time. Their technique does not extend to three dimensions. After a few initial attempts [**CP, AHL**], Matoušek developed a data structure that answers a halfspace reporting query in $\mathbb{R}^d$ in time $O(n^{1-1/\lfloor d/2 \rfloor} \operatorname{polylog} n + k)$. His structure is based on the following two observations. A hyperplane is called λ-*shallow* if one of the halfspaces bounded by h contains at most λ points of S. If the hyperplane bounding a query halfspace is not λ-shallow, for some $\lambda = \Omega(n)$, then a simplex range-reporting data structure can be used to answer a query in time

Problem	d	Size	Query Time	Source
Reporting	$d = 2$	n	$\log n + k$	[CGL]
Emptiness		n	$\log n$	[PrS]
Reporting	$d = 3$	$n \log n$	$\log n + k$	[AHL]
Emptiness		n	$\log n$	[DHKS]
Reporting	$d > 3$	$n \log\log n$	$n^{1-1/\lfloor d/2 \rfloor} \operatorname{polylog} n + k$	[M2]
Emptiness		n	$n^{1-1/\lfloor d/2 \rfloor} 2^{O(\log^* n)}$	[M2]

TABLE 3. Asymptotic upper bounds for halfspace range searching in near-linear space.

$O(n^{1-1/d+\varepsilon} + k) = O(k)$. For shallow hyperplanes, Matoušek proves the following theorem, which is an analog of Theorem 3.

THEOREM 4 (Matoušek [**M2**]). *Let S be a set of n points in $\mathbb{R}^d$ $(d \geq 4)$ and let $1 \leq r < n$ be a given parameter. Then there exists a family of pairs*

$$\Pi = \{(S_1, \Delta_1), \ldots, (S_m, \Delta_m)\}$$

such that each $S_i \subseteq S$ lies inside the simplex Δ_i, $n/r \leq |S_i| \leq 2n/r$, $S_i \cap S_j = \emptyset$ for all $i \neq j$, and every (n/r)-shallow hyperplane crosses $O(r^{1-1/\lfloor d/2 \rfloor})$ simplices of Π. If $r \leq n^\alpha$ for some suitable constant $0 < \alpha < 1$, then Π can be constructed in $O(n \log r)$ time.

Using this theorem, a partition tree for S can be constructed in the same way as for simplex range searching, except that at each node v of the tree, we also preprocess the corresponding canonical subset C_v for simplex range searching and store the resulting data structure as a secondary data structure of v. While answering a query for a halfspace h^+, if h^+ crosses more than $O(r^{1-1/\lfloor d/2 \rfloor})$ simplices of the partition Π_v associated with a node v, then it reports all points of $h^+ \cap C_v$ using the simplex range-reporting data structure stored at v. Otherwise, for each pair $(S_i, \Delta_i) \in \Pi_v$, if $\Delta_i \subseteq h^+$, it reports all points S_i, and if Δ_i is crossed by h, it recursively visits the corresponding child of v.

If we are interested only in determining whether $h^+ \cap S = \emptyset$, we do not have to store simplex range-searching structure at each node of the tree. Consequently, the query time and the size of the data structure can be improved slightly; see Table 3 for a summary of results.

Since the query time of a linear-size simplex range-searching data structure is only a $n^{1/d}$ factor better than the naïve method, researchers have developed practical data structures that work well most of the time. For example, Arya and Mount [**ArM3**] have developed a linear-size data structure for answering approximate range-counting queries, in the sense that the points lying within distance $\delta \cdot \operatorname{diam}(\Delta)$ to the boundary of the query simplex Δ may or may not be counted. Its query time is $O(\log n + 1/\delta^{d-1})$. Overmars and van der Stappen [**OvdS**] developed fast data structures for the special case in which the ranges are "fat" and have bounded size.

In practice, the data structures described in Section 3.4 are used even for simplex range searching. Agarwal *et al.* [**AAEFV**] have described I/O-efficient data structures for halfspace range-reporting queries in two and three dimensions. Recently, Goldstein *et al.* [**GRSY**] presented an algorithm for simplex range searching using R-trees. Although these data structures do not work well in the worst case, they perform reasonably well in practice, for example, when the points are close to uniformly distributed. It is an open question whether simple data structures can be developed for simplex range searching that work well on typical data sets.

4.2. Data structures with logarithmic query time. For the sake of simplicity, we first consider the halfspace range-counting problem. We need a few definitions and concepts before we describe the data structures.

The dual of a point $(a_1, \ldots, a_d) \in \mathbb{R}^d$ is the hyperplane $x_d = -a_1x_1 - \cdots - a_{d-1}x_{d-1} + a_d$, and the dual of a hyperplane $x_d = b_1x_1 + \cdots b_{d-1}x_{d-1} - b_d$ is the point $(-b_1, \ldots, -b_d)$. A nice property of duality is that it preserves the above-below relationship: a point p is above a hyperplane h if and only if the dual hyperplane p^* is above the dual point h^*; see Figure 6.

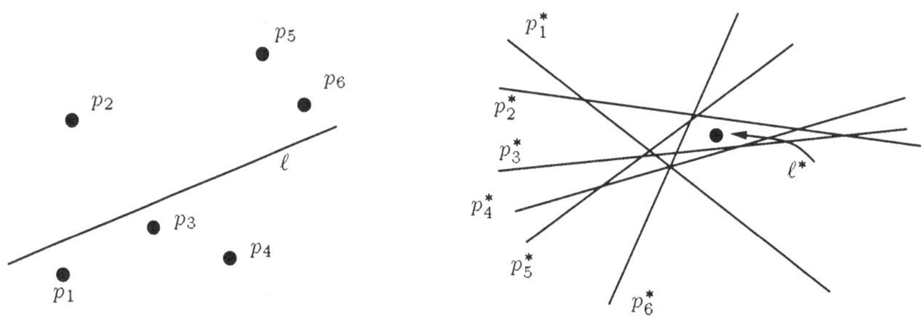

FIGURE 6. A set of points and the arrangement of their dual lines.

The *arrangement* of a set H of hyperplanes in $\mathbb{R}^d$ is the subdivision of $\mathbb{R}^d$ into cells of dimensions k, for $0 \le k \le d$, each cell being a maximal connected set contained in the intersection of a fixed subset of H and not intersecting any other hyperplane of H. The level of a point in $\mathcal{A}(H)$ is the number of hyperplanes lying strictly below the point. Let $\mathcal{A}_{\le k}(H)$ denote the (closure of the) set of points with level at most k. A $(1/r)$-*cutting* of H is a set Ξ of (relatively open) disjoint simplices covering $\mathbb{R}^d$ so that the interior of each simplex intersects at most n/r hyperplanes of H. Clarkson [**Cl**] and Haussler and Welzl [**HW**] were the first to show the existence of a $(1/r)$-cutting of H of size $O(r^d \log^d r)$. Chazelle and Friedman [**CF**] improved the size bound to $O(r^d)$, which is optimal in the worst case. Several efficient algorithms are developed for computing a $(1/r)$-cutting. The best algorithm known for computing a $(1/r)$-cutting was discovered by Chazelle [**Ch8**]; his result is summarized in the following theorem.

THEOREM 5 (Chazelle [**Ch8**]). *Let H be a set of n hyperplanes and $r \le n$ a parameter. Set $k = \lceil \log_2 r \rceil$. There exist k cuttings $\Xi_1, \ldots, \Xi_k$ so that Ξ_i is a $(1/2^i)$-cutting of size $O(2^{id})$, each simplex of Ξ_i is contained in a simplex of Ξ_{i-1}, and each simplex of Ξ_{i-1} contains a constant number of simplices of Ξ_i. Moreover, $\Xi_1, \ldots, \Xi_k$ can be computed in time $O(nr^{d-1})$.*

This theorem has been successfully applied to many geometric divide-and-conquer algorithms; see [**A, Ch8, dBGH, Pe2**] for a few such instances.

Returning to halfspace range searching, suppose that the query halfspace always lies below its bounding hyperplane. Then the halfspace range-counting problem reduces via duality to the following problem: Given a set H of n hyperplanes in

$\mathbb{R}^d$, determine the number of hyperplanes of H that lie above a query point. Since the same subset of hyperplanes lies above all points in a single cell of $\mathcal{A}(H)$, the arrangement of H, we can answer a halfspace range-counting query by locating the cell of $\mathcal{A}(H)$ that contains the point dual to the hyperplane bounding the query halfspace. Theorem 5 can be used in a straightforward manner to obtain a data structure of size $O((n/\log n)^d)$ with $O(\log n)$ query time.

The above approach for halfspace range counting can be extended to the simplex range-counting problem as well. That is, store the solution of every combinatorially distinct simplex (two simplices are combinatorially distinct if they do not contain the same subset of S). Since there are $\Theta(n^{d(d+1)})$ combinatorially distinct simplices, such an approach will require $\Omega(n^{d(d+1)})$ storage; see **[CY, EKM]**.

Cole and Yap **[CY]** were the first to present a near-quadratic size data structure that could answer a triangle range-counting query in the plane in $O(\log n)$ time. They present two data structures: the first one answers a query in time $O(\log n)$ using $O(n^{2+\varepsilon})$ space, and the other in time $O(\log n \log\log n)$ using $O(n^2/\log n)$ space. For $d = 3$, their approach gives a data structure of size $O(n^{7+\varepsilon})$ that can answer a tetrahedron range-counting query in time $O(\log n)$. Chazelle *et al.* **[CSW]** describe a multi-level data structure (see Section 5.1) of size $O(n^{d+\varepsilon})$ that can answer a simplex range-counting query in time $O(\log n)$. The space bound can be reduced to $O(n^d)$ by increasing the query time to $O(\log^{d+1} n)$ **[M6]**. Both data structures can answer simplex range-reporting queries by spending an additional $O(k)$ time.

The size of a data structure can be significantly improved if we want to answer halfspace range-reporting queries. Using random sampling, Clarkson **[Cl]** showed that a halfspace-emptiness query can be answered in $O(\log n)$ time using $O(n^{\lfloor d/2 \rfloor+\varepsilon})$ space. In order to extend his algorithm to halfspace range-reporting queries, we need the following additional idea. Let H be a set of hyperplanes in $\mathbb{R}^d$. For a parameter $1 \le r < n$, we define a $(1/r)$*-cutting* for $\mathcal{A}_{\le l}(H)$ to be a collection Ξ of (relatively open) disjoint simplices that cover $\mathcal{A}_{\le l}(H)$ and the interior of each simplex intersects at most n/r hyperplanes of H. The following theorem by Matoušek **[M2]** leads to a better data structure for answering halfspace range-reporting queries.

THEOREM 6 (Matoušek **[M2]**). *Let H be a collection of n hyperplanes in $\mathbb{R}^d$, let $1 \le l, r < n$ be parameters, and let $q = lr/n + 1$. Then there exists a $(1/r)$-cutting for $\mathcal{A}_{\le l}(H)$, consisting of $O(r^{\lfloor d/2 \rfloor} q^{\lceil d/2 \rceil})$ simplices. If $r \le n^\alpha$ for some suitable constant $0 < \alpha < 1$, then Ξ can be computed in $O(n \log r)$ time.*

Using Theorem 6, a halfspace range-reporting data structure T can be constructed as follows. Each interior node v of T is associated with a canonical subset $C_v \subseteq H$ and a simplex Δ_v; the root of T is associated with H and $\mathbb{R}^d$. Choose r to be a sufficiently large constant. If $|C_v| \le 4r$, then v is a leaf. Otherwise, set $l = |C_v|/r$, compute a $(1/r)$-cutting Ξ_v of size $O(r^{\lfloor d/2 \rfloor})$ for $\mathcal{A}_{\le l}(C_v)$, and create a child w_i for each $\Delta_i \in \Xi_v$. Set C_{w_i} to be the set of hyperplanes that either intersect or lie below Δ_i. We also store C_v at v. The size of the data structure is $O(n^{\lfloor d/2 \rfloor+\varepsilon})$. Let γ be a query point. The goal is to report all points lying above γ. Follow a path of T as follows. Suppose the query procedure is visiting a node v of T. If v is a leaf or γ does not lie in any simplex of Ξ_v (*i.e.*, the level of γ is at least $|C_v|/r$), then report all hyperplanes of C_v lying above γ, by checking each hyperplane explicitly; this step takes $O(|C_v|) = O(kr) = O(k)$ time. Otherwise, recursively visit the node

w_i if Δ_i contains γ. The query time is obviously $O(\log n + k)$. The size of the data structure can be improved to $O(n^{\lfloor d/2 \rfloor} \operatorname{polylog} n)$ without affecting the asymptotic query time.

4.3. Trading space for query time. In the previous two subsections we surveyed data structures for simplex range searching that either use near-linear space or answer a query in polylogarithmic time. By combining these two types of data structures, a tradeoff between the size and the query time can be obtained [**AS, CSW, M6**]. Actually, the approach described in these papers is very general and works for any geometric-searching data structure that can be viewed as a decomposition scheme (described in Section 2), provided it satisfies certain assumptions. We state the general result here, though one can obtain a slightly better bounds (by a polylogarithmic factor) by exploiting special properties of the data structures.

It will be convenient to regard range-searching data structures in the following abstract form, previously described at the end of Section 2. Let $\mathcal{P}$ be a d-dimensional range-searching problem and $\mathcal{D}$ a decomposition scheme for $\mathcal{P}$. That is, for a given set S of n points in $\mathbb{R}^d$, $\mathcal{D}$ constructs a family (multiset) $\mathcal{F} = \mathcal{F}(S)$ of canonical subsets. For a query range γ, the query procedure implicitly computes a sub-family $\mathcal{C}_\gamma = \mathcal{C}(\gamma, S) \subseteq \mathcal{F}$ that partitions $\gamma \cap S$ into canonical subsets, and returns $\sum_{C \in \mathcal{C}_\gamma} w(C)$.

As we mentioned in Section 2, in order to compute $\mathcal{C}_\gamma$ efficiently, $\mathcal{D}$ must be stored in a hierarchical data structure. We call a decomposition scheme *hierarchical* if $\mathcal{F}$ is stored in a tree T. Each node v of T is associated with a canonical subset $C_v \in \mathcal{F}$ and each interior node v satisfies the following property.

(P1) For any query range γ, there exists a subset $Q(v, \gamma) = \{z_1, \ldots, z_a\}$ of children of v so that $\gamma \cap C_{z_1}, \ldots, \gamma \cap C_{z_a}$ partition $\gamma \cap C_v$.

For example, the linear-size partition trees described in Section 4.1 store a simplex Δ_v at each node v. In these partition trees, a child z of a node v is in $Q(v, \gamma)$, for any query halfspace γ, if Δ_z intersects the halfspace γ.

Property (P1) ensures that, for a node v, $w(C_v)$ can be computed by searching only in the subtree rooted at v. The query procedure performs a depth-first search on T to compute $\mathcal{C}_\gamma$. Let $\overline{\mathcal{C}}_\gamma = \overline{\mathcal{C}}(\gamma, S)$ denote the canonical subsets in $\mathcal{F}$ associated with nodes visited by the query procedure; clearly, $\mathcal{C}_\gamma \subseteq \overline{\mathcal{C}}_\gamma$.

Let $r \geq 2$ be a parameter and let $\mathcal{D}$ be a hierarchical decomposition scheme. For any $0 \leq i \leq \lceil \log_r n \rceil$, let $\mathcal{F}_i = \{C \in \mathcal{F} \mid r^i \leq |C| < r^{i+1}\}$. We say that $\mathcal{D}$ is *r-convergent* if there exist constants $\alpha \geq 1$ and $0 \leq \beta < 1$ so that the following three conditions hold for all i.

(C1) The degree of each node in T is $O(r^\alpha)$.

(C2) $|\mathcal{F}_i| = O\left((n/r^i)^\alpha\right)$.

(C3) For any query range γ, $|\overline{\mathcal{C}}_\gamma \cap \mathcal{F}_i| = O\left((n/r^i)^\beta\right)$.

The second and third conditions imply that the number of canonical subsets in $\mathcal{F}$ and the the number of subsets in $\overline{\mathcal{C}}_\gamma$, for any query range γ, decrease exponentially with size.

The size of $\mathcal{D}$ is $O(n^\alpha)$, provided the weight of each canonical subset can be stored in $O(1)$ space, and the query time of $\mathcal{D}$, under the semigroup model, is $O(n^\beta)$

if $\beta > 0$ and $O(\log n)$ if $\beta = 0$. $\mathcal{D}$ is called *efficient* if for any query range γ, each $\mathcal{C}_\gamma \cap \mathcal{F}_i$ can be computed in time $O\left((n/r^i)^\beta\right)$.

THEOREM 7. *Let S be a set of n points in $\mathbb{R}^d$, and let r be a sufficiently large constant. Let $\mathcal{P}$ be a range-searching problem. Let D^1 be a decomposition scheme for $\mathcal{P}$ of size $O(n^\alpha)$ and query time $O(\log n)$, and let $\mathcal{D}^2$ be another decomposition scheme of size $O(n)$ and query time $O(n^\beta)$, for some constants $\alpha > 1$ and $\beta > 0$. If either $\mathcal{D}^1$ or $\mathcal{D}^2$ is hierarchical, efficient, and r-convergent, then for any $n \le m \le n^\alpha$, we can construct a decomposition scheme for $\mathcal{P}$ of size $O(m)$ and query time*

$$O\left(\left(\frac{n^\alpha}{m}\right)^{\beta/(\alpha-1)} + \log\frac{m}{n}\right).$$

PROOF. Suppose $\mathcal{D}^1$ is hierarchical, efficient, and r-convergent. We present a decomposition scheme $\mathcal{D}$ of size $O(m)$. We first define the canonical subsets $\mathcal{F}(S)$ constructed by $\mathcal{D}$ and then define $\mathcal{C}(\gamma, S)$ for each range γ.

Let $\mathcal{F}^1 = \mathcal{F}^1(S)$ be the family of canonical subsets constructed by $\mathcal{D}^1$ on S and T^1 be the corresponding tree. We define a parameter

$$\tau = 1 + \left\lceil \frac{\log_r(n^\alpha/m)}{\alpha - 1} \right\rceil.$$

Informally, to construct $\mathcal{F}$, we discard all nodes in T^1 whose parents are associated with canonical subsets of size less than r^τ. Then we replace the deleted subsets by constructing, for for every leaf z of the pruned tree, the canonical subsets $\mathcal{F}^2(C_z)$ using the second decomposition scheme $\mathcal{D}^2$. See Figure 7.

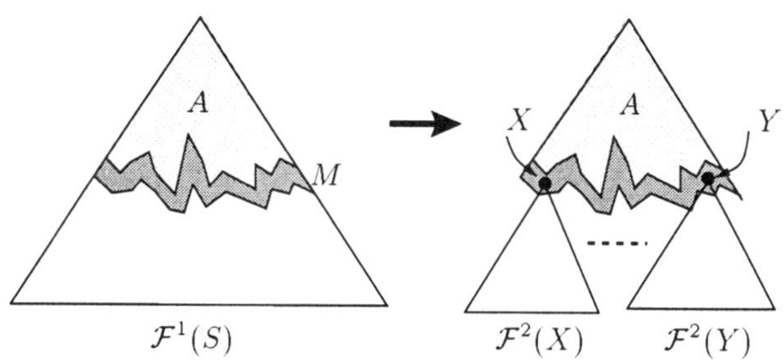

FIGURE 7. The general space query-time tradeoff scheme.

More formally, let $A = \bigcup_{i \ge \tau} \mathcal{F}_i^1$, and let $M \subseteq \mathcal{F}^1 \setminus A$ be the set of canonical subsets whose predecessors lie in A. Since $\mathcal{D}^1$ is r-convergent,

$$|A| = \sum_{i \ge \tau} |\mathcal{F}_i^1| = \sum_{i \ge \tau} O\left(\left(\frac{n}{r^i}\right)^\alpha\right) = O\left(\left(\frac{n}{r^\tau}\right)^\alpha\right).$$

The degree of each node in T^1 is $O(r^\alpha)$, so

$$|M| = O(r^\alpha) \cdot |A| = O\left(\left(\frac{n}{r^{\tau-1}}\right)^\alpha\right).$$

For each canonical subset $C \in M$, we compute $\mathcal{F}^2(C)$ using the second decomposition scheme $\mathcal{D}^2$. The size of each subset in M is at most $r^{\tau-1}$, so $|\mathcal{F}^2(C)| = O(r^{\tau-1})$. Set

$$\mathcal{F}(S) = A \cup \bigcup_{C \in M} \mathcal{F}^2(C).$$

The total number of canonical subsets in $\mathcal{F}(S)$ is

$$|A| + \sum_{C \in M} |\mathcal{F}^2(C)| = O\left(\frac{n^\alpha}{r^{\tau\alpha}}\right) + O\left(\frac{n^\alpha}{r^{(\tau-1)\alpha}}\right) \cdot O(r^{\tau-1})$$
$$= O\left(\frac{n^\alpha}{r^{(\tau-1)(\alpha-1)}}\right) = O(m).$$

For a query range γ, let $M_\gamma = M \cap \overline{\mathcal{C}}^1(\gamma, S)$ and $A_\gamma = A \cap \mathcal{C}^1(\gamma, S)$. We now define $\mathcal{C}(\gamma, S)$ as follows.

$$\mathcal{C}(\gamma, S) = A_\gamma \cup \bigcup_{C \in M_\gamma} \mathcal{C}^2(\gamma, C).$$

It can be shown that $\mathcal{C}(\gamma, S)$ forms a partition of $\gamma \cap S$. Since $\mathcal{D}^1$ is efficient, A_γ and M_γ can be computed in time $O(\log(n/r^\tau)) = O(\log(m/n))$. The size of each canonical subset $C \in M_\gamma$ is at most $r^{\tau-1}$; therefore, each $\mathcal{C}^2(\gamma, C)$ can be computed in time $O(r^{\beta(\tau-1)}) = O((n^\alpha/m)^{\beta/(\alpha-1)})$. By condition (C3), $|M_\gamma| = O(1)$, so the overall query time is

$$O\left(\left(\frac{n^\alpha}{m}\right)^{\beta/(\alpha-1)} + \log\frac{m}{n}\right),$$

as desired.

A similar approach can be used to construct $\mathcal{D}$ if $\mathcal{D}^2$ is r-convergent and efficient. We omit further details. □

For the d-dimensional simplex range-counting problem, for example, we have $\alpha = d + \varepsilon$ and $\beta = 1 - 1/d$. Thus, we immediately obtain the following space query-time tradeoff.

COROLLARY 8. *For any $n \leq m \leq n^{d+\varepsilon}$, a simplex range-counting query can be answered in time $O(n^{1+\varepsilon/d}/m^{1/d} + \log(m/n))$ using $O(m)$ space.*

We conclude this section by making a few remarks on Theorem 7.

(i) Theorem 7 can be refined to balance polylogarithmic factors in the sizes and query times of $\mathcal{D}^1$ and $\mathcal{D}^2$. For example, if the size of $\mathcal{D}^1$ is $O(n^\alpha \operatorname{polylog} n)$ and rest of the parameters are the same as in the theorem, then the query time of the new data structure is

$$O\left(\left(\frac{n^\alpha}{m}\right)^{\beta/(\alpha-1)} \operatorname{polylog}\left(\frac{m}{n}\right)\right).$$

Using a similar argument, Matoušek **[M6]** showed that a simplex range-counting query can be answered in time $O((n/m^{1/d})\log^{d+1}(m/n))$, which improves Corollary 8 whenever $m = O(n^d)$.

(ii) Theorem 7 is quite general and holds for any decomposable geometric searching problem as long as there exists an efficient, r-convergent decomposition scheme for the problem. We will discuss some such results in the next two sections.

(iii) Theorem 7 actually holds under weaker assumptions on $\mathcal{D}^1$ and $\mathcal{D}^2$. For example, even though halfspace range-reporting data structures do not fit in the above framework, they nevertheless admit a tradeoff. In particular, a halfspace reporting query in $\mathbb{R}^d$ can be answered in

$$O((n \operatorname{polylog} n)/m^{1/\lfloor d/2 \rfloor} + k)$$

time using $O(m)$ space.

(iv) Finally, it is not essential for $\mathcal{D}_1$ or $\mathcal{D}_2$ to be tree-based data structures. It is sufficient to have an efficient, r-convergent decomposition scheme with a partial order on the canonical subsets, where each canonical subset satisfies a property similar to (C1).

4.4. Lower bounds. Fredman [**Fr5**] showed that a sequence of n insertions, deletions, and halfplane queries on a set of points in the plane requires $\Omega(n^{4/3})$ time, in the semigroup model. His technique, however, does not extend to static data structures. In a series of papers, Chazelle has proved nontrivial lower bounds on the complexity of online simplex range searching, using various elegant mathematical techniques. The following theorem is perhaps the most interesting result on lower bounds.

THEOREM 9 (Chazelle [**Ch5**]). *Let n, m be positive integers such that $n \le m \le n^d$, and let S be a random set of points in $[0,1]^d$. If only m units of storage are available, then with high probability, the worst-case query time for a simplex range query in S is $\Omega(n/\sqrt{m})$ for $d = 2$, or $\Omega(n/(m^{1/d} \log n))$ for $d \ge 3$, in the semigroup model.*

It should be pointed out that this theorem holds even if the query ranges are wedges or strips, but not if the ranges are hyperplanes. Chazelle and Rosenberg [**CR2**] proved a lower bound of $\Omega(n^{1-\varepsilon}/m + k)$ for simplex range reporting under the pointer-machine model. These lower bounds do not hold for halfspace range searching. A somewhat weaker lower bound for halfspace queries was proved by Brönnimann *et al.*[**BCP**].

As we saw earlier, faster data structures are known for halfspace emptiness queries. A recent series of papers by Erickson established the first nontrivial lower bounds for online and offline emptiness query problems, in the partition-graph model of computation. His techniques were first applied to Hopcroft's problem—Given a set of n points and m lines, does any point lie on a line?—for which he obtained a lower bound of $\Omega(n \log m + n^{2/3}m^{2/3} + m \log n)$ [**Er2**], almost matching the best known upper bound $O(n \log m + n^{2/3}m^{2/3}2^{O(\log^*(n+m))} + m \log n)$, due to Matoušek [**M6**]. Slightly better lower bounds are known for higher-dimensional versions of Hopcroft's problem [**Er2, Er**], but for the special case $n = m$, the best known lower bound is still only $\Omega(n^{4/3})$, which is quite far from the best known upper bound $O(n^{2d/(d+1)}2^{O(\log^* n)})$. More recently, Erickson established lower bounds for the tradeoff between space and query time, or preprocessing and query time, for online hyperplane emptiness queries [**Er3**]. The space-time tradeoffs are established by showing that a partition graph that supports hyperplane emptiness queries also (implicitly) supports halfspace semigroup queries, and then applying the lower bounds of Brönnimann *et al.*[**BCP**]. For d-dimensional hyperplane queries, $\Omega(n^d/\operatorname{polylog} n)$ preprocessing time is required to achieve polylogarithmic query time, and the best possible query time is $\Omega(n^{1/d}/\operatorname{polylog} n)$ if only $O(n \operatorname{polylog} n)$ preprocessing time is allowed. More generally, in two dimensions, if the preprocessing time is p, the query time is $\Omega(n/\sqrt{p})$. Erickson's techniques also imply nontrivial lower bounds for online and offline halfspace emptiness searching, but with a few exceptions, these are quite weak.

Table 4 summarizes the best known lower bounds for online simplex queries, and Table 5 summarizes the best known lower bounds for offline simplex range searching. Lower bounds for emptiness problems apply to counting and reporting problems as well. No nontrivial lower bound was known for any offline range searching problem under the group model until Chazelle's result **[Ch10]**.

Range	Problem	Model	Query Time	Source
Simplex	Semigroup	Semigroup ($d = 2$)	$\frac{n}{\sqrt{m}}$	**[Ch5]**
	Semigroup	Semigroup ($d > 2$)	$\frac{n}{m^{1/d}\log n}$	**[Ch5]**
	Reporting	Pointer machine	$\frac{n^{1-\varepsilon}}{m^{1/d}} + k$	**[CR2]**
Hyperplane	Semigroup	Semigroup	$\left(\frac{n}{m^{1/d}}\right)^{2/(d+1)}$	**[Er3]**
	Emptiness	Partition graph	$\left(\frac{n}{\log n}\right)^{\frac{d^2+1}{d^2+d}} \cdot \frac{1}{m^{1/d}}$	**[Er3]**
Halfspace	Semigroup	Semigroup	$\left(\frac{n}{\log n}\right)^{\frac{d^2+1}{d^2+d}} \cdot \frac{1}{m^{1/d}}$	**[BCP]**
	Emptiness	Partition graph	$\left(\frac{n}{\log n}\right)^{\frac{\delta^2+1}{\delta^2+\delta}} \cdot \frac{1}{m^{1/\delta}}$, where $d \geq \delta(\delta+3)/2$	**[Er3]**

TABLE 4. Asymptotic lower bounds for online simplex range searching using $O(m)$ space.

Range	Problem	Model	Time	Source
Halfspace	Emptiness	Algebraic computation tree	$n \log n$	**[BO]**
		Partition graph ($d \leq 4$)	$n \log n$	**[Er]**
		Partition graph ($d \geq 5$)	$n^{4/3}$	**[Er]**
	Counting	Partition graph	$n^{4/3}$	**[Er2]**
	Group	Group (with $n/2$ help gates)	$n \log n$	**[Ch10]**
Hyperplane	Emptiness	Algebraic computation tree	$n \log n$	**[BO]**
		Partition graph	$n^{4/3}$	**[Er2]**
	Semigroup	Semigroup	$n^{4/3}$	**[Ch11, Er2]**
Simplex	Semigroup	Semigroup	$\frac{n^{2-2/(d-1)}}{\log^{5/2} n}$	**[Ch11]**

TABLE 5. Asymptotic lower bounds for offline simplex range searching.

See the survey papers **[Ch9, M7]** for a more detailed discussion on lower bounds.

5. Variants and extensions

In this section we review some extensions of range-searching data structures, including multi-level data structures, semialgebraic range searching, and dynamization. As in the previous section, the preprocessing time for each of the data structures we describe is at most a polylogarithmic or n^ε factor larger than its size.

5.1. Multi-level data structures. A rather powerful property of data structures based on decomposition schemes (described in Section 2) is that they can be cascaded together to answer more complex queries, at the increase of a logarithmic factor in their performance. This property has been implicitly used for a long time; see, for example, [**EM, LW, Lu, WL, SO**]. The real power of the cascading property was first observed by Dobkin and Edelsbrunner [**DoE**], who used this property to answer several complex geometric queries. Since their result, several papers have exploited and extended this property to solve numerous geometric-searching problems; see [**AS, GOS, vK, M6, Pe**]. In this subsection we briefly sketch the general cascading scheme, as described in [**M6**].

Let S be a set of weighted objects. Recall that a geometric-searching problem $\mathcal{P}$, with underlying relation $\Diamond$, requires computing $\sum_{p \Diamond \gamma} w(p)$ for a query range γ. Let $\mathcal{P}^1$ and $\mathcal{P}^2$ be two geometric-searching problems with the same sets of objects and ranges, and let $\Diamond^1$ and $\Diamond^2$ be the corresponding relations. Then we define $\mathcal{P}^1 \circ \mathcal{P}^2$ to be the conjunction of $\mathcal{P}^1$ and $\mathcal{P}^2$, whose relation is $\Diamond^1 \cap \Diamond^2$. That is, for a query range γ, we want to compute $\sum_{p \Diamond^1 \gamma, p \Diamond^2 \gamma} w(p)$. Suppose we have hierarchical decomposition schemes $\mathcal{D}^1$ and $\mathcal{D}^2$ for problems $\mathcal{P}^1$ and $\mathcal{P}^2$. Let $\mathcal{F}^1 = \mathcal{F}^1(S)$ be the set of canonical subsets constructed by $\mathcal{D}^1$, and for a range γ, let $\mathcal{C}^1_\gamma = \mathcal{C}^1(S, \gamma)$ be the corresponding partition of $\{p \in S \mid p \Diamond^1 \gamma\}$ into canonical subsets. For each canonical subset $C \in \mathcal{F}^1$, let $\mathcal{F}^2(C)$ be the collection of canonical subsets of C constructed by $\mathcal{D}^2$, and let $\mathcal{C}^2(C, \gamma)$ be the corresponding partition of $\{p \in C \mid p \Diamond^2 \gamma\}$ into level-two canonical subsets. The decomposition scheme $\mathcal{D}^1 \circ \mathcal{D}^2$ for the problem $\mathcal{P}^1 \circ \mathcal{P}^2$ consists of the canonical subsets $\mathcal{F} = \bigcup_{C \in \mathcal{F}^1} \mathcal{F}^2(C)$. For a query range γ, the query output is $\mathcal{C}_\gamma = \bigcup_{C \in \mathcal{C}^1_\gamma} \mathcal{C}^2(C, \gamma)$. Note that we can cascade any number of decomposition schemes in this manner.

If we view $\mathcal{D}^1$ and $\mathcal{D}^2$ as tree data structures, then cascading the two decomposition schemes can be regarded as constructing a two-level tree, as follows. We first construct the tree induced by $\mathcal{D}^1$ on S. Each node v of $\mathcal{D}^1$ is associated with a canonical subset C_v. We construct a second-level tree $\mathcal{D}^2_v$ on C_v and store $\mathcal{D}^2_v$ at v as its secondary structure. A query is answered by first identifying the nodes that correspond to the canonical subsets $C_v \in \mathcal{C}^1_\gamma$ and then searching the corresponding secondary trees to compute the second-level canonical subsets $\mathcal{C}^2(C_v, \gamma)$.

The range tree, defined in Section 3.1, fits in this framework. For example, a two dimensional range tree is obtained by cascading two one-dimensional range trees, as follows. Let S be a set of n weighted points and $\mathcal{R}$ the set of all orthogonal rectangles in the plane. We define two binary relations $\Diamond^1$ and $\Diamond^2$, where for any rectangle $\gamma = [\alpha_1, \beta_1] \times [\alpha_2, \beta_2]$, $p \Diamond^i \gamma$ if $x_i(p) \in [\alpha_i, \beta_i]$. Let $\mathcal{P}^i$ be the searching problem associated with $\Diamond^i$, and let $\mathcal{D}^i$ be the data structure corresponding to $\mathcal{P}^i$. Then the two-dimensional orthogonal range-searching problem is the same as $\mathcal{P}^1 \circ \mathcal{P}^2$. We can therefore cascade $\mathcal{D}^1$ and $\mathcal{D}^2$, as described above, to answer a two-dimensional orthogonal range-searching query. Similarly, a data structure for d-dimensional simplex range-searching can be constructed by cascading $d+1$

halfspace range-searching structures, since a d-simplex is an intersection of at most $d+1$ halfspaces. Multi-level data structures were also proposed for *range restriction*, introduced by Willard and Lueker [**WL**] and Scholten and Overmars [**SO**].

The following theorem, whose straightforward proof we omit, states a general result for multi-level data structures.

THEOREM 10. *Let $S, \mathcal{P}^1, \mathcal{P}^2, \mathcal{D}^1, \mathcal{D}^2$ be as defined above, and let r be a constant. Suppose the size and query time of each decomposition scheme are at most $S(n)$ and $Q(n)$, respectively. If $\mathcal{D}^1$ is efficient and r-convergent, then we obtain a hierarchical decomposition scheme $\mathcal{D}$ for $\mathcal{P}^1 \circ \mathcal{P}^2$ whose size and query time are $O(S(n)\log_r n)$ and $O(Q(n)\log_r n)$. If $\mathcal{D}^2$ is also efficient and r-convergent, then $\mathcal{D}$ is also efficient and r-convergent.*

In some cases, the added logarithmic factor in the query time or the space can be saved. The real power of multi-level data structures stems from the fact that there are no restrictions on the relations $\lozenge^1$ and $\lozenge^2$. Hence, any query that can be represented as a conjunction of a constant number of "primitive" queries, each of which admits an efficient, r-convergent decomposition scheme, can be answered by cascading individual decomposition schemes. We will describe a few multi-level data structures in this and the following sections.

5.2. Semialgebraic range searching. So far we assumed that the ranges were bounded by hyperplanes, but many applications involve ranges bounded by nonlinear functions. For example, a query of the form "For a given point p and a real number r, find all points of S lying within distance r from p" is a range-searching problem in which ranges are balls.

As shown below, range searching with balls in $\mathbb{R}^d$ can be formulated as an instance of halfspace range searching in $\mathbb{R}^{d+1}$. So a ball range-reporting (resp. range-counting) query in $\mathbb{R}^d$ can be answered in time $O((n/m^{1/\lceil d/2 \rceil}) \operatorname{polylog} n + k)$ (resp. $O((n/m^{1/(d+1)}) \log(m/n))$), using $O(m)$ space. (Somewhat better performance can be obtained using a more direct approach, which we will describe shortly.) However, relatively little is known about range-searching data structures for more general ranges.

A natural class of more general ranges is the family of Tarski cells. A *Tarski cell* is a real semialgebraic set defined by a constant number of polynomials, each of constant degree. In fact, it suffices to consider the ranges bounded by a single polynomial because the ranges bounded by multiple polynomials can be handled using multi-level data structures. We assume that the ranges are of the form

$$\gamma_f(a) = \{x \in \mathbb{R}^d \mid f(a, x) \geq 0\},$$

where f is a $(d+b)$-variate polynomial specifying the type of range (disks, cylinders, cones, etc.), and a is a b-tuple specifying a specific range of the given type (*e.g.*, a specific disk). Let $\Gamma_f = \{\gamma_f(a) \mid a \in \mathbb{R}^b\}$. We will refer to the range-searching problem in which the ranges are from the set Γ_f as the *Γ_f-range searching*.

One approach to answer Γ_f-range queries is to use *linearization*, originally proposed by Yao and Yao [**YY**]. We represent the polynomial $f(a, x)$ in the form

$$f(a, x) = \psi_0(a)\varphi_0(x) + \psi_1(a)\varphi_1(x) + \cdots + \psi_\ell(a)\varphi_\ell(x)$$

where $\varphi_0, \ldots, \varphi_\ell, \psi_0, \ldots, \psi_\ell$ are polynomials. A point $x \in \mathbb{R}^d$ is mapped to the point

$$\varphi(x) = [\varphi_0(x), \varphi_1(x), \varphi_2(x), \ldots, \varphi_\ell(x)] \in \mathbb{R}^\ell,$$

represented in homogeneous coordinates. Then each range $\gamma_f(a) = \{x \in \mathbb{R}^d \mid f(x,a) \geq 0\}$ is mapped to a halfspace

$$\psi^{\#}(a) : \{y \in \mathbb{R}^\ell \mid \psi_0(a)y_0 + \psi_1(a)y_1 + \cdots + \psi_\ell(a)y_\ell \geq 0\},$$

where, again, $[y_0, y_1, \ldots, y_\ell]$ are homogeneous coordinates.

The constant ℓ is called the *dimension* of the linearization. The following algorithm, based on an algorithm of Agarwal and Matoušek **[AgM2]**, computes a linearization of smallest dimension.[7] Write the polynomial $f(a,x)$ as the sum of monomials

$$f(a,x) = \sum_{\mu \in M} \sum_{\nu \in N} c_{\mu,\nu} a^\mu x^\nu,$$

where $M \subset \mathbb{N}^b$ and $N \subset \mathbb{N}^d$ are finite sets of exponent vectors, $c_{\mu,\nu}$ are real coefficients, and a^μ and x^ν are shorthand for the monomials $a_1^{\mu_1} a_2^{\mu_2} \ldots a_d^{\mu_b}$ and $x_1^{\nu_1} x_2^{\nu_2} \ldots x_d^{\nu_d}$, respectively. Collect the coefficients $c_{\mu,\nu}$ into a matrix C whose rows are indexed by elements of M (*i.e.*, monomials in a) and whose columns are indexed by elements of N (*i.e.*, monomials in x). The minimum dimension of linearization is one less than the rank of this matrix. The polynomials $\varphi_i(x)$ and $\phi_j(a)$ are easily extracted from any basis of the vector space spanned by either the rows or columns of the coefficient matrix C.

For example, a disk with center (a_1, a_2) and radius a_3 in the plane can be regarded as a set of the form $\gamma_f(a)$, where $a = (a_1, a_2, a_3)$ and f is a 5-variate polynomial

$$f(a_1, a_2, a_3; x_1, x_2) = -(x_1 - a_1)^2 - (x_2 - a_2)^2 + a_3^2$$

This polynomial has the following coefficient matrix.

	1	x_1	x_2	x_1^2	x_2^2
1	0	0	0	−1	−1
a_1	0	2	0	0	0
a_2	0	0	2	0	0
a_1^2	−1	0	0	0	0
a_2^2	−1	0	0	0	0
a_3^2	1	0	0	0	0

This matrix has rank 4, so the linearization dimension of f is 3. One possible linearization is given by the following set of polynomials:

$$\begin{aligned} &\psi_0(a) = -a_1^2 - a_2^2 + a_3^2, && \psi_1(a) = 2a_1, && \psi_2(a) = 2a_2, && \psi_3(a) = -1, \\ &\varphi_0(x) = 1, && \varphi_1(x) = x_1, && \varphi_2(x) = x_2, && \varphi_3(x) = x_1^2 + x_2^2. \end{aligned}$$

In general, balls in $\mathbb{R}^d$ admit a linearization of dimension $d+1$; cylinders and other quadrics in $\mathbb{R}^3$ admit a linearization of dimension 9. One of the most widely used linearizations in computational geometry uses the so-called *Plücker coordinates*, which map a line in $\mathbb{R}^3$ to a point in $\mathbb{R}^5$; see **[CEGSS, So, St]** for more details on Plücker coordinates.

A Γ_f-range query can now be answered using a ℓ-dimensional halfspace range-searching data structure. Thus, for counting queries, we immediately obtain a linear-size data structure with query time $O(n^{1-1/\ell})$ **[M6]**, or a data structure of size $O(n^\ell / \log^\ell n)$ with logarithmic query time **[Ch8]**. When $d < \ell$, the performance

[7] In some cases, Agarwal and Matoušek's algorithm returns a dimension one higher than the true minimum, since they consider only linearizations with $\psi_0(a) = 1$.

of the linear-size data structures can be improved by exploiting the fact that the points $\varphi(x)$ have only d degrees of freedom. Using results of Aronov *et al.* [**APS**] on the size of the zone of an algebraic variety in a k-dimensional hyperplane arrangement, Agarwal and Matoušek [**AgM2**] show that the query time for a linear-space data structure can be reduced to $O(n^{1-1/\lfloor (d+\ell)/2 \rfloor+\varepsilon})$. It is an open problem whether one can similarly exploit the fact that the halfspaces $\psi^{\#}(a)$ have only b degrees of freedom to reduce the size of data structures with logarithmic query time when $b < \ell$.

In cases where the linearization dimension is very large, semialgebraic queries can also be answered using the following more direct approach proposed by Agarwal and Matoušek [**AgM2**]. Let S be a set of n points in $\mathbb{R}^d$. For each point p_i, we can define a b-variate polynomial $g_i(a) \equiv f(p_i, a)$. Then $\Gamma_f(a) \cap S$ is the set of points p_i for which $g_i(a) \geq 0$. Hence, the problem reduces to point location in the arrangement of algebraic surfaces $g_i = 0$ in $\mathbb{R}^b$. Let G be the set of resulting surfaces. The following result of Chazelle *et al.* [**CEGS, CEGS2**] leads to a point-location data structure.

THEOREM 11 (Chazelle *et al.* [**CEGS**]). *Let $\mathcal{F} = \{f_1, \ldots, f_n\}$ be a set of n d-variate polynomials, with $d \geq 3$, where each f_i has maximum degree δ in any variable. Then $\mathbb{R}^d$ be partitioned into a set Ξ of $O(n^{2d-3}2^{\alpha(n)^{(2\delta)^{2^d-1}}})$ Tarski cells so that the sign of each f_i remains the same for all points within each cell of Ξ. Moreover, Ξ can be computed in $O(n^{2d-1}\log n)$ time.*

Improving the combinatorial upper bound in Theorem 11 is an open problem. The best known lower bound is $\Omega(n^d)$, and this is generally believed to be the right bound. Any improvement would also improve the bounds for the resulting semialgebraic range searching data structures.

Returning to the original point-location problem for g_i's, using this theorem and results on ε-nets and cuttings, G can be preprocessed into a data structure of size $O(n^{2b-3+\varepsilon})$ if $b \geq 3$, or $O(n^{2+\varepsilon})$ if $b = 2$, so that for a query point $a \in \mathbb{R}^b$, we can compute $\sum_{g_i(a) \geq 0} w(p_i)$ in $O(\log n)$ time.

Using Theorem 11, Agarwal and Matoušek [**AgM2**] also extended Theorem 3 to Tarski cells and showed how to construct partition trees using this extension, obtaining a linear-size data structure with query time $O(n^{1-1/\gamma+\varepsilon})$, where $\gamma = 2$ if $d = 2$ and $\gamma = 2d - 3$ if $d \geq 3$.

As in Section 4.3, the best data structures with linear space and logarithmic query time can be combined to obtain the following tradeoff between space and query time.

THEOREM 12. *Let $f : \mathbb{R}^d \times \mathbb{R}^b \to \mathbb{R}$ be a $(d+b)$-variate polynomial with linearization dimension ℓ. Let $\lambda = \min(2d-3, \lfloor (d+\ell)/2 \rfloor, \ell)$, and let $\gamma = \min(2b-3, \ell)$. For any $n \leq m \leq n^\gamma$, we can build a data structure of size $O(m)$ that supports Γ_f-counting queries in time*

$$O\left(\left(\frac{n^\gamma}{m}\right)^{(\lambda-1)/\lambda(\gamma-1)+\varepsilon} + \log\frac{m}{n}\right).$$

For example, if our ranges are balls in $\mathbb{R}^d$, we have $b = d+1$, $\ell = d+1$, $\lambda = d$, and $\gamma = d+1$, so we can answer queries in time $O((n^{d+1}/m)^{(d-1)/d^2+\varepsilon} + \log(m/n))$ using space $O(m)$.

5.3. Dynamization. All the data structures discussed above assumed S to be fixed, but in many applications one needs to update S dynamically—insert a new point into S or delete a point from S. We cannot hope to perform insert/delete operations on a data structure in less than $P(n)/n$ time, where $P(n)$ is the preprocessing time of the data structure. If we allow only insertions (*i.e.*, a point cannot be deleted from the structure), static data structures can be modified using standard techniques [**BS, Meh, Ov**], so that a point can be inserted in time $O(P(n)\log n/n)$ and a query can be answered in time $O(Q(n)\log n)$, where $Q(n)$ is the query time of the original static data structure. Roughly speaking, these techniques proceed as follows. Choose a parameter $r \geq 2$ and set $t = \lceil \log_r n \rceil$. Maintain a partition of S into t subsets $S_0, \ldots S_{t-1}$ so that $|S_i| \leq (r-1)r^i$, and preprocess each S_i for range searching separately. We call a subset S_i *full* if $|S_i| = (r-1)r^i$. A query is answered by computing $w(S_i \cap \gamma)$ for each subset S_i independently and then summing them up. The total time spent in answering a query is thus $O(t + \sum_{i=1}^{t} Q(r^i))$. Suppose we want to insert a point p. We find the least index i such that the subsets $S_0, \ldots, S_{i-1}$ are full. Then we add the point p and $\bigcup_{j<i} S_j$ to S_i, set $S_j = \emptyset$ for all $j < i$, and preprocess the new S_i for range searching. The amortized insertion time is $O(\sum_{i=0}^{t-1} P(r^i)/r^i)$. We can convert this amortized behavior into a worst-case performance using known techniques [**Sm**]. In some cases the logarithmic overhead in the query or update time can be avoided.

Although the above technique does not handle deletions, many range-searching data structures, such as orthogonal and simplex range-searching structures, can handle deletions at polylogarithmic or n^{ε} overhead in query and update time, by exploiting the fact that a point is stored at roughly $S(n)/n$ nodes [**AS**]. Table 6 summarizes the known results on dynamic 2D orthogonal range-searching data structures; these results can be extended to higher dimensions at a cost of an additional $\log^{d-2} n$ factor in the storage, query time, and update time. Klein *et al.* [**KNOW**] have described an optimal data structure for a special case of 2D range-reporting in which the query ranges are translates of a polygon.

Problem	Size	Query Time	Update Time	Source
Counting	n	$\log^2 n$	$\log^2 n$	[**Ch4**]
Reporting	n	$k\log^2(2n/k)$	$\log^2 n$	[**Ch4**]
	n	$n^{\varepsilon} + k$	$\log^2 n$	[**Sm2**]
	$n\log n$	$\log n \log\log n + k$	$\log n \log\log n$	[**MN**]
	$\frac{n\log n}{\log\log n}$	$\frac{\log^{2+\varepsilon} n}{\log\log n} + k$	$\frac{\log^2 n}{\log\log n}$	[**Sm2**]
Semigroup	n	$\log^4 n$	$\log^4 n$	[**Ch4**]

TABLE 6. Asymptotic upper bounds for dynamic 2D orthogonal range-searching.

Although Matoušek's $O(n\log n)$-size data structure for d-dimensional halfspace range reporting [**M2**] can be dynamized, the logarithmic query time data structure is not easy to dynamize because some of the points may be stored at $\Omega(n^{\lfloor d/2 \rfloor})$ nodes of the tree. Agarwal and Matoušek [**AgM3**] developed a rather sophisticated data structure that can insert or delete a point in time $O(n^{\lfloor d/2 \rfloor - 1 + \varepsilon})$ time and can answer a query in $O(\log n + k)$ time. As in [**CSW**], at each node of the tree, this structure computes a family of partitions (instead of a single partition), each of size

$O(r^{\lfloor d/2 \rfloor})$ for some parameter r. For every shallow hyperplane h, there is at least one partition so that h intersects $O(r^{\lfloor d/2 \rfloor - 1})$ simplices of the partition.

Grossi and Italiano [**GI**], generalizing and improving earlier results of van Kreveld and Overmars [**vKO, vKO2**], describe dynamic d-dimensional orthogonal range searching data structures that also support *split* and *merge* operations, defined as follows. Given a point set S, a point $p \in S$, and an integer i between 1 and d, a split operation divides S into two disjoint subsets S_1, S_2 separated by the hyperplane normal the x_i-axis passing through p, and splits the data structure for S into data structures for S_1 and S_2. Given two point sets S_1 and S_2 separated by a hyperplane normal to some coordinate axis, the merge operation combines the data structures for S_1 and S_2 into a single data structure for their union $S_1 \cup S_2$. Grossi and Italiano's data structure, called a *cross tree*, requires linear space and $O(n \log n)$ preprocessing time and supports insertions and deletions in time $O(\log n)$; splits, merges, and counting queries in time $O(n^{1-1/d})$; and reporting queries in time $O(n^{1-1/d} + k)$. Their technique gives efficient solutions to many other order-decomposable problems involving split and merge operations, including external-memory range searching.

Since an arbitrary sequence of deletions is difficult to handle in general, researchers have examined whether a random sequence of insertions and deletions can be handled efficiently; see [**Mul2, Mul3, Sc**]. Mulmuley [**Mul2**] proposed a reasonably simple data structure for halfspace range reporting that can process a random update sequence of length m in expected time $O(m^{\lfloor d/2 \rfloor + \varepsilon})$ and can answer a query in time $O(k \log n)$. If the sequence of insertions, deletions, and queries is known in advance, the corresponding static data structures can be modified to handle such a sequence of operations by paying a logarithmic overhead in the query time [**EO**]. These techniques work even if the sequence of insertions and queries is not known in advance, but the deletion time of a point is known when it is inserted [**DS**]; see also [**Sm**]. See the survey paper by Chiang and Tamassia [**CT**] for a more detailed review of dynamic geometric data structures.

6. Intersection searching

A general intersection-searching problem can be formulated as follows. Given a set S of objects in $\mathbb{R}^d$, a semigroup $(\mathbf{S}, +)$, and a weight function $w : S \to \mathbf{S}$, we wish to preprocess S into a data structure so that for a query object γ, we can compute the weighted sum $\sum_{p \cap \gamma \neq \emptyset} w(p)$, where the sum is taken over all objects $p \in S$ that intersect γ. Range searching is a special case of intersection searching in which S is a set of points. Just as with range searching, there are several variations on intersection searching: intersection counting ("How many objects in S intersect γ?"), intersection detection ("Does any object in S intersect γ?"), intersection reporting ("Which objects in S intersect γ?"), and so on.

Intersection searching is a central problem in a variety of application areas such as robotics, geographic information systems, VLSI, databases, and computer graphics. For example, the *collision-detection* problem—Given a set O of obstacles and a robot B, determine whether a placement p of B is free—can be formulated as a point intersection-detection query amid a set of regions. If B has k degrees of freedom, then a placement of B can be represented as a point in $\mathbb{R}^k$, and the set of placements of B that intersect an obstacle $O_i \in m$ is a region $K_i \subseteq \mathbb{R}^k$. If B and the obstacles are semialgebraic sets, then each K_i is also a semialgebraic set. A placement p of B is free if and only if p does not intersect any of K_i's. See [**L**] for a survey of known results on the collision-detection problem. Another intersection searching problem that arises quite frequently is the *clipping problem*: Preprocess a given set of polygons into a data structure so that all polygons intersecting a query rectangle can be reported efficiently.

An intersection-searching problem can be formulated as a semialgebraic range-searching problem by mapping each object $p \in S$ to a point $\varphi(p)$ in a parametric space $\mathbb{R}^\ell$ and every query range γ to a semialgebraic set $\psi^\#(\gamma)$ so that p intersects γ if and only if $\varphi(p) \in \psi^\#(\gamma)$. For example, let S be a set of segments in the plane and the query ranges be also segments in the plane. Each segment $e \in S$ with left and right endpoints (p_x, p_y) and (q_x, q_y), respectively, can be mapped to a point $\varphi(e) = (p_x, p_y, q_x, q_y)$ in $\mathbb{R}^4$ and a query segment γ can be mapped to a semialgebraic region $\psi^\#(\gamma)$ so that γ intersects e if and only if $\varphi(e) \in \psi^\#(\gamma)$. Hence, a segment intersection query can be answered by preprocessing the set $\{\varphi(e) \mid e \in S\}$ for semialgebraic searching. A drawback of this approach is that the dimension of the parametric space is typically much larger than d, and, therefore, it does not lead to an efficient data structure.

The efficiency of an intersection-searching structure can be significantly improved by expressing the intersection test as a conjunction of simple primitive tests (in low dimensions) and then using a multi-level data structure to perform these tests. For example, a segment γ intersects another segment e if the endpoints of e lie on the opposite sides of the line containing γ and vice-versa. Suppose we want to report those segments of S whose left endpoints lie below the line supporting a query segment (the other case can be handled in a similar manner). We define three searching problems $\mathcal{P}_1, \mathcal{P}_2$, and $\mathcal{P}_3$, with relations $\lozenge^1, \lozenge^2, \lozenge^3$, as follows:

- $e \lozenge^1 \gamma$: The left endpoint of e lies below the line supporting γ.
- $e \lozenge^2 \gamma$: The right endpoint of e lies above the line supporting γ.
- $e \lozenge^3 \gamma$: The line ℓ_e supporting e intersects γ; equivalently, in the dual plane, the point dual to ℓ_e lies in the double wedge dual to e.

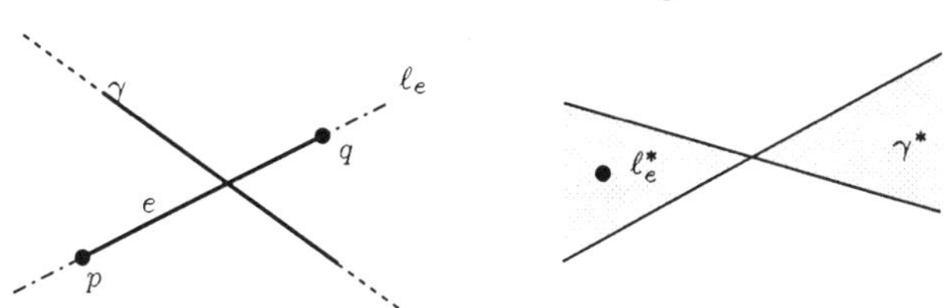

FIGURE 8. Segment intersection searching.

For $1 \le i \le 3$, let $\mathcal{D}^i$ denote a data structure for $\mathcal{P}^i$. Then $\mathcal{D}^1$ (resp. $\mathcal{D}^2$) is a halfplane range-searching structure on the left (resp. right) endpoints of segments in S, and $\mathcal{D}^3$ is (essentially) a triangle range-searching structure for points dual to the lines supporting S. By cascading $\mathcal{D}^1$, $\mathcal{D}^2$, and $\mathcal{D}^3$, we obtain a data structure for segment-intersection queries. Therefore, by Theorem 10, a segment-intersection query can be answered in time $O(n^{1/2+\varepsilon})$ using $O(n \log^3 n)$ space, or in $O(\log^3 n)$ time using $O(n^{2+\varepsilon})$ space; the size in the first data structure and the query time in the second one can be improved to $O(n)$ and $O(\log n)$, respectively. As usual, we can obtain a tradeoff between query time and space using Theorem 7.

It is beyond the scope of this survey paper to cover all intersection-searching problems. Instead, we discuss a few basic problems that have been studied extensively. All intersection-counting data structures described here can also answer intersection-reporting queries at an additional cost that is proportional to the output size. In some cases, an intersection-reporting query can be answered faster. Moreover, using intersection-reporting data structures, intersection-detection queries can be answered in time proportional to their query-search time. Finally, all the data structures described in this section can be dynamized at an expense of $O(n^{\varepsilon})$ factor in the storage and query time.

6.1. Point intersection searching. Preprocess a set S of objects (such as balls, halfspaces, simplices, or Tarski cells) in $\mathbb{R}^d$ into a data structure so that all the objects of S containing a query point can be reported (or counted) efficiently. This is the inverse or dual of the usual range-searching problem. As discussed in Section 4.2, using the duality transformation, a halfspace range-searching problem can be reduced to a point-intersection problem for a set of halfspaces, and vice versa. In general, as mentioned in Section 5.2, a d-dimensional Γ_f-range searching query, where f is $(d+b)$-variate polynomial, can be viewed as a b-dimensional point-intersection searching problem. Therefore, a very close relationship exists between the data structures for range searching (including orthogonal range searching) and for point-intersection searching. Point intersection queries can also be viewed as locating a point in the subdivision of $\mathbb{R}^d$ induced by the objects in S.

Suppose the objects in S are semialgebraic sets of the form $\{x \in \mathbb{R}^d \mid f_1(x) \geq 0, \ldots, f_u(x) \geq 0\}$, where each f_i is a $(d+b)$-variate polynomial of bounded degree that admits a linearization of dimension at most ℓ. Let $\lambda = \min(\ell, 2d-3)$ and $\gamma = \min(2b-3, \lfloor (b+\ell)/2 \rfloor, \ell)$. By constructing a multi-level data structure, point-intersection queries for S can be answered in time $O(\log n)$ using $O(n^{\lambda+\varepsilon})$ space, or in time $O(n^{1-1/\gamma+\varepsilon})$ using $O(n)$ space. Once again, we can obtain a space-time tradeoff, similar to Theorem 12. Table 7 gives some of the specific bounds that can be attained using this general scheme. If S is a set of rectangles in $\mathbb{R}^d$, then the bounds mentioned in Table 1 hold for the point-intersection problem.

Agarwal *et al.* [**AES**] extended the approach for dynamic halfspace range searching to answer point-intersection queries amid the graphs of bivariate algebraic functions, each of bounded degree. Let $\mathbf{F}$ be an infinite family of bivariate polynomials, each of bounded degree, and let $\Lambda(m)$ denote the maximum size of the lower envelope of a subset of $\mathbf{F}$ of size m. Their techniques maintains an n-element subset $\mathcal{F} \subseteq \mathbf{F}$ in a data structure of size $O(\Lambda(n) \cdot n^{\varepsilon})$, so that a polynomial $f \in \mathbf{F}$ can be inserted into or deleted from $\mathcal{F}$ in $O(n^{\varepsilon})$ time and, for a query point p, all functions of $\mathcal{F}$ whose graphs lie below p can be reported in time $O(\log n + k)$.

Besides the motion-planning application discussed above, point location in an arrangement of surfaces, especially determining whether a query point lies above a given set of regions of the form $x_{d+1} \geq f(x_1, \ldots, x_d)$, has many other applications in computational geometry; see [**AST, CEGS2, CEGS3**] for examples. However, most of these applications call for an offline data structure because the query points are known in advance.

6.2. Segment intersection searching. Preprocess a set of objects in $\mathbb{R}^d$ into a data structure so that all the objects of S intersected by a query segment can be reported (or counted) efficiently. We have already given an example of segment intersection-searching in the beginning of this section. See Table 8 for some of the

d	Objects	Problem	Size	Query Time	Source
$d = 2$	Disks	Counting	m	$(n^{4/3}/m^{2/3})\log(m/n)$	**[AgM2]**
	Disks	Reporting	$n \log n$	$\log n + k$	**[AHL]**
	Triangles	Counting	m	$\frac{n}{\sqrt{m}} \log^3 n$	**[AS]**
	Fat triangles	Reporting	$n \log^2 n$	$\log^3 n + k$	**[K]**
	Tarski cells	Counting	$n^{2+\varepsilon}$	$\log n$	**[CEGS2]**
$d = 3$	Functions	Reporting	$n^{1+\varepsilon}$	$\log n + k$	**[AES]**
	Fat tetrahedra	Reporting	m	$\frac{n^{1+\varepsilon}}{\sqrt{m}} + k$	**[EKNS]**
$d \geq 3$	Simplices	Counting	m	$\frac{n}{m^{1/d}} \log^{d+1} n$	
	Balls	Counting	$n^{d+\varepsilon}$	$\log n$	**[AgM2]**
	Balls	Reporting	m	$\frac{n}{m^{1/\lceil d/2 \rceil}} \operatorname{polylog} n + k$	**[M2]**
	Tarski cells	Counting	$n^{2d-3+\varepsilon}$	$\log n$	**[CEGS2]**
			$n^{\gamma+\varepsilon}$	$\log n$	**[AgM2]**

TABLE 7. Asymptotic upper bounds for point intersection searching.

known results on segment intersection searching. For the sake of clarity, we have omitted polylogarithmic factors from the query-search time whenever it is of the form n/m^{α}.

If we are interested in just determining whether a query segment intersects any of the input objects, better bounds can be achieved in some cases. For example, a segment intersection-detection query for a set of balls in $\mathbb{R}^d$, where $d \leq 3$, can be answered in $O(\log n)$ time using $O(n^{d+\varepsilon})$ storage **[AAS]**.

d	Objects	Problem	Size	Query Time	Source
$d = 2$	Simple polygons	Reporting	n	$(k+1)\log n$	**[HS]**
	Lines	Reporting	m	$n/\sqrt{m} + k$	**[AS, CJ2]**
	Segments	Counting	m	$n/\sqrt{m}$	**[AS, CJ2]**
	Circles	Counting	$n^{2+\varepsilon}$	$\log n$	**[AvKO]**
	Circular arcs	Counting	m	$n/m^{1/3}$	**[AvKO]**
$d = 3$	Planes	Counting	m	$n/m^{1/3}$	**[AgM]**
	Halfplanes	Reporting	m	$n/m^{1/3} + k$	**[AgM]**
	Triangles	Counting	m	$n/m^{1/4}$	**[AgM2]**
	Spheres	Counting	m	$n/m^{1/4}$	**[AgM2]**
	Spheres	Reporting	$n^{3+\varepsilon}$	$(k+1)\log^2 n$	**[AAS]**
	Hyperplanes	Counting	m	$n/m^{1/d}$	**[AgM]**

TABLE 8. Asymptotic upper bounds for segment intersection searching, with polylogarithmic factors omitted.

A special case of segment intersection searching, in which the objects are horizontal segments in the plane and query ranges are vertical segments, has been widely studied. In this case a query can be answered in time $O(\log n + k)$ using $O(n \log n)$

space and preprocessing [**VW**]. If we also allow insertions and deletions, the query and update time are respectively $O(\log n \log\log n + k)$ and $O(\log n \log\log n)$ [**MN**], or $O(\log^2 n + k)$ and $O(\log n)$ using only linear space [**CJ**]; if we allow only insertions, the query and update time become $O(\log n + k)$ and $O(\log n)$ [**IA**].

A problem related to segment intersection searching is the *stabbing problem*. Given a set S of objects in $\mathbb{R}^d$, determine whether a query k-flat $(0 < k < d)$ intersects all objects of S. Such queries can also be answered efficiently using semialgebraic range-searching data structures. A line-stabbing query amid a set of triangles in $\mathbb{R}^3$ can be answered in $O(\log n)$ time using $O(n^{2+\varepsilon})$ storage [**PS**]. The paper by Goodman *et al.* [**GPW**] is an excellent survey of this topic.

6.3. Rectangle intersection searching. Given a set S of polygons in the plane, preprocess them into a data structure so that all objects intersecting a query rectangle can be reported efficiently. This problem, also known as the *windowing query* problem, arises in a variety of applications. In many situations, the query output is required to be clipped within the query rectangle. In practice, each polygon in S is approximated by its smallest enclosing rectangle and the resulting rectangles are preprocessed for rectangle-rectangle intersection searching, as discussed in Section 3.6. If the polygons in S are large, then this scheme is not efficient, especially if we want to clip the query output within the query rectangle. A few data structures, for example, strip trees [**B**] and V-trees [**MCD**], have been proposed that store each polygon hierarchically. We can use these data structures to store each polygon and then construct an R-tree or any other orthogonal range-searching data structure on the smallest enclosing rectangles of the polygons. Nievergelt and Widmayer [**NW**] describe another data structure, called a *guard file*, which is suitable if the polygons are fat (have bounded aspect ratio). They place a set of well-chosen points, called *guards*, and associate a subset of polygons with each guard that either contain the guard or lie "near" the guard. For a query rectangle γ, they determine the set of guards that lie inside γ; the lists of polygons associated with these guards give the candidates that intersect γ.

6.4. Colored intersection searching. Preprocess a given set S of colored objects in $\mathbb{R}^d$ (*i.e.*, each object in S is assigned a color) so that the we can report (or count) the colors of the objects that intersect the query range. This problem arises in many contexts where one wants to answer intersection-searching queries for input objects of non-constant size. For example, given a set $P = \{P_1, \ldots, P_m\}$ of m simple polygons, one may wish to report all the simple polygons that intersect a query segment; the goal is to return the indices, and not the descriptions, of these polygons. If we color the edges of P_i by the color i, the problem reduces to colored segment intersection searching in a set of segments.

If an intersection-detection query for S with respect to a range γ can be answered in $Q(n)$ time, then a colored intersection-reporting query with γ can be answered in time $O((k\log(n/k) + 1)Q(n))$. Thus, logarithmic query-time intersection-searching data structures can easily be modified for colored intersection reporting, but very little is known about linear-size colored intersection-searching data structures, except in some special cases [**AvK, BKMT, GJS, GJS2, GJS3, JL**].

Gupta *et al.* [**GJS**] have shown that the colored halfplane-reporting queries in the plane can be answered in $O(\log^2 n + k)$ using $O(n \log n)$ space. Agarwal and van Kreveld [**AvK**] presented a linear-size data structure with $O(n^{1/2+\varepsilon} + k)$ query time for colored segment intersection-reporting queries amid a set of segments in the plane, assuming that the segments of the same color form a connected planar graph, or if they form the boundary of a simple polygon; these data structures can also handle insertions of new segments. Gupta *et al.* [**GJS, GJS3**] present segment intersection-reporting structures for many other special cases.

7. Optimization queries

The goal of an optimization query is to return an object that satisfies a certain condition with respect to a query range. Ray-shooting queries are perhaps the most common example of optimization queries. Other examples include segment-dragging and linear-programming queries.

7.1. Ray-shooting queries. Preprocess a set S of objects in $\mathbb{R}^d$ into a data structure so that the first object (if any) hit by a query ray can be reported efficiently. This problem arises in ray tracing, hidden-surface removal, radiosity, and other graphics problems. Recently, efficient solutions to many other geometric problems have also been developed using ray-shooting data structures.

A general approach to the ray-shooting problem, using segment intersection-detection structures and Megiddo's parametric searching technique [**Me**], was proposed by Agarwal and Matoušek [**AgM**]. Suppose we have a segment intersection-detection data structure for S. Let ρ be a query ray. Their algorithm maintains a segment $\overline{ab} \subseteq \rho$ such that the first intersection point of $\overline{ab}$ with S is the same as that of ρ. If a lies on an object of S, it returns a. Otherwise, it picks a point $c \in \overline{ab}$ and determines, using the segment intersection-detection data structure, whether the interior of the segment $\overline{ac}$ intersects any object of S. If the answer is yes, it recursively finds the first intersection point of $\overline{ac}$ with S; otherwise, it recursively finds the first intersection point of $\overline{cb}$ with S. Using parametric searching, the points c at each stage can be chosen in such a way that the algorithm terminates after $O(\log n)$ steps. Recently, Chan [**C2**] described a randomized reduction that is much simpler than parametric search and usually avoids the added logarithmic factor in the query time. In some cases, by using a more direct approach, we can improve the query time by a polylogarithmic factor. For example, by exploiting some additional properties of input objects and of partition trees, we can modify a segment intersection-searching data structure in some cases to answer ray shooting queries [**A2, CJ2, GOS**].

Another approach for answering ray-shooting queries is based on visibility maps. A ray in $\mathbb{R}^d$ can be represented as a point in $\mathbb{R}^d \times \mathbb{S}^{d-1}$. Given a set S of objects, we can partition the parametric space $\mathbb{R}^d \times \mathbb{S}^{d-1}$ into cells so that all points within each cell correspond to rays that hit the same object first; this partition is called the *visibility map* of S. Using this approach and some other techniques, Chazelle and Guibas [**CG2**] showed that a ray-shooting query in a simple polygon can be answered in $O(\log n)$ time using $O(n)$ space. Simpler data structures were subsequently proposed by Chazelle *et al.* [**CEGG$^+$**] and Hershberger and Suri [**HS**]. Following a similar approach, Pocchiola and Vegter [**PV**] showed that a ray-shooting query amid a set $\mathcal{P}$ of s disjoint convex polygons, with a total of n vertices, can be answered in $O(\log n)$ time, using $O(n + m)$ space, where $m = O(s^2)$ is the size of the visibility graph of $\mathcal{P}$.[8]

---

[8]The vertices of the *visibility graph* are the vertices of the polygons. Besides the polygon edges, there is an edge in the graph between two vertices v_i, v_j of convex polygons P_i and P_j if

d	Objects	Size	Query Time	Source
$d=2$	Simple polygon	n	$\log n$	[HS]
	s disjoint simple polygons	n	$\sqrt{s}$	[AS2, HS]
	s disjoint simple polygons	$(s^2+n)\log s$	$\log s \log n$	[AS2]
	s disjoint convex polygons	s^2+n	$\log n$	[PV]
	s convex polygons	$sn \log s$	$\log s \log n$	[AS2]
	Segments	m	$n/\sqrt{m}$	[AS, CJ2]
	Circular arcs	m	$n/m^{1/3}$	[AvKO]
	Disjoint arcs	n	$\sqrt{n}$	[AvKO]
$d=3$	Convex polytope	n	$\log n$	[DK]
	c-oriented polytopes	n	$\log n$	[dBO]
	s convex polytopes	$s^2 n^{2+\varepsilon}$	$\log^2 n$	[AS3]
	Halfplanes	m	$n/m^{1/3}$	[AgM]
	Terrain	m	$n/\sqrt{m}$	[AgM, CEGSS]
	Triangles	m	$n/m^{1/4}$	[AgM2]
	Spheres	$n^{3+\varepsilon}$	$\log^2 n$	[AAS]
$d>3$	Hyperplanes	m	$n/m^{1/d}$	[AgM]
	Hyperplanes	$\dfrac{n^d}{\log^{d-\varepsilon} n}$	$\log n$	[CF2, M5]
	Convex polytope	m	$n/m^{1/\lfloor d/2 \rfloor}$	[AgM, MS]
	Convex polytope	$\dfrac{n^{\lfloor d/2 \rfloor}}{\log^{\lfloor d/2 \rfloor-\varepsilon} n}$	$\log n$	[MS]

TABLE 9. Asymptotic upper bounds for ray shooting queries, with polylogarithmic factors omitted

Table 9 gives a summary of known ray-shooting results. For the sake of clarity, we have omitted polylogarithmic factors from query times of the form n/m^{α}. The ray-shooting structures for d-dimensional convex polyhedra by Matoušek and Schwarzkopf [**MS**] assume that the source point of the query ray lies inside the polytope. All the ray-shooting data structures mentioned in Table 9 can be dynamized at a cost of polylogarithmic or n^{ε} factor in the query time. Goodrich and Tamassia [**GT**] have developed a dynamic ray-shooting data structure for connected planar subdivisions, with $O(\log^2 n)$ query and update time.

Like range searching, many practical data structures have been proposed that, despite having bad worst-case performance, work well in practice. The books by Foley *et al.* [**FvDFH**] and Glassner [**G**] describe several practical data structures for ray tracing that are used in computer graphics. One common approach is to construct a subdivision of $\mathbb{R}^d$ into constant-size cells so that the interior of each cell does not intersect any object of S. A ray-shooting query can be answered by traversing the query ray through the subdivision until we find an object that intersects the ray. The worst-case query time is proportional to the maximum number

the line segment $\overline{v_i v_j}$ does not intersect any other convex polygon and the line supporting the segment is tangent to both P_i and P_j.

of cells intersected by a segment that does not intersect any object in S; we refer to this quantity as the *crossing number* of the triangulation. Hershberger and Suri [**HS**] showed that if S is the boundary of a simple polygon, then a triangulation (using Steiner points) with $O(\log n)$ crossing number can be constructed in $O(n \log n)$ time. See [**AASu, MMS, DK2, MB, WSCK$^+$**] and the references therein for other ray-shooting results using this approach. Agarwal *et al.* [**AASu**] proved worst-case bounds for many cases on the number of cells in the subdivision that a line can intersect. For example, they show that the crossing number for a set of k disjoint convex polyhedra in $\mathbb{R}^3$ is $\Omega(k + \log n)$, and they present an algorithm that constructs a triangulation of size $O(nk \log n)$ with stabbing number $O(k \log n)$. Aronov and Fortune [**AF**] prove a bound on the average crossing number of set of disjoint triangles in $\mathbb{R}^3$, and present a polynomial-time algorithm to construct a triangulation that achieves this bound. In practice, however, very simple decompositions, such as oct-trees and binary space partitions [**FKN**] are used to trace a ray.

7.2. Nearest-neighbor queries. The *nearest-neighbor query* problem is defined as follows: Preprocess a set S of points in $\mathbb{R}^d$ into a data structure so that a point in S closest to a query point ξ can be reported quickly. This is one of the most widely studied problems in computational geometry because it arises in so many different areas, including pattern recognition [**CH, DH**], data compression [**ArM, RP**], information retrieval [**FRM, Sa**], CAD [**MG**], molecular biology [**SBK**], image analysis [**KK, KSFSP**], data mining [**FL, HaT**], machine learning [**CoS**], and geographic information systems [**RKV, Sp**]. Most applications use so-called *feature vectors* to map a complex object to a point in high dimensions. Examples of feature vectors include color histograms, shape descriptors, Fourier vectors, and text descriptors.

For simplicity, we assume that the distance between points is measured in the Euclidean metric, though a more complicated metric can be used depending on the application. For $d = 2$, one can construct the Voronoi diagram of S and preprocess it for point-location queries in $O(n \log n)$ time [**PrS**]. For higher dimensions, Clarkson [**Cl2**] presented a data structure of size $O(n^{\lceil d/2 \rceil + \varepsilon})$ that can answer a query in $2^{O(d)} \log n$ time. The query time can be improved to $O(d^3 \log n)$, using a technique of Meiser [**Mei**].

A nearest-neighbor query for a set of points under the Euclidean metric can be formulated as an instance of the ray-shooting problem in a convex polyhedron in $\mathbb{R}^{d+1}$, as follows. We map each point $p = (p_1, \ldots, p_d) \in S$ to a hyperplane $\hat{p}$ in $\mathbb{R}^{d+1}$, which is the graph of the function

$$f_p(x_1, \ldots, x_d) = 2p_1x_1 + \cdots + 2p_dx_d - (p_1^2 + \cdots + p_d^2).$$

Then p is a closest neighbor of a point $\xi = (\xi_1, \ldots, \xi_d)$ if and only if

$$f_p(\xi_1, \ldots, \xi_d) = \max_{q \in S} f_q(\xi_1, \ldots, \xi_d).$$

That is, if and only if f_p is the first hyperplane intersected by the vertical ray $\rho(\xi)$ emanating from the point $(\xi_1, \ldots, \xi_d, 0)$ in the negative x_{d+1}-direction. If we define $P = \bigcap_{p \in S} \{(x_1, \ldots, x_{d+1}) \mid x_{d+1} \geq f_p(x_1, \ldots, x_d)\}$, then p is the nearest neighbor of ξ if and only if the intersection point of $\rho(\xi)$ and ∂P lies on the graph of f_p. Thus a nearest-neighbor query can be answered in time roughly $n/m^{1/\lceil d/2 \rceil}$ using

$O(m)$ space. This approach can be extended to answer farthest-neighbor and k-nearest-neighbor queries also. In general, if we have an efficient data structure for answering disk-emptiness queries for disks under a given metric ρ, we can apply parametric searching [**Me**] or Chan's randomized reduction [**C2**] to answer nearest-neighbor queries under the ρ-metric, provided the data structure satisfies certain mild assumptions [**AgM**].

Note that the query time of the above approach is exponential in d, so it is impractical even for moderate values of d (say $d \approx 10$). This has lead to the development of algorithms for finding approximate nearest neighbors [**ArM2, AMN, AMNSW, Cl4, IM, Kl, KSFSP**] or for special cases, such as when the distribution of query points is known in advance [**Cl5, Yi**].

Because of wide applications of nearest-neighbor searching, many heuristics have been developed, especially in higher dimensions. These algorithms use practical data structures described in Section 3, including *kd*-trees, R-trees, R*-trees, and Hilbert R-trees; see *e.g.*, [**FBF, HjS, KSFSP, KK, FL, HaT, RKV, Sp**]. White and Jain [**WJ**] described a variant of R-tree for answering nearest-neighbor queries in which they use spheres instead of rectangles as enclosing regions. This approach was further extended by Katayama and Satoh [**KS**]. Berchtold *et al.* [**BBBKK**] present a parallel algorithm for nearest-neighbor searching. For large input sets, one desires an algorithm that minimizes the number of disk accesses. Many of the heuristics mentioned above try to optimize the I/O efficiency, though none of them gives any performance guarantee. A few recent papers [**AMN, BBKK, PM, Cle**] analyze the efficiency of some of the heuristics, under certain assumptions on the input.

7.3. Linear programming queries. Let S be a set of n halfspaces in $\mathbb{R}^d$. We wish to preprocess S into a data structure so that for a direction vector $\vec{v}$, we can determine the first point of $\bigcap_{h \in S} h$ in the direction $\vec{v}$. For $d \leq 3$, such a query can be answered in $O(\log n)$ time using $O(n)$ storage, by constructing the normal diagram of the convex polytope $\bigcap_{h \in S} h$ and preprocessing it for point-location queries. For higher dimensions, Matoušek [**M4**] showed that, using multidimensional parametric searching and a data structure for answering halfspace emptiness queries, a linear-programming query can be answered in $O((n/m^{1/\lfloor d/2 \rfloor}) \operatorname{polylog} n)$ with $O(m)$ storage. Recently Chan [**C**] has described a randomized procedure whose expected query time is $n^{1-1/\lfloor d/2 \rfloor} 2^{O(\log^* n)}$, using linear space.

7.4. Segment dragging queries. Preprocess a set S of objects in the plane so that for a query segment e and a ray ρ, the first position at which e intersects any object of S as it is translated (dragged) along ρ can be determined quickly. This query can be answered in $O((n/\sqrt{m}) \operatorname{polylog} n)$ time, with $O(m)$ storage, using segment intersection-searching structures and parametric searching. Chazelle [**Ch3**] gave a linear-size, $O(\log n)$ query-time data structure for the special case in which S is a set of points, e is a horizontal segment, and ρ is the vertical direction. Instead of dragging a segment along a ray, one can ask the same question for dragging along a more complex trajectory (along a curve and allowing both translation and rotation). These problems arise quite often in motion planning and manufacturing. See [**Mi, ST**] for a few such examples.

8. Concluding remarks

In this survey paper we reviewed both theoretical and practical data structures for range searching. Theoretically optimal or near-optimal data structures are known for most range searching problems. However, from a practical standpoint, range searching is still largely open; known data structures are rather complicated and do not perform well in practice, especially as the dimension increases. Lower bounds suggest that we cannot hope for data structures that do significantly better than the naïve algorithm in the worst case (and for some problems, even in the average case), but it is still an interesting open question to develop simple data structures that work well on typical inputs, especially in high dimensions.

As we saw in this survey, range-searching data structures are useful for other geometric-searching problems as well. In the quest for efficient range-searching data structures, researchers have discovered several elegant geometric techniques that have enriched computational geometry as a whole. It is impossible to describe in a survey paper all the known techniques and results on range searching and their applications to other geometric problems. We therefore chose a few of these techniques that we thought were most interesting. For further details, we refer the interested reader to the books by Mulmuley [**Mul4**], Preparata and Shamos [**PrS**], and Samet [**Sam2**], and the survey papers by Chazelle [**Ch9**], Güting [**Güt**], Matoušek [**M3, M7**], and Nievergelt and Widmayer [**NW2**].

References

[AM] D. J. Abel and D. M. Mark, *A comparative analysis of some two-dimensional orderings*, Intl. J. Geographic Informations Systems 4 (1990), 21–31.

[A] P. K. Agarwal, *Geometric partitioning and its applications*, Computational Geometry: Papers from the DIMACS Special Year (J. E. Goodman, R. Pollack, and W. Steiger, eds.), American Mathematical Society, Providence, RI, 1991, pp. 1–37.

[A2] P. K. Agarwal, *Ray shooting and other applications of spanning trees with low stabbing number*, SIAM J. Comput. 21 (1992), 540–570.

[AAEFV] P. K. Agarwal, L. Arge, J. Erickson, P. G. Franciosa and J. S. Vitter, *Efficient searching with linear constraints*, Proc. 17th Annu. ACM Sympos. Principles of Database Syst., 1998, to appear.

[AAS] P. K. Agarwal, B. Aronov and M. Sharir, *Computing envelopes in four dimensions with applications*, Proc. 10th Annu. ACM Sympos. Comput. Geom., 1994, pp. 348–358.

[AASu] P. K. Agarwal, B. Aronov and S. Suri, *Line stabbing bounds in three dimensions*, Proc. 11th Annu. ACM Sympos. Comput. Geom., 1995, pp. 267–276.

[AES] P. K. Agarwal, A. Efrat and M. Sharir, *Vertical decomposition of shallow levels in 3-dimensional arrangements and its applications*, Proc. 11th Annu. ACM Sympos. Comput. Geom., 1995, pp. 39–50.

[AgM] P. K. Agarwal and J. Matoušek, *Ray shooting and parametric search*, SIAM J. Comput. 22 (1993), 794–806.

[AgM2] P. K. Agarwal and J. Matoušek, *On range searching with semialgebraic sets*, Discrete Comput. Geom. 11 (1994), 393–418.

[AgM3] P. K. Agarwal and J. Matoušek, *Dynamic half-space range reporting and its applications*, Algorithmica 13 (1995), 325–345.

[AS] P. K. Agarwal and M. Sharir, *Applications of a new space partitioning technique*, Discrete Comput. Geom. 9 (1993), 11–38.

[AS2] P. K. Agarwal and M. Sharir, *Ray shooting amidst convex polygons in 2D*, J. Algorithms 21 (1996), 508–519.

[AS3] P. K. Agarwal and M. Sharir, *Ray shooting amidst convex polyhedra and polyhedral terrains in three dimensions*, SIAM J. Comput. 25 (1996), 100–116.

[AST] P. K. Agarwal, M. Sharir and S. Toledo, *Applications of parametric searching in geometric optimization*, J. Algorithms 17 (1994), 292–318.

[AvK] P. K. Agarwal and M. van Kreveld, *Polygon and connected component intersection searching*, Algorithmica 15 (1996), 626–660.

[AvKO] P. K. Agarwal, M. van Kreveld and M. Overmars, *Intersection queries in curved objects*, J. Algorithms 15 (1993), 229–266.

[AHL] A. Aggarwal, M. Hansen and T. Leighton, *Solving query-retrieval problems by compacting Voronoi diagrams*, Proc. 22nd Annu. ACM Sympos. Theory Comput., 1990, pp. 331–340.

[AHU] A. V. Aho, J. E. Hopcroft and J. D. Ullman, The Design and Analysis of Computer Algorithms, Addison-Wesley, Reading, MA, 1974.

[AnS] A. Andersson and K. Swanson, *On the difficulty of range searching*, Comput. Geom. Theory Appl. 8 (1997), 115–122.

[AB] M. Anthony and N. Biggs, Computational Learning Theory, Cambridge University Press, Cambridge, UK, 1992.

[Ar] L. Arge, *The Buffer Tree: A new technique for optimal I/O-algorithms*, Proc. 4th Workshop Algorithms Data Struct., Lecture Notes Comput. Sci., vol. 955, Springer-Verlag, 1995, pp. 334–345.

[AV] L. Arge and J. S. Vitter, *Optimal interval management in external memory*, Proc. 37th Annu. IEEE Sympos. Found. Comput. Sci., 1996, pp. 560–569.

[AF] B. Aronov and S. Fortune, *Average-case ray shooting and minimum weight triangulation*, Proc. 13th Annu. Sympos. Comput. Geom., 1997, pp. 203–212.

[APS] B. Aronov, M. Pellegrini and M. Sharir, *On the zone of a surface in a hyperplane arrangement*, Discrete Comput. Geom. 9 (1993), 177–186.

[ArM] S. Arya and D. M. Mount, *Algorithms for fast vector quantization*, Data Compression Conference, IEEE Press, 1993, pp. 381–390.

[ArM2] S. Arya and D. M. Mount, *Approximate nearest neighbor queries in fixed dimensions*, Proc. 4th ACM-SIAM Sympos. Discrete Algorithms, 1993, pp. 271–280.

[ArM3] S. Arya and D. M. Mount, *Approximate range searching*, Proc. 11th Annu. ACM Sympos. Comput. Geom., 1995, pp. 172–181.

[AMN] S. Arya, D. M. Mount and O. Narayan, *Accounting for boundary effects in nearest neighbor searching*, Discrete Comput. Geom. 16 (1996), 155–176.

[AMNSW] S. Arya, D. M. Mount, N. S. Netanyahu, R. Silverman and A. Wu, *An optimal algorithm for approximate nearest neighbor searching*, Proc. 5th ACM-SIAM Sympos. Discrete Algorithms, 1994, pp. 573–582.

[ARWW] T. Asano, T. Roos, P. Widmayer and E. Welzl, *Space filling curves and their use in the design of geometric data structures*, Proc. 2nd Latin Amer. Sympos. Theoret. Informatics, Lecture Notes Comput. Sci., vol. 911, Springer-Verlag, 1995, pp. 36–48.

[Av] D. Avis, *Non-partitionable point sets*, Inform. Process. Lett. 19 (1984), 125–129.

[B] D. H. Ballard, *Strip trees: A hierarchical representation for curves*, Commun. ACM, 24 (1981), 310–321.

[BYF] R. Bar-Yehuda and S. Fogel, *Partitioning a sequence into few monotone subsequences*, Technical Report 640, Technion IIT, Haifa, Israel, 1990.

[BYF2] R. Bar-Yehuda and S. Fogel, *Variations on ray shooting*, Algorithmica 11 (1994), 133–145.

[BM] R. Bayer and McCreight, *Organization of large ordered indexes*, Acta Inform. 1 (1972), 173–189.

[BSW] B. Becker, H. W. Six and P. Widmayer, *Spatial priority search: An access technique for scaleless maps*, Proc. ACM SIGMOD Conf. on Management of Data, 1991, pp. 128–138.

[BKSS] N. Beckmann, H.-P. Kriegel, R. Schneider and B. Seeger, *The R^*-tree: An efficient and robust access method for points and rectangles*, Proc. ACM SIGMOD Conf. on Management of Data, 1990, pp. 322–331.

[BA] A. M. Ben-Amram, *Lower bounds on algebraic random access machines*, Proc. 22nd Internat. Colloq. Automata Lang. Prog., Lecture Notes Comput. Sci., vol. 944, Springer-Verlag, 1995, pp. 360–371.

[BA2] A. M. Ben-Amram, *What is a "pointer machine"?*, SIGACT News 26 (1995), 88–95.

[BO] M. Ben-Or, *Lower bounds for algebraic computation trees*, Proc. 15th Annu. ACM Sympos. Theory Comput., 1983, pp. 80–86.

[Be] J. L. Bentley, *Multidimensional binary search trees used for associative searching*, Commun. ACM 18 (1975), 509–517.

[Be2] J. L. Bentley, *Multidimensional divide-and-conquer*, Commun. ACM 23 (1980), 214–229.

[BF] J. L. Bentley and J. H. Friedman, *Data structures for range searching*, ACM Comput. Surv. 11 (1979), 397–409.

[BS] J. L. Bentley and J. B. Saxe, *Decomposable searching problems I: Static-to-dynamic transformation*, J. Algorithms 1 (1980), 301–358.

[BBBKK] S. Berchtold, C. Böhm, B. Barunmüller, D. A. Keim and H.-P. Kriegel, *Fast parallel similarity search in multimedia databases*, Proc. ACM SIGMOD Conf. on Management of Data, 1997, pp. 1–12.

[BBKK] S. Berchtold, C. Böhm, D. A. Keim and H.-P. Kriegel, *A cost model for nearest neighbor search in high-dimensional data space*, Proc. 16th Annu. ACM Sympos. Principles of Database Syst., 1997, pp. 78–86.

[BKK] S. Berchtold, D. A. Keim and H.-P. Kriegel, *The X-tree: An index structure for higher dimensional data*, Proc. 22th VLDB Conference, 1996, pp. 28–39.

[Bi] T. Bially, *Space-filling curves: Their generation and their application to bandwidth reduction*, IEEE Trans. Inform. Theory 15 (1969), pp. 658–664.

[BEHW] A. Blumer, A. Ehrenfeucht, D. Haussler and M. Warmuth, *Learnablility and the Vapnik-Chervonenkis dimension*, J. ACM 36 (1989), 929–965.

[BKMT] P. Bozanis, N. Ktsios, C. Makris and A. Tsakalidis, *New upper bounds for generalized intersection searching problems*, Proc. 22nd Internat. Colloq. Auto. Lang. Program., Lecture Notes in Comput. Sci., vol. 944, 1995, pp. 464–475.

[BKMT2] P. Bozanis, N. Ktsios, C. Makris and A. Tsakalidis, *New results on intersection query problems*, unpublished manuscript, 1996.

[BCP] H. Brönnimann, B. Chazelle and J. Pach, *How hard is halfspace range searching?*, Discrete Comput. Geom. 10 (1993), 143–155.

[BCS] P. Bürgisser, M. Clausen and M. A. Shokrollahi, Algebraic Complexity Theory, Grundlehren der mathematischen Wissenschaften, vol. 315, Springer-Verlag, 1996.

[C] T. M. Chan, *Fixed-dimensional linear programming queries made easy*, Proc. 12th Annu. ACM Sympos. Comput. Geom., 1996, pp. 284–290.

[C2] T. M. Chan, *Geometric applications of a randomized optimization technique*, Proc. 14th Annu. ACM Sympos. Comput. Geom., 1998, to appear.

[Ch] B. Chazelle, *Filtering search: A new approach to query-answering*, SIAM J. Comput. 15 (1986), 703–724.

[Ch2] B. Chazelle, *Computing on a free tree via complexity-preserving mappings*, Algorithmica 2 (1987), 337–361.

[Ch3] B. Chazelle, *An algorithm for segment-dragging and its implementation*, Algorithmica 3 (1988), 205–221.

[Ch4] B. Chazelle, *A functional approach to data structures and its use in multidimensional searching*, SIAM J. Comput. 17 (1988), 427–462.

[Ch5] B. Chazelle, *Lower bounds on the complexity of polytope range searching*, J. Amer. Math. Soc. 2 (1989), 637–666.

[Ch6] B. Chazelle, *Lower bounds for orthogonal range searching, I: The reporting case*, J. ACM 37 (1990), 200–212.

[Ch7] B. Chazelle, *Lower bounds for orthogonal range searching, II: The arithmetic model*, J. ACM 37 (1990), 439–463.

[Ch8] B. Chazelle, *Cutting hyperplanes for divide-and-conquer*, Discrete Comput. Geom. 9 (1993), 145–158.

[Ch9] B. Chazelle, *Computational geometry: A retrospective*, Proc. 26th Annu. ACM Sympos. Theory Comput., 1994, pp. 75–94.

[Ch10] B. Chazelle, *A spectral approach to lower bounds*, Proc. 35th Annu. IEEE Sympos. Found. Comput. Sci., 1994, pp. 674–682.

[Ch11] B. Chazelle, *Lower bounds for off-line range searching*, Discrete Comput. Geom. 17 (1997), pp. 53–65.

[CCPY] B. Chazelle, R. Cole, F. P. Preparata and C. K. Yap, *New upper bounds for neighbor searching*, Inform. Control 68 (1986), 105–124.

[CE] B. Chazelle and H. Edelsbrunner, *Optimal solutions for a class of point retrieval problems*, J. Symbolic Comput. 1 (1985), 47–56.

[CE2] B. Chazelle and H. Edelsbrunner, *Linear space data structures for two types of range search*, Discrete Comput. Geom. 2 (1987), 113–126.

[CEGG+] B. Chazelle, H. Edelsbrunner, M. Grigni, L. J. Guibas, J. Hershberger, M. Sharir and J. Snoeyink, *Ray shooting in polygons using geodesic triangulations*, Algorithmica 12 (1994), 54–68.

[CEGS] B. Chazelle, H. Edelsbrunner, L. J. Guibas and M. Sharir, *A singly-exponential stratification scheme for real semi-algebraic varieties and its applications*, Proc. 16th Internat. Colloq. Automata Lang. Program., Lecture Notes Comput. Sci., vol. 372, Springer-Verlag, 1989, pp. 179–192.

[CEGS2] B. Chazelle, H. Edelsbrunner, L. J. Guibas and M. Sharir, *A singly-exponential stratification scheme for real semi-algebraic varieties and its applications*, Theoret. Comput. Sci. 84 (1991), 77–105.

[CEGS3] B. Chazelle, H. Edelsbrunner, L. J. Guibas and M. Sharir, *Diameter, width, closest line pair and parametric searching*, Discrete Comput. Geom. 10 (1993), 183–196.

[CEGSS] B. Chazelle, H. Edelsbrunner, L. J. Guibas, M. Sharir and J. Stolfi, *Lines in space: Combinatorics and algorithms*, Algorithmica 15 (1996), 428–447.

[CF] B. Chazelle and J. Friedman, *A deterministic view of random sampling and its use in geometry*, Combinatorica 10 (1990), 229–249.

[CF2] B. Chazelle and J. Friedman, *Point location among hyperplanes and unidirectional ray-shooting*, Comput. Geom. Theory Appl. 4 (1994), 53–62.

[CG] B. Chazelle and L. J. Guibas, *Fractional cascading: I. A data structuring technique*, Algorithmica 1 (1986), 133–162.

[CG2] B. Chazelle and L. J. Guibas, *Visibility and intersection problems in plane geometry*, Discrete Comput. Geom. 4 (1989), 551–581.

[CGL] B. Chazelle, L. J. Guibas and D. T. Lee, *The power of geometric duality*, BIT, 25 (1985), 76–90.

[CP] B. Chazelle and F. P. Preparata, *Halfspace range search: An algorithmic application of k-sets*, Discrete Comput. Geom. 1 (1986), 83–93.

[CR] B. Chazelle and B. Rosenberg, *Computing partial sums in multidimensional arrays*, Proc. 5th Annu. ACM Sympos. Comput. Geom., 1989, pp. 131–139.

[CR2] B. Chazelle and B. Rosenberg, *Simplex range reporting on a pointer machine*, Comput. Geom. Theory Appl. 5 (1996), 237–247.

[CSW] B. Chazelle, M. Sharir and E. Welzl, *Quasi-optimal upper bounds for simplex range searching and new zone theorems*, Algorithmica 8 (1992), 407–429.

[CW] B. Chazelle and E. Welzl, *Quasi-optimal range searching in spaces of finite VC-dimension*, Discrete Comput. Geom. 4 (1989), 467–489.

[CJ] S. W. Cheng and R. Janardan, *Efficient dynamic algorithms for some geometric intersection problems*, Inform. Process. Lett. 36 (1990), 251–258.

[CJ2] S. W. Cheng and R. Janardan, *Algorithms for ray-shooting and intersection searching*, J. Algorithms 13 (1992), 670–692.

[CT] Y.-J. Chiang and R. Tamassia, *Dynamic algorithms in computational geometry*, Proc. IEEE 80 (1992), 1412–1434.

[Cl] K. L. Clarkson, *New applications of random sampling in computational geometry*, Discrete Comput. Geom. 2 (1987), 195–222.

[Cl2] K. L. Clarkson, *A randomized algorithm for closest-point queries*, SIAM J. Comput. 17 (1988), 830–847.

[Cl3] K. L. Clarkson, *Randomized geometric algorithms*, Computing in Euclidean Geometry (D.-Z. Du and F. K. Hwang, eds.), Lecture Notes Series on Computing, vol. 1, World Scientific, Singapore, 1992, pp. 117–162.

[Cl4] K. L. Clarkson, *An algorithm for approximate closest-point queries*, Proc. 10th Annu. ACM Sympos. Comput. Geom., (1994), pp. 160–164.

[Cl5] K. L. Clarkson, *Nearest neighbor queries in metric spaces*, Proc. 29th Annu. ACM Sympos. Theory Comput., (1997), pp. 609–617.

[CS] K. L. Clarkson and P. W. Shor, *Applications of random sampling in computational geometry, II*, Discrete Comput. Geom. 4 (1989), 387–421.

[Cle] J. G. Cleary, *Analysis of an algorithm for finding nearest neighbors in Euclidean space*, ACM Trans. Math. Softw. 5 (1979), 183–192.

[Co] R. Cole, *Partitioning point sets in 4 dimensions*, Proc. 12th Internat. Colloq. Automata Lang. Program., Lecture Notes Comput. Sci., vol. 194, Springer-Verlag, 1985, pp. 111–119.

[CY] R. Cole and C. K. Yap, *Geometric retrieval problems*, Inform. Control 63 (1984), 39–57.

[Com] D. Comer, *The ubiquitous B-tree*, ACM Comput. Surv. 11 (1979), 121–137.

[CoS] S. Cost and S. Salzberg, *A weighted nearest neighbor algorithm for learning with symbolic features*, Machine Learning 10 (1993), 57–67.

[CH] T. Cover and P. E. Hart, *Nearest neighbor pattern classification*, IEEE Trans. Inform. Theory 13 (1967), 21–27.

[dBGH] M. de Berg, L. J. Guibas and D. Halperin, *Vertical decompositions for triangles in 3-space*, Discrete Comput. Geom. 15 (1996), 35–61.

[dBO] M. de Berg and M. Overmars, *Hidden surface removal for c-oriented polyhedra*, Comput. Geom. Theory Appl. 1 (1992), 247–268.

[DE] D. Dobkin and H. Edelsbrunner, *Organizing point sets in two and three dimensions*, Tech. Report F130, Inst. Informationsverarb., Tech. Univ. Graz, Graz, Austria, 1984.

[DHKS] D. Dobkin, J. Hershberger, D. Kirkpatrick and S. Suri, *Implicitly searching convolutions and computing depth of collision*, Proc. 1st Annu. SIGAL Internat. Sympos. Algorithms, Lecture Notes Comput. Sci., vol. 450, Springer-Verlag, 1990, pp. 165–180.

[DS] D. Dobkin and S. Suri, *Maintenance of geometric extrema*, J. ACM 38 (1991), 275–298.

[DoE] D. P. Dobkin and H. Edelsbrunner, *Space searching for intersecting objects*, J. Algorithms 8 (1987), 348–361.

[DK] D. P. Dobkin and D. G. Kirkpatrick, *A linear algorithm for determining the separation of convex polyhedra*, J. Algorithms 6 (1985), 381–392.

[DK2] D. P. Dobkin and D. G. Kirkpatrick, *Determining the separation of preprocessed polyhedra – A unified approach*, Proc. 17th Internat. Colloq. Automata Lang. Program., Lecture Notes Comput. Sci., vol. 443, Springer-Verlag, 1990, pp. 400–413.

[DH] R. O. Duda and P. E. Hart, Pattern Classification and Scene Analysis, Wiley Interscience, New York, NY, 1973.

[E] H. Edelsbrunner, Algorithms in Combinatorial Geometry, EATCS Monographs on Theoretical Computer Science, vol. 10, Springer-Verlag, Heidelberg, West Germany, 1987.

[EGHS+] H. Edelsbrunner, L. J. Guibas, J. Hershberger, R. Seidel, M. Sharir, J. Snoeyink and E. Welzl, *Implicitly representing arrangements of lines or segments*, Discrete Comput. Geom. 4 (1989), 433–466.

[EH] H. Edelsbrunner and F. Huber, *Dissecting sets of points in two and three dimensions*, Report F138, Inst. Informationsverarb., Tech. Univ. Graz, Graz, Austria, 1984.

[EKM] H. Edelsbrunner, D. G. Kirkpatrick and H. A. Maurer, *Polygonal intersection searching*, Inform. Process. Lett. 14 (1982), 74–79.

[EM] H. Edelsbrunner and H. A. Maurer, *A space-optimal solution of general region location*, Theoret. Comput. Sci. 16 (1981), 329–336.

[EO] H. Edelsbrunner and M. H. Overmars, *Batched dynamic solutions to decomposable searching problems*, J. Algorithms 6 (1985), 515–542.

[EW] H. Edelsbrunner and E. Welzl, *Halfplanar range search in linear space and $O(n^{0.695})$ query time*, Inform. Process. Lett. 23 (1986), 289–293.

[EKNS] A. Efrat, M. J. Katz, F. Nielsen and M. Sharir, *Dynamic data structures for fat objects and their applications*, Proc. 5th Workshop Algorithms Data Struct., 1997, pp. 297–306.

[ES] P. Erdős and G. Szekeres, *A combinatorial problem in geometry*, Compositio Math. 2 (1935), 463–470.

[Er] J. Erickson, *New lower bounds for halfspace emptiness*, 37th Annu. IEEE Sympos. Found. Comput. Sci., 1996, pp. 472–481.

[Er2] J. Erickson, *New lower bounds for Hopcroft's problem*, Discrete Comput. Geom. 16 (1996), 389–418.

[Er3] J. Erickson, *Space-time tradeoffs for emptiness queries*, Proc. 13th Annu. ACM Sympos. Comput. Geom., 1997, pp. 304–313.

[ELS] G. Evangelidis, D. B. Lomet, and B. Salzberg, *The* hB$^{\Pi}$*-tree: A multi-attribute index supporting concurrency, recovery and node consolidation*, VLDB Journal 6 (1997), 1–25.

[F] C. Faloutsos, *Gray codes for partial match and range queries*, IEEE Trans. Softw. Eng. 44 (1988), 1381–1393.

[FG] C. Faloutsos and V. Gaede, *Analysis of n-dimensional quadtrees using the Hausdorff fractal dimension*, Proc. 22nd VLDB Conference, 1996, pp. 40–50.

[FL] C. Faloutsos and K.-I. Lin, *FastMap: A fast algorithm for indexing, data-mining and visualization of traditional and multimedia databases*, Proc. ACM SIGMOD Conf. on Management of Data, 1995, pp. 163–173.

[FRM] C. Faloutsos, M. Ranganathan and Y. Manolopoulos, *Fast subsequence matching in time-series databases*, Proc. ACM SIGMOD Conf. on Management of Data, 1994, pp. 86–93.

[FR] C. Faloutsos and Y. Rong, *DOT: A spatial access method using fractals*, Proc. 7th IEEE Internat. Conf. on Data Engineering, 1991, pp. 152–158.

[FRo] C. Faloutsos and S. Roseman, *Fractals for secondary key retrieval*, Proc. ACM SIGMOD Conf. on Management of Data, 1989, pp. 247–252.

[FB] R. A. Finkel and J. L. Bentley, *Quad trees: a data structure for retrieval on composite keys*, Acta Inform. 4 (1974), 1–9.

[FvDFH] J. D. Foley, A. van Dam, S. K. Feiner and J. F. Hughes, Computer Graphics: Principles and Practice, Addison-Wesley, Reading, MA, 1990.

[FW] F. W. Fredman and D. E. Willard, *Surpassing the information theoretic bound with fusion trees*, J. Comput. Syst. Sci. 48 (1993), 424–436.

[Fr] M. L. Fredman, *The complexity of maintaining an array and computing its partial sums*, J. ACM 29 (1979), 250–260.

[Fr2] M. L. Fredman, *Inherent complexity of range query problems*, Proc. 17th Allerton Conf. Commun. Control Comput., 1979, pp. 231–240.

[Fr3] M. L. Fredman, *The inherent complexity of dynamic data structures which accommodate range queries*, Proc. 21st Annu. IEEE Sympos. Found. Comput. Sci., 1980, pp. 191–199.

[Fr4] M. L. Fredman, *A lower bound on the complexity of orthogonal range queries*, J. ACM 28 (1981), 696–705.

[Fr5] M. L. Fredman, *Lower bounds on the complexity of some optimal data structures*, SIAM J. Comput. 10 (1981), 1–10.

[FV] M. L. Fredman and D. J. Volper, *The complexity of partial match retrieval in a dynamic setting*, J. Algorithms 3 (1982), 68–78.

[Fre] H. Freeston, *The BANG file: a new kind of grid file*, Proc. ACM SIGMOD Conf. on Management of Data, 1987, pp. 260–269.

[Free] M. Freestone, *A general solution of the n-dimensional B-tree problem*, Proc. ACM SIGMOD Conf. on Management of Data, 1995, pp. 80–91.

[FBF] J. H. Friedman, J. L. Bentley and R. A. Finkel, *An algorithm for finding best matches in logarithmic expected time*, ACM Trans. Math. Softw. 3 (1977), 209–226.

[FKN] H. Fuchs, Z. M. Kedem and B. Naylor, *On visible surface generation by a priori tree structures*, Comput Graph. 14 (1980), 124–133. Proc. SIGGRAPH '80.

[G] A. S. Glassner (ed.), An Introduction to Ray Tracing, Academic Press, 1989.

[GRSY] J. Goldstein, R. Ramakrishnan, U. Shaft and J.-B. Yu, *Processing queries by linear constraints*, Proc. 16th Annu. ACM Sympos. Principles Database Syst., 1997, pp. 257–267.

[GPW] J. E. Goodman, R. Pollack and R. Wenger, *Geometric transversal theory*, New Trends in Discrete and Computational Geometry (J. Pach, ed.), Springer-Verlag, Heidelberg, Germany, 1993, pp. 163–198.

[GT] M. T. Goodrich and R. Tamassia, *Dynamic ray shooting and shortest paths via balanced geodesic triangulations*, Proc. 9th Annu. ACM Sympos. Comput. Geom., 1993, pp. 318–327.

[GBLP] J. Gray, A. Bosworth, A. Layman and H. Patel, *Data cube: A relational aggregation operator generalizing group-by, cross-tab, and sub-totals*, Proc. 12th IEEE Internat. Conf. on Data Engineering, 1996, pp. 152–159.

[Gr] D. Greene, *An implementation and performance analysis of spatial data access methods*, Proc. 5th IEEE Internat. Conf. on Data Engineering, 1989, pp. 606–615.

[Gre] L. Greengard, The Rapid Evaluation of Potential Fields in Particle Systems, MIT Press, Cambridge, MA, 1988.

[GV] D. Grigoriev and N. Vorobjov, *Complexity lower bounds for computation trees with elementary transcendental function gates*, Theoret. Comput. Sci. 157 (1996), 185–214.

[GI] R. Grossi and G. F. Italiano, *Efficient splitting and merging algorithms for order decomposable problems*, Proc. 24th Internat. Colloq. Automata, Lang. Prog., Lecture Notes Comput. Sci., vol. 1256, Springer-Verlag, 1997, pp. 605–615.

[GOS] L. J. Guibas, M. H. Overmars and M. Sharir, *Ray shooting, implicit point location, and related queries in arrangements of segments*, Report 433, Dept. Comput. Sci., New York Univ., New York, NY, Mar. 1989.

[Gü] O. Günther, *The design of the cell tree: An object oriented index structure for geometric data bases*, Proc. 5th IEEE Internat. Conf. on Data Engineering, 1989, pp. 598–605.

[GB] O. Günther and J. Bilmes, *Tree based access methods for spatial databases: Implementation and performance evaluation*, IEEE Trans. Knowledge and Data Engineering 3 (1991), 342–356.

[GHRU] H. Gupta, V. Harinarayan, A. Rajaraman and J. D. Ullman, *Index selection for OLAP*, Proc. ACM SIGMOD Conf. on Management of Data, 1996, pp. 205–216.

[GJS] P. Gupta, R. Janardan and M. Smid, *Efficient algorithms for generalized intersection searching on non-iso-oriented objects*, Proc. 10th Annu. ACM Sympos. Comput. Geom., 1994, pp. 369–378.

[GJS2] P. Gupta, R. Janardan and M. Smid, *On intersection searching problems involving curved objects*, Proc. 4th Scand. Workshop Algorithm Theory, Lecture Notes Comput. Sci., vol. 824, Springer-Verlag, 1994, pp. 183–194.

[GJS3] P. Gupta, R. Janardan and M. Smid, *Further results on generalized intersection searching problems: counting, reporting and dynamization*, J. Algorithms, 19 (1995), 282–317.

[Güt] R. Güting, *An introduction to spatial database systems*, VLDB Journal 4 (1994), 357–399.

[Gu] A. Guttman, *R-trees: A dynamic index structure for spatial searching*, Proc. ACM SIGACT-SIGMOD Conf. Principles Database Systems, 1984, pp. 47–57.

[HT] D. Harel and R. E. Tarjan, *Fast algorithms for finding nearest common ancestors*, SIAM J. Comput. 13 (1984), 338–355.

[HRU] V. Harinarayan, A. Rajaraman and J. D. Ullman, *Implementing data cubes efficiently*, Proc. 13th IEEE Internat. Conf. on Data Engineering, 1997, pp. 208–219.

[HaT] T. Hastie and R. Tibshirani, *Discriminant adaptive nearest neighbor classification*, IEEE Trans. Pattern Anal. Mach. Intell. 18 (1996), 607–616.

[HW] D. Haussler and E. Welzl, *ε-nets and simplex range queries*, Discrete Comput. Geom. 2 (1987), 127–151.

[HKP] J. H. Hellerstein, E. Koustsoupias and C. H. Papadimtriou, *On the analysis of indexing schemes*, Proc. 16th Annu. ACM Sympos. Principles of Database Syst., 1997, pp. 249–256.

[H] A. Henrich, *Improving the performance of multi-dimensional access structures based on kd-trees*, Proc. 12th IEEE Intl. Conf. on Data Engineering, 1996, pp. 68–74.

[HS] J. Hershberger and S. Suri, *A pedestrian approach to ray shooting: Shoot a ray, take a walk*, J. Algorithms 18 (1995), 403–431.

[Hi] K. H. Hinrichs, *The grid file system: Implementation and case studies of applications*, Ph.D. thesis, Dept. Information Science, ETH, Zürich, 1985.

[HjS] G. R. Hjaltason and H. Samet, *Ranking in spatial databases*, Advances in Spatial Databases - Fourth International Symposium, Lecture Notes Comput. Sci., no. 951, Aug. 1995, pp. 83–95.

[HAMS] C.-T. Ho, R. Agrawal, N. Megiddo and R. Srikant, *Range queries in OLAP data cubes*, Proc. ACM SIGMOD Conf. on Management of Data, 1997, pp. 73–88.

[HBA] C.-T. Ho, J. Bruck and R. Agrawal, *Partial-sum queries in OLAP data cubes using covering codes*, Proc. Annu. ACM Sympos. Principles Database Syst., 1997, pp. 228–237.

[HoS] E. G. Hoel and H. Samet, *A qualitative comparison study of data structures for large line segment databases*, Proc. ACM SIGMOD Conf. on Management of Data, 1992, pp. 205–214.

[Ho] P. Houthuys, *Box sort, a multidimensional binary sorting method for rectangular boxes, used for quick range searching*, Visual Comput. 3 (1987), pp. 236–249.

[HSW] A. Hutflesz, H.-W. Six and P. Widmayer, *Globally order preserving multidimensional linear hashing*, Proc. 4th Intl. Conf. on Data Engineering, 1988, pp. 572–579.

[HSW2] A. Hutflesz, H.-W. Six and P. Widmayer, *Twin grid files: Space optimizing access schemes*, Proc. ACM SIGMOD Conf. on Management of Data, 1988, pp. 183–190.

[IA] H. Imai and T. Asano, *Dynamic orthogonal segment intersection search*, J. Algorithms 8 (1987), 1–18.

[IM] P. Indyk and R. Motwani, *Approximate nearest neighbors: Towards removing the curse of dimensionality*, Proc. 30th Annu. ACM Symps. Theory of Comput., 1998, to appear.

[IMRV] P. Indyk, R. Motwani, P. Raghavan and S. Vempala, *Locality-preserving hashing in multidimensional space*, Proc. 29th Annu. ACM Sympos. Theory Comput., 1997, pp. 618–625.

[J] H. V. Jagdish, *Linear clustering of objects with multiple attributes*, Proc. ACM SIGMOD Conf. on Management of Data, 1990, pp. 332–342.

[JL] R. Janardan and M. Lopez, *Generalized intersection searching problems*, Internat. J. Comput. Geom. Appl. 3 (1993), 39–69.

[KF] I. Kamel and C. Faloutsos, *Parallel R-trees*, Proc. ACM SIGMOD Conf. on Management of Data, 1992, pp. 195–204.

[KF2] I. Kamel and C. Faloutsos, *On packing R-trees*, Proc. 2nd Internat. Conf. on Information and Knowledge Management, 1993, pp. 490–499.

[KF3] I. Kamel and C. Faloutsos, *Hilbert R-tree: An improved R-tree using fractals*, Proc. 20th VLDB Conference, 1994, pp. 500–510.

[KRVV] P. C. Kanellakis, S. Ramaswamy, D. E. Vengroff and J. S. Vitter, *Indexing for data models with constraints and classes*, Proc. 12th ACM SIGACT-SIGMOD-SIGART Conf. Princ. Database Sys., 1993, pp. 233–243.

[KS] N. Katayama and S. Satoh, *The SR-tree: An index structure for high-dimensional nearest-neighbor queries*, Proc. ACM SIGMOD Conf. on Management of Data, 1997, pp. 369–380.

[KV] M. D. Katz and D. J. Volper, *Data structures for retrieval on square grids*, SIAM J. Comput. 15 (1986), 919–931.

[K] M. J. Katz, *3-D vertical ray shooting and 2-D point enclosure, range searching, and arc shooting amidst convex fat objects*, Research Report 2583, INRIA, BP93, 06902 Sophia-Antipolis, France, 1995.

[KNOW] R. Klein, O. Nurmi, T. Ottmann and D. Wood, *A dynamic fixed windowing problem*, Algorithmica 4 (1989), 535–550.

[Kl] J. Kleinberg, *Two algorithms for nearest-neighbor search in high dimension*, Proc. 29th Annu. ACM Sympos. Theory Comput., 1997, pp. 599–608.

[KK] V. Koivune and S. Kassam, *Nearest neighbor filters for multivariate data*, IEEE Workshop on Nonlinear Signal and Image Processing, 1995.

[KPW] J. Komlós, J. Pach and G. Woeginger, *Almost tight bounds for ϵ-nets*, Discrete Comput. Geom. 7 (1992), 163–173.

[KSFSP] F. Korn, N. Sidiropoulos, C. Faloutsos, E. Siegel and Z. Protopapa, *Fast nearest neighbor search in medical image database*, Proc. 22nd VLDB Conference, 1996, pp. 215–226.

[KrS] H.-P. Kriegel and B. Seeger, *Multidimensional order preserving linear hashing with partial expansions*, Proc. Intl. Conf. on Database Theory, Lecture Notes Comput. Sci., vol. 243, Springer-Verlag, 1986, pp. 203–220.

[KW] R. Krishnamurthy and K.-Y. Wang, *Multilevel grid files*, Tech. report, IBM T. J. Watson Center, Yorktown Heights, NY, 1985.

[L] J.-C. Latombe, Robot Motion Planning, Kluwer Academic Publishers, Boston, MA, 1991.

[LW] D. T. Lee and C. K. Wong, *Finding intersections of rectangles by range search*, J. Algorithms 2 (1981), 337–347.

[LLE] S. Leutenegger, M. A. Lopez and J. Edington, *STR: A simple and efficient algorithm for R-tree packing*, Proc. 13th IEEE Internat. Conf. on Data Engineering, 1997, pp. 497–506.

[LJF] K. Lin, H. V. Jagdish and C. Faloutsos, *The TV-tree: An index structure for higher dimensional data*, VLDB Journal 4 (1994), 517–542.

[LS] D. Lomet and B. Salzberg, *The hB-tree: A multiattribute indexing method with good guaranteed performance*, ACM Trans. Database Syst., 15 (1990), 625–658.

[Lu] G. S. Lueker, *A data structure for orthogonal range queries*, Proc. 19th Annu. IEEE Sympos. Found. Comput. Sci., 1978, pp. 28–34.

[MB] J. MacDanold and K. Booth, *Heuristics for ray tracing using space subdivision*, Visual Comput. 6 (1990), 153–166.

[M] J. Matoušek, *Efficient partition trees*, Discrete Comput. Geom. 8 (1992), 315–334.

[M2] J. Matoušek, *Reporting points in halfspaces*, Comput. Geom. Theory Appl. 2 (1992), 169–186.

[M3] J. Matoušek, *Epsilon-nets and computational geometry*, New Trends in Discrete and Computational Geometry (J. Pach, ed.), Springer-Verlag, Heidelberg, Germany, 1993, pp. 69–89.

[M4] J. Matoušek, *Linear optimization queries*, J. Algorithms, 14 (1993), 432–448.

[M5] J. Matoušek, *On vertical ray shooting in arrangements*, Comput. Geom. Theory Appl. 2 (1993), 279–285.

[M6] J. Matoušek, *Range searching with efficient hierarchical cuttings*, Discrete Comput. Geom. 10 (1993), 157–182.

[M7] J. Matoušek, *Geometric range searching*, ACM Comput. Surv. 26 (1994), 421–461.

[MS] J. Matoušek and O. Schwarzkopf, *On ray shooting in convex polytopes*, Discrete Comput. Geom. 10 (1993), 215–232.

[MW] J. Matoušek and E. Welzl, *Good splitters for counting points in triangles*, J. Algorithms 13 (1992), 307–319.

[MWW] J. Matoušek, E. Welzl and L. Wernisch,*Discrepancy and ε-approximations for bounded VC-dimension*, Combinatorica 13 (1993), 455–466.

[Mc] E. M. McCreight, *Priority search trees*, SIAM J. Comput. 14 (1985), 257–276.

[MCD] M. R. Mediano, M. A. Casanova and M. Dreux, *V-trees: A storage method for long vector data*, Proc. 20th VLDB Conference, 1994, pp. 321–329.

[Me] N. Megiddo, *Applying parallel computation algorithms in the design of serial algorithms*, J. ACM 30 (1983), 852–865.

[Meh] K. Mehlhorn, Multi-dimensional Searching and Computational Geometry, Data Structures and Algorithms, vol. 3, Springer-Verlag, Heidelberg, West Germany, 1984.

[MN] K. Mehlhorn and S. Näher, *Dynamic fractional cascading*, Algorithmica 5 (1990), 215–241.

[MG] H. Mehrotra and J. E. Gary, *Feature-based retrieval of similar shapes*, Proc. 9th IEEE Intl. Conf. on Data Engineering, 1996, pp. 108–115.

[Mei] S. Meiser, *Point location in arrangements of hyperplanes*, Inform. Comput. 106 (1993), 286–303.

[Mi] J. S. B. Mitchell, *Shortest paths among obstacles in the plane*, Proc. 9th Annu. ACM Sympos. Comput. Geom., 1993, pp. 308–317.

[MMS] J. S. B. Mitchell, D. M. Mount and S. Suri, *Query-sensitive ray shooting*, Proc. 10th Annu. ACM Sympos. Comput. Geom., 1994, pp. 359–368.

[Mo] G. Morton, A computer oriented geodetic data base and a new technique in file sequencing, Tech. report, IBM Ltd., Ottawa, Canada, 1966.

[MR] R. Motwani and P. Raghavan, Randomized Algorithms, Cambridge University Press, New York, NY, 1995.

[Mu] J. Mullin, *Spiral storage: Efficient dynamic hashing with constant-performance*, Comput. J. 28 (1985), 330–334.

[Mul] K. Mulmuley, *A fast planar partition algorithm, I*, J. Symbolic Comput. 10 (1990), 253–280.

[Mul2] K. Mulmuley, *Randomized multidimensional search trees: Dynamic sampling*, Proc. 7th Annu. ACM Sympos. Comput. Geom., 1991, pp. 121–131.

[Mul3] K. Mulmuley, *Randomized multidimensional search trees: Further results in dynamic sampling*, Proc. 32nd Annu. IEEE Sympos. Found. Comput. Sci., 1991, pp. 216–227.

[Mul4] K. Mulmuley, Computational Geometry: An Introduction Through Randomized Algorithms, Prentice Hall, Englewood Cliffs, NJ, 1994.

[NHS] J. Nievergelt, H. Hinterberger and K. C. Sevcik, *The grid file: An adaptable, symmetric multi-key file structure*, ACM Trans. Database Syst., 9 (1984), 38–71.

[NW] J. Nievergelt and P. Widmayer, *Guard files: Stabbing and intersection queries on fat spatial objects*, Comput. J. 36 (1993), 107–116.

[NW2] J. Nievergelt and P. Widmayer, *Spatial data structures: Concepts and design choices*, Algorithmic Foundations of Geographic Information Systems, Lecture Notes Comput. Sci., vol. 1340, Springer-Verlag, 1997, pp. 153–197.

[O] J. Orenstein, *A comparison of spatial query processing techniques for native and parameter spaces*, Proc. ACM SIGMOD Conf. on Management of Data, 1990, pp. 343–352.

[OM] J. Orenstein and T. H. Merrett, *A class of data structures for associative searching*, Proc. ACM SIGMOD Conf. on Management of Data, 1984, pp. 181–190.

[Ou] M. Ouksel, *The interpolation-based grid file*, Proc. 4th Annu. ACM Sympos. Principles of Database Syst., 1985, pp. 20–27.

[Ov] M. H. Overmars, The Design of Dynamic Data Structures, Lecture Notes Comput. Sci., vol. 156, Springer-Verlag, Heidelberg, West Germany, 1983.

[Ov2] M. H. Overmars, *Efficient data structures for range searching on a grid*, J. Algorithms 9 (1988), 254–275.

[OSdBvK] M. H. Overmars, M. H. M. Smid, M. T. de Berg and M. J. van Kreveld, *Maintaining range trees in secondary memory, part I: Partitions*, Acta Inform. 27 (1990), 423–452.

[OvdS] M. H. Overmars and A. F. van der Stappen, *Range searching and point location among fat objects*, Algorithms – ESA'94, Lecture Notes Comput. Sci., vol. 855, Springer-Verlag, 1994, pp. 240–253.

[P] J. Pach, *Notes on geometric graph theory*, Discrete and Computational Geometry: Papers from the DIMACS Special Year (J. E. Goodman, R. Pollack and W. Steiger, eds.), American Mathematical Society, Providence, RI, 1991, pp. 273–285.

[PST] B.-U. Pagel, H.-W. Six and H. Toben, *The transformation technique for spatial objects revisited*, Proc. 3rd Intl. Symp. on Large Spatial Databases, Lecture Notes Comput. Sci., vol. 692, Springer-Verlag, 1993, pp. 73–88.

[PM] A. Papadpoulos and Y. Manolopoulos, *Performance of nearest neighbor queries in R-trees*, Database Theory—ICDT'97, 6th International Conference, Lecture Notes Comput. Sci., vol. 1186, Springer, 1997, pp. 394–408.

[PY] M. S. Paterson and F. F. Yao, *Efficient binary space partitions for hidden-surface removal and solid modeling*, Discrete Comput. Geom. 5 (1990), 485–503.

[Pe] M. Pellegrini, *Ray shooting on triangles in 3-space*, Algorithmica 9 (1993), 471–494.

[Pe2] M. Pellegrini, *On point location and motion planning among simplices*, Proc. 25th Annu. ACM Sympos. Theory Comput., 1994, pp. 95–104.

[PS] M. Pellegrini and P. Shor, *Finding stabbing lines in 3-space*, Discrete Comput. Geom. 8 (1992), 191–208.

[PV] M. Pocchiola and G. Vegter, *Pseudo-triangulations: Theory and applications*, Proc. 12th Annu. ACM Sympos. Comput. Geom., 1996, pp. 291–300.

[PrS] F. P. Preparata and M. I. Shamos, Computational Geometry: An Introduction, Springer-Verlag, New York, NY, 1985.

[RP] V. Ramasubramanian and K. K. Paliwal, *Fast k-dimensional tree algorithms for nearest neighbor search with applications to vector quantization encoding*, IEEE Trans. Signal Processing, 40 (1992), 518–531.

[RS] S. Ramaswamy and S. Subramanian, *Path caching: A technique for optimal external searching*, Proc. 13th Annu. ACM Sympos. Principles Database Syst., 1994, pp. 25–35

[R] J. T. Robinson, *The k-d-B-tree: A search structure for large multidimensional dynamic indexes*, Proc. ACM SIGACT-SIGMOD Conf. Principles Database Systems, 1981, pp. 10–18.

[RKV] N. Roussopoulos, S. Kelley and F. Vincent, *Nearest neighbor queries*, Proc. ACM SIGMOD Conf. on Management of Data, 1995, pp. 71–79.

[RKR] N. Roussopoulos, Y. Kotidis and M. Roussopoulos, *Cubetree: Organization of and bulk incremental updates on the data cube*, Proc. ACM SIGMOD Conf. on Management of Data, 1997, pp. 89–99.

[S] H. Sagan, Space-filling Curves, Springer-Verlag, New York, NY, 1994.

[Sa] G. Salton, Automatic Text Processing, Addison-Wesley, Reading, MA, 1989.

[Sam] H. Samet, Applications of Spatial Data Structures: Computer Graphics, Image Processing, and GIS, Addison-Wesley, Reading, MA, 1990.

[Sam2] H. Samet, The Design and Analysis of Spatial Data Structures, Addison-Wesley, Reading, MA, 1990.

[SO] H. W. Scholten and M. H. Overmars, *General methods for adding range restrictions to decomposable searching problems*, J. Symbolic Comput. 7 (1989), 1–10.

[ST] E. Schömer and C. Thiel, *Efficient collision detection for moving polyhedra*, Proc. 11th Annu. ACM Sympos. Comput. Geom., 1995, pp. 51–60.

[Sc] O. Schwarzkopf, *Dynamic maintenance of geometric structures made easy*, Proc. 32nd Annu. IEEE Sympos. Found. Comput. Sci., 1991, pp. 197–206.

[SK] B. Seeger and H.-P. Kriegel, *The buddy-tree: An efficient and robust access method for spatial data base systems*, Proc. 16th VLDB Conference, 1990, pp. 590–601.

[Se] R. Seidel, *Backwards analysis of randomized geometric algorithms*, New Trends in Discrete and Computational Geometry (J. Pach, ed.), Springer-Verlag, Heidelberg, Germany, 1993, pp. 37–68.

[SRF] T. Sellis, N. Roussopoulos and C. Faloutsos, *The* R^+*-tree: A dynamic index for multi-dimensional objects*, Proc. 13th VLDB Conference, 1987, pp. 507–517.

[SeK] K. Sevcik and N. Koudas, *Filter trees for managing spatial data over a range of size granularities*, Proc. 22nd VLDB Conference, 1996, pp. 16–27.

[SA] M. Sharir and P. K. Agarwal, Davenport-Schinzel Sequences and Their Geometric Applications, Cambridge University Press, New York, 1995.

[SSA] K. Shim, R. Srikant and R. Agrawal, *High-dimensional similarity joins*, Proc. 13th IEEE Internat. Conf. on Data Engineering, 1997, pp. 301–311.

[SBK] B. K. Shoichet, D. L. Bodian and I. D. Kuntz, *Molecular docking using shape descriptors*, J. Computational Chemistry 13 (1992), 380–397.

[Sh] A. Shoshani, *OLAP and statistical databases: Similarities and differences*, Proc. ACM Sympos. Principles of Database Syst., 1997, pp. 185–196.

[Sm] M. Smid, *Algorithms for semi-online updates on decomposable problems*, Proc. 2nd Canad. Conf. Comput. Geom., 1990, pp. 347–350.

[Sm2] M. Smid, *Maintaining the minimal distance of a point set in less than linear time*, Algorithms Rev. 2 (1991), 33–44.

[So] D. M. Y. Sommerville, Analytical Geometry in Three Dimensions, Cambridge University Press, Cambridge, UK, 1951.

[Sp] R. F. Sproull, *Refinements to nearest-neighbor searching*, Algorithmica 6 (1991), 579–589.

[SY] J. M. Steele and A. C. Yao, *Lower bounds for algebraic decision trees*, J. Algorithms 3 (1982), 1–8.

[St] J. Stolfi, Oriented Projective Geometry: A Framework for Geometric Computations, Academic Press, New York, NY, 1991.

[Str] V. Strassen, *Algebraic complexity theory*, Algorithms and Complexity (J. van Leeuwen, ed.), Handbook of Theoretical Computer Science, vol. A, MIT Press, Cambridge, MA, 1990, pp. 633–672.

[SR] S. Subramanian and S. Ramaswamy, *The P-range tree: A new data structure for range searching in secondary memory*, Proc. 6th ACM-SIAM Sympos. Discrete Algorithms, 1995, pp. 378–387.

[T] R. E. Tarjan, *A class of algorithms which require nonlinear time to maintain disjoint sets*, J. Comput. Syst. Sci. 18 (1979), 110–127.

[V] P. M. Vaidya, *Space-time tradeoffs for orthogonal range queries*, SIAM J. Comput. 18 (1989), 748–758.

[VW] V. K. Vaishnavi and D. Wood, *Rectilinear line segment intersection, layered segment trees and dynamization*, J. Algorithms 3 (1982), 160–176.

[vK] M. J. van Kreveld, *New results on data structures in computational geometry*, Ph.D. dissertation, Dept. Comput. Sci., Utrecht Univ., Utrecht, Netherlands, 1992.

[vKO] M. J. van Kreveld and M. H. Overmars, *Divided k-d trees*, Algorithmica 6 (1991), 840–858.

[vKO2] M. J. van Kreveld and M. H. Overmars, *Concatenable structures for decomposable problems*, Inform. Comput. 110 (1994), 130–148.

[VV] D. E. Vengroff and J. S. Vitter, *Efficient 3-d range searching in external memory*, Proc. 28th Annu. ACM Sympos. Theory Comput., 1996, pp. 192–201.

[vzG] J. von zur Gathen, *Algebraic complexity theory*, Annu. Rev. Comput. Sci., vol. 3, Annual Reviews, Palo Alto, CA, 1988, pp. 317–347.

[Vu] J. Vuillemin, *A unifying look at data structures*, Commun. ACM 23 (1980), 229–239.

[W] E. Welzl, *Partition trees for triangle counting and other range searching problems*, Proc. 4th Annu. ACM Sympos. Comput. Geom., 1988, pp. 23–33.

[W2] E. Welzl, *On spanning trees with low crossing numbers*, Data Structures and Efficient Algorithms, Final Report on the DFG Special Joint Initiative, Lecture Notes Comput. Sci., vol. 594, Springer-Verlag, 1992, pp. 233–249.

[We] H. Weyl, *Randbemerkungen zu Hauptproblemen der Mathematik, II, Fundamentalsatz der Alegbra und Grundlagen der Mathematik*, Math. Z. 20 (1924), 131–151.

[WSCK+] K. Y. Whang, J. W. Song, J. W. Chang, J. Y. Kim, W. S. Cho, C. M. Park and I. Y. Song, *Octree-R: An adaptible octree for efficient ray tracing*, IEEE Trans. Visual. and Comp. Graphics 1 (1995), 343–349.

[WJ] D. A. White and R. Jain, *Similarity indexing with the SS-tree*, Proc. 12th IEEE Internat. Conf. on Data Engineering, 1996, pp. 516–523.

[Wh] M. White, *N-trees: Large ordered indexes for multidimensional space*, Tech. report, US Bureau of the Census, Statistical Research Division, Washington, DC, 1982.

[Wi] D. E. Willard, *Polygon retrieval*, SIAM J. Comput. 11 (1982), 149–165.

[Wi2] D. E. Willard, *Lower bounds for the addition-subtraction operations in orthogonal range queries and related problems*, Inform. Comput. 82 (1989), 45–64.

[Wi3] D. E. Willard, *Applications of the fusion tree method to computational geometry and searching*, Proc. 3rd ACM-SIAM Sympos. Discrete Algorithms, 1992, pp. 286–296.

[Wi4] D. E. Willard, *Applications of range query theory to relational data base join and selection operations*, J. Comput. Syst. Sci. 52 (1996), 157–169.

[WL] D. E. Willard and G. S. Lueker, *Adding range restriction capability to dynamic data structures*, J. ACM 32 (1985), 597–617.

[Y] A. C. Yao, *Space-time trade-off for answering range queries*, Proc. 14th Annu. ACM Sympos. Theory Comput., (1982), pp. 128–136.

[Y2] A. C. Yao, *On the complexity of maintaining partial sums*, SIAM J. Comput. 14 (1985), 277–288.

[YY] A. C. Yao and F. F. Yao, *A general approach to D-dimensional geometric queries*, Proc. 17th Annu. ACM Sympos. Theory Comput., (1985), pp. 163–168.

[Ya] F. F. Yao, *A 3-space partition and its applications*, Proc. 15th Annu. ACM Sympos. Theory Comput., 1983, pp. 258–263.

[YDEP] F. F. Yao, D. P. Dobkin, H. Edelsbrunner and M. S. Paterson, *Partitioning space for range queries*, SIAM J. Comput. 18 (1989), 371–384.

[Yi] P. N. Yianilos, *Data structures and algorithms for nearest neighbor search in general metric spaces*, Proc. 4th ACM-SIAM Sympos. Discrete Algorithms, 1993, pp. 311–321.

Center for Geometric Computing, Department of Computer Science, Duke University, Box 90129, Durham, NC 27708-0129

E-mail address: pankaj@cs.duke.edu, *URL*: **http://www.cs.duke.edu/~pankaj**

Center for Geometric Computing, Department of Computer Science, Duke University, Box 90129, Durham, NC 27708-0129

E-mail address: jeffe@cs.duke.edu, *URL*: **http://www.cs.duke.edu/~jeffe**

Contemporary Mathematics
Volume **223**, 1999

Deformed Products and Maximal Shadows of Polytopes

Nina Amenta and Günter M. Ziegler

ABSTRACT. We present a construction of *deformed products* of polytopes that has as special cases all the known constructions of linear programs with "many pivots," starting with the famous Klee-Minty cubes from 1972.

Thus we obtain sharp estimates for the following geometric quantities for d-dimensional simple polytopes with at most n facets:

- the maximal number of vertices on an increasing path,
- the maximal number of vertices on a "greedy" greatest increase path, and
- the maximal number of vertices of a 2-dimensional projection.

This, equivalently, provides good estimates for the worst-case behaviour of the simplex algorithm on linear programs with these parameters with the worst-possible, the greatest increase, and the shadow vertex pivot rules.

The bounds on the maximal number of vertices on an increasing path or a greatest increase path unify and slightly improve a number of known results. The bound on the maximal number of vertices of a 2-dimensional projection is new: we show that a 2-dimensional projection of a d-dimensional polytope with n facets may have as many as $\Theta(n^{\lfloor d/2 \rfloor})$ vertices for fixed d. This provides the same bound for the worst-case behaviour of the simplex algorithm with the shadow vertex pivot rule. The maximal complexity of shadows in fixed dimension is also relevant for problems of Computational Geometry. We give a new algorithm for the construction of the shadow of a d-dimensional polytope.

However, we find that for even $d \geq 4$ the polars of cyclic polytopes, $C_d(n)^{\Delta}$, which have the maximal number of vertices for any given n, do not maximize the shadow: for example, any 2-dimensional projection of $C_4(n)^{\Delta}$ has not more than $3n$ vertices.

1991 *Mathematics Subject Classification.* Primary 90C05, 52Bxx, 68U05.

Key words and phrases. Deformed products of polytopes, linear programs with many pivots, increasing paths, Klee-Minty cubes, upper bound theorem, pivot rules in linear programming, shadow vertex rule, 2-dimensional projections, complexity of shadows, cyclic polytopes.

Most of this work was done at The Geometry Center, supported by NSF/DMS-8920161, at the Freie Universität Berlin, supported by the DAAD, and at Xerox PARC, partially supported by NSF/CCR-9404113.

Supported by a DFG Gerhard-Hess-Forschungsförderungspreis (Zi 475/1-1) and by the German-Israeli Foundation grant I-0309-146.06/93.

© 1999 American Mathematical Society

1. Introduction

In 1965, Victor Klee wrote:

> "Before attacking a linear programming problem with the simplex algorithm, it can be very comforting to have a good estimate of the number of iterations which may be required in order to reach a solution. [...] In addition to the primary interest in the expected number of iterations, there is a strong secondary interest in the *maximum number* of iterations for the problem of a given size. Here [...] the still unsolved problem of determining the maximum number has been of central interest since the inception of the simplex method." (Klee [**Kl**, p. 313])

In other words, Klee asked for the largest possible number $H_{\text{Da}}(d,n)$ of bases that might be visited by the simplex algorithm (with Dantzig's largest coefficient rule) on a linear programming problem that defines a (simple, bounded) d-dimensional polytope with at most n facets. This problem is still not solved. However, obvious upper bounds are given by

$$H_{\text{Da}}(d,n) \;\leq\; H(d,n) \;\leq\; M(d,n), \tag{1}$$

where $H(d,n)$ is the maximal number of vertices in an increasing path on such a polytope, and $M(d,n)$ is the maximal number of vertices that such a polytope may have.

Quite a different line of study, questions of Computational Geometry that arise, for example, in Robotics, lead one to ask for the maximal number of vertices that the k-dimensional projection ("shadow") of a d-dimensional polytope with n facets may have. If the maximal quantity is denoted by $H_{k\text{-sh}}(d,n)$, then again it is clear that one has upper bounds given by

$$H_{2\text{-sh}}(d,n) \;\leq\; H_{k\text{-sh}}(d,n) \;\leq\; M(d,n). \tag{2}$$

This also provides an upper bound for the worst-case behaviour of the simplex algorithm with the shadow vertex pivot rule, which is also given by $H_{2\text{-sh}}(d,n)$. Thus it complements Borgwardt's (much lower) linear bounds for the average case behaviour of this pivot rule.

Already in 1957, T. S. Motzkin had formulated that $M(d,n)$ is given by the number of vertices of the polars (or duals) of cyclic polytopes $C_d(n)$; this conjecture, which became known as the Upper Bound Conjecture, was proved in 1970 by McMullen.

THEOREM 1.1. **(Upper Bound Theorem,** McMullen [**Mc**] [**Zi**, Sect. 8.4]) *A d-dimensional polytope with n facets has no more than*

$$M(d,n) \;:=\; \binom{n-\lceil\frac{d}{2}\rceil}{\lfloor\frac{d}{2}\rfloor}+\binom{n-1-\lceil\frac{d-1}{2}\rceil}{\lfloor\frac{d-1}{2}\rfloor} \;=\; \begin{cases} \frac{n}{n-k}\binom{n-k}{k} & \text{for } d=2k \text{ even,}\\ 2\binom{n-k-1}{k} & \text{for } d=2k+1 \text{ odd.}\end{cases}$$

vertices, where equality is attained by all polars of neighborly polytopes and only by those (in particular, by the polars of cyclic polytopes).

This upper bound is a polynomial in n of degree $\lfloor\frac{d}{2}\rfloor$ in the case of *constant dimension*:

$$M(d,n) = O(n^{\lfloor\frac{d}{2}\rfloor}) \qquad \text{for fixed } d.$$

It is exponential in the *diagonal case* (for constant n/d); for example, for $n = 2d$ it grows like

$$M(d, 2d) \approx \left(\tfrac{27}{4}\right)^{\lfloor d/2 \rfloor}.$$

The Upper Bound Theorem does not, however, answer the original questions: it only provides a (large) upper bound. Klee & Minty's classic 1972 paper [**KM**] established that the quantities $H_{\mathrm{Da}}(d,n)$ and $H(d,n)$, at least asymptotically, grow as badly as $M(d,n)$—both in the *diagonal case* where n/d is constant, and in the case of *constant dimension*.

THEOREM 1.2. (**Long increasing paths,** Klee & Minty [**KM**]) *Increasing paths, even those followed by the simplex algorithm with Dantzig's rule, on a d-dimensional polytope with 2d facets may visit as many as*

$$2^d \quad \leq \quad H_{\mathrm{Da}}(d, 2d) \quad \leq \quad M(d, 2d)$$

distinct vertices (that is, exponentially many in the diagonal case), and they have a bound

$$\frac{1}{2^{\lfloor d/2 \rfloor^2}} \quad \leq \quad \liminf_{n \longrightarrow \infty} \frac{H_{\mathrm{Da}}(d,n)}{n^{\lfloor \frac{d}{2} \rfloor}} \quad \leq \quad \liminf_{n \longrightarrow \infty} \frac{M(d,n)}{n^{\lfloor \frac{d}{2} \rfloor}}$$

for the case of constant dimension d.

Analogous results were subsequently obtained by various authors, in particular

- by Jeroslow [**J**] and Blair [**B**] for the greatest increase rule,
- by Goldfarb & Sit [**GSi**] for the diagonal case of the steepest increase rule,
- by Avis & Chvátal [**AC**] for the diagonal case of Bland's least index rule,
- and by Murty [**Mu**] and Goldfarb [**G, G2**] for the diagonal case of the shadow vertex (Gass-Saaty) rule.

It turns out that *all* these "bad examples" were constructed by (clever) variations of the Klee-Minty method of "tilting facets" or of "deforming products":

> "Then Q is obtained from the prism $[0,1] \times P$ by tilting the left base $\{0\} \times P$ in one way and the right base $\{1\} \times P$ in the opposite way." (Klee & Minty [**KM**, p. 162])

> "the main difficulty to overcome is to insure that the deformed polytope is combinatorially equivalent to $V \times P$." (Jeroslow [**J**, p. 370])

In view of this, it is quite surprising that a proper definition of "deformed products" apparently has never been given. The main point of the present article is to provide such a definition—see Section 3—and to show that it subsumes *all* the main constructions of linear programs on which the simplex algorithm with various pivot rules needs an exponential number of steps: they *are* deformed products. In the process of recreating these "old examples" as deformed products, we also sharpen the analysis, obtaining somewhat better lower bounds than were previously available (Section 4).

As a new result, we produce, also as deformed products, d-dimensional polytopes for which some 2-dimensional projection has as many as

$$H_{2\text{-sh}}(d,n) = \Theta(n^{\lfloor d/2 \rfloor})$$

vertices, for fixed d. In Section 5.2 we also outline the relevance of this for questions of Computational Geometry. The computation of k-shadows is a very natural problem also because it interpolates between linear programming and the convex hull problem, where linear programming is concerned with computing a 1-shadow, while a convex hull computation determines the d-shadow.

As a surprise in the investigation of shadows we find that the 2-shadows of the polars $C_d(n)^\Delta$ of cyclic polytopes have $O(n^{\lfloor (d-1)/2 \rfloor})$ vertices. Thus, in even dimensions, the polars of cyclic polytopes, which provide the maximal numbers $M(d,n)$ of vertices for any given n, do not maximize the shadow. This suggests that the answer to Klee's initial question will not be quite as simple as stating $H_{\mathrm{Da}}(d,n) = M(d,n)$ for all $n > d \geq 0$, although that has not been disproved.

2. Some preliminaries

2.1. Polytopes. A *polytope* is a finite intersection of closed half-spaces in $\mathbb{R}^d$, that is, a bounded set of the form $P = \{x \in \mathbb{R}^d\colon Ax \leq b\}$ with $b \in \mathbb{R}^n$ and $A \in \mathbb{R}^{n \times d}$, where $Ax \leq b$ is the matrix notation for a system of n linear inequalities. Equivalently [**Zi**, Chap. 1], a polytope is the convex hull of a finite set of points in $\mathbb{R}^d$, written $P = \operatorname{conv}\{p_1, \dots, p_m\}$ for points $p_i \in \mathbb{R}^d$. The *dimension* of a polytope is the dimension of its affine hull. A *d-polytope* is a polytope of dimension d.

An inequality "$a^t x \leq \alpha$" (where a and x are column vectors in $\mathbb{R}^d$, a^t is the transpose of a, and α is a scalar) is *valid* for $P \subseteq \mathbb{R}^d$ if it is satisfied for all $x \in P$. The *faces* of P are all the subsets of the form $F = \{x \in P\colon a^t x = \alpha\}$, where $a^t x \leq \alpha$ is a valid inequality. The faces of a polytope are themselves polytopes. (This includes $\emptyset$ and P as the *trivial* faces of P.) The faces of a d-polytope of dimensions 0, 1, $d-2$ and $d-1$ are called the *vertices*, *edges*, *ridges* and *facets* of P, respectively. A *k-face* is a k-dimensional face.

LEMMA 2.1. *Let $P = \operatorname{conv}\{p_i, \dots, p_m\} \subseteq \mathbb{R}^d$. Then for any $I \subseteq \{1, \dots, m\}$, the subset $F := \operatorname{conv}\{p_i : i \in I\}$ is a face of P if and only if there is a linear inequality $a^t x \leq \alpha$ such that $a^t p_i \leq \alpha$ holds for all i, with equality if and only if $i \in I$.*

PROOF. See e.g., [**Zi**, Prop. 2.3]. □

A d-polytope is *simple* if every vertex lies on exactly d facets, or (equivalently) on exactly d edges. A *polar* of a d-polytope $P \subseteq \mathbb{R}^d$ is given by

$$P^\Delta \ := \ \{a \in \mathbb{R}^d\colon a^t x \leq 1 \text{ for all } x \in P'\},$$

where P' denotes a translate of P that has the origin 0 in its interior. Polars of simple polytopes are *simplicial*, that is, all their proper faces are simplices.

In the following, a polytope *of type* (d,n) is a simple d-dimensional polytope with at most n facets. We will usually write $n > d \geq 0$, where it is understood that the parameters n and d satisfy $n > d > 1$, or $(d,n) = (1,2)$, or $(d,n) = (0,0)$. We refer, for example, to [**Zi**] for other basic combinatorial properties of convex polytopes.

DEFINITION 2.2. **(Combinatorially equivalent polytopes)** Two polytopes P and Q are *combinatorially equivalent* if there is a bijection between their vertex sets $\operatorname{vert}(P) = \{p_1, \dots, p_m\}$ and $\operatorname{vert}(Q) = \{q_1, \dots, q_m\}$ such that for any subset

$I \subseteq \{1, \dots, m\}$, the convex hull $\operatorname{conv}\{p_i : i \in I\}$ is a face of P if and only if the convex hull $\operatorname{conv}\{q_i : i \in I\}$ is a face of Q.

DEFINITION 2.3. **(Normally equivalent polytopes)** Two combinatorially equivalent d-polytopes $P, Q \subseteq \mathbb{R}^d$ are *normally equivalent* if additionally the unit facet normals of their corresponding facets coincide, that is, if each facet $\operatorname{conv}\{p_i : i \in I\}$ of P is parallel to the corresponding facet $\operatorname{conv}\{q_i : i \in I\}$ of Q.

An Isomorphism Lemma of the type below was already used by Klee & Minty [**KM**, p. 167] [**J**, p. 371] [**GSi**, p. 278]. Our version adds a dimension condition (which is essential: otherwise see [**Zi**, p. 71] for a counterexample!), but slightly weakens the conditions otherwise.

LEMMA 2.4. **(Isomorphism Lemma)** *Let* $P = \operatorname{conv}\{p_1, \dots, p_m\}$ *and* $Q = \operatorname{conv}\{q_1 \dots, q_m\}$ *be two polytopes with* $\dim(Q) \leq \dim(P)$ *such that for each index set* $I \subseteq \{1, 2, \dots, m\}$ *one has*

$$\operatorname{conv}\{p_i \colon i \in I\} \text{ is a facet of } P \quad \Longrightarrow \quad \operatorname{conv}\{q_i \colon i \in I\} \text{ is a face of } Q.$$

Then the correspondence $p_i \longleftrightarrow q_i$ *defines a combinatorial isomorphism between* P *and* Q*, that is,* $\operatorname{conv}\{p_i \colon i \in I\}$ *is a face (resp. facet) of* P if and only if $\operatorname{conv}\{q_i \colon i \in I\}$ *is a face (resp. facet) of* Q.

PROOF. The faces are exactly the intersections of facets. Hence

$$\operatorname{conv}\{p_i \colon i \in I\} \text{ is a face of } P \quad \Longrightarrow \quad \operatorname{conv}\{q_i \colon i \in I\} \text{ is a face of } Q.$$

Since the dimension of a face G of Q is the largest l such that there is a chain of faces of Q of the form $\emptyset \subset F_0 \subset F_1 \subset \dots \subset F_{l-1} \subset F_l = G$, this implies $\dim(Q) \geq \dim(P)$. Hence $\dim(P) = \dim(Q)$, and

$$\operatorname{conv}\{p_i \colon i \in I\} \text{ is a } k\text{-face of } P \quad \Longrightarrow \quad \operatorname{conv}\{q_i \colon i \in I\} \text{ is a } k\text{-face of } Q.$$

In particular, all facets of P correspond to facets of Q.

Now we argue by induction on the dimension, where the situation is trivial for $\dim(P) = \dim(Q) \leq 1$. Assume that some facets of Q are induced by P, some are not. Since the facet graph is connected, we can find adjacent facets G, G' of Q, with $G = \operatorname{conv}\{q_j : j \in J\}$ and $G' = \operatorname{conv}\{q_j : j \in J'\}$, such that $F := \operatorname{conv}\{p_j : j \in J\}$ is a facet of P, but $\operatorname{conv}\{p_j : j \in J'\}$ is not. Then the ridge

$$H := G \cap G' = \operatorname{conv}\{q_j : j \in J \cap J'\}$$

is a facet of G, hence (by induction applied to G) we know that $K := \operatorname{conv}\{p_j : j \in J \cap J'\}$ is a $(\dim(P) - 2)$-face of P, a facet of F. Let $F'' = \operatorname{conv}\{p_j : j \in J''\}$ be the second facet of P that contains K. Thus $G'' := \operatorname{conv}\{q_j : j \in J''\}$ is different from G and G', and it contains the ridge H: contradiction. □

A *cyclic polytope* $C_d(n)$ is a polytope that is combinatorially equivalent to the polytope

$$\operatorname{conv}\{\gamma(1), \gamma(2), \dots, \gamma(n)\} \subseteq \mathbb{R}^d,$$

where $\gamma(t) := (t, t^2, \dots, t^d)$ denotes the *moment curve* in $\mathbb{R}^d$. The cyclic polytopes $C_d(n)$ are simplicial, and they are *neighborly*: any subset of $\lfloor \frac{d}{2} \rfloor$ of its vertices is the vertex set of a face.

The *product* of two polytopes $P \subseteq \mathbb{R}^d$ and $Q \subseteq \mathbb{R}^e$ is given by

$$P \times Q \quad = \quad \left\{ \begin{pmatrix} x \\ u \end{pmatrix} : \begin{array}{c} x \in P \\ u \in Q \end{array} \right\}.$$

The vertex set of a product is given by

$$\operatorname{vert}(P \times Q) \quad = \quad \left\{ \begin{pmatrix} p_i \\ q_j \end{pmatrix} : \begin{array}{c} p_i \in \operatorname{vert}(P) \\ q_j \in \operatorname{vert}(Q) \end{array} \right\}$$

while the facet-defining inequalities for $P \times Q$ are just the inequalities of P together with those of Q. (Thus products of simple polytopes are simple as well.) Here $\binom{x}{u}$ denotes the concatenation of the column vectors x and u, and $\operatorname{vert}(P)$ is the set of vertices of P. Thus taking products of polytopes the numbers of vertices are multiplied, while the numbers of facets are just added.

Example 2.5. **(Polytopes with many vertices)** Polytopes with "many vertices and few facets" can be constructed by taking products of 2-polytopes, as follows.

If d is even, assume that n is a multiple of $\frac{d}{2}$. Then the product of $\frac{d}{2}$ copies of a $(\frac{2n}{d})$-gon $C_2(\frac{2n}{d})$ is a simple d-polytope $C_2(\frac{2n}{d})^{d/2}$ with n facets and $(\frac{2n}{d})^{d/2}$ vertices.

If d is odd, assume that $n-2$ is a multiple of $\frac{d-1}{2}$. Then the product of an interval I with $\frac{d-1}{2}$ copies of a $(\frac{2(n-2)}{d-1})$-gon $C_2(\frac{2(n-2)}{d-1})$ is a simple d-polytope $I \times C_2(\frac{2(n-2)}{d-1})^{(d-1)/2}$ with n facets and $2(\frac{2(n-2)}{d-1})^{(d-1)/2}$ vertices.

For instance, for $n = 2d$ the two constructions in Example 2.5 yield the d-cubes C_d: simple polytopes of type $(d, 2d)$, with 2^d vertices. Similarly, we get an exponential number of vertices (in d) if n is any larger fixed multiple of d.

Although for fixed dimension d the products in Example 2.5 have $O(n^{\lfloor d/2 \rfloor})$ vertices, equality for the upper bound theorem is attained only for the polar polytopes of neighborly polytopes.

2.2. Linear programming. In geometric terms, linear programming (via the simplex algorithm) deals with edge paths on polyhedra that are strictly increasing with respect to some linear function $c\colon \mathbb{R}^d \longrightarrow \mathbb{R}$. Thus we study linear programming problems of the type

$$\max c^t x\colon \quad Ax \leq b,$$

where $c \in \mathbb{R}^d$, $b \in \mathbb{R}^n$ and $A \in \mathbb{R}^{n \times d}$. Writing $a_1^t, \ldots, a_n^t$ for the rows of A, the i-th inequality of the system is given by $a_i^t x \leq b_i$. The assignment $x \longmapsto c^t x$ represents a linear objective function. Our constructions below maintain all the following nondegeneracy assumptions, although none of them is essential for what we do.

Feasibility: P is non-empty.

Full dimension: $P := \{x \in \mathbb{R}^d\colon Ax \leq b\}$ is a d-dimensional polyhedron with n (distinct) facets (that is, each of the inequalities in the system $Ax \leq b$ defines a facet of P, and all these facets are distinct).

Boundedness: P is a polytope. Hence $c^t x$ is bounded on P for *every* choice of $c \in \mathbb{R}^d$.

Phase I: A start vertex is given. Usually, we furthermore assume that the start vertex is the vertex of P that *minimizes* the objective function.

Primal nondegeneracy: The polytope P is *simple*, that is, every vertex of P lies on exactly d facets (equivalently, every vertex is adjacent to exactly d edges).

Dual nondegeneracy: No two vertices have the same objective function value $c^t v$. In particular, there are no horizontal edges with respect to the objective function.

With the primal nondegeneracy assumption, a *vertex* of the linear program is a point $v \in \mathbb{R}^d$ that satisfies $a_i^t v = b_i$ for exactly d indices $i \in I(v) := \{i_1, i_2, \ldots, i_d\}_<$, while $a_i^t v < b_i$ holds for all $i \in \{1, 2, \ldots, n\} \backslash I(v)$.

The (geometric) simplex algorithm [**Zi**, Sect. 3.2] proceeds from any vertex v to a *better adjacent* vertex, that is, to a vertex v' such that $I(v)$ and $I(v')$ differ in exactly one element, and such that $c^t v < c^t v'$. A *pivot rule* determines which vertex to choose. Here we mostly concentrate on geometrically defined rules. We do not consider randomized pivot rules, although they are very interesting [**K**] [**MSW**]. The same applies for "affirmative action" pivot rules, where exponential lower bounds in the worst case are not available [**Z**] [**FT**]. We do consider the following deterministic and non-deterministic pivot rules.

Greatest increase rule: Choose the vertex v' that gives the greatest increase in the objective function, that is, such that $c^t v' - c^t v$ is maximal.

Dantzig's largest coefficient rule: Choose the vertex v' such that

$$I(v) \backslash I(v') = \{i\},$$

where γ_i is the largest (positive!) coefficient in the (unique) representation of the objective function as $c^t x = c^t v + \sum_{i \in I(v)} \gamma_i a_i^t x$. (The coefficients γ_i in this representation are known as the *reduced costs* or *nonbasic gradients*. The positive reduced cost coefficients correspond to the legal pivots at v [**Kl**, Sect. I] [**KM**, Sect. 7].)

Bland's least index rule: Choose v' such that the (unique) index in

$$I(v) \backslash I(v')$$

is minimal, that is, choose the edge that leaves the facet with the smallest number (according to the fixed numbering of the facets resp. inequalities) [**Bl**].

The Gass-Saaty shadow vertex rule: For the starting vertex one constructs a linear functional $\widetilde{c} \in \mathbb{R}^d$ such that v_0 (uniquely) maximizes $\widetilde{c}^t x$ over P. (E. g., take $\widetilde{c} := \sum_{i \in I(v_0)} a_i$). Then given that v_{i-1} maximizes $(\widetilde{c} + \lambda c)^t x$ exactly for $\lambda \in [\lambda_{i-1}, \lambda_i]$, the next vertex v_i is chosen to maximize $(\widetilde{c} + \lambda c)^t x$ for all $\lambda \in [\lambda_i, \lambda_{i+1}]$. This is equivalent to the condition that v_i must maximize the ratio $\frac{\widetilde{c}^t (v_i - v_{i-1})}{c^t (v_i - v_{i-1})}$.
Thus, in geometric terms, the shadow vertex rule finds a path that projects (under $x \longmapsto (c^t x, \widetilde{c}^t x)$) to a path on the boundary of a 2-dimensional shadow [**GS**] [**Bo**].

OBSERVATION 2.6. **(A relation between Dantzig's and Bland's rules)** If we multiply the i-th inequality $a_i^t x \leq b_i$ in a linear program by δ^{1-i} for some $\delta > 0$, then a_i is replaced by $a_i' = \delta^{1-i} a_i$, b_i is replaced by $b_i' = \delta^{1-i} b_i$, the polytope and the objective function do not change at all, and the reduced costs are scaled by

$\gamma_i' := \delta^{i-1}\gamma_i$. Hence, if we choose $\delta > 0$ small enough, then at every vertex we obtain $\gamma_i' > \gamma_j'$ whenever $i, j \in I(v)$ and $i < j$.

In other words, Dantzig's rule follows the same sequence of pivots that is chosen by Bland's rule, provided that we suitably scale the inequalities.

General theory **[S]** **[GLS]** implies that all the constructions in the following can be realized with small coefficients. Thus every time we exhibit for some pivot rule paths whose length is exponential in the parameters d and n, this also provides examples for which the number of steps is exponential if measured in the (bit) encoding length of the linear programs in question.

2.3. Some basic observations.

DEFINITION 2.7. **(Basic counting functions)** For d-dimensional polytopes with at most n facets, and linear functions φ on them, we study and relate the following quantities:

$M(d,n)$: the maximal number of vertices,

$H(d,n)$: the maximal number of vertices on an increasing path,

$H_{\text{gi}}(d,n)$: the maximal number of vertices on a *reversible* greatest increase path, that is, the maximal number l such that the greatest increase paths with respect to φ and to $-\varphi$, starting at the vertices that minimize φ respectively $-\varphi$, both contain at least l vertices,

$H_{\text{Da}}(d,n)$: the maximal number of vertices on a reversible path for Dantzig's largest coefficient rule,

$H_{\text{Bl}}(d,n)$: the maximal number of vertices on a reversible path for Bland's least index rule,

$H_{\text{2-sh}}(d,n)$: the maximal number of vertices on a 2-dimensional projection (a "shadow").

(Counting the vertices rather than the edges on monotone paths amounts to a shift of 1: this simplifies some of the formulas in the following.)

LEMMA 2.8. *For all $n > d \geq 0$,*

$$H_{\text{gi}}(d,n) \leq H(d,n) \leq M(d,n) \tag{3}$$

$$H_{\text{gi}}(d,n) + (d-1) \leq M(d,n) \tag{4}$$

$$H_{\text{Bl}}(d,n) \leq H_{\text{Da}}(d,n) \leq H(d,n) \leq M(d,n) \tag{5}$$

$$H_{\text{2-sh}}(d,n) \leq H(d,n) \leq M(d,n) \tag{6}$$

PROOF. Inequality (4) follows from the observation that among the (at least) d neighbors of the starting vertex, the greatest increase rule will visit only one: the one which provides the greatest increase. Inequality (5) is a consequence of Observation 2.6. The first inequality in (6), on the other side, follows from the following lemma. □

LEMMA 2.9. *For all $n > d \geq 0$, $H_{\text{2-sh}}(d,n)$ is the maximal number of vertices on a simplex path according to the shadow vertex rule on a nondegenerate linear program of type (d,n).*

PROOF. Every 2-dimensional k-gon Q is projectively equivalent to a k-gon Q' such that all k vertices lie on a path that is increasing with respect to a given linear function $\alpha : \mathbb{R}^2 \longrightarrow \mathbb{R}$. If Q is the shadow of a d-polytope P (under the projection to the last two coordinates, say), then the projective transformation extended to $\mathbb{R}^d$ provides us with P' such that Q' is the shadow of P'. Thus there are k vertices in an increasing edge path of Q', and hence also on a shadow vertex path of P'. □

A key question, which remains open in our study, is whether we really have $H(d,n) = M(d,n)$ for all $n > d \geq 0$, that is, whether the only upper bound on the length of monotone paths is restricted only by the Upper Bound Theorem, which bounds the total number of vertices. The answer is positive for $d \leq 3$.

In fact, for $d = 3$ let $m := 2n-5$ and $k = 0, 1, \dots, m$ and define

$$v_k \quad := \quad \begin{cases} (\cos(k\pi/m), \sin(k\pi/m), \sin(k\pi/m))^t & \text{if } k \equiv 0 \text{ or } k \equiv 1 \pmod 4, \\ (\cos(k\pi/m), \sin(k\pi/m), 0)^t & \text{if } k \equiv 2 \text{ or } k \equiv 3 \pmod 4) \end{cases}$$

This yields a simple polytope with n facets, whose projection to the first two coordinates is "half a regular $2m$-gon," with all its $m+1$ vertices on a semicircular arc. Our figure represents this for the case $m = 7$, $n = 6$.

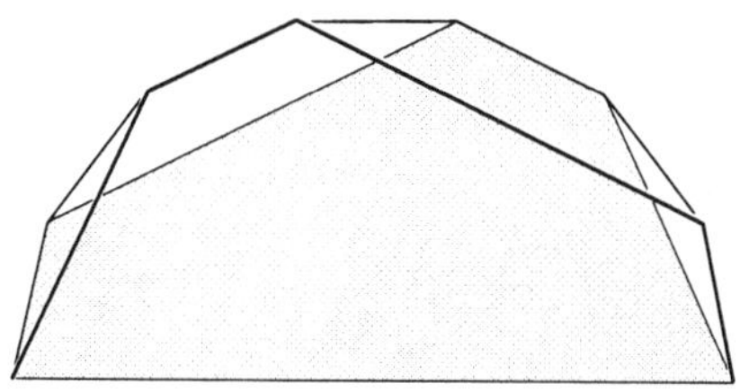

The combinatorial type of the resulting 3-polytope is that of the polar of a cyclic polytope $C_3(n)$. Hence for $d = 3$, the maximum of $H_{2\text{-sh}}(3,n) = H(3,n)$ is achieved by the polar of a cyclic polytope. We will, however, see in Theorem 5.15 that the corresponding statement fails for $d = 4$. Thus it is not at all clear whether $H(d,n) = M(d,n)$ "deserves to be true" for $d > 3$. Nevertheless, we prove that the quantities $H_{2\text{-sh}}(d,n)$ and $H(d,n)$ have the same asymptotic growth for any fixed dimension d, in Theorem 5.3.

3. Deformed products

Here is a simple basic version of the deformed products construction.

DEFINITION 3.1. **(Deformed products)** Let $P \subseteq \mathbb{R}^d$ be a convex polytope, and φ: $P \longrightarrow \mathbb{R}$ a linear functional with $\varphi(P) \subseteq [0,1]$. Let $V, W \subseteq \mathbb{R}^e$ be convex polytopes. Then the *deformed product* of (P,φ) and of (V,W) is

$$(P,\varphi) \bowtie (V,W) \quad := \quad \left\{ \begin{pmatrix} x \\ v + \varphi(x)(w - v) \end{pmatrix} : \begin{array}{c} x \in P \\ v \in V,\ w \in W \end{array} \right\} \quad \subseteq \quad \mathbb{R}^{d+e}.$$

EXAMPLES 3.2. (i) If $V = W$, then the deformed product is the "standard product":

$$(P,\varphi) \bowtie (V,V) \quad = \quad P \times V.$$

(ii) If φ is constant ($\varphi(x) = \lambda$ for all $x \in P$), then the deformed product is a standard product with an appropriate Minkowski sum:

$$(P, \varphi) \bowtie (V, W) \quad = \quad P \times \left(\lambda W + (1-\lambda)V\right).$$

(iii) If $P = Q \times [0,1]$ with $\varphi(q,t) = t$, then

$$(Q \times I, \varphi) \bowtie (V, W) \quad = \quad Q \times \operatorname{conv}\left((V \times \{0\}) \cup (W \times \{1\})\right).$$

This includes the case where $Q = \{0\}$ and $P = [0,1]$ with $\varphi(x) = x$, as illustrated in our figure. Here the left drawing depicts a "general" case, where for the right drawing V and W are normally equivalent, so the resulting polytope turns out to be (combinatorially equivalent to) a prism over V.

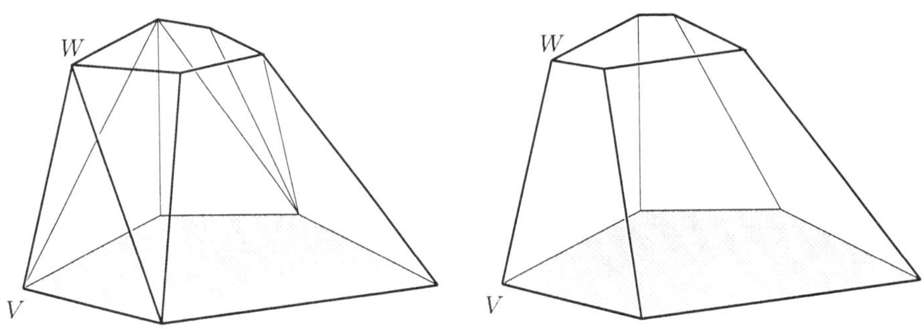

LEMMA 3.3. *Deformed products are convex.*

PROOF. Define

$$f(x, v, w) := \begin{pmatrix} x \\ v + \varphi(x)(w - v) \end{pmatrix},$$

and let $f(x', v', w')$ and $f(x'', v'', w'')$ be two points in the deformed product. We may restrict P to the line segment $[x', x'']$, and verify that the deformed product $([x', x''], \varphi) \bowtie (V, W)$ is convex.

Now if φ is constant on $[x', x'']$ (that is, if $\varphi(x') = \varphi(x'') = \lambda_0$), then we get that

$$([x', x''], \varphi) \bowtie (V, W) = [x', x''] \times (\lambda_0 W + (1-\lambda_0)V),$$

which is convex. (This is the situation of Example 3.2(ii).)

If φ is not constant, take $\lambda' := \varphi(x')$ and $\lambda'' := \varphi(x'')$ and define $V' := \lambda' W + (1-\lambda')V$ and $W'' := \lambda'' W + (1-\lambda'')V$. Both V' and W'' are convex. With this notation the map $\binom{x}{u} \longmapsto \binom{\varphi(x)}{u}$ provides us with an isomorphism

$$([x', x''], \varphi) \bowtie (V, W) \quad \cong \quad \operatorname{conv}\left((V' \times \{\lambda'\}) \cup (W'' \times \{\lambda''\})\right).$$

(This is the situation of Example 3.2(iii).) □

We now show that a deformed product taken with a pair of *normally equivalent* polytopes is combinatorially equivalent to a regular (not deformed) product.

THEOREM 3.4. **(Vertices and facets of deformed products with normally equivalent polytopes)** *Let $P \subseteq \mathbb{R}^d$ be a d-polytope, φ: $\mathbb{R}^d \longrightarrow \mathbb{R}$ a linear function with $\varphi(P) = [0,1]$, and let $V, W \subseteq \mathbb{R}^e$ be normally equivalent e-polytopes.*

(i) *For* $P = \operatorname{conv}\{p_1, \dots, p_m\}$, $V = \operatorname{conv}\{v_1, \dots, v_n\}$ *and*

$$W = \operatorname{conv}\{w_1, \dots, w_n\},$$

the deformed product is given by

$$(P,\varphi) \bowtie (V,W) \quad = \quad \operatorname{conv}\left\{ \begin{pmatrix} p_i \\ v_j + \varphi(p_i)(w_j - v_j) \end{pmatrix} : \begin{array}{c} 1 \le i \le m \\ 1 \le j \le n \end{array} \right\}. \tag{7}$$

In particular, the deformed product is a polytope.

(ii) *If* P, V *and* W *are given by*

$$\begin{aligned} P &= \{x \in \mathbb{R}^d \colon a_k^t x \le \alpha_k \text{ for } 1 \le k \le s\}, \\ V &= \{u \in \mathbb{R}^e \colon b_l^t u \le \beta_l \text{ for } 1 \le l \le t\}, \qquad \text{and} \\ W &= \{u \in \mathbb{R}^e \colon b_l^t u \le \beta_l' \text{ for } 1 \le l \le t\}, \end{aligned}$$

then the deformed product is given by

(8)

$$(P,\varphi) \bowtie (V,W) \;=\; \left\{ \begin{pmatrix} x \\ u \end{pmatrix} \in \mathbb{R}^{d+e} \colon \begin{array}{ll} (a_k^t, 0^t)\begin{pmatrix} x \\ u \end{pmatrix} \le \alpha_k & \text{for } 1 \le k \le s, \\ ((\beta_l - \beta_l')\varphi\,,\, b_l^t)\begin{pmatrix} x \\ u \end{pmatrix} \le \beta_l & \text{for } 1 \le l \le t. \end{array} \right\}$$

(iii) *The deformed product* $(P,\varphi) \bowtie (V,W)$ *is combinatorially equivalent to* $P \times V$ *and to* $P \times W$. *In particular, if* P *is a* d*-polytope with* n *vertices and* s *facets, and if* V *and* W *are* e*-polytopes with* m *vertices and* t *facets (that is, if the descriptions given in* (i) *and* (ii) *were irredundant), then* $Q := (P,\varphi) \bowtie (V,W)$ *is a* $(d{+}e)$*-polytope with* nm *vertices and* $s{+}t$ *facets, and the descriptions of* Q *are also irredundant: the points in the convex hull description* (7) *are the vertices, and the inequalities in* (8) *determine the facets of* Q.

PROOF.

(i) Denote by $Q \subseteq \mathbb{R}^{d+e}$ the deformed product, by $\mathbb{R} \subseteq \mathbb{R}^{d+e}$ the right-hand-side of equation (7). Then clearly $Q \supseteq R$. Take an arbitrary linear functional $\ell = (\ell^1, \ell^2)\colon \mathbb{R}^{d+e} \longrightarrow \mathbb{R}$. The maximum of ℓ on Q is attained by some $\binom{x}{\varphi(x)w + (1-\varphi(x))v}$. Since $1 \ge \varphi(x) \ge 0$ the maximum is attained in vertices $w \in W$ and $v \in V$. Since W, V are normally equivalent the maximum of $\ell^2(\varphi(x)w + (1-\varphi(x))v)$, for fixed x, is attained by corresponding vertices v_j and w_j. Hence we may assume that $v = v_j$ and $w = w_j$ for some $1 \le j \le n$. Fixing those, we get that ℓ is linear in x, and hence the maximum of ℓ on Q is attained for some $x = p_i$, where $1 \le i \le m$. Thus the maximum of ℓ over Q is attained for some point of $\mathbb{R}$. Since $\mathbb{R}$ is convex, we conclude $Q \subseteq R$.

(ii) Let $U \subseteq \mathbb{R}^{d+e}$ denote the solution set of the inequality system in (8), that is, the right-hand side of the equation (8). First, a simple computation shows that the inequalities in (8) are valid for Q:

$$(a_k^t, 0^t)\begin{pmatrix} x \\ v + \varphi(x)(w - v) \end{pmatrix} \quad = \quad a_k^t x \quad \le \quad \alpha_k$$

and

$$\begin{array}{rcl} ((\beta_l - \beta_l')\varphi, b_l^t)\begin{pmatrix} x \\ v + \varphi(x)(w-v) \end{pmatrix} & = & \\ (\beta_l - \beta_l')\varphi(x) + b_l^t v + \varphi(x)(b_l^t w - b_l^t v) & = & \\ (1-\varphi(x))(b_l^t v - \beta_l) + \beta_l + \varphi(x)(b_l^t w - \beta_l) & \leq & \beta_l \end{array}$$

follow easily from the corresponding inequalities for P, V and W. This implies $Q \subseteq U$. Furthermore, we have equality for a point of the form $\binom{p_i}{v_j+\varphi(p_i)(w_j-v_j)}$ in the first case if and only if $a_k^t p_i = \alpha_k$, and in the second case if and only if $b_l^t v_j = \beta_l$ and $b_l^t w_j = \beta_l'$: to see this, note that in case $\varphi(x) > 0$ we get $b_l v_j = \beta_l$, while in case $\varphi(x) < 1$ we get $b_l w_j = \beta_l'$. Finally, we know that the two conditions $b_l^t v_j = \beta_l$ and $b_l^t w_j = \beta_l'$ are equivalent.

At this point we invoke the Isomorphism Lemma 2.4, applied to $P \times V$ as the first polytope and the deformed product $(P, \varphi) \bowtie (V, W)$ as the second polytope. This shows us that the description of the deformed product in terms of inequalities (8) is complete, and it also establishes part (iii). □

REMARK 3.5. From now on we will maintain the assumptions of Theorem 3.4, namely that

- $P \subseteq \mathbb{R}^d$ is d-dimensional (although this is not essential),
- $\varphi(P) = [0,1]$ (it is only important that $\varphi(P) \subseteq [0,1]$: without this assumption the deformed product is not usually convex, as may be seen for $P = [0,1] \subseteq \mathbb{R}^1$, $\varphi(x) = 2x$, $V = [0,1]$, $W = \{0\}$),
- $V, W \subseteq \mathbb{R}^e$ are e-dimensional and normally equivalent.

COROLLARY 3.6. *Let $Q := (P, \varphi) \bowtie (V, W) \subseteq \mathbb{R}^{d+e}$ be a deformed product as above.*

(i) *The projection $\pi_1 \colon \mathbb{R}^{d+e} \longrightarrow \mathbb{R}^d$ to the first d components maps Q to $\pi_1(Q) = P$.*

(ii) *The projection $\pi_2 \colon \mathbb{R}^{d+e} \longrightarrow \mathbb{R}^e$ to the last e components maps Q to $\pi_2(Q) = \operatorname{conv}(V \cup W)$.*

COROLLARY 3.7. *The deformed product $(P, \varphi) \bowtie (V, W)$ is connected to the "standard" product $P \times V$ in the realization space of the product.*

PROOF. Consider

$$(P, t\varphi) \bowtie (V, W) \quad := \quad \operatorname{conv}\left\{ \begin{pmatrix} p_i \\ u \end{pmatrix} : \begin{array}{l} p_i \in \operatorname{vert}(P), \\ u = v_j + t\varphi(p_i)(w_j - v_j),\ 1 \leq j \leq n \end{array} \right\}$$

for $0 \leq t \leq 1$. For $t = 0$ this realizes the standard product, for $t = 1$ one gets the deformed product. □

REMARK 3.8. R. Seidel has pointed out the following alternative way to view the construction of deformed products $(P, \varphi) \bowtie (V, W)$. For this again assume that $\varphi(P) = [0,1]$. If V and W are normally equivalent, then $\operatorname{conv}(V \times \{0\} \cup W \times \{1\}) \subseteq \mathbb{R}^{e+1}$ is combinatorially equivalent to $V \times I$ (as in Example 3.2(iii)). Hence

$$\begin{array}{rcl} C & := & P \times \operatorname{conv}(V \times \{0\} \cup W \times \{1\}) \\ & = & \operatorname{conv}(P \times V \times \{0\} \cup P \times W \times \{1\}) \ \subseteq \ \mathbb{R}^{d+e+1} \end{array}$$

is combinatorially equivalent to $P \times V \times I$ (it is a "deformed prism" over $P \times V$), where

$$C = \left\{ \begin{pmatrix} x \\ u \\ t \end{pmatrix} : \begin{array}{l} x \in P \\ u = v + t(w - v), \ v \in V, \ w \in W \\ t \in I \end{array} \right\}.$$

Now cut C with the $(d + e)$-dimensional hyperplane $\varphi(x) = t$, and project the intersection to $\mathbb{R}^{d+e}$ by deleting the last coordinate. Call the image S. We get

$$S = \left\{ \begin{pmatrix} x \\ u \end{pmatrix} : \begin{array}{l} x \in P \\ u = v + \varphi(x)(w - v), \ v \in V, \ w \in W \end{array} \right\},$$

which is the deformed product. S is a convex polytope. It is combinatorially equivalent to $P \times V$ or $P \times W$, since it is a slice of a (combinatorial) prism that has the top on one side and the bottom on the other side.

DEFINITION 3.9. **(Increasing edges)** If p', p'' are adjacent vertices of a polytope $P \subseteq \mathbb{R}^d$, then we write $[p', p'']$ for the edge between p' and p''. We say that $[p', p'']$ is *α-increasing* if $\alpha(p'') > \alpha(p')$ for a linear function $\alpha\colon \mathbb{R}^d \longrightarrow \mathbb{R}$. (The ordering of the vertices p', p'' is relevant in this notation.)

DEFINITION 3.10. **(Deformed product programs)** With the notation and assumptions of Theorem 3.4, let

$$\max \varphi(x): \quad x \in P$$

be a linear program in $\mathbb{R}^d$, and let

$$\max \alpha(u): \ u \in V \qquad \text{resp.} \qquad \max \alpha(u): \ u \in W$$

be two normally equivalent linear programs (that is, one arises from the other by changing the right hand side, without change of the combinatorial type).

The linear function $\alpha\colon \mathbb{R}^e \longrightarrow \mathbb{R}$ induces a linear function $\widehat{\alpha}\colon \mathbb{R}^{d+e} \longrightarrow \mathbb{R}$ via $\widehat{\alpha}\binom{x}{u} := \alpha(u)$. Using this function, the *deformed product program* is given by

$$\max \widehat{\alpha}\begin{pmatrix} x \\ u \end{pmatrix} = \alpha(u): \quad \begin{pmatrix} x \\ u \end{pmatrix} \in Q = (P, \varphi) \bowtie (V, W).$$

For the deformed product $Q = (P, \varphi) \bowtie (V, W)$ of Theorem 3.4 use the notation

$$v(i, j) := \begin{pmatrix} p_i \\ v_j + \varphi(p_i)(w_j - v_j) \end{pmatrix}$$

for the vertices $(1 \leq i \leq m, \ 1 \leq j \leq n)$ of Q. By Theorem 3.4(iii) the edges of a deformed product just correspond to those of the regular product. Thus Q has two types of edges:

- edges of the form $[v(i', j), v(i'', j)]$ (called *P-edges*) for $1 \leq j \leq n$ and for any $1 \leq i', i'' \leq m$ such that $p_{i'}$ and $p_{i''}$ are adjacent vertices of P, and
- edges of the form $[v(i, j'), v(i, j'')]$ (named *(V, W)-edges*) for any $1 \leq i \leq m$ and for $1 \leq j', j'' \leq n$ such that $v_{j'}$ and $v_{j''}$ are adjacent vertices of V (equivalently, $w_{j'}$ and $w_{j''}$ are adjacent vertices of W).

PROPOSITION 3.11. **(Increasing edges of deformed products)** *For any deformed product program* $\max \widehat{\alpha}\binom{x}{u} = \alpha(u)$: $\binom{x}{u} \in Q = (P, \varphi) \bowtie (V, W)$,

- *a P-edge $[v(i',j),v(i'',j)]$ is $\widehat{\alpha}$-increasing if and only if either $[p_{i'},p_{i''}]$ is φ-increasing and $\alpha(w_j)>\alpha(v_j)$, or $[p_{i'},p_{i''}]$ is φ-decreasing and $\alpha(w_j)<\alpha(v_j)$,*
- *a (V,W)-edge $[v(i,j'),v(i,j'')]$ is $\widehat{\alpha}$-increasing if and only if $[v_{j'},v_{j''}]$ is α-increasing.*

PROOF. We compute for a P-edge that

$$\begin{array}{rcll} & [v(i',j),v(i'',j)] \text{ is } \widehat{\alpha}\text{-increasing} & & \iff \\ \alpha(v_j+\varphi(p_{i''})(w_j-v_j)) & - \quad \alpha(v_j+\varphi(p_{i'})(w_j-v_j)) & >0 & \iff \\ & \big(\varphi(p_{i''})-\varphi(p_{i'})\big)\big(\alpha(w_j)-\alpha(v_j)\big) & >0 & \end{array}$$

which yields the claim. For a (V,W)-edge, we get

$$\begin{array}{rcll} & [v(i,j'),v(i,j'')] \text{ is } \widehat{\alpha}\text{-increasing} & & \iff \\ \alpha(v_{j''}+\varphi(p_i)(w_{j''}-v_{j''})) & - \quad \alpha(v_{j'}+\varphi(p_i)(w_{j'}-v_{j'})) & >0 & \iff \\ \varphi(p_i)\big(w_{j''}-w_{j'}\big) & + \quad (1-\varphi(p_i))\big(v_{j''}-v_{j'}\big) & >0 & \end{array}$$

which implies the claim since $[v_{j'},v_{j''}]$ is α-increasing if and only if $[w_{j'},w_{j''}]$ is α-increasing. □

COROLLARY 3.12. **(Bland's rule on deformed product programs)** *Let p_1 and p_m be the vertices of P that minimize resp. maximize φ, with $\varphi(p_1)=0$ and $\varphi(p_m)=1$. On the deformed product program $\max \widehat{\alpha}\binom{x}{u}$: $\binom{x}{u}\in(P,\varphi)\bowtie(V,W)$, according to* Definition 3.10, *we have the following behavior.*

If the inequalities of Q are numbered such that all the inequalities of P get smaller indexes than all the inequalities corresponding to (V,W), then the simplex algorithm with Bland's rule prefers P-edges over (V,W)-edges.

Hence, if Bland's rule on P for the objective function φ takes a path with k vertices from p_1 to p_m, and for $-\varphi$ takes a path with k' vertices from p_m to p_1, then for $(Q,\widehat{\alpha})$ it will follow a path with k vertices from $v(1,j)$ to $v(m,j)$ if $\alpha(v_j)<\alpha(w_j)$, and a path with k' vertices from $v(m,j)$ to $v(1,j)$ if $\alpha(v_j)>\alpha(w_j)$.

Analogous results are easy to derive for many other pivot rules—see the examples in the next section. This is the basis for virtually *all* known constructions of classes linear programs on which variants of the simplex algorithm are exponential.

4. Some (old) examples

4.1. The Klee-Minty cubes. The Klee-Minty cubes [**KM**, Sect. 4] are classical: they can be found in any standard text on linear programming, in Schrijver [**S**], in Chvátal [**Chv**], and, reluctantly, in Padberg [**P**, Sect. 5.7.1].

$$\max x_d\colon\ x\in \mathrm{KM}_d$$

where KM_d is, for some ε with $0<\varepsilon<\frac{1}{2}$, given by the inequality system

$$\begin{array}{rcccl} 0 & \le & x_1 & \le & 1 \\ \varepsilon x_{j-1} & \le & x_j & \le & 1-\varepsilon x_{j-1} \text{ for } 2\le j\le d \end{array}$$

As deformed products, we obtain the Klee-Minty cubes recursively as

$$\begin{array}{lcll} \mathrm{KM}_0 & = & \{0\} & \subseteq \mathbb{R}^0 \\ \mathrm{KM}_1 & = & [0,1] & \subseteq \mathbb{R}^1 \\ \mathrm{KM}_{d+1} & = & (\mathrm{KM}_d,x_d)\bowtie([0,1],[\varepsilon,1-\varepsilon]) & \subseteq \mathbb{R}^{d+1}. \end{array}$$

The boundary case $\varepsilon = \frac{1}{2}$ is analyzed in Clausen [**Cl2**]. It is not difficult to explicitly determine both vertex coordinates and the combinatorial structure of the Klee-Minty cubes, see Avis & Chvátal [**AC**], Clausen [**Cl**], and Gärtner & Ziegler [**GZ**]. However, this is not necessary for us, since the basic extremal properties follow already from the construction as deformed products.

THEOREM 4.1. **(The Klee-Minty cubes)** *Number the inequalities of* KM_d *in exactly the order in which they appear in the system, that is, the i-th inequality is* $x_j \geq \varepsilon x_{j-1}$ *for* $i = 2j-1$ *and* $x_j \leq 1 - \varepsilon x_{j-1}$ *for* $i = 2j$. *Then the simplex algorithm with Bland's least index rule follows an* x_d*-increasing path through all* 2^d *vertices of the d-dimensional Klee-Minty cube* KM_d.

If we multiply the i-th inequality by δ^{1-i} *for some small enough* $\delta > 0$, *then the same is true for Dantzig's largest coefficient rule.*

PROOF. Take $V = \mathrm{conv}\{v_1, v_2\}$ with $v_1 = 0$, $v_2 = 1$ and $W = \mathrm{conv}\{w_1, w_2\}$ with $w_1 = \varepsilon$, $w_2 = 1 - \varepsilon$. The linear functional we use on KM_d is $\varphi(x) = x_d$, so we get $\widehat{\alpha}\binom{x}{u} = u = x_{d+1}$ for $\binom{x}{u} \in \mathbb{R}^{d+1}$. Among the 2^d vertices of KM_d we find $p_1 = 0$ and $p_{2^d} = e_d$.

By Corollary 3.12 and induction on d, we find that starting at $v(1,1) = 0$ Bland's rule on KM_{d+1} will take a path through all the 2^d vertices of the form $v(i,1)$, ending at $v(2^d,1) = \varepsilon e_{d+1}$. At this point all P-edges are x_{d+1}-decreasing, so the simplex algorithm takes the (unique) (V,W)-edge to $v(2^d,2) = (1-\varepsilon)e_{d+1}$. After that, Bland's rule takes a path through all the 2^d vertices $v(i,2)$, ending at $v(1,2) = e_{d+1}$. □

Avis & Chvátal [**AC**] proved that if we take the i-th inequality to be $x_i \geq \varepsilon x_{i-1}$ for $i \leq d$ and to be $x_{i-d} \leq 1 - \varepsilon x_{i-d-1}$ for $i > d$, then the Bland's rule will follow a shorter path, whose number of vertices is, however, still exponential. By Observation 2.6, the same is true for the worst case of the Dantzig rule for this ordering of the inequalities.

OBSERVATION 4.2. The Klee-Minty cubes are *not* projectively equivalent to "standard" cubes, as was recently stated by Gritzmann & Klee [**GK**, p. 646]. To see this, analyze the following picture, which represents a parallel projection of the 3-dimensional Klee-Minty cube (for $\varepsilon = \frac{1}{3}$). If the Klee-Minty cube was a projective image of a regular cube, then the four "horizontal" edges would have to meet in one point (or be parallel), which is clearly not the case.

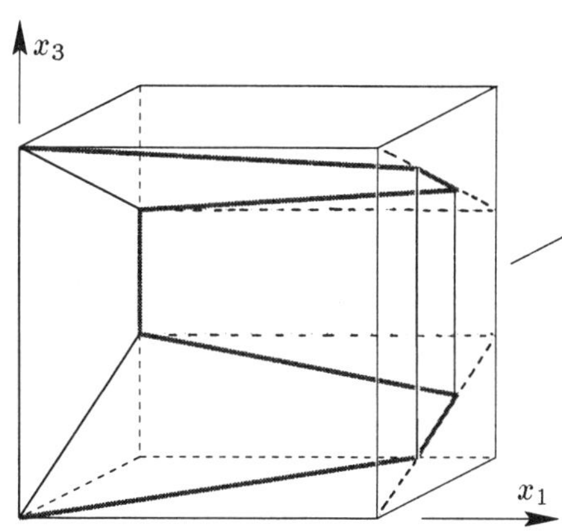

THEOREM 4.3. *For $n > d \geq 0$:*

$$H_{\mathrm{Bl}}(d{+}1, n{+}2) \quad \geq \quad 2\, H_{\mathrm{Bl}}(d, n).$$

Analogous inequalities are true for the maximal numbers of vertices $H(d,n)$ and $H_{\mathrm{Da}}(d,n)$ on arbitrary increasing paths resp. for paths according to Dantzig's rule. Hence for $d \geq 0$ we have

$$H(d, 2d) \quad \geq \quad H_{\mathrm{Da}}(d, 2d) \quad \geq \quad H_{\mathrm{Bl}}(d, 2d) \quad \geq \quad 2^d.$$

PROOF. Let $P \subseteq \mathbb{R}^d$ be a simple (d,n)-polytope which achieves the maximum length for an arbitrary increasing path (respectively of a path according to Bland's or Dantzig's rule), and let $\varphi : P \longrightarrow [0,1]$ be a corresponding objective function. Then for some $0 < \varepsilon < \frac{1}{2}$ form the deformed product

$$(P, \varphi) \bowtie ([0,1], [\varepsilon, 1{-}\varepsilon])$$

and use the same proof as for the Klee-Minty cubes. □

4.2. The Goldfarb-Sit cubes. Goldfarb & Sit [**GSi**] constructed linear programs—rescaled Klee-Minty cubes tailored to fool the steepest increase rule, see also Clausen [**Cl**]—as follows. They analyzed the programs

$$\max \sum_{i=1}^{d} \beta^{i-1} x_i \colon\ x \in \mathrm{GS}_d$$

where $\mathrm{GS}_d \subseteq \mathbb{R}^d$ is the polytope given by

$$\begin{array}{rcccl} 0 & \leq & x_1 & \leq & 1 \\ \beta x_{j-1} & \leq & x_j & \leq & \delta_j - \beta x_{j-1} \text{ for } 2 \leq j \leq d, \end{array}$$

where $\beta \geq 2$, $\theta > 2$ and $\delta_i := (\theta\beta)^{i-1}$. It is immediate from Theorem 3.4 that we may construct the *Goldfarb-Sit cubes* as deformed products

$$\begin{array}{lcll} \mathrm{GS}_0 & = & \{0\} & \subseteq \mathbb{R}^0 \\ \mathrm{GS}_1 & = & [0,1] & \subseteq \mathbb{R}^1 \\ \mathrm{GS}_{d+1} & = & (\mathrm{GS}_d, x_d) \bowtie ([0, \delta_{d+1}], [\beta, \delta_{d+1} - \beta]) & \subseteq \mathbb{R}^{d+1}. \end{array}$$

and thus that they are combinatorial d-cubes.

4.3. The Goldfarb cubes. Murty [**Mu**] was the first to construct (in the framework of linear complementarity problems) linear programs of type $(d, 2d)$ that are exponential for the shadow vertex algorithm. Essentially equivalent, but much more explicit examples (at least from a linear programming point of view) were given by Goldfarb [**G, G2**]. Goldfarb's programs are

$$\max \xi_d\colon \quad \xi \in \mathrm{Gol}_d$$

where the polytopes Gol_d are given by

$$\begin{array}{rcccl} 0 & \leq & \xi_1 & \leq & 1 \\ \beta\xi_1 & \leq & \xi_2 & \leq & \delta - \beta\xi_1 \\ \beta\xi_j - \xi_{j-1} & \leq & \xi_{j+1} & \leq & \delta^j - \beta\xi_j + \xi_{j-1} \text{ for } 2 \leq j < d \end{array}$$

with $\beta \geq 2$ and $\delta > 2\beta$. They are deformed products! The "additional rotation" that Goldfarb talks about [**G**, p. 2] is represented by a different functional φ:

$$\begin{array}{lcll} \mathrm{Gol}_0 & = & \{0\} & \subseteq \mathbb{R}^0 \\ \mathrm{Gol}_1 & = & [0,1] & \subseteq \mathbb{R}^1 \\ \mathrm{Gol}_{d+1} & = & (\mathrm{Gol}_d, \xi_d - \frac{1}{\beta}\xi_{d-1}) \bowtie ([0,\delta_d],[\beta,\delta_d-\beta]) & \subseteq \mathbb{R}^{d+1}. \end{array}$$

The β-δ-scaling in these examples seems, however, to be an artifact that remained from the previous work by Goldfarb & Sit. Setting $\varepsilon := \frac{\beta}{\delta} < \frac{1}{2}$, $\gamma := \frac{1}{\beta\delta} \leq \frac{1}{4}\varepsilon$, and substituting $x_j := \frac{\xi_j}{\delta^{j-1}}$ we get

$$\max x_d: \quad x \in \mathrm{Gol}'_d$$

where the polytopes Gol'_d are given by:

$$\begin{array}{rcccl} 0 & \leq & x_1 & \leq & 1 \\ \varepsilon x_1 & \leq & x_2 & \leq & 1-\varepsilon x_1 \\ \varepsilon(x_j - \gamma x_{j-1}) & \leq & x_{j+1} & \leq & 1-\varepsilon(x_j-\gamma x_{j-1}) \text{ for } 2 \leq j \leq d \end{array}$$

which for $\gamma \longrightarrow 0$ yields the Klee-Minty cube (and for $\varepsilon, \gamma \longrightarrow 0$ yields the usual 0/1-cube).

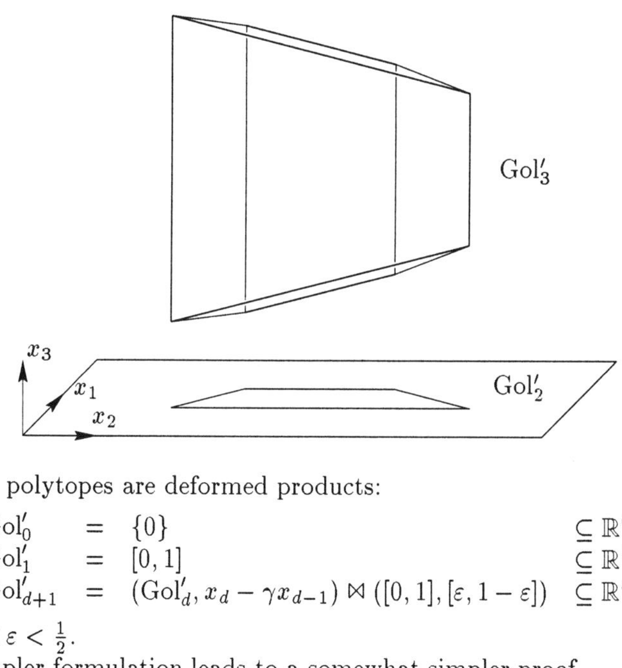

Again, these polytopes are deformed products:

$$\begin{array}{lcll} \mathrm{Gol}'_0 & = & \{0\} & \subseteq \mathbb{R}^0 \\ \mathrm{Gol}'_1 & = & [0,1] & \subseteq \mathbb{R}^1 \\ \mathrm{Gol}'_{d+1} & = & (\mathrm{Gol}'_d, x_d - \gamma x_{d-1}) \bowtie ([0,1],[\varepsilon,1-\varepsilon]) & \subseteq \mathbb{R}^{d+1} \end{array}$$

for $0 < 4\gamma < \varepsilon < \frac{1}{2}$.

This simpler formulation leads to a somewhat simpler proof.

THEOREM 4.4. (Goldfarb [**G**, **G2**]) *The projection* π: $\mathrm{Gol}'_d \longrightarrow \mathbb{R}^2$ *given by* $\pi(x) = (x_{d-1}, x_d)$ *has* 2^d *vertices.*

PROOF. Our proof follows [**G**]. For any vertex v of Gol'_d, let A_v be the $(d \times d)$-matrix whose rows are the facet normals at v:

$$A_v = \begin{pmatrix} \sigma_1 & & & & & & \\ \varepsilon & \sigma_2 & & & & & \\ -\varepsilon\gamma & \varepsilon & \sigma_3 & & & & \\ & -\varepsilon\gamma & \varepsilon & \sigma_4 & & & \\ & & -\varepsilon\gamma & \ddots & \ddots & & \\ & & & \ddots & \ddots & \sigma_{d-1} & \\ & & & & -\varepsilon\gamma & \varepsilon & \sigma_d \end{pmatrix} =: \begin{pmatrix} a_1^t \\ a_2^t \\ a_3^t \\ a_4^t \\ \vdots \\ \vdots \\ a_d^t \end{pmatrix}$$

where $\sigma_i = -1$ if the i-th inequality at v is "$\varepsilon(x_j - \gamma x_{j-1}) \leq x_{j+1}$," and $\sigma_i = +1$ otherwise. A_v can be taken in this form since Gol'_d is combinatorially equivalent to a d-cube. We get that

$$A_v\, x \;\leq\; A_v\, v$$

is a system of inequalities that is valid for all $x \in \mathrm{Gol}'_d$, with equality in all components if and only if $x = v$.

Now define a row vector $\alpha^t = (\alpha_1, \dots, \alpha_d)$ recursively by $\alpha_1 := 1$, $\alpha_2 := \frac{2}{\varepsilon}$ and $\alpha_{i+2} := \frac{1}{\gamma\varepsilon}(\varepsilon\alpha_{i+1} + \sigma_i\alpha_i)$. We verify $\alpha_{i+1} \geq \frac{2}{\varepsilon}\alpha_i > 0$ for all $i \geq 1$. This is true for $i = 1, 2$, and for $i \geq 3$ we get

$$\alpha_{i+2} \;\geq\; \frac{1}{\gamma\varepsilon}(\varepsilon\alpha_{i+1} - \alpha_i) \;\geq\; \frac{1}{\gamma\varepsilon}(\varepsilon\alpha_{i+1} - \frac{\varepsilon}{2}\alpha_{i+1}) \;=\; \frac{1}{2\gamma}\alpha_{i+1} \;\geq\; \frac{2}{\varepsilon}\alpha_{i+1} > 0,$$

using induction for the first and for the second inequality. Since α^t is a positive row vector, we deduce that

$$\alpha^t A_v\, x \;\leq\; \alpha^t A_v\, v$$

holds for all $x \in \mathrm{Gol}'_d$, with equality if and only if $x = v$.

From $\alpha^t A_v = (0, 0, \dots, 0, \alpha_{d-1}, \alpha_d)$ we get

$$(\alpha_{d-1}, \alpha_d)\pi(x) \;=\; \alpha^t A_v\, x \;\leq\; \alpha^t A_v\, v$$

with equality if and only if $x = v$: hence $\pi(v)$ is a vertex of $\pi(\mathrm{Gol}'_d)$, and v is the only point of Gol'_d that projects to $\pi(v)$, so all the 2^d vertices $\pi(v)$ of $\pi(\mathrm{Gol}'_d)$ are distinct. □

The condition $\gamma \leq \frac{\varepsilon}{4}$ that we have used in the proof (equivalent to Goldfarb's $\beta \geq 2$) is not the best one could do, but it is the condition that works without problems for the proofs.

The sequence in which the shadow boundary algorithm visits the vertices of the Goldfarb cubes is the "same" as for the Klee-Minty cubes. Using that for $\gamma \to 0$ we get the Klee-Minty cubes from the Goldfarb cubes, this follows from the fact that the shadow vertex rule visits the vertices in increasing order with respect to the objective function $\varphi(x) = x_d$.

COROLLARY 4.5. *For $d \geq 0$, $H_{2\text{-sh}}(d, 2d) \geq 2^d$.*

4.4. The Klee-Minty products. In view of the deformed product construction, it is not hard to construct polytopes with "many" vertices on increasing paths in fixed dimension [**KM**, Sect. 5]: we iteratively construct deformed products with polygons (that is, we take $e = 2$). The single extra ingredient needed here—and in the following constructions—is a construction of suitable polygons $V, W \subseteq \mathbb{R}^2$. Again, this was first done by Klee and Minty; we replace their lemma [**KM**, p. 165] by the following.

LEMMA 4.6. *Let $\alpha(x)$ denote a linear function on the plane ($x \in \mathbb{R}^2$). For each $k \geq 4$, and for each $\lambda > 0$, there exist normally equivalent k-gons*

$$V = \operatorname{conv}\{v_1, \dots, v_k\} \qquad \textit{and} \qquad W = \operatorname{conv}\{w_1, \dots, w_k\}$$

(both labeled in clockwise order), such that

$$0 = \alpha(v_1) < \alpha(w_1) < \alpha(w_2) < \alpha(v_2) < \alpha(v_3) < \alpha(w_3) < \alpha(w_4) < \dots$$

(The sequence ends with $\dots < \alpha(v_k) < \alpha(w_k) = 1$ if k is odd, and with $\dots < \alpha(w_k) < \alpha(v_k) = 1$ if k is even.)

PROOF. Take $2k-4 \geq 4$ points equally spaced on a semicircular arc ("half a $2(2k-5)$-gon") and label them $w_2, v_2, v_3, w_3, w_4 \dots$, ending with $\dots, v_{k-1}, w_{k-1}$ if k is odd and with $\dots, w_{k-1}, v_{k-1}$ if k is even. The additional points v_1, w_1, v_k and w_k are then chosen appropriately on a line that is parallel to the diameter on which the semicircular arc was based.

Our figures illustrate this in the cases $k = 4$ and $k = 7$.

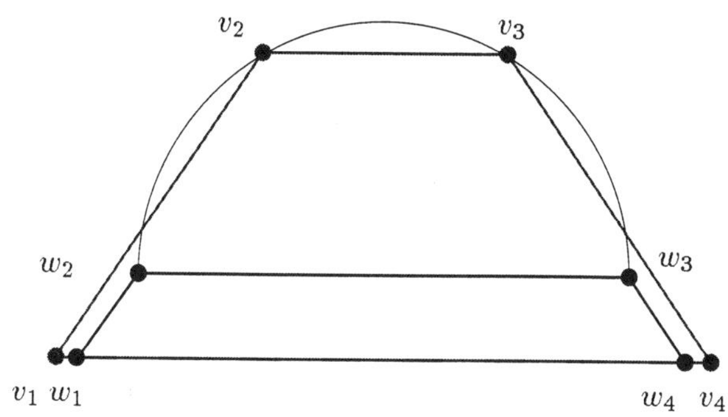

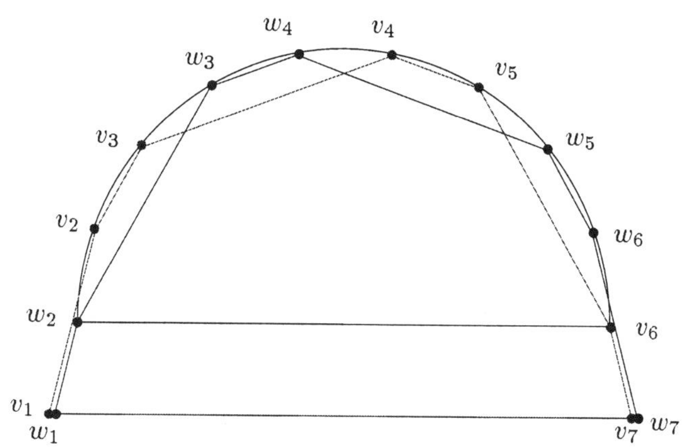

Note: this lemma is false for $k = 3$. □

THEOREM 4.7. *For $k \geq 4$ and $n > d \geq 0$,*

$$H_{\mathrm{Bl}}(d{+}2, n{+}k) \quad \geq \quad k\, H_{\mathrm{Bl}}(d,n). \tag{9}$$

The same recursion is valid for $H(d,n)$ and for $H_{\mathrm{Da}}(d,n)$.

PROOF. After rescaling the objective function and relabeling the vertices we may take a polytope P of type (d,n) for which Bland's rule, for a functional $c^t x = \varphi(x)$ with $\varphi(P) = [0,1]$, follows an increasing path on $l := H_{\mathrm{Bl}}(d,n)$ vertices starting at the vertex p_1 with $\varphi(p_1) = 0$ and ending at the vertex p_m with $\varphi(p_m) = 1$. Now construct the deformed product

$$Q \quad := \quad (P, \varphi) \bowtie (V, W),$$

where $V, W \subseteq \mathbb{R}^2$ are convex k-gons according to Lemma 4.6.

By Corollary 3.12 we get that Bland's rule, applied to Q with objective function $\widehat{\alpha}\binom{x}{u} = \alpha(u)$, first follows a P-path with l vertices from $v(1,1)$ to $v(m,1)$, then after one (V,W)-pivot it follows a P-path with l vertices from $v(m,2)$ to $v(1,2)$, then after one (V,W)-pivot it first follows a P-path with l vertices from $v(1,3)$ to $v(m,3)$, etc.. The complete path will thus visit kl vertices, arriving at $v(1,k)$ or at $v(m,k)$, depending on whether k is even or odd. □

COROLLARY 4.8. *For $n \geq 2d \geq 2$:*

$$H(d,n) \;\geq\; H_{\mathrm{Da}}(d,n) \;\geq\; H_{\mathrm{Bl}}(d,n) \;\geq\; \left\lfloor \frac{2n}{d} \right\rfloor^{d/2} \qquad \text{if } d \text{ is even,}$$

and

$$H(d,n) \;\geq\; H_{\mathrm{Da}}(d,n) \;\geq\; H_{\mathrm{Bl}}(d,n) \;\geq\; 2\left\lfloor \frac{2n-4}{d-1} \right\rfloor^{\lfloor d/2 \rfloor} \qquad \text{if } d \text{ is odd.}$$

PROOF. Starting at $H_{\mathrm{Bl}}(0,0) = 1$ and $H_{\mathrm{Bl}}(1,2) = 2$, we get

$$H_{\mathrm{Bl}}(2m, km) \quad \geq \quad k^m \tag{10}$$

and

$$H_{\mathrm{Bl}}(2m{+}1, km{+}2) \quad \geq \quad 2\,k^m \tag{11}$$

for all $m \geq 0$, from (9) together with induction on m.

For even d, we use (10), and substitute $m := \frac{d}{2}$ and $k := \lfloor \frac{n}{m} \rfloor = \lfloor \frac{2n}{d} \rfloor$, where $n \geq 2d$ guarantees $k \geq 4$. For odd d, we use (11), and substitute $m := \frac{d-1}{2}$ and $k := \lfloor \frac{n-2}{m} \rfloor = \lfloor \frac{2n-4}{d-1} \rfloor$, where $n \geq 2d$ guarantees $k \geq 4$. □

COROLLARY 4.9. *For constant dimension d, the function $H(d,n)$ grows like a polynomial of degree $\lfloor \frac{d}{2} \rfloor$ in n. Furthermore, for even d*

$$\frac{1}{(d/2)!} \;=\; \lim_{n \longrightarrow \infty} \frac{M(d,n)}{n^{d/2}} \;\geq\; \liminf_{n \longrightarrow \infty} \frac{H(d,n)}{n^{d/2}} \;\geq\; \frac{1}{(d/2)^{d/2}},$$

and for odd d

$$\frac{2}{\lfloor d/2 \rfloor !} \;=\; \lim_{n \longrightarrow \infty} \frac{M(d,n)}{n^{\lfloor d/2 \rfloor}} \;\geq\; \liminf_{n \longrightarrow \infty} \frac{H(d,n)}{n^{\lfloor d/2 \rfloor}} \;\geq\; \frac{2}{\lfloor d/2 \rfloor^{\lfloor d/2 \rfloor}}.$$

PROOF. While the upper bounds for $H(d,n)$ are from the Upper Bound Theorem 1.1, the lower bounds are easily derived from Corollary 4.8: for example, in the case where d is even we have $H(d,n) \geq (\frac{2n}{d} - 1)^{d/2}$, and hence

$$\frac{H(d,n)}{n^{(d-1)/2}} \;\geq\; \left(\frac{1}{d/2} - \frac{2}{n} \right)^{d/2},$$

which converges nicely for $n \longrightarrow \infty$ if d is fixed. $\square$

(This amounts to a certain improvement in comparison with the analysis in Klee & Minty [**KM**, p. 170].)

4.5. The Jeroslow construction. Jeroslow [**J**] gave a construction of polytopes for which the greatest increase rule makes many pivots, which we can restate as deformed products. (Later Blair [**B**] proposed different programs for a diagonal case, which are also essentially deformed products.)

In the following, we keep close to Jeroslow's construction and notation. However, we complement his remark "The reader may wish to follow our construction with paper and pencil, since we shall refer to geometrical aspects of it." [**J**, p. 372] by a suitable figure.

LEMMA 4.10. *Let $\alpha(x)$ denote the first coordinate for points $x \in \mathbb{R}^2$ in the plane.*

For each $k \geq 1$, and for each $\lambda > 0$, there exist normally equivalent $(4k+2)$-gons $V = \operatorname{conv}\{v_0, v_1, \ldots, v_{4k+1}\}$ and $W = \operatorname{conv}\{w_0, w_1, \ldots, w_{4k+1}\}$, such that, for $1 \leq l < 4k$,

$$\alpha(v_{l+1} - v_l) = \begin{cases} \lambda & \text{if } l \equiv 1 \bmod 4 \\ 3 & \text{if } l \equiv 2 \bmod 4 \\ \lambda & \text{if } l \equiv 3 \bmod 4 \\ 1 & \text{if } l \equiv 0 \bmod 4 \end{cases} \quad \text{and } \alpha(w_{l+1} - w_l) = \begin{cases} \lambda & \text{if } l \equiv 1 \bmod 4 \\ 1 & \text{if } l \equiv 2 \bmod 4 \\ \lambda & \text{if } l \equiv 3 \bmod 4 \\ 3 & \text{if } l \equiv 0 \bmod 4 \end{cases}$$

and

$$\alpha(w_l - v_l) = \begin{cases} +1 & \text{if } l \equiv 1 \text{ or } 2 \bmod 4, \\ -1 & \text{if } l \equiv 3 \text{ or } 0 \bmod 4. \end{cases}$$

Furthermore, we require that $\alpha(v_1) < \alpha(v_0) < \alpha(v_2)$ and $\alpha(v_{4k-1}) < \alpha(v_{4k+1}) < \alpha(v_{4k})$.

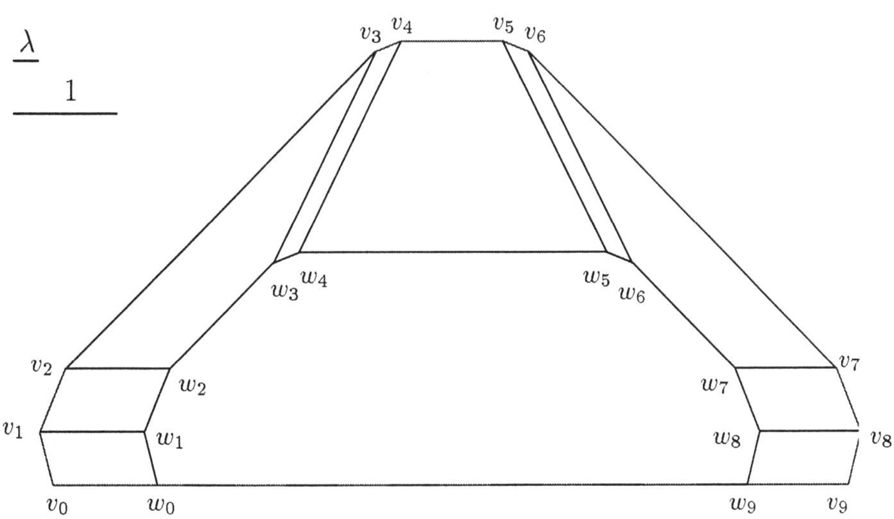

PROOF. If we place the vertex v_1 at the origin, then the polygons V and W are, in fact, determined by k and λ as soon as we prescribe the slopes m_i of the the edges $[v_l v_{l+1}]$ and $[w_l w_{l+1}]$ against the x-axis.

One possible choice is to take the slopes of a regular $8k$-gon, with $m_i = \cot(\pi/4k)$: this is what we (roughly) took for our figure, which represents the situation for $k = 2$ and $\lambda = 1/4$. (In the upper lefthand corner we represent the scale of the drawing.)

Alternatively one could for example take $m_l := 2k - l$: this yields "small" rational coordinates for the vertices, as is required in complexity arguments for the greatest increase pivot rule. □

With these polygons, a recursive deformed products construction can be used to establish the asymptotic lower bound on the length of greatest increase paths, as indicated in the next figure. To improve the bound, Jerowlow uses the fact that the resulting polytopes are "reversible":

DEFINITION 4.11. **(Reversible pairs,** Jeroslow **[J])** Let $P \subseteq \mathbb{R}^n$ be a polytope, and $\varphi\colon \mathbb{R}^d \longrightarrow \mathbb{R}$ a linear function with $\varphi(P) = [0,1]$ in general position on P. Let p_1 denote the unique minimal vertex of P with $\varphi(p_1) = 0$, and p_m the unique maximal vertex with $\varphi(p_m) = 1$.

Then (P, φ) is a *(reversible) greatest increase pair of length* l if the greatest increase path with respect to the objective function $c^t x := \varphi(x)$ (from p_1 to p_m) and the greatest increase path with respect to $c^t x := -\varphi(x)$ (from p_m to p_1) both have at least l vertices.

Thus $H_{\mathrm{gi}}(d,n)$ is the greatest length of a reversible pair (P, φ) of type (d,n).

THEOREM 4.12. *For all* $k \geq 1$,

$$H_{\mathrm{gi}}(d{+}2, n{+}4k{+}2) \quad \geq 2k\, H_{\mathrm{gi}}(d,n) \; + \; 2k.$$

PROOF. For any reversible pair (P, φ) of length l of dimension d we can choose a number $\lambda > 0$ such that the progress along each edge in the two greatest increase paths for φ and for $-\varphi$ on P is larger than λ. (In particular, this condition implies that $\lambda < \frac{1}{l} \leq 1$.) Corresponding to this value of λ, construct the $(4k{+}2)$-gons $V, W \subseteq \mathbb{R}^2$ described in Lemma 4.10. With these one obtains a deformed product

$$Q \;:=\; (P, \varphi) \bowtie (V, W)$$

of dimension $d+2$, with $n{+}4k{+}2$ facets. The linear function $\widehat{\alpha}\binom{x}{u} = \alpha(u)$ defines an objective function on Q that is in general position.

We have constructed V and W such that on every P-edge that the simplex algorithm may encounter the progress is more than λ, while the progress along a V-edge or W-edge of the deformed product is at most λ. Thus, once again, the simplex algorithm with the greatest increase rule prefers P-edges over (V, W)-edges. Thus the greatest increase rule applied to the objective function $\alpha \circ \pi_2$ on Q will trace a greatest increase path with l vertices from $v(1,1)$ to $v(m,1)$, then two (V, W)-steps from $v(m,1)$ via $v(m,2)$ to $v(m,3)$, then trace a greatest increase path of length l to $v(1,3)$, then do two steps from $v(1,3)$ via $v(1,4)$ to $v(1,5)$, etc. The vertices v_0, w_0, v_{4k+1} and w_{4k+1} are used to force the initial step of the path from v_1 resp. from v_{4k} to choose the P-edge. Our figure illustrates this for the case $l = 3$ and $\lambda = \frac{1}{4}$.

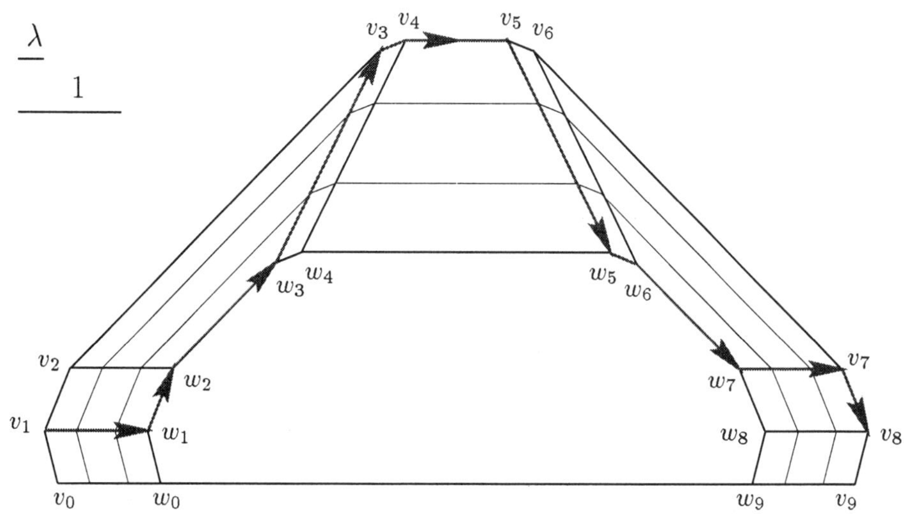

All in all, this determines a greatest increase path that has $2k$ sequences of P-edges that visit l vertices each, and touches $2k$ additional vertices, namely $w_2, v_4, w_2, \ldots, v_{4k}$. Thus the greatest increase path on (P, α) has $2kl + 2k$ vertices.

Analogously (or by symmetry), the same is true for the greatest increase path on $(Q, -\alpha)$. Thus (Q, α) is a reversible pair of length $2k(l+1)$. □

We use this lemma to derive both an exponential lower bound for $H_{\mathrm{gi}}(d, n)$ in the diagonal case and to show that it grows like $n^{\lfloor d/2 \rfloor}$ in the case of constant dimension.

COROLLARY 4.13. *The greatest increase rule traces, in the worst case, paths of exponential length:*

$$H_{\mathrm{gi}}(d, 3d-1) \;\geq\; 3 \cdot 2^{\lfloor d/2 \rfloor} - 2$$

holds for $d \geq 1$. *Furthermore,*

$$H_{\mathrm{gi}}(d, n) \;\geq\; 2\left(2\left\lfloor \frac{n-d}{2d} \right\rfloor\right)^{d/2} - 2 \qquad \text{if } d > 0 \text{ is even,}$$

and

$$H_{\mathrm{gi}}(d, n) \;\geq\; 3\left(2\left\lfloor \frac{n-d-1}{2(d-1)} \right\rfloor\right)^{\lfloor d/2 \rfloor} - 2 \qquad \text{if } d \text{ is odd.}$$

PROOF. The case $k = 1$ of Theorem 4.12, the initial conditions $H_{\mathrm{gi}}(0,0) = 1$ and $H_{\mathrm{gi}}(1,2) = 2$ and induction on m provide

$$H_{\mathrm{gi}}(2m, 6m) \;\geq\; 3 \cdot 2^m - 2 \qquad \text{and} \qquad H_{\mathrm{gi}}(2m{+}1, 6m{+}2) \;\geq\; 4 \cdot 2^m - 2$$

for all $m \geq 0$. Clearly $H_{\mathrm{gi}}(d, n) \leq H_{\mathrm{gi}}(d, n{+}1)$, from which we derive the first claim.

For the second claim, we start at $H_{\mathrm{gi}}(0,0) = 1$ and $H_{\mathrm{gi}}(1,2) = 2$, and get by induction that $H_{\mathrm{gi}}(2m, (4k{+}2)m) \geq 2\,(2k)^m - 2$ and

$$H_{\mathrm{gi}}(2m{+}1, (4k{+}2)m{+}2) \geq 3\,(2k)^m - 2$$

for all $m \geq 0$, from which the result follows via suitable substitutions (as in the proof of Corollary 4.8). □

With the method of Corollary 4.9 one also gets a lower bound for

$$\liminf_{n \longrightarrow \infty} H_{\text{gi}}(d,n)/n^{\lfloor d/2 \rfloor}$$

that improves somewhat on that by Jeroslow [**J**, p. 370].

5. Shadows

To the previous constructions that we have surveyed, we now add a new deformed product construction, which—via Lemma 2.9—gives a lower bound for the number of simplex pivots required by the shadow vertex rule. We do this, of course, by constructing polytopes whose shadows have many vertices.

5.1. Shadows of d-polytopes.

LEMMA 5.1. *Let $\alpha(x)$ denote a non-constant linear function on the plane ($x \in \mathbb{R}^2$). For each even $k \geq 4$, and for each $\lambda > 0$, there exist normally equivalent k-gons*

$$V = \text{conv}\{v_1, \ldots, v_k\} \qquad \textit{and} \qquad W = \text{conv}\{w_1, \ldots, w_k\}$$

(both labeled in clockwise order), such that

$$\begin{array}{rcl} 0 & = & \alpha(v_1) < \alpha(w_1) < \alpha(w_2) < \alpha(v_2) < \alpha(v_3) \\ & < & \alpha(w_3) < \ldots < \alpha(w_k) < \alpha(v_k) = 1 \end{array}$$

and such that $\text{conv}(V \cup W)$ *is a convex $2k$-gon whose vertices appear on its boundary in the order*

$$v_1, w_1, w_2, v_2, v_3, w_3, w_4, \ldots, w_k, v_k.$$

PROOF. Take $2k$ points equally spaced on a semicircular arc of radius $\frac{1}{2}$ ("half a $2(2k-1)$-gon") and label them according to the order given by this lemma.

Our figure illustrates the case $k = 4$. $\square$

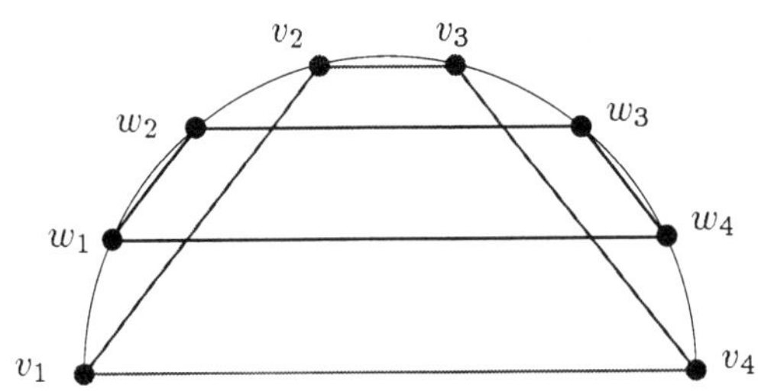

Note: this lemma is false for all odd k.

THEOREM 5.2. *For all even $k \geq 4$,*

$$H_{2\text{-sh}}(d+2, n+k) \quad \geq \quad k\, H_{2\text{-sh}}(d,n).$$

PROOF. Let P be a convex polytope with a 2-dimensional projection that has $m := H_{2\text{-sh}}(d,n)$ vertices. After a projective transformation, we may assume that the projection is given by

$$\pi \colon \mathbb{R}^d \longrightarrow \mathbb{R}^2, \qquad x \longmapsto (\varphi(x), \varphi'(x)),$$

such that the image of this projection is a planar m-gon $\operatorname{conv}\{u_1, \dots, u_m\}$ (labeled clockwise) with $u_1 = (0,0)$, $u_m = (1,0)$, and $0 = \varphi(u_1) < \varphi(u_2) < \dots < \varphi(u_m) = 1$, while $\varphi'(i) \geq 0$ (with equality only for $i = 1$ and $i = m$), as in the figure:

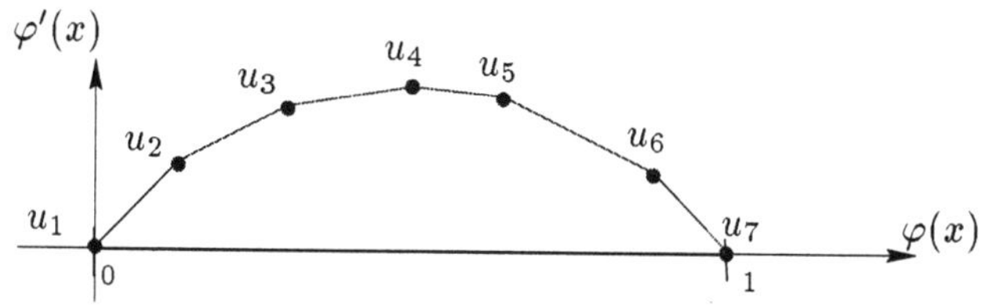

With this preparation, we construct polygons $V, W \subseteq \mathbb{R}^2$ as in Lemma 5.1, and with these the deformed product

$$Q \ := \ (P, \varphi) \bowtie (V, W).$$

The projection of this polytope to the last two coordinates given by Corollary 3.6(ii) produces the following picture:

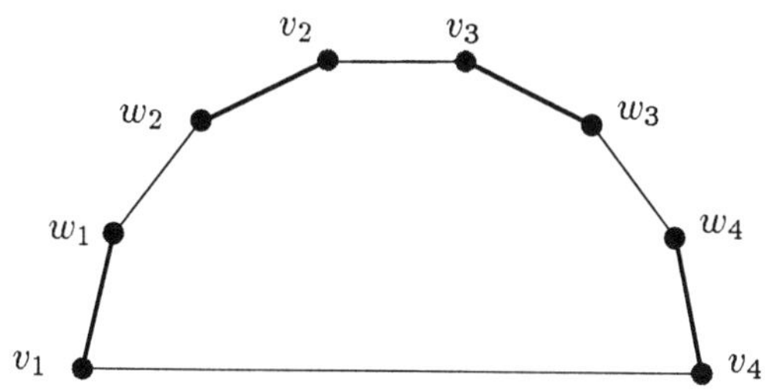

where the fat edges represent the images of the copies P_i of the polytope P. In fact, if instead we use the projection to $\mathbb{R}^3$ given by

$$\widehat{\pi}\colon \mathbb{R}^d \longrightarrow \mathbb{R}^3, \qquad x \longmapsto (\pi_2(x), \varphi'(x)),$$

then the image is a 3-dimensional polytope whose "base" is still given by this figure, while there is a little "arc" erected above each of the fat edges. The graph of the resulting polytope,

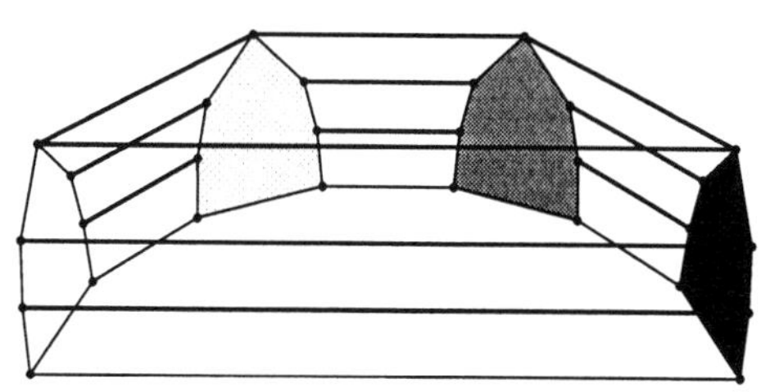

in the top view (for $m = 6$), may look somewhat like the following:

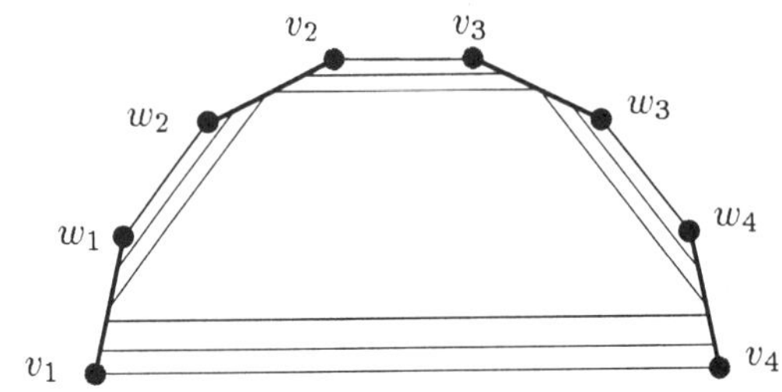

This top view corresponds to deleting the last coordinate from $\mathbb{R}^3$, and thus to taking the projection π_2 applied to Q. Now we take a slightly shifted projection, namely the projection

$$\begin{array}{rcccccc} \widehat{\pi}^\varepsilon\colon & \mathbb{R}^d & \xrightarrow{\widehat{\pi}} & \mathbb{R}^3 & \longrightarrow & \mathbb{R}^2 \\ & x & \longmapsto & (\pi_2(x), \varphi'(x)) & \longmapsto & \pi_2(x) + \varepsilon\varphi'(x)\, e_2 \end{array}$$

If $\varepsilon > 0$ is chosen small enough, then all the vertices of the "arcs above the fat edges" will appear in the shadow boundary; thus instead of the $2k$ vertices of the image of π_2 (two for each of the fat edges $[v_i, w_i]$), the image of the shifted projection will have mk vertices (very flat arcs with m vertices each for each of the k fat edges).

Hence the projection $\widehat{\pi}^\varepsilon$ of Q has $km = k\, H_{2\text{-sh}}(d, m)$ vertices. □

COROLLARY 5.3. *For $n \geq 2d \geq 2$,*

$$H_{2\text{-sh}}(d,n) \;\geq\; \left(2\left\lfloor \frac{n}{d} \right\rfloor\right)^{d/2} \qquad \text{if } d \text{ is even,}$$

and

$$H_{2\text{-sh}}(d,n) \;\geq\; 2\left(2\left\lfloor \frac{n-2}{d-1} \right\rfloor\right)^{\lfloor d/2 \rfloor} \qquad \text{if } d \text{ is odd.}$$

PROOF. This is (nearly) the same analysis as for Corollary 4.8. □

For $\liminf\limits_{n\longrightarrow\infty} H_{2\text{-sh}}(d,n)/n^{\lfloor d/2 \rfloor}$ we also get the *same* bounds that were derived in Corollary 4.9 for $\liminf\limits_{n\longrightarrow\infty} H(d,n)/n^{\lfloor d/2 \rfloor}$, with the same proof.

5.2. k-shadows and applications. The deformed product of polytopes with "large shadows" described in the last section has a number of applications outside of linear programming. The lower bound applies to more general geometric constructions, thanks to the following observations.

OBSERVATION 5.4. The complexity of the shadow of a polytope on the plane (the 2-shadow) is a lower bound on the complexity of a k-shadow (the projection of a d-dimensional polytope to a k-dimensional subspace), since projections do not increase the number of vertices of a polytope.

A *k-section* of a polytope P in $\mathbb{R}^d$ is the intersection of P with a k-dimensional hyperplane.

OBSERVATION 5.5. The polar of a k-shadow of P is a k-section of the polar of P. (For instance the polar of a projection of P from a point x is the intersection of the polar hyperplane of x with the polar of P.)

Analogous to $H_{k\text{-sh}}(d,n)$, which denotes the maximal number of vertices of any k-shadow of a (d,n)-polytope, let $H_{k\text{-sec}}(d,n)$ be the maximum, over all d polytopes with n vertices, of the number of facets in a k-section.

COROLLARY 5.6. $H_{k\text{-sec}}(d,n) = H_{k\text{-sh}}(d,n) \in O(n^{\lfloor d/2 \rfloor})$.

The computation of shadows and sections are natural problems in Computational Geometry; we can think of the computation of k-shadows as an intermediate problem between Linear Programming (the 1-shadow) and Halfspace Intersection (the d-shadow). One natural application is the maximization of a convex function of k variables over a polytope with n facets; the maximum is achieved at a vertex of the k-shadow.

Several researchers have studied the construction of k-shadows with respect to particular geometric search and optimization problems. Ponce & Faverjon [**PF**] compute the set of stable three-finger grasps, with friction, of a polygon in the plane, as the 3-shadow of a 5-dimensional polytope. This result is extended in [**PSSBM**], where a subset of the stable four-finger grasps, with friction, of a polyhedron in $\mathbb{R}^3$ is computed the as 8-shadow of an 11-dimensional polytope. Agarwal, Amenta & Sharir [**AAS**] reduce the problem of finding the largest similar copy of one convex polygon contained in another to the maximization of a convex function in two dimensions over a polytope in $\mathbb{R}^4$; they use the particular structure of this polytope to get an efficient algorithm for the computation of the shadow. Algorithms for the computation of general shadows were also developed in the context of Logic Programming by Huynh, Lassez & Lassez [**HLL**].

5.3. k-shadow algorithms. In this section we consider some old and new algorithms for the problem:

PROBLEM 5.7. **(k-Shadow)**
Input: A set $\mathcal{H}$ of n linear halfspaces in $\mathbb{R}^d$
Output: The projection of the polytope $P = \bigcap \mathcal{H}$ to the subspace spanned by $e^1, \ldots, e^k$.

Corollary 5.6 settles the question of the worst-case complexity of the problem when P is a simple polytope. But the problem remains interesting because one expects that, in most situations, the k-shadow, $k < d$, will be significantly less complex than P itself. (For $k = 2$ this intuition is proved, in a probabilistic model, by Borgwardt's analysis of the Shadow Vertex Algorithm [**Bo**].) The goal therefore is an optimal output-sensitive algorithm. An algorithm is *output-sensitive* when the running time depends on both the size of the input and the size of the output.

A hyperplane of the form $\{x \in \mathbb{R}^d : a_1x_1 + a_2x_2 + \ldots + a_kx_k + b = 0\}$ in $\mathbb{R}^d$ is called *k-vertical* (a k-vertical hyperplane is parallel to the $x_{k+1}, \ldots, x_d$ axes). We say that a face f of P is *supported by* a k-vertical hyperplane H if f is contained in H and P is entirely contained in one of the closed halfspaces bounded by H. The faces of P supported by k-vertical hyperplanes are the faces of the k-shadow.

We begin with the obvious worst-case optimal algorithm for computing the k-shadow. This algorithm returns a list F of all faces of the k-shadow.

ALGORITHM 5.8. **(Worst-case Algorithm)**

```
    Compute P
    For every face f of P
        if f is supported by a k-vertical hyperplane
```

```
4           add f to F
```

For Step **1**, there is an asymptotically worst-case optimal algorithm, due to Chazelle [**Ch**], for computing a simple polytope as an intersection of halfspaces, which runs in time $O(n^{\lfloor d/2 \rfloor})$. The test in Step **3** can be performed in constant time (when d is taken to be constant). Together with Corollary 5.6 we get the following.

THEOREM 5.9. *The k-shadow of a simple polytope P can be computed in*

$$O(n^{\lfloor d/2 \rfloor})$$

time, and this is optimal in the worst case.

We now turn to algorithms that achieve some degree of output-sensitivity by computing the k-shadow of P without computing P itself. We compare the running times in terms of both the number s of shadow facets and the number v of shadow vertices. We shall consider only the case which P is simple. Since any vertex of a simple polytope is adjacent to exactly $\binom{d}{k-1}$ faces of dimension $k-1$, taking d and k to be constants implies that, in the k-shadow, $s = O(v)$. But the only bound we have on v with respect to s is that $v = O(s^{\lfloor k/2 \rfloor})$, from the Upper Bound Theorem. So it is possible, for large k, that $v \gg s$.

Ponce et al. [**PSSBM**] described and tested the following pivoting algorithm, analogous, for instance, to the convex hull algorithm of Avis & Fukuda [**AF**]. It uses two data structures, a stack S and a dictionary D, and returns a list V of the vertices of the k-shadow.

ALGORITHM 5.10. **(Shadow Vertex Pivoting Algorithm)**

```
1     Find the minimum vertex p_0 of P w.r.t. the x_1 coordinate
2     Insert p_0 in S
3     Insert p_0 in D
4     Add p_0 to V
5     While S ≠ ∅
6           remove a vertex p from S
7           for each edge e adjacent to p
8              if e is supported by a k-vertical hyperplane
9                 find the other endpoint p' of e
10                if p' ∉ D
11                   insert p' in D
12                   insert p' in S
13                   add p' to V
14    Return V
```

Step **1** can be done by linear programming, which, when d is taken to be a constant, requires time $O(n)$ (although exponential in d). Step **9** is done by intersecting the ray supporting e anchored at p with each of the input halfspaces in turn. The intersection nearest p is p'. Since the degree of each vertex in the k-shadow is constant, the entire face structure can be derived from the list of vertices in time $O(v)$.

THEOREM 5.11. (Ponce et. al., [**PSSBM**]) *The k-shadow of a simple polytope given as an intersection of n halfspaces in fixed dimension can be computed in time $O(vn)$.*

We give a different algorithm for computing the k-shadow, which is more efficient when $v \gg s$. The algorithm operates in two phases. In the first phase, we find all the shadow facets, rather than the shadow vertices, by pivoting. In the second phase we construct the k-shadow as the intersection of the shadow facets using a halfspace intersection algorithm.

In a pivoting step in the first phase, we find all shadow facets adjacent to the current one. Let f be the current shadow facet, a $(k-1)$-face of P that projects to a facet of the shadow. The face f lies in the intersection G of some set $\mathcal{G}$ of $d-k+1$ input hyperplanes. For each input hyperplane F not in $\mathcal{G}$, we replace each of the $d-k+1$ hyperplanes of $\mathcal{G}$ in turn with F, to get a set $\mathcal{G}'$ which intersects in a different $(k-1)$-flat G'. G' contains a shadow facet if, first, G' contains a $(k-1)$-face of P, and second, if that $(k-1)$-face supported by a vertical hyperplane.

The input halfspaces corresponding to $\mathcal{G}'$ define an unbounded polytope P' which contains P. If G', as a face of P', is not supported by a k-vertical hyperplane, then any face of P contained in G' will not be supported by a k-vertical hyperplane either. In that case we know that G' cannot contain an adjacent shadow facet.

Otherwise we run a $(k-1)$-dimensional linear program in the affine subspace G', minimizing x_1, to find out if $G' \bigcap P = \emptyset$. If the linear program is infeasible, G' fails to intersect P; otherwise it returns a point p which minimizes x_1 on an adjacent shadow facet f' induced by G'.

For each f, we run at most $(d-k)n$ such linear programs, each requiring $O(n)$ time. So finding all shadow facets adjacent to f requires $O(n^2)$ time.

The algorithm for the pivoting phase again requires a stack S and a dictionary D, and returns a list F of shadow facets.

ALGORITHM 5.12. **(Shadow Facet Pivoting Algorithm)**

```
1     Find the minimum vertex p_0 of P w.r.t. the x_1 coordinate
2     For each (k−1)-face f adjacent to p_0
3         if f is supported by a k-vertical hyperplane
4           insert f in S
5           insert f in D
6           add f to F
7     While S ≠ ∅
8           remove a shadow facet f from S
9           find all shadow facets adjacent to f
10          for every shadow facet f' adjacent to f
11              if f' ∉ D
12                insert f' in D
13                insert f' in S
14                add f' to F
15    Return F
```

Step **9** is executed once for every shadow facet and requires (by the previous discussion) $O(n^2)$ time, so the running time of the pivoting phase is $O(sn^2)$.

From the list F of shadow facets, the entire k-shadow can be computed by a convex hull algorithm in dimension k. Using Chazelle's algorithm this takes time $O(s^{\lfloor k/2 \rfloor})$, for an overall running time of $O(sn^2 + s^{\lfloor k/2 \rfloor})$. When $s^{\lfloor k/2 \rfloor} \gg v \gg s$, we can do better by using Seidel's output-sensitive convex hull algorithm, which takes time $O(s^2 + v \log s)$. The s^2 term in the running time for Seidel's algorithm comes from a first phase which finds the minimal vertex, with respect to x_1, in each facet of the final halfspace intersection. Since the Shadow Facet Pivoting Algorithm produces these minimal vertices as a by-product of finding the shadow facets, we can skip the first phase, for an overall running time of $O(sn^2 + v \log s)$.

THEOREM 5.13. *A k-shadow of a polytope P, given as the intersection of an input set $\mathcal{H}$ of halfspaces, can be computed in time $O(sn^2 + \min(s^{\lfloor k/2 \rfloor}, v \log s))$, where $n = |\mathcal{H}|$, s is the number shadow facets, and v is the number shadow vertices.*

Both pivoting algorithms run a lot of linear programs using almost identical sets of constraints. The running times can therefore be improved slightly using the ray-shooting data structure of Matoušek & Schwarzkopf [**M**, **MS**] (a similar observation is made in [**PSSBM**], and by Chan [**C**]).

5.4. Shadows of cyclic polytopes. The polar of a cyclic polytope has the maximum number of faces of all dimensions among polytopes with n facets. We show, however, that in even dimension the polars of cyclic polytopes do not maximize the complexity of the shadow among all polytopes with n facets. This is equivalent to the conclusion that for large n a monotone path through all vertices cannot be found by the "shadow vertex algorithm," for any realization of $C_d(n)^\Delta$. Equivalently, a Bruggesser-Mani shelling of $C_d(n)$ [**Zi**, Sect. 8.2] cannot in general be generated by a 2-dimensional section of $C_d(n)$ that would cut all the facets. In fact, we show that in even dimensions the shadow of a polar of a cyclic polytope must have asymptotically fewer vertices than the polytope itself.

Let us review some properties of cyclic polytopes; for more details, see [**Zi**, Example 0.6]. A curve C in $\mathbb{R}^d$ is *of order d* if every hyperplane intersects C in at most d points. The convex hull of any set of $n > d$ points that lie on a curve of order d is a cyclic polytope $C_d(n)$. Sturmfels [**St**] has shown that if the dimension d is even, then any polytope P that is combinatorially isomorphic to a polytope $C_d(n)$ has its vertices on some curve of order d.

A facet of P is supported by a $(d-1)$-plane H which contains d vertices. Let H^+ denote the positive halfspace of H, which is determined by $P \subseteq H^+$. Index the vertices $v_1, \dots, v_n$ along a suitable directed d-order curve C. If C leaves H^+ at a vertex v_i, then it must reenter H^+ at v_{i+1}, since otherwise v_{i+1} would be outside H^+ and hence outside P. Thus the vertex set of any facet of P is a union of adjacent pairs v_i, v_{i+1} ("Gale's evenness criterion" [**Zi**, Thm. 0.7]).

Every face of smaller dimension is determined by a subset of the vertex set of a facet. This means that the $(d-2)$-faces of P are of the form

$$T(i_1, \dots, i_{d/2}) := \operatorname{conv}\{v_{i_1}, v_{i_1+1}, v_{i_2}, v_{i_2+1}, \dots, v_{i_{d/2}}\},$$

with $i_k \notin \{i_l, i_l+1\}$ (and indices taken mod n). That is, there are $d/2 - 1$ adjacent pairs of vertices, and then one vertex, $v_{i_{d/2}}$, which needs not be adjacent to any other. We choose our numbering in such a way that $v_{i_{d/2}+1}$ is not a vertex of $T(i_1, \dots, i_{d/2})$.

THEOREM 5.14. *Let $P \subseteq \mathbb{R}^d$ be a cyclic d-polytope with n vertices, where $d \geq 4$ is even, and let P^Δ be a polar polytope of P. Then any 2-dimensional shadow of P^Δ has at most*

$$\frac{6n}{d-2}\binom{n-d/2}{d/2-2} \quad = \quad O(n^{d/2-1})$$

vertices.

PROOF. Using the duality between projections and sections, we consider a 2-dimensional section $F \cap P$ of the cyclic polytope P, and verify that it cannot have more than $\frac{6n}{d-2}\binom{n-d/2}{d/2-2}$ vertices.

The maximum is clearly achieved for a 2-plane that meets the interior of P, and that is in general position with respect to P. In fact, given any 2-plane F that meets the interior of P, the number of vertices of $F \cap P$ cannot decrease by a sufficiently small perturbation of F.

Under these conditions, every vertex p of $F \cap P$ is determined by a unique $(d-2)$-dimensional face $T(i_1, \dots, i_{d/2})$. Consider the $(d-1)$-dimensional hyperplane

$$H(i_2, \dots, i_{d/2}) \quad := \quad \text{aff}(F \cup \{v_{i_2}, \dots, v_{i_{d/2}}\}).$$

This hyperplane intersects the $(d-2)$-flat supporting $T(i_1, \dots, i_{d/2})$ in a $(d-3)$-flat which contains the vertices $v_{i_2}, \dots, v_{i_{d/2}}$. There is a one-dimensional family of such $(d-3)$-flats, one of which supports the $(d-3)$-face $\text{conv}\{v_{i_1}, v_{i_2}, \dots, v_{i_{d/2}}\}$ and another supporting the $(d-3)$-face $\text{conv}\{v_{i_1+1}, \dots, v_{i_{d/2}}\}$. The $(d-3)$-flats that intersect the interior of $T(i_1, \dots, i_{d/2})$ separate the points v_{i_1} and v_{i_1+1}. If F hits $T(i_1, \dots, i_{d/2})$ in the interior point p, then $H(i_2, \dots, i_{d/2})$ separates the points v_{i_1} and v_{i_1+1} on the moment curve. The figure represents the four-dimensional case.

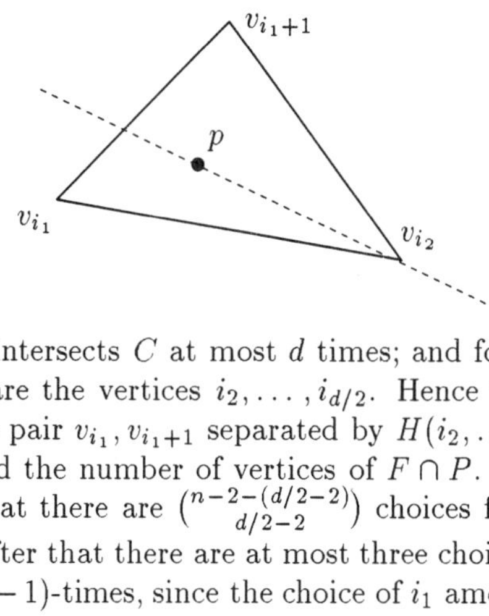

But every hyperplane intersects C at most d times; and for $H(i_2, \dots, i_{d/2})$, $d-3$ of those intersections are the vertices $i_2, \dots, i_{d/2}$. Hence there are at most three possible choices for the pair v_{i_1}, v_{i_1+1} separated by $H(i_2, \dots, i_{d/2})$.

Thus we can bound the number of vertices of $F \cap P$. There are n choices for the index $i_{d/2}$, after that there are $\binom{n-2-(d/2-2)}{d/2-2}$ choices for the set set of indices $\{i_2, \dots, i_{d/2-1}\}$, and after that there are at most three choices for i_1. In this count, each face appears $(d/2-1)$-times, since the choice of i_1 among $\{i_1, \dots, i_{d/2-1}\}$ was arbitrary. Thus there are at most

$$\frac{1}{d/2-1} 3n \binom{n-d/2}{d/2-2} \quad = \quad \frac{6n}{d-2}\binom{n-d/2}{d/2-2} \quad = \quad O(n^{d/2-1})$$

vertices p of $F \cap P$. □

In dimension $d = 4$ this means that the 2-shadow cannot have more than $3n$ vertices. For this we have an (almost) matching lower bound.

THEOREM 5.15. *Let P be a cyclic polytope in $\mathbb{R}^4$ with n vertices, with 0 in its interior, and let P^Δ be the polar polytope of P. The shadow of P^Δ in $\mathbb{R}^2$ can have at many as $3n-10$ vertices.*

PROOF. For the lower bound, construct a curve of degree 4, a flat F and points $v_1, \dots, v_n$ on the curve so that if you project the curve to an orthogonal complement to F, which maps F itself to a point, you obtain the situation in our figure.

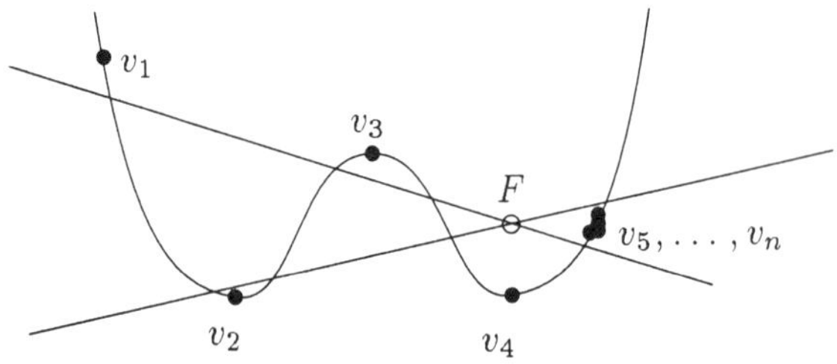

This can, for example, be achieved very explicitly by taking the curve C given by $\gamma(t) := (t, \binom{t}{2}, \binom{t}{3}, \binom{t}{4})$, taking π_{14} to be the projection to the first and fourth coordinate, and taking $F := \{(\frac{5}{2}, x, y, 0)\}$, and choosing the points on C as $v_1 := \gamma(-1)$, $v_2 := \gamma(\frac{1}{2})$, $v_3 := \gamma(\frac{3}{2})$, $v_4 := \gamma(\frac{5}{2})$, and $v_i := \gamma(t_i)$ with t_i very close to 4, for all i $(5 \le i \le n)$.

$F \cap T(i,j) \neq \emptyset$ holds if and only if in the 2-dimensional projection the point $\pi_{14}(F)$ is in the interior of the triangle $\pi_{14}(T(i,j))$, which is spanned by $\pi_{14}(v_i)$, $\pi_{14}(v_j)$ and $\pi_{14}(v_{j+1})$.

Every 2-face of P formed by $\{v_1, v_2\}$, $\{v_2, v_3\}$ or $\{v_3, v_4\}$, together with any one of the vertices $v_5, \dots, v_n$ (that is, the triangles $T(i,1)$, $T(i,2)$ and $T(i,3)$ for $5 \le i \le n$) intersect F. This gives $3(n-4)$ vertices in $P \cap F$. In addition, the 2-faces $\mathrm{conv}\{v_1, v_4, v_5\}$ and $\mathrm{conv}\{v_1, v_4, v_n\}$ (i.e., the triangles $T(1,4)$ and $T(4,n)$) also intersect F, for a total of $3n - 10$. □

Acknowledgments. We are grateful to Bernd Gärtner and to Raimund Seidel for helpful discussions.

References

[AAS] P. Agarwal, N. Amenta, & M. Sharir, *Largest placement of one convex polygon inside another,* Discrete Comput. Geometry **19** (1998), 95–104.

[AZ] N. Amenta and G. M. Ziegler, *Shadows and slices of polytopes,* in: Proceedings of the 12th Annual ACM Symposium on Computational Geometry, May 1996, pp. 10–19.

[ABS] D. Avis, D. Bremner and R. Seidel, *How good are convex hull algorithms?* Computational Geometry: Theory and Applications **7** (1997), 265–301.

[AC] D. Avis and V. Chvátal, *Notes on Bland's pivoting rule,* in: "Polyhedral Combinatorics" (M. L. Balinski, A. J. Hoffmann, eds.), Math. Programming Study **8** (1978), 24–34.

[AF] D. Avis and K. Fukuda, *A pivoting algorithm for convex hulls and vertex enumeration of arrangements and polyhedra,* Discrete Comput. Geometry **8** (1992) 295–314.

[B] C. E. Blair, Some linear programs requiring many pivots, Working Paper No. 867, College of Commerce and Business Administration, University of Illinois, May 1982.

[Bl] R. G. Bland, *New finite pivoting rules for the simplex method,* Math. Operations Research **2** (1977), 103–107.

[Bo] K. H. Borgwardt, The Simplex Method. A Probabilistic Analysis, Algorithms and Combinatorics **1**, Springer-Verlag, Heidelberg 1987.

[C] T. Chan, *Output-sensitive results on convex hulls, extreme points and related problems,* in: Proc. "11th Annual Symposium on Computational Geometry" 1995, 10–19.

[Ch] B. Chazelle, *An optimal convex hull algorithm for point sets in any fixed dimension,* Discrete Comput. Geometry **9** (1993), 145–158.

[CEG] B. Chazelle, H. Edelsbrunner and L. J. Guibas, *The complexity of cutting complexes,* Discrete Comput. Geometry **4** (1989), 139–182.

[Chv] V. Chvátal, Linear Programming, W. H. Freeman, New York 1983.

[Cl] J. Clausen, *Worst-case behaviour of simplex-algorithms: From theory to numerical examples,* Report 82/1, DIKU – Institute of Datology, University of Copenhagen, October 1981, revised February 1985, 35 pages.

[Cl2] J. Clausen, *A new family of exponential LP problems,* European J. Operational Research **32** (1987), 130–139.

[FT] Y. Fathi and C. Tovey, *Affirmative action algorithms,* Math. Programming **34** (1986), 292–301.

[GS] S. I. Gass and T. Saaty, *The computational algorithm for the parametric objective function,* Naval Research Logistics Quarterly **2** (1955), 39–45.

[GZ] B. Gärtner and G. M. Ziegler, *Randomized simplex algorithms on Klee-Minty cubes,* in: Proc. 35th Annual "Symposium on Foundations of Computer Science" (FOCS), November 20–22, 1994, Santa Fe NM, IEEE Computer Society Press, Los Alamitos CA, 1994, pp. 502–510.

[G] D. Goldfarb, *Worst case complexity of the shadow vertex simplex algorithm,* Preprint, Columbia University 1983, 11 pages.

[G2] D. Goldfarb, *On the complexity of the simplex algorithm,* in: Advances in optimization and numerical analysis, Proc. 6th Workshop on Optimization and Numerical Analysis, Oaxaca, Mexico, January 1992; Kluwer, Dordrecht 1994, 25–38.

[GSi] D. Goldfarb and W. T. Sit, *Worst case behaviour of the steepest edge simplex method,* Discrete Applied Math. **1** (1979), 277–285.

[GK] P. Gritzmann and V. Klee, *Mathematical programming and convex geometry,* in: Handbook of Convex Geometry (P. Gruber, J. Wills, eds.), Vol. B, North Holland, Amsterdam 1993, pp. 627-674.

[GLS] M. Grötschel, L. Lovász and A. Schrijver, Geometric Algorithms and Combinatorial Optimization, Algorithms and Combinatorics **2**, Springer-Verlag, Berlin Heidelberg 1988; second edition 1994.

[HLL] T. Huynh, C. Lassez and J.-L. Lassez, *Practical issues on the projection of polyhedral sets,* Annals of Mathematics and Artificial Intelligence **6** (1992), 295–316.

[J] R. G. Jeroslow, *The simplex algorithm with the pivot rule of maximizing improvement criterion,* Discrete Math. **4** (1973), 367–377.

[K] G. Kalai, *A subexponential randomized simplex algorithm,* in: "Proc. 24th ACM Symposium on the Theory of Computing (STOC)", ACM Press 1992, pp. 475–482.

[Kl] V. Klee, *A class of linear programs requiring a large number of iterations,* Numerical Math. **7** (1965), 313–321.

[KM] V. Klee and G. J. Minty, *How good is the simplex algorithm?,* in: "Inequalities III," (O. Shisha, ed.), Academic Press, New York, (1972), 159–175.

[M] J. Matoušek, *Linear optimization queries,* Journal of Algorithms **14** (1993), 432–448.

[MS] J. Matoušek and O. Schwarzkopf, *Linear optimization queries,* in: "Proc. 8th Annual Symposium on Computational Geometry" (Berlin 1992), ACM Press 1992, 16–25.

[MSW] J. Matoušek, M. Sharir and E. Welzl, *A subexponential bound for linear programming,* in: "Proc. 8th Annual ACM Symp. Computational Geometry" (Berlin 1992), ACM Press 1992, 1–8.

[Mc] P. McMullen, *The maximum numbers of faces of a convex polytope,* Mathematika **17** (1970), 179–184.

[Mu] K. G. Murty, *Computational complexity of parametric linear programming,* Math. Programming **19** (1980), 213–219.

[P] M. Padberg, Linear Programming and Extensions, Algorithms and Combinatorics **12**, Springer-Verlag, Heidelberg 1995.

[PF] J. Ponce and B. Faverjon, *On computing three-finger force-closure grasps of polygonal objects,* IEEE Transactions on Robotics and Automation **11** (1995), 868–881.

[PSSBM] J. Ponce, S. Sullivan, A. Sudsang, J.-D. Boissonnat and J.-P. Merlet, *On computing four-finger equilibrium and force-closure grasps of polyhedral objects,* International Journal of Robotics Research **16** (1996), 11–35.

[S] A. Schrijver, Theory of Linear and Integer Programming, Wiley-Interscience Series in Discrete Mathematics and Optimization, John Wiley & Sons, Chichester New York 1986.

[Se] R. Seidel, *Constructing higher dimensional convex hulls at logarithmic cost per face,* Proc. 18th Annual Symposium on the Theory of Computing, (1986), 404–413.

[St] B. Sturmfels, *Cyclic polytopes and d-order curves,* Geometriae Dedicata **24** (1987), 103–107.

[Z] N. Zadeh, *What is the worst case behavior of the simplex algorithm?*, Technical Report No. 27, Dept. Operations Research, Stanford 1980, 26 pages.

[Zi] G. M. Ziegler, Lectures on Polytopes, Graduate Texts in Mathematics **152**, Springer-Verlag, New York 1995; Revised edition 1998.
(Updates, corrections, and more: `http://www.math.tu-berlin.de/~ziegler`)

Computer Sciences Department, University of Texas, Austin, TX 78712,USA
E-mail address: `amenta@cs.utexas.edu`

Dept. Mathematics 7-1, Technische Universität Berlin, 10623 Berlin, Germany
E-mail address: `ziegler@math.tu-berlin.de`

Contemporary Mathematics
Volume **223**, 1999

Flag Complexes, Labelled Rooted Trees, and Star Shellings

Louis J. Billera, Clara S. Chan, and Niandong Liu

ABSTRACT. Flag complexes, also known as clique complexes, are studied in various guises. In this paper we demonstrate that labeled rooted trees form flag complexes when given a natural partial order. Consequently we derive enumerative facts about certain complexes of cubes, which we call star-shellable, by using a correspondence between the number of faces of each dimension in such complexes with the number of labeled rooted plane trees having certain properties.

1. Introduction

In this paper we will construct flag complexes from sets of labeled rooted plane trees and show a surprising correspondence to a class of cubical complexes which we call star-shellable. This leads to enumerative results about star-shellable cubical complexes, which we conjecture to hold more generally.

In §2, we will show that for each $d \geq 2$, the set of all rooted plane trees on $d+1$ nodes forms a flag complex in a natural way. We show in §3 that this flag complex has as its f-vector the h-vector of the d-dimensional cube. This observation gives a "strong g-theorem" for cubes.

In section 4, we define star-shellable complexes and show that every cubical star-shellable complex of dimension less than 5 has for its h-vector the f-vector of a flag complex corresponding to a set of labeled rooted plane trees. We conjecture that this holds for all star-shellable complexes, and thus that the h-vectors of all star-shellable complexes satisfy the Kruskal-Katona inequalities for simplicial complexes.

In section 5, we observe that the h-vectors of cubical zonotopes and cuboids also satisfy these conditions, for dimensions up to several thousand.

2. Flag complexes of rooted plane trees

A flag complex is a simplicial complex whose minimal non-faces contain exactly 2 vertices. Equivalently, any set of vertices forming a complete graph in the 1-skeleton of the complex is the set of vertices of some face. Thus, flag complexes are

1991 *Mathematics Subject Classification.* Primary 52B05; Secondary 05A15.

Key words and phrases. cubical complex, h-vector, labelled rooted tree, flag complex, star shelling.

Partially supported by NSF grant DMS-9500581.

Partially supported by NSF grant DMS-9410679.

© 1999 American Mathematical Society

also known as clique complexes. For each $d \geq 2$, we construct a flag complex Δ_d whose simplices correspond to the rooted plane trees on $d+1$ nodes.

We order the nodes of each rooted plane tree recursively by root first, and then subtrees of the root, from left to right. This is called *preorder* (or *depth-first* order). We will refer to rooted plane trees with preorder simply as trees, and vertex i will mean the ith vertex in preorder.

If two nodes of a tree share an edge, we call the lower node a *child* of the upper. A subgraph consisting of a node with more than one child, together with all of its children (and connecting edges) is called a *fork*. More formally, we define a *fork* to be a graph $G = (V, E)$, where $V = \{n_0 < n_1 < \ldots < n_k\}$ (with $k \geq 2$, $n_1 = n_0 + 1$) and $E = \{ \{n_0, n_j\} \mid j = 1, \ldots, k\}$. The node n_0 will be called the *root* of this fork. For ease of notation, we will denote G by x_A, where $A = \{n_1, n_2, \ldots, n_k\}$ and $n_1 < n_2 < \cdots < n_k$. It is easy to see that there is a unique tree (with a given number of nodes) having x_A as its only fork. (See Figure 1).

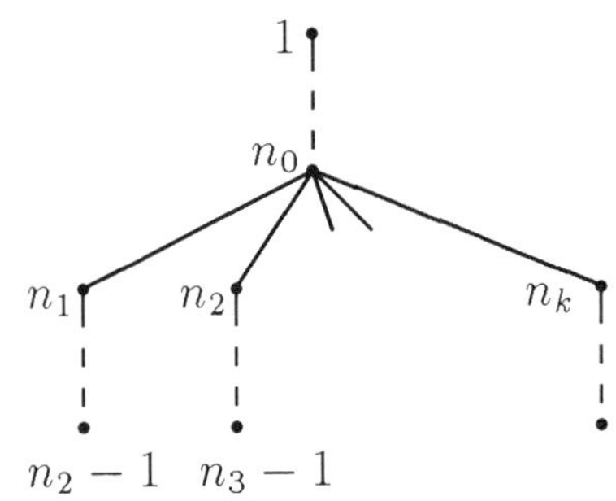

FIGURE 1: The unique tree with $x_{\{n_1, n_2, \ldots, n_k\}}$ as its only fork.

DEFINITION 1. A *fork system* is the set of forks of a tree. Forks in a fork system are said to be *consistent* with each other.

We define a simplicial complex Δ_d whose vertex set is the set of all forks x_A, where $A = \{n_1, \ldots, n_k\}$ and $n_k \leq d+1$. A set of forks forms a face of Δ_d if and only if it is a fork system, that is, if and only if they are consistent with each other. That faces of Δ_d correspond directly to $(d+1)$-node trees is a consequence of the following.

PROPOSITION 2.1. *Each fork system arises from a unique tree. In other words, if two n-node trees T_1 and T_2 have the same fork system S, then $T_1 = T_2$.*

PROOF. If $n = 1$, this is clear. We proceed by induction on n. Suppose it is true for $n < k$. For $n = k$, let l be the parent of k in T_1. If l is the root of some fork in S, then it is a fork with the same children in T_1 as in T_2, so l is also the parent of k in T_2. If l is not a root in S, then k does not belong to any fork in T_1 or T_2, so $l = k - 1$ is also the parent of k in T_2. In either case, (l, k) is an edge in both T_1 and T_2. Delete the edge (l, k) and the node k from both trees T_1 and T_2. The new trees still have the same fork system, so by induction they are identical. $\square$

We show now that the collection Δ_d is a simplicial complex, in fact a flag complex. Let x_A and x_B be forks with children $A = \{m_1 < m_2 < \ldots < m_k\}$ and $B = \{n_1 < n_2 < \ldots < n_l\}$. We assume $m_1 < n_1$. For convenience, we set $m_{k+1} = d + 2$.

LEMMA 2.2. *The forks x_A and x_B are consistent if and only if they are disjoint and non-interlacing,* i.e.,

1. $A \cap B = \emptyset$, *and*
2. *if p is such that $m_p < n_1 < m_{p+1}$, then $n_l < m_{p+1}$.*

PROOF. Suppose these conditions hold. For $i \leq j$, let $P(i,j)$ denote the path of edges $\{i, i+1\}, \{i+1, i+2\}, \ldots, \{j-1, j\}$. Now let T be the plane tree with edges $P(1, m_1 - 1) \cup x_A \cup (\cup_{i \neq p} P(m_i, m_{i+1} - 1)) \cup P(m_p, n_1 - 1) \cup x_B \cup (\cup_{i<l} P(n_i, n_{i+1} - 1)) \cup P(n_l, m_{p+1})$. Then T has fork-system $\{x_A, x_B\}$.

On the other hand, if x_A and x_B are forks in some tree, then $A \cap B = \emptyset$, since x_A and x_B can not share a child. Thus $m_p < n_1 < m_{p+1}$ for some p, and so the root $n_0 = n_1 - 1$ of x_B must satisfy $m_p \leq n_0 < m_{p+1}$. By the nature of preorder, it follows that $m_p < n_l < m_{p+1}$. □

THEOREM 2.3. *S is a fork system if and only if each pair in S is a fork system. Thus Δ_d is a flag complex.*

PROOF. It follows from the proof of Lemma 2.2 that if two forks are consistent (that is, appear together in a fork system), then they themselves *form* a fork system. This shows the necessity. To show sufficiency, suppose each pair in S is a fork system. We show that S is a fork system by induction on $|S|$. If $|S| = 2$, this is clear, so assume $|S| = k > 2$. Let $S = \{x_1, x_2, \ldots, x_k\}$ where n_0, the root of x_k, is greater than the root of x_i for any $i < k$. By induction on k, there exists a $(d+1)$-node tree T with fork system $\{x_1, \ldots, x_{k-1}\}$. Let n_* denote the largest descendant of n_0 in T. Then the subtree of T at n_0 is simply the path $P(n_0, n_*)$, since the roots of all forks in T are less than n_0.

Let x_k have set of children $\{n_1, n_2, \ldots, n_l\}$. We claim that $n_* \geq n_l$: if $n_* = d+1$, this is clear, so assume $n_* < d+1$. Since n_* has no child, $n_* + 1$ is a child of some fork x_p with $p < k$. Since x_p and x_k are consistent, we have $n_l < n_* + 1$, by Lemma 2.2.

Now we construct a new tree T' from T by replacing $P(n_0, n_*)$ with $x_k \cup (\cup_{i<l} P(n_i, n_{i+1} - 1)) \cup P(n_l, n_*)$. Then T' has fork system $S = \{x_1, x_2, \ldots, x_k\}$.

Thus S is a fork system if and only if each pair in S forms a fork system, and thus if and only if each subset of S is a fork system. It follows that Δ_d is a flag complex. □

3. A strong g-theorem for cubes

The fact that rooted plane trees form a flag complex is interesting in itself, but furthermore we can apply this fact to prove an intriguing enumerative result about cubes, analogous to the g-theorem for simplicial polytopes.

The enumerative invariant we will focus on is the toric h-vector, defined for an arbitrary d-dimensional polyhedral complex P as follows [**S3**]:

DEFINITION 2. For any face $F \in P$ let ∂F denote the boundary complex of F. Now define $\overline{h}(P, x)$ and $g(P, x)$ inductively by

1. $\overline{h}(\phi, x) = g(\phi, x) = 1$
2. $\overline{h}(P, x) = \sum_{F \in P} g(\partial F, x)(x-1)^{d - dim\, F}$
3. $g(P, x) = \sum_{i=0}^{\lfloor (d+1)/2 \rfloor} (k_i - k_{i-1}) x^i$, where k_i is the coefficient of x^i in $\overline{h}(P, x)$

Then $h(P,x) = \sum h_i x^i := x^{d+1}\overline{h}(P, \frac{1}{x})$ is a polynomial of degree $d+1$, and the *toric h-vector* of P is $h(P) = (h_0, h_1, \ldots, h_{d+1})$. We also define the *toric g-vector* $g(P) = (h_0, h_1 - h_0, \ldots, h_m - h_{m-1})$ where $m = \lfloor \frac{d+1}{2} \rfloor$. From here on we will refer to toric h- and g-vectors simply as h- and g-vectors.

Note that since all equi-dimensional faces of a cubical complex P are combinatorially equivalent, $h(P)$ is just an invertible linear transformation of $f(P)$. We can use Gessel's formulation for h-vectors of cubes to write this down explicitly [**S3**]. Thus, the study of h-vectors of cubical complexes is equivalent to the study of f-vectors, under this invertible linear transformation. (An alternative linear transformation is used in [**A**].)

Let C_d denote the d-dimensional cube. As a result of the Dehn-Sommerville equations $h_i = h_{d-i}$ for its boundary ∂C_d (see, for example, [**S3**]), the h-vector of C_d is simply the g-vector of its boundary complex, which has the following combinatorial interpretation, due to Shapiro. Let $a_n(i)$ denote the number of n-node rooted plane trees with exactly i forks. A proof of the following result can be found in [**C**].

LEMMA 3.1. *We have*

$$g(\partial C_d, x) = \sum_{i=0}^{\lfloor d/2 \rfloor} a_{d+1}(i) x^i.$$

Now we can establish the following.

THEOREM 3.2. *For any $d \geq 2$, $g(\partial C_d, x) = h(C_d, x) = f(\Delta, x)$ for some simplicial flag complex Δ.*

PROOF. Let Δ be the flag complex Δ_d from Section 2. By the previous lemma, it is clear that $f(\Delta_d, x) = g(\partial C_d, x)$. □

Thus $g(\partial C_d)$ satisfies the well-known Kruskal-Katona inequalities, which characterize the f-vectors of simplicial complexes [**S**]. Very similar to the Kruskal-Katona conditions are the weaker Macaulay conditions, which characterize the f-vectors of order ideals of monomials. The so-called g-theorem for simplicial polytopes states that the g-vector of the boundary of a simplicial polytope satisfies the Macaulay conditions [**S**]. Thus we have what might be termed a "strong g-theorem for cubes" [**S2**].

COROLLARY 3.3. *For any $d \geq 2$, $g(\partial C_d)$ satisfies the Kruskal-Katona inequalities.*

It now follows easily that the weaker conditions hold for the h-vectors of cubes.

COROLLARY 3.4. *For any $d \geq 2$, we have that $h(\partial C_d)$ satisfies Macaulay's conditions,* i.e., *there exists an order ideal of monomials M such that $h(\partial C_d) = f(M)$.*

PROOF. Since $h_i = g_0 + g_1 + \ldots + g_i$, the required order ideal M can be obtained by adding a new variable z to the square-free order ideal $\overline{M}$ corresponding to the simplicial complex Δ_d and defining $M_i = \cup_{k \leq i, d-i} \overline{M}_k \cdot z^{i-k}$ for $0 \leq i \leq d$. □

We note finally that one can also describe Δ as follows. Let $\Delta_0 = \{X \in \{0,1\}^d \mid X \text{ contains at least two 1's }\}$, so $|\Delta_0| = 2^d - (d+1)$. Let $\{X_i, X_j\} \in \Delta$

if and only if the 1's of one fit into consecutive 0's of the other. For $k > 2$, let $\Delta_{k-1} = \{\{X_{i_1}, X_{i_2}, ..., X_{i_k}\} \mid \{X_{i_j}, X_{i_{j'}}\} \in \Delta \text{ for all } j \neq j'\}$. Then let

$$\Delta =: \Delta_0 \cup \Delta_1 \cup ... \cup \Delta_{[\frac{d}{2}]-1}.$$

To see that $\Delta = \Delta_d$, consider the correspondence

$$f = \{n_1, ..., n_k\} \longleftrightarrow \chi_{\{n_1,...,n_k\}}$$

where $k \geq 2$, $n_i \in \{1, 2, ..., d\}$, and χ_S is the characteristic function of S. As an example, for the case of $d = 4$, $\Delta_1 = \{\{(1001), (0110)\}, \{(1100), (0011)\}\}$, and so $g_2(\partial C_4) = f_1(\Delta) = |\Delta_1| = 2$.

4. Star-shellable complexes

One can view the strong g-theorem as a result that holds for cubical complexes consisting of a single cube. It is natural to ask whether this result extends to more general cubical complexes. We now extend the construction from Section 2 to achieve the analogous result for *star-shellable* complexes of dimension less than 5, and certain star-shellable complexes of arbitrary dimension.

DEFINITION 3. A pure polyhedral complex is *star-shellable* if its facets can be ordered $F_1, F_2, \ldots$ so that for each $k \geq 2$, $\langle F_k \rangle \cap \langle F_1, F_2, \ldots, F_{k-1} \rangle$ is the star of some nonempty face in ∂F_k.

Here, $\langle F_k \rangle$ denotes the complex of faces of F_k and $\langle F_1, F_2, \ldots, F_k \rangle$ denotes the complex generated by $F_1, \ldots, F_k$. Note that for cubical complexes, the condition that $\langle F_k \rangle \cap \langle F_1, F_2, \ldots, F_{k-1} \rangle$ is the star of some nonempty face in ∂F_k is equivalent to saying that this subcomplex of $\langle F_k \rangle$ is pure and codimension 1 and contains no pair of antipodal facets of F_k. Note the difference between star-shellable and shellable cubical complexes, the latter requiring only that each facet intersect previous facets in the shelling in any codimension 1 ball or sphere. Prototypes of star-shellable cubical complexes lie on the boundaries of cubical zonotopes. (See [**Zi**] for background on zonotopes.)

PROPOSITION 4.1. *For any zonotope Z and any fixed point p outside of Z, the region of Z visible from p is star-shellable.*

PROOF. We can use a Bruggesser-Mani line shelling [**BM**]. Let F_0, F_1, F_2 be facets of Z visible from p such that $F_1 \cap F_0$ and $F_2 \cap F_0$ are antipodal facets of F_0, and for each i, let $\mathcal{F}_i$ denote the hyperplane spanned by F_i. We just need to show that there is no way that we will hit both $\mathcal{F}_1$ and $\mathcal{F}_2$ before $\mathcal{F}_0$ as we walk out of Z toward p along a generic line L. Notice that the intersections of any two of $\mathcal{F}_0$, $\mathcal{F}_1$ and $\mathcal{F}_2$ are parallel to some $d-2$ dimensional linear subspace W, so projecting onto the plane $W^\perp$ reduces the problem to the case $dim(Z) = 2$, which is clear. (See Figure 2). □

For the remainder of this section, let C denote a star-shellable cubical complex of dimension $d \geq 2$, and $F_1, F_2, \ldots, F_r$ a star shelling for C. We will call F_k an *(i,0)-face* if $\langle F_k \rangle \cap \langle F_1, \ldots, F_{k-1} \rangle$ contains exactly i (antipodally unpaired) facets of F_k. (The significance of the second label will be made clear in the last section.) It is useful to note that star-shellings have the following property.

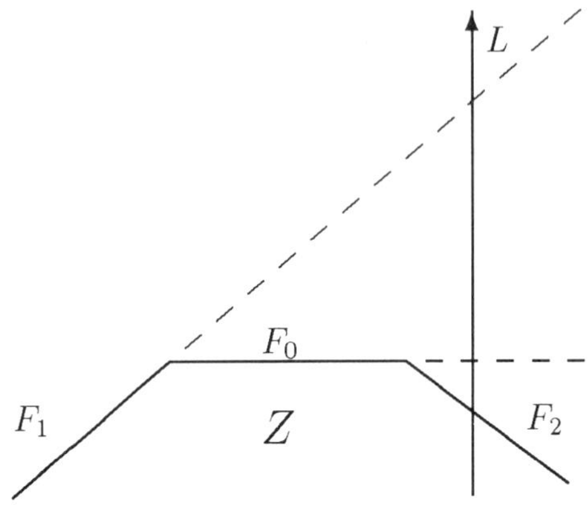

FIGURE 2: L hits $\mathcal{F}_0$ before $\mathcal{F}_1$.

LEMMA 4.2. *For every $(i,0)$-face with $i \geq 1$, there exists an earlier $(i-1,0)$-face in $F_1, \ldots, F_r$.*

PROOF. Suppose the Lemma is false. Let k be the least integer such that F_k is an $(i,0)$ face for some $i \geq 1$ yet there is no $(i-1,0)$ face among $F_1, F_2, \ldots, F_{k-1}$. Then for $j < k$, F_j must be an $(l,0)$ face for some $l < i-1$.

Let $\langle F_k \rangle \cap \langle F_1, \ldots, F_{k-1} \rangle = \langle G_1, \ldots, G_i \rangle$ where $G_1, \ldots, G_i$ are facets of F_k and if n_j denotes the least integer such that $G_j \subset F_{n_j}$ then $n_l < n_m$ only if $l < m$, *i.e.*, they are listed in the order in which they appear in the shelling. Note that $dim(G_i \cap G_j) = d-2$ for $j < i$ since G_i and G_j are not opposite in the cube F_k.

Let $n = n_i$ be the least integer such that $G_i \subset F_n$; necessarily $n < k$. For each $j < i$, since we have a shelling and since $G_i \cap G_j \in \langle F_n \rangle \cap \langle F_1, \ldots, F_{n-1} \rangle$, there exists a $(d-1)$-face H_j in $\langle F_n \rangle \cap \langle F_1, \ldots, F_{n-1} \rangle$ such that $H_j \supset G_i \cap G_j$.

Suppose $H_{j_1} = H_{j_2}$ for some $j_1 \neq j_2$. Then H_{j_1} contains $G_i \cap G_{j_1}$ and $G_i \cap G_{j_2}$, two different facets of G_i, so $G_i = H_{j_1}$. Since F_n is the first facet in the shelling containing G_i, G_i is not in $\langle F_1, \ldots, F_{n-1} \rangle$ and this contradicts $G_i = H_{j_1}$; we conclude that $H_{j_1} \neq H_{j_2}$ for all $j_1 \neq j_2$ $(j_1, j_2 < i)$. Thus F_n is an $(l,0)$ face with $l \geq i-1$, which is a contradiction. □

To generalize the work of the last section from a single cube to a star-shellable cubical complex C, we will use a generalization of Lemma 3.1. For any facet F_k of C, let $h(F_k, C, x)$ denote the contribution of F_k to $h(C,x)$ with respect to the given shelling order, so that $h(C,x) = \sum_{k=1}^{r} h(F_k, C, x)$. In particular, $h(F_k, C, x) = x^{d+1}\overline{h}(F_k, C, \frac{1}{x})$, where

$$\overline{h}(F_k, C, x) = \sum_{F \in \langle F_k \rangle \setminus \langle F_1, \ldots, F_{k-1} \rangle} g(\partial F, x)(x-1)^{d - dim\, F}.$$

A *leafnode* in a rooted tree is a node with no successor. Let $c_d(i,k)$ be the number of $(d+1)$-node trees with exactly k nodes which are either roots of forks or non-leafnodes at most i in preorder. See [C] for the full generality of the following result as well as for its proof.

THEOREM 4.3. *If F_k is an $(i,0)$-face, then $h(F_k, C, x) = \sum_j c_d(i,j)x^j$. Thus if C has a star-shelling $F_1, F_2, \ldots, F_r$, then $h(C,x) = \sum_i s_i (\sum_j c_d(i,j)x^j)$ where s_i is the number of $(i,0)$-faces in $F_1, F_2, \ldots, F_r$.*

Figure 3 illustrates this result for a star-shellable complex of dimension 2.

$\sum c_2(0,j)x^j = 1 + x$ $\sum c_2(1,j)x^j = 2x$ $\sum c_2(2,j)x^j = x + x^2$

$C =$ 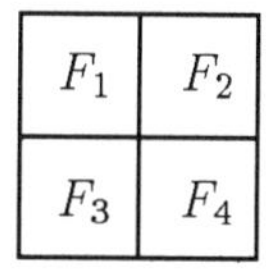

$s_0 = 1$
$s_1 = 2$
$s_2 = 1$

FIGURE 3: $h(C, x) = (1 + x) + 2(2x) + (x + x^2) = 1 + 6x + x^2$.

Now we can generalize the strong g-theorem for cubes.

THEOREM 4.4. *If there exists a star-shelling $F_1, F_2, \dots, F_r$ of C such that each F_k is an $(s, 0)$-face with $s \leq 3$, then $h(C) = f(\Delta)$ for some flag complex Δ; hence $h(C)$ satisfies the Kruskal-Katona inequalities.*

PROOF. Let m be the maximum s such that an $(s, 0)$-face exists. Let k_s be the smallest k such that F_k is an $(s, 0)$ face. By Lemma 4.2, we know that $k_s < k_r$ for $s < r \leq m$.

Now let $\Delta^1 = \Delta_d$ from Section 2, so $f(\Delta^1) = h(C_d) = h(F_1)$. It is convenient to think of $x_{c(i)}$ as a label of node i in T, where $c(i)$ denotes the set of children of i. We add faces to the simplicial complex inductively, so that at step k we have $f(\Delta^k) = h(\langle F_1, \dots, F_k \rangle)$, and finally $f(\Delta) = h(\langle F_1, \dots, F_r \rangle) = h(C)$.

The inductive process is as follows. Suppose $f(\Delta^{k-1}) = h(\langle F_1, \dots, F_{k-1} \rangle)$, and F_k is an $(s, 0)$-face. We want to add faces to Δ^{k-1} to get Δ^k with $f(\Delta^k) = h(\langle F_1, \dots, F_k \rangle)$. The new vertices will be denoted x^k_A where A is some subset of $\{2, \dots, d+1\}$ containing $\{2, \dots, s+1\}$. The new faces will be in one-to-one correspondence with the $(d+1)$-node trees. Let $c([n])$ denote the set of nodes whose parents are in $[n] = \{1, 2, \dots, n\}$. The rule is that for each tree T, we add to Δ^{k-1} a face $\sigma = \sigma^k_T$ with the following vertices:

1. $x^k_{c([s])}$;
2. $x^{k_{n-1}}_{c([n-1])}$ if n is a non-leafnode with $2 \leq n \leq s$;
3. $x_{c(i)}$ where $i > s$ is a fork in T.

It is convenient to think of the vertices of σ^k_T as labels of certain nodes in T; in particular, the root node always gets the label described in (1), any non-leafnode $2 \leq n \leq s$ gets the label described in (2), and any fork $i > s$ gets the label described in (3). Figure 4 illustrates the labeling of one particular tree with $d = 5$ and $s = 0, 1, 2$.

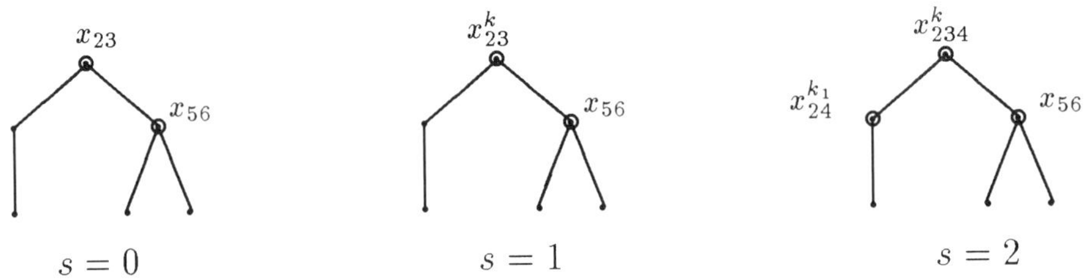

FIGURE 4: Example of labeling with $d = 5$ and $s = 0, 1, 2$.

We now show that for any $\sigma^k_T \in \Delta^k$, all of its facets are in Δ^k. Each facet results from removing a single vertex from σ^k_T. Removing vertices of type (1), (2), and (3) respectively results in the following facets:

1. $\sigma^{k_{s-1}}_{T'} = \sigma^k_T \setminus x^k_{c([s])}$, where T' is derived from T by collapsing the largest type (2) node in T, if it is a fork (otherwise $T' = T$).
2. $\sigma^k_{T'} = \sigma^k_T \setminus x^{k_{n-1}}_{c([n-1])}$, where T' is derived from T by moving the children of node n of T to the parent of node n. (Since $m \leq 3$, moving the children from a node to its parent does not change any other label.)
3. $\sigma^k_{T'} = \sigma^k_T \setminus x_{c(\{i\})}$, where $i > s$ is a fork in T and T' is derived from T by collapsing i.

Thus each facet of σ^k_T belongs to Δ^k. Thus $\Delta = \Delta^r$ is a simplicial complex with $f(\Delta) = h(C)$.

Figure 5 illustrates a case where F_k is a (2,0)-face and σ^k_T has 3 vertices, one of each type.

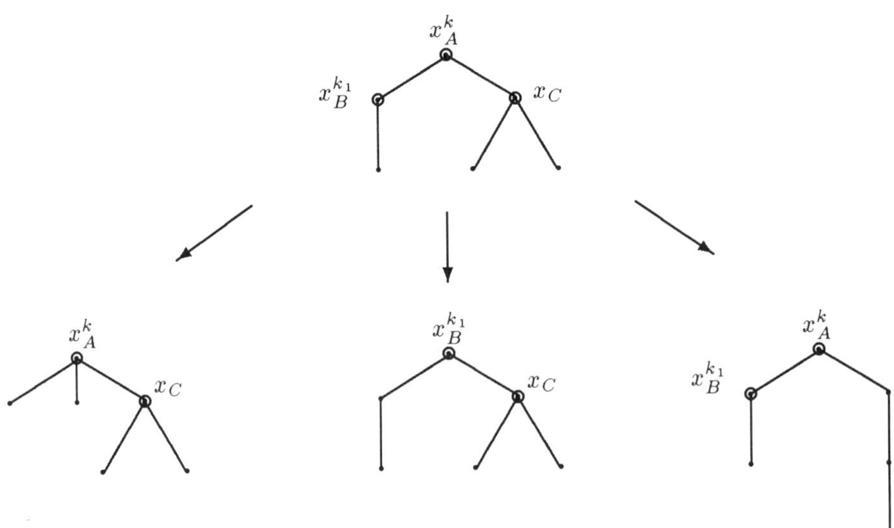

FIGURE 5: Each facet of σ^k_T belongs to Δ.

We now show that Δ is a flag complex. Let σ be a set of $k \geq 3$ vertices of Δ such that $\sigma \setminus x$ forms a face in Δ for each $x \in \sigma$. We need to show that $\sigma \in \Delta$.

Let $\sigma = \{x^{i_1}_{A_1}, \dots, x^{i_j}_{A_j}, x_{A_{j+1}}, \dots, x_{A_k}\}$, where F_{i_v} is an $(s_v, 0)$-face for each v, and $s_1 < \dots < s_j$. Thus we need to show there exists a tree T with fork system $S = \{x_{A_1}, x_{A_2 \setminus A_1}, \dots, x_{A_j \setminus A_{j-1}}, x_{A_{j+1}}, \dots, x_{A_k}\}$. (Some of these "forks" may have only one child, but the analysis of fork systems still goes through.) We may assume $j \geq 1$, otherwise Theorem 2.3 applies. By hypothesis, $\sigma \setminus x^{i_j}_{A_j} \in \Delta$. Thus there exists a tree T_0 with fork system $S_0 = \{x_{A_1}, x_{A_2 \setminus A_1}, \dots, x_{A_{j-1} \setminus A_{j-2}}, x_{A_{j+1}}, \dots, x_{A_k}\}$. Suppose $j \geq 3$. Then by hypothesis and the fact that Δ is a simplicial complex, we know that $\{x^{i_{j-1}}_{A_{j-1}}, x^{i_j}_{A_j}, x_{A_{j+1}}, \dots, x_{A_k}\} \in \Delta$, so there exists a tree with fork system $\{x_{A_{j-1}}, x_{A_j \setminus A_{j-1}}, x_{A_{j+1}}, \dots, x_{A_k}\}$. In particular this means that $x_{A_j \setminus A_{j-1}}$ is consistent with every fork in S_0, since x_A consistent with x_B implies x_A consistent with x_C for any $C \subset B$. Thus there exists a tree T with fork system S, as desired. The cases where $j = 1, 2$ can be dealt with similarly. □

Note that the tree in Figure 6 shows that our construction fails if F_k is a (4,0)-face for some k, since then $\sigma^k_T = \{x^{k_1}_{24}, x^{k}_{2345}, x^{k_3}_{234}\} \in \Delta$, but $\sigma^k_T \setminus x^{k_3}_{234} \notin \Delta$.

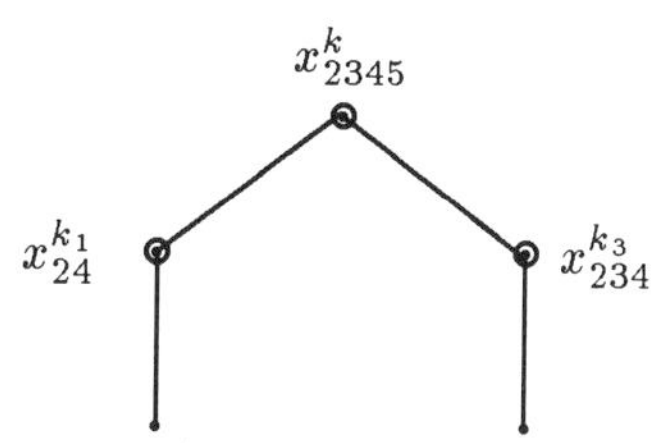

FIGURE 6: $\sigma^k_T \setminus x^{k_3}_{234} \notin \Delta$

If $d = 4$, however, we can modify the construction easily so that we obtain the following corollary.

COROLLARY 4.5. *If C is a star-shellable cubical complex of dimension less than 5, then $h(C) = f(\Delta)$ for some flag complex Δ. Hence $h(C)$ satisfies the Kruskal-Katona inequalities.*

PROOF. The construction of Δ is almost exactly as it was above, except that for an $(s, 0)$-face F_k we now add, for each $(d+1)$-node tree T, a new face F with vertices

1. $x^k_{c([s])}$;
2. $x^{k_{n-1}}_{c([n-1])}$ if n is a non-leafnode with $2 \leq n \leq s$, and $c([n-1])$ is determined after the children of nodes $n < t \leq s+1$ have been moved to their parents.
3. $x_{c(i)}$ where $i > s$ is a fork in T.

Again, it follows that for each $x \in \sigma$, $\sigma \setminus x$ is the set of labels for a tree derived from T by labeling T with respect to a previous facet in the shelling or collapsing out a fork, or moving the children of one node to its parent. Thus Δ is a simplicial complex with $f(\Delta) = h(C)$; moreover the same argument given above shows that Δ is a flag complex. (Figure 7 shows the new labeling of the tree in Figure 6.) □

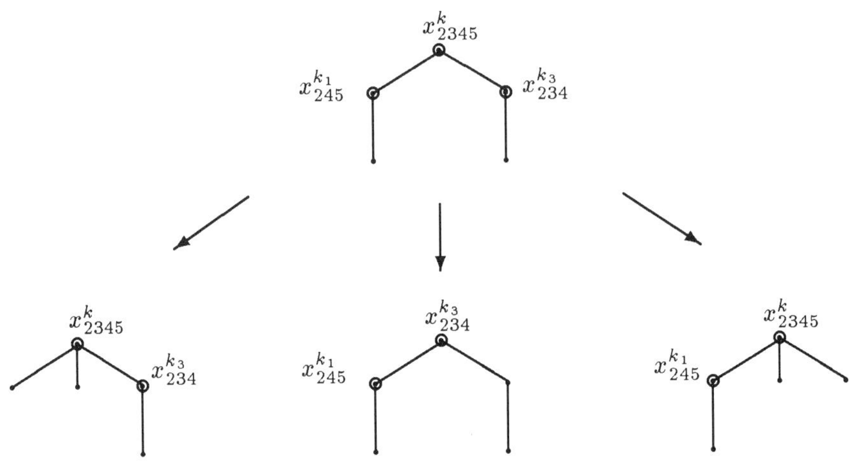

FIGURE 7: Labelling for $d = 4$.

CONJECTURE 4.6. *Let C be a star-shellable cubical complex of dimension $d \geq 5$. Then $h(C) = f(\Delta)$ for some flag complex Δ.*

5. Speculations and computations

Motivated by our g-theorem for cubes, we consider the following question. Let C be the boundary complex of any cubical polytope of dimension $d \geq 2$. Do $g(C)$ and $h(C)$ both satisfy the Kruskal-Katona inequalities? It seems likely to be true for cubical zonotopes, though the general situation is quite unclear.

In the special case where $C = Z$, a cubical zonotope, recall that S_Z, the visible half of the boundary of Z, is star-shellable. Given a shelling order, we define an *(i,j)-face* to be any facet F whose intersection with the previous facets in the shelling consists of i unpaired and j antipodal pairs of facets of F. We can extend a star shelling of S_Z to a shelling of $\partial(Z)$ with one $(i, d-i)$-facet for each $(i, 0)$-facet. Thus, using a reciprocity theorem of Stanley [**S3**], we have

$$h(C) = h(S_Z) + \overline{h}(S_Z),$$

where $\overline{h}(S_Z)$ denotes the reverse of $h(S_Z)$.

For a cubical d-zonotope Z with n zones, it is easy to see that

$$f_i(Z) = \sum_{j=i}^{d}(1 - (-1)^{d-j})\binom{n}{j}\binom{j}{i},$$

from the corresponding equations for generic hyperplane arrangements [**Za**]. Using Gessel's formula for the g-vector of a cube, it is easy to go from $f(Z)$ to $h(Z)$ and $g(Z)$, and then to check Kruskal-Katona directly. In this way we have verified by computer that the g- and h-vectors of cubical zonotopes satisfy Kruskal-Katona for the following values of the dimension d and the number of generators n: $d \leq 8$, $n \leq 5000$; $d \leq 10$, $n \leq 1000$; $d \leq 16$, $n \leq 100$; $d \leq 21$, $n \leq 50$.

Recall that if $n = d$ then we have the d-cube, for which we proved that the g-vector satisfies the Kruskal-Katona inequalities and the h-vector satisfies the weaker Macaulay inequalities, for all $d \geq 2$. In general, one cannot conclude that h satisfies

Kruskal-Katona when g does. (For example, $g = (1,2,1)$ satisfies Kruskal-Katona, yet the corresponding $h = (1,3,4)$ does not.)

However, for $n \geq d$, $d \leq 5$, it is easy to see that the g- and h-vectors of cubical zonotopes satisfy Kruskal-Katona, since in this case the Dehn-Sommerville Equations reduce the space of possible f-vectors to two dimensions. For example, for cubical 5-polytopes, the relation $h_2 \leq h_1^{(1)}$ can be expressed as

$$f_4 \leq \frac{1}{14}\left(f_0 - \frac{9}{2}\right)^2 + \frac{1}{8},$$

and by the formula above, any cubical 5-zonotope has $f_0 > f_4$ and $f_0 \geq 32$, so it easily satisfies $h_2 \leq h_1^{(1)}$.

Does an analogous proof exist for arbitrary $n \geq d$?

Another standard class of cubical polytopes is the *cuboids,* which are successive cuts of the cube [**G**]. For each d, there are $(d+1)$ distinct d-dimensional cuboids (including the d-cube, which is the 0th cut.) Again it is easy to check the Kruskal-Katona inequalities for the g- and h-vectors of any d-cuboid with $k \leq d$ cuts. In this way we have verified by computer that the conditions do hold for all cuboids of dimension $d \leq 26$. Again we ask, is there a proof, for arbitrary d?

Further experiments suggest that these conditions hold in possibly greater generality. We make the following definition as a generalization of star-shellable complexes.

DEFINITION 4. A shellable cubical complex C is *inductive* if it has a shelling with $c_{i,j}$ (i,j)-faces for each (i,j), such that

1. $c_{i,0} > 0$ implies $c_{i-1,0} > 0$ for all $i \geq 1$;
2. $c_{i,j} \geq c_{i,j+1}$ for all $i \geq 1$ and all j.

EXAMPLE: For any d, the boundary of the d-cube has a shelling with coefficients $c_{0,0} = c_{0,d-1} = 1$ and $c_{i,1} = 2$ for all $1 \leq i \leq d-1$, and $c_{i,j} = 0$ for all other (i,j). Thus the boundary of any cube is inductive.

Various computer experiments indicate that the h-vector of a shellable cubical complex C satisfies the Kruskal-Katona inequalities if C is inductive. (Thanks to David Robbins' convenient model for plane trees.)

References

[A] R. Adin, *A new cubical h-vector*, Discrete Math. **157** (1996), 3–14.

[BM] M. Bruggesser and P. Mani, *Shellable decompositions of cells and spheres*, Math. Scand. **25** (1971), 197–205.

[C] C. Chan, *Plane trees and h-vectors of shellable cubical complexes*, SIAM Journal of Discrete Mathematics **4** (1991), 568–574.

[G] B. Grünbaum, Convex Polytopes, Interscience Publishers, John Wiley & Sons, London, 1967.

[S] R. Stanley, *Combinatorics and Commutative Algebra*, Progress in Mathematics **41** (1983), Birkhäuser, Boston/Basel/Stuttgart.

[S2] R. Stanley, *The number of faces of simplicial polytopes and spheres*, Discrete Geometry and Convexity (J. E. Goodman *et al.*, eds.), Annals of the N.Y. Acad. Sci. **440** (1985), 212–223.

[S3] R. Stanley, *Generalized h-vectors, intersection cohomology of toric varieties, and related results*, Adv. Stud. Pure Math. 11 (1987), Commutative Algebra and combinatorics, 187–213.

[Za] T. Zaslavsky, *Facing up to Arrangements: Face Count Formulas for Partitions of Space by Hyperplanes*, Memoirs of the AMS **1** # 154 (1975).

[Zi] G. M. Ziegler, *Lectures on Polytopes*, Graduate Texts in Mathematics **152**, Springer-Verlag, New York, 1995.

DEPARTMENT OF MATHEMATICS, CORNELL UNIVERSITY, ITHACA, NY 14853
E-mail address: `billera@math.cornell.edu`

IDA CENTER FOR COMMUNICATIONS RESEARCH, 29 THANET RD., PRINCETON, NJ 08540
E-mail address: `clara@ccr-p.ida.org`

DEPARTMENT OF MATHEMATICS, CORNELL UNIVERSITY, ITHACA, NY 14853

Contemporary Mathematics
Volume **223**, 1999

Discrepancy Bounds for Geometric Set Systems with Square Incidence Matrices

Bernard Chazelle

ABSTRACT. Alexander has proven the existence of a set of n points in the plane such that, given any two-coloring of the points, there exists a halfplane within which one color outnumbers the other by $\Omega(n^{1/4})$. We strengthen this result by showing that the halfplane can be chosen among n fixed ones. In other words, we build a point/halfplane set system of discrepancy $\Omega(n^{1/4})$, whose incidence matrix is $n \times n$. By a result of Matoušek, this lower bound is tight.

The second result is an $n \times n$ variant of a classical lower bound of Roth on the discrepancy of arithmetic progressions. Stated in dual form, our result asserts the existence of a set system of discrepancy $\Omega(n^{1/4})$, whose $n \times n$ incidence matrix $(a_{i,j})$ is formed as follows: each row (resp. column) corresponds to a line segment (resp. horizontal line) in the plane, and $a_{i,j} = 1$ if segment i and line j intersect in an integer point. Matoušek and Spencer have shown the lower bound to be tight.

1. Introduction

Schmidt [S] has shown the existence of n points in the plane such that, given any two-coloring of the points, there is always an axis-parallel box within which one color outnumbers the other by $\Omega(\log n)$. It is natural to ask the question: can the box be chosen among a small set of prespecified boxes? Given an incidence matrix A of a set system, we define the discrepancy of A,

$$D(A) = \Big\{ \min \|Ax\|_\infty \, : \, x \in \{-1, 1\} \Big\}.$$

Is there an $n \times n$ incidence matrix for boxes, $A = (a_{i,j})$, of discrepancy $\Omega(\log n)$, such that $a_{i,j} = 1$ if the box associated with row i contains the point associated with column j? As it happens, the answer is trivially affirmative. Here is why: the set of candidate boxes for Schmidt's bound can obviously be restricted to $O(n^4)$. So, by applying the bound for $n' = O(n^{1/4})$ points, we can make the number of candidates equal to n. Next, we form an $n \times n$ incidence matrix A by filling in $n - n'$ columns of zeroes derived from the addition of $n - n'$ dummy points. The

1991 *Mathematics Subject Classification.* Primary 68U05, 68Q15.

Key words and phrases. Discrepancy theory, geometric set systems, eigenvalue bounds.

Supported in part by NSF Grant CCR-96-23768 and by the US Army Research Office under Grant DAAH04-96-1-0181.

© 1999 American Mathematical Society

discrepancy of the resulting set system is $\Omega(\log n') = \Omega(\log n)$. Of course, this trick does not work if the discrepancy is of the form n^α, for some constant α. By using a more subtle argument, we prove the following:

THEOREM 1.1. *There exist n points and n halfplanes in $\mathbf{R}^2$, such that the $n \times n$ incidence matrix $A = (a_{i,j})$ has discrepancy $D(A) = \Omega(n^{1/4})$, where $a_{i,j} = 1$ if and only if the i-th halfplane contains the j-th point.*

THEOREM 1.2. *There exist n horizontal lines and n segments in $\mathbf{R}^2$, such that the $n \times n$ incidence matrix $A = (a_{i,j})$ has discrepancy $D(A) = \Omega(n^{1/4})$, where each row (resp. column) corresponds to a segment (resp. horizontal line), and $a_{i,j} = 1$ if and only if segment i and line j intersect in an integer point.*

Both theorems are optimal. This follows from results by Matoušek [**M**] and Matoušek and Spencer [**MS**], respectively. It remains an open question whether the same bounds hold for $m \times n$ matrices, where m is asymptotically smaller than n.

2. The proof of Theorem 1.1

Let P be the set of n integer points in $[1, \sqrt{n}]^2$ (assume that n is a large square), and let x be a vector in $\mathbf{R}^n$ whose i-th coordinate x_i is associated with $p_i \in P$. Given a closed halfplane h bounded above by a nonvertical line, we define $f(h) = \sum_{p_i \in h} x_i$. Let ω be the motion-invariant measure for lines, normalized so as to provide a probability measure for the lines crossing the square $[1, \sqrt{n}]^2$. Alexander [**A**] has proven that if $x_1 + \cdots + x_n = 0$, then

$$\int f^2(h)\, d\omega(h) = \Omega(1/\sqrt{n})\, \|x\|_2^2.$$

For completeness, we repeat (with only a few modifications) the argument we used in [**C**] to discretize Alexander's result and derive Lemma 2.1 below. We subdivide the space of lines crossing $[1, \sqrt{n}]^2$ into $N + O(n^2)$ regions within which $f(h)$ remains invariant. By choosing n and N large enough, say, $N = 2^n$, we can easily ensure that the ω-area σ of N of these regions is exactly the same, i.e., about $1/N$, while the other $O(n^2)$ regions have smaller areas. Thus, the error produced in computing $\int f^2(h)\, d\omega(h)$ by integrating f^2 only over the equal-area regions is $O(n^2/N) \sup f^2$. Because $|f|$ cannot exceed

$$|x_1| + \cdots + |x_n| \le \sqrt{n}\, \|x\|_2,$$

this error is bounded by $O(n^3 \|x\|_2^2/N)$. Let B be the $N \times n$ matrix whose rows are indexed by the N equal-area regions $\hat{\sigma}$ and are the characteristic vectors of the set of x_i's appearing in (the unique form) $f(h)$, for $h \in \hat{\sigma}$. We have

$$\left| \|Bx\|_2^2 - \frac{1}{\sigma} \int f^2(h)\, d\omega(h) \right| = O(n^3) \frac{\|x\|_2^2}{N\sigma},$$

and because $\sigma = 1/N \pm O(n^2/N^2)$,

$$\left| \|Bx\|_2^2 - N \int f^2(h)\, d\omega(h) \right| = O(n^3 \|x\|_2^2).$$

LEMMA 2.1. [**C**]

$$\det B^T B = \Omega\left(N \big/ \sqrt{n}\right)^{n-1}.$$

PROOF. Let $\mu_1 \geq \cdots \geq \mu_n \geq 0$ be the eigenvalues of $B^T B$ and let $\{v_i\}$ be an orthonormal eigenbasis, with μ_i corresponding to v_i. Let $(\xi_1, \ldots, \xi_n)$ be the coordinates of x in the basis $\{v_i\}$. The solution space of the system of equations, $x_1 + \cdots + x_n = 0$ and $\xi_j = 0$ $(j < n-1)$, is of dimension at least 1. It lies in the (ξ_{n-1}, ξ_n)-plane, so it intersects the cylinder $\xi_{n-1}^2 + \xi_n^2 = 1$. For any point x of the intersection,

$$\|Bx\|_2^2 = \sum_{i=1}^{n} \mu_i \xi_i^2 = \mu_{n-1}\xi_{n-1}^2 + \mu_n \xi_n^2 \leq \mu_{n-1}.$$

This implies that for the unit vector x,

$$\mu_{n-1} \geq N \int f^2(h)\, d\omega(h) - O(n^3 \|x\|_2^2) \geq \Omega(N/\sqrt{n}) - O(n^3),$$

and hence,

$$\mu_{n-1} \geq \Omega(N/\sqrt{n}). \tag{1}$$

We need a lower bound on the smallest eigenvalue. With N being large enough, we can always assume that, for each point p_i, there exist two lines (adding them on, if necessary, and updating N accordingly), each represented by a distinct row of B, that pass right above and below p_i. The contribution of these two rows to $\|Bx\|_2^2$ is of the form $\Phi^2 + (\Phi + x_i)^2$, which is always at least $x_i^2/2$. It follows that $\|Bx\|_2^2 \geq \frac{1}{2}\|x\|_2^2$, and hence, $\mu_n \geq 1/2$. The lemma follows from (1) and the fact that $\det B^T B$ is the product of the eigenvalues.

By the Binet-Cauchy formula,[1]

$$\det B^T B = \sum_{1 \leq j_1 < \cdots < j_n \leq N} \left| \det B \begin{pmatrix} j_1 & j_2 & \cdots & j_n \\ 1 & 2 & \cdots & n \end{pmatrix} \right|^2 .$$

Therefore, there exists an $n \times n$ submatrix A of B such that

$$\begin{aligned} (\det A)^2 &= \left| \det B \begin{pmatrix} j_1 & j_2 & \cdots & j_n \\ 1 & 2 & \cdots & n \end{pmatrix} \right|^2 \\ &\geq \binom{N}{n}^{-1} \det B^T B = \Omega(1)^n \left(\frac{n}{eN}\right)^n \left(\frac{N}{\sqrt{n}}\right)^{n-1} \\ &\geq (cn)^{n/2}, \end{aligned}$$

for some fixed $c > 0$. Lovász et al. [**LSV**] define the hereditary discrepancy, $D^H(A)$, of the incidence matrix A to be the maximum value of $D(A')$ over all matrices A' formed by subsets of the columns of A. They prove that

$$D^H(A) = \Omega(|\det A|^{1/n}).$$

In our case, this implies that

$$D^H(A) = \Omega(n^{1/4}).$$

Let A' be the (or any) submatrix of A that achieves the hereditary discrepancy, and let M be the matrix derived from A by zeroing out the columns not in the submatrix A'. By introducing artificial points if necessary, we can make M the incidence matrix of a point/halfplane set system, whose discrepancy is thus $\Omega(n^{1/4})$. This completes the proof of Theorem 1.1. □

[1] The notation following $\det B$ refers to the matrix obtained by picking the rows indexed $j_1, \ldots, j_n$ in B.

For reference it might be useful to make a general lemma out of the technique we just used.

LEMMA 2.2. *If B is an $N \times n$ incidence matrix of a set system, then there exists an $n \times n$ matrix A formed by n rows of B, such that*

$$D^H(A) \geq c\sqrt{\frac{n}{N}}\left(\det B^T B\right)^{1/2n},$$

for some constant $c > 0$.

3. The proof of Theorem 1.2

PROOF. Let B be the $N \times n$ incidence matrix of the following set system: each set is an arithmetic progression modulo m, of length k and difference at most $6k$, where $k = \lfloor\sqrt{n/6}\rfloor$; note that $N = O(n\sqrt{n})$. By adapting an argument of Roth [**R**], Beck and Sós [**BS**] have shown that the matrix $B^T B$ has all its eigenvalues in $\Omega(n)$. This implies that

$$\det B^T B = \Omega(n)^n.$$

By Lemma 2.2, we derive the existence of an $n \times n$ submatrix A, such that

$$D^H(A) = \Omega(n^{1/4}).$$

This result can be interpreted in terms of arithmetic progressions, but it is perhaps better grasped in dual space. Since arithmetic progressions are considered modulo n, a row of A might consist of two distinct progressions. By doubling the number of rows if necessary we can make them into regular arithmetic progressions. Let n denote the new number of rows. If $b, a+b, \ldots, ka+b$ is the progression associated with row i, let us now associate with that row the segment on line $Y = aX + b$ running from $X = 0$ to $X = k$. Column j is associated with line $Y = j$. The lower bound on the hereditary discrepancy implies that the restriction of the set system to a certain subset of lines has discrepancy $\Omega(n^{1/4})$. By zeroing out the leftover columns and adding dummy column lines, we thus create an $n \times n$ incidence matrix of discrepancy $\Omega(n^{1/4})$, where element $a_{i,j}$ is 1 if and only if segment i and line j intersect in an integer point. This proves Theorem 1.2. □

Acknowledgments. I wish to thank the referees for their constructive comments.

References

[A] R. Alexander, *Geometric methods in the study of irregularities of distribution*, Combinatorica, 10 (1990), 115–136.

[BC] J. Beck and W. W. L. Chen, *Irregularities of Distribution*, Cambridge Tracts in Mathematics, 89, Cambridge Univ. Press, Cambridge, 1987.

[BS] J. Beck and V. T. Sós, *Discrepancy theory*, in "Handbook of Combinatorics," eds. R. L. Graham, M. Grötschel and L. Lovász, MIT Press (1995), chap. 26, 1405–1446.

[C] B. Chazelle, *A spectral approach to lower bounds with applications to geometric searching*, SIAM J. Comput., to appear. Preliminary version in Proc. 35th IEEE Symp. Found. Comp. Sci. (1994), 674–682.

[LSV] L. Lovász, J. Spencer and K. Vesztergombi, *Discrepancy of set systems and matrices*, European Journal of Combinatorics, 7 (1986), 151–160.

[M] J. Matoušek, *Tight upper bounds for the discrepancy of halfspaces*, Disc. Comput. Geom., 13 (1995), 593–601.

[MS] J. Matoušek and J. Spencer, *Discrepancy in arithmetic progressions*, J. AMS, 9 (1996), 195–204.

[S] W. M. Schmidt, *Irregularities of distribution, VII*, Acta Arithmetica, 21 (1972), 45–50.
[R] K. F. Roth, *Remark concerning integer sequences*, Acta Arithmetica, 9 (1964), 257–260.

DEPARTMENT OF COMPUTER SCIENCE, PRINCETON UNIVERSITY PRINCETON, NJ 08544
E-mail address: chazelle@cs.princeton.edu

Contemporary Mathematics
Volume **223**, 1999

Computational Topology

Tamal K. Dey, Herbert Edelsbrunner, and Sumanta Guha

ABSTRACT. The authors of this article believe there is or should be a research area appropriately referred to as computational topology. Its agenda includes the identification and formalization of topological questions in computer applications and the study of algorithms for topological problems. It is hoped this article can contribute to the creation of a computational branch of topology with a unifying influence on computing and computer applications.

Introduction

The title of this article combines *computation* with *topology*, suggesting a general research activity that studies the computational aspects of problems with topological flavor. What we have in mind is distinctly different from studying the *topology of computing* or the *computer animation of topology*. Computational studies of topological questions can be found in the mathematics and the computer science literature, but no concerted effort is apparent. The authors hope that together with the like-minded survey paper by Vegter [**V**] this article can contribute to the general awareness of the pervasive presence of topological notions in computer applications.

What is it about a problem that makes it topological? The standard answer mentions a type of geometry devoid of concrete spatial notions such as straightness, convexity, distance, and the like. The emphasis lies on connectivity, continuity, on space, and on maps. This does not mean that the more geometric notions have to be absent but rather that their role is deemphasized benefiting the focused attention on topological properties. The authors believe in a synergy between geometry and topology, and in the value of intuition and visualization as a general path to understanding. They also believe in the general scientific method of studying aspects in deliberate isolation. In other words, it is worthwhile to occasionally focus on topological problems. The connection to geometry can be forgotten temporarily but needs to be remembered eventually.

Many computer application areas deal with geometric data and problems, and there we find a rich collection of questions with topological flavor. Typically, these questions are not well defined and it is part of the task to find the most meaningful

1991 *Mathematics Subject Classification.* Primary 68-02, Secondary 51-02, 55-02, 57-02, 68Q25.

Key words and phrases. Survey; topology, geometry, algorithms, computer applications.

The authors of this paper are grateful to the advise and suggestions offered by three referees and many friends and collegues in the mathematics and computer science communities.

© 1999 American Mathematical Society

formalization in the given context. Once a concrete problem has been formulated we can study its computational complexity. It is important to notice this is an over-simplification of the actual situation: the problem formalization cannot be independent of studying the computational complexity, or else we are likely to generate many computationally infeasible problems. First we need to be convinced there really are essentially topological questions in computer application areas. Part I of this article provides sufficient evidence for their existence by discussing problems in image processing, cartography, computer graphics, solid modeling, mesh generation, and molecular modeling. Second, we need the necessary background in topology. Part II surveys topological methods categorized under decompositions, fixed-points, surfaces, embedding, three-manifolds, and homology computation.

It has been observed that a good fraction of the literature in topology is not or only barely intelligible to readers with interest but without proper topological education. The first barrier is the occasionally complex notation and the usual assumption of accumulated concepts and definitions. Indeed, theory building is the main purpose of topology, and as a consequence definitions are more important than in some other areas of mathematics. The authors attempt to produce a document accessible to non-specialists by collecting formal and standard definitions in an appendix. The main parts of the paper replace technical detail by intuitive examples and arguments. However, it seems impossible to avoid technical definitions altogether and we hope the appendix will be helpful in recalling their meaning.

I. Application Problems

We consider problems in six applications areas of computer science: image processing, cartography, computer graphics, solid modeling, mesh generation, and molecular modeling.

1. Image processing

A *pixel* is a unit square and an *image* is a rectangle decomposed into pixels. Images are the most common means for representing pictorial information in a computer, which explains the enormous literature in the area of image processing. The papers collected in [**OTF**] cover a fairly wide range of the topics. The related area of mathematical morphology is treated in Serra [**Ser**].

We restrict our attention to *binary images* where each pixel is either black or white. The *foreground* consists of all black pixels, and the *background* consists of all white pixels.

Connectivity. Each pixel shares an edge each with 4 other pixels, and it shares a vertex each with 4 additional pixels. The numbers are smaller for pixels at the image boundary. Questions of proximity and connectivity can be based on the two corresponding notions of adjacency: pixels $p \neq q$ are 4-*adjacent* if they share an edge and they are 8-*adjacent* if they share an edge or a vertex. 4-adjacency implies 8-adjacency. A *path* is a sequence of pixels so that any two contiguous ones are adjacent. It is *simple* if two pixels are adjacent only if they are contiguous, and it is a *closed curve* if the first is adjacent to the last pixel. Depending on the notion of adjacency we get 4-*connected* and 8-*connected* paths and curves. For $i = 4, 8$ two pixels are i-*connected* if they belong to a common i-connected path.

A basic result in plane topology is the Jordan curve theorem which states that every simple closed curve in $\mathbb{R}^2$ decomposes the plane into two disjoint open connected components. Neither notion of connectivity among pixels supports an analoguous result for images. The following version of the Jordan curve theorem uses different notions of connectivity for the curve and the components [**KR**]:

> a simple closed 8-connected curve partitions an image into the path and two 4-connected components,

see figure 1. Symmetrically, if the closed curve is 4-connected then the two components are 8-connected. A necessary assumption for both results is that the curve touches the image boundary at most once.

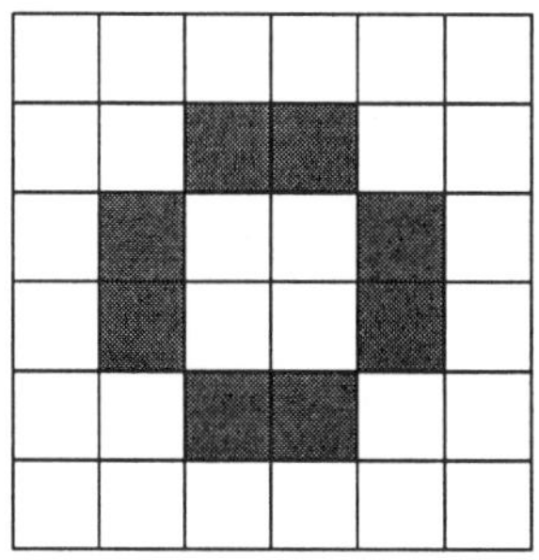

FIGURE 1: A simple closed 8-connected curve with two 4-connected components in the complement.

The image version of the Jordan curve theorem can be generalized to 3- and higher-dimensional arrays of unit cubes and hypercubes, see Herman [**Her**]. This should be compared to the situation in Euclidean space. Already $\mathbb{R}^3$ permits a counterexample to the straightforward extension of the Jordan curve theorem: there is a *wild* 2-sphere in $\mathbb{R}^3$ so that the bounded component of $\mathbb{R}^3$ minus the 2-sphere is not homeomorphic to the open 3-ball, see e.g. [**BEf**, chapter 2.9]. Such counterexamples do not exist for *tame* embeddings and the Jordan curve theorem generalizes as expected.

Skeletonizing. A common operation in image processing replaces the foreground by its skeleton. There is no commonly accepted definition of what exactly this is. One expects a thin subset that reflects the connectivity of the foreground. Often the skeleton is stored together with additional information that permits the exact or approximate reconstruction of the foreground.

Algorithms for constructing skeletons are abundant in the image processing literature. One class of algorithms is based on the medial axis transform. For each foreground pixel p consider a shortest path that connects p to a background pixel q. The length of this path is $\delta(p)$. Boundary pixels are characterized by $\delta = 1$. The medial axis consists of all pixels p with at least two shortest paths to the background that intersect in no pixel other than p.

Another class of algorithms constructs a skeleton by repeatedly removing pixels of the boundary [**NS**], compare this with the discussion on collapsing in section 7. The foreground is eroded layer by layer, and pixels are removed as long as the connectivity of the foreground does not change. The skeleton is the collection of foreground pixels that remain. Although this collection retains the connectivity of the foreground it often does not preserve its overall shape.

Inner and outer boundaries. Consider a collection of disjoint 8-connected simple closed curves that bound the foreground of a binary image. Each curve C in this collection defines two 4-connected components: an outer and an inner region. C is an *outer boundary* if the foreground component bounded by C is part of its inner region. Otherwise, C is an *inner boundary*. How can we distinguish outer from inner boundaries?

Lee, Poston and Rosenfeld [**LPR**] propose an algorithm that can be explained using the topological concept of winding number, which counts for an oriented closed curve how often the normal vector winds around the circle of directions. The number is signed counting counterclockwise order positive and clockwise order negative. Orient each curve C so that the foreground lies to its left. Then the winding number of any outer boundary is $+1$ and that of any inner boundary is -1.

The algorithm traverses C and adds the supplementary angles at the pixel centers. This angle is positive if C makes a left turn and negative if C makes a right turn. Because C is simple the sum of angles can only be 360 or -360 degrees, which is 360 times the winding number.

2. Cartography

The purposeful deformation of a geographic map is a common operation in cartography. We consider two types of deformations, one bringing two maps into correspondence and the other reflecting quantities other than geographic distance and area.

Rubber sheets. As a common practice in cartography, different information about a geographic region is represented in different maps. We can imagine a map M showing mineral distribution as obtained from a geological survey, and another map N displaying agricultural land distribution as obtained from satellite images. The two geographic maps need to be brought into correspondence so that mineral and agricultural land distributions can be related.

To model this problem let $P \subseteq M$ and $Q \subseteq N$ be two sets of n points each together with a bijection $b : P \to Q$. The construction of a homeomorphism $h : M \to N$ that agrees with b at all points of P is popularly known as *rubber sheeting*, see for example [**G, WG**].

Suppose $\mathcal{K}$ and $\mathcal{L}$ are simplicial complexes whose simplices cover M and N: $M = |\mathcal{K}|$ and $N = |\mathcal{L}|$. Suppose also that the points in P and Q are vertices of $\mathcal{K}$ and $\mathcal{L}$: $P \subseteq \operatorname{Vert}\mathcal{K}$ and $Q \subseteq \operatorname{Vert}\mathcal{L}$, and that there is a bijective vertex map

$$v : \operatorname{Vert}\mathcal{K} \to \operatorname{Vert}\mathcal{L}$$

that agrees with b at all points of P. The extension of v to a simplicial map $f : M \to N$ is a simplicial homeomorphism effectively solving the rubber sheet problem.

Several variations of the construction of such complexes $\mathcal{K}$ and $\mathcal{L}$ have been tackled in the recent past. Aronov, Seidel and Souvaine [**ASS**] consider simply-connected polygons M and N with n vertices each. They show there are always isomorphic complexes $|\mathcal{K}| = M$ and $|\mathcal{L}| = N$ with at most $O(n^2)$ vertices each. They also prove that sometimes $\Omega(n^2)$ vertices are necessary and they show how to construct the complexes in $O(n^2)$ time. Gupta and Wenger [**GW**] solve the same problem with at most $O(n + m \log n)$ vertices, where m is the minimum number of extra points required in any particular problem instance.

Cartograms. A *cartogram* is a geographic map that is deliberately deformed so that area in the map reflects a quantity other than geographic area, such as for example population density [**DCN**]. Figure 2 shows a deformation of the United States so that the area of each state is proportional to the number of electoral votes in the 1992 presidential election. The cartogram is related to the usual geographic representation of the United States through a homeomorphism displayed in figure 3.

FIGURE 2: The shaded states have a majority of votes for the elected president.

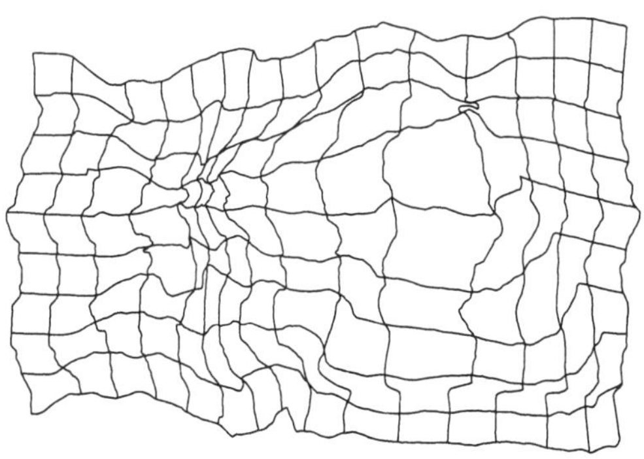

FIGURE 3: The image of a regular square grid under the homeomorphism that generates the cartogram of the United States shown in figure 2.

This particular example has been generated by an iterative algorithm described in [**EW**]. It deforms the Euclidean plane through a series of elementary piecewise linear homeomorphisms. Each elementary homeomorphism leaves the area measure invariant everywhere except in a local region. The approach is based on a regular triangular grid tiling and can be extended to three dimensions using the Delaunay tiling of the body-centered cube lattice.

3. Computer graphics

In computer graphics we find an unexpected connection between computer hardware and topology. A *texture map* is the hardware implementation of a continuous function, which is a basic concept in topology. It requires a triangulated surface and a range or texture space, which is a k-dimensional unit cube of colored pixels or voxels that can be accessed through k texture coordinates. Typical values of k are 1, 2 and 3. The map is explicitly stored at vertices and computed through barycentric combination at all other points of the surface.

Warping. *Warping* refers to the distortion or deformation of a two-dimensional image. A popular approach is based on Coons patches which are bivariate parametric functions [**He**]. They warp a source image bounded by four curves to a rectangular target image. Given the left, right, bottom, top boundary curves $L(v) = C(0, v)$, $R(v) = C(1, v)$, $B(u) = C(u, 0)$, $T(u) = C(u, 1)$, the Coons patch adds an interpolation between L and R to an interpolation between B and T and subtracts an interpolation between the four corner points:

$$\begin{aligned} C(u,v) &= (1-u)\cdot L(v) + u\cdot R(v) \\ &+ (1-v)\cdot B(u) + v\cdot T(u) \\ &- (1-u)(1-v)\cdot C(0,0) - uv\cdot C(1,1) \\ &- u(1-v)\cdot C(1,0) - (1-u)v\cdot C(0,1). \end{aligned}$$

The inverse, C^{-1}, maps the source image bounded by L, R, B, T to the unit square.

A more general but little explored approach to warping can be based on simplicial maps carrying the contents of one triangulated image to another. This is similar to the approach to cartograms mentioned in section 2. We can use a simplicial map to warp a rectangular image onto a torus or any other surface. An important question in this context is the formulation of algorithms that automatically generate interesting simplicial maps between two given 2-dimensional domains.

Morphing. The transformation of one geometrical or graphical object into another is referred to as *metamorphosis* or *morphing*. The general objective is to compute a transformation that is smooth, in a vague and intuitive sense, and visually pleasing.

The morphing of a digital image $\mathbb{X}$ into another such image $\mathbb{Y}$ has found broad commercial applications. A common step in the generation of the morphing sequence is the construction of isomorphic simplicial complexes $|\mathcal{K}| = \mathbb{X}$ and $|\mathcal{L}| = \mathbb{Y}$. The transformation can be performed by slowly moving each vertex in $\mathcal{K}$ to the corresponding vertex in $\mathcal{L}$ and letting this process carry the entire image along. The construction of $\mathcal{K}$ and $\mathcal{L}$ is made difficult by external requirements that have to do with features that need to be matched across the two images.

A more challenging version of the morphing problem concerns geometric shapes and in particular shapes in $\mathbb{R}^3$. An early paper on this subject is Kent, Carlson and Parent [**KCP**]. It presents a method that takes a topological 2-sphere $\mathbb{X}$ to another such sphere $\mathbb{Y}$ by first constructing isomorphic triangulations and then moving parts of $\mathbb{X}$ towards the corresponding parts of $\mathbb{Y}$. This approach is limited to surfaces that share the same topology. The morphing of a sphere to a torus is difficult because it is not clear when and how the sphere surface should be pinched and punctured to change its topology.

An entirely different approach based on the representation of shapes as envelopes of spheres is described in [**CEF**]. Both $\mathbb{X}$ and $\mathbb{Y}$ are given in terms of finitely many spheres, and the deformation is created by moving the spheres of $\mathbb{X}$ to the spheres of $\mathbb{Y}$. Duplication and merging of spheres is permitted. At any moment in time we have a finite set of spheres, and the shape at that time is the envelope of an infinite family of shrunken spheres [**E5**]. The infinite family makes the envelope smooth and guarantees that any motion translates into a smooth deformation. This method changes topology automatically and extends to morphs between three or more given shapes.

4. Solid modeling

This section discusses three topological questions which arise in the area of solid modeling: how to represent a model in the computer, how to construct or reconstruct a surface from discrete point data, and how to extract features of a model.

Model representation. Models are continuous domains and they need finite representations in the computer. Most commonly we find complexes of one kind or another used for that purpose. The model or its boundary is decomposed into simple pieces or cells, and each cell is stored along with adjacencies to neighboring cells.

The *boundary representation*, for short *b-rep*, describes a 3-dimensional model through its boundary complex, which consists of vertices, edges, and 2-dimensional faces. The winged-edge structure of Baumgart [**Bau**] and the quad-edge structure of Guibas and Stolfi [**GS**] are examples of b-rep data structures that assume the boundary is a manifold. This means every point of the boundary has a local neighborhood homeomorphic to $\mathbb{R}^2$. If the boundary is triangulated then every edge belongs to two triangles and every vertex belongs to a simple cycle of triangles and edges. Non-manifold b-reps have been proposed by Wesley [**Wes**] and a survey of b-rep schemes can be found in Weiler [**We**].

A data structure for 3-dimensional complexes is described in Dobkin and Laszlo [**DL**]. The basic element is the oriented edge-face pair. All edge-faces sharing the same edge belong to a cycle or ring around the edge. Symmetrically, all edge-faces sharing the same face are arranged in a cycle describing the face boundary. The two types of rings connect all edge-faces into a single topological data structure. More general complexes of arbitrary fixed dimension are discussed in Brisson [**Br**].

A common concern is the topological validity of a given complex data structure. Topologically invalid structures may be caused by numerical errors inevitable in algorithms that perform geometric calculations in floating-point arithmetic. To check a 2-manifold we would include tests that check orientability and the Euler relation, see section 7. A popular approach to building b-reps manipulates the data structure exclusively through Euler operators, see Mäntylä [**M**]. The purpose of the restriction is to avoid any side-effects that may compromise the topological validity of the data structure.

Surface reconstruction. Given a finite set of points in $\mathbb{R}^3$, the *surface reconstruction* problem aims at producing a surface that passes through the points. Versions of the problem differ in their assumptions on the point distribution and their requirements on the surface. In all cases the surface is supposed to adapt to the general shape of the point set and surfaces with fewer wrinkles and less area are generally preferred.

The most common version of the problem assumes the points represent slices through the object and thus lie on a series of parallel planes in space. CAT scans in medical imaging are a prime example of data that falls into this category. Within each plane the points are to be connected by edges forming a closed polygon, and polygons in contiguous planes are to be connected by triangles forming a cylindrical surface, see figure 4. The collection of cylinders makes up the reconstructed surface.

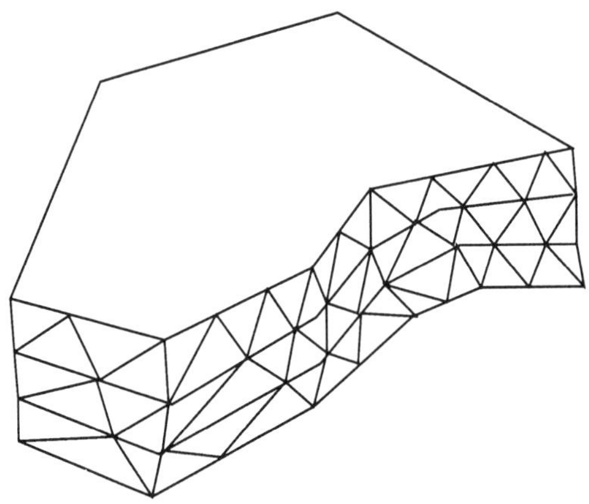

FIGURE 4: A triangulated cylinder connecting polygons in four parallel planes.

The operation that connects polygons in two parallel planes is often referred to as *lofting*. Let P and Q be two polygons and let n be the total number of vertices. In one of the first papers on the topic, Fuchs, Kedem and Uselton [**FKU**] describe a dynamic programming algorithm that finds the minimum area cylinder connecting P and Q in time $O(n^2 \log n)$. The cylinder is not guaranteed to be free of self-intersections, and it is generally not clear how to modify the construction so that a guarantee can be given. In fact, there are cases where all lofting cylinders that avoid self-intersections have at least one vertex in addition to the n vertices of P and Q [**GOS**].

The surface reconstruction problem seems to get considerably more difficult if no assumptions on the point distribution are made. An approach to this general problem based on 3-dimensional Delaunay complexes is described in Boissonnat [**Bo**]. The initial surface is the convex hull, which is the boundary of the Delaunay complex. The surface is moved inwards by successively removing tetrahedra from the complex. The removal heuristic of Boissonnat is replaced in [**E4**] by a rational and unambiguous collapsing rule sensitive to local and global distribution patters. Figure 5 shows a sample surface reconstructed with this method.

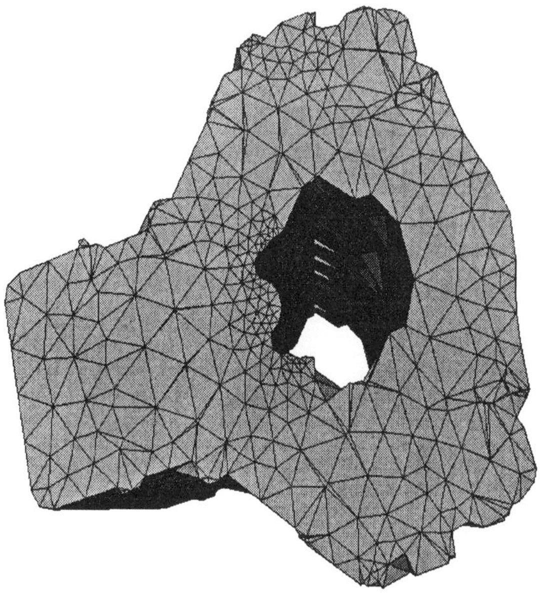

FIGURE 5: An engine block bounded by a 2-manifold of genus 1.

Feature extraction. There is no common agreement on what exactly a feature of a geometric model is or should be. If the model is a solid shape in $\mathbb{R}^3$ then the three types of holes characterizing its connectivity belong to almost everybody's list: the *gaps* that separate components, the *tunnels* that pass through the shape, and the *voids* that are components of the complement space inaccessible from the outside. These features are counted by the Betti numbers $\beta_0, \beta_1, \beta_2$.

It is easier to define components than gaps between them and they are counted by β_0. The voids are similarly unambiguous and are counted by β_2. The concept of a tunnel and how to count tunnels is somewhat more delicate. For example consider the 1-skeleton of a tetrahedron, which consists of 4 vertices and 6 edges. There are 4 triangle entries connecting the outside with the inside, but β_1 is only 3. The reason is that only 3 of the tunnels are independent, and all other paths around and through the 1-skeleton are combinations of the independent tunnels: the 3 independent tunnels form a basis of the group of all tunnels. The basis is not unique and in general there is no canonical or best choice. See section 12 for more information about Betti numbers and how to compute them.

Another type of commonly sought features are pronounced protrusions and intrusions. Intrusions are similar to the three types of holes except that they are not holes in any topological sense and require a geometric characterization. A protrusion of the shape is an intrusion of the complement space, and an intrusion of the shape is a protrusion of the complement space. A more detailed discussion of these features in the context of macromolecules can be found in section 6.

5. Mesh generation

Finite element and other numerical methods for physical simulation rely on the availability of complexes, called meshes or grids, that decompose shapes and space of interest [**KS**]. The two dominant types of complexes in $\mathbb{R}^3$ are the structured mesh [**Cos, TWM**] and the tetrahedral mesh [**BE**].

Structured meshes. The prototypical structured mesh in $\mathbb{R}^2$ is the decomposition of a recangular region into smaller rectangles by drawing vertical and horizontal lines. Each rectangle is referred to as an *element*. Each element has 4 neighbor elements and each vertex is surrounded by 4 elements. Exceptions to this rule are found only on the boundary of the region where the numbers are less than 4. There is flexibility in the shape of elements but not in the structure of adjacencies, which is rigid and motivates the name of this kind of mesh. For example the drawing in figure 3 qualifies as a structured mesh. One of the reasons for insisting on the regular adjacency structure is the convenient representation with two-dimensional matrices. Each matrix entry stores an element and adjacencies are represented by index increments and decrements.

A structured mesh in $\mathbb{R}^3$ consists of hexahedral elements adjacent to 6 other elements. Each edge is surrounded by 4 elements and each vertex is surrounded by 8 elements. Exceptions are found only at the boundary where the numbers are smaller. Such a mesh can again be stored in a matrix with adjacencies represented through index increments and decrements. Even for only moderately interesting 3-dimensional shapes the problem of constructing a structured mesh can be challenging. In many cases it amounts to constructing a homeomorphism between the shape and the unit cube. Meshes with hexahedral elements that relax the degree requirements for edges and vertices are considered in [**Ep, SB**].

Tetrahedral meshes. The most common un-structured mesh is the triangular mesh in $\mathbb{R}^2$ and the tetrahedral mesh in $\mathbb{R}^3$, see figure 5. By far the most popular such mesh is the Delaunay complex named after Boris Delaunay [**De**]. For a finite point set $S \subseteq \mathbb{R}^3$ the Delaunay complex is unique and decomposes the convex hull of the set. In the non-degenerate case all elements are tetrahedra, and a tetrahedron spanned by 4 points in S belongs to the complex iff its circumsphere encloses no point of S.

An important but seemingly difficult problem is the generation of quality meshes. The precise meaning of this term depends on the application and sometimes is not clear altogether. In many cases it means the angles of the triangles and tetrahedra are not too small and not too large, the edges are not too long and not too short, etc. Ideally the shape and size of the elements should adapt to the solution of whatever numerical problem is considered. The resulting chicken-and-egg paradox suggests that both the mesh and the solution are to be refined simultaneously or at least in alternating steps.

It has been observed that Delaunay complexes in $\mathbb{R}^3$ tend to contain many "slivers", which are fairly flat tetrahedra of quadrilateral shape [**CFF**]. In contrast to a triangle shaped one, a quadrilateral shaped flat tetrahedron can have a fairly small circumsphere and can therefore exist in the middle of the Delaunay complex. It would be interesting to find a point insertion strategy that can eliminate slivers:

> is there always a point p so that the Delaunay complex of $S \cup \{p\}$ contains fewer slivers than the Delaunay complex of S?

Slivers close to the boundary can just be taken off from the outside, but for slivers in the middle it would be important to have other means of elimination, which an answer to the above question would provide.

Mesh partitioning. The primary motivation for partitioning a mesh are parallel analysis algorithms that assign processors to work concurrently in different parts of the mesh. The partition defines *blocks* which should be roughly the same size and there should be few adjacencies across blocks.

For a planar graph such a partition can be constructed by removing a small number of vertices. Let the *size* of a graph be the number of vertices. Lipton and Tarjan [**LT**] prove that

> every planar graph of size n contains a subset of at most $\sqrt{8n}$ vertices whose removal decomposes the graph into components of size at most $2n/3$ each.

Call the removed vertex set a *separator*. Lipton and Tarjan also give an algorithm that constructs a separator in time $O(n)$. For any given size bound the graph can be decomposed into blocks of at most that size by repeated removal of separators.

A geometric proof of the existence of separators for planar graphs is described in [**MTTV**]. It is based on Koebe's theorem, see section 10, and the existence of centerpoints in $\mathbb{R}^3$, see section 8. The geometric proof extends to tetrahedral meshes in $\mathbb{R}^3$ provided the aspect ratios are bounded from above by some constant. In this case the mesh contains $O(n^{2/3})$ vertices whose removal leaves components of size at most $c \cdot n$, for any positive constant $c < 1$.

6. Molecular modeling

We are interested in the study of molecules as 3-dimensional shapes and conformations. A big motivation for research in this area is the protein folding problem that aims at a computer simulation of the natural folding process. That process takes an amorphous chain of amino acid components into a folded state where shape and geometry are determining factors of the protein's function. We quote from a recent article by Rose [**Ros**]:

> "What role a protein takes in the grand biological opera depends on exactly one thing: its shape. For a protein molecule, function follows form."

At the current state of the research no successful computer simulation of the folding process is in sight.

Molecular surfaces. A basic question concerns the geometric definition of a folded protein. Even if we make the unrealistic assumption that the protein is a static assembly of its atoms, it is not clear which part of space exactly the protein occupies. What is its interior and where is its boundary?

A rigorous but computationally demanding answer to these questions is described in a recent book by Bader [**B**]. Based on Morse theoretic ideas of flow and critical points, the 3-dimensional space is partitioned into catchment regions or unstable manifolds of various dimensions. The protein itself is a union of such regions and its shape changes sometimes catastrophically as small changes in the flow field imply big changes in the catchment regions.

A more pragmatic answer to the shape definition question starts from the geometric idea of modeling an atom as a spherical ball. The atom kernel is at the center and the electrons populate the region between the center and the sphere boundary, see [**LR, Ri**]. Chemically bound atoms correspond to overlapping balls. The protein itself is the union of all the balls.

This geometric idea can be extended to modeling the shape of a protein as it appears to a solvent such as water. If we model the solvent as an omnipresent sphere of radius ϱ, we can just increase the atom balls by that radius and shrink the solvent to a point. The resulting model is sometimes referred to as the *solvent accessible* model of the protein.

Dual complexes. The union of balls model has a simplicial complex as its dual. The complex is defined in geometric terms and topological language is useful in understanding its relation to the model.

Let B be the finite set of balls defining the model $\bigcup B$ of some protein. We decompose $\bigcup B$ into convex cells using weighted Voronoi cells. The *weighted distance* of a point $x \in \mathbb{R}^3$ from a ball $b \in B$ is the square distance to the center minus the square radius. The *weighted Voronoi cell* V_b of b is the set of points x for which b minimizes the weighted distance, see figure 6.

The weighted Voronoi cells decompose $\bigcup B$ into convex cells of the form $b \cap V_b$. Any two cells have disjoint interiors but they may overlap along their boundary. The boundary overlap figures in the definition of the dual complex. To describe this let T' be the set of centers of the balls in T. We are interested in the subsets $T \subseteq B$ whose cells have non-empty common intersection. The *dual complex* of B consists of all simplices spanned by the centers of balls in such subsets:

$$\mathcal{K} = \{\sigma = \operatorname{conv} T' \mid \bigcap_{b \in T} (b \cap V_b) \neq \emptyset\}.$$

Assuming general position, at most 4 Voronoi cells can meet in a point. In this case all elements of $\mathcal{K}$ are simplices in $\mathbb{R}^3$ and altogether they form a simplicial complex, see figure 6.

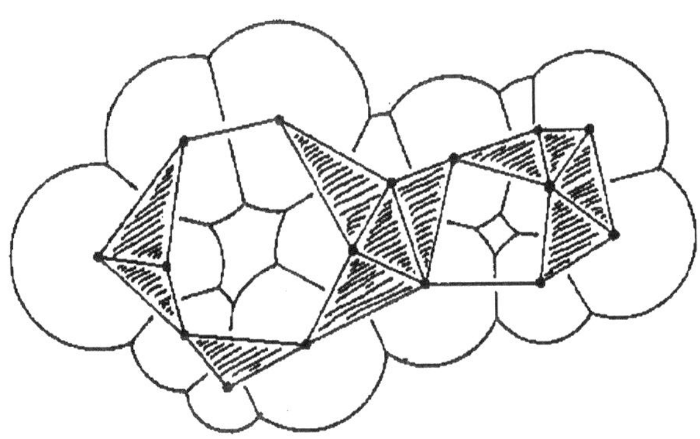

FIGURE 6: The union of disks is decomposed by the weighted Voronoi cells. The dual complex represents the overlap structure of the convex cells in the decomposition.

The original definition of the dual complex dates back to [**EKS**] where unit-disks in the plane are considered. Algorithms for the 3-dimensional case are described in [**EM**]. The general definition for mixed radii and arbitrary fixed dimension can be found in [**E2**]. The latter reference also contains a proof of the homotopy equivalence between the ball union and the dual complex and short inclusion-exclusion formulas for measuring the volume and surfaces area of a ball union model.

Pockets and protrusions. The dual complex of a ball union model is a combinatorial structure consisting of vertices and of pairs, triplets, and quadruplets of vertices. It allows fast combinatorial algorithms for computing the connectivity in terms of Betti numbers, see section 15 of this paper. Because the ball union and the dual complex are homotopy equivalent, we know that the two have the same Betti numbers. The second Betti number counts the *voids*, which are the bounded components of the complement of $\bigcup B$. These are interesting to biologists because they may contain water molecules caught at some moment during the folding process and influencing the further evolution of that process.

Of even greater significance than voids are cavities or depressions that extend to the surface of a protein. These are not holes in a topological sense although common language refers to them often as such [**CV**]. To avoid any confusion we refer to such cavities as *pockets*, see figure 7. The significance of pockets lies in their prominent role in molecular interactions. Proteins communicate with other molecules via surface contact along partial albeit imperfect shape complementarity. We wish to give a geometrically rigorous definition of a pocket as a part of space that can be computed and analyzed. Such a definition using concepts in Morse theory is given in [**EFL**]. We present a slightly simplified version of that definition.

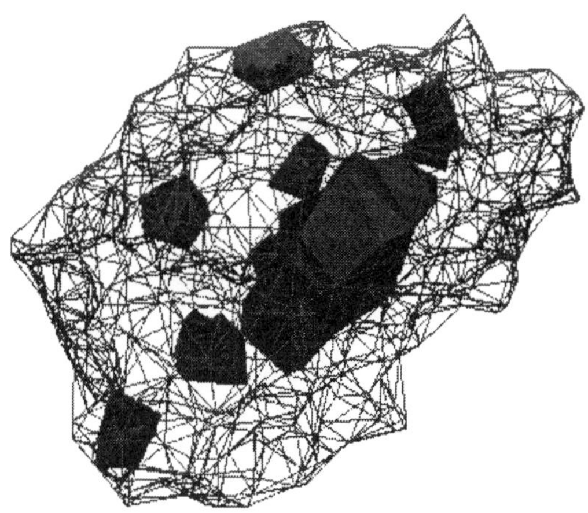

FIGURE 7: Visualization of voids and pockets of the dual complex of HIV-1 proteise.

For each point $x \in \mathbb{R}^3$ let $f(x)$ be the minimum weighted distance to any ball $b \in B$. We can make f smooth using convenient cut-off functions close to the boundary of Voronoi cells. Assuming smoothness we take the gradient of f,

$$\nabla f = \left(\frac{\partial f}{\partial x_1}, \frac{\partial f}{\partial x_2}, \frac{\partial f}{\partial x_3} \right),$$

which is a smooth vector field $\nabla f : \mathbb{R}^3 \to \mathbb{R}^3$. From a point x we follow the flow of ∇f, which either takes us to infinity or we get caught at a local maximum. A *pocket* is defined as a maximal connected piece of space outside $\bigcup B$ whose points do not flow to infinity.

It appears that a pocket is an inherently continuous concept whose computation requires following flow lines for as many points as the desired accuracy of the construction requires. Fortunately, this is not the case and a close relationship between pockets and Delaunay simplices can be used to obtain a fast combinatorial algorithm, see [**EFL**].

II. Topological Methods

We consider six topics in topology: decompositions, fixed points, surfaces, embedding, three-manifolds, and homology computation. This leaves large domains untouched, such as the entire fields of Morse theory [**Mi2**] and differential topology [**GP**]. The deliberate omission of differentiable topics is motivated by the size and depth of that area and the attempt to preserve the elementary character of this paper.

7. Decompositions

Questions about decomposing a shape into simple pieces can have geometric as well as topological flavor. Restrictions on the type of pieces that are allowed and how they can overlap give rise to different types of complexes. This section considers a few decomposition questions for the simplest kind of complex where each cell is convex and any two cells intersect in another cell.

Euler characteristic. The Euler relation for the boundary complex of a polyhedron is possibly *the* oldest topological result in all of mathematics. If the polyhedron is convex and bounded in $\mathbb{R}^3$ the relation states that the number of vertices and 2-dimensional faces exceeds the number of edges by 2:

$$v - e + f \quad = \quad 2.$$

This result was know already to René Descartes. Initial attempts to extend this result to polyhedra with holes and to polyhedra bounded by faces with holes were only partially successful, see Lakatos [**L**]. Eventually, Poincaré settled the issue [**P, P2**]. He observed that the result of the alternating face sum is a property of the space the faces cover, and that it equals the alternating sum of Betti numbers counting the holes in that space, see section 15. For example, the boundary of a 3-dimensional convex bounded polyhedron is homeomorphic to the 2-sphere, $\mathbb{S}^2$, and 2 is the characteristic number for $\mathbb{S}^2$.

The *Euler characteristic* of a space $\mathbb{X}$ is the alternating simplex count of any triangulation $\mathcal{K}$ of $\mathbb{X}$:

$$\chi(\mathbb{X}) \quad = \quad \sum_{\emptyset \neq \sigma \in \mathcal{K}} (-1)^{\dim \sigma}.$$

For the torus we have $\chi = 0$, and more generally for orientable closed surfaces with genus g we have $\chi = 2 - 2g$. For non-orientable surfaces with genus g we have $\chi = 2 - g$, see section 9. It is thus easy to determine the type of a closed surface given as a triangulation. First check whether or not the triangles can be consistently oriented and then computing the Euler characteristic by counting vertices, edges, and triangles. The same approach fails for compact 3-manifolds because they all have vanishing Euler characteristic.

The Euler characteristic has found a multitude of applications in geometry and topology, including volume formulas for unions of spherical balls [**E2, NW**], see also section 6. A survey of extensions and generalizations can be found in the booklet by Shashkin [**Sha**].

Covers and nerves. A *cover* of a topological space $\mathbb{X}$ is a collection $\mathcal{C}$ of subsets whose union is $\mathbb{X}$: $\mathbb{X} = \bigcup \mathcal{C}$. For example the set of d-faces of a convex polyhedron of dimension $d+1$ is a cover of the polyhedron boundary. The *nerve* of a cover $\mathcal{C}$ is the system of subcollections with non-empty common intersection:

$$\mathrm{Nrv}\,\mathcal{C} \quad = \quad \{X \subseteq \mathcal{C} \mid \bigcap X \neq \emptyset\}.$$

The nerve is an abstract simplicial complex because it is closed under taking subsets: $X \in \mathrm{Nrv}\,\mathcal{C}$ and $Y \subseteq X$ implies $Y \in \mathrm{Nrv}\,\mathcal{C}$. We get a geometric realization of the abstract complex by mapping each $C \in \mathcal{C}$ to a point $p(C)$ in $\mathbb{R}^k$. Every collection $X \in \mathrm{Nrv}\,\mathcal{C}$ is represented by the convex hull of the points $p(C)$, $C \in X$. If $\mathcal{C}$ is finite then $\mathrm{Nrv}\,\mathcal{C}$ is finite and if k is sufficiently large then the points can be chosen so that the resulting simplices form a genuine simplicial complex. Call this complex $\mathcal{K}$ and consider its underlying space, $|\mathcal{K}| = \bigcup_{\sigma \in \mathcal{K}} \sigma$, which is again a topological space. The nerve theorem originally due to Leray [**Le**] states that under reasonable assumptions about the sets in the cover and about their common intersections the underlying space of $\mathcal{K}$ has the same homotopy type as $\mathbb{X}$:

> if $\mathcal{C}$ is finite and $\bigcap X$ is either empty or contractible for every $X \subseteq \mathcal{C}$ then $|\mathcal{K}|$ and $\mathbb{X} = \bigcup \mathcal{C}$ are homotopy equivalent.

The literature contains versions of this theorem that differ in the requirements on the sets in the cover, the allowed size of the cover, and the resulting topological relation between $|\mathcal{K}|$ and $\mathbb{X}$, see e.g. [**Bj, Wu**].

The homotopy equivalence implies that the Euler characteristics of $\mathcal{K}$ and $\mathbb{X}$ are the same. This fact has been used in [**E3**] to derive boolean formulas for non-convex polyhedra. The nerve theorem is also useful in computational settings because covers are relatively easy to get and through the nerve operation large and meaningful complexes can be automatically generated. An example of this idea is the dual complex of a union of balls discussed in section 6. Another example is the automatic triangulation of a topological space developed in [**C, ES, MS**].

Collapsing. If a simplicial complex can be collapsed then its underlying space is contractible. Unfortunately, the reverse is not correct: there are contractible spaces that have non-collapsible triangulations. An example of such a space is the dunce cap [**Mu**]. Although the cap is contractible every edge in its triangulation belongs to at least two triangles.

What exactly is a collapse? Given a simplicial complex $\mathcal{K}$, an *elementary collapse* removes a pair of simplices $\tau \subseteq \sigma$ provided they are the only simplices in the star of τ: $\mathrm{St}\,\tau = \{\tau, \sigma\}$. For example the right complex in figure 8 is obtained from the left complex by removing a vertex τ and an edge σ. More generally, a *collapse* removes the entire star of a simplex τ provided there is a unique highest-dimensional simplex in this star. For example the rightmost vertex of the right complex in figure 8 can be removed together with the two edges and one triangle that contains it. The topmost vertex cannot be removed through a single collapse.

A collapse removing a star of 2^k simplices can be expressed as the composition of 2^{k-1} elementary collapses.

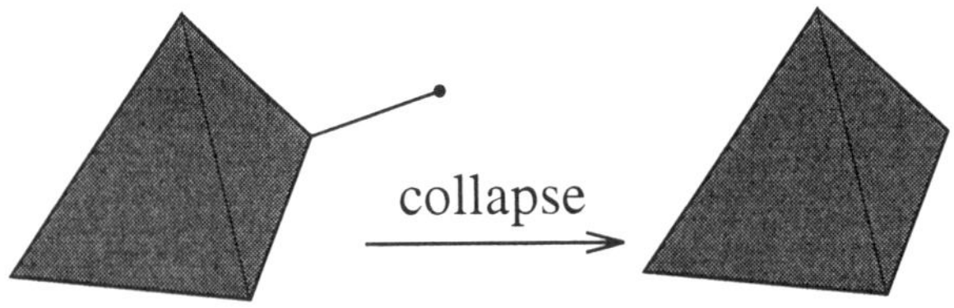

FIGURE 8: Collapsing a vertex together with the only edge that contains it.

It is fairly easy to see that for each collapse there is a corresponding deformation retraction that modifies the underlying space of the complex. This implies that a collapse retains the homotopy type. If $\mathcal{K}$ can be reduced by a sequence of collapses all the way down to a single vertex then $|\mathcal{K}|$ is contractible.

An interesting result due to Wegner [**W**] relates the collapsibility of a simplicial complex with its representability by convex sets. Wegner calls an abstract simplicial complex $\mathcal{A}$ *d-representable* if it is the nerve of a collection of convex sets in $\mathbb{R}^d$, and he shows that $\mathcal{A}$ is d-representable iff it is d-collapsible. The notion of d-collapsibility used here is slightly different from collapsibility as it allows the removal of principal simplices of dimension less than d.

Shelling. A concept similar to collapsing that also applies to non-simplicial complexes is shelling. A superficial difference is that shelling is applied in forward direction constructing a complex by adding one cell at a time. Suppose $\mathcal{K}$ is a complex of convex cells so the union of cells is a topological ball: $|\mathcal{K}| \approx \mathbb{B}^d$. A *shelling* is an ordering of the d-dimensional cells, $\sigma_1, \sigma_2, \ldots, \sigma_n$, so that every prefix defines a d-ball:

$$\bigcup_{i=1}^{j} \sigma_i \quad \approx \quad \mathbb{B}^d$$

for every $1 \leq j \leq n$. If $|\mathcal{K}| \approx \mathbb{S}^d$, as is the case for the boundary complex of every $(d+1)$-dimensional convex polytope, then the requirement is that all *proper* prefixes define a d-ball. $\mathcal{K}$ is *shellable* if it has a shelling.

Shelling is important in the theory of convex polytopes. Schläfli [**Sc**] proves the Euler relation for convex polytopes of any dimension assuming their boundary complexes are shellable. Many years later, Bruggesser and Mani [**BM**] establish that these complexes are indeed shellable. Seidel [**Se**] uses the same shelling order to compute the convex hull of a set of n points in time $\mathrm{O}(\log n)$ per face of the constructed polytope. The algorithm requires an initial preprocessing phase of time $\mathrm{O}(n^2)$ before the faces can be enumerated.

Given a complex $\mathcal{K}$, Danaraj and Klee [**DK**] study algorithms that compute a shelling, if it exists. If $|\mathcal{K}| \approx \mathbb{B}^2$ there always is a shelling and it can be constructed in time proportional to the number of cells. Already for $|\mathcal{K}| \approx \mathbb{B}^3$ it is possible that $\mathcal{K}$ is not shellable [**Rud**] and no polynomial time algorithm deciding the case is known. The problem is made difficult by the existence of partial shellings that cannot be extended to a complete shelling, see Ziegler [**Z**].

8. Fixed points

A point is *fixed* by a function if the point is its own image. In many circumstances, the continuity of the function implies the existence of a fixed point. A nice and elementary treatment of this subject can be found in the booklet by Shashkin [**Sha2**].

A metric fixed point theorem. A map $f : \mathbb{R}^d \to \mathbb{R}^d$ *contracts* if there exists a constant γ, $0 < \gamma < 1$, such that the Euclidean distance between image points is less than γ times the Euclidean distance between arguments:

$$\|f(x) - f(y)\| \quad \leq \quad \gamma \cdot \|x - y\|.$$

For example if you consult a geographic map, you can think of each point on the map as the image of a point in the real world. The geographic map is the image of a map or continuous function, and usually this map is contracting. A classical result on contracting maps is that they have a unique fixed point. In the case of a posted geographic map this fixed point is typically indicated by the words "you are here" or a similar label.

Brouwer's theorem. Brouwer's fixed-point theorem is one of the most basic facts about topological spaces and generalizes the metric fixed point theorem. Originally formulated for the d-dimensional ball, the theorem is stated here for a d-simplex σ^d homeomorphic to $\mathbb{B}^d$:

> every map $f : \sigma^d \to \sigma^d$ has a fixed point.

Observe that the only assumption on f is that it be continuous. Brouwer [**Bro**] provides an existential proof for this theorem. Cohen [**Co**] gives a constructive proof based on a combinatorial result of Sperner [**Spe**]. We describe the combinatorial result and comment on its connection to Brouwer's theorem.

Let σ^d be a d-dimensional simplex with vertex set $T = \{v_0, v_1, \dots, v_d\}$ and triangulation $\mathcal{K}$. A *labeling* is a function $\ell : \mathrm{Vert}\,\mathcal{K} \to T$. Each vertex of $\mathcal{K}$ either belongs to the interior of σ or to the interior of a proper face of σ. The labeling is *proper* if $\ell(u)$ is one of the vertices of the face containing u in its interior. A d-simplex in $\mathcal{K}$ is *complete* if its vertices have $d+1$ different labels. Sperner's lemma guarantees the existence of a complete d-simplex in every properly labeled triangulation. More precisely:

> every properly labeled triangulation of σ has an odd number of complete d-simplices.

How does this combinatorial result relate to the continuous Brouwer theorem? Given a map $f : \sigma^d \to \sigma^d$, we can define a labeling of a triangulation based on the vectors $f(x) - x$ so that a complete d-simplex indicates a possible fixed point in its interior. If the triangulation is chosen sufficiently fine then the possibility becomes certainty and the existence of complete d-simplices implies the existence of fixed points.

Borsuk-Ulam theorem. Intuitively it seems clear that the d-dimensional sphere, $\mathbb{S}^d$, cannot be embedded in $\mathbb{R}^d$ without overlap. A theorem conjectured by Ulam and proved by Borsuk [**Bor**] is a stronger version of this statement.

Two points $x, y \in \mathbb{S}^d$ with coordinates x_i and y_i in $\mathbb{R}^{d+1}$ are *antipodal* if $x_i = -y_i$ for $1 \leq i \leq d+1$. We write $y = -x$ if x and y are antipodal. The Borsuk-Ulam theorem states that if $\mathbb{S}^d$ is mapped continuously to $\mathbb{R}^d$ then there are two antipodal points with identical image:

every map $f : \mathbb{S}^d \to \mathbb{R}^d$ has a point $x \in \mathbb{S}^d$ with $f(x) = f(-x)$.

Ramos [**R**] extends the result to maps from a Cartesian products of spheres to the real space of same dimension. Not all such maps have antipodal points with identical images and Ramos describes a combinatorial criterion that determines when the map has such a point.

Geometric applications. The Bosuk-Ulam theorem has many interesting geometric consequences, see for example the survey by Bárány [**Ba**]. Consider d finite sets, $S_1, S_2, \ldots, S_d$, in $\mathbb{R}^d$. A hyperplane h *bisects* S_i if at most half of the points in S_i lie in each open half-space defined by h. A *ham sandwich cut* is a hyperplane that bisects all d sets. Using the Borsuk-Ulam theorem it is fairly straightforward to prove that a ham sandwich cut always exists. A linear time algorithm for finding ham sandwich cut in two dimensions can be found in [**LMS**].

Consider a single set $S \subseteq \mathbb{R}^d$ and several hyperplanes cutting S. $k \leq d$ hyperplanes $h_1, h_2, \ldots, h_k$ form an *equipartition* if each of the 2^k open d-dimensional cells contains at most $n/2^k$ points, $n = \text{card}\, S$. Using the ham sandwich cut theorem it is easy to prove that every $S \subseteq \mathbb{R}^2$ admits an equipartion formed by 2 lines. It is also true that every $S \subseteq \mathbb{R}^3$ admits an equipartition formed by 3 planes, but the argument requires more than just the 3-dimensional ham sandwich cut theorem, see Hadwiger [**H**] for a proof and [**YDEP**] for applications to range searching. Five hyperplanes cut the moment curve in $\mathbb{R}^5$ into at most $5 \cdot 5 + 1 < 2^5$ pieces, which implies that for $d \geq 5$ not every $S \subseteq \mathbb{R}^d$ admits a d-partition [**A**]. The question for $d = 4$ is currently undecided. Ramos [**R**] studies extensions of these equipartition results to values $k \leq d$.

A related geometric concept is a point so all hyperplanes passing through this point cut a given set in reasonably balanced fractions. A point $x \in \mathbb{R}^d$ is a *centerpoint* of $S \subseteq \mathbb{R}^d$ if every closed halfspace that contains x contains at least $\frac{n}{d+1}$ points of S. The centerpoint theorem states that every finite set has a centerpoint. The classic proof of this theorem is based on Helly's result on overlapping convex sets, see e.g. [**E**]. Using a new approach using Brouwer's fixed point theorem, Živaljević and Vrećica [**ZV**] prove a more general result that contains the ham sandwich and the centerpoint theorem as special cases. A linear time algorithm for finding a centerpoint of a set of n points in $\mathbb{R}^2$ is given in [**JM**].

9. Surfaces

The notion of homotopy formalizes the idea of a continuous deformation of a topological space. Since the fundamental group of a space is invariant under such deformations, this group can sometimes be used to prove the non-existence of a continuous deformation of one space to another.

Classification. A classic result due to Brahana and to Dehn and Heegard says that two surfaces or 2-manifolds are topologically different iff their fundamental groups are non-isomorphic. It follows that surfaces can be classified just by considering their fundamental groups. Descriptions of this result can be found in many textbooks in topology [**Mas, Sti**].

The fundamental group $\pi(\mathbb{M})$ of a surface $\mathbb{M}$ can be finitely presented as $\pi(\mathbb{M}) = (G : R)$, where G is a set of generators and R is a set of relations. Specifically, for an orientable surface of genus g, there are $2g$ generators in G, which are typically represented as g pairs where one member is called 'latitudinal' and the other is 'longitudinal'. For a non-orientable surface of genus g there are g generators. Each relation identifies a product consisting of generators and inverses as a trivial path. For example, $(a, b : aba^{-1}b^{-1})$ is a finite representation of the fundamental group of the torus, see figure 9.

Alternatively, each surface can be combinatorially represented as a polygon P with an even number of edges. Each edge is oriented and labeled. Such a representation is called a *polygonal schema*. The surface can be obtained from P by gluing edges with equal labels consistent with their orientations. Among all polygonal schema of a given surface it is possible to identify a canonical one. We use a superscribed '$-$' to indicate orientation reversal. The canonical polygonal schema of an orientable surface of genus $g \geq 1$ has the form

$$x_1 y_1 x_1^- y_1^- \ldots x_g y_g x_g^- y_g^- .$$

Similarly, the canonical polygonal schema of a non-orientable surface of genus $g \geq 1$ has the form

$$x_1 x_1 x_2 x_2 \ldots x_g x_g .$$

The only remaining case is the 2-sphere, which has genus 0 and canonical schema xx^-.

Algorithms. The algorithmic problem of computing polygonal schemas has been considered in the computational geometry literature. Let $\mathcal{K}$ be a simplicial complex triangulating a surface $\mathbb{M}$, and let $n = \operatorname{card} \mathcal{K}$ be the number of simplices. Vegter and Yap [**VY**] describe an algorithm that takes time $O(n \log n)$ to construct the canonical schema of $\mathbb{M}$. A not necessarily canonical polygonal schema that has the same number of edges can be constructed in optimal $O(n)$ time [**DS**].

At the beginning of this century, Dehn posed two problems related to homotopy on surfaces. Recall that $\mathbb{M}$ is a surface and $\mathcal{K}$ is a triangulation of $\mathbb{M}$.

(i) Given a closed curve in terms of a cycle of edges in $\mathcal{K}$, determine whether or not the curve is contractible.

(ii) Given two closed curves, determine whether or not they are homotopic.

Optimal algorithms for both problems are given in [**DG**]. The algorithms are based on observing that deciding the homotopy equivalence of two closed curves is equivalent to deciding whether their representatives in the fundamental group of the surface are conjugate. The conjugacy problem for this fundamental group is then solved by methods from combinatorial group theory [**Gr**, **LS**]. The group theoretic results allow a reduction to Dehn-like algorithms in all cases except for orientable surfaces of genus 2 and for non-orientable surfaces of genus 3 and 4. The problem of finding an optimal algorithm that decides the homotopy between two curves on such surfaces is still open.

Algorithms for deciding the contractibility of a closed curve can also be based on universal covering spaces, see figure 9. Let $\mathbb{Y}$ be a universal covering space of $\mathbb{X}$ with projection $p : \mathbb{Y} \to \mathbb{X}$, and let $y_0 \in \mathbb{Y}$ and $x_0 = p(y_0)$. Every path on $\mathbb{X}$ with initial point x_0 has a unique lifting on $\mathbb{Y}$ with initial point y_0. Furthermore, a closed curve is contractible in $\mathbb{X}$ if and only if its lift on $\mathbb{Y}$ is closed. This was first discovered by Dehn [**D**].

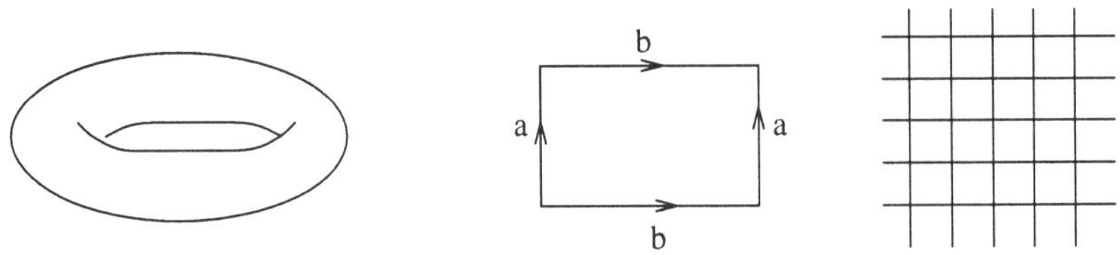

FIGURE 9: The polygonal schema of a torus is a rectangle with opposite edges identified. The torus can be constructed from a rectangular piece of paper by gluing edges as shown. The universal covering space of the torus is the plane decomposed into rectangles each a copy of the polygonal schema.

Let n be the number of simplices in the given triangulation of $\mathbb{M}$ and m the number of edges forming the closed curve. The algorithm of Schipper [**S**] takes time $O(gn + g^2m)$ to construct a finite portion of the covering space to check whether the lifted curve is indeed closed. The algorithm has been improved by Dey and Schipper [**DS**] to run in time $O(n + m \log g)$ and memory $O(n + m)$.

Surface homology. It may be remarked that by the Hurewicz isomorphism theorem [**Sp**] the generators for the homotopy group of a surface are also generators for its homology group. In fact, for an orientable surface of genus g the 1-st homology group is a free abelian goup on the set of g latitudinal and g longitudinal generators. In the case of a non-orientable surface of genus g the 1-st homology group is a direct sum of a $(g-1)$-dimensional free abelian group and the torsion group $\mathbb{Z}_2$.

10. Embedding

An embedding of one topological space in another is an injection whose restriction to the image is a homeomorphism. Using a geometric realization as intermediate step we can stretch the concept and talk about embeddings of an abstract simplicial complex $\mathcal{A}$. To do this we first get a simplicial complex $\mathcal{K}$ that geometrically realizes $\mathcal{A}$ and then embed $|\mathcal{K}|$.

Geometric realization. Every abstract simplicial complex $\mathcal{A}$ has a geometric realization in $\mathbb{R}^d$, provided d is large enough. We can for example use one dimension per vertex of $\mathcal{A}$ and place that vertex at unit distance from the origin on the corresponding coordinate axis. This is clearly a waste of real space and we should expect that the minimum d depends on $k = \dim \mathcal{A}$ but not on the cardinality of $\operatorname{Vert} \mathcal{A}$.

How large does d have to be? As it turns out $d = 2k + 1$ always suffices. To see this map the vertices $u \in \operatorname{Vert} \mathcal{A}$ to points $p(u)$ in general position in $\mathbb{R}^d$. Every subset of $d + 1 = 2k + 2$ or fewer points is affinely independent. It follows that the intersection of two geometric simplices is the realization of the intersection of the two corresponding abstract simplices:

$$\operatorname{conv} p(\alpha) \cap \operatorname{conv} p(\beta) \quad = \quad \operatorname{conv} p(\alpha \cap \beta),$$

which is exactly what is required from a geometric realization. To prove that the bound is tight, Flores [**Fl**] and van Kampen [**vK**] independently construct k-dimensional abstract simplicial complexes without geometric realization in $\mathbb{R}^{2k}$.

One of their examples generalizes K_5, the complete graph of 5 vertices. Think of this graph as the 1-skeleton of a 4-simplex; it is the case $k = 1$ of the following construction. Take a $(2k+2)$-simplex σ^{2k+2}, which has $2k+3$ vertices and a face for every subset of the vertices. The k-skeleton of σ^{2k+2} is obtained after removing all faces of dimension larger than k. As proved in [**Fl, vK**] the k-skeleton of σ^{2k+2} cannot be embedded in $\mathbb{R}^{2k}$.

Planar graphs. The question of embedding a graph in the Euclidean plane is one of the very few embedding problems that has been studied from a mathematical as well as computational point of view. A graph is a 1-dimensional abstract simplicial complex and can always be embedded in $\mathbb{R}^3$. A graph is *planar* if it can be embedded in $\mathbb{R}^2$. Planar graphs are completely characterized by a classic result due to Kuratowski [**Ku**]:

> a graph is planar iff it does not contain a subgraph homeomorphic to K_5 or to $K_{3,3}$.

K_5 is the complete graph with 5 vertices and $K_{3,3}$ is the complete bipartite graph with $3+3$ vertices. It is not entirely clear how to turn this result into an efficient algorithm for testing whether or not a given graph is planar. Indeed, the fastest algorithms for planarity testing are based on rather different principles and construct a concrete embedding, if it exists. By Euler's relation, a graph with $n \geq 3$ vertices can only be planar if it has $3n-6$ or fewer edges. It thus seems possible to test the planarity of such a graph in time $O(n)$. The first such algorithm due to Hopcroft and Tarjan [**HT**] is based on fast graph traversal methods.

Fáry [**F**] proves that every planar graph has a straight line embedding in $\mathbb{R}^2$. In other words, every planar graph has a geometric realization in $\mathbb{R}^2$, see also Tutte [**Tu**]. This result also follows from an older result by Steinitz which states that every 3-connected planar graph is the 1-skeleton of the boundary complex of a 3-dimensional convex polytope [**SR**]. The existence of a straight line embedding also follows from a considerably stronger result by Koebe [**K**]:

> for every planar graph with n vertices there are n non-overlapping disks in the plane and a bijection between the vertices and the disks so that two vertices are connected by an edge iff the two corresponding disks touch.

In other words, the graph is the nerve of a set of n possibly touching but not otherwise overlapping disks. Numerical algorithms for constructing such sets of disks are studied by Stephenson [**St**].

Beyond graphs. Embedding problems for abstract complexes of dimension 2 or higher are considerably more difficult than for graphs. For example the problem to decide whether or not a 2-complex can be embedded in $\mathbb{R}^3$ seems to be related to the recognition problem for the 3-sphere for which algorithms have been found only recently, see section 11.

A few results concerning the extension of Kuratowski's theorem beyond graphs are described in [**Sh**]. For dimension $k = 3$ or higher there is an obstruction so that a k-dimensional complex $\mathcal{A}$ can be embedded in $\mathbb{R}^{2k}$ iff the obstruction vanishes. The obstruction itself belongs to the cohomology groups of a complex related to $\mathcal{A}$. For $k \geq 5$ similar obstructions are known for embeddings in $\mathbb{R}^{2k-1}$.

For a graph the questions whether it has an embedding or a geometric realization in the plane are the same: every planar graph has a geometric realization in $\mathbb{R}^2$. This is no longer true for complexes of dimension higher than 1: there are

2-dimensional complexes with embeddings in $\mathbb{R}^3$ that have no geometric realization in $\mathbb{R}^3$. Examples can be found in Bing [**Bi**] and Grünbaum [**Gru**, chapter 11].

Instead of general 2-dimensional complexes consider triangulations of 2-manifolds. Such triangulations are characterized by having every edge shared by exactly two triangles and the link of every vertex being a simple cycle of edges. A 2-manifold can be embedded in $\mathbb{R}^3$ iff it is orientable. Examples of 2-manifolds that are non-orientable and thus have no embedding in $\mathbb{R}^3$ are the projective plane and the Klein bottle.

11. Three-manifolds

A complete classification of all 2-manifolds is mentioned at the beginning of section 9. Such a thing is not possible for d-manifolds if $d \geq 4$. The case of 3-manifolds is still open and Thurston [**Th**] sketches a program that is hoped to eventually settle the problem. A central problem in this context is the Poincaré conjecture which has not been settled in spite of a considerable amount of effort in the mathematics community.

Poincaré conjecture. Poincaré originally conjectured that every 3-manifold with betti numbers $\beta_0 = 1$, $\beta_1 = \beta_2 = 0$, and $\beta_3 = 1$ is a 3-sphere, or more accurately is homeomorphic to $\mathbb{S}^3$. He found a counterexample himself constructed from a dodecahedron whose pentagonal faces are pairwise identified with a twist. Pointcaré modified his original conjecture to

every simply-connected 3-manifold is a 3-sphere.

It is still open whether or not this is the case. The specialization of the conjecture to 2-manifolds is correct. The generalization to d-manifolds has been settled by Smale [**Sm**] for $d \geq 5$ and by Freedman and Quinn [**FQ**] for $d = 4$. See Bing [**Bi**] for a very accessible survey of topological problems and issues related to the Poincaré conjecture.

An algorithm that decides whether or not a given complex is a triangulation of the 3-sphere has recently been discovered by Rubinstein [**Ru**], see also [**T**]. The existence of such an algorithm is possibly surprising considering the unsettling results of past attempts to characterize the 3-sphere algebraically.

Knots. A *knot* is an embedding of a closed curve in space:

$$K : \mathbb{S}^1 \to \mathbb{R}^3.$$

It is *polygonal* if it consists of a finite number of line segments. The *standard knot* is the unit circle in the x_1x_2-plane, and a knot is *trivial* if there is an isotopy that connects it to the standard knot. The computational complexity of deciding the triviality of a knot is as yet unresolved. For a polygonal knot with n line segments an algorithm with running time at most doubly-exponential in n is given by Haken [**Ha**].

A remarkable result due to Seifert states that every knot is the boundary of an orientable 2-manifold embedded in $\mathbb{R}^3$. Such a 2-manifold with boundary is called a *Seifert surface* for the knot.

The Seifert surface can be doubled to form a 2-manifold without boundary, see figure 10. The *genus* of a knot is the least genus of any such 2-manifold. It is known that a knot is trivial iff its genus vanishes or equivalently if it has a Seifert surface that is a disk. It seems difficult to use this observation in the design of a polynomial time algorithm for testing knot triviality because there exist polygonal knots K with n edges so that every piecewise linear Seifert disk whose boundary is K has at least exponentially many vertices, see Snoeyink [**Sn**].

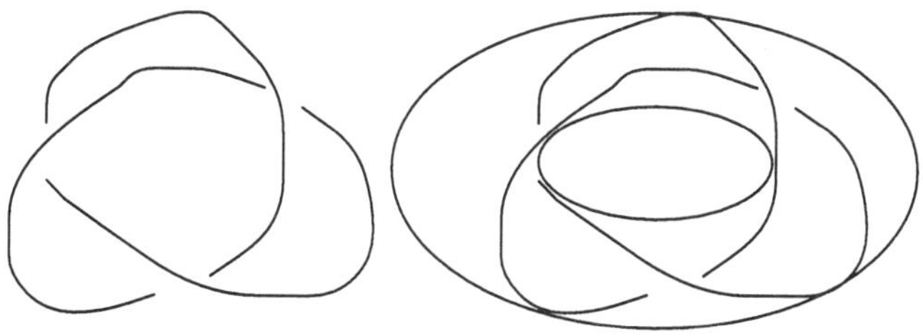

FIGURE 10: A trefoil knot and its embedding on a torus.

A useful invariant of knots is the *knot group*, that is the fundamental group $\pi(\mathbb{R}^3 - K)$ of the complement space of the knot. Although the knot group is invariant under isotopy it does not classify knots: there exist non-isotopic knots with isomorphic knot groups [**Ro**].

Knots can be classified by classifying a special class of 3-manifolds [**Hem**]. Consider a regular neighborhood R of K in $\mathbb{R}^3$. The space $\mathbb{R}^3 - R$ when compacted with a single point at infinity forms an orientable compact 3-manifold with boundary. These manifolds are often referred to as *Haken manifolds* named after Wolfgang Haken who initiated their study [**Ha**].

Computational complexity. Markov [**Ma**] proves that the question whether two finitely represented groups are isomorphic is undecidable. Since every such group is the fundamental group of a 4-manifold, it follows that the question whether two 4-manifolds are homeomorphic is undecidable. The undecidability result extends to dimensions higher than 4. This is why we said earlier that the classification of 4- and higher-dimensional manifolds is out the question.

Manifolds can be specified in terms of triangulations. Triangulations are finite combinatorial objects and the existence of an isomorphism between two can be decided by trying all possibilities. Probably, this is the best one can do. Even for graphs or 1-complexes no polynomial time algorithm is known for deciding isomorphism, see Garey and Johnson [**GJ**].

12. Homology computation

Homology groups offer a formal algebraic framework for studying and counting holes in a topological space. The situation is fairly intuitive for spaces or shapes $\mathbb{X} \subseteq \mathbb{R}^3$. The ranks or Betti numbers $\beta_0, \beta_1, \beta_2$ of the three possibly non-trivial homology groups count the number of gaps, tunnels, and voids. By Alexander duality these correspond to the number of components, loops, and shells in $\mathbb{X}$. The corresponding algebraic notions are non-bounding 0-, 1-, and 2-cycles. They form cosets in the homology groups counted by $\beta_0, \beta_1, \beta_2$.

This section considers several approaches to computing Betti numbers. Since Betti numbers are invariant over all triangulations of a topological space we may assume without loss of generality that $\mathbb{X}$ is given in terms of a triangulating simplicial complex $\mathcal{K}$.

Smith normal form. The classic algorithm for Betti numbers manipulates the incidence matrices of simplices of contiguous dimensions. For complexes in $\mathbb{R}^3$ we have vertices, edges, triangles, and tetrahedra and therefore 3 incidence matrices for dimensions 0 and 1, 1 and 2, 2 and 3. Using standard row and column operations, the matrices are brought into a normal form, from which the Betti numbers can be derived. The normal forms also encode the torsion subgroups and thus contain complete information about all homology groups of $\mathcal{K}$.

If implemented as described in textbooks like Munkres [**Mu**], the running time is not known to be polynomial in $n = \text{card}\,\mathcal{K}$. Part of the reason for the slow running time is that the number of row and column operations depends not only on the number of simplices but also on the size of the integer coefficients describing chains of simplices computed during the algorithm. An improvement of the worst-case running time to $O(n^5)$ can be found in [**KB**]. The complexes that arise in the study of macromolecules, see section 8, are usually too large to even keep the incidence matrices in internal memory. We need new ideas to work on these applications.

Incremental algorithm. A combinatorial approach to computing Betti numbers is described in [**DE**]. Let $\sigma_1, \sigma_2, \ldots, \sigma_n$ be an ordering of the simplices in $\mathcal{K}$ so that every prefix $\mathcal{K}_i = \{\sigma_1, \sigma_2, \ldots, \sigma_i\}$ is a complex. For example the sequence of simplices sorted in order of non-decreasing dimension satisfies this requirement. The sequence of σ_i is a *filter* and the sequence of complexes $\mathcal{K}_i$ is a *filtration* of $\mathcal{K}$.

The Betti numbers of $\mathcal{K}_i$ can be computed from σ_i and the Betti numbers of $\mathcal{K}_{i-1}$. Let k be the dimension of σ_i. If σ_i completes a k-cycle in $\mathcal{K}_i$ then

$$\beta_k(\mathcal{K}_i) = \beta_k(\mathcal{K}_{i-1}) + 1.$$

Indeed, none of the simplices in the star of σ_i have been added yet so σ_i does not bound in $\mathcal{K}_i$. In the other case

$$\beta_{k-1}(\mathcal{K}_i) = \beta_{k-1}(\mathcal{K}_{i-1}) - 1$$

because σ_i fills a non-bounding $(k-1)$-cycle in $\mathcal{K}_{i-1}$, which now bounds in $\mathcal{K}_i$. A formal proof of this case analysis can be based on Mayer-Vietoris sequences as described in any textbook in algebraic topology.

If $\mathcal{K}$ is a complex in $\mathbb{R}^3$ then there are fast algorithms that distinguish between the two possible cases. Given the filter, all questions can be decided in time $O(n\alpha(n))$, where $\alpha(n)$ is the extremely slowly growing inverse of the Ackermann function, see [**CLR**]. In the same time the Betti numbers of all complexes in the filtration can be computed. Already in $d = 4$ dimensions it is not clear how to efficiently decide whether or not a simplex σ_i belongs to a cycle. It would be interesting to find fast data structures for the cycle test in all dimensions.

Solids. Call a connected 3-manifold with boundary embedded in $\mathbb{R}^3$ a *solid.* Suppose $\mathbb{M}$ is a solid and $\mathcal{K}$ is a triangulation of $\mathbb{M}$. By definition $\mathcal{K}$ is a pure 3-complex which means that every $\sigma \in \mathcal{K}$ is face of a tetrahedron. The boundary of $\mathbb{M}$ is a 2-manifold and the boundary complex of $\mathcal{K}$ triangulates it. Let $\mathbb{S}_1, \mathbb{S}_2, \ldots, \mathbb{S}_m$ be the components of the boundary of $\mathbb{M}$. Each $\mathbb{S}_i$ is a connected orientable closed surface. Assume $\mathbb{S}_1$ through $\mathbb{S}_{m-1}$ are 'inner' surfaces and $\mathbb{S}_m$ is the 'outer' surface. Note that the dispositions of the surfaces in $\mathbb{R}^3$ can form exotic configurations that may be knotted or linked, see figure 11.

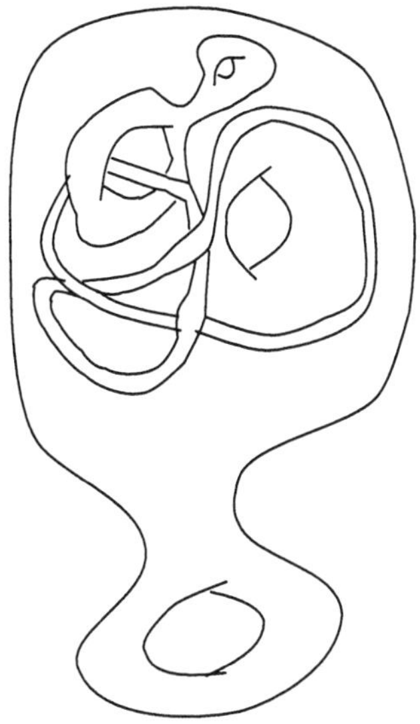

FIGURE 11: A solid with knotted and linked boundaries.

By assumption $\mathbb{M}$ is connected and we have $\beta_0 = 1$. The number of loops and tunnels is the sum of surface tunnels:

$$\beta_1 = \sum_{i=1}^{m} g_i,$$

where g_i is the genus of $\mathbb{S}_i$. Furthermore, there is a void for each inner surface and therefore $\beta_2 = m - 1$. In other words the Betti numbers of $\mathbb{M}$ can be computed directly from the genuses of the $\mathbb{S}_i$. As explained in section 7, the genus of $\mathbb{S}_i$ can be calculated from the triangulation:

$$g_i = 1 - \frac{1}{2}(v_i - e_i + f_i),$$

where v_i, e_i, f_i are the number of vertices, edges, triangles in the triangulation of $\mathbb{S}_i$.

The above algorithm applies to all simplicial complexes in $\mathbb{R}^3$, not just to triangulations of solids. This is because every such complex $\mathcal{K}$ has a homotopy equivalent solid that can be obtained by thickening $\mathcal{K}$. Delfinado and Edelsbrunner [**DE**] explain how the thickening process can be avoided. Dey and Guha [**DG2**] show that the latitudinal generators of the inner surfaces together with the longitudinal generators of the outer surface form a basis of the first homology group of $\mathbb{M}$. This basis can be computed in time $O(gn^2)$, where g is the maximum genus of any surface [**VY, DS**].

Combinatorial Laplacians. Friedman [**Fr**] proposes the use of Laplacians for computing Betti numbers of simplicial complexes of arbitrary dimension. His approach appears to be more efficient than computing Smith normal forms, at least for some types of complexes.

Consider the homology groups of a simplicial complex $\mathcal{K}$ with real rather than integer coefficients; the Betti numbers remain the same. More precisely let

$$\ldots \xrightarrow{\partial_{j+2}} \mathsf{C}_{j+1} \xrightarrow{\partial_{j+1}} \mathsf{C}_j \xrightarrow{\partial_j} \mathsf{C}_{j-1} \xrightarrow{\partial_{j-1}} \ldots$$

be the chain complex where C_j is the vector space of j-chains with coefficients in $\mathbb{R}$. $\mathsf{H}_j(\mathbb{R}) = \mathrm{Ker}\,\partial_j | \mathrm{Im}\,\partial_{j+1}$ is the j-th homology group with real coefficients, and β_j is the dimension of H_j. With the usual inner product $\langle \cdot, \cdot \rangle$ on vector spaces over $\mathbb{R}$, we get maps

$$\partial_j^* : \mathsf{C}_{j-1} \to \mathsf{C}_j$$

defined by $\langle \partial_j c, c' \rangle = \langle c, \partial_j^* c' \rangle$, where $c \in \mathsf{C}_j$ and $c' \in \mathsf{C}_{j-1}$. In other words, ∂_j^* is the transpose of ∂_j. Finally for each j define the *Laplacian* $\triangle_j : \mathsf{C}_j \to \mathsf{C}_j$:

$$\triangle_j = \partial_{j+1}\partial_{j+1}^* + \partial_j^*\partial_j,$$

and the *harmonic j-forms*:

$$\mathcal{H}_j = \{c \in \mathsf{C}_j \mid \triangle_j c = 0\}.$$

By the elementary linear algebra behind Hodge theory, $\mathcal{H}_j$ is isomorphic to $\mathsf{H}_j(\mathbb{R})$ so β_j of $\mathcal{K}$ is the dimension of $\mathcal{H}_j$, see **[Fr]**. By definition, this is the dimension of the null space of $\triangle_j$, which is the multiplicity of the eigenvalue 0 of $\triangle_j$. Friedman uses matrix-theoretic methods to probabilistically estimate the eigenvalues of the Laplacian $\triangle_j$.

References

[ASS] B. Aronov, R. Seidel and D. Souvaine, *On compatible triangulations of simple polygons*, Comput. Geom. Theory Appl. **3** (1993), 27–35.

[A] D. Avis, *On the partitionability of point sets in space*, In "Proc. 1st Ann. Sympos. Comput. Geom. 1985," 116–120.

[B] R. F. W. Bader, Atoms in Molecules. A Quantum Theory. Clanrendon Press, Oxford, England, 1994.

[Ba] I. Bárány, *Geometric and combinatorial applications of Borsuk's theorem*. Chapter IX in New Trends in Discrete Geometry, J. Pach (ed.), Springer-Verlag, Berlin, 1993, 235–249.

[Bau] B. Baumgart, A polyhedron representation for computer vision. In "Proc. AFIPS Conf. 1975," 589–596.

[BE] M. Bern and D. Eppstein, *Mesh generation and optimal triangulation*. In Computing in Euclidean Geometry, F. K. Hwang and D.-Z. Du (eds.), World Scientific, 1992, 47–123.

[Bi] R. Bing, *Some aspects of the topology of 3-manifolds related to the Poincaré conjecture*. In Lectures on Modern Mathematics, vol. II, T. L. Saaty (ed.), Wiley, New York, 1964, 93–128.

[Bj] A. Björner, *Topological methods*. In Handbook of Combinatorics, vol. I and II, R. L. Graham, M. Grötschel and L. Lovász (ed.), Elsevier, Amsterdam, 1995, 1819–1872.

[Bo] J.-D. Boissonnat, *Geometric structures for three-dimensional shape representation*. ACM Trans. Graphics **3** (1984), 266–286.

[BEf] V. G. Boltjanskij and V. A. Efermovič, Anschauliche kombinatorische Topologie. Vieweg, Braunschweig/Wiesbaden, Germany, 1986.

[Bor] K. Borsuk, *Drei Sätze über die n-dimensionale euklidische Sphäre*. Fund. Math. **20** (1933), 177–190.

[Br] E. Brisson, *Representing geometric structures in d dimensions: topology and order*. Discr. Comput. Geom. **9** (1993), 387–426.

[Bro] L. E. J. Brouwer, *Über Abbildung von Mannigfaltigkeiten*. Math. Ann. **71** (1912), 97–115.

[BM] H. Bruggesser and P. Mani, *Shellable decompositions of cells and spheres*. Math. Scand. **29** (1971), 197–205.

[CV] R. Casati and A. C. Varzi, Holes and Other Superficialities. MIT Press, Cambridge, Massachusetts, 1994.

[CFF] J. C. Cavendish, D. A. Field and W. H. Frey, *An approach to automatic three-dimensional finite element mesh generation*. Internat. J. Numer. Methods Engrg. **21** (1985), 329–347.

[CEF] H.-L. Cheng, H. Edelsbrunner and P. Fu, Shape space from deformation, Rept. rgi-tech-98-011, Raindrop Geomagic, Champaign, Illinois, 1998.

[C] L. P. Chew, *Guaranteed-quality mesh generation for curved surfaces.* In "Proc. 9th Ann. Sympos. Comput. Geom. 1993," 274–280.

[Co] D. I. A. Cohen, *On the Sperner lemma.* J. Combin. Theory **2** (1967), 585–587.

[CLR] T. H. Cormen, Ch. E. Leiserson and R. L. Rivest, Introduction to Algorithms. MIT Press, Cambridge, Massachusetts, 1990.

[Cos] J. E. Costillo (ed.), Mathematical Aspects of Numerical Grid Generation. Frontiers in Applied Math. **8**, SIAM, Philadelphia, 1991.

[DK] G. Danaraj and V. Klee, *Which spheres are shellable?* Ann. Discrete Math. **2** (1978), 33–52.

[D] M. Dehn, *Über unendliche diskontinuierliche Gruppen.* Math. Ann. **71** (1912), 116–144.

[De] B. Delaunay, *Sur la sphère vide.* Izv. Akad. Nauk SSSR, Otdelenie Matematicheskii i Estestvennyka Nauk **7** (1934), 793–800.

[DG] T. K. Dey and S. Guha, *Optimal algorithms for curves on surfaces.* In "Proc. 35th IEEE Ann. Sympos. Found. Comput. Sci. 1995," 266–274.

[DG2] T. K. Dey and S. Guha, Algorithms for manifolds and simplicial complexes in Euclidean 3-space. In "Proc. 28th ACM Sympos. Theory Comput. 1996," 398–407.

[DS] T. K. Dey and H. Schipper, *A new technique to compute polygonal schema for 2-manifolds with application to null-homotopy detection.* Discrete Comput. Geom. **14** (1995), 93–110.

[DE] C. J. A. Delfinado and H. Edelsbrunner, *An incremental algorithm for Betti numbers of simplicial complexes on the 3-sphere.* Comput. Aided Geom. Design **12** (1995), 771–784.

[DL] D. P. Dobkin and M. J. Laszlo, *Primitives for the manipulation of three-dimensional subdivisions.* Algorithmica **4** (1989), 3–32.

[DCN] J. A. Dougenik, N. R. Chrisman and D. R. Niemeyer, *An algorithm to construct continuous area cartograms.* Professional Geographer **37** (1985), 75–81.

[E] H. Edelsbrunner, Algorithms in Combinatorial Geometry. Springer-Verlag, Heidelberg, Germany, 1987.

[E2] H. Edelsbrunner, *The union of balls and its dual shape.* Discrete Comput. Geom. **13** (1995), 415–440.

[E3] H. Edelsbrunner, *Algebraic decomposition of non-convex polyhedra.* In "Proc. 36th Ann. IEEE Sympos. Found. Comput. Sci. 1995," 248–257.

[E4] H. Edelsbrunner, Surface reconstruction by wrapping finite sets in space. Rept. rgi-tech-96-001, Raindrop Geomagic, Champaign, Illinois, 1996.

[E5] H. Edelsbrunner, Deformable smooth surface design. Rept. rgi-tech-96-002, Discrete Comput. Geom., to appear.

[EFL] H. Edelsbrunner, M. A. Facello and J. Liang, *On the definition and the construction of pockets in macromolecules.* In "Proc. Pacific Sympos. Biocomputing 1996," World Scientific, Singapore, 272–281.

[EKS] H. Edelsbrunner, D. G. Kirkpatrick and R. Seidel, *On the shape of a set of points in the plane.* IEEE Trans. Inform. Theory **IT-29** (1983), 551–559.

[EM] H. Edelsbrunner and E. P. Mücke. *Three-dimensional alpha shapes.* ACM Trans. Graphics **13** (1994), 43–72.

[ES] H. Edelsbrunner and N. R. Shah, *Triangulating topological spaces.* Internat. J. Comput. Geom. Appl. **7** (1997), 365–378.

[EW] H. Edelsbrunner and R. Waupotitsch, *A combinatorial approach to cartograms.* J. Comput. Geom. Theory Appl. **7** (1997), 343–360.

[Ep] D. Eppstein, *Linear complexity hexahedral mesh generation.* In "Proc. 12th Ann. Sympos. Comput. Geom. 1996," 58–67.

[F] I. Fáry, *On straight line representation of planar graphs.* Acta Sci. Math. (Szeged) **11** (1948), 229–233.

[Fl] A. Flores, *Über n-dimensionale Komplexe die in* R_{2n+1} *selbstverschlungen sind.* Ergeb. Math. Koll. **6** (1933/34), 4–7.

[FQ] M. H. Freedman and F. Quinn. Topology of 4-manifolds. Princeton Univ. Press, Princeton, New Jersey, 1990.

[Fr] J. Friedman, *Computing Betti numbers via combinatorial Laplacians.* In "Proc. 28th ACM Sympos. Theory Comput. 1996," 386–391.

[FKU] H. Fuchs, Z. M. Kedem and S. P. Uselton, *Optimal surface reconstruction from planar contours.* Commun. ACM **20** (1977), 693–702.

[GJ] M. R. Garey and D. S. Johnson, Computers and Intractibility: A Guide to the Theory of NP-Completeness. Freeman, New York, 1979.

[G] D. Gillman, *Triangulations for rubber-sheeting.* Auto Cartography **7** (1985), 191–197.

[GOS] C. Gitlin, J. O'Rourke and V. Subramanian, *On reconstructing polyhedra from parallel slices.* Intern. J. Comput. Geom. Appl. **6** (1996), 103–122.

[Gr] M. D. Greendlinger, *On Dehn's algorithm for the conjugacy and word problems with applications.* Comm. Pure Appl. Math. **13** (1960), 641–677.

[Gru] B. Grünbaum, Convex Polytopes. John Wiley & Sons, London, 1967.

[GS] L. Guibas and J. Stolfi, *Primitives for the manipulation of general subdivisions and the computation of Voronoi diagrams.* ACM Trans. Graphics **4** (1985), 74–123.

[GP] V. Guillemin and A. Pollack, Differential Topology. Prentice Hall, Englewood Cliffs, New Jersey, 1974.

[GW] H. Gupta and R. Wenger, *Constructing piecewise linear homeomorphisms of simple polygons.* J. Algorithms **22** (1997), 142–157.

[H] H. Hadwiger, *Simultane Vierteilung zweier Körper.* Arch. Math. (Basel) **17** (1966), 274–278.

[Ha] W. Haken, *Theorie der Normalflächen,* Acta Math. **105** (1961), 245–375.

[Ha2] W. Haken, *Various aspects of the three-dimensional Poincaré problem.* In "Topology of Manifolds," Proc. Univ. Georgia Topology of Manifolds Institute, J. C. Cantrella and C. H. Edwards Jr. (eds.), Markham, Chicago, 1970, 140–152.

[He] P. S. Heckbert, *Bilinear Coons patch image warping.* In Graphics Gems IV, P. S. Heckbert (ed.), Academic Press, Boston, 1994, 438–446.

[Hem] G. Hemion, The Classification of Knots and 3-dimensional Spaces. Oxford Univ. Press, Oxford, 1992.

[Her] G. T. Herman, Oriented surfaces in digital spaces. CVGIP: Graphical Models Image Process. **55** (1993), 381–396.

[HT] J. E. Hopcroft and R. E. Tarjan, *Efficient planarity testing.* J. Assoc. Mach. Soc. **21** (1974), 549–568.

[JM] S. Jadhav and A. Mukhopadhyay, *Computing a centerpoint of a finite planar set of points in linear time.* In "Proc. 9th Ann. Sympos. Comput. Geom. 1993," 83–90.

[vK] E. R. van Kampen, *Komplexe in euklidischen Räumen.* Abh. Math. Sem. **9** (1932), 72–78 and 152–153.

[KB] R. Kannan and A. Bachem, *Polynomial algorithms for computing the Smith and Hermite normal forms of an integer matrix.* SIAM J. Comput. **8** (1979), 499–507.

[KS] W. J. Kaufmann III and L. L. Smarr, Supercomputing and the Transformation of Science. Scientific American Library, New York, 1993.

[KCP] J. R. Kent, W. E. Carlson and R. E. Parent, *Shape transformation for polyhedral objects.* Computer Graphics **26** (1992), 47–54.

[K] P. Koebe, *Kontaktprobleme der konformen Abbildung.* Ber. Verh. Sächs. Akad. Wiss., Leipzig, Math.-Phys. Kl. **88** (1936), 141–164.

[KR] T. Y. Kong and A. Rosenfeld, *Digital topology: introduction and survey.* Comput. Vision Graph. Image Process. **48** (1989), 357–393.

[Ku] K. Kuratowski, *Sur le problème des courbes en topologie.* Fund. Math. **15** (1930), 271–283.

[L] I. Lakatos, Proofs and Refutations. The Logic of Mathematical Discovery. Cambridge Univ. Press, England, 1976.

[LR] B. Lee and F. M. Richards, *The interpretation of protein structures: estimation of static accessibility.* J. Mol. Biol. **55** (1971), 379–400.

[LPR] C. N. Lee, T. Poston and A. Rosenfeld, *Winding and Euler numbers for 2D and 3D digital images.* CVGIP: Graphical Models Image Process. **53** (1991), 522–537.

[Le] J. Leray, *Sur la forme des espaces topologiques et sur les point fixes des représentations.* J. Math. Pures Appl. **24** (1945), 95–167.

[LT] R. J. Lipton and R. E. Tarjan, *A separator theorem for planar graphs.* SIAM J. Appl. Math. **36** (1979), 177–189.

[LMS] C.-Y. Lo, J. Matoušek and W. Steiger, *Algorithms for ham-sandwich cuts.* Discrete Comput. Geom. **11** (1994), 433–452..

[LS] R. C. Lyndon and P. E. Schupp, Combinatorial Group Theory. Springer-Verlag, New York, 1977.

[M] M. Mäntylä, An Introduction to Solid Modeling. Comput. Sci. Press, Rockville, Maryland, 1988.

[Ma] A. A. Markov, *Insolvability of the problem of homeomorphy.* In "Proc. Intern. Cong. Math. 1958," Cambridge, 300–306.

[MS] T. Martinetz and K. Schulten, *Topology representing networks.* Neural Networks **7** (1994), 507–522.

[Mas] W. S. Massey, Algebraic Topology: An Introduction. Springer-Verlag, New York, 1967.

[MTTV] G. L. Miller, S.-H. Teng, W. P. Thurston and S. A. Vavasis, *Separators for sphere-packings and nearest neighbor graphs.* J. Assoc. Comput. Mach. **44** (1997), 1–29.

[Mi] J. Milnor, *Two complexes which are homeomorphic but combinatorially distinct.* Ann. Math. (1961), 575–590.

[Mi2] J. Milnor, Morse Theory. Annals Math. Studies, Princeton Univ. Press, New Jersey, 1963.

[Mo] E. Moise, Geometric Topology in Dimensions 2 and 3. Springer-Verlag, New York, 1977.

[Mu] J. R. Munkres, Elements of Algebraic Topology, Addison-Wesley, Redwood City, 1984.

[NS] N. J. Naccache and R. Shinghal, *SPTA: A proposed algorithm for thinning binary patterns.* IEEE Trans. System, Man, Cybernet. **14** (1984), 409–418.

[NW] D. Q. Naiman and H. P. Wynn, *Inclusion-exclusion-Bonferroni identities and inequalities for discrete tube-like problems via Euler characteristics.* Ann. Statist. **20** (1992), 43–76.

[OTF] Y.-L. O, A. Toet, D. Foster, H. J. A. M. Heijmans and P. Meer (eds.) Shape in Picture. Mathematical Description of Shape in Grey-level Images. NATO ASI Series, Springer-Verlag, Berlin, 1992.

[P] H. Poincaré, *Sur la généralisation d'un théorème d'Euler relatif aux polyèdres.* C. R. Acad. Sci. Paris **117** (1893), 144–145.

[P2] H. Poincaré, *Complément à l'analysis situs.* Rend. Circ. Mat. Palermo **13** (1899), 285–343.

[R] E. A. Ramos, *Equipartition of mass distributions by hyperplanes.* Discrete Comput. Geom. **15** (1996), 147–167.

[Ri] F. M. Richards, *Areas, volumes, packing, and protein structure.* Ann. Rev. Biophys. Bioeng. **6** (1977), 151–176.

[Ro] D. Rolfsen, Knots and Links. Publish or Perish, Berkeley, 1976.

[Ros] G. D. Rose, *No assembly required.* The Sciences **36** (1996), 26–31.

[Ru] J. H. Rubinstein, An algorithm to recognise the 3-sphere. Manuscript, Univ. Melbourne, Australia, 1995.

[Rud] M. E. Rudin, *An unshellable triangulation of a tetrahedron.* Bull. Amer. Math. Soc. **64** (1958), 90–91.

[S] H. Schipper, *Determining contractibility of curves.* In "Proc. 8th Sympos. Comput. Geom. 1992," 358–367.

[Sc] L. Schläfli, *Theorie der vielfachen Kontinuität.* Denkschr. Schweiz. naturf. Ges. **38** (1901), 1–237.

[SB] R. Schneiders and R. Bunton, Automatic generation of hexahedral finite element meshes. Comput. Aided Geom. Design **12** (1995).

[Se] R. Seidel, *Constructing higher dimensional convex hulls at logarithmic cost per face.* In "Proc. ACM 18th Ann. Sympos. Theory Comput. 1986," 404–413.

[Ser] J. P. Serra, Image Analysis and Mathematical Morphology. Academic Press, London, 1982.

[Sh] A. Shapiro, *Obstructions to the imbedding of a complex in a Euclidean space. I. The first obstruction.* Ann. Math. **66** (1957), 256–269.

[Sha] Yu. A. Shashkin, The Euler Characteristic. Little Math. Lib., Mir Publ., Moscow, 1989.

[Sha2] Yu. A. Shashkin, Fixed Points. Amer. Math. Soc. and Math. Assoc. Amer., 1991.

[Sm] S. Smale, *Generalized Poincaré's conjecture in dimensions greater than 4.* Ann. Math. **74** (1961), 391–406.

[Sn] J. Snoeyink, *A trivial knot whose spanning disks have exponential size.* In "Proc. 6th Sympos. Comput. Geom. 1990," 139–147.

[Sp] E. H. Spanier, Algebraic Topology. Springer-Verlag, New York, 1966.

[Spe] E. Sperner, *Neuer Beweis für die Invarianz der Dimensionszahl und des Gebietes.* Hamburger Abhand. **6** (1928), 265–272.

[SR] E. Steinitz and H. Rademacher, Vorlesungen über die Theorie der Polyeder. Springer-Verlag, Berlin, 1934.

[St] K. Stephenson, *Circle packings in the approximation of conformal mappings*. Bull. Amer. Math. Soc. **23** (1990), 407–415.

[Sti] J. Stillwell, Classical Topology and Combinatorial Group Theory. Springer-Verlag, New York, 1980.

[T] A. Thompson, *This position and the recognition problem for* S^3. Math. Res. Lett. **1** (1994), 613–630.

[TWM] J. F. Thompson, Z. U. A. Warsi and C. W. Mastin, Numerical Grid Generation. Foundations and Applications. North-Holland, New York, 1985.

[Th] W. P. Thurston, Three-dimensional Geometry and Topology, vol. 1. S. Levy (ed.), Princeton Univ. Press, New Jersey, 1985.

[Tu] W. Tutte, *Convex representation of graphs*. Proc. London Math. Soc. **10** (1960), 304–320.

[V] G. Vegter, *Computational topology*. In The Handbook of Discrete and Computational Geometry, J. E. Goodman and J. O'Rourke (eds.), CRC Press LLC, Boca Raton, Florida, 1997, 517–536.

[VY] G. Vegter and C. K. Yap, *Computational complexity of combinatorial surfaces*. In "Proc. 6th Sympos. Comput. Geom. 1990," 102.–111.

[W] G. Wegner, *d-collapsing and nerves of families of convex sets*. Arch. Math. **26** (1975), 317–321.

[We] K. J. Weiler, *Topological structures for geometric modeling*. Ph.D. Thesis, Comput. Syst. Engin., Rensselaer Polytechnic Inst., 1986.

[Wes] M. A. Wesley, Construction and use of geometric modeling systems. In Computer Aided Design - Modeling, Systems Engineering CAD Systems, J. Encarnacao (ed.), Springer-Verlag, 1980.

[WG] M. S. White Jr. and P. Griffin, *Piecewise linear rubber-sheet map transformation*. Amer. Cartographer **12** (1985), 123–131.

[Wu] W.-T. Wu, *On a theorem of Leray*. Chinese Math. **2** (1962), 398–410.

[YDEP] F. F. Yao, D. P. Dobkin, H. Edelsbrunner and M. S. Paterson, *Partitioning space for range queries*. SIAM J. Comput. **18** (1989), 371–384.

[Z] G. M. Ziegler, *Shelling polyhedral 3-balls and 4-polytopes*. Discrete Comput. Geom. **19** (1998), 159–174.

[ZV] R. T. Živaljević and S. Vrećica, *An extension of the ham sandwich theorem*. Bull. London Math. Soc. **22** (1990), 183–186.

A. Definitions

The appendix presents definitions from topology grouped under topology and space, affine and convex hull, simplicial complexes, maps and equivalences, fundamental groups and covering spaces, manifolds, and homology groups and Betti numbers.

Topology and space. The mathematical definition of a topology is rather general and abstract. We present the definition and immediately specialize it assuming an embedding in Euclidean space.

A **topology** on a set $\mathbb{X}$ is a system $\mathcal{X}$ of subsets with the following properties:

(i) $\emptyset, \mathbb{X} \in \mathcal{X}$,

(ii) if $\mathcal{Z} \subseteq \mathcal{X}$ then $\bigcup \mathcal{Z} \in \mathcal{X}$, and

(iii) if $\mathcal{Z} \subseteq \mathcal{X}$ is finite then $\bigcap \mathcal{Z} \in \mathcal{X}$.

A set $\mathbb{X}$ with a topology is called a **topological space**. The sets in $\mathcal{X}$ are the **open** sets and the complements of the open sets are the **closed** sets of $\mathbb{X}$. A **neighborhood** of a point $x \in \mathbb{X}$ is an open set that contains x. A **cover** is a collection of sets whose union is $\mathbb{X}$. $\mathbb{X}$ is **compact** if every cover of $\mathbb{X}$ with open sets has a finite subcover. $\mathbb{X}$ is **connected** if the only subsets of $\mathbb{X}$ that are both open and closed are $\emptyset$ and $\mathbb{X}$. The **subspace topology** of $\mathbb{Y} \subseteq \mathbb{X}$ is the system $\mathcal{Y} = \{\mathbb{Y} \cap X \mid X \in \mathcal{X}\}$. With this topology $\mathbb{Y}$ is called a **subspace** of $\mathbb{X}$.

The d**-dimensional Euclidean space** is the set of real d-tuples, $\mathbb{R}^d = \{x = (x_1, x_2, \ldots, x_d) \mid x_i \in \mathbb{R}\}$. The **norm** of $x \in \mathbb{R}^d$ is $\|x\| = (\sum_{i=1}^d x_i^2)^{\frac{1}{2}}$, and the **distance** between points $x, y \in \mathbb{R}^d$ is $|xy| = \|x - y\|$. The **open ball** with center $z \in \mathbb{R}^d$ and radius $\varepsilon > 0$ is $b_{z,\varepsilon} = \{x \mid |xz| < \varepsilon\}$. The set of open balls generates what we call the **Euclidean topology** of $\mathbb{R}^d$, namely the system whose sets are unions of open balls. Every subset $\mathbb{Y} \subseteq \mathbb{R}^d$ defines a subspace topology and thus a subspace of $\mathbb{R}^d$. Particularly useful subspaces of $\mathbb{R}^d$ are the d**-halfspace**, the **(closed)** d**-ball**, and the $(d-1)$**-sphere**:

$$\begin{aligned} \mathbb{H}^d &= \{x \in \mathbb{R}^d \mid x_1 \geq 0\}, \\ \mathbb{B}^d &= \{x \in \mathbb{R}^d \mid \|x\| \leq 1\}, \text{ and} \\ \mathbb{S}^{d-1} &= \{x \in \mathbb{R}^d \mid \|x\| = 1\}. \end{aligned}$$

Affine and convex hull. We assume the usual vector space with componentwise addition of two vectors and multiplication of a vector with a scalar. The terms vector and point are used synonymously.

Given a set of points $T = \{p_0, p_1, \ldots, p_n\}$ in $\mathbb{R}^d$, the **affine hull** is

$$\operatorname{aff} T = \{\sum_{i=0}^{n} \phi_i p_i \mid \sum_{i=0}^{n} \phi_i = 1\}.$$

T is **affinely independent** if $\operatorname{aff} T$ is different from the affine hull of every proper subset of T. A k**-flat** is the affine hull of $k+1$ affinely independent points. The **convex hull** of T is

$$\operatorname{conv} T = \{x \in \operatorname{aff} T \mid \phi_i \geq 0 \text{ for all } i\}.$$

A k**-simplex** is the convex hull of $k+1$ affinely independent points. We call a 0-simplex a **vertex**, a 1-simplex an **edge**, a 2-simplex a **triangle**, and a 3-simplex a **tetrahedron**. Let $T = \{p_0, p_1, \ldots, p_k\}$ be affinely independent. Then $\sigma = \operatorname{conv} T$ is a k-simplex and its **dimension** is $\dim \sigma = k = \operatorname{card} T - 1$. A **face** of σ is a simplex $\tau = \operatorname{conv} U$ with $U \subseteq T$; it is **proper** if U is a proper subset of T. The **barycentric coordinates** of a point $x \in \sigma$ are the real numbers ϕ_i with

$$\sum_{i=0}^{k} \phi_i p_i = x \text{ and } \sum_{i=0}^{k} \phi_i = 1.$$

The **barycenter** of σ is the point $b(\sigma)$ with barycentric coordinates $\phi_i = \frac{1}{k+1}$ for all i.

Simplicial complexes. A **simplicial complex** is a finite collection $\mathcal{K}$ of simplices with the following two properties:

(i) if $\sigma \in \mathcal{K}$ and τ is a face of σ then $\tau \in \mathcal{K}$, and

(ii) if $\sigma, \sigma' \in \mathcal{K}$ then $\sigma \cap \sigma'$ is empty or a face of both.

The **vertex set** is $\operatorname{Vert} \mathcal{K} = \{\sigma \in \mathcal{K} \mid \dim \sigma = 0\}$. The **underlying space** is the part of space covered by simplices in $\mathcal{K}$: $|\mathcal{K}| = \bigcup \mathcal{K}$. The **dimension** is $\dim \mathcal{K} = \max\{\dim \sigma \mid \sigma \in \mathcal{K}\}$. A **subcomplex** is a simplicial complex $\mathcal{L} \subseteq \mathcal{K}$. The

closure of a subset $L \subseteq \mathcal{K}$ is the smallest subcomplex that contains L: $\mathrm{Cl}\, L = \{\sigma \in \mathcal{K} \mid \sigma \subseteq \tau \text{ for some } \tau \in L\}$. The **star** and **link** of $\tau \in \mathcal{K}$ are

$$\begin{aligned} \mathrm{St}\, \tau &= \{\sigma \in \mathcal{K} \mid \tau \text{ is a face of } \sigma\} \text{ and} \\ \mathrm{Lk}\, \tau &= \{\sigma \in \mathrm{Cl}\,\mathrm{St}\, \tau \mid \sigma \cap \tau = \emptyset\}. \end{aligned}$$

A simplicial complex $\mathcal{L}$ with $|\mathcal{L}| = |\mathcal{K}|$ is a **subdivision** of $\mathcal{K}$ if every simplex in $\mathcal{L}$ is contained in a simplex in $\mathcal{K}$.

An **abstract simplicial complex** is a finite system $\mathcal{A}$ of sets so $X \in \mathcal{A}$ and $Y \subseteq X$ implies $Y \in \mathcal{A}$. The sets in $\mathcal{A}$ are called **abstract simplices**. The **dimensions** of X and $\mathcal{A}$ are $\dim X = \mathrm{card}\, X - 1$ and $\dim \mathcal{A} = \max\{\dim X \mid X \in \mathcal{A}\}$. The **vertex set** of $\mathcal{A}$ is $\mathrm{Vert}\, \mathcal{A} = \bigcup \mathcal{A}$. A **graph** is a 1-dimensional abstract simplicial complex. The **nerve** of a finite system A is

$$\mathrm{Nrv}\, A = \{X \subseteq A \mid \bigcap X \neq \emptyset\}.$$

$\mathrm{Nrv}\, A$ is an abstract simplicial complex with vertex set A. A **geometric realization** is a map $f : \mathrm{Vert}\, \mathcal{A} \to \mathbb{R}^d$ for which $\mathcal{K} = \{\mathrm{conv}\, f(X) \mid X \in \mathcal{A}\}$ is a simplicial complex. A useful result is that every abstract simplicial complex of dimension k has a geometric realization in $\mathbb{R}^d$ for some large enough d.

A **chain** is a collection of abstract simplices that can be ordered so that $X_0 \subset X_1 \subset \ldots \subset X_j$. The **order complex** is the system $\mathcal{C}$ of chains in $\mathcal{A}$; it is an abstract simplicial complex because every subset of a chain is again a chain. Let $\mathcal{K}$ be the simplicial complex defined by a geometric realization $f : \mathrm{Vert}\, \mathcal{A} \to \mathbb{R}^d$. Extend f to $g : \mathrm{Vert}\, \mathcal{C} \to \mathbb{R}^d$ by sending every abstract simplex $X \in \mathcal{A} = \mathrm{Vert}\, \mathcal{C}$ to its barycenter: $g(X) = b(\mathrm{conv}\, f(X))$. The **barycentric subdivision** of $\mathcal{K}$ is

$$\mathrm{Sd}\, \mathcal{K} = \{\mathrm{conv}\, g(C) \mid C \in \mathcal{C}\}.$$

$\mathrm{Sd}\, \mathcal{K}$ is indeed a simplicial complex and g is a geometric realization of the order complex.

Maps and equivalences. A function $f : \mathbb{X} \to \mathbb{Y}$ is **continuous** if the preimage of every open set in $\mathbb{Y}$ is open in $\mathbb{X}$. A **map** is a continuous function. A **homeomorphism** is a bijective map whose inverse is also continuous. $\mathbb{X}$ and $\mathbb{Y}$ are **homeomorphic** or **topologically equivalent**, denoted $\mathbb{X} \approx \mathbb{Y}$, if there is a homeomorphism between them. A **triangulation** of $\mathbb{X}$ is a simplicial complex $\mathcal{K}$ together with a homeomorphism $h : |\mathcal{K}| \to \mathbb{X}$. $\mathbb{X}$ is **triangulable** if it has a triangulation. An **embedding** is an injection $e : \mathbb{X} \to \mathbb{Y}$ whose restriction to the image, $e(\mathbb{X}) \subseteq \mathbb{Y}$, is a homeomorphism. A **retraction** of a topological space $\mathbb{X}$ to a subspace $\mathbb{Y}$ is a map $r : \mathbb{X} \to \mathbb{Y}$ with $r(y) = y$ for all $y \in \mathbb{Y}$. An **immersion** is a map $f : \mathbb{X} \to \mathbb{Y}$ such that every $x \in \mathbb{X}$ has a neighborhood on which f is injective.

A **homotopy** between two maps $f, g : \mathbb{X} \to \mathbb{Y}$ is a map

$$F : \mathbb{X} \times [0, 1] \to \mathbb{Y}$$

for which $F(x, 0) = f(x)$ and $F(x, 1) = g(x)$ for all $x \in \mathbb{X}$. f and g are **homotopic** if there exists a homotopy F between f and g. F **fixes** x if $F(x, t)$ is the same for all $t \in [0, 1]$. An **isotopy** between f and g is a homotopy F so that every $f_t : \mathbb{X} \to \mathbb{Y}$ defined by $f_t(x) = F(x, t)$ is a homeomorphism between $\mathbb{X}$ and its image. A **deformation retraction** of $\mathbb{X}$ to $\mathbb{Y} \subseteq \mathbb{X}$ is a homotopy between the identity $1_{\mathbb{X}} : \mathbb{X} \to \mathbb{X}$ and a retraction $r : \mathbb{X} \to \mathbb{Y}$ that fixes all $y \in \mathbb{Y}$.

Two topological spaces, $\mathbb{X}$ and $\mathbb{Y}$, are **homotopy equivalent**, denoted $\mathbb{X} \simeq \mathbb{Y}$, if there are maps $a : \mathbb{X} \to \mathbb{Y}$ and $b : \mathbb{Y} \to \mathbb{X}$ with $b \circ a$ homotopic to $1_{\mathbb{X}}$ and $a \circ b$

homotopic to $1_{\mathbb{Y}}$. For example, $\mathbb{X} \simeq \mathbb{Y}$ if there is a deformation retraction of $\mathbb{X}$ to $\mathbb{Y}$. Indeed, define $a = r$ and $b = 1_{\mathbb{Y}}$. Then $a \circ b = 1_{\mathbb{Y}}$ and the deformation retraction itself is a homotopy between $b \circ a$ and $1_{\mathbb{X}}$. $\mathbb{X}$ is **contractible** if it is homotopy equivalent to a point.

Topological spaces given in terms of simplicial complexes can be related via piecewise linear maps. Let $\mathcal{K}, \mathcal{L}$ be simplicial complexes. A **vertex map** is a function $v : \text{Vert}\,\mathcal{K} \to \text{Vert}\,\mathcal{L}$ with the property that if $\sigma = \text{conv}\,T$ is a simplex in $\mathcal{K}$ then $\text{conv}\, v(T)$ is a simplex in $\mathcal{L}$. A **simplicial map** $f : |\mathcal{K}| \to |\mathcal{L}|$ is the linear extension of a vertex map v:

$$f(x) = \sum_{p_i \in T} \phi_i \cdot v(p_i),$$

where $\sigma = \text{conv}\,T$ is a simplex in $\mathcal{K}$ that contains x and ϕ_i is the barycentric coordinate of x that corresponds to $p_i \in T$. $\mathcal{K}$ and $\mathcal{L}$ are **isomorphic** or **simplicially equivalent** if they permit a bijective vertex map v. In this case the linear extension is a simplicial homeomorphism between $|\mathcal{K}|$ and $|\mathcal{L}|$. $\mathcal{K}$ and $\mathcal{L}$ are **PL-equivalent** if there are simplicially equivalent subdivisions $\mathcal{K}'$ of $\mathcal{K}$ and $\mathcal{L}'$ of $\mathcal{L}$. It has been conjectured that two simplicial complexes are PL-equivalent if they triangulate the same topological space. This is known as the **Hauptvermutung** in combinatorial topology and has since been disproved for complexes of dimension 3 and higher [**Mi**]. The conjecture is correct for 3-dimensional manifolds and 2-dimensional triangulable spaces [**Mo**].

Fundamental group and covering spaces. Let $\mathbb{X}$ be a topological space. A **path** in $\mathbb{X}$ is a continuous map $p : [0,1] \to \mathbb{X}$. The path **joins** the **initial** point, $p(0)$, to the **terminal** point, $p(1)$. $\mathbb{X}$ is **path-connected** if every pair of points in $\mathbb{X}$ can be joined by a path. Two paths are **equivalent** if they are connected by a homotopy that leaves the common initial and terminal points fixed. The **inverse** of p is $p^{-1}(t) = p(1-t)$. The **product** of two paths p and q is defined if $p(1) = q(0)$:

$$(p \cdot q)(t) = \begin{cases} p(2t) & \text{if } 0 \le t \le \frac{1}{2}, \text{ and} \\ q(2t-1) & \text{if } \frac{1}{2} \le t \le 1 . \end{cases}$$

The equivalence relation and the product are compatible, that is, if p_1, q_1 and p_2, q_2 are two pairs of equivalent paths then $p_1 \cdot q_1$ and $p_2 \cdot q_2$ are equivalent. A path p is a **loop** if $p(0) = p(1) = x_0$, and the loop is **based** at x_0. For example, $p \cdot p^{-1}$ is a loop equivalent to the constant map $[0,1] \to x_0$. The equivalence classes of loops based at x_0 together with the product operation form a group called the **fundamental group** of $\mathbb{X}$ at x_0 and denoted $\pi(\mathbb{X}, x_0)$. For a path-connected space $\mathbb{X}$ any two groups $\pi(\mathbb{X}, x_0)$ and $\pi(\mathbb{X}, y_0)$ are isomorphic. Therefore, for a path-connected space $\mathbb{X}$ we have a unique fundamental group for the entire space, denoted $\pi(\mathbb{X})$. We note that for finite simplicial complexes and for manifolds the notions of path-connectedness and connectedness are the same and do not have to be distinguished. The fundamental group is invariant over homotopy equivalent spaces. $\mathbb{X}$ is **simply connected** if $\pi(\mathbb{X})$ is trivial. If $\mathbb{X}$ is contractible then it is simply connected. The reverse is not correct: for $d \ge 2$ the d-sphere is simply connected but not contractible.

Let $\mathbb{X}$ be a topological space. A **covering space** of $\mathbb{X}$ is a topological space $\mathbb{Y}$ together with a map $p : \mathbb{Y} \to \mathbb{X}$. p is called a **projection** and satisfies the following property: for each point $x \in \mathbb{X}$ there is a path-connected neighborhood U so that for each path-connected component V of $p^{-1}(U)$ the restriction of p to V

is a homeomorphism. A covering space $\mathbb{Y}$ of $\mathbb{X}$ is **universal** if $\mathbb{Y}$ is connected. Any two universal covering spaces of $\mathbb{X}$ are topologically equivalent. For example, $\mathbb{R}^1$ is a universal covering space of $\mathbb{S}^1$ and $\mathbb{R}^2$ is a universal covering space of the torus. An important property of covering spaces is that they lift maps. For example, if α is a path in a space $\mathbb{X}$ with covering space $\mathbb{Y}$ and projection p, and if $y_0 \in \mathbb{Y}$ is such that $p(y_0) = \alpha(0)$, then there exists a unique path α' in $\mathbb{Y}$ with initial point $\alpha'(0) = y_0$ and such that $p \circ \alpha' = \alpha$. We say the path α' is a **lifting** of the path α.

Manifolds. A topological space $\mathbb{M}$ is a d**-manifold** if every point $x \in \mathbb{M}$ has a neighborhood homeomorphic to $\mathbb{R}^d$. A **closed surface** is a compact 2-manifold. The long-standing **Poincaré conjecture** states that every simply connected 3-manifold is homeomorphic to the 3-sphere. In spite of a tremendous amount of effort there is still no proof and no counterexample [**Bi, Ha2**]. $\mathbb{M}$ is a d**-manifold with boundary** if every $x \in \mathbb{M}$ has a neighborhood homeomorphic to $\mathbb{R}^d$ or to the d-dimensional half-space, $\mathbb{H}^d$. The **boundary** of $\mathbb{M}$ is the set of points with neighborhoods homeomorphic to $\mathbb{H}^d$. The boundary of $\mathbb{M}$ is either empty or a $(d-1)$-manifold. All manifolds of dimension $d \leq 3$ are triangulable.

An **ordered** k**-simplex** is a k-simplex together with a permutation of its vertices, and we write

$$\sigma = [p_0, p_1, \ldots, p_k].$$

Two orderings have the same **orientation** if they differ by an even permutation. All simplices of dimension 1 or higher have two orientations. The orientation of a $(k-1)$-face **induced** by σ is

$$\tau = (-1)^i [p_0, \ldots, p_{i-1}, p_{i+1}, \ldots, p_k],$$

where a leading minus reverses orientation. Two k-simplices sharing a $(k-1)$-face τ are **consistently oriented** if they induce *different* orientations of τ. A triangulable d-manifold is **orientable** if all d-simplices in a triangulation can be ordered so all adjacent pairs are consistently oriented. Otherwise, the d-manifold is **non-orientable**. For example, the projective plane and the Klein bottle are two non-orientable closed surfaces, and the Möbius strip is a non-orientable 2-manifold with boundary. Non-orientable closed surfaces cannot be embedded in $\mathbb{R}^3$.

Homology groups and Betti numbers. We consider only simplicial homology with integer coefficients. Let $\mathcal{K}$ be a simplicial complex and assign to each simplex an arbitrary but fixed ordering of its vertices. $\mathcal{K}^j \subseteq \mathcal{K}$ is the set of ordered j-simplices. A j**-chain** is a function $c : \mathcal{K}^j \to \mathbb{Z}$. We write the chain as a formal sum:

$$c = \sum_{\sigma \in \mathcal{K}^j} c(\sigma) \cdot \sigma.$$

Two j-chains are added componentwise. The j-chains together with addition form the **group of** j**-chains**, C_j. The **boundary operator** maps an ordered j-simplex $\sigma = [p_0, p_1, \ldots, p_j]$ to a $(j-1)$-chain

$$\partial_j \sigma = \sum_{i=0}^{j} (-1)^i [p_0, \ldots, p_{i-1}, p_{i+1}, \ldots, p_j].$$

The **boundary homomorphism**, $\partial_j : \mathsf{C}_j \to \mathsf{C}_{j-1}$, is defined by $\partial_j c = \sum c(\sigma) \cdot \partial_j \sigma$. The **chain complex** is the sequence of chain groups connected by boundary homomorphisms:

$$\dots \xrightarrow{\partial_{j+2}} \mathsf{C}_{j+1} \xrightarrow{\partial_{j+1}} \mathsf{C}_j \xrightarrow{\partial_j} \mathsf{C}_{j-1} \xrightarrow{\partial_{j-1}} \dots$$

The **image** and **kernel** of the boundary homomorphism are

$$\begin{aligned} \operatorname{Im} \partial_j &= \{\partial_j c \mid c \in \mathsf{C}_j\}, \\ \operatorname{Ker} \partial_j &= \{c \in \mathsf{C}_j \mid \partial_j c = 0\}. \end{aligned}$$

$c \in \mathsf{C}_j$ is a ***j*-cycle** if $c \in \operatorname{Ker} \partial_j$ and it is a ***j*-boundary** if $c \in \operatorname{Im} \partial_{j+1}$. A j-boundary is sometimes called a **null-homologous** j-cycle. The j-boundaries form a subgroup of the group of j-cycles which form a subgroup of the group of j-chains: $\mathsf{B}_j \subseteq \mathsf{Z}_j \subseteq \mathsf{C}_j$. The fact that every j-boundary is also a j-cycle is important and follows from $\partial_j \partial_{j+1} c = 0$ for every $c \in \mathsf{C}_{j+1}$. The ***j*-th homology group** is the quotient of the cycle group over the boundary group:

$$\mathsf{H}_j = \mathsf{Z}_j | \mathsf{B}_j.$$

Homology groups are finitely generated and abelian. The fundamental theorem on such groups implies H_j can be written as the direct sum of two subgroups, $\mathbb{Z}^{\beta_j}$ and T. β_j is the **rank** of H_j. T is the direct sum of finitely many cyclic groups, $\mathbb{Z}_k$, and is referred to as the **torsion subgroup** of H_j.

The above seemingly roundabout construction is justified by the fact that H_j is invariant over all simplicial complexes triangulating the same topological space $\mathbb{X} \approx |\mathcal{K}|$. It is therefore meaningful to call H_j the j-th homology group of $\mathbb{X}$. Similarly, the ***j*-th Betti number**, β_j, and the **Euler characteristic**,

$$\chi = \sum (-1)^j \beta_j,$$

are invariant over all triangulations of $\mathbb{X}$. For example the Euler characteristics of the d-sphere and the d-ball are

$$\begin{aligned} \chi &= 1 + (-1)^d, \\ \chi &= 1. \end{aligned}$$

Indeed, the only non-zero Betti numbers of $\mathbb{S}^d$ are $\beta_0 = \beta_d = 1$ and the only non-zero Betti number of $\mathbb{B}^d$ is $\beta_0 = 1$. The Euler characteristic of every compact d-manifold with odd d is 0.

DEPARTMENT OF COMPUTER SCIENCE AND ENGINEERING, INDIAN INSTITUTE OF TECHNOLOGY, KHARAGPUR 721302, INDIA.

E-mail address: dey@cse.iitkgp.ernet.in

DEPARTMENT OF COMPUTER SCIENCE, UNIVERSITY OF ILLINOIS AT URBANA-CHAMPAIGN, URBANA, ILLINOIS 61801, USA.

E-mail address: edels@cs.uiuc.edu

DEPARTMENT OF ELECTRICAL ENGINEERING AND COMPUTER SCIENCE, UNIVERSITY OF WISCONSIN AT MILWAUKEE, MILWAUKEE, WISCONSIN 53201, USA.

E-mail address: guha@sprecher.cs.uwm.edu

Contemporary Mathematics
Volume **223**, 1999

Recent Progress on Packing and Covering

Gábor Fejes Tóth

ABSTRACT. We survey the theory of packing and covering with special emphasis on the developments in the last ten years.

Introduction

In this survey we give an account of the progress of the theory of packing and covering in the past ten years. This theory constitutes the oldest chapter of Discrete and Computational Geometry. It originated as an offspring of number theory and crystallography at the turn of the century and developed into an independent theory in the fifties. It was then that the term Discrete Geometry was introduced as a synonym for problems concerning packing, covering and their special case, tiling. About ten years ago, the words Discrete Geometry gained a new, broader meaning in the compound Discrete and Computational Geometry which includes besides Computational Geometry not just problems concerning packing, covering and tiling, but also the theory of polytopes, as well as combinatorial geometry both in the sense of Hadwiger and Debrunner, and Erdős. The new terminology reflects new trends: besides the classical theory of packing and covering densities, which we discuss in Part I, there has been an increasing amount of research on packing and covering influenced by combinatorics or computer science. We mention some of these new aspects in Part II. Our treatment is by no means comprehensive. Readers who are interested in a more thorough treatment of the subject are referred to [**CS, FK2, FK3, Fe, Fe2, PA, R5**].

I Classical theory

1. Basic concepts

A *packing* is a family of sets whose interiors are mutually disjoint, and a *covering* is a family of sets whose union is the whole space. Packings and coverings can be defined in various spaces. In this paper we restrict our attention to the d-dimensional Euclidean space E^d. Moreover, we shall consider only packings and coverings of *convex bodies*, that is compact convex sets with nonempty interior. The volume,

1991 *Mathematics Subject Classification.* Primary 52C15, 52C17; Secondary 5202, 52B55, 52C07, 52C10.

Key words and phrases. Packing, covering, lattice-arrangements, density bounds, geometric algorithms, combinatorial geometry.

© 1999 American Mathematical Society

boundary, interior and convex hull of K are denoted by $V(K)$, $\mathrm{bd}(K)$, $\mathrm{int}(K)$ and $\mathrm{conv}(K)$, respectively. When $d = 2$, that is on the plane, we use the terms *convex disc* instead of convex body and *area*, denoted by $a(K)$, instead of volume. The *Minkowski sum* $K + L$ of the sets K and L consists of all points $x + y$, $x \in K$, $y \in L$. The set $K - K = K + (-K)$ is called the *difference body* of K. B^d denotes the unit ball centered at the origin in E^d.

The basic problems in the theory of packing and covering are the determination of the most efficient packings and coverings of congruent replicas of a given set K in the space. The usual measure of the efficiency of an arrangement is density.

Let $\mathcal{A}$ be an arrangement (family of sets with finite volume) and D a set with finite volume. The *inner density* $d_{\mathrm{inn}}(\mathcal{A}|D)$ and *outer density* $d_{\mathrm{out}}(\mathcal{A}|D)$ *of* $\mathcal{A}$ *relative to* D are defined by

$$d_{\mathrm{inn}}(\mathcal{A}|D) = \frac{1}{V(D)} \sum_{A \in \mathcal{A}, A \subset D} V(A)$$

and

$$d_{\mathrm{out}}(\mathcal{A}|D) = \frac{1}{V(D)} \sum_{A \in \mathcal{A},\, A \cap D \neq \emptyset} V(A).$$

The *lower* and *upper density of* $\mathcal{A}$ are given by the limits

$$d_-(\mathcal{A}) = \liminf_{\lambda \to \infty} d_{\mathrm{inn}}(\mathcal{A}|\lambda B^d)$$

and

$$d_+(\mathcal{A}) = \limsup_{\lambda \to \infty} d_{\mathrm{out}}(\mathcal{A}|\lambda B^d).$$

If $d_-(\mathcal{A}) = d_+(\mathcal{A})$, then we call the common value *density of* $\mathcal{A}$ and denote it by $d(\mathcal{A})$. It is easily seen that these quantities are independent of the choice of the origin.

The *packing density* $\delta(K)$ and the *covering density* $\vartheta(K)$ of a convex body (or more generally of a measurable set) K are defined by

$$\delta(K) = \sup\{d_-(\mathcal{P});\ \mathcal{P} \text{ is a packing of } E^d \text{ with congruent copies of } K\}$$

and

$$\vartheta(K) = \inf\{d_+(\mathcal{C});\ \mathcal{C} \text{ is a covering of } E^d \text{ with congruent copies of } K\}.$$

If we restrict the arrangements to those which consist of parallel translates of K, then we obtain the similarly defined notions of *translational packing density* and *translational covering density* of K denoted by $\delta_T(K)$ and $\vartheta_T(K)$. By further restriction to lattice arrangement we obtain the *lattice packing density* $\delta_L(K)$ and *lattice covering density* $\vartheta_L(K)$ of K. A *lattice* is the set of all integer linear combinations of a particular basis of E^d and a *lattice arrangement* is the set of translates of a given set in E^d by all vectors of a lattice. It is easy to see that in the definition of $\delta_L(K)$ and $\vartheta_L(K)$ we can take maximum and minimum instead of supremum and infimum. By a theorem of Groemer **[Gro2]** the same holds for the translational and for the general packing and covering densities.

2. The sphere packing problem

Indubitably, of all the problems concerning packing or covering it was the sphere packing problem, that is the determination of the packing density of the three-dimensional ball B^3, which attracted the most attention in the past decade. This is one of the oldest unsolved mathematical problems, which can be traced back as far as to a work of Kepler. In his treatise "On the Six-Cornered Snowflake" [**Ke**] Kepler described a packing of equal balls which in modern terminology is called the face centered cubic lattice. It consists of layers of balls arranged in a square lattice so that balls of each layer fit in the holes of the neighboring layers. Kepler also observed that this arrangement can be built also from hexagonal layers. Then he wrote: "The packing will be the tightest possible, so that in no other arrangement could more pellets be stuffed into the same container". This packing has density $\pi/\sqrt{18} = 0.74048\ldots$ and Kepler's statement quoted above has been interpreted as the conjecture that this is the maximum possible density of a packing of equal balls, that is, $\delta(B^3) = \pi/\sqrt{18}$. The sphere packing problem is mentioned as part of Problem 18 in Hilbert's famous collection of unsolved problems [**Hi**].

Experts consider the sphere packing problem to be extremely difficult, but they agree that the difficulties are mostly of a technical nature. Already in the early fifties, L. Fejes Tóth [**Fe**] gave a strategy which, if it can be carried out successfully, might reduce the problem to establishing a finite number of inequalities of the form $f(x) \geq c$, where f is a function satisfying Lipschitz conditions on a compact set, and some years later [**Fe2**] he pointed out that this job could be done by computers. But considering the great number of variables of the functions involved, even modern computers are unable to carry out his program.

In 1990 W.Y. Hsiang announced the solution of the sphere packing problem. His first preprint of about 20 pages [**Hs**] contained serious errors and difficult statements without proof. Disregarding criticism, Hsiang maintained his claim of having a proof. He produced subsequent preprints and an extended abstract [**Hs2, Hs3, Hs4, Hs5**] giving more details, modifying slightly his method and correcting some errors. But, unfortunately, even his final publication [**Hs6**] cannot be accepted as a proof: It needs further modification and much more detail. His work has to be considered one of the most detailed strategies for a proof published so far. We should note that Hsiang's strategy is very similar to the one proposed by L. Fejes Tóth. The reader interested in the controversy about Hsiang's paper is referred to [**Bez, C, CHMS, F2, H3, Hs7, Mu2, St**].

Different strategies, without the claim of being a proof, were given by Hales [**H, H2, H4, H5**]. While Hsiang investigates the Dirichlet cells of the balls, Hales' first strategy is based on the analysis of the Delone decomposition which is dual to the decomposition by Dirichlet cells. In his latest papers Hales combined Dirichlet and Delone cells. He divided the problem into five parts which he thinks are approximately equally difficult. He solved two of the subproblems [**H4**] and expected that the rest of the work can be done within some years. After further modification of the method [**FH**], two additional sub-cases could be settled by Hales [**H6**] and by his student Ferguson [**Fer**].

Recapitulating, the sphere packing problem is still unsolved, but there is hope that it will be settled in the near future. We cannot foresee whether it will be Hsiang's, Hales' or a third strategy which leads to a solution. On the other hand, we can be almost sure that the solution will be several hundreds pages long and will rely on an essential amount of computer work. Rogers wrote in [**R3**] that "many mathematicians believe, and all physicists know" that $\delta(B^3) = \pi/\sqrt{18}$. It is doubtful if the physicists' knowledge will be consolidated after the publication of such a proof.

We conclude this section by summarizing the known results about the sphere packing problem. Already Gauss [**G**] showed that $\delta_L(B^3) = \pi/\sqrt{18}$. The strongest result establishing Kepler's conjecture for a restricted class of packings is due to A. Bezdek, W. Kuperberg and E. Makai Jr. [**BKM**]. They proved the conjecture for packings consisting of parallel infinite strings of balls. (A *string of balls* is a collection of congruent balls whose centers are collinear and such that each of them touches two others). The best known upper bound on $\delta(B^3)$ is

$$\delta(B^3) \leq 0.773055$$

given by Muder [**Mu**]. We also know that the face centered cubic lattice is a locally optimal arrangement [**BBC, DZ, H**].

3. Some exact results

A convex body is called a *space-filler* if there exists a tiling with congruent copies of it. Obviously, we have $\delta(K) = \vartheta(K) = 1$ for any space-filler. Until recently, for $d \geq 3$, no other convex body in E^d was known for which the packing or covering density could be determined. The first three-dimensional non space-filler for which the packing density could be determined was a non-compact set. A. Bezdek and W. Kuperberg [**BK**] showed that circular cylinders of infinite length in both directions cannot be packed more densely than their bases. The obvious conjecture that this statement holds for circular cylinders of any finite length, is still open. We note that the definitions in Section 1 do not apply for non-compact sets, but they readily can be modified so as to include such sets as well.

When Pach took notice of the result of A. Bezdek and W. Kuperberg he posed the following provocative problem [**MP**, page 44]:

Give an example of a bounded convex body K which is not a space-filler but whose packing density $\delta(C)$ can be exactly determined.

A candidate for such a body appears to be a cube with one corner chopped off. It has been a widespread conjecture that if Q is a cube and $\widetilde{Q}$ is the body obtained from Q by chopping off a sufficiently small tetrahedron at a vertex of Q, then $\delta(\widetilde{Q}) = V(\widetilde{Q})/V(Q)$. As was pointed out by Makai [**M**], this is not the case. For, it follows from results of Ollerenshaw and Mahler (see Theorem 7 and the note after Theorem 4 in Chapter 4, §27 in [**GL**]), that a cube is irreducible, that is, it does not contain any proper subset with the same critical determinant. On the other hand, A. Bezdek [**Be**] proved a similar statement for the rhombic dodecahedron: Let D be a rhombic dodecahedron and $\widetilde{D}$ the body obtained from D by chopping off a sufficiently small tetrahedron at a trihedral vertex of D. Then $\delta(\widetilde{D}) = V(\widetilde{D})/V(D)$.

The packing density is not determined for any set which is not a space-filler in dimensions greater than 3. More surprisingly, in dimensions greater than 1, the circle is the only set which is not a space-filler and for which the covering density is known.

4. Density bounds

In high dimensions the only realistic goal is to give bounds for the packing and covering densities. This is the subject of Rogers' classical book "Packing and Covering" [**R5**]. In the introduction to his book Rogers wrote: "I have the impression that most of the simplest general results of the subject have already been discovered, and further progress may be rather slow, depending on detailed and complicated technical developments". He added: "If I am proved wrong in this I shall be happy". Unfortunately, Rogers' forecast turned out to be very accurate. Thus, when giving an account of the presently known best results we shall mention several ones which were known already in 1964, the year in which Rogers' book appeared.

Concerning the existence of reasonably economical packings the Theorem of Minkowski-Hlawka states that

$$\delta_L(K) \geq \zeta(d)/2^{d-1}$$

for every centrally symmetric convex body K in E^d, where $\zeta(d) = \sum\limits_{k=1}^{\infty} k^{-d}$.

There are slight refinements of the Theorem of Minkowski-Hlawka, the best due to Schmidt [**S2**] who established the inequality

$$\delta_L(K) \geq \frac{cd}{2^d}$$

for every centrally symmetric convex body K in E^d and sufficiently large d, provided $c < \ln 2$.

We recall the observation of Minkowski according to which an arrangement of translates of a convex body K is a packing if and only if the corresponding translates of $\frac{1}{2}(K-K)$ form a packing. It follows that

$$\delta_T(K) = 2^d \delta_T(K-K)\frac{V(K)}{V(K-K)} \qquad \text{and} \qquad \delta_L(K) = 2^d \delta_L(K-K)\frac{V(K)}{V(K-K)}$$

for every convex body K in E^d. Combining this with the bound

$$\frac{V(K-K)}{V(K)} \leq \binom{2d}{d}$$

by Rogers and Shephard (see [**R3**, p. 37]), it easily follows that

$$\delta_T(K) \geq \delta_L(K) \geq c\frac{d^{3/2}}{4^d}$$

for each convex body K in E^d, where c is a suitable constant, independent of K and d.

There has been some important recent progress concerning the existence of economical packings. The known proofs of the Theorem of Minkowski-Hlawka, as well as of its refinements, use some mean value argument. In 1987 Rush and Sloane [**RS**] gave a construction based on codes which produces lattice packings of balls whose density comes very close to the Minkowski-Hlawka bound and improves on that bound for d-dimensional *superballs* defined by $\sum_{i=1}^{d} |x_i|^\sigma \leq 1$ when σ is an integer greater than 2. Later, using codes fitted to the shape of the body to be packed, Rush [**Ru**] constructed for every d-dimensional convex body K which is symmetric through each of the coordinate hyperplanes a lattice packing whose density is at least $2^{-d+o(d)}$. In [**Ru2**] he gets an essential improvement of the Minkowski-Hlawka bound for a large class of convex bodies.

Ball [**B**] used a completely new method for proving the existence of dense lattice ball packings. Through a variational argument he established

$$\delta_L(B^d) \geq 2\zeta(d)\frac{d-1}{2^d},$$

which is the presently best known lower bound for $\delta_L(B^d)$.

Most of the best known upper bounds for covering densities are due to Rogers [**R2, R4**]. He proved

$$\vartheta_T(K) \leq d \ln d + d \ln \ln d + 5d$$

and

$$\vartheta_L(K) \leq d^{\log_2 \ln d + c}$$

for an arbitrary convex body K in E^d, while for $\vartheta_L(B^d)$ he established the upper bound

$$\vartheta_L(B^d) \leq cd(\ln d)^{\log_2 \sqrt{2\pi e}}.$$

Gritzmann [**Gr**] obtained a similar bound for a large class of convex bodies:

$$\vartheta_L(K) \leq cd(\ln d)^{1+\log_2 e}$$

for every convex body K in E^d which has an affine image symmetric about at least $\log_2 \ln d + 4$ coordinate hyperplanes.

All the methods establishing the existence of dense packings in E^d for $d \geq 3$ rely on the theory of lattices and thus the best lower bound for $\delta(K)$ and $\delta_T(K)$ are the same as for $\delta_L(K)$. Only in the plane could the apparent advantage of taking arbitrary congruent copies instead of translates or lattice translates be exploited. G. Kuperberg and W. Kuperberg [**KK**] showed that

$$\delta(K) \geq \frac{1}{2}\sqrt{3}$$

for every convex disk K. Refining a method of W. Kuperberg [**Ku**] Ismailescu [**I**] established the upper bound

$$\vartheta(K) \leq 1.2281771\ldots$$

for every convex disk K.

The lower bound

$$\vartheta(B^d) \geq \frac{d}{e^{3/2}} + o(d)$$

by Coxeter, Few and Rogers [**CFR**] compares favorably with the opposite bound by Rogers. The situation is less satisfactory in the case of packing. In 1929 Blichfeldt proved the inequality

$$\delta(B^d) \leq \frac{d+2}{2}2^{-d/2}.$$

Rankin and Rogers improved Blichfeldt's bound by constant factors, but almost fifty years had to elapse until the bound was improved in the exponential order of magnitude first by Sidelnikov, later by Levenštein and finally by Kabatjanskiĭ and Levenštein [**KL**] who established the presently best known asymptotic bound

$$\delta(B^d) \leq 2^{-(0.599+o(1))d} \qquad (\text{as} \quad d \to \infty).$$

A central problem, is to find the limits

$$\lim_{d\to\infty} \frac{1}{d} \log_2 \delta(B^d)$$

and

$$\lim_{d\to\infty} \frac{1}{d} \log_2 \delta_L(B^d),$$

or more precisely, as we don't even know whether the limits exist, to find

$$\liminf_{d\to\infty} \frac{1}{d} \log_2 \delta(B^d), \quad \limsup_{d\to\infty} \frac{1}{d} \log_2 \delta(B^d), \quad \liminf_{d\to\infty} \frac{1}{d} \log_2 \delta_L(B^d)$$

and $\limsup_{d\to\infty} \frac{1}{d} \log_2 \delta_L(B^d)$. It is also a fundamental question, whether the asymptotic behavior of $\delta(B^d)$ as $d \to \infty$ is the same as that of $\delta_L(B^d)$.

According to a result of Schmidt [**S**] we have $\delta(K) < 1$ and $\vartheta(K) > 1$ for every smooth convex body K. However, Schmidt's method does not allow the derivation of explicit bounds for any given body. We mention the following tantalizing questions:

One can conjecture that there are constants $c < 1$ and $C > 1$, independent of d, such that $\delta(E) \leq c$ and $\vartheta(E) \geq C$ for every ellipsoid E in E^d. Find such constants.

Let T denote a regular tetrahedron in E^3. Since the face angle of T is not commensurable with π, T is not a space-filler. As we will see in the next section, this implies that $\delta(T) < 1 < \vartheta(T)$. Give explicit non-trivial bounds for $\delta(T)$ and $\vartheta(T)$.

By a modification of Blichfeldt's method we can get a quite reasonable upper bound for the packing density of some special bodies K, *e.g* of cylinders, [**FK**] but no non-trivial bound is known for $\vartheta(K)$ for any K other than a ball.

5. Connection between the different packing and covering densities

We have obviously

$$\delta_L(K) \leq \delta_T(K) \leq \delta(K) \leq 1 \leq \vartheta(K) \leq \vartheta_T(K) \leq \vartheta_L(K).$$

It is a fundamental problem to characterize the case of equality in these inequalities. Only the characterization of convex bodies with packing and covering density equal 1, is settled. Schmidt [**S**] showed, without explicitly stating, that the three statements, $\delta(K) = 1$, $\vartheta(K) = 1$ and K is a space-filler, are equivalent. We recall a theorem of Venkov [**V**] (see also [**Mc**]) according to which any convex body which tiles E^d by translates is a polytope and admits a face-to-face lattice tiling of E^d. Such sets are called *parallelohedra*. Thus the five statements, $\delta_L(K) = 1$, $\delta_T(K) = 1$, $\vartheta_L(K) = 1$, $\vartheta_T(K) = 1$ and K is a parallelohedron, are also equivalent.

Rogers [**R**] proved that $\delta_L(K) = \delta_T(K)$ for every convex disc in the plane. It is conjectured that an analogous statement holds for the corresponding covering densities. It is also conjectured that the situation is different in high dimensions and convex bodies K in E^d for which $\delta_T(K) = \delta_L(K)$ or $\vartheta_T(K) = \vartheta_L(K)$ are rather exceptional. Specifically, Rogers [**R5**, p. 15] conjectures that $\delta_L(B^d) < \delta_T(B^d)$ for all sufficiently large d. Non-lattice packings of balls in E^{10}, E^{11} and E^{13} that are denser than any known lattice packing, give some support to Rogers' conjecture (see [**CS**, Chapter 5]). But, quite surprisingly, no example of a convex body K is known for which $\delta_L(K) < \delta_T(K)$ or $\vartheta_L(K) > \vartheta_T(K)$. On the other hand, there are some recent results which could exploit the advantage of taking arbitrary arrangements of congruent copies of a body instead of just lattice arrangements.

A. Bezdek and W. Kuperberg [**BK2**] showed that for $d \geq 3$ there are ellipsoids E in E^d for which

$$\delta_L(E) < \delta(E).$$

A similar, but more general theorem holds for coverings [**FK4**]: For $d \geq 3$ every strictly convex body K in E^d has an affine image K' such that

$$\vartheta(K') < \vartheta_L(K').$$

In particular, there is an ellipsoid E in E^3 for which $\vartheta(E) < 1.394$. We note that according to the result of Coxeter, Few and Rogers mentioned in the previous section we have

$$\vartheta_T(E) = \vartheta(B^3) \geq \frac{3\sqrt{3}}{2}(3\,\mathrm{arcsec}\ 3 - \pi) > 1.394.$$

Thus, we have

$$\vartheta(E) < \vartheta_T(E)$$

for this ellipsoid.

The discs K in the plane for which $\delta(K) = \delta_L(K)$ or $\vartheta(K) = \vartheta_L(K)$ are exceptional in the following sense [**F, FZ**]: Let $\mathcal{K}$ be the space of all convex discs in the plane equipped with the Hausdorff metric. Let $\mathcal{L}_p$ and $\mathcal{L}_c$ be the subsets of $\mathcal{K}$ consisting of all convex discs K for which $\delta(K) = \delta_L(K)$ and $\delta(K) = \delta_L(K)$, respectively. Then $\mathcal{L}_p$ and $\mathcal{L}_c$ are nowhere dense in $\mathcal{K}$. (A set is *nowhere dense*, if its closure has empty interior.) It is conjectured that analogous statements hold for higher dimensions as well.

It is an interesting question whether among the arrangements realizing the optimal densities $\delta(K)$ and $\vartheta(K)$ there exist some with specific regularity conditions. In particular, can the optimal densities be realized by *periodic arrangements*, that is by arrangements arising from a finite set of congruent copies of K by translations through the vectors of a full dimensional lattice? A body K is called *packing aperiodic* if there is no periodic packing of congruent copies of K with density $\delta(K)$, and K is *covering aperiodic* if no periodic covering with congruent copies of K has density $\vartheta(K)$. We do not know whether a convex disc exists which is packing aperiodic or covering aperiodic. On the other hand, Schmitt [**Sc2**] constructed a packing aperiodic star-shaped disc, which is the union of two convex discs. With the same method one can construct a covering aperiodic star-shaped disc as well. Schmitt [**Sc**] also constructed a star-shaped polytope in E^3 which tiles E^3 but no tiling with its replicas is periodic. With a slight modification of Schmitt's construction, Conway produced a convex space-filler in E^3 which admits only non-periodic tilings if no mirror-image copies of the tiles are allowed (see [**SS**]).

II New trends

6. Algorithmic aspects

We mentioned in Section 2 that with high probability the sphere packing problem can be solved by a computer. Several other questions concerning packing and covering can be reduced to a finite optimization problem. It seems that it was Linhart who first adopted the algorithmic point of view for problems on packing and covering. An arrangement $\mathcal{A}$ is called *k-fold packing* if each point of the space belongs to the interior of at most k members of $\mathcal{A}$ and it is called *k-fold covering* if each point of the space belongs to the closure of at least k members of $\mathcal{A}$. Linhart [**L**] gave algorithms approximating the maximum density $\delta_L^k(B^2)$ of a k-fold lattice packing of circles and the minimum density $\vartheta_L^k(B^2)$ of a k-fold lattice covering by circles with arbitrary accuracy. He implemented his algorithms for $k \leq 20$. His method did not determine the exact values of $\delta_L^k(B^2)$ and $\vartheta_L^k(B^2)$, nor the corresponding optimal arrangements. Other methods, which fulfill this job as well, were developed by Temesvári, Horváth and Yakovlev [**THY**] (for k-fold packings) and by Temesvári [**T**] (for k-fold coverings). These methods reduce both problems to the determination of the optima of finitely many well defined functions of one variable. The proofs readily offer algorithms finding the best arrangements, however the authors did not try to implement them.

Mount and Silverman [**MS**] gave algorithms determining certain packing and covering densities of convex n-gons. Their results are based on the following results of L. Fejes Tóth (see [**Fe2**, page 163]):

$$\delta(K) \leq \frac{a(K)}{H(K)}$$

and

$$\vartheta_T(K) \geq \frac{a(K)}{h(K)},$$

where $H(K)$ denotes the minimum area of a hexagon containing a convex disc K and $h(K)$ denotes the maximum area of a hexagon contained in K. Both bounds are best possible for centrally symmetric discs, and they imply for such discs that

$$\delta(K) = \delta_T(K) = \delta_L(K) = \frac{a(K)}{H(K)}$$

and

$$\vartheta_T(K) = \vartheta_L(K) = \frac{a(K)}{h(K)}.$$

Thus the problems of determining $\delta(K)$ and $\vartheta_L(K)$ for a centrally symmetric convex disc are reduced to the determination of $H(K)$ and $h(K)$, respectively.

Let K be a centrally symmetric convex disc. A *central parallelogram* in K is a parallelogram one of whose vertices is the center of symmetry of K and whose remaining vertices lie on the boundary of K. One can show that the minimum of the area of a central parallelogram in K is $H(K)/4$. Mount and Silverman gave algorithms determining the minimum area central parallelogram in a centrally symmetric convex n-gon P, and thereby a densest packing of P, in $O(n)$ time. We recall from Section 4 that

$$\delta_T(K) = \frac{4a(K)}{a(K-K)}\delta_T(K-K)$$

for all convex discs. Since the number of sides of the difference body of an n-gon is at most $2n$, the translational packing density of an n-gon can also be determined in $O(n)$ time.

Turning to covering, we observe that, for a centrally symmetric convex disc K, we have $h(K) = 2t(K)$, where $t(K)$ denotes the maximum area of a triangle contained in K. Since there exists an $O(n)$ time algorithm determining $t(P)$ for every n-gon P [**DS**], the translational covering density of a centrally symmetric n-gon can be found in linear time as well.

We mention some further results where the algorithmic aspect plays an important role. Recently Peikert [**Pe**] described a heuristic algorithm for the determination of the densest packings of n congruent circles in a square. He did not prove that his algorithm always provides the optimal arrangement in finite time, but he implemented the method successfully for $n \leq 20$.

Let r_n be the maximum number such that n caps of radius r_n can form a packing on S^2 and let R_n be the minimum number such that n caps of radius R_n can cover S^2. The value of r_n is known for $n \leq 12$ and $n = 24$ and R_n has been determined for $n \leq 10$, $n = 12$ and $n = 14$ (see [**Fe**]). There is a great amount of literature dealing with the construction of good circle packings and circle coverings on the sphere. The ad hoc methods of the earlier papers have recently been replaced by different computer algorithms, but none of them has been shown to give the optimum. In [**HSS**] conjecturally best circle packings and circle coverings are given for $n \leq 130$. In addition, for $n \leq 55000$ good arrangements with icosahedral symmetry have been constructed.

7. Combinatorial aspects

There are several problems on packing and covering with a distinct combinatorial flavor. Very often, the solution of such problems relies on combinatorial methods. A typical example is the following nice theorem of Bárány, Füredi and Pach [**BFP**]:

In a packing of circles in which each circle has at least 6 neighbors either all circles are congruent or arbitrarily small circles must occur.

Two circles in a packing are *neighbors* if their boundaries have a common point. Of course, if in a packing with the above property all the circles are congruent, then the packing is uniquely determined: it is the well known hexagonal lattice packing.

We sketch how this theorem follows from a combinatorial theorem stating that a convex function on a locally finite graph is either constant or not bounded above. A real valued function defined on the vertices of a graph is *convex* if the value of the function at any vertex does not exceed the average of the values at the neighboring vertices. A graph is *locally finite* if the valence of each vertex is finite. The *nerve graph* of a circle packing is the graph whose vertices are the centers of the circles and two vertices are joined by an edge if the corresponding circles touch each other.

Let $\mathcal{P}$ be a packing of circles such that each circle in $\mathcal{P}$ has at least 6 neighbors. Let $r(C)$ denote the radius of the circle $C \in \mathcal{P}$. Besides the combinatorial theorem mentioned above, we need the following simple geometric inequality. If $C_0, C_1, \dots, C_k$ are circles from $\mathcal{P}$ such that $C_1, \dots, C_k$ ($k \geq 6$) are tangent to C_0, then

$$\frac{1}{k}\sum_{i=1}^{k}\frac{1}{r(C_i)} \geq \frac{1}{r(C_0)}.$$

Now, if all the circles of $\mathcal{P}$ are congruent, or if the nerve of $\mathcal{P}$ is not locally finite, then there is nothing to be proved. Otherwise, there are two neighboring elements of $\mathcal{P}$, say C_1 and C_2, such that $r(C_1) > r(C_2)$. Using the inequality above, it is easily seen that the function

$$f(C) = \left(\frac{1}{r(C)} - \frac{1}{r(C_1)}\right) \Big/ \left(\frac{1}{r(C_2)} - \frac{1}{r(C_1)}\right)$$

defined on the vertices of the nerve of $\mathcal{P}$ is convex. Since it is not constant, it is not bounded above. It follows that there are arbitrarily small circles in $\mathcal{P}$.

Multiple arrangements give rise to another combinatorial type problem: Decompose a k-fold packing or a k-fold covering into possibly few simple arrangements. Pach [**P**] showed that any double packing with positively homothetic copies of a convex disc can be decomposed into 4 simple packings. Further, if $\mathcal{P}$ is a k-fold packing with convex discs such that for some integer L the inradius $r(K)$ and the area $a(K)$ of each member K of $\mathcal{P}$ satisfies the inequality $9k\pi r^2(K)/a(K) \leq L$, then $\mathcal{P}$ can be decomposed into L simple packings.

Concerning the decomposition of multiple coverings we know that to any centrally symmetric polygon P and positive integer r there exists an integer $k = k(P, r)$ such that every k-fold covering with translates of P can be decomposed into r coverings [**P2**]. Also any 33-fold covering with congruent circles can be decomposed into two coverings [**M-LP**]. In 3-space results analogous to the above two theorems do not hold.

Recall the classical theorem of Erdős and Szekeres [**PA**, Theorem 9.20] stating for every natural number $n \geq 4$ the existence of a smallest natural number $f(n)$ with the property that any set of points in general position in the plane with cardinality greater than $f(n)$ contains the vertices of a convex n-gon. Erdős and Szekeres showed that

$$2^{n-2} \leq f(n) \leq \binom{2n-4}{n-2}.$$

It is conjectured that the lower bound is sharp for all n. The upper bound was not improved for more than sixty years. Recently, Chung and Graham [**CG**] managed to improve it by 1. Shortly after, Kleitman and Pachter [**KP**] showed that $f(n) \leq \binom{2n-4}{n-2} + 6 - 2n$. A further improvement was achieved by Tóth and Valtr [**TV**] who presented the presently known best upper bound, namely

$$2^{n-2} \leq f(n) \leq \binom{2n-5}{n-2} + 1.$$

We call a family $\mathcal{F}$ of sets in the plane *convexly independent* if no member of $\mathcal{F}$ is contained in the convex hull of some other members of $\mathcal{F}$. With this terminology, the theorem of Erdős and Szekeres can be formulated as follows.

A set of points in which every three, but no n, points are convexly independent contains at most $f(n)$ elements.

It is natural to ask whether this reformulation of the theorem of Erdős and Szekeres remains true instead of sets of points for families of sets. Concerning this problem the following theorem holds [**BF**]:

For every integer $n \geq 4$ there exists a smallest natural number $g(n)$ such that the cardinality of every family of mutually disjoint convex discs in the plane with the property that any three, but no n, members of the family are convexly independent, is at most $g(n)$.

Obviously, we have $f(n) \leq g(n)$. The value of $f(n)$, as well as that of $g(n)$, is known only for $n = 4$ and $n = 5$. We have $f(4) = g(4) = 4$ and $f(5) = g(5) = 8$ (see [**BF2**]). It is an open question, whether $f(n) = g(n)$ for all $n \geq 4$.

We note that the assumption in the theorem that the members of the family constitute a packing is essential. The following simple example shows that for every natural number n there exist n (intersecting) convex sets $C_1, \ldots, C_n$ such that any three, but no four of them, are convexly independent.

We consider $m \cdot (n+1)$ points on the boundary of a circle K, where $m = \binom{n}{3}$. We denote these points, in their cyclic order, by $a_1 = b_0^1, b_1^1, \dots, b_n^1, a_2 = b_0^2, b_1^2, \dots, b_n^2, \dots, a_m = b_0^m, b_1^m, \dots, b_n^m, a_{m+1} = a_1$. Further let $\mathcal{H}_1, \dots, \mathcal{H}_m$ denote the triples from $\{1, \dots, n\}$. We define the sets $C_1, \dots, C_n$ as follows. First of all, we agree that the points $a_1, \dots, a_m$ belong to $\operatorname{bd} C_i$ for $i = 1, \dots, n$. We shall ensure that the sets C_i, C_j and C_k with $= \{i, j, k\} = \mathcal{H}_h$ are convexly independent by appropriately defining their boundaries between the points a_h and a_{h+1}.

Let A_i^h denote the arc of $\operatorname{bd} C_i$ between a_h and a_{h+1}. If $i \in \mathcal{H}_h = \{i, j, k\}$ and $1 \le i < j < k \le n$, then let A_i^h be the union of the arc of $\operatorname{bd} K$ between a_h and b_i^h and the segment $b_i^h a_{h+1}$. This will be the definition of A_i^h also in the case when $i \notin \mathcal{H}_h$. If $i \in \mathcal{H}_h = \{i, j, k\}$, $1 \le j < i < k \le n$, then let A_i^h consist of the arcs $a_h b_{j-1}^h$ and $b_{j+1}^h b_i^h$ of $\operatorname{bd} K$ and the segments $b_{j-1}^h b_{j+1}^h$ and $b_i^h a_{h+1}$. Finally, if $i \in \mathcal{H}_h = \{i, j, k\}$, $1 \le j < k < i \le n$, then let A_i^h consist of the arcs $a_h b_{j-1}^h$, $b_{j+1}^h b_{k-1}^h$ and $b_{k+1}^h b_i^h$ of $\operatorname{bd} K$ and of the segments $b_{j-1}^h b_{j+1}^h$, $b_{k-1}^h b_{k+1}^h$ and $b_i^h a_{h+1}$.

Now, it is obvious from the construction that any three sets C_i, C_j, C_k are convexly independent; however, $C_i \subset \operatorname{conv}(C_j \cup C_k \cup C_l)$ for $i < j < k < l$.

We note that in the above example the boundaries of two sets of the family have many points in common. Pach and Tóth [**PT**] constructed an infinite family of straight-line segments such that any three of them are convexly independent, but no four are. However, they also showed that there is a function $q(n)$ such that every family of at least $q(n)$ segments, any four of which are convexly independent, has n members in convex position. Further, there is a function $p(n)$ with the property that if $\mathcal{F}$ is a family of at least $p(n)$ plane convex sets with non-empty interiors, any two members of which have at most two boundary points in common, and any three of which are convexly independent, then $\mathcal{F}$ has n convexly independent members.

A set of points in which every four points are convexly independent is itself convexly independent. This is not the case for families of sets. Thus, it makes sense to ask for the maximum cardinality $h(k, n)$ of a family of convex bodies with the property that any k but no n members of the family are convexly independent $(3 \le k \le n)$.This question was raised in [**BF3**], where an exponential upper bound was given for $h(4, n)$ and an $O(n^2)$ upper bound was given for $h(k, n)$ when $k \ge 5$. From these, using Ramsey's theorem, a super exponential upper bound was derived for $h(3, n) = g(n)$. These bounds were recently improved by Pach and Tóth [**PT2**] and Tóth [**T**] who showed that

$$h(3, n) = g(n) \le \binom{2n-4}{n-2}^2,$$

$$2\left\lfloor \frac{n+1}{4} \right\rfloor^2 \le h(4, n) \le n^3,$$

$$h(5, n) \le 6n - 12$$

and

$$h(k, n) \le \frac{k-4}{k-5} n \quad \text{for } 6 \le k \le n.$$

The last two bounds come very close to the lower bound

$$n - 1 + \left\lfloor \frac{n-1}{k-1} \right\rfloor \le h(k, n) \qquad (n \ge 5)$$

given in [**BF3**].

8. Finite packings and coverings

There are two main types of problems concerning finite packings and coverings. In both cases we arrange a finite number n of congruent copies of a given object C. While in the first case the container we try to pack, or the body we try to cover, is of a given shape K, for the second type of problem we assume of this set only convexity.

Investigations about the first type of problems were restricted essentially to the case when $C = B^2$. For a set K and a positive integer n we define the quantities

$$M_p(K,n) = \inf\{\lambda | n \text{ congruent copies of } B^2 \text{ can be packed in } \lambda K\}$$

and

$$M_c(K,n) = \sup\{\lambda | n \text{ congruent copies of } B^2 \text{ can cover } \lambda K\}.$$

$M_p(K,n)$ has been determined in the following cases: K is a square, $n \leq 20$, $n = 25$ and $n = 36$; K is a circle, $n \leq 11$ and $n = 19$; and, finally, when K is a regular triangle for $n \leq 12$ as well as for all triangular numbers $n = k(k+1)/2$. $M_c(K,n)$ is known when K is a square and $n \leq 7$, when K is a circle and $n \leq 9$, and also when K is a regular triangle and $n \leq 6$, $n = 9$ or $n = 10$. Most of these results were obtained in the last few years (see [**CFG**, sections D1, D2, D3] for a partial survey and [**F3, Fo, HM, Me, Me2, Me3, Me4, Me5, MSHGT, NO, Pe**] for the most recent results).

We mentioned in Section 6 the heuristic algorithm given by Peikert for the solution of the problem of densest packing of congruent circles in a square. Other algorithms have been created as well to find good packings of circles [**GLu, GLu2, GLNO, LG, MFP, NO2**]. It is a challenging problem to find methods to prove the optimality of the arrangements so found.

Now we turn to finite packing and covering problems of the second type. Following are the basic problems here:

Find the minimum volume $M_p(n,d)$ of a convex body in E^d that can accommodate k non-overlapping unit balls.

Find the maximum volume $M_c(n,d)$ of a convex body in E^d that can be covered by k unit balls.

We note that such problems have been considered already at a very early stage of the history of packings and coverings. As an example we mention the following theorems [**Fe**, page 67]:

If $n \geq 2$ unit circles are packed in a convex domain K, then $a(K) \geq n\sqrt{12}$.

If $n \geq 2$ unit circles cover a convex domain K, then $a(K) \leq n\sqrt{27}/2$.

These results, and the generalization of the first one by Groemer [**Gro**], show that the arrangements realizing the extreme areas $M_p(n,2)$ and $M_c(n,2)$ in the plane are very close to an appropriate cluster chosen from the densest lattice packing and thinnest lattice covering of circles, respectively. Moreover, it is conjectured that a packing of n unit circles contained in a convex disc of area $M_p(n,2)$ is always part of a densest lattice packing of unit circles. This conjecture has been verified for infinitely many integers n by Wegner [**G**]. The first number for which the conjecture is not settled is 121.

The situation is completely different in high dimensions. Let $\mathcal{P}$ be a packing of unit balls in E^d with density $\delta(B^d)$ and let $\mathcal{C}$ be a covering with unit balls in E^d with density $\vartheta(B^d)$. If K is a large "round" convex body, that is one expanding in all directions, then $d_{\text{inn}}(\mathcal{P}|K)$ is approximately equal to $\delta(B^d)$ and $d_{\text{out}}(\mathcal{C}|K)$ is approximately equal to $\vartheta(B^d)$. We recall from Section 4 that $\delta(B^d)$ approaches 0 in the exponential order of magnitude, and $\vartheta(B^d)$ approaches ∞ at least as fast as order d when $d \to \infty$. On the other hand, an easy computation shows that the inner density of n unit balls in E^d whose centers are placed equidistantly at distance 2 on a straight line relative to the "sausage-like" convex hull of the balls is at least $\sqrt{\pi/2(d+1)}$. This density also approaches 0 as $d \to \infty$, however at a much slower pace than $\delta(B^d)$. Similarly, a linear arrangement of balls in E^d can cover a sausage-like convex body K so that the outer density of the balls relative to K is less than $\sqrt{\pi e/2}$. These examples and some numerical computation led L. Fejes Tóth **[Fe3]** and Wills **[W]** to the following *sausage conjectures*:

If, for $d \geq 5$, n unit balls in E^d are packed in a convex body of volume $M_p(n, d)$, then the centers of the balls are equally placed on a line segment.

If, for $d \geq 5$, n unit balls in E^d cover a convex body of volume $M_c(n, d)$, then the centers of the balls are equally placed on a line segment.

These conjectures stimulated intensive research. A detailed survey of partial results concerning these problems and related questions is given in **[GW]**. A breakthrough concerning the sausage conjecture for ball packings was recently achieved by Betke, Henk and Wills **[BHW]**: they proved that the conjecture holds for dimensions $d \geq 13387$. Later Betke and Henk **[BH]** improved the bound on d to $d \geq 42$.

References

[B] K. M. Ball, *A lower bound for the optimal density of lattice packings,* Duke J. Mathematics **68** (1993), 217–221.

[BFP] I. Bárány, Z. Füredi and J. Pach, *Discrete convex functions and proof of the six circle conjecture of Fejes Tóth,* Can. J. Math. **36** (1984), 569–576.

[BH] U. Betke and M. Henk, *Finite packings of spheres,* Discrete Comput. Geom. **19** (1998).

[BHW] U. Betke, M. Henk and J. M. Wills, *Finite and infinite packings,* J. reine angew. Math. **453** (1994), 165–191.

[Be] A. Bezdek, *A remark on the packing density in the 3-space,* in: Intuitive Geometry, ed. by K. Böröczky and G. Fejes Tóth, Colloquia Math. Soc. János Bolyai **63**, North-Holland, Amsterdam–New York (1994), 17–22.

[BBC] A. Bezdek, K. Bezdek and R. Connelly, *Finite and uniform stability of sphere packings,* Discrete Comput. Geom., to appear.

[BK] A. Bezdek and W. Kuperberg, *Maximum density space packing with congruent circular cylinders of infinite length,* Mathematika **37** (1990), 74–80.

[BK2] A. Bezdek and W. Kuperberg, *Packing Euclidean space with congruent cylinders and with congruent ellipsoids,* in: Applied Geometry and Discrete Mathematics: the "Victor Klee Festschrift," ed. by P. Gritzmann and B. Sturmfels, DIMACS Series on Discrete Mathematics and Computer Science **4**, Amer. Math. Soc. & Assoc. Comput. Math. (1991), 71–80.

[BKM] A. Bezdek, W. Kuperberg and E. Makai Jr., *Maximum density space packing with parallel strings of balls,* Discrete Comput. Geom. **6** (1991), 227–283.

[Bez] K. Bezdek, *Isoperimetric inequalities and the dodecahedral conjecture,* International J. Math. **8** (1997), 759–880.

[BF] T. Bisztriczky and G. Fejes Tóth, *A generalization of the Erdős–Szekeres convex n-gon theorem,* J. reine angew. Math. **395** (1989), 167–170.

[BF2] T. Bisztriczky and G. Fejes Tóth, *Nine convex sets determine a pentagon with convex sets as vertices,* Geometriae Dedicata **31** (1989), 89–104.

[BF3] T. Bisztriczky and G. Fejes Tóth, *Convexly independent sets,* Combinatorica **10** (1990), 195–202.

[CG] F. R. K. Chung and R. L. Graham, *Forced convex n-gons in the plane,* Discrete Comput. Geom. **19** (1998), 365–369.

[C] B. Cipra, *Gaps in the sphere packing proof?* Science **259** (1993), 895.

[CHMS] J. H. Conway, T. C. Hales, D. J. Muder and N. J. A. Sloane, *On the Kepler conjecture,* Math. Intelligencer **16** no. 2 (1994), 5.

[CS] J. H. Conway and N. J. A. Sloane, Sphere Packings, Lattices and Groups, Second edition, Springer-Verlag, New York, 1993.

[CFR] H. S. M. Coxeter, L. Few and C. A. Rogers, *Covering space with equal spheres,* Mathematika **6** (1959), 147–157.

[CFG] H. T. Croft, K. J. Falconer and R. K. Guy, Unsolved Problems in Geometry, Springer-Verlag, New York, 1991.

[DZ] M. H. Dauenhauer and H. J. Zassenhaus, *Local optimality of critical lattice sphere-packing of regular tetrahedra,* Discrete Math. **64** (1987), 129–146.

[DS] D. P. Dobkin and L. Snyder, *On a general method for maximizing and minimizing among certain geometric problems,* Proc. 20th IEEE Sympon Foundation of Computer Science (1979), 9–17.

[EGH] P. Erdős, P. M. Gruber and J. Hammer, Lattice Points, Pitman Monograph 39, Longman Scientific & Technical and John Wiley & Sons, New York, 1989.

[F] G. Fejes Tóth, *Densest packings of typical convex discs are not lattice-like,* Discrete Comput. Geom. **14** (1995), 1–8.

[F2] G. Fejes Tóth, *Review of W.Y. Hsiang, On the sphere packing problem and the proof of Kepler's conjecture,* Math. Rev. 95g:52032.

[F3] G. Fejes Tóth, *Thinnest covering of a circle by eight or nine congruent circles*, preprint, 1997.

[FK] G. Fejes Tóth and W. Kuperberg, *Blichfeldt's density bound revisited,* Math. Annalen. **295** (1993), 721–727.

[FK2] G. Fejes Tóth and W. Kuperberg, *Recent results in the theory of packing and covering,* in: New Trends in Discrete and Computational Geometry, ed. by J. Pach, Springer (1993), 251–279.

[FK3] G. Fejes Tóth and W. Kuperberg, *Packing and covering with convex sets,* in: Handbook of Convex Geometry, ed. P. M. Gruber and J. M. Wills, North-Holland, Amsterdam (1993), 799–860.

[FK4] G. Fejes Tóth and W. Kuperberg, *Thin non-lattice covering with an affine image of a strictly convex body,* Mathematika **42** (1995), 239–250.

[FZ] G. Fejes Tóth and T. Zamfirescu, *For most convex discs thinnest covering is not lattice-like,* in: Intuitive Geometry, Colloquia Math. Soc. János Bolyai **63**, ed. by K. Böröczky and G. Fejes Tóth, North-Holland, Amsterdam–New York (1994), 105–108.

[Fe] L. Fejes Tóth, Lagerungen in der Ebene auf der Kugel und im Raum, Zweite verbesserte und erweiterte Auflage, Springer-Verlag, Berlin–New York, 1972.

[Fe2] L. Fejes Tóth, Regular Figures, Pergamon Press, Oxford–London–New York, 1964.

[Fe3] L. Fejes Tóth, *Research Problem no. 13,* Periodica Math. Hungar. **6** (1975), 197–199.

[Fer] S. P. Ferguson, *Sphere packings V,* Ph.D. thesis, University of Michigan, 1997.

[FH] S. P. Ferguson and T. C. Hales, *A formulation of the Kepler conjecture*, preprint, 1997.

[Fo] F. Fodor, *The densest packing of 19 congruent circles in a circle*, preprint, 1997.

[G] C. F. Gauss, *Untersuchungen über die Eigenschaften der positiven ternären quadratischen Formen von Ludwig August Seber,* Göttingische gelehrte Anzeigen, 1831 Juli 9 = J. reine angew. Math. **20** (1840), 312-320 = Werke II, Königliche Gesellschaft der Wissenschaften, Göttingen, 1876, 188–196.

[GLu] R. L. Graham and B. D. Lubachevsky, *Dense packings of equal disks in an equilateral triangle: from 22 to 34 and beyond,* Electronic J. Combinatorics **2** (1995), A1.

[GLu2] R. L. Graham and B. D. Lubachevsky, *Repeated patterns of dense packings of equal disks in a square,* Electronic J. Combinatorics **3** (1996), R16.

[GLNO] R. L. Graham, B. D. Lubachevsky, K. J. Nurmela and P. R. J. Östergård, *Dense packings of congruent circles in a circle,* Discrete Math, to appear.

[Gr] P. Gritzmann, *Lattice covering of space with symmetric convex bodies,* Mathematika **32** (1985), 311–315.

[GW] P. Gritzmann and J. M. Wills, *Finite packing and covering,* in: Handbook of Convex Geometry, ed. P. M. Gruber and J. M. Wills, North-Holland, Amsterdam (1993), 861–897.

[Gro] H. Groemer, *Über die Einlagerung von Kreisen in einem konvexen Bereich,* Math. Zeitschr. **73** (1960), 285–294.

[Gro2] H. Groemer, *Existenzsätze für Lagerungen im Euklidischen Raum,* Math. Zeitschr. **81** (1963), 260–278.

[GL] P. M. Gruber and C. G. Lekkerkerker, Geometry of Numbers, Elsevier, North-Holland, Amsterdam, 1987.

[H] T. C. Hales, *The sphere packing problem,* Journal of Computational and Applied Math. **44** (1992), 41–76.

[H2] T. C. Hales, *Remarks on the density of sphere packings in three dimensions,* Combinatorica **13** (1993), 181–197.

[H3] T. C. Hales, *The status of the Kepler conjecture,* Math. Intelligencer **16** no. 3 (1994), 47–58.

[H4] T. C. Hales, *Sphere packings I,* Discrete Comput. Geom. **17** (1997), 1–51.

[H5] T. C. Hales, *Sphere packings II,* Discrete Comput. Geom. **18** (1997), 135-149.

[H6] T. C. Hales, *Sphere packings III,* preprint, 1997.

[HSS] R. J. Hardin, M. J. A. Sloane and W. D. Smith, Spherical Codes, Book in preparation.

[HM] A. Heppes and J. B. M. Melissen, *Covering a rectangle with equal circles,* Period. Math. Hung. **34** (1997), 63–79.

[Hi] D. Hilbert, *Mathematische Probleme,* Archiv Math. Phys. **1** (1901), 44–63 and 213–237=Gesammelte Abhandlungen III, 290–329=Mathematical problems, Bull. Amer. Math. Soc. **8** (1902), 437–497= Proc. Symp. Pure Math. **28** (1976), 1–34.

[Hs] W. Y. Hsiang, *On the density of sphere packings in* E^3, I, preprint, 1990.

[Hs2] W. Y. Hsiang, *On the density of sphere packings in* E^3*-Kepler's conjecture and Hilbert's 18th problem,* preprint, 1990.

[Hs3] W. Y. Hsiang, *Sphere packings and spherical geometry – Kepler's conjecture and beyond,* Center for Pure and Appl. Math. University of California, Berkeley, preprint PAM-528, July 1991.

[Hs4] W. Y. Hsiang, *On the density of sphere packings in* E^3, I, Center for Pure and Appl. Math. University of California, Berkeley, preprint PAM-530, August 1991.

[Hs5] W. Y. Hsiang, *On the sphere packing problem and the proof of Kepler's conjecture,* in: Differential Geometry and Topology, Alghero (1992), 117–127.

[Hs6] W. Y. Hsiang, *On the sphere packing problem and the proof of Kepler's conjecture,* International J. Math. **4** (1993), 739–831.

[Hs7] W. Y. Hsiang, *A rejoinder to Hales's article,* Math. Intelligencer **17** no. 1 (1995), 35-42.

[I] D. Ismailescu, *Covering the plane with copies of a convex disc,* Discrete Comput. Geom., to appear.

[KL] G. A. Kabatjanskiĭ and V. I. Levenštein, *Bounds for packings on a sphere and in space (in Russian),* Problemy Peredachy Informatsii **14** (1978), 3–25; English translation in Problems of Information Transmission **14** (1978), 1–17.

[Ke] J. Kepler, The Six-cornered Snowflake, Oxford Clarendon Press, Oxford, 1966.

[KP] D. J. Kleitman and L. Pachter, *Finding convex sets among points in the plane,* Discrete Comput. Geom. **19** (1998).

[KK] G. Kuperberg and W. Kuperberg, *Double-lattice packings of convex bodies in the plane,* Discrete Comput. Geom. **5** (1990), 389–397.

[Ku] W. Kuperberg, *Covering the plane with congruent copies of a convex body,* Bull. London Math. Soc. **21** (1989), 82–86.

[L] J. Linhart, *Eine Methode zur Bestimmung der Dichte einer dichtesten gitterförmigen k-fachen Kreispackung,* Arbeitsberichte Math. Inst. Univ. Salzburg, 1983.

[LG] B. D. Lubachevsky and R. L. Graham, *Curved hexagonal packings of equal disks in a circle,* Discrete Comput. Geom. **18** (1997), 179–194.

[M] E. Makai Jr., Oral communication, 1997.

[M-LP] P. Mani-Levitska and J. Pach, *Decomposition problems for multiple coverings of unit balls,* preprint, 1990.

[MFP] C. D. Maranas, C. A. Floudas and P. M. Pardalos, *New results in the packing of equal circles in a square,* Discrete Math. **128** (1995), 187–193.

[Me] J. B. M. Melissen, *Densest packings of congruent circles in an equilateral triangle,* Amer. Math. Monthly **100** (1993), 916–925.

[Me2] J. B. M. Melissen, *Densest packings of eleven congruent circles in a circle,* Geom. Dedicata **50** (1994), 15–25.

[Me3] J. B. M. Melissen, *Optimal packings of eleven circles in an equilateral triangle,* Acta Math. Hungar. **65** (1994), 389–393.

[Me4] J. B. M. Melissen, *Loosest circle coverings of an equilateral triangle,* Math. Mag. **70** (1997), 119–125.

[Me5] J. B. M. Melissen, *Packing and covering with circles,* Ph.D. thesis, University of Utrecht, 1997.

[MSHGT] J. B. M. Melissen, P. C. Schuur, A. Heppes, Zs. Gáspár and T. Tarnai, *New circle coverings of an equilateral triangle,* preprint, 1997.

[Mc] P. McMullen, *Convex bodies which tile space by translation,* Mathematika **27** (1980), 113–121; **28** (1981), 191.

[MP] W. Moser and J. Pach, *Research Problems in Discrete Geometry,* DIMACS Technical Report 93-32, 1993.

[MS] D. M. Mount and R. Silverman, *Packing and covering the plane with translates of a convex polygon,* J. Algorithms **11** (1990), 564–580.

[Mu] D. J. Muder, *A new bound on the local density of sphere packings,* Discrete Comput. Geom. **10** (1993), 351–375.

[Mu2] D. J. Muder, *Letter about the Hsiang controversy,* in: S. Sight, Fermat's Last Theorem, Fourth Estate, London, (1997), 316–318.

[NO] K. J. Nurmela and P. R. J. Östergård, *Optimal packings of equal circles in a square,* Proc. 8th Quadrennial International Conference on Graph Theory, Combinatorics and Appications, 1996.

[NO2] K. J. Nurmela and P. R. J. Östergård, *Packing up to 50 equal circles in a square,* Discrete Comput. Geom. **18** (1997), 111–120.

[P] J. Pach, *Decomposition of multiple packing and covering,* Diskrete Geometrie, 2. Kolloq. Inst. Math. Univ. Salzburg (1980), 169–178.

[P2] J. Pach, *Covering the plane with convex polygons,* Discrete Comput. Geom. **1** (1986), 73–81.

[PA] J. Pach and P. K. Agarwal, Combinatorial Geometry, John Wiley, New York 1995.

[PT] J. Pach and G. Tóth, *Erdős-Szekeres type theorems for segments and non-crossing convex sets,* preprint, 1997.

[PT2] J. Pach and G. Tóth, *A generalization of the Erdős-Szekeres theorem to disjoint convex sets,* Discrete Comput. Geom. **19** (1998).

[Pe] R. Peikert, *Dichteste Packung von gleichen Kreisen in einem Quadrat,* El. Math. **49** (1994), 16–26.

[R] C. A. Rogers, *The closest packing of convex two-dimensional domains,* Acta Math. **86** (1951), 309–321.

[R2] C. A. Rogers, *A note on coverings,* Mathematika **4** (1957), 1–6.

[R3] C. A. Rogers, *The packing of equal spheres,* J. London Math. Soc. (3) **8** (1958), 609–620.

[R4] C. A. Rogers, *Lattice coverings of space,* Mathematika **6** (1959), 33–39.

[R5] C. A. Rogers, Packing and Covering, Cambridge University Press, Cambridge, 1964.

[RS] J. A. Rush and N. J. A. Sloane, *An improvement to the Minkowski-Hlawka bound for packing superballs,* Mathematika **34** (1987), 8–18.

[Ru] J. A. Rush, *A lower bound on packing density,* Invent. Math. **98** (1989), 499–509.

[Ru2] J. A. Rush, *A bound, and a conjecture on the maximum lattice-packing density of a superball,* Mathematika **40** (1993), 137–143.

[SS] D. Schattschneider and M. Senechal, *Tilings,* in Handbook of Discrete and Computational Geometry, ed. by J. E. Goodman and J. O'Rourke, CRC Press, Boca Raton (1997), 43–62.

[S] W. M. Schmidt, *Zur Lagerung kongruenter Körper im Raum,* Monatsh. Math. **65** (1961), 154–158.

[S2] W. M. Schmidt, *On the Minkowski-Hlawka theorem,* Illinois J. Math. **7** (1963), 18–23.

[Sc] P. Schmitt, *An aperiodic prototile in space,* preprint, 1988.

[Sc2] P. Schmitt, *Discs with special properties of densest packings,* Discrete Comput. Geom. **6** (1991), 181–190.

[St] I. Stewart, *Has the sphere packing problem been solved?,* New Scientist **134** (2 May 1992), 16.

[T] Á. H. Temesvári, *Eine Methode zur Bestimmung der dünnsten gitterförmigen k-fachen Kreisüberdeckungen,* Studia Sci. Math. Hungar. **23** (1988), 23–35. English translation in Math. Notes **41** (1987), 349–355.

[THY] Á. H. Temesvári, J. Horváth and N. N. Yakovlev, *A method for finding the densest lattice k-fold packing of circles* (Russian), Math. Zametki **41** (1987), 625–636, 764;

[T] G. Tóth, *Finding convex sets in convex position,* preprint, 1997.

[TV] G. Tóth and P. Valtr, *Note on the Erdős-Szekeres theorem,* Discrete Comput. Geom., to appear.

[V] B. A. Venkov, *On a class of euclidean polytopes* (in Russian), Vestnik Leningrad. Univ. Ser. Mat. Fiz. Him. **9** (1954), 11–31.

[G] G. Wegner, *Über endliche Kreispackungen in der Ebene,* Studia Sci. Math. Hungar. **21** (1986), 1–28.

[W] J. M. Wills, *Research Problem no. 33,* Periodica Math. Hungar. **14** (1983), 189–191.

Mathematical Institute of the Hungarian Academy of Sciences, P.O. Box 127, H-1364 Budapest, Hungary.

E-mail address: gfejes@math-inst.hu

Contemporary Mathematics
Volume **223**, 1999

Acoptic Polyhedra

Branko Grünbaum

ABSTRACT. Acoptic polyhedra are polyhedra in 3-space, with simple polygons as faces and with no selfintersections. These polyhedra are generalizations of convex polyhedra, and present a variety of interesting properties and open problems. Among the most challenging is the "general realizability conjecture," according to which every cell-complex decomposition of an orientable 2-manifold (satisfying some natural conditions) is isomorphic to an acoptic polyhedron. The known partial results on this conjecture are given. Definitions and concepts that may be useful in future studies are presented together with a variety of illustrative examples and additional open questions.

1. General introduction

The theory of convex polytopes has had a phenomenal flowering during the last fifty or so years, and is at present a mature field.[1] Hence it seems to be the appropriate time to start the systematic study of more general, not necessarily convex, polyhedra and polytopes.

There are many reasons for such activity:

(i) The collection of objects to be studied is vastly greater and more interesting if not restricted by requiring convexity. In particular, such polyhedra can be used to model a variety of orientable and non-orientable maps in a visually accessible manner.

(ii) Convexity is not essential for many results in the formulation of which it is assumed. But regardless of whether a certain property characterizes convex polyhedra or not, the investigation of its range of applicability is bound to produce new insights. Is convexity just a convenient assumption, which makes it possible to carry out a certain proof, or do there exist some limitations on the validity of the theorem in the nonconvex case; if so—what are the limitations, and what happens if they are exceeded?

1991 *Mathematics Subject Classification.* Primary 52B10, 52B15.

Key words and phrases. Polyhedron, convex, acoptic, duality, rigidity, selfintersecting, enumeration realization, regular, toroidal, Eberhard type theorems, isogonal, isohedral, monohedral, dynatogram, panorama.

Research supported in part by NSF grant DMS-9300657. Many of the new results and insights presented were obtained in long-term collaboration with G. C. Shephard, but the author alone is responsible for the views and statements formulated in the paper.

[1]An attractive and up-to-date introduction to this topic is Ziegler's book [**Z**].

© 1999 American Mathematical Society

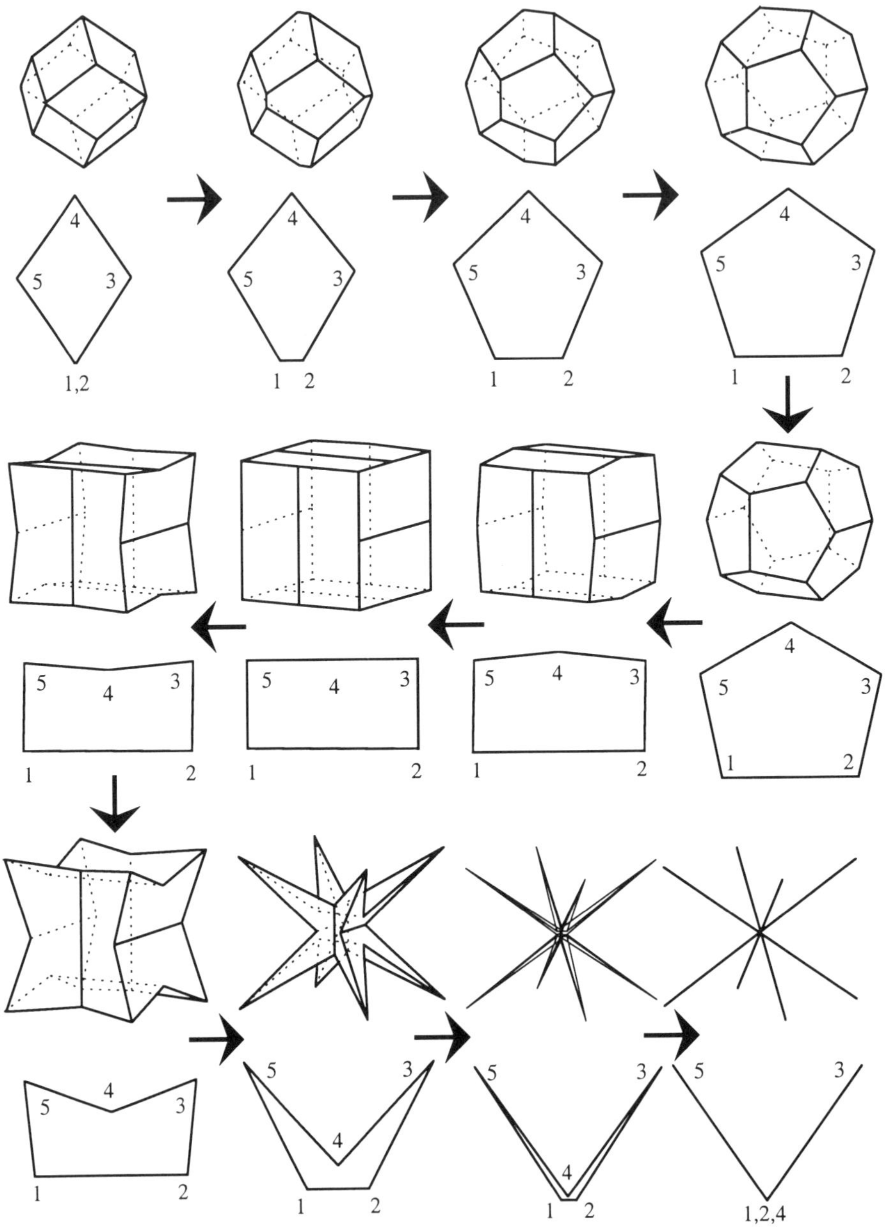

FIGURE 1. The continuum of *isohedral acoptic* polyhedra of *type* [3.3.3.3.3] with a plane of symmetry, starting with a *weakly convex* polyhedron *isomegethic* with the rhombic dodecahedron, and ending with a *weakly acoptic* polyhedron. The *protoface* of each polyhedron is shown below a perspective view of the polyhedron itself. Unfamiliar terms (italicized) are explained later in the text.

A good example of such developments is Cauchy's uniqueness theorem for convex polyhedra, which implies the rigidity of cardboard models.[2] It has been known for more than two centuries that the uniqueness theorem does not hold if nonconvex polyhedra are admitted. However, in the class of polyhedra without selfintersections, Cauchy's theorem is valid not only for convex polyhedra, but for some nonconvex ones as well; how can these polyhedra be characterized? How can polyhedra with only a finite number of realizations be characterized? What is the relation between the maximal number of distinct realizations and the number of faces or vertices? Continuously movable polyhedra without selfintersections are known,[3] but there is no characterization or even a general way of generating such polyhedra. The exciting "bellows conjecture" that the volume enclosed by a movable polyhedron is constant has just been decided.[4] Still unsolved is the question whether the space of realizations of a movable polyhedron with no selfintersections is always simply connected.[5]

(iii) Many interesting families of polyhedra form continua which are not inherently limited to convex polyhedra. A simple example of this possibility is shown in Figure 1; additional illustrations are furnished by the isogonal polyhedra discussed in Section 7.

(iv) If duality or polarity of polyhedra are to be concepts applicable to even mildly nonconvex polyhedra (such as those with convex faces and no selfintersections), one has to admit polyhedra with selfintersections, and even polyhedra with selfintersecting faces.[6] In fact, one is forced to go to the very general definitions in order to have consistent and usable concepts. This is the topic of a paper in preparation.

In each of these directions, any results that are obtained yield also additional understanding of convex polytopes or polyhedra.

In recent years papers discussing various aspects of polyhedra more general than the convex ones have started appearing in increasing numbers. However, no general framework is available, and the last attempt to give a comprehensive survey of such more general polyhedra was in Brückner's well known book [**Bru**], almost a whole century ago! As any critical reading shows, Brückner failed to make the basic definitions consistent,[7] is woefully incomplete concerning the topics he

---

[2] Cauchy's proof appeared in [**C**]; it was reproduced in some editions of Hadamard's well-known geometry text [**HG**], but excluded from other editions because of flaws in the proof as noted by Hadamard [**H**] and Steinitz [**St**]. Partial correction of a minor flaw in the proof of the combinatorial lemma can be found in Lebesgue [**Le**]. A complete proof appears in Steinitz-Rademacher [**SR**], as well as in the books of Aleksandrov [**A**] and Lyusternik [**Ly**]. For other proofs see [**Ho**], [**Sto**]. Simplifications of Steinitz's proof of the geometric lemma are given in [**Eg**] and [**SZ**].

[3] See the descriptions of such polyhedra in Connelly [**Co**], [**Co2**], Kuiper [**K**], Aleksandrov [**A2**]; selfintersecting movable polyhedra have been known for more than a century, see Bricard [**Bri**], Lebesgue [**Le2**]. Various aspects of rigidity and uniqueness are discussed in Crapo-Whiteley [**CW**], Connelly [**Co4**], Sabitov [**S**], Maksimov [**M**].

[4] See the discussions of the conjectures in Connelly [**Co3**], Sabitov [**S2**], Aleksandrov [**A2**]; the solution appears in Connelly *et al.* [**CSW**].

[5] It is known that in the case of the selfintersecting movable octahedra of Bricard the realizations space is not simply connected, see [**BS**].

[6] For a discussion of problems in this context, and the inadequacy of the "folk wisdom" regarding duality, see [**GS4**].

[7] Details concerning the inconsistencies and other shortcomings of Brückner's book [**Bru**] are given in [**Grü6**], [**Grü7**].

discusses, and has probably greatly contributed to the long-lasting neglect of the whole field by the mathematical community. The present paper overlaps several recent surveys,[8] which the reader may wish to consult for additional references and for different views on various questions.

When attempting to give an overview of the available information on these polyhedra, two basic facts need to be taken into account at the outset:

(1) At this time there is no experience, and no experimental material, to guide the formulation of a reasonable program of investigation in dimensions higher than 3.[9] Hence the thrust of the present exposition is confined to polyhedra in Euclidean 3-space $\mathbb{E}^3$.

(2) There are many different classes of polyhedra that make reasonable topics of investigation.[10] They differ not only by the hierarchical level of generality considered, but also by the point of view regarding what is a polyhedron. In some contexts polyhedra are best interpreted as "solids"; this is also the traditional approach, going back to antiquity. Examples of such interpretations can be found in works on the *stellation* and *facetting* of various regular or other polyhedral,[11] in the studies of polyhedral scenes,[12] and in investigations of the "convex ring."[13] In other situations, it is most appropriate to consider polyhedra as surfaces built up from simple planar polygons, or else, to consider them as built up from planar polygons but interpreting these not as patches of the plane but as collections of straight-line segments some of which may intersect each other.[14] For convex polyhedra these different points of view essentially coalesce, due to the famous Steinitz Theorem.[15]

We are concentrating on the latter type of interpretations, and in Figure 9 we illustrate the differences that arise from this distinction. As we go along, we introduce several specific classes of polyhedra, with appropriate terminology and notation. We hope that this will facilitate the development of the field, and that the various open problems and conjectures will present tempting challenges.

[8]Particularly near to the topics discussed here is the work of Brehm and Wills [**BWi**]. Their polyhedral manifolds are essentially the same as our AC-polyhedra. See also Martini [**Ma**], Brehm and Schulte [**BrS**].

[9]Nearly the only exception are investigations concerning regular polytopes of various degrees of generality and abstraction; for a guide to this literature see, for example, Coxeter [**Cox**], Johnson [**Jo**], McMullen-Schulte [**MSc**].

[10]See, for example, Cromwell [**Cr**] for a survey of the various kinds of polyhedra and their history.

[11]Stellations of polyhedra are discussed, among others, by Wenninger [**We**], [**We2**], Hudson and Kingston [**HK**], and Messer [**Me**]; the latter contains many references to other relevant books and papers. For facetting see, for example, Bridge [**Brid**].

[12]See, for example, Shirai [**Sh**].

[13]Introduced by Hadwiger [**Ha**], for more recent accounts see McMullen-Schneider [**MS**]; see also [**GS6**] for related material.

[14]Even more general types of polyhedra, in which the faces are polygons that need not be planar, have been considered (and may find applications in crystallography and other fields); we shall not deal with them here. For some points of view and various results about these polyhedra see, for example, Grünbaum [**Grü4**], Dress [**D**], [**D2**], Burt [**Bu**], Molnár [**Mol**], Farris [**F**], [**F2**], McMullen and Schulte [**MSc2**]. In the terminology of Grünbaum [**Grü6**, p. 50] the present exposition is restricted to *epipedal* polyhedra, that is, polyhedra in which each face is a planar polygon.

[15]First announced in [**St**, p. 77] with several different detailed proofs in [**SR**]. However, these expositions are written in quite cumbersome form, and the result has remained not widely known for a long time. It appears in Lyusternik [**Ly**], but the proof there is deeply flawed, ignoring the depth of the required arguments. One of Steinitz's proofs appears in [**Grü**]; additional proofs of the result are given in [**BG**] and [**Z**].

In order to avoid excessive length, we discuss here only polyhedra we call **acoptic**, that is polyhedra with no selfintersections (including no selfintersecting faces). Precise definitions are given below. Certain aspects of more general polyhedra, with selfintersections allowed, have been discussed in [**Grü6**] and [**Grü7**], but much additional work remains to be done.

One of the obstacles encountered already in the study of acoptic polyhedra, and with even stronger effect if the polyhedra have selfintersections, is the problems of visualization. A perusal of this paper shows the use of several methods of presentation, but it must be admitted that none is quite satisfactory for any but the simplest situations. It is to be hoped that computer-based modes of presentation will alleviate this difficulty in the near future.

Even with the limitation to acoptic polyhedra, it has been necessary to exclude many topics which fit the aim of the paper. These will be discussed in other venues. The paper is organized as follows. Acoptic polyhedra, and some other concepts, are defined and illustrated in Section 2, which proposes a terminology for the subject. Sections 3, 4 and 5 are devoted to results on the realization of various maps by polyhedra, while more general questions of realizability are considered in Section 6. Isogonal polyhedra and their dynatograms and panoramas are the subject of Section 7. In Section 8 we discuss the known results on monohedral polyhedra, while the concluding Section 9 is concerned with spanning trees in the graphs of polyhedra.

2. Acoptic polyhedra, and some general considerations

To judge from all appearances and experience, there is no single class of polyhedra that deserves to be considered as *the* general kind deserving investigation. Different levels of generality are appropriate in different situations. This is no problem in itself, but in order to avoid difficulties and misunderstandings, precise definitions and terminology are required. Without pretending that the following are more than possibly suitable categories of polyhedra, here are several definitions.

To begin with, we consider polyhedra as certain collections of **planar, compact, simply-connected polygonal regions**; the boundary of such a region is called a **simple polygon** and the region itself is referred to as a **face** of the polyhedron. A simple polygon consists of a finite number of line segments of positive length (the **edges** of the polygon); the endpoints of the edges are the **vertices** of the polygon. Each vertex of a polygon belongs to precisely two edges (said to be mutually adjacent), and the edges form a simple circuit (Jordan polygon). It should be noted that we made no assumption of convexity, and that edges (adjacent or nonadjacent) may be collinear; examples may be seen below, in Figure 2. For brevity, we call faces, edges and vertices of a polyhedron its **elements**.

In general, the conditions under which a finite collection of polygons (as described above, or of some other type) is called a **polyhedron** are:

(P1) each edge is shared by precisely two faces;

(P2) all faces containing a given vertex form a single circuit of at least three faces; and

(P3) no proper subfamily has both these properties.

In most cases, we find it convenient to restrict the polyhedra considered by imposing additional conditions. The following are examples of such restrictions. They

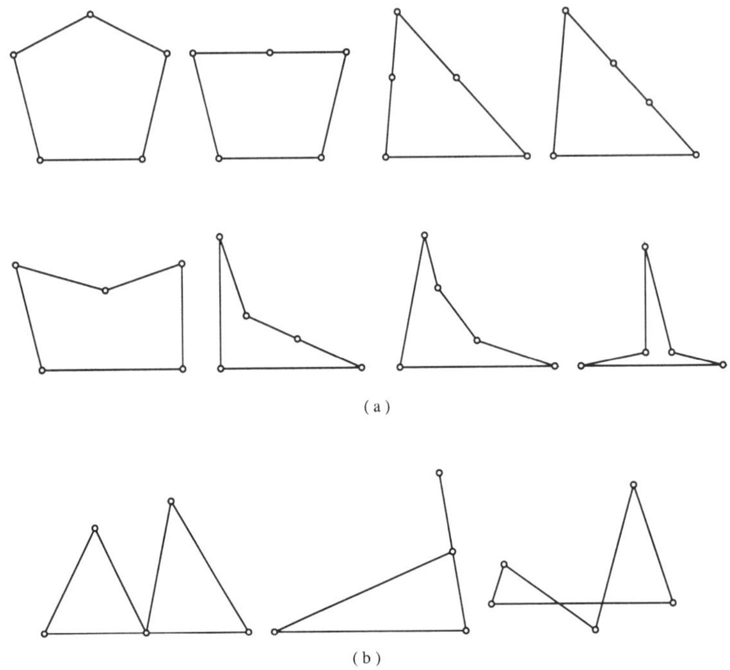

FIGURE 2. Examples of (a) simple polygons, and (b) polygons which are not simple. All polygons shown are pentagons.

determine the **acoptic** polyhedra (from Greek *acoptos*, uncut) which are the main topic of this report, and several of their subclasses.

A polyhedron is **acoptic**[16] if

(P4) the faces are simple polygonal regions; and

(P5) the relative interiors of its elements are disjoint.

This definition does not preclude the possibility that distinct faces (contiguous or not) are coplanar, and that two faces have several common edges and/or vertices. It implies that the polyhedron has a well-defined bounded **interior**, and an unbounded **exterior**. It also implies that the polyhedron is homeomorphic to a **cell-decomposition**[17] (or **map**) $\mathcal{C}(P)$ of a compact, orientable 2-manifold $M(P)$; the

---

[16]The meaning of the word "acoptic" as used here differs from the usage in [**Grü6**], [**Grü7**]. The earlier concept corresponds to "weakly acoptic," as defined below.

[17]A cell-complex decomposition of a 2-manifold M is a finite collection of closed topological disks ("faces") whose union is M, with each two faces meeting in at most a finite number of arcs ("edges") and singleton points, each disk meeting at least three other disks along edges, and no two edges having coinciding endpoints ("vertices"). This definition is more general than the one in [**Grü**], where the disks were assumed to be convex; this latter has been used by many other authors. However, the present definition is more restrictive than the usual topological concept. The prohibition of multiple edges between the same pair of vertices excludes some maps which are acceptable in the topological literature. This condition also shows that (with this interpretation) there is no duality possible among the cell-decompositions of a 2-manifold, since there is no prohibition of two faces having more than one edge in common. Duality can be restored only by allowing much more general cell complexes. In this context, the examples and discussion in [**GS4**] are instructive. It should be noted that some authors use "*polyhedral map*" for the different concept of a cell-decomposition of a 2-manifold in which there is no assumption that the faces are

vertices, edges and faces (countries) of the map $\mathcal{C}(P)$ are in bijective correspondence with those of P, hence we may say that P and $\mathcal{C}(P)$ are **combinatorially equivalent**. (Two polyhedra or maps are called *combinatorially equivalent*, or of the same *combinatorial type*, provided there exists an incidence-preserving bijection between their sets of vertices, edges, and faces.) We also say that P is a **realization** of $\mathcal{C}(P)$; in fact, it is an imbedding of $\mathcal{C}(P)$ in the Euclidean 3-space $\mathbb{E}^3$.

We come now to several other conditions imposed frequently on polyhedra. For obvious reasons, a polyhedron is said to be **convex-faced** if

(P6) each face is a convex polygon.

A polyhedron is called **conservative** if

(P7) distinct elements have distinct affine hulls.

An acoptic polyhedron is **strictly acoptic** if

(P8) there are no overarching elements.[18]

The class of convex-faced conservative acoptic polyhedra is the most-studied type of polyhedra more general than the convex ones.[19] Obviously, convex polyhedra are acoptic polyhedra which are convex-faced, conservative, and for which the associated 2-manifold is a sphere. Conversely, by Steinitz's Theorem, acoptic polyhedra with these properties are combinatorially equivalent to convex polyhedra. While the combinatorial types of "small" convex polyhedra have been enumerated to the limits of feasibility (see, for example, Federico [**Fe**]), essentially nothing is known regarding the number $t(f)$ of combinatorial types of acoptic polyhedra with f faces that are not realizable by convex polyhedra. It is easy to see[20] that $t(f) = 0$ for $f \leq 5$, and $t(6) = 3$ (see Figure 3), but already $t(7)$ is not known with certainty. The only publication I am aware of with some information on the numbers $t(n)$ is Gardner [**G2**, pp. 32, 77]. Gardner states[21] that $t(7) = 26$ and $t(8) = 277$. It is not clear what these numbers are expressing; no class I could think of yields these values. With the definitions adopted here, I find that $t(7) = 17$; the 17 types are shown in Figure 4. On the other hand, for other classes of polyhedra, or under some other classification principles, the number of types changes, see Figures 5 to 9. A decision concerning what constitutes convenient distinctions between polyhedra seems to depend in an essential way on the context. However, it is appropriate to formulate here as a major open question the following:

Enumeration problem. Determine—at least for small n—the possible combinatorial types of acoptic polyhedra with n faces (or with n vertices). More specialized enumerations, such as those of conservative, or simple, or simplicial polyhedra, or by the genus of the associated map, would also be of interest.

---

polygons, but only that any two meet in either a common arc, or a common point or not at all. See, for example, Pulapaka and Vince [**PV**].

[18] A cell complex or polyhedron has *overarching elements* if it contains two vertices and two faces that are mutually incident but are not all indicent with one edge. The three polyhedra in Figure 3 have overarching elements; the Kepler-Poinsot regular polyhedra $\{5, 5/1\}$ and $\{5/2, 5\}$ have overarching elements as well. Convex-faced acoptic polyhedra are automatically free of overarching elements. Other formulations of this condition can be found in [**B4**].

[19] Hajós and Heppes [**HH**] describe the construction of conservative polyhedra P with the following startling property: every vertex of P is exposed (that is, there is a supporting plane of P which intersects P at this vertex only) and every supporting plane of P meets P at vertices only.

[20] See Norgate [**N**], where nonacoptic pentahedra are investigated as well.

[21] Gardner [**G2**] gives no reference for these results, nor a definition of the polyhedra whose types he counts.

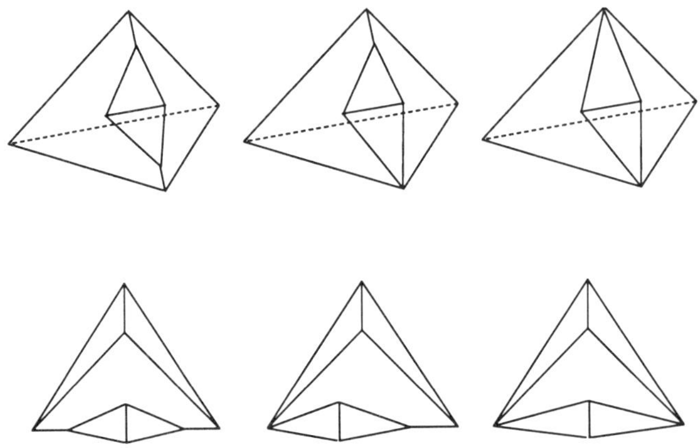

FIGURE 3. Examples of the three combinatorial types of acoptic hexahedra that cannot be represented by convex polyhedra. One polyhedron of each type is shown in perspective view in the top row. Beneath each is shown a "pseudo-Schlegel" diagram, which gives a planar view of the combinatorial structure of the polyhedron. Pseudo-Schlegel diagrams of this kind are possible for all acoptic polyhedra of genus 0.

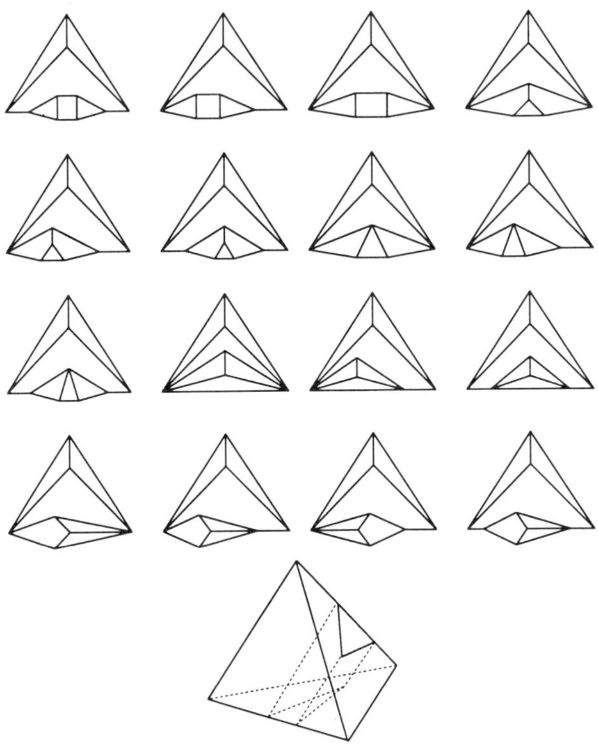

FIGURE 4. Representatives of the 17 types of acoptic heptahedra known to the author; they may be conjectured to be the only possible types. The first 16 have genus 0, and are shown by pseudo-Schlegel diagrams; all are easily derivable from the tetrahedron. The last polyhedron is toroidal, and is shown in a perspective view.

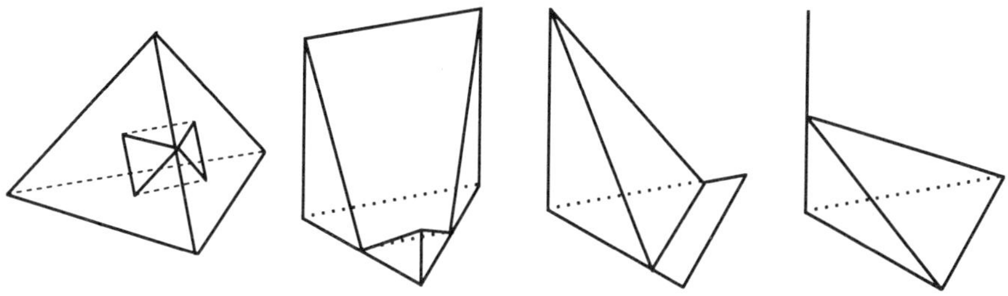

FIGURE 5. Examples of weakly acoptic polyhedra which are not acoptic.

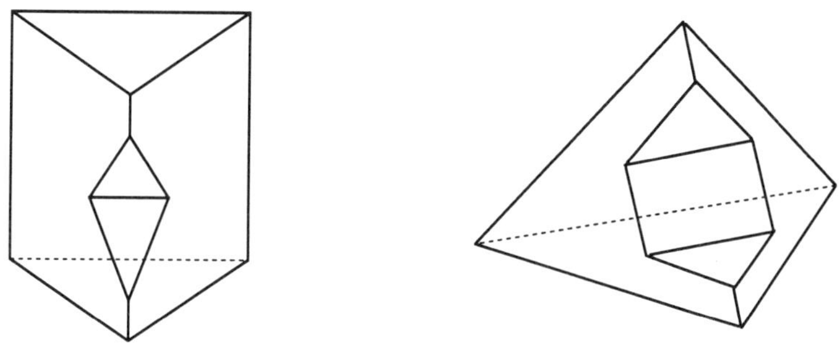

FIGURE 6. Two combinatorially equivalent acoptic heptahedra. In many geometric respects they are sufficiently different to be distinguished in a classification finer than the one by combinatorial type—they are not isomorphic.

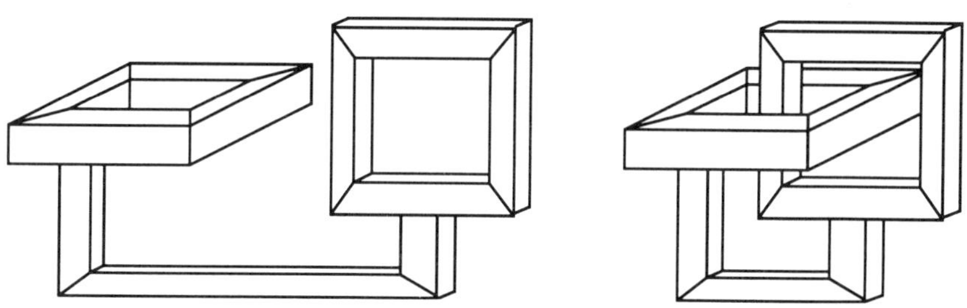

FIGURE 7. Two isomorphic but not equiform acoptic polyhedra of genus 2.

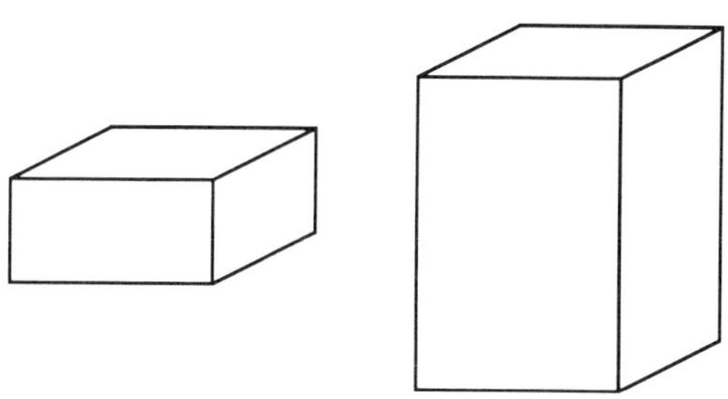

FIGURE 8. Two convex polyhedra that have the same symmetry group cannot be connected by a continuous family of isomorphic polyhedra with the same symmetry group.

A convex-faced acoptic polyhedron in which all vertices are convex is itself convex. It follows that every acoptic polyhedron of positive genus must have a nonconvex vertex.[22] The question how many such vertices must be present in convex-faced polyhedra of various kinds was investigated by several authors; the most inclusive results are those of Betke and Gritzmann [**BeG**]. In this paper it is shown that every convex-faced acoptic polyhedron of positive genus must have at least five nonconvex vertices, and that for every positive genus there is such a polyhedron with precisely five nonconvex vertices.

For classifications of polyhedra finer than by combinatorial type it is convenient to distinguish three possible types of edges. An edge E of an acoptic polyhedron P is said to be **convex** (resp. **flat, concave**) if the dihedral angle of the two faces of P that contain E, measured in the interior of P, is less than π (resp. equals π, is greater than π). Edges E, E' of polyhedra P, P' have the same **convexity character** if both are convex, or both flat, or both concave. Polyhedra P, P' are **isomorphic** (or have the same **isomorphism type**) if there is a map establishing P and P' as combinatorially equivalent, such that corresponding edges have the same convexity character. An example of combinatorially equivalent but nonisomorphic polyhedra is shown in Figure 6. Polyhedra P, P' are **equiform** if they are isomorphic and there exists a continuous family $P(t)$, $0 \leq t \leq 1$, of polyhedra isomorphic to P, such that $P(0) = P$ and $P(1)$ is congruent to P' or the mirror image of P'. An example of isomorphic but not equiform polyhedra is shown in Figure 7. Analogous examples can easily be constructed with acoptic polyhedra of genus 1, but we conjecture that isomorphic acoptic polyhedra of genus 0 are equiform. The analogous result is known to be valid for polygons, see [**Grü3**]. These concepts can be refined further; we say that two equiform polyhedra are *symmetrically equiform* if they, and all the polyhedra involved in the isotopy connecting them, have the same symmetry group. The polyhedra in the continuous family in Figure 1 are symmetrically equiform, but the example in Figure 8 shows that symmetric equiformity is more restrictive than equiformity.

When attention is restricted to convex polyhedra, combinatorial equivalence is a useful equivalence relation. Although it is too coarse an equivalence to be very useful in the more general context of acoptic polyhedra, it is sufficient in certain contexts which occupy us first. The more detailed classifications just described, various levels of which may be appropriate in different situations, are considered in some specific cases later.

For polyhedra of any particular class, two kinds of general constructions are often used. In one construction, limits of sequences of polyhedra of the same combinatorial type are taken, with convergence in the Hausdorff metric. This can be interpreted in two ways. If—as is often customary, see Figure 9—the polyhedra are considered just as polyhedral sets of points, ignoring the combinatorial structure they possess, then the class of convex polyhedra is closed under such limits (if subdimensional limit sets are included).[23] However, a much more interesting concept arises if the convergence is understood as requiring that the mutually corresponding elements form convergent sequences, and the limit is considered as having the same

[22] A vertex V of an acoptic polyhedron P is **convex** provided there is a neighborhood N of V such that the plane of every face incident with V is a supporting plane of $N \cap V$; in other words, if the 2-manifold P is locally convex at V.

[23] The numbers of elements of the limit polyhedron may be smaller than for the terms of the sequence; see [**EGK**] for a discussion of the lower semicontinuity of the numbers of elements.

combinatorial type as the terms of the sequence. In case the class under consideration is that of convex polyhedra, the resulting limits form the **weakly convex polyhedra**. An example of such a polyhedron is shown in Figure 10; other examples appear in Figure 15 and in Figure 18 (for $a = 0$, or $b = 0$, or $a = b$). In case the point-sets of two polyhedra coincide, we shall say that they are **isomegethic**.[24]

If the class of polyhedra whose limits are taken is that of acoptic polyhedra, the resulting limit polyhedra are **weakly acoptic**. Examples of weakly acoptic but not acoptic polyhedra are shown in Figure 5. Weakly acoptic polyhedra can be interpreted as images of cell decompositions in which distinct vertices may be represented by the same point, edges may have zero length, relative interiors of elements may have nonempty intersection, and various other coincidences are possible. However, if the relative interiors of two faces (or a face and an edge, or two edges) of a weakly acoptic polyhedron intersect, the two elements must be coplanar. A different, direct definition of weakly acoptic polyhedra (called there "acoptic") was given in [**Grü6**] and [**Grü7**].

The second general question of interest in connection with any given class of polyhedra is whether the class admits **duality**.[25] It is well known that the class of convex polyhedra is, in fact, closed under duality. For a satisfactory duality among more general acoptic polyhedra one would wish to have corresponding edges (or all corresponding elements) exhibit the same convexity character. For example, isogonal icosahedra (such as the one shown in Figure 11(a), which has been considered by Jessen [**Je**]) are dual in this sense to the isohedral dodecahedra (such as the one in Figure 11(b), described in [**O**] and [**Ste**], and more generally in [**GS5**]). Many pairs of dual acoptic polyhedra can be found such that a dual correspondence between their elements is more or less satisfactory from the point of view of their convexity character. However, no general framework for examples of this kind has been proposed, and it is doubtful that it exists. The only consistent approach to duality of (acoptic, or more general) polyhedra is via **polarity**. However, as shown in Grünbaum-Shephard [**GS4**], with this construction the duals of acoptic polyhedra are often nonacoptic. This fact is one of the reasons for the desirability of considering more general concepts of polyhedra. We shall expand on this topic in a separate note.

One other problem that arises in connection with the concepts introduced above is finding characterizations of each class of polyhedra through orientable maps $\mathcal{C}$ homeomorphic to $\mathcal{C}(P)$ for some polyhedron P in the class considered. For example, the famous theorem of Steinitz characterizes convex polyhedra as combinatorially equivalent to spherical maps with 3-connected graphs. Many of the results discussed below can be considered as giving support to the following conjecture,[26] which seems rather preposterous, but nevertheless is still undefeated.

---

[24] From Greek $\mu\varepsilon\gamma\varepsilon\theta o\sigma$—extent, bulk.

[25] Two polyhedra are *duals* of each other if there exists an incidence-reversing bijection between their sets of elements.

[26] Variants of this conjecture were first proposed in several talks I gave in the mid-1980's. A formulation was also included in a manuscript [**GS3**], several copies of which were privately circulated; however, this paper was never submitted for publication since major parts of it were rendered obsolete by other developments. The conjecture was also mentioned by Barnette *et al.* [**BGH**], and Ljubić [**Lj**].

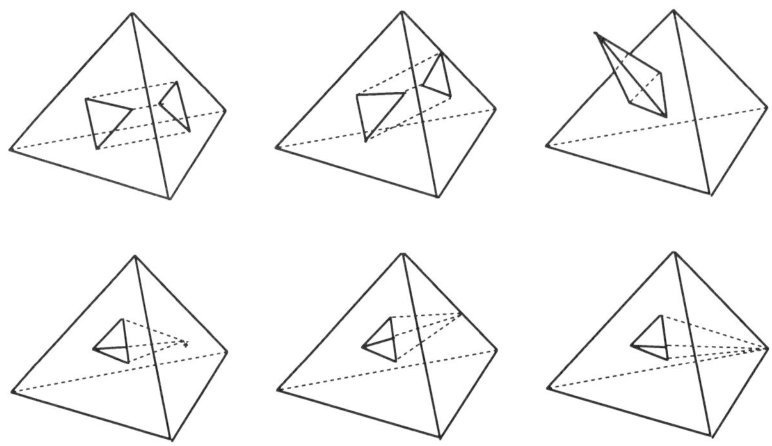

FIGURE 9. Examples of "polyhedral solids" which are not polyhedra under our definitions, since not all their "faces" are simply-connected or limits of simply connected polygons. However, they can be interpreted as limits of simply-connected "polyhedral solids."

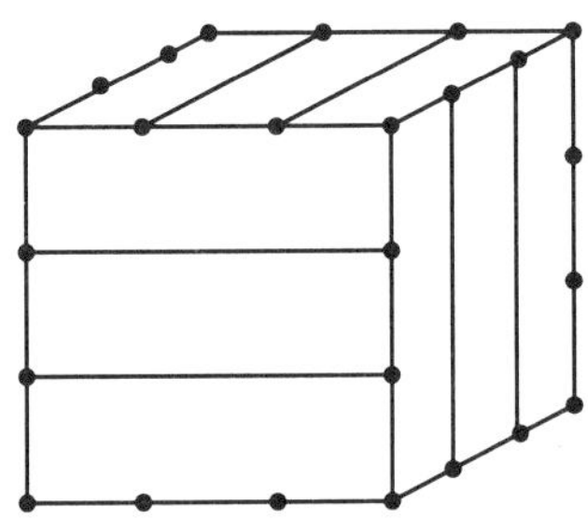

FIGURE 10. An example of a weakly convex polyhedron. The point set determined by the polyhedron and its interior is convex, but the facial structure of this **convex** polyhedron does not coincide with the given weakly convex one. The polyhedron is isomegethic to a cube.

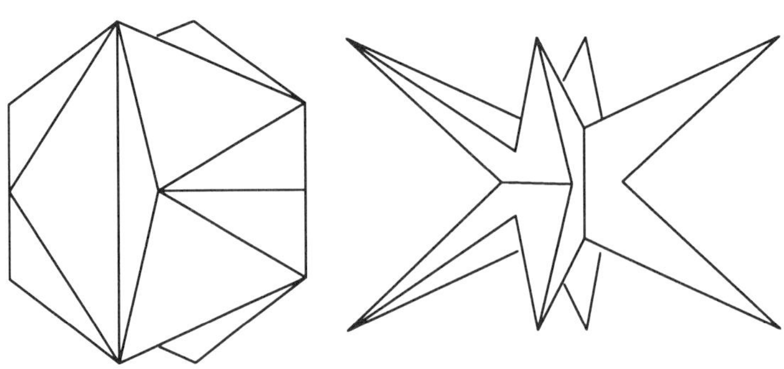

FIGURE 11. An acoptic isogonal icosahedron and an acoptic isohedral dodecahedron which can be considered dual to each other in a correspondence under which corresponding elements have analogous convexity character. However, no general duality theory exists which acts on all acoptic polyhedra under preservation of the convexity character of corresponding elements.

General Realizability Conjecture. Every cell-complex decomposition, without overarching elements, of any compact orientable 2-manifold is realizable by a strictly acoptic polyhedron.

While a proof of the General Realizability Conjecture would characterize strictly acoptic polyhedra, at present time there is no analogous conjecture attempting to characterize the class of all acoptic polyhedra, or the subclasses consisting of those that are convex-faced, or convex-faced and conservative. Also, there is no proposed characterization of weakly acoptic polyhedra, or of weakly convex ones.

3. Realizations of triangulations by acoptic polyhedra

Restricted to triangulations of 2-manifolds, the General Realizability Conjecture appears to have been first formulated in [**Grü**, p. 253] [27] Clearly, in this case any acoptic polyhedron realizing the triangulation is automatically convex-faced and, using admissible perturbations, may be assumed conservative as well. It should be stressed that despite several attempts to establish the conjecture at least for triangulations of the torus, even this special case is still undecided. The first examples of triangulations of the torus seem to go back to Möbius [**Mö**] and Reinhardt [**Re**], but it is not clear exactly what kinds of polyhedra they had in mind—whether combinatorial, topological or geometric. Császár [**Cs**] described a realization by an acoptic polyhedron of the 7-vertex triangulation of the torus (which is a neighborly polyhedron). Other triangulated tori realized by acoptic polyhedra were described by Altshuler [**Al**], [**Al2**]. Various toroidal acoptic polyhedra were described by Alaoglu and Giese [**AG**]; their results and other works concerning polyhedra of positive genus that have certain regularity properties are discussed in more detail in Section 8.

One approach to the proof of the realizability conjecture for triangulations[28] of the torus is by first showing that there is a finite number of "irreducible" triangulations, then finding acoptic realizations of these maps, and finally showing that if one can realize by an acoptic polyhedron a triangulation resulting by a "reduction" of a given triangulation, then the given triangulation can be realized by an acoptic polyhedron as well. The first step, in which a triangulation is called **irreducible** if the contraction of any edge results in a complex which is not a map in the sense used here, has been carried out repeatedly. Independently of each other, this enumeration was carried out in the early 1970's by R. A. Duke and the present author, by D. W. Barnette, and at a later time by S. A. Lavrenchenko; only Lavrenchenko published the list of the 21 irreducible triangulations, see [**L**]. These irreducible triangulations are shown in Figure 12.[29] The feasibility of the second step is also known. However, the last step has so far resisted all attempts at proof.

---

[27] Various aspects of the triangulations case of the General Realizability Conjecture have been discussed, among others, by Altshuler [**Al**], [**Al2**], Barnette [**B4**], Duke [**Du**], and Reay [**R**]; many additional references can be found in these papers. Certain cases in which triangulations of the torus can be realized by an acoptic polyhedron are presented in [**Al**], [**Al2**].

[28] Another approach, mentioned in [**Grü**, p. 253] in connection with the triangulation of the torus with 7 vertices, finds the given map as a subcomplex of the 2-skeleton of a convex 4-polytope, and obtains an imbedding in 3-space by considering the Schlegel diagram of the polytope. For elaboration of this approach see [**Du**], [**AB**] and the references given there.

[29] The statements in Barnette [**B4**] and Schipper [**Sc**] that there are 24 resp. 22 such triangulations are erroneous.

FIGURE 12. The 21 irreducible triangulations of the torus. In each part, the standard identification of the top and bottom, and of the side margins, yields the toroidal triangulation.

In analogy to the first step above, Barnette and Edelson [**BE**], [**BE2**] showed that **every** 2-manifold has only finitely many irreducible triangulations; Nakamoto and Ota [**NO**] showed that the number of vertices in any irreducible triangulation of a 2-manifold M is bounded from above by a linear function of the Euler characteristic of M. Algorithmic aspects of this result have been considered by Schipper [**Sc**].

For triangulations of 2-manifolds of genus greater than 1 only a few specific instances have been decided, by the construction of appropriate polyhedra. In particular, acoptic polyhedra that realize triangulations of orientable 2-manifolds of genus 2, 3, and 4 having the minimal possible number of vertices[30] (10, 10, and 11, respectively) have been obtained by Brehm [**Br2**], [**Br3**] and by Bokowski and Brehm [**BB**].[31] The minimal triangulations of the 2-manifolds of genus 5 and 6 have 12 vertices (in particular, for genus 6 the graph is the complete 12-vertex graph), but so far no realizations by acoptic polyhedra have been reported. Since the number of vertices of minimal triangulations of 2-manifolds of genus g grows only as $\sqrt{g}$, for sufficiently high genus if a realizations by an acoptic polyhedron exists it will have more "holes" than vertices.[32] This was often deemed as an unlikely situation, leading to expressions of disbelief in the existence of such realizations.[33] However, this argument is invalid; in fact, McMullen, Schulz and Wills [**MSW**] constructed acoptic polyhedra of genus g with $O(g/\log g)$ vertices. Their smallest example showing the possibility of genus exceeding the number of vertices occurs for $g = 577$, with 576 vertices.

4. Realizations of toroidal quadrangulations and other maps by acoptic polyhedra

Realizations by acoptic polyhedra of certain cell-complex decompositions of the torus into quadrangles, or into hexagons, have been described already by Becker [**Bec**] (see also Brückner [**Bru**, p. 221]. Realizations of toroidal maps without overarching elements by convex-faced conservative acoptic polyhedra were considered by Simutis [**Si**]; see also [**B4**]. Among other results she showed that various classes of toroidal maps can be realized by such polyhedra, while other maps cannot be realized under these restrictions. However, many of Simutis' unrealizable examples are obtained by face-splitting from realizable maps, and hence can be realized if the polyhedra are not required to be conservative.[34] In other cases in which Simutis shows that the map is not realizable, no convex-faced acoptic polyhedron—conservative or not—can realize the map. This includes, in particular, the well-known 3-valent toroidal map of 7 hexagons,[35] and the 9-quadrangles map shown in Figure 13 (the "twisted triangular picture frame"); however, both maps are realizable by acoptic polyhedra which have some nonconvex faces. For

---

[30] According to Ringel and Youngs [**RY**], the minimal number of vertices in a triangulation of an orientable 2-manifold of genus g equals the least integer $\geq (7 + \sqrt{48g + 1})/2$ if $g \neq 2$, and 10 if $g = 2$.

[31] For information about maps having complete graphs as their 1-skeleton see [**AB2**].

[32] The first case in which this happens is $g = 20 > 19 =$ number of vertices in a minimal triangulation of the manifold of genus 20.

[33] For example, in Schulz [**Sch**].

[34] This applies, in particular, to the examples in Figures 16 and 18 of [**Si**].

[35] 3-valent maps can be realized by convex-faced acoptic polyhedra only if they are of genus 0, see [**Grü**, p. 206] [**B4**]. For far-reaching generalizations of this observation see [**BGH**] and the references given there.

realizations of the former map, found by Szilassi, there are illustrations in several publications.[36] In contrast, it seems that no illustration is available in the literature of the Ljubić torus [**Lj**], which is a realization of the latter map and was described by Ljubić at a regional meeting of the AMS in 1987; one version of the Ljubić torus is shown in Figure 13.

There are many quadrangulations and other cell-decompositions of the torus for which the existence of a realization by an acoptic polyhedron is undecided. Three small maps of this kind, which are prime candidates for counterexamples for the General Realization Conjecture, are shown in Figure 14.

5. Realizations of regular maps by acoptic polyhedra

For **regular maps**[37] on the torus it was shown by Schwörbel [**Schw**] that every such map is realizable by an acoptic polyhedron; this is especially remarkable for the maps consisting of quadrangles, and of hexagons. Two special cases are presented in [**SWi**].

It is known that for every $p \geq 3$ and $q \geq 3$ there exist regular maps with p-gonal faces and q-valent vertices.[38] According to a private communication from Prof. S. E. Wilson, such regular maps exist even if one requires them to be without overarching elements. The General Realizability Conjecture implies that all such maps can be realized by acoptic polyhedra. Unfortunately, a proof of such a far-reaching generalization of Schwörbel's results appears unlikely in the near future.

For an account of the known other cases in which a regular map is realized by a convex-faced acoptic polyhedron, or by an acoptic polyhedron, and for references to the original papers, see Bokowski and Wills [**BW**], Brehm and Wills [**BWi**], Schulte and Wills [**SW**], and their references. The list of such maps is quite short. Specifically, convex-faced acoptic realizations are known for:

(i) The triangulations of genus 3 usually denoted $\{3,7\}_8$ and $\{3,8\}_6$ and named after Klein and Dyck, respectively.

(ii) Quadrangulations (of valences 6 and 8) usually denoted $\{4,6\,|\,3\}$ and $\{4,8\,|\,3\}$, which are most easily visualized as "thickened" versions of the 1-skeleton of the Schlegel diagram in 3-space of the regular 4-simplex and the regular 24-cell; their genera are 6 and 73. In suitable realizations these have the full tetrahedral resp. octahedral groups as symmetry group.[39]

(iii) The two maps $\{6,4\,|\,3\}$ and $\{8,4\,|\,3\}$ dual (as maps) to the ones in (ii).

(iv) An infinite family of maps $\{4,k\,|\,4^{[p/2]-1}\}$ for $k \geq 4$, of genus $1+(k-4)2^{p-3}$, constructed in [**MSW**] and [**MSWi**], and their dual maps.

(v) A realization of the well-known Fricke-Klein map $(3,8)_{12}$ of genus 5 by an acoptic isogonal polyhedron is described below and shown in Figure 19.

---

[36] See, for example, Gardner [G], Stewart [Ste], Szilassi [Sz].

[37] A map (cell decomposition) or polyhedron is **regular** if the group of its combinatorial automorphisms acts transitivity on its **flags**, where each flag is a triplet consisting of a vertex, an edge, and a face, all mutually incident.

[38] This conjecture from [Grü5] has been proved by Vince [V] and Wilson [GW].

[39] As pointed out by Prof. J. M. Wills, acoptic realizations with full icosahedral symmetry group have not been found for any regular map. Possibly, none exist.

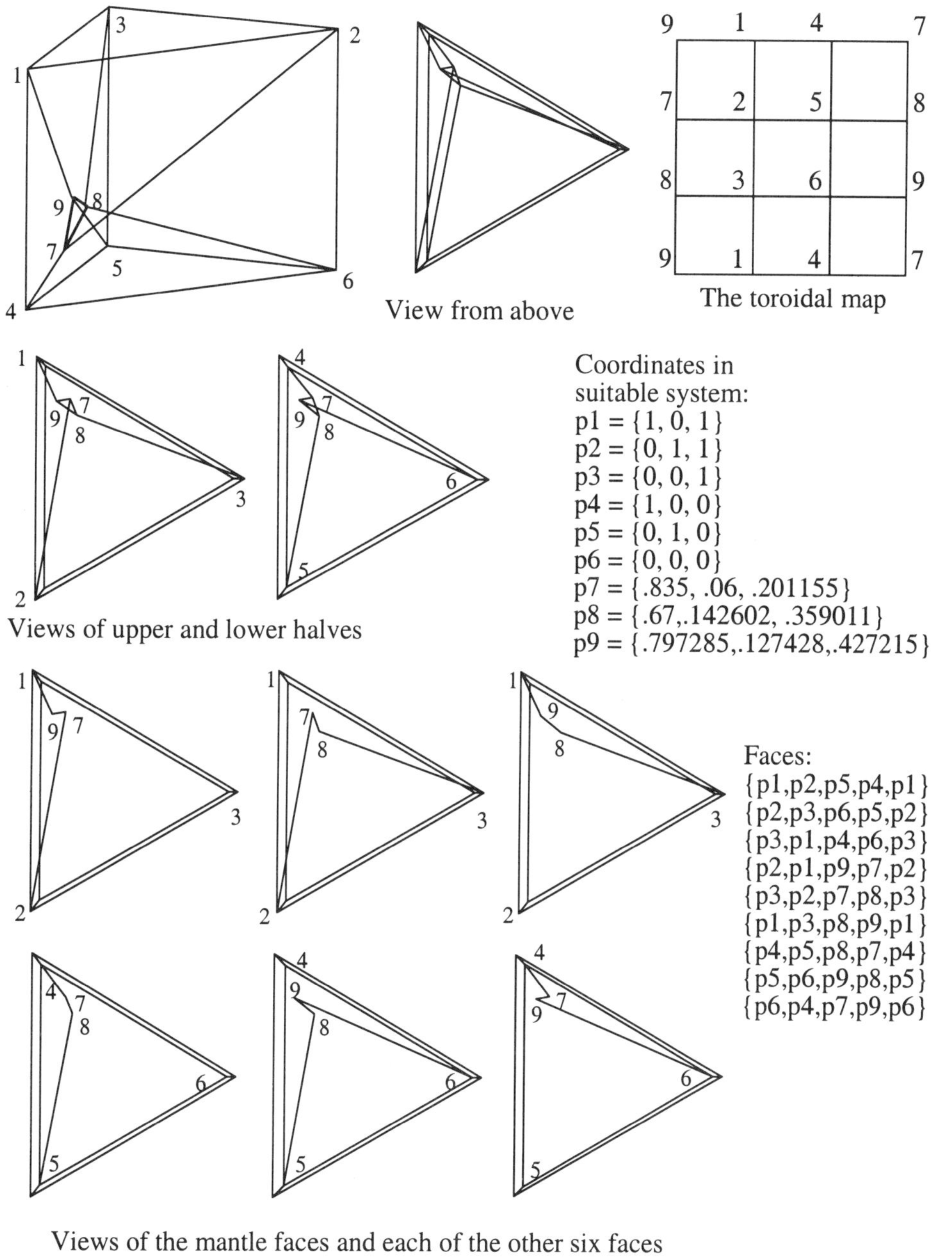

FIGURE 13. The Ljubić torus, an acoptic realization of the 9-face quadrangulation shown top right, that is not realizable by convex-faced acoptic polyhedra.

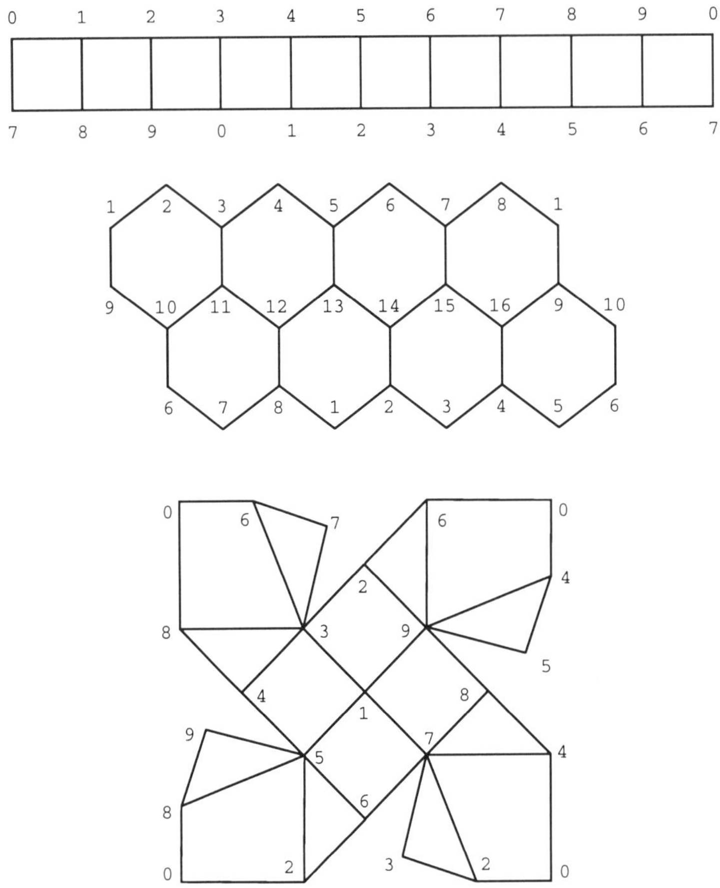

FIGURE 14. Three maps without overarching elements, for which it is not known whether they can be realized by acoptic polyhedra. The first two are toroidal, the last has genus 2 and appears in [**PV**] as example in a different context.

6. Acoptic polyhedra with prescribed face-vectors or f-vectors

The General Realization Conjecture and related approaches to other classes of polyhedra aim at a very detailed description of possible combinatorial types. A coarser classification is provided by Eberhard-type theorems, which have been investigated for convex polyhedra for over a century.[40] These results deal with realizations by polyhedra of a given class of sequences that specify the number of k-gonal faces (or k-valent vertices) for various values of k. The basic restrictions on the sequences involved usually stem from the Euler relation (and, in some cases,

[40]The first results are those of Eberhard [**E**], dealing with convex polyhedra. For a more accessible presentation of Eberhard's results see [**Grü**]. The topic has had (and continues to have) many extensions, specializations and related developments; for some of these see [**GZ**], [**J**], [**JO**] and the references given there. It is interesting to note that the constructions used by Eberhard have recently been rediscovered by chemists investigating the "Fullerene" forms of carbon; see, for example, Fowler and Redmond [**FR**] and the references given there.

from more specific constraints); in many cases, the difficult part is the construction of the polyhedra.

In more recent times the topic also attracted interesting extensions to convex-faced acoptic polyhedra of positive genus, see [**JJ**], [**Ju**], [**JJ2**], [**Gr**]. There seem to have been no corresponding examinations of the case in which other classes of acoptic polyhedra are admitted. However, there has been considerable work done on the question of realization of even coarser data sets. In particular, investigations that dealt with the realization by convex-faced acoptic polyhedra of the **f-vectors** (whose components are the numbers of elements of different dimensions of the polyhedron in question) or some other global characteristic (such as the **valence-functional**)[41] have led to very interesting results. Concerning the latter, we mention here only the following result from [**BGH**]:

If P is a convex-faced acoptic polyhedron of genus $g \geq 1$ then $\delta(P) \geq 2 + 2g$. This is best possible for $g = 1$.

Many other results and open problems on valence functionals of convex-faced acoptic polyhedra, dealing also with polyhedra realizable in higher-dimensional spaces and with non-orientable polyhedra, are presented in [**BGH**] together with references to earlier literature.

7. Isogonal acoptic polyhedra

Another direction of investigation concerns the acoptic polyhedra which possess a high degree of symmetry. If the group of (isometric) symmetries of a polyhedron acts transitively of its vertices (or edges, or faces) then the polyhedron is said to be **isogonal** (or **isotoxal**, or **isohedral**, respectively). For convex polyhedra and for the analogously defined maps on the sphere, the possibilities have been studied in some detail for close to two centuries.[42] Results on one aspect of this topic, dealing with maps on the sphere and convex polyhedra appear in Grünbaum and Shephard [**GS**], while the existence of isogonal acoptic polyhedra of higher genus was established in [**GS2**].[43] The construction of isogonal acoptic tori is discussed in some detail in [**Grü7**]; there are two main kinds, with vertex symbols (3.3.3.3.3.3) and (3.3.3.4.4). For each kind there are distinct combinatorial types that depend on two discrete parameters; each combinatorial type depends (up to similarity) on a continuous parameter. As shown by Schwörbel, the regular toroidal map $\{3,6\}_{2,2}$ (in the notation of Coxeter and Moser [**CM**]) can be realized by an acoptic torus. Slightly different realizations of the map $\{3,6\}_{2,2}$ are provided by members of the continuous family of isogonal tori shown in Figure 15. Another type of highly symmetric polyhedra is described by Wills [**Wi**].

---

[41] Following [**BGH**], the *valence functional* $\delta(P)$ of a polyhedron P is $\delta(P) = \Sigma$ (valence(V)-3), where the summation is over all the vertices of V. It is a way of measuring the departure of the polyhedron or map from being simple (3-valent). It is likely that analogous results may hold for the similarly defined functionals Σ(valence(V)-4) and Σ(valence(V)-6.

[42] For the early history see Brückner [**Bru**]; a recent account is given by Martini [**Ma**].

[43] According to [**Ku**] and [**BWi**, p. 547] the "flat tori" mentioned in Brehm and Kühnel [**BK**], which were described by Brehm [**Br**] in 1978, are isogonal; however, the published accounts do not mention any symmetry properties. According to Kurth [**Ku**], these polyhedra have been independently discovered by H. Leitzke.

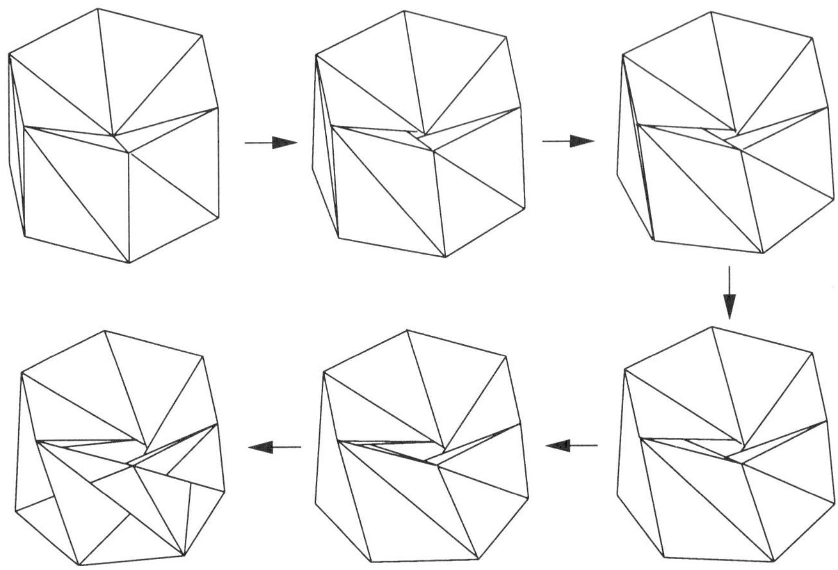

FIGURE 15. Several isogonal tori that realize the regular map $\{3,6\}_{2,2}$. All are acoptic except the first and the last, which are weakly acoptic.

In order to present information about isogonal polyhedra of a given combinatorial type in a convenient manner, we use **dynatograms**[44] and **panoramas**. Both representations are possible because each such family depends on at most two parameters. In a dynatogram we show the regions (or arcs) that correspond to parameter values that yield isomorphic polyhedra, while in a panorama small copies of views of polyhedra are placed in positions that correspond (approximately) to the parameter values that determine representative examples of the type. As an illustration, we show in Figure 16 a parametrization of isogonal polyhedra of the combinatorial type of the snub cube (3.3.3.3.4). The dynatogram of this family is shown in Figure 17, while a panorama appears in Figure 18. These tools lead to various insights; the following are some examples.

(i) Despite appearances, the weakly acoptic polyhedron that corresponds to $(a, b) = (1, 0)$ is not isomegethic with the cuboctahedron; in the notation of Figure 16, pairs of vertices such as D and F coincide, hence (proper) triangles ADE and ADF also coincide, but extend into the *interior* of the cuboctahedron!

(ii) It is obvious on contemplating Figure 18 that the nonconvex acoptic polyhedra with $a > b$ are all isomorphic, and that the nonconvex ones with $a < b$ are also all isomorphic—but the polyhedra in two families are not isomorphic to each other.

(iii) The Boolean sum[45] of a polyhedron with $b < 0$ with its convex hull yields an acoptic isogonal polyhedron of genus 5, while the same construction on a polyhedron with $b > 1$ leads to an acoptic isogonal polyhedron of genus 7; these polyhedra have been described in [**GS2**].

---

[44]From the Greek δυνατοσ—possible.

[45]The Boolean sum of two acoptic polyhedra is obtained from the union of the elements of both by deleting the faces that occur in both polyhedra.

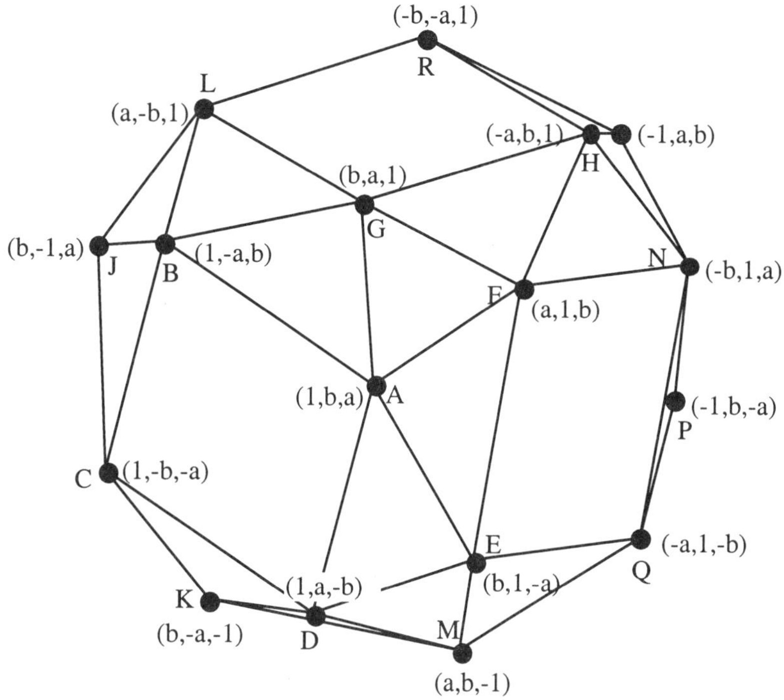

FIGURE 16. A snub cube (3.3.3.3.4), with notation for its vertices, and with indication of a parametrization of the vertices which is convenient for isogonal polyhedra of this combinatorial type.

(iv) The Boolean sum of the polyhedron that corresponds to $(a, b) = (1/3, 1/6)$ with the mirror image of the polyhedron with $(a, b) = (1/3, -1/6)$ is the new realization of the regular map $(3, 8)_{12}$ of genus 5 mentioned earlier (see Figure 19). For more details concerning polyhedral realizations of this map see [**Grü8**].

As another illustration we show in Figure 21 a dynatogram of the acoptic isogonal icosahedra (3.3.3.3.3), using the notation of Figure 20, and with explanations analogous to those in the caption of Figure 18. Comments similar to (i), (ii) and (iii) above apply in this case as well.

Analogous dynatograms and panoramas can be presented for many other families of polyhedra. A panorama of those acoptic polyhedra of the combinatorial type of the regular dodecahedron, which have a plane of symmetry, is shown in Figure 1. A systematic exposition of dynatograms and panoramas for acoptic polyhedra that are isogonal, or isotoxal, or isohedral is being prepared.

8. Monohedral polyhedra

Several different directions of investigation concern polyhedra that are **monohedral**, that is, have all faces mutually congruent but not necessarily equivalent under symmetries of the polyhedron. The common shape of faces of a monohedral polyhedron is called its **protoface**. A well-known example of a convex monohedral but not isohedral polyhedron is shown in Figure 22(a). It is remarkable that the following question, widespread in the folklore, is still open:

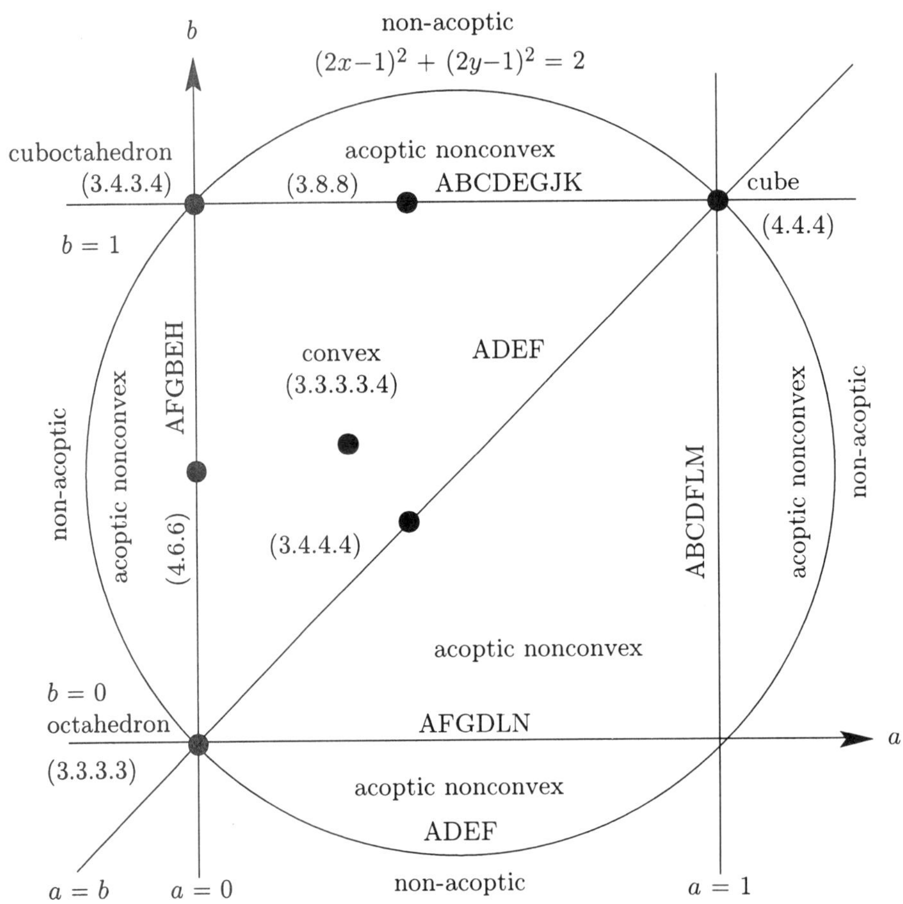

FIGURE 17. Dynatogram for acoptic isogonal polyhedra of combinatorial type of the snub cube, (3.3.3.3.4). The parameters a and b determine the coordinates of the vertices, as shown it Figure 16. All regions outside of the large circle correspond to nonacoptic polyhedra, while the polyhedra with parameter values on the circle are weakly acoptic. The strings of capital letters near the lines and the circle indicate (in the notation of Figure 16) examples of sets of coplanar vertices. The symbols for isogonal polyhedra near the lines and near three of the vertices of the square indicate to which isogonal polyhedra the polyhedra of type (3.3.3.3.4) are isomegethic (that is, consist of the same set of points). The solid dots indicate the parameter values for which the polyhedra of type (3.3.3.3.4) are isomegethic to a uniform (Archimedean) polyhedra. The uniform (Archimedean) snub cube (3.3.3.3.4) corresponds to the parameter values $a = 0.29955977$, $b = 0.5436890$.

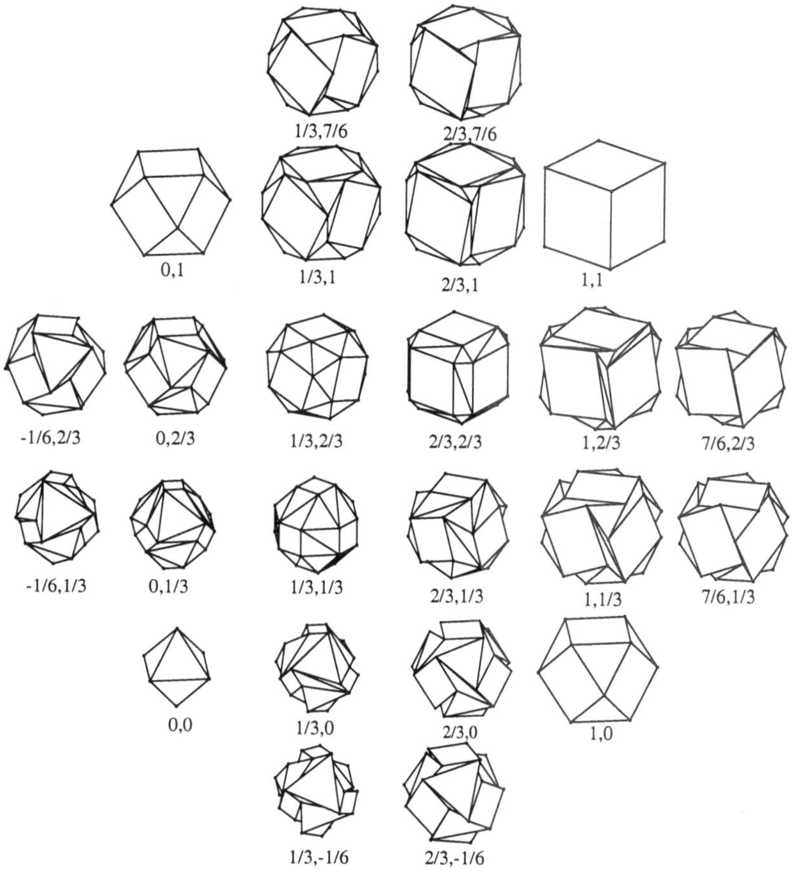

FIGURE 18. A panorama of acoptic isogonal polyhedra combinatorially equivalent to the snub cube (3.3.3.3.4). This panorama corresponds to the dynatogram in Figure 17. The values of the parameters a and b are given under each polyhedron.

Conjecture. If the simple polygon F is the protoface of a strictly convex monohedral polyhedron then F is also the protoface of an isohedral polyhedron.[46]

The requirement of strict convexity is essential for the validity of the conjecture, as shown be the example of the monohedral polyhedron in Figure 10; its protoface is a 3 by 1 rectangle, which is not the protoface of any isohedral polyhedron. The conjecture is analogous to one part of Problem 18 of Hilbert's famous collection [**Hi**] concerning monohedral tilings of the plane. The tiling analog of the conjecture is known to fail, but the complete determination of strictly convex protofaces of monohedral tilings is still open; see [**GS5**] for details and references.[47]

[46] It seems that the protoface of the polyhedron in Figure 22 is the only quadrangle which is the protoface of a strictly convex monohedral but not isohedral polyhedron. No pentagon with a similar property is known.

[47] One particularly attractive open question concerning monohedral tilings is whether the prototile of some such tiling is *aperiodic*, that is, *every* monohedral tiling with this prototile has no symmetries besides the identity. Analogously, it is not known whether there exists a polygon F such that every acoptic monohedral polyhedron with protoface F has only the identity symmetry.

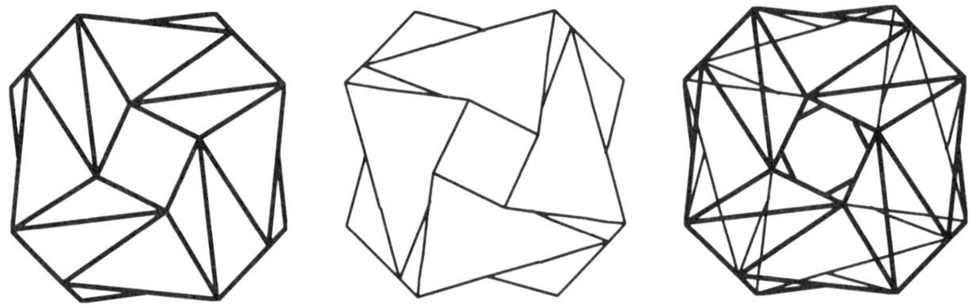

FIGURE 19. The Boolean sum of the polyhedron in (a) which corresponds to $(a, b) = (1/3, 1/6)$, with the polyhedron in (b) for which $(a, b) = (1/3, -1/6)$, yields the isogonal acoptic polyhedron in (c); the elimination of the square faces makes the bottom part of the polyhedron visible in (c). The polyhedron in (c) is a realization of the regular map $(3.8)_{12}$ (which is map #96.162 in Wilson's catalog [**Wil**] of regular maps).

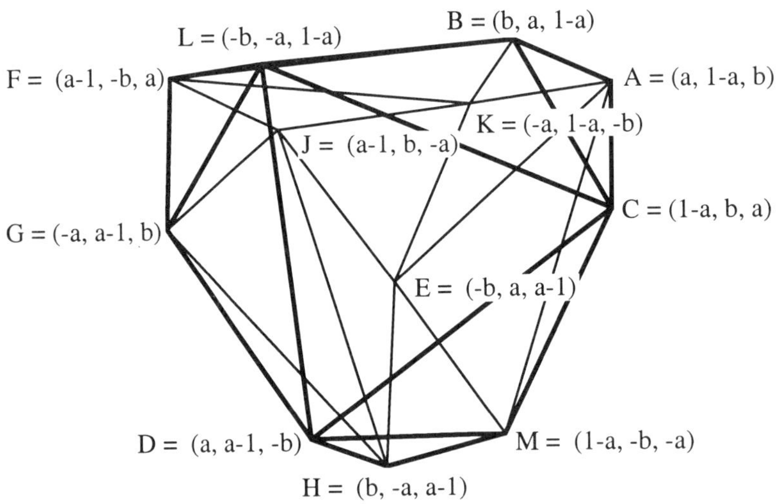

FIGURE 20. An icosahedron (3.3.3.3.3), with notation for its vertices, and with indication of a parametrization of its vertices which is convenient for isogonal polyhedra of this combinatorial type.

A different type of question, first raised long ago by Tom Banchoff[48] for icosahedra, is the following: For any given polyhedron P, all faces of which have the same number of sides, determine all monohedral acoptic polyhedra Q combinatorially equivalent to P. In particular, one may ask what are the different isomorphism types of such polyhedra Q, what polygons F may serve as protofaces for the Q's, and how many distinct polyhedra Q there are for a given protoface F. Various results on these and similar questions were presented by the author in talks and courses at the University of Washington since 1990; detailed expositions are planned. These results include the following.

[48]Private communication, some twenty years ago.

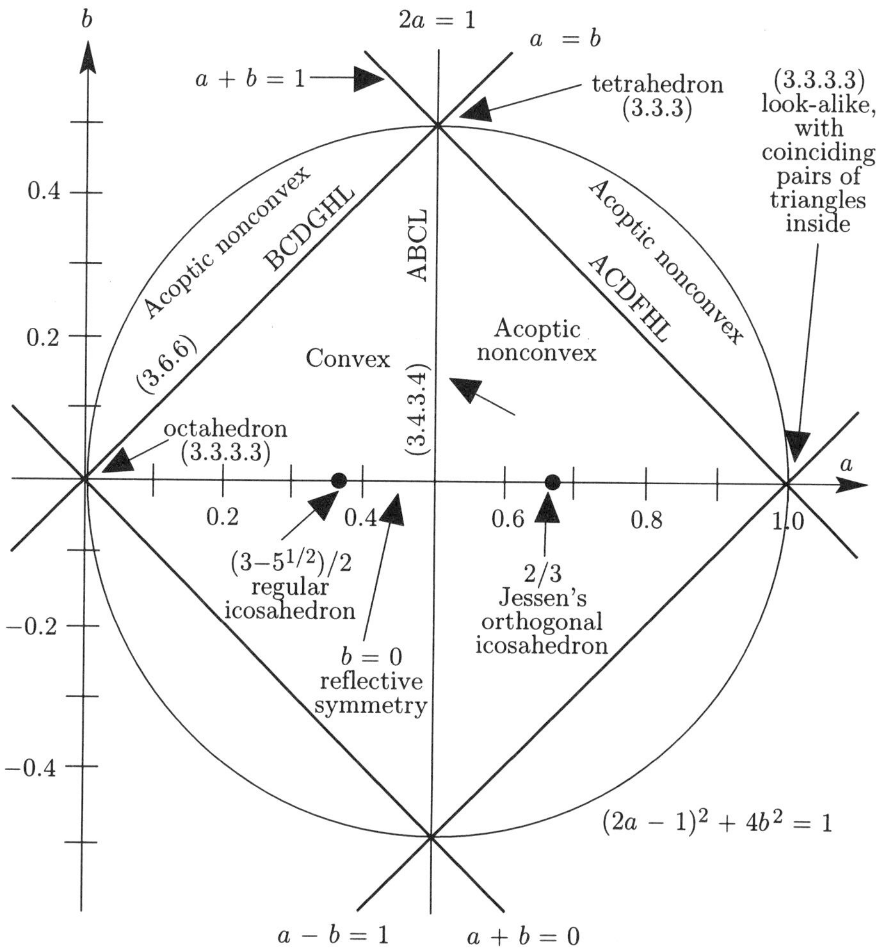

FIGURE 21. Dynatogram for acoptic polyhedra of combinatorial type of the icosahedron, (3.3.3.3.3). As shown in Figure 20, the parameters a and b determine the coordinates of the vertices; they are given (in Cartesian coordinates) by all cyclic permutations of $(\pm a, \pm(1-a), \pm b)$ with an even number of minus signs. The similarity of many features of this dynatogram with the dynatogram of snub cubes in Figure 17 is not surprising since icosahedra can be interpreted as "snub tetrahedra." However, since triangles here play the roles played by both triangles and squares in the case of snub cubes, polyhedra corresponding to the bottom part of the dynatogram are congruent mirror images of the polyhedra corresponding to the upper part; polyhedra with $b = 0$ have mirror-symmetry.

(1) Every triangle is the protoface of a convex monohedral polyhedron combinatorially equivalent to the regular octahedron. Moreover, every monohedral acoptic polyhedron combinatorially equivalent to the regular octahedron is isohedral. Each scalene acute triangle is the protoface of four distinct monohedral octahedra; if symmetry groups are taken into account, these four are of two distinct isomorphism types. Additional information is contained in the dynatogram in Figure 23, in which the parameters a and b are the sides of the triangular protoface, with $a \leq b \leq c = 1$, and in the "panorama" in Figure 24. For parameter values in E

and G, one of the polyhedra is weakly acoptic but not acoptic. If P is a simplicial convex polyhedron such that every triangle is the protoface of a monohedral convex polyhedron combinatorially equivalent with P, then P is combinatorially equivalent to the regular octahedron.[49]

(2) Banchoff's original question about monohedral acoptic polyhedra of the combinatorial type of the regular icosahedron is much more complicated, and is not fully answered.[50] It is known that there are many distinct isomorphism types, even if only those polyhedra are considered which have three mutually orthogonal axes of halfturn symmetry. As indicated in Figure 25, these polyhedra have nine transitivity classes of edges, but all faces are equivalent under (combinatorial) automorphisms compatible with the orbits of the edges. A dynatogram for such polyhedra, obtained by numerical calculations, is shown in Figure 26. It shows, among other facts, the possibility of two distinct convex polyhedra with the same scalene protoface; this shows that the formulation of Cauchy's rigidity theorem has to be formulated more carefully than is often done. Many of the polyhedra of this kind are visually quite attractive, as shown by the examples in Figure 27. However, many additional types of monohedral icosahedra with other symmetries are possible; so far these have been explored only superficially. Several examples with isosceles triangles as protofaces are shown in Figure 28.

(3) There exist convex polyhedra for which no isomorphic polyhedron is monohedral. An example is shown in Figure 29(a). It is realizable by a weakly convex monohedral polyhedron, but the polyhedron in Figure 29(b) admits no such realization. The polyhedron in Figure 30 admits no realization as a monohedral acoptic polyhedron. All these are the smallest examples of their kind known to me.

(4) For any given triangle T the number of monohedral convex polyhedra with protoface T is finite, but for suitable T it can be arbitrarily large.

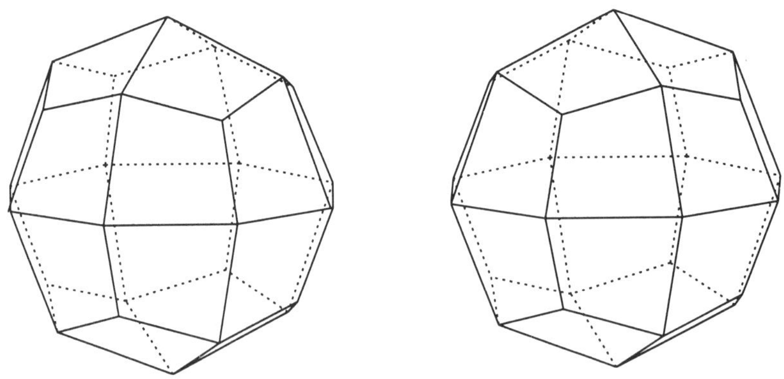

FIGURE 22. (a) A monohedral polyhedron which is not isohedral. It is the polar of the pseudorhombicuboctahedron. (b) An isohedral polyhedron with the same protoface; it is the polar of the rhombicuboctahedron.

---

[49]For some of these results see also Webber [**W**].

[50]The convex ones among them were investigated by E. Miller, in an undergraduate research project directed by Banchoff. Among Miller's results is the discovery that a unique convex icosahedron has a right-angled protoface. A publication is being prepared [**Mi**].

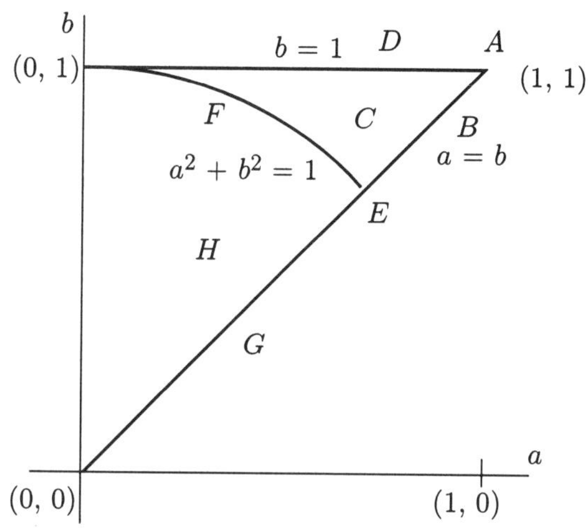

FIGURE 23. A dynatogram of monohedral (and isohedral) octahedra that have as prototile a triangle with sides $a \leq b \leq c = 1$.

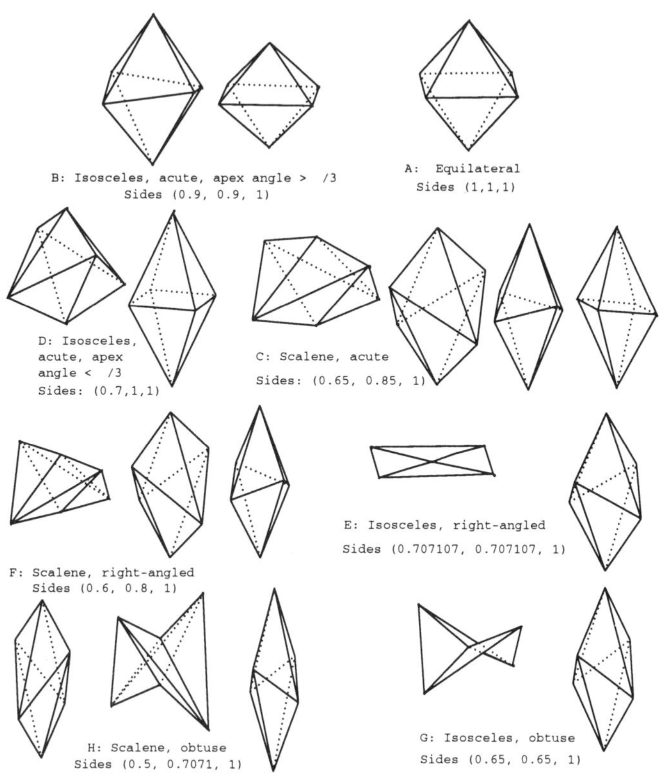

FIGURE 24. A panorama of the monohedral octahedra, corresponding to the dynatogram in Figure 23. The panorama is greatly distorted, due to the necessity of accommodating up to four polyhedra at one location.

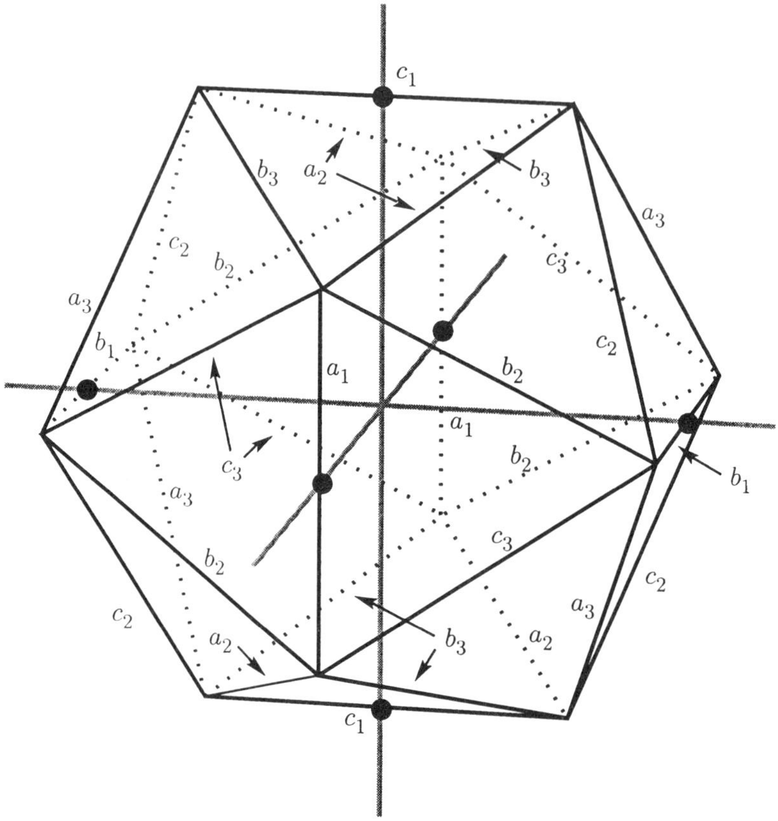

FIGURE 25. Notation used in Figure 26, for the edges of one type of monohedral acoptic isocahedra combinatorially equivalent to the regular icosahedron. The polyhedra in question have three mutually perpendicular axes of halfturn symmetry, indicated by the shaded lines.

In this context we also have the following

Conjecture. If $f > 120$ there are at most three convex monohedral polyhedra with precisely f faces.

A different direction concerns the study of monohedral tori, which was started by Alaogly and Giese [**AG**]. They describe 6-valent triangle-faced as well as 4-valent quadrangle-faced monohedral tori. Additional examples can be found in [**Ste**]. Recently Webber [**W**] has developed methods of construction of many interesting kinds of 6-valent triangle-faced monohedral tori. Webber's tori can have arbitrarily large numbers of faces, and even can be knotted. It is interesting to note that the question whether there exist monohedral tori with hexagonal protoface is still open.[51]

Conjecture. There exist no monohedral acoptic polyhedra of genus greater than 1 in which all vertices have the same valence.

---

[51]It may be mentioned that the isogonal acoptic tori described in [**GS2**] and [**Grü7**], with vertex symbols (3.3.3.3.3.3), have polars which are 3-valent monohedral (even isohedral) realizations of hexagon-faced toroidal maps; however, these polar polyhedra are not acoptic, see [**GS4**].

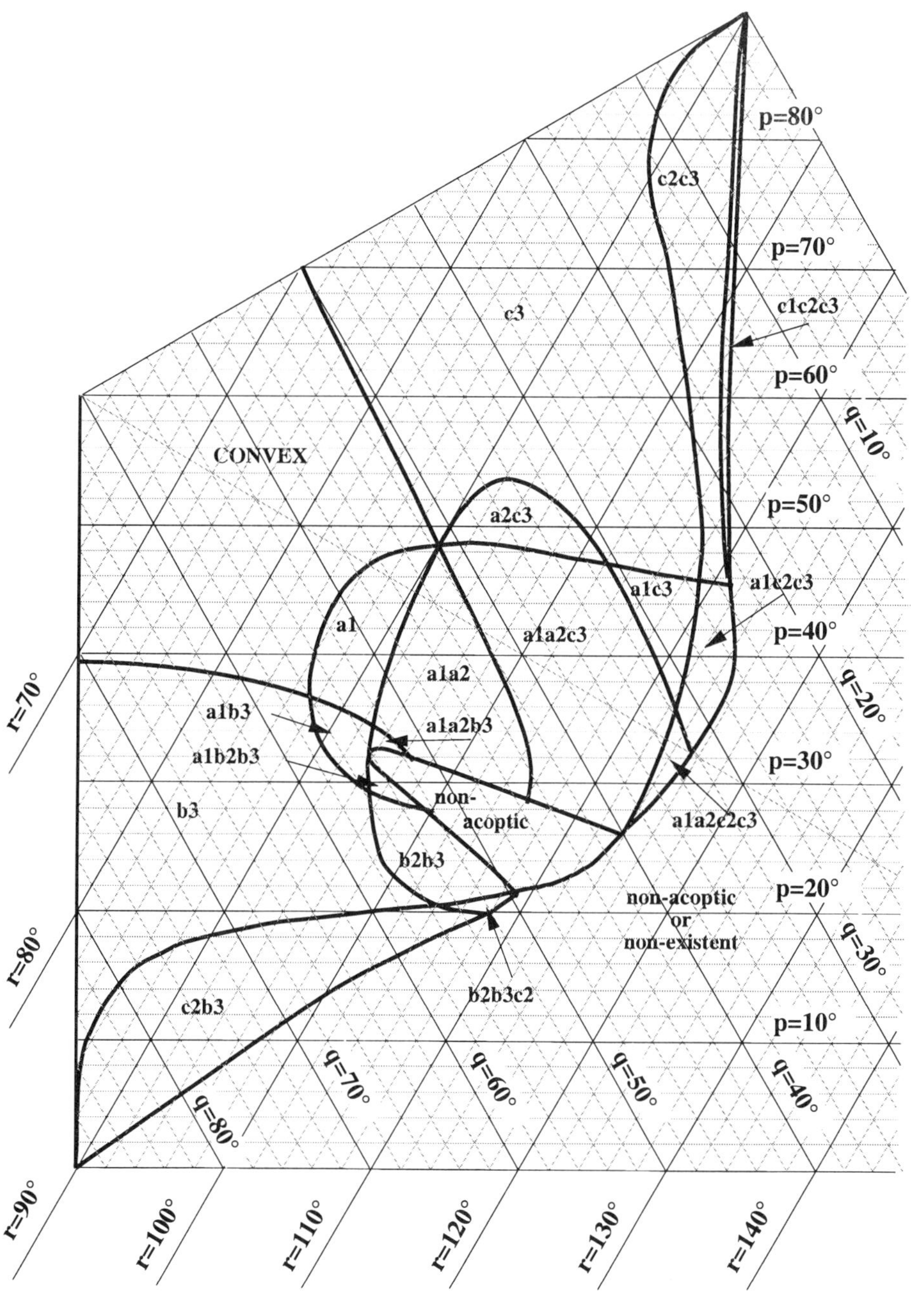

FIGURE 26. An empirical dynatogram of the convexity types of the kind of acoptic polyhedra isomorphic to the regular icosahedron described in Figures 24 and 25. The symbols designate the nonconvex edges, in the notation of Figure 25. The parameters are the angles of the protoface, and the presentation of the dynatogram is in trilinear coordinates.

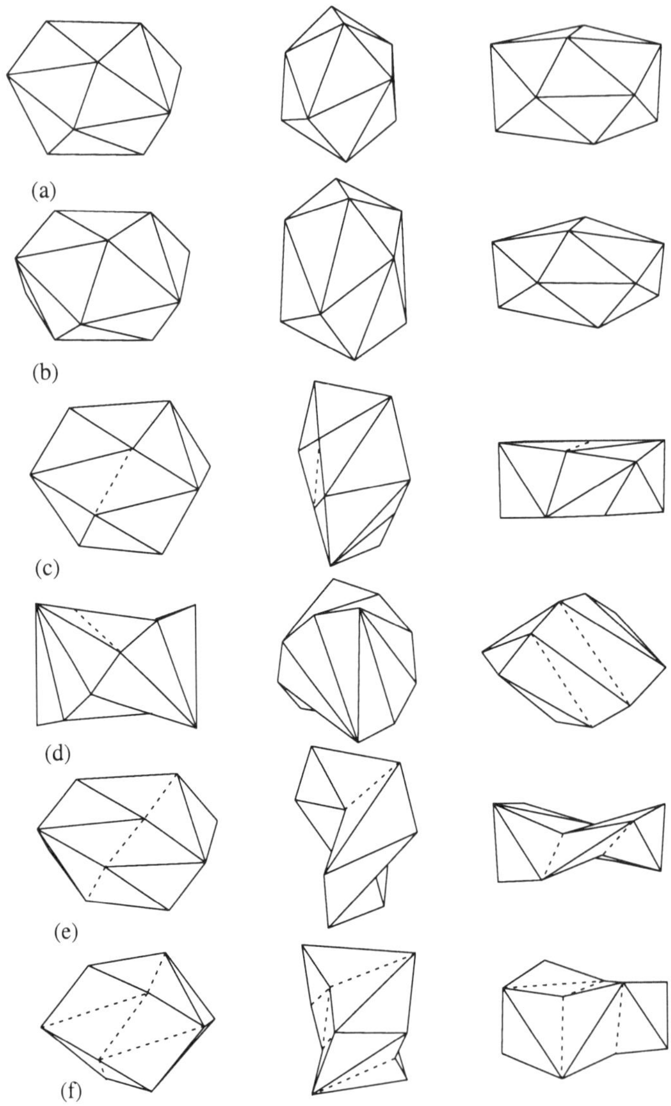

FIGURE 27. Examples of monohedral icosahedra with three mutually perpendicular axes of 2-fold symmetry. Each is shown in perspective views from three different directions (close to—but not coinciding with—the three axes of symmetry); concave edges are indicated by dashed lines. (a) and (b) show the two distinct convex icosahedra having as protoface a 45°-60°-75° triangle. The protoface in (c) is a 45°-45°-90° triangle, and in (d) a 24.4°-77.8°-77.8° triangle. In both only the edges of one orbit are concave, but it is a 2-edge orbit in (c) and a 4-edge orbit in (d). The two icosahedra with a 34°-46°-100° triangle as protoface are shown in (e) and (f); the former has edges of two orbits concave, while the latter has edges of three orbits concave.

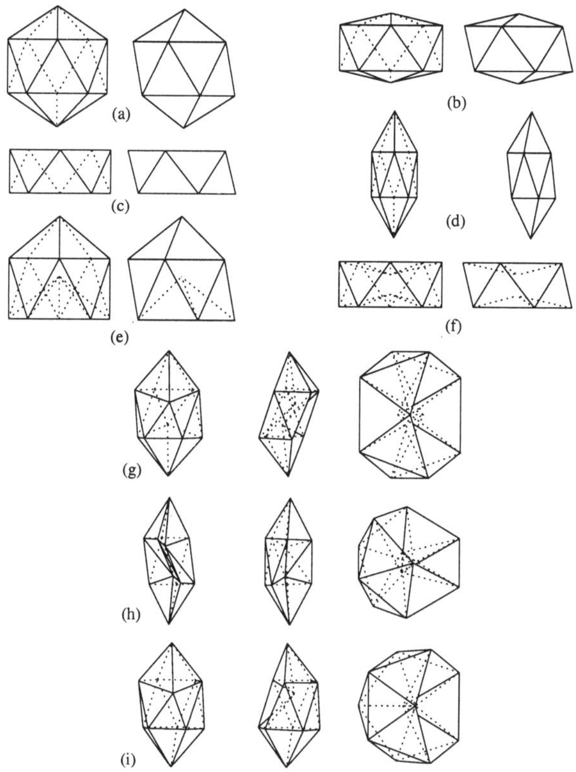

FIGURE 28. Examples of monohedral polyhedra with isosceles triangles as protofaces, and combinatorially equivalent to the regular icosahedron. The first six have an axis of 5-fold rotational symmetry; the axis is taken as vertical, and each polyhedron is presented by two views in horizontal, mutually perpendicular directions. The last three polyhedra are shown in views from three mutually perpendicular directions. The symmetry groups are: (a) [3,5]; (b), (c), (d), (f) [2,5]; (e) [5]; (g) [2,1]; (h) $[2]^+$; (i) [1]. If symmetry groups are taken into account, all polyhedra shown are of different isomorphism types, except for (b) and (d); these two have the same isomorphism type and are equiform but not symmetrically equiform.

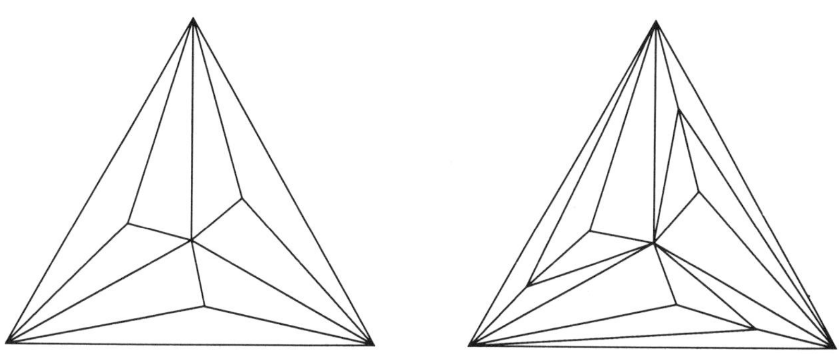

FIGURE 29. Schlegel diagrams of two convex polyhedra; the polyhedron in (a) does not have a monohedral realization as a convex polyhedron, while the one in (b) cannot be monohedrally realized even with a weakly convex polyhedron.

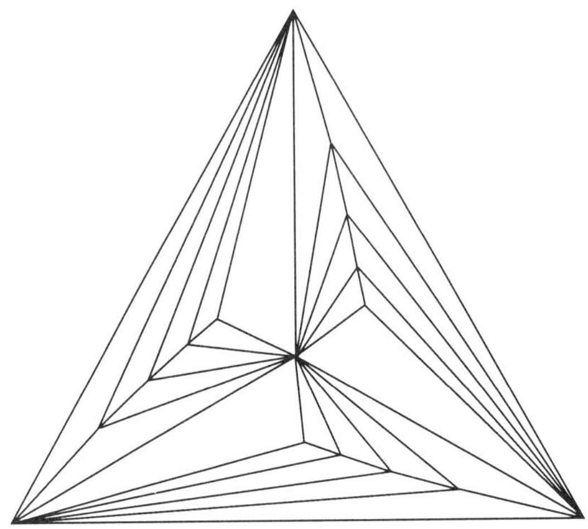

FIGURE 30. A Schlegel diagram of a convex polyhedron for which no combinatorially equivalent acoptic polyhedron is monohedral.

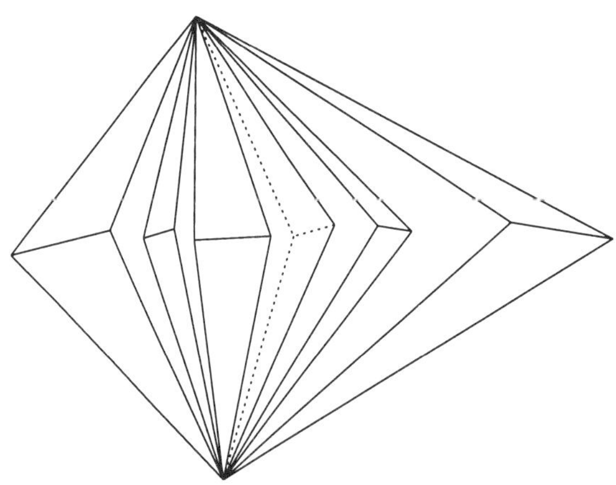

FIGURE 31. An acoptic polyhedron of genus 0 with no spanning tree of valence ≤ 3.

9. Spanning trees

Barnette [**B3**] proved that each toroidal convex-faced acoptic polyhedron has a spanning tree of maximal valence 3. Does this extend to polyhedra of higher genus, or to more general acoptic polyhedra without overarching elements? As can be seen from Figure 31, an extension is not possible without some conditions on overarching elements. Earlier, in [**B**], Barnette established the existence of such trees for convex polyhedra. In this context, an old question from [**Grü2**, p. 1148], which would extend Barnette's result, may be mentioned:

Conjecture. Every convex polyhedron admits a spanning tree of maximal valence 3 such that in the dual polyhedron there is a spanning tree of maximal valence 3 which uses only edges that correspond to edges not used in the spanning tree of the starting polyhedron.

With obvious definitions, this can be reformulated as saying that each convex polyhedron admits simultaneously a spanning vertex-tree and a spanning face-tree, both of maximal valence 3 and with no edge used in both. In this version it is possible to inquire what happens for convex-faced toroidal polyhedra, or for other classes of polyhedra.

* * * * *

Many additional topics, results and problems dealing with acoptic polyhedra could have been included in the above discussion. Although limitations of time and space require us to stop, it is our hope that the presentation will arouse the curiosity of the reader concerning the many questions about acoptic polyhedra—whether posed above or not—and lead to interesting new developments.

References

[AG] L. Alaoglu and J. H. Giese, *Uniform isohedral tori.* Amer. Math. Monthly, **53** (1946), 14–17.

[A] A. D. Aleksandrov, *Convex Polyhedra.* [In Russian]. Moscow 1950. German translation: A. D. Alexandrow, Konvexe Polyeder, Akademie-Verlag, Berlin, 1958.

[A2] V. A. Aleksandrov, *A new example of a flexible polyhedron.* Siberian Math. J. **36** (1995), 1049–1057.

[Al] A. Altshuler, *Polyhedral realizations in* R^3 *of triangulations of the torus and 2-manifolds in cyclic 4-polytopes.* Discrete Math. **1** (1971), 211–238.

[Al2] A. Altshuler, *Construction and enumeration of regular maps on the torus.* Discrete Math. **4** (1973), 201–217.

[AB] A. Altshuler and U. Brehm, *A non-Schlegelian polyhedral map on the torus.* Mathematika, **31** (1984), 83–88.

[AB2] A. Altshuler and U. Brehm, *Neighborly maps with few vertices.* Discrete Comput. Geom. **8** (1992), 93–104.

[B] D. W. Barnette, *Trees in polyhedral graphs.* Canad. J. Math. **18** (1966), 731–736.

[B2] D. W. Barnette, *Graph theorems for manifolds.* Israel J. Math. **16** (1973), 62–72.

[B3] D. W. Barnette, *3-trees in polyhedral maps.* Israel J. Math. **79** (1992), 251–256.

[B4] D. W. Barnette, *Polyhedral maps of 2-manifolds.* Convexity and Related Combinatorial Geometry Proc. Second University of Oklahoma Conference, D. C. Kay and M. Breen, eds. Dekker, New York 1982, 7–19.

[BE] D. W. Barnette and A. L. Edelson, *All orientable 2-manifolds have finitely many minimal triangulations.* Israel J. Math. **62** (1988), 90–98.

[BE2] D. W. Barnette and A. L. Edelson, *All 2-manifolds have finitely many minimal triangulations.* Israel J. Math. **67** (1989), 123–128.

[BGH] D. W. Barnette, P. Gritzmann and R. Höhne, *On valences of polyhedra.* J. Combinat. Theory A **58** (1991), 279–300.

[BG] D. W. Barnette and B. Grünbaum, *On Steinitz's theorem concerning convex 3-polytopes and on some properties of planar graphs.* In "The Many Facets of Graph Theory," G. Chartrand and S. F. Kapoor, eds. Lecture Notes in Math, **110** (1969), 27–40. Springer, Berlin-Heidelberg-New York 1969.

[Bec] J. K. Becker, *Über Polyeder.* Z. Math. Phys. **14** (1869), 65 ff.

[BeG] U. Betke and P. Gritzmann, *Polyedrische 2-Mannigfaltigkeiten mit wenigen nicht-convexen Ecken.* Monatshefte Math. **97** (1984), 1–21.

[BB] J. Bokowski and U. Brehm, *A polyhedron of genus 4 with minimal number of vertices and maximal symmetry.* Geom. Dedicata **29** (1989), 53–64.

[BW] J. Bokowski and J. M. Wills, *Regular polyhedra with hidden symmetries.* Math. Intelligencer **10** (1988), 27–32.

[Br] U. Brehm, *Polyeder gegebenen Geschlechtes mit minimaler Eckenzahl.* Tagungsbericht Oberwolfach **19** (1978), 5.

[Br2] U. Brehm, *Polyeder mit zehn Ecken vom Geschlecht 3.* Geom. Dedicata **11** (1981), 119–124.

[Br3] U. Brehm, *A maximally symmetric polyhedron of genus 3 with 10 vertices.* Mathematika **34** (1987), 237–242.

[BK] U. Brehm and W. Kühnel, *Smooth approximation of polyhedral surfaces regarding curvatures.* Geometriae Dedicata **12** (1982), 435–461.

[BrS] U. Brehm and E. Schulte, *Polyhedral maps.* In "Handbook of Discrete and Computational Geometry," J. E. Goodman and J. O'Rourke, eds. CRC Press 1997, pp. 345–358.

[BWi] U. Brehm and J. M. Wills, *Polyhedral manifolds.* Chapter 2.4 in "Handbook of Convex Geometry," P. M. Gruber and J. M. Wills, eds. North-Holland, Amsterdam 1993. Vol A, 535–554.

[Bri] R. Bricard, *Mémoire sur la théorie de l'octaèdre articulé.* J. Math. Pures Appl. Ser. 5, **3** (1897), 113–148.

[Brid] N. J. Bridge, *Facetting the dodecahedron.* Acta Crystallographica **A30** (1974), 548–552.

[Bru] M. Brückner, *Vielecke und Vielflache.* Theorie und Geschichte. Teubner, Leipzig 1900.

[Bu] M. Burt, *Spatial arrangement and polyhedra with curved surfaces and their architectural applications.* M.Sc. Thesis, Technion, Haifa, Israel, 1966.

[BS] A. V. Bushmelev and I. H. Sabitov, *Configuration spaces of the Bricard octahedra.* [In Russian] Ukrain. Geom. Sbornik **33** (1990), 36–41.

[C] A. L. Cauchy, *Sur les polygones et les polyèdres.* Second Mémoire. J. École Polytechn. 9 (1813) pp. 87 ff. (= Oeuvres Complètes d'Augustin Cauchy, 2e Série, Tome 1 (1905), 26–38.)

[Co] R. Connelly, *A counterexample to the rigidity conjecture for polyhedra.* Publ. Math. I. H. E. S. **47** (1978), 333–338.

[Co2] R. Connelly, *A flexible sphere.* Math. Intelligencer **1** (1978), 130–131.

[Co3] R. Connelly, *Conjectures and open questions in rigidity.* Proc. Internat. Congress of Math. Helsinki 1978, 407–414.

[Co4] R. Connelly, *Rigidity.* In "Handbook of Convex Geometry" P. M. Gruber and J. M. Wills, eds. North-Holland, Amsterdam 1993. Vol A, 223–271.

[CSW] R. Connelly, I. Sabitov and A. Walz, *The bellows conjecture.* Beiträge zur Algebra und Geometrie, to appear.

[Cox] H. S. M. Coxeter, Regular Polytopes. 3rd ed., Dover, New York 1973.

[CM] H. S. M. Coxeter and W. O. J. Moser, Generators and Relations for Discrete Groups. 4th ed. Springer, Berlin 1980.

[CW] H. Crapo and W. Whiteley, *The Geometry of Rigid Structure*, in preparation.

[Cr] P. Cromwell, *Polyhedra.* Cambridge Univ. Press 1997.

[Cs] A. Császár, *A polyhedron without diagonals.* Acta Sci. Math. Szeged **13** (1949/50), 140–142.

[D] A. W. M. Dress, *A combinatorial theory of Grünbaum's new regular polyhedra, Part I: Grünbaum's new regular polyhedra and their automorphism group.* Aequationes Math. **23** (1981), 252–265.

[D2] A. W. M. Dress, *A combinatorial theory of Grünbaum's new regular polyhedra, Part II: Complete enumeration.* Aequationes Math. **29** (1985), 222–243.

[Du] R. A. Duke, *Geometric embedding of complexes.* Amer. Math. Monthly **77** (1970), 597–603.

[E] V. Eberhard, *Zur Morphologie der Polyeder.* Teubner, Leipzig 1891.

[Eg] W. Egloff, *Eine Bemerkung zu Cauchy's Satz über die Starrheit konvexer Vielflache.* Abh. Math. Sem. Hamburg **20** (1956), 253–256.

[EGK] H. G. Eggleston, B. Grünbaum and V. Klee, *Some semicontinuity theorems for convex polytopes and cell complexes.* Comment. Math. Helvet. **39** (1964), 165–188.

[F] S. L. Farris, *Completely classifying all vertex-transitive and edge-transitive polyhedra, part I: necessary class conditions.* Geometriae Dedicata **26** (1988), 111–124.

[F2] S. L. Farris, *Completely classifying all vertex-transitive and edge-transitive polyhedra, part II: finite, fully-transitive polyhedra*, in preparation.

[Fe] P. J. Federico, *The number of polyhedra.* Philips Res. Reports **30** (1975), 220*–231*.

[FR] P. W. Fowler and D. B. Redmond, *Symmetry aspects of leapfrog and truncated polyhedra.* MATCH Communications in Mathematical and Computer Chemistry **33** (1996), 101–119.

[G] M. Gardner, *Mathematical Games.* Scientific American **239** (1978), No. 5, 22–32.

[G2] M. Gardner, My Best Mathematical and Logic Puzzles. Dover, New York 1994.

[GW] A. Gray and S. Wilson, *A more elementary proof of Grünbaum's conjecture.* Congressus Numerantium **72** (1990), 25–32.

[Gr] P. Gritzmann, *The toroidal analogue to Eberhard's theorem.* Mathematika **30** (1983), 274–290.

[Grü] B. Grünbaum, *Convex Polytopes.* Interscience Monographs in Pure and Applied Mathematics, Vol. XVI. John Wiley & Sons, London-New York-Sydney, 1967.

[Grü2] B. Grünbaum, *Polytopes, graphs, and complexes.* Bull. Amer. Math. Soc. **76** (1970), 1131–1201.

[Grü3] B. Grünbaum, *Polygons.* In "The Geometry of Metric and Linear Spaces," L. M. Kelly, ed. Lecture Notes in Mathematics Number 490, 147–184. Springer-Verlag, Berlin-Heidelberg-New York 1975.

[Grü4] B. Grünbaum, *Regular polyhedra—old and new.* Aequationes Math. **16** (1977), 1–20.

[Grü5] B. Grünbaum, *Regularity of graphs, complexes and designs.* Problèmes Combinatoires et Théorie des Graphes. Colloq. Internat. C.N.R.S. No. 260, Paris 1978, 191–197.

[Grü6] B. Grünbaum, *Polyhedra with hollow faces.* In "POLYTOPES: Abstract, Convex and Computational," Proc. NATO - ASI Conference, Toronto 1993. T. Bisztriczky, P. McMullen, R. Schneider and A. Ivic' Weiss, eds. Kluwer Acad. Publ., Dordrecht 1994, 43–70.

[Grü7] B. Grünbaum, *Isogonal prismatoids.* Discrete Comput. Geom. 18 (1997), 13–52.

[Grü8] B. Grünbaum, *Polyhedral realizations of a regular map of genus 5*, in preparation.

[GS] B. Grünbaum and G. C. Shephard, *Spherical tilings with transitivity properties.* In "The Geometric Vein - The Coxeter Festschrift," Chandler Davis, Branko Grünbaum and F. A. Sherk, eds. Springer-Verlag, New York 1982, 65–98.

[GS2] B. Grünbaum and G. C. Shephard, *Polyhedra with transitivity properties.* Math. Reports Acad. Sci. Canada **6** (1984), 61–66.

[GS3] B. Grünbaum and G. C. Shephard, *On the geometry of polygons and polyhedra,* unpublished manuscript. 1985-1988.

[GS4] B. Grünbaum and G. C. Shephard, *Duality of polyhedra.* In "Shaping Space: A Polyhedral Approach," Proc. "Shaping Space"Conference, Smith College, April 1984. M. Senechal and G. Fleck, eds. Birkhäuser, Boston 1988, 205–211.

[GS5] B. Grünbaum and G. C. Shephard, *Tilings and Patterns.* Freeman, New York 1987.

[GS6] B. Grünbaum and G. C. Shephard, *A new look at Euler's theorem for polyhedra.* Amer. Math. Monthly **10** (1994), 109–128. Response to comments by P. Hilton and J. Pedersen, *ibid.*, 961–962.

[GZ] B. Grünbaum and J. Zaks, *The existence of certain planar maps.* Discrete Math. **10** (1974), 93–115.

[H] J. Hadamard, *Erreurs des mathématiciens.* Interméd. Math. **14** (1907), 31. See also: Problème 3456, Interméd. Math. **15** (1908), 222.

[HG] J. Hadamard, Lecons de Géométrie élémentaire. Vol. II. Géométrie dans l'éspace. 2nd ed. Paris 1910; 4th ed. Paris 1921; 7th ed. Paris 1931; 8th ed. Paris 1949.

[Ha] H. Hadwiger, *Eulers Characteristik und kombinatorische Geometrie.* J. reine angew. Math. **194** (1955), 101–110.

[HH] G. Hajós and A. Heppes, *On a special kind of polyhedra.* Acta Math. Acad. Sci. Hungar. **21** (1970), 101–103.

[Hi] D. Hilbert, *Mathematische Probleme.* Göttinger Nachrichten, 1900, 253–297; English translation: Mathematical problems. Bull. Amer. Math. Soc. **8** (1902), 437–479; reprinted in F. E. Browder, ed., Mathematical Developments Arising From Hilbert's Problems, Proc. Sympos. Pure Math. vol. 28, Amer. Math. Soc., Providence RI, 1876, 1–34.

[Ho] H. Hopf, *Selected Topics in Geometry.* Lecture Notes, New York University 1946.

[HK] J. L. Hudson and J. G. Kingston, *Stellating polyhedra.* Math. Intelligencer **10** (1988), 50–61.

[J] S. Jendrol, *On face vectors and vertex vectors of convex polyhedra.* Discrete Math. **118** (1993), 119–144.

[JJ] S. Jendrol and E. Jucovič, *On the toroidal analogue of Eberhard's theorem.* Proc. London Math. Soc. (3) **25** (1972), 385–398.

[JJ2] S. Jendrol and E. Jucovič, *Generalization of a theorem by V. Eberhard.* Math. Slovaca **27** (1977), 383–407.

[JO] S. Jendrol and P. J. Owens, *Pentagonal 3-polytopal graphs with edges of only two types and shortness parameters.* Discrete Math. **137** (1995), 251–263.

[Je] B. Jessen, *Orthogonal icosahedra.* Nordisk Matem. Tidsk. **15** (1967), 90–96.

[Jo] N. W. Johnson, *Uniform Polytopes.* Cambridge University Press, to appear.

[Ju] E. Jucovic, *On the face-vectors and vertex-vectors of cell-decompositions of orientable 2-manifold.* Math. Nachrichten **73** (1976), 285–295.

[K] N. H. Kuiper, *Sphères polyédriques flexibles dans E^3, d'après Robert Connelly.* Séminaire Bourbaki **30** (1977/78) No. 514. Lecture Notes in Mathematics #710, Springer, New York 1979, 147–168.

[Ku] W. Kurth, *Enumeration of Platonic maps on the torus.* Discrete Math. **61** (1986), 71–83.

[L] S. A. Lavrenchenko, *Irreducible triangulations of the torus.* [In Russian] Ukrain. Geom. Sbornik **30** (1987), 52–62.

[Le] H. Lebesgue, *Démonstration complète du théorème du Cauchy sur l'égalité des polyèdres convexes.* Interméd. Math. **16** (1909), 113–120.

[Le2] H. Lebesgue, *Octaèdres articulés de Bricard.* Enseign. Math., (2) **13** (1967), 175–185.

[Lj] D. Ljubić, *A twisted quadrangulated torus.* Abstract 835-52-08, Abstracts Amer. Math. Soc. **8** (1987), 302.

[Ly] L. A. Lyusternik, *Convex Figures and Polyhedra.* [In Russian]. GITTL, Moscow 1956. English translation by T. J. Smith, Dover, New York 1963. English translation by D. L. Barnett, Heath, Boston 1966.

[M] I. G. Maksimov, *Polyhedra with bendings and Riemann surfaces.* Russian Math. Surveys **50** (1995), 821–823.

[Ma] H. Martini, *A hierarchical classification of Euclidean polytopes with regularity properties.* In "POLYTOPES: Abstract, Convex and Computational", Proc. NATO-ASI Conference, Toronto 1993. T. Bisztriczky, P. McMullen, R. Schneider and A. Ivic' Weiss, eds. Kluwer Acad. Publ., Dordrecht 1994, 71–96.

[MS] P. McMullen and R. Schneider, *Valuations on convex bodies.* In "Convexity and its Applications," P. M. Gruber and J. M. Wills, eds. Birkhäuser, Basel 1983, 170–247.

[MSc] P. McMullen and E. Schulte, *Abstract Regular Polytopes*, in preparation.

[MSc2] P. McMullen and E. Schulte, *Regular polytopes in ordinary space.* Discrete Comput. Geom. 17 (1997), 449–478.

[MSW] P. McMullen, C. Schulz and J. M. Wills, *Polyhedral 2-manifolds in E^3 with unusually large genus.* Israel J. Math. **46** (1983), 127–144. Correction in: *Two remarks on equivelar manifolds*, ibid. **52** (1985), 28–32.

[MSWi] P. McMullen, E. Schulte and J. M. Wills, *Infinite series of combinatorially regular polyhedra in three-space.* Geometriae Dedicata **2** (1988), 299–307.

[Me] P. W. Messer, *Stellation of the rhombic triacontahedron and beyond.* Structural Topology **21** (1995), 25–46.

[Mi] E. N. Miller, *Icosahedra constructed with congruent triangles*, in preparation.

[Mö] A. F. Möbius, *Über die Bestimmung des Inhalts eines Polyëders.* Ber. Verh. Sächs. Ges. Wiss., Math.-Phys. Klasse **17** (1985), 31–68. (= Gesammelte Werke, Vol. 2, 473–512. Hirzel, Leipzig 1886.)

[Mol] J. Molnár, *Diamond polyhedra I. Simple diamond polyhedra.* Ann. Univ. Sci. Budapest. Eötvös Sect. Math. **19** (1976), 129–139.

[NO] A. Nakamoto and K. Ota, *Note on irreducible triangulations of surfaces.* J. Graph Theory **20** (1995), 227–233.

[N] M. Norgate, *Non-convex pentahedra.* Math. Gazette **54** (1970), 115–124.

[O] J. Ounsted, *An unfamiliar dodecahedron.* Mathematics Teaching **83** (1978), 46–47.

[PV] H. Pulapaka and A. Vince, *Nonrevisiting paths on surfaces.* Discrete Comput. Geom. **15** (1996), 353–357.

[R] J. Reay, *Can neighborly polyhedra be realized geometrically?* In "Shaping Space: A Polyhedral Approach," Proc. "Shaping Space" Conference, Smith College, April 1984. M. Senechal and G. Fleck, eds. Birkhäuser, Boston 1988, 251–253, 259.

[Re] C. Reinhardt, *Zu Möbius' Polyedertheorie.* Ber. Verh. sächs. Gesellschaft der Wissenschaften, Math.-phys. Cl. **37** (1885), 106–125.

[RY] G. Ringel and J. W. T. Youngs, *Solution of the Heawood map-coloring problem.* Proc. Nat. Acad. Sci. USA **60** (1968), 438–445.

[S] I. K. Sabitov, *Local theory of bending of surfaces.* In "Geometry III. Theory of Surfaces," Y. D. Burago and V. A. Zalgaller, eds. Encyclop. of Math. Sci. vol. 48. Springer, Berlin 1992.

[S2] I. K. Sabitov, *On the problem of invariance of the volume of a flexible polyhedron.* Russian Math. Surveys **50** (1995), 452–452.

[Sc] H. Schipper, *Generating triangulations of 2-manifolds.* In "Computational Geometry-Methods, Algorithms and Applications" Proc. Internat. Workshop on Comput. Geom CG '91, Bern 1991. Springer, New York 1991, 237–248.

[SZ] I. J. Schoenberg and S. K. Zaremba, *On Cauchy's lemma concerning convex polygons.* Canad. J. Math. **19** (1967), 1062–1071.

[SW] E. Schulte and J. M. Wills, *Combinatorially regular polyhedra.* In: Symmetry of Discrete Mathematical Structures and Their Symmetry Groups, ed. by K. H. Hoffman and R. Wille, Heldermann, Berlin 1990, 49–88.

[Sch] C. Schulz, *Geometrische Realisierungen geschlossener Flächen mit wenigen Ecken.* Geom. Dedicata **11** (1981), 309–314.

[Schw] J. Schwörbel, *Die kombinatorisch regulären Tori.* Thesis, Siegen University, 1988.

[SWi] J. Schwörbel and J. M. Wills, *The two Pappus-tori* Geom. Dedicata **28** (1988), 359–362.

[Sh] Y. Shirai, *Three-Dimensional Computer Vision.* Springer, New York 1987.

[Si] J. S. Simutis, *Geometric realization of toroidal maps.* Ph.D. thesis, University of California, Davis, CA, 1977.

[St] E. Steinitz, *Polyeder und Raumteilungen.* Encykl. Math. Wissenschaften **3** (1922), Geometrie, Part 3AB12, 1–139.

[SR] E. Steinitz and H. Rademacher, *Vorlesungen über die Theorie der Polyeder.* Springer, Berlin 1934.

[Ste] B. M. Stewart, *Adventures Among the Toroids.* The author, Okemos, MI, 1980.

[Sto] J. J. Stoker, *Geometrical problems concerning polyhedra in the large.* Comm. Pure Appl. Math. **21** (1968), 119–168.

[Sz] L. Szilassi, *Regular toroids.* Structural Topology **13** (1986), 69–80.

[V] A. Vince, *Regular combinatorial maps.* J. Combinat. Theory **35** (1983), 256–277.

[W] W. T. Webber, *Monohedral idemvalent toroidal polyhedra.* Ph.D. thesis, University of Washington, Seattle 1994.

[We] M. J. Wenninger, *Polyhedron Models.* Cambridge Univ. Press 1971.

[We2] M. J. Wenninger, *Dual Models.* Cambridge Univ. Press 1983.

[Wi] J. M. Wills, *On polyhedra with transitivity properties.* Discrete Comput. Geom. **1** (1986), 195–199.

[Wil] S. E. Wilson, *New techniques for the construction of regular maps.* Ph.D. thesis, University of Washington, Seattle 1976.

[Z] G. M. Ziegler, *Lectures on Polytopes.* Graduate Texts in Mathematics #152, Springer, New York 1995.

Department of Mathematics, University of Washington, Seattle, Washington 98195

E-mail address: grunbaum@math.washington.edu

Contemporary Mathematics
Volume **223**, 1999

A Proof of the Strict Monotone 4-step Conjecture

Fred Holt and Victor Klee

Dedicated to Branko Grünbaum for the thirtieth anniversary of his book on Convex Polytopes

ABSTRACT. With $\Delta(d, n)$ denoting the maximum diameter attained by (the graphs of) d-polytopes having n facets, the still unsettled *Hirsch conjecture* asserts that $\Delta(d, n) \leq n - d$ whenever $n > d \geq 2$. Its special case, the *d-step conjecture*, asserts that $\Delta(d, 2d) = d$. The present note deals with two related functions, Δ_{sm} and Δ_m, which involve paths along which a given linear objective function is steadily increasing. This note was motivated by Ziegler's *strict monotone Hirsch conjecture*, asserting that always $\Delta_{sm}(d, n) \leq n - d$. (Since $\Delta \leq \Delta_{sm} \leq \Delta_m$, this implies the Hirsch conjecture.) When Γ is any of the functions Δ, Δ_{sm}, and Δ_m, the numbers of the form $\Gamma(k, 2k)$ are of special interest because of the fact that $\Gamma(d, n) = \Gamma(n - d, 2(n - d))$ for $d < n \leq 2d$. (In particular, $\Gamma(d, n) \leq n - d$ for all d and n if and only if $\Gamma(d, 2d) = d$ for all d.) This note summarizes the present knowledge concerning the functions Δ_{sm} and Δ_m, and proves the *strict monotone 4-step conjecture* asserting that $\Delta_{sm}(4, 8) = 4$.

Introduction

As the term is used here, a linear functional φ is *admissible* for a (convex) polytope P provided that φ does not attain the same value at any two vertices of P. The three functions defined in the next paragraph are all of interest in connection with the behavior of the simplex method of linear programming and the open problem of whether there exists a pivot rule that turns the simplex method into an LP algorithm whose worst-case behavior is polynomially bounded.

Consider all triples (P, x, y) consisting of a d-polytope with n facets ($(d-1)$-faces) and two vertices x and y of P. Let $\Delta(d, n)$ denote the smallest integer k such that for each such triple, x and y can be joined by a path of length $\leq k$ (i.e., one

---

1991 *Mathematics Subject Classification.* Primary 90C05, 52B05, 52B11, 52B55.

Key words and phrases. d-step conjecture, Dantzig figure, diameter, Hirsch conjecture, linear programming, LP orientation, monotone diameter, monotone path, monotone orientation, strict monotone diameter.

The research of both authors was supported in part by the National Science Foundation. They are also indebted to a referee for some questions and suggestions that helped to improve the exposition.

© 1999 American Mathematical Society

formed from k or fewer edges of P). Let $\Delta_{sm}(d,n)$ denote the smallest integer k such that whenever (P,x,y) is such a triple and φ is an admissible functional for which

$$\min\varphi(P)=\varphi(x)<\varphi(y)=\max\varphi(P),$$

then x can be joined to y by a path of length $\leq k$ along which the objective function φ is steadily increasing. The function $\Delta_m(d,n)$ is defined similarly, omitting the requirement that $\min\varphi(P)=\varphi(x)$. Evidently,

$$\Delta(d,n)\leq\Delta_{sm}(d,n)\leq\Delta_m(d,n),$$

where the first inequality is proved by an easy application of a projective transformation. Also, the d-cubes demonstrate that $\Delta(d,2d)\geq d$ for all d.

An important property of certain diameter functions Γ is that whenever $1<d<n$,

$$\Gamma(d,n)\leq\Gamma(d+1,n+1),$$

with equality when $n\leq 2d$. For the choice $\Gamma=\Delta$, that was established in [**KW**] with the aid of perturbation and wedging. In Section 1 of the present note, it is established for the functions Δ_{sm} and Δ_m. For each such Γ, it is true that for $n>2d$

$$\Gamma(d,n)\leq\Gamma(n-d,2(n-d)),$$

and that for all $k\geq 0$

$$\Gamma(d,2d)=\Gamma(d+k,2d+k).$$

Thus it is clear for each such Γ that the d-step conjecture implies the Hirsch conjecture (though not necessarily on a dimension-for-dimension basis), and that, whether the conjectures are true or false, a large share of the information concerning the numbers $\Gamma(d,n)$ is carried by those of the form $\Gamma(d,2d)$. That is the reason for our emphasis here on the numbers $\Delta_{sm}(d,2d)$ and $\Delta_m(d,2d)$.

As reported in [**D, D2**], W. M. Hirsch conjectured that always $\Delta(d,n)\leq n-d$. Ziegler [**Z**] strengthened this conjecture by stating what he called the *strict monotone Hirsch conjecture*: $\Delta_{sm}(d,n)\leq n-d$. Its special case, the *strict monotone d-step conjecture*, asserts that $\Delta_{sm}(d,2d)=d$.

It was shown in [**Kl**] that $\Delta(3,n)=\lfloor\frac{2}{3}n\rfloor-1$ for all n; in [**KW**] that $\Delta(d,2d)=d$ for all $d\leq 5$; in [**Kl2**] that $\Delta_m(3,n)\leq n-3$; and in [**T**] that $\Delta_m(4,8)\geq 5$. We note in this paper that $\Delta_{sm}(3,n)=\Delta(3,n)$ and $\Delta_m(3,n)=n-3$ for all n, and we show that $\Delta_{sm}(4,8)=4$ and $\Delta_m(4,8)=5$. The precise value of $\Delta(d,2d)$ is unknown when $d\geq 6$, and of $\Delta_{sm}(d,2d)$ or $\Delta_m(d,2d)$ when $d\geq 5$.

Thus the d-step conjecture holds for $d\leq 5$. The Hirsch conjecture holds for $d\leq 3$. The strict monotone d-step conjecture holds for $d\leq 4$, and the strict monotone Hirsch conjecture for $d\leq 3$. The monotone Hirsch conjecture holds for $d\leq 3$; and although the monotone d-step conjecture fails for $d\geq 4$, the monotone Hirsch conjecture has been established without dimensional restrictions for several classes of polytopes that arise in connection with practical optimization problems [**BR, G, M, R, R2**].

The functions Δ, Δ_m, and Δ_{sm} are all of interest in connection with edge-following algorithms for linear programming. Just as Δ describes, in a sense, the worst possible behavior of the best possible edge-following LP algorithm (see [**KK**] and [**KW**]), Δ_m does the same for monotone edge-following algorithms. Similarly, Δ_{sm} applies to monotone edge-following algorithms for those LP problems in which an admissible linear objective function φ is to be maximized and there is a natural starting vertex at which φ attains its minimum. For example, that would be the situation in any problem whose (nonempty, bounded) feasible region P is defined by constraints of the form

$$Ax \leq b, \quad x \geq 0$$

where all entries of A are nonnegative and the admissible objective function is of the form $\varphi(x) = c^T x$ with $c \geq 0$. Here the origin is a vertex of P and φ attains its P-minimum at the origin.

Beyond this admittedly limited class of LP problems, the function Δ_{sm} is of additional interest because of its relevance to the geometry of polytopes. In particular, the known counterexamples to the monotone Hirsch conjectures have "back doors" —very short routes from the starting vertex to the maximum vertex provided the beginning move(s) can be against the gradient. In these counterexamples, the minimum vertex is often a neighbor of both the maximum vertex and the starting vertex for the counterexample.

1. Equivalence of Hirsch conjectures to d-step conjectures

When P is a polytope, x and y are vertices of P, and φ is an admissible functional such that $\varphi(y) = \max \varphi(P)$, we use $\delta_P^{\varphi}(x, y)$ to denote the smallest integer k such that x is joined to y by a path of length k along which φ is increasing. The *monotone diameter* of P is the maximum of $\delta_P^{\varphi}(x, y)$ over all (x, y, φ) of the indicated sort, and the *strict monotone diameter* of P is similarly defined with respect to (x, y, φ) such that $\varphi(x) = \min \varphi(P)$ and $\varphi(y) = \max \varphi(P)$. Thus $\Delta_m(d, n)$ and $\Delta_{sm}(d, n)$ are respectively the maximum of the monotone diameter and the maximum of the strict monotone diameter as P ranges over all d-polytopes with n facets.

Just as for the diameter function Δ, an easy perturbation argument shows that Δ_m and Δ_{sm} are unchanged when the d-polytopes in question are restricted to those that are *simple* (i.e., each vertex is incident to precisely d edges, or, equivalently, to precisely d facets). As in **[HK]**, we use the term *(d, n)-polytope* to denote a simple d-polytope with n facets.

Now we want to prove the following result, whose consequences were described in the Introduction.

1.1. PROPOSITION. *If Γ is Δ_m or Δ_{sm}, then $\Gamma(d, n) \leq \Gamma(d + 1, n + 1)$ for all $1 < d < n$, with equality when $n \leq 2d$.*

PROOF. With $1 < d < n$, consider a (d, n)-polytope $P \subset \mathbb{R}^d$ and a linear functional φ on $\mathbb{R}^d$ such that

$$|\varphi(u) - \varphi(v)| > 1 \text{ for each pair } u, v \text{ of distinct vertices of } P.$$

Choose any facet F of P, and let ψ be a linear functional on $\mathbb{R}^d$ such that

$$\psi(F) = \{0\} \text{ and } \psi(P \setminus F) \subset]0, 1[.$$

Using ψ, construct the $(d + 1, n + 1)$-polytope

$$W = \{(p, \alpha) : p \in P, 0 \leq \alpha \leq \psi(p)\} \subset \mathbb{R}^{d+1},$$

a wedge over P with foot F, and define the linear functional

$$\eta(u, \alpha) = \varphi(u) + \alpha$$

for all $(u, \alpha) \in \mathbb{R}^{d+1} = \mathbb{R}^d \times \mathbb{R}$.

For any vertex u of P, denote the vertices $(u, 0)$ and $(u, \psi(u))$ of W by u_b and u^t respectively. Then each edge of W is of the form $[u_b, v_b]$ or $[u^t, v^t]$ where $[u, v]$ is an edge of P, or of the form $[u_b, u^t]$ where u is a vertex of $P \setminus F$ (these are called *vertical edges*). Note that any two vertices of P have φ-values that differ by more than 1, while on P the range of ψ (and hence of α in the definition of η) is contained in $[0, 1[$. Thus for any two vertices u and v of P, the condition that $\varphi(u) < \varphi(v)$ is equivalent to the condition that

$$\eta(u_b) \leq \eta(u^t) < \eta(v_b) \leq \eta(v^t).$$

Since $\eta(u_b) = \eta(u^t)$ if and only if u_b and u^t coincide, it follows that the linear functional η is admissible for W and that if the minimum and maximum of φ on P are attained at vertices x and y respectively, then the minimum and maximum of η on W are attained at x_b and y^t respectively. Moreover, any η-monotone edge-path in W projects onto a φ-monotone edge-path in P, and the projection of the edges in an η-monotone path is one-to-one except that the vertical edges of W produce vertices rather than edges in P. In any case, the path in W has at least as many edges as its projection in P. Thus we may conclude, for both of the mentioned choices of Γ, that

$$\Gamma(d, n) \leq \Gamma(d + 1, n + 1).$$

To complete the proof, we assume that $d < n \leq 2d$ and show that then

$$\Gamma(d + 1, n + 1) \leq \Gamma(d, n).$$

Since $n + 1 < 2(d + 1)$, any two vertices of a $(d + 1, n + 1)$-polytope P must lie on a common facet Q of P, and Q is a (d, m)-polytope for some $m \leq n$. For any linear functional φ that is admissible for P, the restriction of φ to Q is admissible for Q. For any pair of vertices x and y of Q, each φ-monotone path from x to y on Q is by inclusion a φ-monotone path on P, and hence the minimum length of such paths on Q is no smaller than the minimum on P. From this it follows that $\Gamma(d + 1, n + 1) \leq \Gamma(d, m)$. Successive truncation shows that $\Gamma(d, m) \leq \Gamma(d, n)$ and thus completes the proof. □

We also need the following, known from **[KW]** when Γ is Δ.

1.2. PROPOSITION. *Suppose that* Γ is Δ_m *or* Δ_{sm}, *and that* $n \geq 2d$. *Then the value of* $\Gamma(d, n)$ *is unaltered if the relevant maximum is restricted to pairs of vertices that do not share a facet.*

PROOF. Suppose that $n \geq 2d$. Among the 4-tuples (P, x, y, φ) which instantiate the bound $\Gamma(d, n)$, choose one that maximizes the dimension of the smallest face G of P that is incident to both x and y. We want to show that $G = P$. Suppose, to the contrary, that G is contained in some facet F of P, and note that F is a $(d - 1, m)$-polytope for some $m \leq n - 1$. Note also that

$$\delta_F^{\varphi}(x, y) \geq \delta_P^{\varphi}(x, y).$$

Truncating as necessary, we produce a contradiction by taking the prism $F \times I$ if $m < n - 1$ or by taking the wedge over F if $m = n - 1$. The details are omitted, for they are virtually identical with those in the proof of 2.8 in **[KW]** and in the proof of 1.1 above. □

The above facts motivate our focus on the numbers $\Gamma(d, 2d)$, and on the triples known as *Dantzig figures*. A d-dimensional Dantzig figure is a triple (P, x, y) such that P is a $(d, 2d)$-polytope and x and y are vertices of P that are *estranged* in the sense that they do not share a facet. Such figures play an essential role in all that follows.

2. A strictly monotone version of d-connectedness

We need a strictly monotone variant of Balinski's observation [**B**] that the graph of a d-polytope is d-connected. The proof below is based on a comment made by David Walkup in the 1960's, and it is somewhat similar in spirit to Barnette's short proof [**Ba**] of Balinski's theorem.

2.1. PROPOSITION. *Suppose that φ is an admissible functional for a d-polytope P, and x and y are vertices of P such that $\varphi(x) = \min\varphi(P)$ and $\varphi(y) = \max\varphi(y)$. Then among the φ-monotone paths from x to y in P's edge-graph, there are d that are pairwise vertex-disjoint except for having x and y in common.*

PROOF. The assertion is obvious when $d = 2$, and we proceed by induction on the dimension. When $d \geq 3$, we apply the lower-dimensional results in conjunction with a directed and vertex-oriented version of Menger's connectivity theorem (e.g., Theorem 11.6 of [**BM**]).

For each point $q = (q_1, ..., q_d) \in \mathbb{R}^d$, let $\Phi(q) = (q_1, ..., q_{d-1}, 0)$ and $\varphi(q) = q_d$. Then $q = \Phi(q) + \varphi(q)z$, where $z = (0, ..., 0, 1)$. In treating the theorem's d-dimensional case, we may assume that this φ is the admissible functional in question, that P's vertex x is the origin 0, and that $\varphi(y) = 1$. Since φ's level sets are preserved by the linear transformation that takes q into $q - \varphi(q)y$, we may assume also that $y = z$. Now turn each edge of P into a directed edge (u, v) such that $\varphi(u) < \varphi(v)$. To show there are d independent monotone paths from x to y, it suffices, in view of the version of Menger's theorem mentioned above, to show that for each set S of $d - 1$ vertices other than x or y, there is a φ-monotone path from x to y that misses the set S.

Let $J = \{q \in \mathbb{R}^d : q_{d-1} = 0\}$, $J^+ = \{q \in \mathbb{R}^d : q_{d-1} > 0\}$, and $J^- = -J^+$. The orthogonal projection $\Phi(S)$ of S on the hyperplane $H = \{q \in \mathbb{R}^d : q_d = 0\}$ is contained in a $(d-2)$-flat G in H. With G_0 denoting the $(d-2)$-subspace of H that is parallel to G, we may assume (with the aid of a suitable rotation about the line $\mathbb{R}z$, and, if necessary, a reflection across the hyperplane J) that $G_0 = \{q \in \mathbb{R}^d : q_{d-1} = q_d = 0\}$ and that either $S \subset J$ or $S \subset J^-$.

If $S \subset J^-$ and P misses J^+, then the intersection $J \cap P$ is a face of P that misses S and includes the vertices x and y. By the inductive hypothesis it must contain a φ-monotone path from x to y. In the remaining cases, either (a) $S \subset J^-$ and P intersects J^+ or (b) $S \subset J$ and (since P is not contained in J) we may assume that P intersects J^+.

Now let Π denote the transformation that projects P orthogonally onto the 2-dimensional plane $\{q \in \mathbb{R}^d : q_1 = ... = q_{d-2} = 0\}$. The projection $\Pi(P)$ is a convex polygon K that intersects J^+, and it follows from the 2-dimensional result that the boundary of K contains a φ-monotone path that goes from x to y and lies, except for its endpoints, in J^+. This path can be "lifted" to a φ-monotone path that goes from x to y in P and that misses the set S. It remains only to describe the lifting.

With $x = v_0$ and $v_m = y$, let $[v_0, v_1], [v_1, v_2], \ldots, [v_{m-1}, v_m]$ be the successive edges of the mentioned φ-monotone path from x to y in $K \cap J^+$. Then each of the edges $[v_{j-1}, v_j]$ is of the form $\Pi(F_j)$ for some face F_j of P. Since the projection Π preserves φ-values, there are unique vertices w_{j-1} and w_j of F_j that project into v_{j-1} and v_j respectively, and these are respectively the unique minimizer and the unique maximizer of φ's restriction to F_j. In each F_j there is a φ-monotone path T_j from w_{j-1} to w_j, and stringing these paths together produces a φ-monotone path

$$T = T_1 \cup T_2 \cup \cdots \cup T_m$$

from x to y in P. Except for its end vertices, T is contained in the open halfspace J^+ and hence it misses the set S. That completes the proof. □

Balinski's original d-connectedness theorem is an easy consequence of Proposition 2.1. Consider any two vertices x and y of a d-polytope P in $\mathbb{R}^d$, and let H_x and H_y be hyperplanes whose intersections with P are respectively $\{x\}$ and $\{y\}$. With the aid of a projective transformation (as in [**Ba**]), we may assume that H_x and H_y are parallel, and then a slight perturbation turns them into the level sets of an admissible function φ whose P-minimum is attained uniquely at x and P-maximum is attained uniquely at y. Then apply Proposition 2.1.

3. Monotone paths on 3-polytopes

Now consider the d independent monotone paths whose existence is asserted by Proposition 2.1. If the shortest of these d paths uses k edges, then each path has at least $k-1$ internal vertices and hence the total number of vertices of P is at least $d(k-1)+2$. When P is a $(3,n)$-polytope, P has precisely $2n-4$ vertices, so $3k-1 \leq 2n-4$ andhence $k \leq \lfloor(2n/3)\rfloor - 1 \leq n-3$. Thus the strict monotone Hirsch conjecture is correct when $d=3$. On the other hand, this bound is achieved as the diameter of a family of 3-polytopes identified in [**Kl**]. Consideration of prisms over $(n-2)$-gons, in the spirit of the d_2 shown in Figure 1, shows that $\Delta_m(3,n) \geq n-3$, and equality results from the arguments in [**Kl2**]. In fact,

$$\Delta(3,n) = \Delta_{sm}(3,n) = \lfloor(2n/3)\rfloor - 1 \leq n-3 = \Delta_m(3,n)$$

for all $n \geq 4$. The inequality "$\leq$" is strict for all $n > 6$. The facts that $\Delta_m(3,6) = 3$ and $\Delta_m(3,7) = 4$ are of particular importance in the work below.

There are five combinatorial types of $(3,7)$-polytopes, denoted in [**GS**] by d_1, d_2, d_3, d_4, d_5. Each of the d_k has diameter 3 but monotone diameter 4. Moreover, except for the pentagonal prism (d_2), there is an embedding of d_k in $\mathbb{R}^3$ and an admissible linear functional $\varphi : \mathbb{R}^3 \longrightarrow \mathbb{R}$ such that more than one vertex is at monotone distance 4 from the top vertex. The following figures illustrate these conclusions.

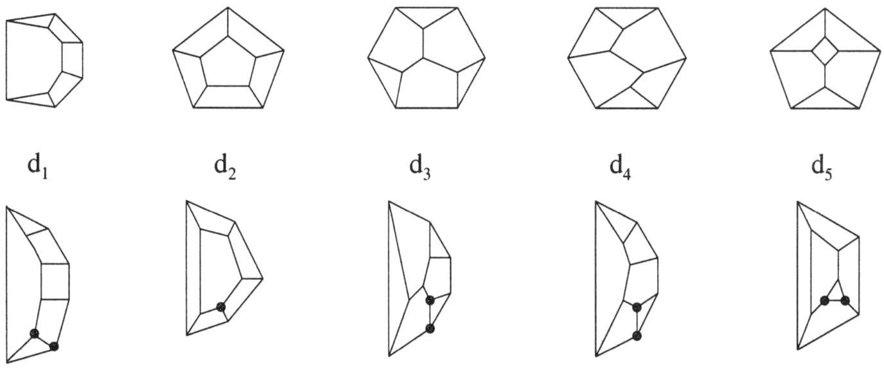

FIGURE 1: The 2-diagrams in the upper row are those given in [**GS**] for the $(3,7)$-polytopes $d_1, ..., d_5$. The alternate 2-diagrams in the lower row illustrate the fact that each d_k admits a realization that has monotone diameter 4; in fact, each of the highlighted vertices is at monotone distance 4 from the top vertex.

There is danger in jumping to d-dimensional conclusions on the basis of plausible $(d-1)$-dimensional diagrams. That is illustrated by the "Brückner sphere" discussed in [**GS**]—a 3-diagram that is not combinatorially equivalent to the Schlegel diagram of any 4-polytope. Thus we should state explicitly that, while the diagrams in the lower row of Figure 1 *illustrate* the fact that each of the d_k admits a realization with monotone diameter 4, they do not *prove* this fact. A proof would in each case require an algebraic representation P of d_k in $\mathbb{R}^3$, a specification of an admissible linear functional φ on $\mathbb{R}^3$, and a specification of two vertices x and y of P such that $\varphi(y) = \max \varphi(P)$ and each φ-increasing path from x to y uses at least 4 edges of P. We do not supply such details here because they are not needed for the proof of our main result. The fact that each of the d_k can be geometrically realized so as to have monotone diameter 4 shows that none of the d_k can be neglected in our analysis. However, our analysis does not depend on geometric realizations of the d_k, but only on properties of certain orientations of the graphs of the d_k that are introduced in Section 4 as combinatorial generalizations of the orientations that can be realized geometrically.

4. Proof of the strict monotone 4-step conjecture

A complete catalog of the 37 combinatorial types of $(4,8)$-polytopes was produced by Grünbaum and Sreedharan [**GS**]. We use their tables to exhaust the possibilities for a counterexample to the strict monotone 4-step conjecture. Let us start, then, to develop a profile for a counterexample to the strict monotone 4-step conjecture.

4.1. LEMMA. *Suppose that x and y are two vertices of a $(4,8)$-polytope P in $\mathbb{R}^4$, and that $\varphi : \mathbb{R}^4 \longrightarrow \mathbb{R}$ is an admissible functional such that $\varphi(y) = \max \varphi(P)$ and $\delta_P^{\varphi}(x,y) \geq 5$. Then x and y are estranged vertices and hence (P,x,y) is a Dantzig figure. Now suppose also that $[x,v]$ is an edge of P, and that F is the (unique) facet of P that is incident to both v and y. If $\varphi(x) < \varphi(v)$, then F is a $(3,7)$-polytope and $\delta_F^{\varphi}(v,y) = 4$.*

PROOF. If x and y are incident to a common facet, then that facet is at most a $(3, 7)$-polytope and hence

$$\delta_P^\varphi(x, y) \leq \Delta_m(3, 7) = 4,$$

contradicting the assumption that $\delta_P^\varphi(x, y) \geq 5$. Hence x and y are estranged, and since P is a $(4, 8)$-polytope, (P, x, y) is a Dantzig figure.

Now consider any edge $[x, v]$ such that $\varphi(x) \leq \varphi(v)$. Since (P, x, y) is a Dantzig figure, v and y are incident to a unique common facet F, and we have

$$5 \leq \delta_P^\varphi(x, y) \leq 1 + \delta_F^\varphi(v, y).$$

Of course, v need not be the bottom vertex in F, but in any case the term $\delta_F^\varphi(v, y)$ cannot exceed the monotone diameter of F. From the facts that $\delta_F^\varphi(v, y) \geq 4$, that F has at most 7 2-faces, and that $\Delta_m(3, 6) = 3$, it follows that F has exactly 7 2-faces and $\delta_F^\varphi(v, y) = 4$. □

From the final statement of Lemma 4.1 it follows that $\delta_P^\varphi(x, y) \leq 5$. In conjunction with Todd's example [**T**], this shows that $\Delta_m(4, 8) = 5$.

4.2. LEMMA. *With hypotheses as in Lemma* 4.1, *assume in addition that* $\varphi(x) = \min \varphi(P)$. *Then the number of vertices of P is* 18, 19, *or* 20.

PROOF. Since P is a simple 4-polytope with 8 facets, the number m of vertices of P is between 14 and 20. Under the additional assumption that $\varphi(x) = \min \varphi(P)$, there are (by Proposition 2.1) four independent monotone paths from x to y. It follows that the shortest such path in any such set of paths is of length $k \leq \lfloor \frac{m+2}{4} \rfloor$. When $m < 18$, this yields $k \leq 4$, so only the cases $m = 18$, $m = 19$, and $m = 20$ remain. □

It suffices, then, to consider 4-dimensional Dantzig figures (P, x, y) such that P has 18, 19, or 20 vertices, and such that every edge incident to x terminates on a $(3, 7)$-facet incident to y. In [**HK**] we worked with the duals of the polytopes described in the Grünbaum-Sreedharan [**GS**] catalog of simplicial 4-polytopes with 8 vertices, and for each of these we found the pairs of vertices that turned these polytopes into Dantzig figures. Only 21 of those Dantzig figures, in 7 of the $(4, 8)$-polytopes, satisfy the profile, provided in Lemmas 4.1 and 4.2, of a counterexample to the strict monotone 4-step conjecture.

The following table summarizes our study of these 21 Dantzig figures. Let us describe the row corresponding to the first Dantzig figure, D_1.

	$N=$	3	4	5	6	$Q=$	1	2	7	8	
$D_1 = (P_{25}, N, Q)$	↓	O	K	M	J	↓	T	R	P	B	$2(d_1)JKAEDPSTBQ$
		7	2	7	2		3	6	6	3	

The first column indicates that the Dantzig figure D_1 is found in the $(4, 8)$-polytope P_{25} from the [**GS**] catalog (our P_k is their P_k^8), with estranged vertices N and Q. The next two columns summarize the edge-facet intersections in D_1; the second column records that the vertex N is the intersection of the four facets indexed 3, 4, 5, 6, that the edge from N not incident to facet 3 terminates on facet 7 in vertex O, and similarly that the edge from N not incident to facet 4 terminates on facet 2 in vertex K. The third column contains the complementary information about the vertex Q. Similar comments apply to the remaining 20 rows. Thus the first three columns describe, for each of the Dantzig figures $D_1, \ldots, D_{21}$, certain combinatorial aspects that are especially relevant to our analysis; the remaining details of combinatorial structure can be found in **[GS]** and **[HK]**.

The fourth column in the table outlines the proofs that the various D_i cannot have strict monotone diameter > 4. These proofs are based on properties of the $(3,7)$-polytopes d_k with respect to orientations that generalize those orientations which have geometric realizations. For example, the fourth column for D_1 indicates that facet 2 is of type d_1, and under the listed combinatorial equivalence, the lemma and corollary for d_1 polytopes (4.3 and 4.4) applies to D_1. The detailed arguments appear in the proofs of Corollaries 4.4, 4.6, 4.8, and 4.10, and those proofs contain more detailed explanations of the table's fourth column.

We list only one combinatorial equivalence for each Dantzig figure, although there may be others. For example, for D_1, we could also list $3(d_3)$BTAFMUSKJN and $7(d_4)$OMFGHRUTBQ, demonstrating that D_1 has strict monotone diameter 4 via Lemmas 4.5 and 4.7 respectively. We could establish the strict monotone diameter of D_1 by considering facet 6, of type d_5, as well; however, the analogous lemma for types d_5 is not necessary for our present purposes and so is omitted.

Dantzig figures	Edge-facet intersections										Facet for lemmas no.(type)comb.equiv.
$D=(P,x,y)$	$x=$	$h_1\cap$	$h_2\cap$	$h_3\cap$	h_4						$h_i(d_j)\alpha\beta\gamma\delta\epsilon\zeta\eta\theta\iota\omega$
	$\downarrow$	v_1	v_2	v_3	v_4						
		h_{i_1}	h_{i_2}	h_{i_3}	h_{i_4}						
$D_1=(P_{25},N,Q)$	$N=$	3	4	5	6	$Q=$	1	2	7	8	$2(d_1)$JKAEDPSTBQ
	$\downarrow$	O	K	M	J	$\downarrow$	T	R	P	B	
		7	2	7	2		3	6	6	3	
$D_2=(P_{29},V,K)$	$V=$	4	5	6	8	$K=$	1	2	3	7	$2(d_3)$EOMFBQPJNK
	$\downarrow$	U	O	T	E	$\downarrow$	B	C	J	N	
		1	2	1	2		6	6	4	4	
$D_3=(P_{31},N,S)$	$N=$	3	4	5	6	$S=$	1	2	7	8	$7(d_1)$OMHGFTWVCS
	$\downarrow$	O	K	M	J	$\downarrow$	V	T	P	C	
		7	2	7	2		6	3	4	6	
$D_4=(P_{33},M,P)$	$M=$	3	4	6	7	$P=$	1	2	5	8	$5(d_3)$ONWTRKJEUP
	$\downarrow$	O	L	F	N	$\downarrow$	R	U	Q	E	
		5	2	1	5		6	4	6	4	
$D_5=(P_{33},J,S)$	$J=$	2	3	4	5	$S=$	1	6	7	8	$1(d_3)$EAPUVFBCQS
	$\downarrow$	N	E	K	A	$\downarrow$	T	V	Q	C	
		6	1	6	1		5	4	2	2	
$D_6=(P_{33},T,A)$	$T=$	5	6	7	8	$A=$	1	2	3	4	$4(d_3)$OWNMFVUEJA
	$\downarrow$	S	W	R	O	$\downarrow$	J	F	E	B	
		1	4	2	4		5	7	5	7	
$D_7=(P_{33},L,U)$	$L=$	2	3	6	7	$U=$	1	4	5	8	$1(d_3)$BCAFVSQPEU
	$\downarrow$	M	C	B	K	$\downarrow$	W	P	V	E	
		4	1	1	5		7	2	7	2	
$D_8=(P_{35},Y,A)$	$Y=$	5	6	7	8	$A=$	1	2	3	4	$1(d_1)$XWQCDEUSPA
	$\downarrow$	X	V	W	O	$\downarrow$	J	S	E	P	
		1	4	1	4		5	8	5	8	
$D_9=(P_{35},T,D)$	$T=$	3	4	7	8	$D=$	1	2	5	6	$1\cup 2(d_1\cup d_1)$
	$\downarrow$	V	R	S	M	$\downarrow$	K	W	C	E	RLKJEAPQCD
		5	2	1	6		3	8	7	4	SUWX

TABLE 1: Candidate Dantzig figures from P_{25}, P_{29}, P_{31}, P_{33}, and P_{35}.

Dantzig figures	Edge-facet intersections										Facet for lemmas no.(type)comb.equiv.
$D = (P, x, y)$	$x =$	$h_1 \cap$	$h_2 \cap$	$h_3 \cap$	h_4						$h_i(d_j)\alpha\beta\gamma\delta\epsilon\zeta\eta\theta\iota\omega$
	$\downarrow$	v_1	v_2	v_3	v_4						
		h_{i_1}	h_{i_2}	h_{i_3}	h_{i_4}						
$D_{10} = (P_{35}, J, X)$	$J =$	2	3	4	5	$X =$	1	6	7	8	$1(d_1)AEPSUWDCQX$
	$\downarrow$	N	E	K	A	$\downarrow$	Y	Q	W	C	
		6	1	6	1		5	2	5	2	
$D_{11} = (P_{35}, L, U)$	$L =$	2	3	6	7	$U =$	1	4	5	8	$1 \cup 8(d_1 \cup d_1)$
	$\downarrow$	M	C	R	K	$\downarrow$	V	W	S	E	$RTVYWXQPSU$
		4	1	8	5		7	6	3	2	$CDEA$
$D_{12} = (P_{35}, N, Q)$	$N =$	3	4	5	6	$Q =$	1	2	7	8	$2(d_1)JKAEDCLRPQ$
	$\downarrow$	O	K	M	J	$\downarrow$	R	X	P	C	
		7	2	7	2		3	6	3	6	
$D_{13} = (P_{35}, O, P)$	$O =$	4	5	6	7	$P =$	1	2	3	8	$3(d_1)NMJKLRTSAP$
	$\downarrow$	Y	M	V	N	$\downarrow$	R	S	Q	A	
		8	3	8	3		7	4	7	4	
$D_{14} = (P_{36}, J, Q)$	$J =$	2	3	4	5	$Q =$	1	6	7	8	$1(d_4)AEWUSPDHXQ$
	$\downarrow$	N	E	K	A	$\downarrow$	R	X	P	H	
		6	1	6	1		2	5	2	5	
$D_{15} = (P_{36}, O, S)$	$O =$	4	5	6	7	$S =$	1	2	3	8	$3(d_1)NMJKLTVUAS$
	$\downarrow$	H	M	Y	N	$\downarrow$	T	U	P	A	
		1	3	8	3		7	4	6	4	
$D_{16} = (P_{36}, D, V)$	$D =$	1	2	5	6	$V =$	3	4	7	8	$7 \cup 8(d_1 \cup d_1)$
	$\downarrow$	K	H	P	E	$\downarrow$	Y	T	U	M	$PSUWYXQRTV$
		3	7	8	4		5	2	1	6	$HOML$
$D_{17} = (P_{36}, L, W)$	$L =$	2	3	6	7	$W =$	1	4	5	8	$8(d_1)RTQPSUVYXW$
	$\downarrow$	M	R	T	K	$\downarrow$	Y	X	U	E	
		4	8	8	5		7	7	3	2	
$D_{18} = (P_{37}, A, R)$	$A =$	1	2	3	4	$R =$	5	6	7	8	$8(d_1)WUXYVTSPQR$
	$\downarrow$	J	W	E	U	$\downarrow$	T	Q	P	O	
		5	8	5	8		2	1	1	4	
$D_{19} = (P_{37}, O, U)$	$O =$	4	5	6	7	$U =$	1	2	3	8	$3(d_1)NMJKLVYWAU$
	$\downarrow$	R	M	G	N	$\downarrow$	V	W	S	A	
		8	3	1	3		7	4	6	4	
$D_{20} = (P_{37}, X, K)$	$X =$	1	4	7	8	$K =$	2	3	5	6	$3(d_1)WYAUVLMNJK$
	$\downarrow$	Y	Q	W	G	$\downarrow$	N	D	L	J	
		3	5	3	5		4	1	7	4	
$D_{21} = (P_{37}, D, Y)$	$D =$	1	2	5	6	$Y =$	3	4	7	8	$8(d_1)PSQRTVUWXY$
	$\downarrow$	K	P	S	E	$\downarrow$	X	V	W	M	
		3	8	8	4		1	2	1	6	

TABLE 1 (cont'd): Candidate Dantzig figures from P_{36} and P_{37}. The first column in the table identifies the Dantzig figure; $D = (P, x, y)$ indicates that the Dantzig figure is found in the $(4, 8)$-polytope P from the **[GS]** catalog, with estranged vertices x and y. The next two columns summarize the edge-facet intersections in the Dantzig figure; for example, the vertex $v_1 = h_{i_1} \cap h_2 \cap h_3 \cap h_4$ is the neighbor of x along the edge not incident to h_1. The last column in the table provides a specific combinatorial equivalence between a facet or pair of facets in the Dantzig figure and one of the graphs considered in the following lemmas.

Let G be the graph of a d-polytope P. By a *monotone orientation* of G (or of P) we mean a way of directing all of G's edges so that the resulting digraph satisfies the following conditions:

the digraph is acyclic;

in each k-face P $(1 \leq k \leq d)$ there is a unique source and a unique sink;

in each k-face, there are k independent paths from source to sink.

Such orientations are called *good orientations* by Kalai [**K**].

A monotone orientation is a combinatorial phenomenon. In contrast, if P is a d-polytope with m vertices, an *LP-imbedding* (X, ϕ) of P is an assignment $X_{d\times m}$ of coordinates to the vertices of P and of a linear functional φ such that $\operatorname{conv} X$ is combinatorially equivalent to P and ϕ is admissible for X. An LP-imbedding induces an orientation on the edges of P; an edge is directed $v_i \to v_j$ iff $\phi(v_i) < \phi(v_j)$. An *LP-orientation* of P is an orientation of the edges of P that is induced by some LP-imbedding (X, ϕ) of P. Every LP-orientation is a monotone orientation of P.

In the proofs of the following four lemmas, we orient edges as necessary to avoid cycles, to preserve one-source/one-sink per face, and to maintain the required monotone distances. In each case, we arrive at a contradiction; none of the four graphs considered in the following lemmas can be given a monotone orientation that preserves the required monotone distances. Each of the Dantzig figures listed above is covered by at least one of these lemmas, as indicated by the fourth column in the table.

4.3. LEMMA. *In the d_1* polytope $\alpha\beta\gamma\delta\epsilon\zeta\eta\theta\iota\omega$ *of Figure 2, with sink ω, either α* or *β is at monotone distance 3 from ω.*

PROOF. Since ω is the sink, we have $\iota \to \omega$, $\zeta \to \omega$, and $\theta \to \omega$. To keep β at monotone distance 4, we must take $\eta \to \beta$.

If $\delta \to \epsilon$, then $\epsilon \to \zeta$, $\epsilon \to \beta$, $\beta \to \alpha$, $\delta \to \alpha$, $\alpha \to \gamma$, $\delta \to \gamma$, $\gamma \to \iota$; but now $[\alpha, \gamma, \iota, \omega]$ is a monotone path from α to ω of length 3 .

On the other hand, if $\epsilon \to \delta$, then $\delta \to \gamma$, $\gamma \to \iota$; $\gamma \to \alpha$, $\delta \to \alpha$, $\alpha \to \beta$; $\epsilon \to \beta$, and β is a second sink. □

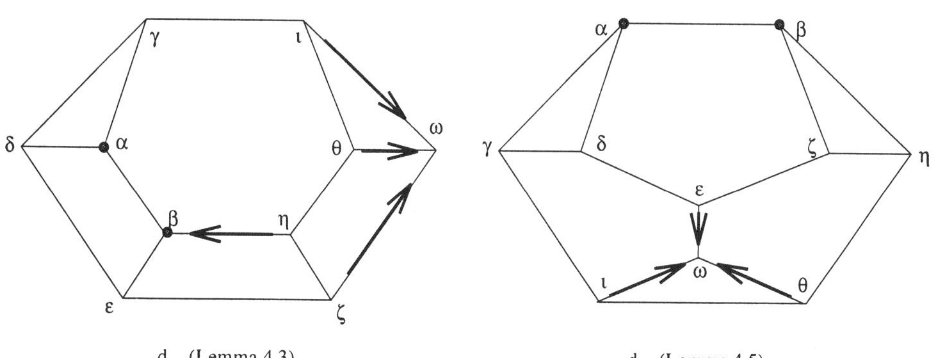

FIGURE 2: The diagrams for Lemmas 4.3 and 4.5. In each, the vertex ω is the sink, and for any monotone orientation, the monotone distance from one of the highlighted vertices to the sink is less than 4.

4.4. COROLLARY. *The Dantzig figures $D_1, D_3, D_8, D_{10}, D_{12}, D_{13}, D_{15}, D_{17}, D_{18}, D_{19}, D_{20}, D_{21}$ have strict monotone diameter* 4.

PROOF. For each of these Dantzig figures, the last column in its entry in the table above gives a combinatorial equivalence between a d_1-facet of the Dantzig figure and the d_1 polytope $\alpha\beta\gamma\delta\epsilon\zeta\eta\theta\iota\omega$ of Lemma 4.3.

For example, in the table entry for D_1, the last column indicates that facet 2 is of type d_1, and the listed map of vertices yields a combinatorial equivalence between facet 2 in D_1 and $\alpha\beta\gamma\delta\epsilon\zeta\eta\theta\iota\omega$. Note that under this equivalence $Q(\Leftrightarrow \omega)$ is the sink, and by Lemma 4.3 either $J(\Leftrightarrow \alpha)$ or $K(\Leftrightarrow \beta)$ is at monotone distance 3 from Q. In D_1, if Q is the sink, then N is the source by Lemma 4.1. Since either J or K is at monotone distance 3 from Q, at least one of these directed edges $N \to K$ and $N \to J$ starts a monotone path from N to Q of length 4. Thus the strict monotone diameter of D_1 is 4. □

4.5. LEMMA. *In the d_3 polytope $\alpha\beta\gamma\delta\epsilon\zeta\eta\theta\iota\omega$ of Figure 2, with sink ω, either α or β is at monotone distance 3 from ω.*

PROOF. Since ω is the sink, we have $\iota \to \omega$, $\epsilon \to \omega$, and $\theta \to \omega$. We derive contradictions from all four possible orientations of the pair of edges $\gamma\delta$ and $\eta\zeta$.

If $\eta \to \zeta$ and $\gamma \to \delta$, then $\delta \to \epsilon$, $\zeta \to \epsilon$; $\delta \to \alpha$, $\zeta \to \beta$, but now the face $\alpha\beta\zeta\epsilon\delta$ has two sources.

If $\zeta \to \eta$ and $\delta \to \gamma$, then $\gamma \to \iota$, $\eta \to \theta$; $\eta \to \beta$, $\gamma \to \alpha$, and the face $\alpha\beta\eta\theta\iota\gamma$ has two sources.

If $\zeta \to \eta$ and $\gamma \to \delta$ (or $\eta \to \zeta$ and $\delta \to \gamma$ by symmetry), then $\delta \to \epsilon$, $\eta \to \theta$; $\eta \to \beta$, $\zeta \to \beta$; $\delta \to \alpha$, $\gamma \to \alpha$, and either β or α is a sink. □

4.6. COROLLARY. *The Dantzig figures D_2, D_4, D_5, D_6, D_7 have strict monotone diameter 4.*

PROOF. For each of these Dantzig figures, the last column in its entry in the table above gives a combinatorial equivalence between a d_3-facet of the Dantzig figure and $\alpha\beta\gamma\delta\epsilon\zeta\eta\theta\iota\omega$.

For example, in the table entry for D_2, we see that facet 2 is of type d_3. A combinatorial equivalence between facet 2 of D_2 and the d_3 polytope of Lemma 4.5 is listed by the images of the vertices; we see that $K(\Leftrightarrow \omega)$ is the sink, and either $E(\Leftrightarrow \alpha)$ or $O(\Leftrightarrow \beta)$ is at monotone distance 3 from K. However, V is the source, and at least one of the directed edges $V \to O$ and $V \to E$ initiates a monotone path of length 4 from V to K. Hence, the strict monotone diameter of D_2 is 4. □

4.7. LEMMA. *In the d_4 polytope $\alpha\beta\gamma\delta\epsilon\zeta\eta\theta\iota\omega$ of Figure 3, with sink ω, either α or β is at monotone distance 3 from ω.*

PROOF. Since ω is the sink, we have $\iota \to \omega$, $\zeta \to \omega$, and $\theta \to \omega$. To keep β at monotone distance 4, we must take $\eta \to \beta$.

If $\delta \to \epsilon$, then $\epsilon \to \zeta$; $\epsilon \to \alpha$, $\delta \to \alpha$, $\alpha \to \beta$, $\beta \to \gamma$, $\delta \to \gamma$, $\gamma \to \iota$; but now $[\beta, \gamma, \iota, \omega]$ is a monotone path from β to ω of length 3.

On the other hand, if $\epsilon \to \delta$, then $\delta \to \gamma$, $\gamma \to \iota$; $\gamma \to \beta$, $\beta \to \alpha$, $\delta \to \alpha$, $\epsilon \to \alpha$, and α is a second sink. □

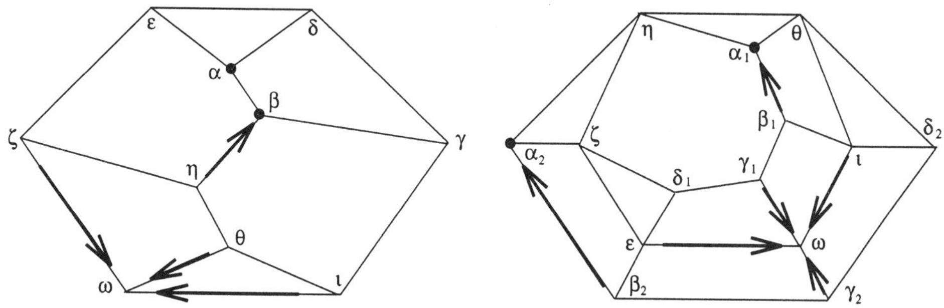

d_4 (Lemma 4.7) $\qquad$ $d_1 \cup d_1$ (Lemma 4.9)

FIGURE 3: The diagrams for Lemmas 4.7 and 4.9. In each, the vertex ω is the sink, and for any monotone orientation, the monotone distance from one of the highlighted vertices to the sink is less than 4.

4.8. COROLLARY. *The Dantzig figure D_{14} has strict monotone diameter* 4.

PROOF. In the table entry for D_{14}, the last column indicates that facet 1 has type d_4. Under the listed combinatorial equivalence between facet 1 and the d_4 of Lemma 4.7, $Q(\Leftrightarrow \omega)$ is the sink, and either $A(\Leftrightarrow \alpha)$ or $E(\Leftrightarrow \beta)$ is at monotone distance 3 from Q. However J is the source, and at least one of the directed edges $J \rightarrow A$ or $J \rightarrow E$ initiates a monotone path from J to Q of length 4. Hence the strict monotone diameter of D_{14} is 4. □

4.9. LEMMA. *For any consistent monotone orientation of the two adjacent d_1 polytopes*

$$\alpha_1\beta_1\gamma_1\delta_1\epsilon\zeta\eta\theta\iota\omega \quad \text{and} \quad \alpha_2\beta_2\gamma_2\delta_2\epsilon\zeta\eta\theta\iota\omega$$

of Figure 3, *with ω as sink, at least one of α_1 and α_2 is at monotone distance* 3 *from ω.*

PROOF. Since ω is the sink, we have $\gamma_1 \rightarrow \omega$, $\gamma_2 \rightarrow \omega$, $\epsilon \rightarrow \omega$, and $\iota \rightarrow \omega$. To keep α_1 and α_2 at monotone distance 4, we must take $\beta_1 \rightarrow \alpha_1$ and $\beta_2 \rightarrow \alpha_2$.

If $\eta \rightarrow \zeta$, then $\zeta \rightarrow \epsilon$, $\zeta \rightarrow \alpha_2$, $\eta \rightarrow \alpha_2$, and α_2 is a sink.

If $\zeta \rightarrow \eta$, then $\eta \rightarrow \theta$, $\theta \rightarrow \iota$, $\theta \rightarrow \alpha_1$, $\eta \rightarrow \alpha_1$, and α_1 is a sink. □

4.10. COROLLARY. *The Dantzig figures D_9, D_{11}, D_{16} have strict monotone diameter* 4.

PROOF. The proof here is similar to those of the previous corollaries, with the exception that here we must exhibit a combinatorial equivalence between a union of two d_1 facets of the Dantzig figure and the graph considered in Lemma 4.9. These equivalences are listed in the fourth column of the table entries for D_9, D_{11}, D_{16}, with the first line identifying the two facets, the second line listing the images of $\alpha_1\beta_1\gamma_1\delta_1\epsilon\zeta\eta\theta\iota\omega$ under this equivalence, and the third line listing the images of $\alpha_2\beta_2\gamma_2\delta_2$.

For example, from the table entry for D_9, we see that facets 1 and 2 are both of type d_1, and their union is combinatorially equivalent to the union of two d_1 polytopes as in Lemma 4.9. Under the listed equivalence, $D(\Leftrightarrow \omega)$ is the sink, and either $R(\Leftrightarrow \alpha_1)$ or $S(\Leftrightarrow \alpha_2)$ is at monotone distance 3 from D. In D_9, T is the source, and so at least one of the directed edges $T \to R$ or $T \to S$ initiates a monotone path from T to D of length 4. □

The above sequence of lemmas and corollaries leads to the following, our main result.

4.11. THEOREM. $\Delta_{sm}(4,8) = 4$.

PROOF. The 37 combinatorial types of $(4,8)$-polytopes are listed in [**GS**]. The strict monotone diameter of the 4-cube is equal to 4. The purpose of the lemmas and corollaries is to show that no $(4,8)$-polytope has strict monotone diameter greater than 4. This involves geometric realizations and LP-orientations of the various combinatorial types.

First, lemmas 4.1 and 4.2 show that it suffices to consider LP-orientations of Dantzig figures $D = (P, x, y)$ in which

the estranged vertices x and y are source and sink in the LP-orientation;

every edge from from x terminates on a $(3,7)$-facet incident to y and *vice versa*;

the $(4,8)$-polytope P has at least 18 vertices.

These requirements rule out all but seven of the 37 $(4,8)$-polytopes, leaving as candidates only the 21 Dantzig figures listed in Table 1.

For each candidate Dantzig figure $D = (P, x, y)$, Lemmas 4.3, 4.5, 4.7, and 4.9 served to identify a pair of edge-neighbors of x such that any monotone orientation of P with x as source and y as sink leaves at least one of these two neighbors at monotone distance 3 from y. The fourth column of Table 1 provides a key to our specific arguments along these lines, showing that each of the 21 Dantzig figures is covered by one of the lemmas. It follows that under any monotone orientation of P with x as source and y as sink, the monotone distance from x to y is 4.

The strict monotone 4-step conjecture is a statement about LP-orientations. Since every LP-orientation is a monotone orientation, we conclude that the strict monotone diameter of each of these 21 Dantzig figures is 4, and since these 21 Dantzig figures exhaust the possibilities for a higher strict monotone diameter, $\Delta_{sm}(4,8) = 4$. □

5. Comment

We suspect that the strict monotone d-step conjecture is false when d is sufficiently large, and that it may therefore eventually be added to [**KK**]'s list of strengthenings of the d-step conjecture that hold for $d \leq 3$ but fail for some larger d. More specifically, we suspect that the failure of the strict monotone d-step conjecture can be shown by means of a polytope combinatorially equivalent to the polytope P_d that was used in [**HK**] to provide a counterexample to the Lagarias-Prabhu-Reeds strengthening [**LPR**] of the d-step conjecture. For the $(5,10)$-polytope P_5 of [**HK**], we have produced a monotone orientation for which the strict monotone diameter is 6 (see Figure 4). Thus the purely combinatorial methods used here cannot be extended beyond $d = 4$. Moreover, if this monotone orientation of P_5 is an LP-orientation, then $\Delta_{sm}(5,10) \geq 6$.

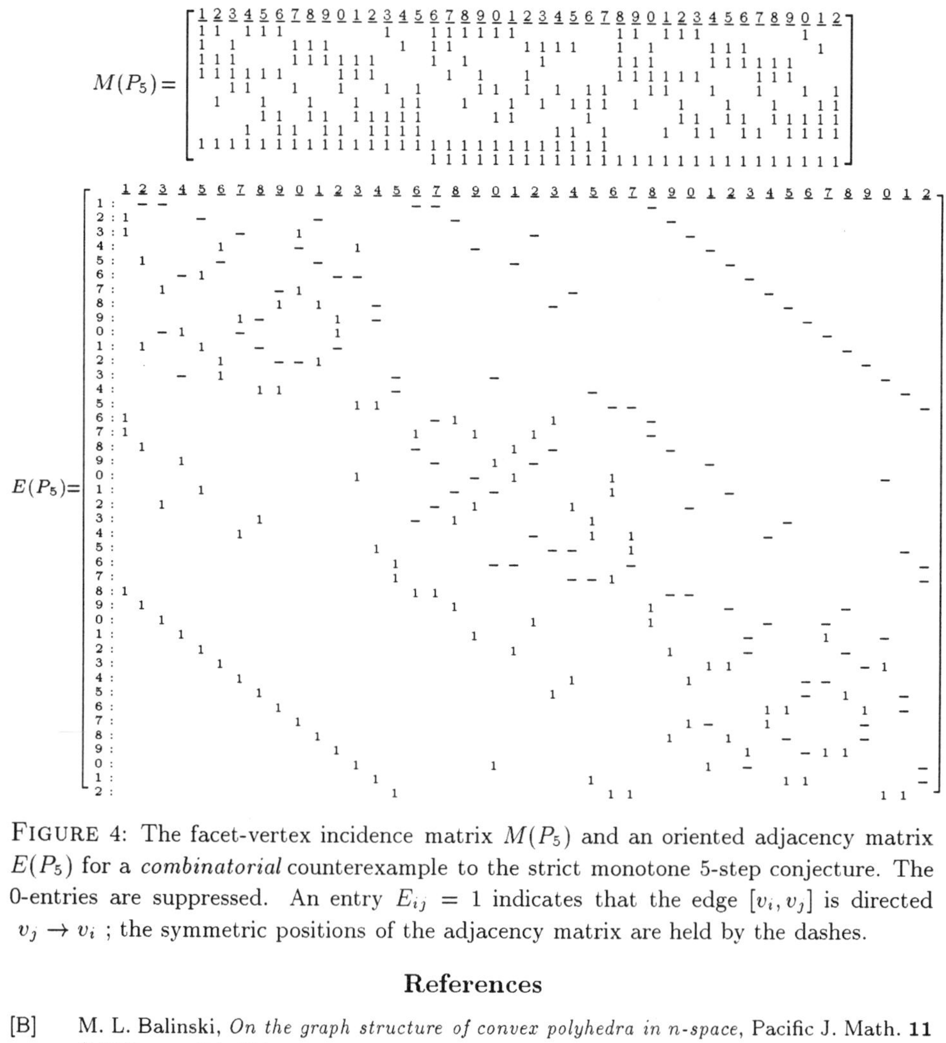

FIGURE 4: The facet-vertex incidence matrix $M(P_5)$ and an oriented adjacency matrix $E(P_5)$ for a *combinatorial* counterexample to the strict monotone 5-step conjecture. The 0-entries are suppressed. An entry $E_{ij} = 1$ indicates that the edge $[v_i, v_j]$ is directed $v_j \to v_i$; the symmetric positions of the adjacency matrix are held by the dashes.

References

[B] M. L. Balinski, *On the graph structure of convex polyhedra in n-space*, Pacific J. Math. **11** (1961), pp. 431–434.

[BR] M. L. Balinski and F. J. Rispoli, *Signature classes of transportation polytopes*, Math. Programming **60** (1993), pp. 127–144.

[Ba] D. Barnette, *A short proof of the d-connectedness of d-polytopes*, Discrete Math. **137** (1995), pp. 351–352.

[BM] J. A. Bondy and U. S. R. Murty, *Graph Theory with Applications*, American Elsevier, New York, 1976.

[D] G. B. Dantzig, *Linear Programming and Extensions*, Princeton Univ. Press, Princeton, N. J. 1963.

[D2] G. B. Dantzig, *Eight unsolved problems from mathematical programming*, Bull. Amer. Math. Soc. **70** (1964), pp. 499–500.

[G] R. C. Grinold, *The Hirsch conjecture in Leontief substitution systems*, SIAM J. Appl. Math. **21** (1970), pp. 483–485.

[GS] B. Grünbaum and V. Sreedharan, *An enumeration of simplicial 4-polytopes with 8 vertices*, J. Combin. Theory **2** (1967), pp. 437–465.

[HK] F. Holt and V. Klee, *Counterexamples to the strong d-step conjecture for $d \geq 5$*, Discrete Comput. Geom. **19** (1988), pp. 33–46.

[K] G. Kalai, *A simple way to tell a simple polytope from its graph*, J. Combin. Theory A **49**, (1988), pp. 381–383.

[Kl] V. Klee, *Diameters of polyhedral graphs*, Canad. Math. J. **16** (1964), pp. 602–614.

[Kl2] V. Klee, *Paths on polytopes. I*, SIAM J. Appl. Math. **13**, (1965), pp. 946–956.

[KK] V. Klee and P. Kleinschmidt, *The d-step conjecture and its relatives*, Math. of Operations Research **12** (1987), 718–755.

[KW] V. Klee and D. W. Walkup, *The d-step conjecture for polyhedra of dimension* $d < 6$, Acta Math. **133** (1967), pp. 53–78.

[LPR] J. C. Lagarias, N. Prabhu and J. A. Reeds, *The d-step conjecture and Gaussian elimination*, Discrete Comput. Geom. **18** (1997), pp. 53–82.

[M] T. Matsui, *The monotone Hirsch conjecture for matching polytopes*, Research Report B-211, Department of Information Sciences, Tokyo Institute of Technology (1988).

[R] F. Rispoli, *The monotonic diameter of the perfect matching and shortest path polytopes*, Operations Research Letters **12** (1992), pp. 23–27.

[R2] F. Rispoli, *The monotonic diameter of the perfect 2-matching polytope*, SIAM J. Optimization **4** (1994), pp. 455–460.

[T] M. Todd, *The monotonic bounded Hirsch conjecture is false for dimension at least* 4, Math. Operations Research **5** (1980), pp. 599–601.

[Z] G. Ziegler, *Lectures on Polytopes*, Springer-Verlag, New York, 1994.

THE BOEING COMPANY, P.O. BOX 3707, M/S 7L-21, SEATTLE, WA 98124-2207
E-mail address: `fred.b.holt@boeing.com`.

UNIVERSITY OF WASHINGTON, DEPARTMENT OF MATHEMATICS, BOX 354350, SEATTLE, WA 98195-4350
E-mail address: `klee@math.washington.edu`

Contemporary Mathematics
Volume **223**, 1999

Interactions between Real Algebraic Geometry and Discrete and Computational Geometry

Ilia Itenberg and Marie-Françoise Roy

ABSTRACT. We give a short introduction to real algebraic geometry, and show how some of its results are connected with discrete and computational geometry. Several applications of the theory of semi-algebraic sets to algorithmic problems arising in geometry are discussed. We also describe a combinatorial patchworking which makes it possible to construct real algebraic varieties in a completely combinatorial fashion.

1. Introduction

One of the most basic results in real algebraic geometry is Descartes' rule [**Des**]. Given a univariate polynomial

$$P = a_1 X^{d(1)} + \ldots + a_\ell X^{d(\ell)}$$

($d(1) > \ldots > d(\ell)$, $a_i \neq 0$), the number of its positive real roots counted with multiplicities is bounded by the number of sign changes in the coefficients of P (for a proof see [**BCR**]). The difference between the number of positive real roots counted with multiplicities and the number of sign changes in the coefficients is even. Moreover, there are polynomials such that the maximum of positive real roots predicted by Descartes' rule is reached. This result can be seen as a prototype of the interaction between real algebraic geometry and combinatorics: information about the real roots can be extracted from the knowledge of a combinatorial object, the *signed Newton diagram* of P: which monomials appear in P and with which sign.

What is real algebraic geometry? Real algebraic geometry is the study of real solutions of polynomial systems of equations. As we all know from the univariate polynomials of degree 2, the number of real solutions depends on sign conditions on polynomial expressions of the coefficients, and so real algebraic geometry also includes the study of systems of polynomial inequalities.

Descartes' rule does not provide a complete answer to the problem of real root counting in the univariate case. A complete answer was discovered by Sturm in the 19th century [**Stu**]. The famous Tarski-Seidenberg principle [**Tar, Sei**] can be seen as a vast generalization of Sturm's theorem. The *Existential Theory of the*

1991 *Mathematics Subject Classification.* Primary 14P10, 68Q25, 14P25.
Key words and phrases. Semi-algebraic sets, complexity of algorithms, Viro method.

© 1999 American Mathematical Society

Reals (ETR) is concerned with whether or not a system of polynomial equalities and inequalities has a common solution. Sturm's theorem, or more precisely its generalizations due to Hermite and Sylvester [**Her, Syl**], provides the answer in the univariate case. Tarski proved that over the reals ETR is decidable. This is quite remarkable since, for polynomials with integer coefficients, the similar problem for integer solutions is known to be undecidable [**Mat**]. Tarski's work on this problem is the starting point of model theory, a whole branch of mathematical logic. He defined formulae in a first order language and formulated his main result in this new terminology.

A typical example of a first order theory is the theory of real closed fields, which had already been studied by Artin and Schreier [**Art, ArS**]. Tarski gave first order axioms for real closed fields which express the algebraic aspect of the theory of the reals.

DEFINITION 1. A real closed field is an ordered field whose positive elements are squares and such that every polynomial of odd degree with coefficients in the field has a root in the field.

The quantifier elimination theorem of Tarski is the following.

THEOREM 1. *The first order theory of real closed fields admits quantifier elimination in the language of ordered fields.*

All of us eliminated quantifiers without knowing it when studying second degree equations in high school. If $a \neq 0$, the formula

$$\exists x \; ax^2 + bx + c = 0$$

is equivalent to

$$b^2 - 4ac \geq 0$$

over the reals.

Tarski's effective proof of quantifier elimination establishes the decidability of elementary problems in real geometry. It took many years for these logical results to be translated in a more geometric language. This required the concept of a semi-algebraic set.

DEFINITION 2. Semi-algebraic sets are the sets of solutions of boolean combinations of polynomial inequalities. Semi-algebraic maps are maps whose graph is semi-algebraic.

Then quantifier elimination can be formulated as follows.

THEOREM 2. *Linear projections of semi-algebraic sets are semi-algebraic sets.*

The equivalence between this result and the quantifier elimination theorem is easy, since to make a linear projection is to eliminate an existential quantifier.

Various extensions of Tarski's results are studied nowadays. In the case of polynomials, as we shall see later in this text, the main problems studied are complexity and implementation. Generalizations to the exponential function [**Wilk**], and more generally to o-minimal structures [**vD**] have been developed recently. The complexity of Pfaffian functions has also been considered [**GaV**].

Hilbert's 16th problem is another important source of development in real algebraic geometry. Consider a real algebraic hypersurface H of degree d in $\mathbb{R}P^k$. Such a hypersurface is, by definition, a real homogeneous polynomial in $k+1$ variables of degree d considered up to multiplication by a nonzero real number. We suppose the hypersurface is nonsingular, which means that the polynomial has no critical points in $\mathbb{R}^{k+1} \setminus \{0\}$. This polynomial has a well defined zero locus $\mathbb{R}H$ in the real projective space $\mathbb{R}P^k$. The set $\mathbb{R}H$ is called *the real point set* of the hypersurface H. The first part of Hilbert's 16-th problem concerns the possible topological types of the manifolds $\mathbb{R}H$ or of the pairs $(\mathbb{R}P^k, \mathbb{R}H)$ (for a given degree d). The complete answer for curves in $\mathbb{R}P^2$ is known only for curves of degree at most 7. For surfaces in $\mathbb{R}P^3$, the answer is known up to degree 4.

To solve the problem, it is necessary to work in two main directions: first, find the restrictions on the topological types of $\mathbb{R}H$ or of pairs $(\mathbb{R}P^k, \mathbb{R}H)$, and, second, give constructions of hypersurfaces for realizable topological types. Many deep and important results were obtained in the first direction using algebraic and differential topology (see, for example, the survey papers [**Vi3, Wils**]). However, the methods of constructions had not been seriously changed since the XIXth century until 1980, when Viro proposed a new method to construct real algebraic varieties. We will present a particular case of this method in what follows.

Hilbert's 17th problem plays an important role in real algebraic geometry as well. Hilbert asked whether positive polynomials are sums of squares of rational functions. This problem was solved quickly and beautifully by Artin in 1925. Abstract real algebra starts here, and real closed fields were defined on that occasion. Though this problem was solved early, it is still the center of an important activity in various geometric situations: for example, how many sums of squares occur?

What about the interactions with discrete and computational geometry? Several basic problems in discrete and computational geometry are formulated starting with n points in the plane. For instance, two ordered lists $[A_1, \dots, A_n]$ and $[B_1, \dots, B_n]$ of n points in the plane have the same *order* (or *orientation*) *type* if every two triangles $[A_i, A_j, A_k]$ and $[B_i, B_j, B_k]$ have the same orientation. The isotopy problem asks, given two such lists of points with the same order type, if it possible to move continuously from $[A_1, \dots, A_n]$ to $[B_1, \dots, B_n]$ keeping the same order type?

The set of configurations of an order type (the n-tuples of points of the plane with a given order type) is a basic semi-algebraic set of $\mathbb{R}^{2n}$ (given by a finite conjunction of sign conditions on polynomials). It looks like a very special semi-algebraic set. But according to Mnev's universality theorem [**Mne**], every basic semi-algebraic set is (up to cartesian product by an affine space) homeomorphic to the configuration space of an order type. Thus, the answer to the isotopy problem is no in a very strong sense. Moreover, it means that the general study of semi-algebraic sets does not bring us too far from basic problems of discrete and computational geometry [**HRR**].

2. Qualitative properties of semi-algebraic sets

2.1. Stability and finiteness. We have seen that linear projections of semi-algebraic sets are semi-algebraic. Due to the quantifier elimination property, many sets constructed from semi-algebraic sets are again semi-algebraic: the closure, the interior, the set of singular points, and the boundary of a semi-algebraic set are semi-algebraic. The image of a semi-algebraic set under a semi-algebraic map is semi-algebraic. Logic is a powerful tool for showing that many sets are semi-algebraic.

A more detailed geometric study is needed to prove that semi-algebraic sets have a finite number of connected components. These connected components are also semi-algebraic sets. The real dimension and local real dimension of semi-algebraic sets can be defined for any real closed field [**BCR**], and the subset of points of a semi-algebraic set with a fixed local real dimension is semi-algebraic.

Semi-algebraic sets can be decomposed into a finite number of smooth semi-algebraic pieces. This is called a stratification. Additional regularity conditions can be imposed as well. The boundary condition asks that the boundary of a stratum of any dimension is the union of strata of smaller dimensions. Other regularity conditions are often considered, for example, Whitney conditions on the limits of tangent spaces [**BCR**]. Open simplices and finite simplicial complexes are obviously semi-algebraic. The triangulation theorem [**BCR**] states that a semi-algebraic set is semi-algebraically homeomorphic to a simplicial complex.

An important property states that in a semi-algebraic family of semi-algebraic sets, there is a finite number of topological types.

Let R be a real closed field.

DEFINITION 3. A semi-algebraic family of semi-algebraic sets is a semi-algebraic subset S of $R^{\ell+k}$. The set S is considered as a family of fibers $S_p = \{x \in R^k \mid (p, x) \in S\} \subset R^k$ depending on $p \in R^\ell$. Two sets S_p and $S_{p'}$ have the same topological type when they are homeomorphic.

THEOREM 3. *Given a semi-algebraic family of semi-algebraic sets $S \subset R^{\ell+k}$, there exist a finite number of simplicial complexes such that every S_p is semi-algebraically homeomorphic to one of these simplicial complexes.*

This is a consequence of Hardt's triviality theorem [**Hard**], [**BCR**].

THEOREM 4. *Given a semi-algebraic family of semi-algebraic sets $S \subset R^{\ell+k}$, there exists a partition of R^ℓ into a finite number of semi-algebraic sets F_i such that*

$$S_{F_i} = \{(p, x) \in S \mid p \in F_i\}$$

is semi-algebraically homeomorphic to $S_p \times F_i$ for any $p \in F_i$.

This theorem relies on geometric results, such as the triangulation theorem. Another consequence of Hardt's triviality theorem is that for a semi-algebraic family of sets, the set of parameters where the topological type is fixed is a semi-algebraic set.

Another finiteness property should be noticed here: families of semi-algebraic sets with fixed complexity (defined by a fixed number of inequalities of a fixed degree) have finite VC dimension: a family of sets is of VC dimension n if n is the greatest integer such that for any set M of n points in the space, any subset of M can be separated from its complement in M by a member of that family.

All these quantitative results are valid over arbitrary real closed fields. The consideration of general real closed fields is natural in the theoretical development of real algebraic geometry [**BCR**]. More surprisingly, real closed fields different from the field of real numbers play an important role for complexity questions: in the recent study of complexity of algorithms in real algebraic geometry, various intermediate nonarchimedean real closed fields are considered (for example [**GV1, Ca1, HRS1, Re1, BPR1**]). In the study of lower bounds, abstract real algebra is important as well (see [**LR1**]).

2.2. Semi-algebraic modeling. Semi-algebraic sets are encountered naturally in many algorithmic problems arising in geometry. For example, a problem of floodlights [**ORo**] can be solved at least theoretically with semi-algebraic sets. Given a line segment (the stage) and a set of floodlights of fixed apex and apertures, there is an algorithm to decide if the floodlights can be rotated to illuminate the stage, since this problem can be easily translated into a first order formula. Of course, what is asked for is an efficient algorithm which is not provided by the general tools described here.

2.2.1. *Visual aspects.* An object has in general an infinite number of different images, depending on the viewpoint (the point from which we project the surface on an image plane). However it is possible using semi-algebraic geometry to prove the intuitively clear fact that the number of different topological types of the *image contours* of a 3D object is finite when the object is semi-algebraic [**RvE**].

From there comes the idea of representing a 3D object with a graph: its *aspect graph*, whose nodes are its different (topologically equivalent) image contours and whose arcs are the *visual events* where the topological aspect of the image contour changes [**KV1, KV2, Mal**]. The position of the eye where the contour seen has a fixed topological type is semi-algebraic.

2.2.2. *The Piano Mover's Problem.* One famous problem where semi-algebraic modeling is used is the *"piano mover's problem"*.

Given a room with doors and other openings and an object (a robot, often called "the piano"), we want to determine if there is a permissible rigid motion of the piano from its initial position p_0 inside the room to a desired final position p_1 outside the room, while the piano avoids the obstacles, defined by the walls of the room [**SS**]. Here, the set O of obstacles and the piano P are presumed to be semi-algebraic subsets of $\mathbb{R}^k$.

Let C denote the configuration space and F the free space, that is, the subset of configurations such that the piano does not run across the walls. It is not difficult to see that the set C can be represented as a semi-algebraic subset of some $\mathbb{R}^N$. If $k = 2$, that is, if the problem is planar, a frame is fixed to the piano and the positions of the piano are given by the image of this frame. Two coordinates are needed to determine the origin of the frame and it is possible to add two more coordinates (for example a rotation matrix) to determine the position completely. The group of planar isometries is of dimension 3, and is a semi-algebraic subset of $\mathbb{R}^4$. So one can take $N = 4$.

If p is a position, that is, an element of C, we shall denote by P_p the subset of $\mathbb{R}^k$ corresponding to the piano in the position p. Since P is semi-algebraic, P_p is semi-algebraic too. We have $F = \{p \in C \mid P_p \cap O = \emptyset\}$. This set, defined by a first order formula of the language of ordered fields, is therefore semi-algebraic, and a quantifier elimination algorithm will give an explicit description of it.

Given two allowed positions p_0 and p_1, the problem we started with reduces to the following: do p_0 and p_1 belong to the same connected component of F? If they do, the piano can be moved from p_0 to p_1, and if not, such a motion is impossible. So we are led to consider the following general problem: given a semi-algebraic set S explicitly described, decide whether two elements x_0 and x_1 of S belong to the same connected component of S. If they do, explicitly construct a path between them lying in S. We will return to this problem later in the text.

2.3. Is the world semi-algebraic? The answer is obviously no. Besides fractal geometry, we need to integrate differential equations such as

$$y' = y$$

and add the exponential function.

A natural class of sets extending semi-algebraic sets is the class of semi-pfaffian sets. They are sets defined by a finite number of nested first order algebraic differential equations. These sets enjoy finiteness properties very similar to those of semi-algebraic sets [**Kh2**]. It is essential that the differential equations considered are first order. For example, the equation $y'' = -y$ should not be considered, since $\sin x$ has infinite oscillations, and hence does not have finiteness properties. It is possible to define the sine function by a first order differential equation, but only on a bounded region, and this restricted function enjoys finiteness properties.

Interestingly, Pfaffian functions lead us back to special kind of polynomials. Fewnomials, polynomials with few monomials, can be considered as a particular case of Pfaffian functions. This is because the monomial $y = x^M$ can be defined by the differential equation $y' = (M/x)y$.

The number of connected components and the topological types of sets defined by fewnomials depend only on the number of monomials and not on the degree. These results can be seen as a generalization of Descartes' rule described in the introduction: the number of real solutions of a polynomial equation is bounded by the number of monomials, not by the degree [**Kh2**].

3. Quantitative aspects of semi-algebraic sets

3.1. Explicit bounds in semi-algebraic geometry. According to Petrowsky/Oleinik/Thom/Milnor [**Pet, Ole, Tho, Mil**], the number of connected components of an algebraic set of degree d in $\mathbb{R}^k$ is $O(d)^k$.

The proof relies on the *critical point method*: in the case of a bounded smooth algebraic hypersurface defined by a polynomial Q, in a well chosen coordinate system the sweep hyperlane $X_k = c$ will touch a connected component for the first time at an isolated critical point. This point satisfies the equations

$$Q = \frac{\partial Q}{\partial X_1} = \ldots = \frac{\partial Q}{\partial X_{k-1}} = 0.$$

Since the point where the connected component is touched for the last time is critical too, the number of connected components is smaller than half of the number of critical points. According to Bezout's theorem, the number of these critical points is of order $O(d)^k$. The same method proves as well that the sum of all Betti numbers of an algebraic set is of order $O(d)^k$.

If we consider a real algebraic plane curve C defined by a polynomial $P(X,Y)$ of degree d, the critical values in the X direction correspond to zeroes of the discriminant of P with respect to Y, which is of degree $d(d-1)$. Under general position assumptions, this corresponds also to the number of critical points of X on C, and implies that the maximum number of compact connected components of C is $d(d-1)/2$. This bound is not sharp: the maximal number of connected components of an algebraic curve of degree d in $\mathbb{R}P^2$ is by Harnack's theorem [**Harn**] at most $(d-1)(d-2)/2+1$, which is sharp.These results on the number of connected components of an algebraic set have been extended to semi-algebraic sets in various directions. In particular, the number of nonempty sign conditions defined by a family of polynomials has been considered [**Alo**]. It can be proved [**BPR3**] that the number of connected components of sign conditions defined by s polynomials in k variables of degree at most d on an algebraic set of dimension k' defined by an equation of degree at most d is

$$\binom{O(s)}{k'}O(d)^k.$$

The key idea is to use a general position argument. Intuitively, general position corresponds to the maximum number of possible nonempty connected components and it is enough to compute the number of possible connected components in this case. Perturbations by various infinitesimals are needed to go to general position. Since $k'+1$ polynomials in general position do not intersect a set of dimension k', we have $\binom{O(s)}{k'}$ algebraic cases to consider, and apply the critical point method for each case.

The last result has interesting consequences in discrete and computational geometry. Two ordered lists $[A_1, \dots, A_n]$ and $[B_1, \dots, B_n]$ of n points in $\mathbb{R}^d$ have the same order type if for any $\phi : [1, \dots, d+1] \mapsto [1, \dots, n]$, the simplexes $[A_{\phi(1)}, \dots, A_{\phi(d+1)}]$ and $[B_{\phi(1)}, \dots, B_{\phi(d+1)}]$ have the same orientation. Ordered lists of n points having the same order type form the configuration of the order type, which is a basic semi-algebraic subset of $\mathbb{R}^{nd}$. Two ordered lists of n points in $\mathbb{R}^d$ have the same isotopy type if they have the same order type and belong to the same connected component of the configuration of their order type. Using the preceding results, it can be proved [**PR**] that the number of isotopy types of n points in $\mathbb{R}^d$ is

$$O(1)^{nd}n^{nd^2} = n^{nd^2+O(nd/\log n)}.$$

This extends a bound of the same nature known for the number of order types [**GoP**]. Another application, where the algebric set considered is the Grassmanian, is the following: a family of $k+1$ separated n compact bodies in $\mathbb{R}^d$ can be intersected by k transversals in at most

$$O(n^{k(k+1)(d-k)})$$

order types [**GPW**].

3.2. Complexity of algorithms. Starting from [**GV1**], many algorithms based on the critical point method have been obtained in the last decade. They have a complexity single exponential in the number of variables (see for example [**Ca1, Gri, Ca2, HRS1, HRS2, HRS3, Re1, Re2, GRi, HRS4, GV2, Ca4, BPR1, BPR6**]).

We describe some recent results where the algebraic part (dependence on the degree of polynomials) and the combinatorial part (dependence on the number of polynomials) are separated. The degrees of the polynomials in the output are also independent of the number of polynomials. These features are important for discrete and computational geometry where natural assumptions are $d=1$, or d, k fixed.

For example, finding a point in every connected component of sign conditions defined by a list of s polynomials in k variables of degree d on a variety of dimension k' can be done in time

$$\binom{O(s)}{k'}sd^{O(k)}$$

according to [**BPR7**].

For the complexity of Tarski-Seidenberg principle, the following result holds. Given

$$\Phi(Y) = (Q_\omega X^{[\omega]}) \dots (Q_1 X^{[1]}) F(P_1, \dots, P_s),$$

where Q_i is $\exists$ or $\forall$, an equivalent quantifier-free formula can be computed in

$$s^{(\ell+1)\Pi(k_i+1)} d^{(\ell+1)\Pi O(k_i)}$$

arithmetic operations, where ℓ is the number of free variables of the formula and k_i the number of variables of the block $X^{[i]}$. The degrees in the output are bounded by

$$d^{\Pi_i O(k_i)}$$

independently of s [**BPR4, BPR5**].

Let us come back to the piano mover's problem. Let R be a real closed field. A *roadmap* [**Ca1**] of $S \subset R \times R^{k-1}$ is a set $R(S) \subset S$ which is a semi-algebraic set of dimension at most one, satisfying

RM1 for every semi-algebraically connected component C of S, the intersection $C \cap R(S)$ is nonempty and semi-algebraically connected,

RM2 for every $p \in R$, and for every semi-algebraically connected component C of S_p, the intersection $C \cap R(S)$ is nonempty.

According to [**BPR8**], the road map of a semi-algebraic set S contained in a variety $Z(Q)$ of dimension k' can be constructed in time

$$\binom{O(s)}{k'} s d^{O(k^2)}.$$

It is clear that once a roadmap is constructed the piano mover's problem is solved in principle.

The main ideas of these results are the following. The combinatorial complexity part is based on the fact that it is possible to study only equalities, with many parallel algebraic subcases. Taking sums of squares does not increase the degree very much and makes the problem independent of the number of polynomials. Infinitesimal deformations allow us to go to general position in order to avoid considering too many parallel algebraic cases. Techniques of approximating varieties [**RV**] are used in order to make deformations with a fixed number of infinitesimals without changing the dimension of the variety being deformed. The idea is to find a finite number of approximating varieties of dimension k' defined by $k - k'$ equations covering $Z(Q)$ in the sense that every point of $Z(Q)$ is close to one of these approximating varieties, to perform the construction on these varieties and to come back to the original variety by a limit process. Finally, subroutines for algebraic sets give a point in every connected component through the critical point method. The algebraic complexity is concentrated in this part of the work.

Let us quote another result [**Bas**]: the sum of Betti numbers for a general semi-algebraic set is $\binom{O(s)}{k'} O(d)^k$ and the complexity of computing the Euler characteristic of a semi-algebraic set is $\binom{O(s)}{k'} d^{O(k^2)}$.

Here are some open problems on the complexity of algorithms for semi-algebraic sets.

1) Compute the real dimension of an algebraic set, or of a semi-algebraic set. A doubly exponential algorithm in the number of variables follows easily from Collins' Cylindrical Algebraic Decomposition (CAD), and an algorithm with complexity

$d^{O(k^2)}$ can be found in [**RV**]. In the complex case, the complexity of computing dimension is $d^{O(k)}$ [**Chi**] and the same bound is hoped for in the real case.

2) Compute the connected components of a semi-algebraic set. Single exponential algorithms are known [**CGV, HRS5**], the improvement to $s^{k'+1}d^{O(k^3)}$ is expected.

3) Decompose a semi-algebraic set into a finite number of smooth pieces. The problem can be solved with doubly exponential complexity in the number of variables using Collins' CAD. According to [**MoRan**] and [**Ran**] an improved algorithm gives complexity doubly exponential in the number of dimensions where nonempty strata appear. Several unsuccessful attempts have been made to prove that the complexity of this problem is single exponential in the number of variables. Another approach could be to deduce lower bounds from the universality theorem [**Mne**].

3.3. A computational challenge. Take a small systems of polynomial equalities and inequalities (written on a half page) and decide if there exist solutions.

Based on the recent progress mentioned in the complexity of algorithms through the critical point method, software projects are developed in this direction such as [**Ca3, CaR**] and Real Solving [**Ro2**] (inside European projects POSSO and FRISCO). Efficient methods based on [**PRS, ABRW**] are known for systems with a finite number of solutions [**GRR, GRRT**]. For general systems of inequalities, an efficient arithmetic for infinitesimals is needed.

3.4. Lower bounds. Semi-algebraic sets give lower bounds for discrete and computational geometry problems in the following manner: small semi-algebraic decision or computation trees cannot define semi-algebraic sets possessing any of the following features: many connected components [**Ben**], big sum of Betti numbers, complicated intersection with low degree sets [**MPR**], high geometric degree [**LR1**] or high curvature [**GKV**]. Hence, if we can prove that one of these quantities is high for a given problem, we know also that no quick algorithm can be designed for solving it.

One of the most classical results in this direction is the fact that an algorithm for sorting n points has complexity at least $n\log(n)$ [**Ben**]. We quote a more recent result, inspired by [**Str1**] and [**Str2**]. For two nonzero polynomials P, Q of degree d, lower bounds of order $d\log(d)$ for deciding the signs of the sequence of remainders of P and Q in a point $a \in R$ are found [**LR1**], while an upper bound of $d(\log(d))^2$ is known (see [**LR2**]).

4. Combinatorial patchworking

4.1. Real algebraic hypersurfaces in projective space. In this section, we discuss some combinatorial aspects appearing in the topology of real algebraic varieties. We describe first *combinatorial patchworking* (see [**It1, It2, It3, ItVi**]) which is a special case of the Viro method. This construction is very useful and has some fruitful applications. Combinatorial patchworking enables the construction of hypersurfaces using a simple combinatorial procedure: one can patchwork them from the pieces which are essentially hyperplanes. Slight modifications of this construction yield different objects such as, for example, real polynomials in two variables with prescribed collections of types of critical points [**Sh**] and real polynomial vector fields in $\mathbb{R}^2$ with prescribed collections of types of nondegenerate singular points **[ItSh]**.

The construction of combinatorial patchworking (or T-construction) works in any dimension. For simplicity, we restrict ourselves to the case of curves in $\mathbb{R}P^2$. The general T-construction can be formulated in a completely similar way.

4.2. Construction of an algebraic curve. To construct an algebraic curve in $\mathbb{R}P^2$, we start from the following combinatorial data.

Let d be a positive integer, and T be the triangle

$$\{(x,\ y) \in \mathbb{R}^2\ :\ x \geq 0,\ \ y \geq 0,\ \ x + y \leq d\}.$$

The number d will be the degree of a curve under construction, and the triangle T will be the Newton polygon of the curve.

Suppose that T is triangulated so that the vertices of all the triangles of the triangulation have integer coordinates. Suppose also that some distribution of signs, $a_{i,j} = \pm$ at the vertices of the triangulation, is given (see Figure 1). Then, there arises a naturally associated piecewise-linear curve L in $\mathbb{R}P^2$.

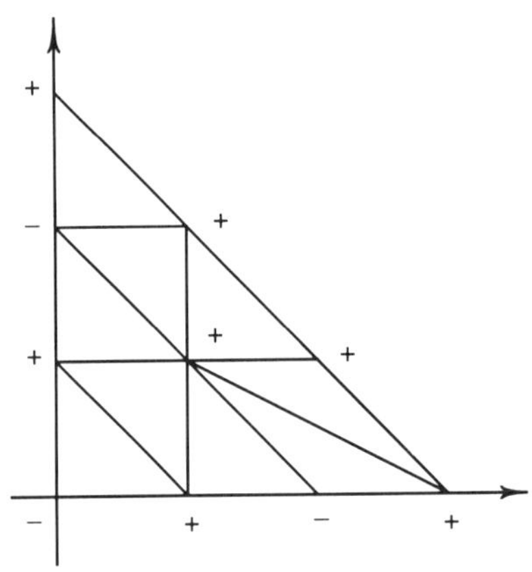

Figure 1

The construction of L is the following. Take copies $T_x = s_x(T)$, $T_y = s_y(T)$, $T_{xy} = s(T)$ of T, where $s = s_x \circ s_y$ and s_x, s_y are reflections with respect to the coordinate axes. Extend the triangulation of T to a symmetric triangulation of $T \cup T_x \cup T_y \cup T_{xy}$, and extend the distribution of signs to a distribution at the vertices of the extended triangulation using the following rule: passing from a vertex to its mirror image with respect to a coordinate axis, we preserve its sign if the distance from the vertex to the axis is even, and change the sign if the distance is odd (see Figure 2).

If a triangle of the triangulation has vertices of different signs, select a midline separating them. Denote by L' the union of the selected midlines. It is contained in $T \cup T_x \cup T_y \cup T_{xy}$ (see Figure 3). Glue by s the sides of $T \cup T_x \cup T_y \cup T_{xy}$. The resulting space T_* is homeomorphic to $\mathbb{R}P^2$. Let us take the curve L to be the image of L' in T_*.

A pair (T_*, L) is called *a chart* of a real algebraic curve C in $\mathbb{R}P^2$, if there exists a homeomorphism of pairs $(T_*, L) \longrightarrow (\mathbb{R}P^2, \mathbb{R}C)$.

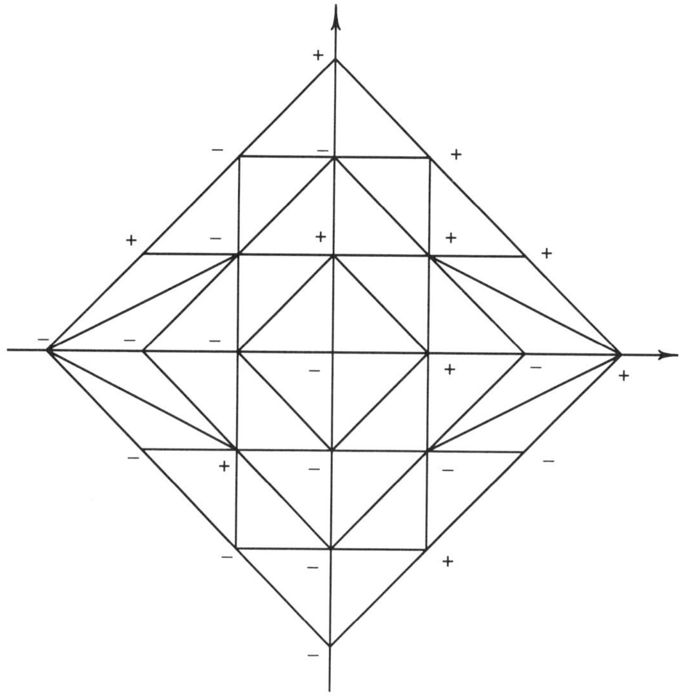

Figure 2

Suppose now that the triangulation of T is *convex*. This means that there exists a piecewise-linear convex function $\nu : T \longrightarrow \mathbb{R}$ which is linear on each triangle of the triangulation and not linear on the union of any two different triangles. Sometimes, such triangulations are also called regular or coherent.

The following statement is a special case of Viro's theorem (see, for example, [**Vi2**]).

THEOREM 5. *If the triangulation of T is convex, then there exists a nonsingular real algebraic curve C of degree d in $\mathbb{R}P^2$ with the chart (T_*, L).*

A curve having the chart (T_*, L) is called *a T-curve.* This notion was introduced by Orevkov [**Ore**].

It is easy to verify that the triangulation of the triangle

$$\{(x,\ y) \in \mathbb{R}^2\ :\ \ x \geq 0,\ \ y \geq 0,\ \ x + y \leq 3\}$$

shown in Figure 1 is convex. Thus, Figure 3 represents a "topological picture" of a curve of degree 3 in $\mathbb{R}P^2$. The real point set of this curve has two connected components in $\mathbb{R}P^2$.

In fact, a polynomial defining the curve C in Theorem 5 can be written down explicitly. If t is positive and sufficiently small, the polynomial

$$\sum_{(i,j)\in V} a_{i,j} x_0^i x_1^j x_2^{d-i-j} t^{\nu(i,j)}$$

(where V is the set of vertices of the triangulation of T, the coefficient $a_{i,j}$ is the sign associated to (i,j), and ν is a convex function $\nu : T \to \mathbb{R}$ certifying that the chosen triangulation of T is convex) defines a curve with the properties described in Theorem 5.

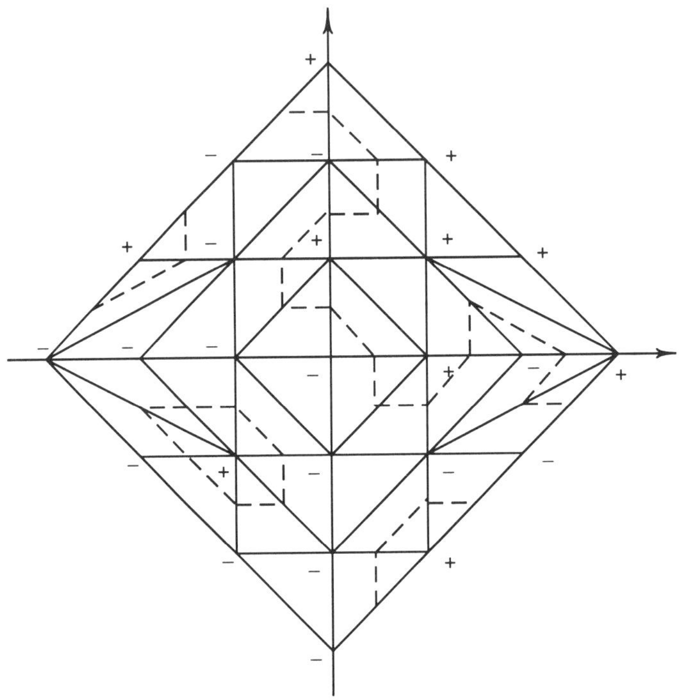

Figure 3

4.3. Topology of T-curves and T-surfaces. The class of T-curves is quite rich. For example, all the curves in $\mathbb{R}P^2$ up to degree 6 are T-curves. In the study of the topology of real algebraic hypersurfaces, special attention is traditionally devoted to *M-hypersurfaces*. An *M-curve* is a curve in $\mathbb{R}P^2$ whose real point set has the maximal possible number of connected components for a given degree; this number is equal to $(d-1)(d-2)/2+1$ for the degree d according to Harnack's theorem. For any degree, there exists a T-curve which is an M-curve.

The counter-examples to the oldest and most important conjecture in the topology of real algebraic curves, the Ragsdale conjecture, were constructed as T-curves (see [**It1, It2, It3, ItVi**]).

While there are many interesting examples of T-curves, there exist real algebraic curves in $\mathbb{R}P^2$ which are not T-curves. Let us first give necessary definitions. Each connected component of the real point set $\mathbb{R}C$ of a curve of even degree is called *an oval* and it divides $\mathbb{R}P^2$ in two parts. We call the part homeomorphic to a disk *the interior* of the oval. A pair of ovals is *injective* if one oval of this pair lies inside of the other one. Denote by J the number of ovals of a curve containing inside of them at least one injective pair.

It is shown in [**It2**] that for any T-curve of degree d which is an M-curve, the following inequality holds

$$J \leq 3d/2.$$

This gives a strong restriction on the topology of T-curves being M-curves. One can easily construct a family of M-curves of increasing degrees such that the numbers J of the curves of this family depend quadratically in the degree.

The situation with T-surfaces is rather curious. If we suppose that all the tetrahedra of a triangulation are primitive, *i.e.* have the volume 1/6, then there are very strong restrictions on the topology of the resulting T-surface. Namely, the Euler characteristic of the real point set of a T-surface constructed from a primitive triangulation is completely determined by the degree d (and does not depend on triangulation and distribution of signs !). It implies, in particular, that the number of connected components of a T-surface of degree d constructed using a primitive triangulation is at most $d^3/6 - d^2 + 11d/6$ (this result on the number of connected components was first obtained by Shustin [**Sh**]).

There exist T-surfaces of any degree in $\mathbb{R}P^3$ which are M-surfaces (it means that the total $\mathbb{Z}/2\mathbb{Z}$-homology group of the real point set has the same rank as that of the complexification).

4.4. Viro method for complete intersections. Viro method can be used in many situations. As shown by Sturmfels [**St1**], the construction of combinatorial patchworking can be extended to the case of complete intersections.

We again restrict ourselves to a particular case, namely, that of complete intersections of dimension 0. Even in this situation, combinatorial patchworking is an interesting and nontrivial construction.

Suppose that we are given the following combinatorial data. Let A_i ($i = 1, \dots, k$) be k finite collections of points in $(\mathbb{R}_+)^k$ with integer coordinates (where $\mathbb{R}_+$ is the set of nonnegative real numbers). Denote by $\Delta_1, \dots, \Delta_k$ the convex hulls of $A_1, \dots, A_k$, respectively. Let δ_i be a distribution of signs ("+" or "−") at the points of A_i. A polytope Δ_i with a distribution of signs δ_i is called a *signed Newton diagram* and is denoted by $\tilde{\Delta}_i$.

We say that a real polynomial P_i in k variables has the signed Newton diagram $\tilde{\Delta}_i$ if the Newton polytope of P_i coincides with Δ_i, the monomials of P_i correspond to the points of A_i, and the signs of the monomials of P_i coincide with the signs attributed to the corresponding points of A_i by the distribution δ_i.

We perform the following construction (*cf.* [**St1**]). Let ω_i be a real-valued function defined on the set A_i. By taking the lower convex hull of the graph of ω_i and then projecting each facet to $\mathbb{R}^k \times \{0\}$, the function ω_i defines a polyhedral subdivision τ_i of Δ_i. Denote by Δ_M the Minkowski sum of the polytopes $\Delta_1, \dots, \Delta_n$ and by A the set

$$\{a \in \Delta_M \mid a = a_1 + \ldots + a_k, \text{ where } a_i \in A_i\}.$$

Let us define a function $\omega : A \longrightarrow \mathbb{R}$ as follows:

$$\omega(a) = \min\{\omega_1(a_1) + \ldots + \omega_k(a_k) \mid a_1 + \ldots + a_k = a\}.$$

Such a function ω defines a polyhedral subdivision τ_ω of Δ_M. The vertices of τ_ω belong to A. Each facet V of τ_ω has a unique representation

$$V = v_1 + \ldots + v_k,$$

where v_i is a face of τ_i. Suppose that the functions $\omega_1, \dots, \omega_k$ are *generic*: this means that for any facet V of τ_ω equal to $v_1 + \ldots + v_k$, we have

$$\dim(V) = \dim(v_1) + \ldots + \dim(v_k).$$

A facet $V = v_1 + \ldots + v_k$ of τ_ω is called a *mixed cell* if $\dim(v_1) = \ldots = \dim(v_k) = 1$. Let v_i be one of the edges defining a mixed cell V. We call the edge v_i *alternating* if the distribution of signs δ_i associates different signs to the endpoints of v_i.

To each open orthant m of $\mathbb{R}^k$ we associate a vector $\vec{m}$ of $(\mathbb{Z}/2\mathbb{Z})^k$ in the following way: let $(x_1, \ldots, x_k)$ belong to m; then the i-th coordinate of $\vec{m}$ is 0 if $x_i > 0$ and 1 otherwise. We also associate to any integer point $a_i \in A_i$ a vector $\vec{a}_i$ of $(\mathbb{Z}/2\mathbb{Z})^k$ by replacing each coordinate of a_i by its parity (0 or 1). Consider a composition of symmetries with respect to coordinate hyperplanes sending the points of the positive orthant to the points of an orthant m. Let $a_i^{(m)}$ be the image of a point a_i under this composition. Now define the sign of $a_i^{(m)}$ as follows:

$$\mathrm{sign}(a_i^{(m)}) = (-1)^{\vec{a}_i \cdot \vec{m}} \delta_i(a_i),$$

where $\vec{a} \cdot \vec{m}$ is the scalar product (over $\mathbb{Z}/2\mathbb{Z}$) of the vectors $\vec{a}$ and $\vec{m}$ and $\delta_i(a_i)$ is the sign associated by δ_i to a_i.

Let $V = v_1 + \ldots + v_k$ be a mixed cell of the subdivision τ_ω, and let $V^{(m)}$ be the symmetric copy of V in the orthant m. Denote by a_i and b_i the endpoints of v_i $(i = 1, \ldots, k)$. The cell $V^{(m)}$ is called *alternating*, if for any $i = 1, \ldots, k$ the signs $sign(a_i^{(m)})$ and $sign(b_i^{(m)})$ are opposite.

The following theorem is a particular case of Viro's theorem adapted for complete intersections by Sturmfels [**St1**] (see also [**PSt, ItR**]).

THEOREM 6. *If t is positive and sufficiently small, then the number of common real zeroes of the polynomials*

$$P_i^{(t)}(X) = \sum_{a_i \in \Delta_i} \delta_i(a_i) X^{a_i} t^{\omega_i(a_i)}, \quad i = 1, \ldots, k$$

in an orthant m is equal to the total number of alternating copies in m of all the mixed cells of the subdivision τ_ω.

Now, we can associate a number $n(\tilde{\Delta})$ to the collection $\tilde{\Delta}$ of signed Newton diagrams $\tilde{\Delta}_1, \ldots, \tilde{\Delta}_k$ in the following way.

First, we define the number $n(\tilde{\Delta}, \omega)$ as the number of alternating copies of all the mixed cells of τ_ω. Then, we define $n(\tilde{\Delta})$ to be the maximal number $n(\tilde{\Delta}, \omega)$ for all possible choices of generic functions $\omega_1, \ldots, \omega_k$. Similarly, for any orthant m, we define the number $n(\tilde{\Delta}, \omega; m)$ as the number of alternating copies in the orthant m of all the mixed cells of τ_ω, and we define $n(\tilde{\Delta}; m)$ to be the maximal number $n(\tilde{\Delta}, \omega; m)$ for all possible choices of generic functions $\omega_1, \ldots, \omega_k$.

In fact, these choices are grouped into finitely many equivalence classes: we say that two choices of generic functions $\omega_1, \ldots, \omega_k$ are equivalent if they give the same subdivision τ_ω of Δ_M. There exists a one-to-one correspondence between the subdivisions τ_ω and convex triangulations of the Cayley polytope associated with $\tilde{\Delta}_1, \ldots, \tilde{\Delta}_k$ (see [**St2**]). Thus, the calculation of the numbers $n(\tilde{\Delta}; m)$ and $n(\tilde{\Delta})$ can be carried out once we have a list of all the convex triangulations of the Cayley polytope. To get such a list, one can use the algorithm due to De Loera [**DLo**] which enumerates all the convex triangulations of a point configuration in an affine space.

In the case of a univariate polynomial there are two orthants, numbered 0 (corresponding to $X > 0$) and 1 (corresponding to $X < 0$). A signed Newton diagram is simply a list of pairs of integers and associated signs, listed by increasing order of the integers. A mixed cell is given by two consecutive integers in this list. The number $n(\tilde{\Delta}; 0)$ is the number of sign changes in the list of associated signs. Descartes' rule [**Des**] states that the number of positive real zeroes of a polynomial is not greater than the number of sign changes in the list of its coefficients.

It is natural to ask the following question: what can be the number of common real zeroes in $(\mathbb{R}^*)^k$ of polynomials $P_1, \dots, P_k$ having Newton signed diagrams $\tilde{\Delta}_1, \dots, \tilde{\Delta}_k$, respectively? The multivariate Descartes' rule we proposed in [**ItR**] conjectures that the maximal number of isolated zeroes in $(\mathbb{R}^*)^k$ is realized by the T-construction described above:

Suppose that k real polynomials in k variables with signed Newton diagrams $\tilde{\Delta}_1, \dots, \tilde{\Delta}_k$ have a finite number of common real zeroes in $(\mathbb{R}^*)^k$ (where $\mathbb{R}^* = \mathbb{R} \setminus \{0\}$). Then

(1) *the number of common zeroes of these polynomials in an open orthant m of $\mathbb{R}^k$ is at most $n(\tilde{\Delta}; m)$;*
(2) *the number of common zeroes of these polynomials in $(\mathbb{R}^*)^k$ is at most $n(\tilde{\Delta})$.*

There is an important case when the second part of the conjecture stated above is true. Namely, if one can find such functions $\omega_1, \dots, \omega_k$ that all the mixed cells of the subdivision τ_ω of Δ_M are of volume 1, then the number $n(\tilde{\Delta}, \omega)$ is equal to the mixed volume of the polytopes $\Delta_1, \dots, \Delta_k$. Thus, the second statement of the conjecture in this case follows from Bernstein's theorem [**Ber**].

A counter-example due to T. Y. Li and X. Wang [**LiW**] shows that the conjecture is not true in general. Consider the system

$$y - x - 1 = 0$$
$$y^3 - 2 - 9x^3 + 0.01x^3y^3 = 0.$$

This system has 3 positive solutions:

$$x \approx 0.317659, \; y = x + 1$$
$$x \approx 0.659995, \; y = x + 1$$
$$x \approx 8.12058, \; y = x + 1$$

while the combinatorial bound is 1.

The problem of finding a correct generalization of Descartes's rule in the multivariate case is wide open.

References

[Alo] N. Alon, *The number of polytopes, configurations, and real matroids.* Mathematika, 33, 62–71 (1986).

[ABRW] M. E. Alonso, E. Becker, M.-F. Roy and T. Wormann, *Zeroes, multiplicities, and idempotents for zero dimensional systems.* In: Algorithms in Algebraic Geometry and Applications, L. Gonzalez Vega and T. Recio eds., Progress in Math., 143, 6–16. Birkhäuser (1996).

[Art] E. Artin, *Uber die Zerlegung defniter Funktionen in Quadrate.* Hamb. Abh. 5, 100–115 (1927). In: The collected papers of E. Artin, 273–288. Addison Wesley (1963).

[ArS] E. Artin and O. Schreier, *Algebraische Konstruktionen reeller Korper.* Ham. Abh., 5, 85–99 (1926). In: The collected papers of E. Artin, 258–272. Addison Wesley (1963).

[Bas] S. Basu, *On bounding the Betti numbers and computing the Euler characteristic of semi-algebraic sets.* Proc. 28th ACM Annual Symp. on Theory of Comp., 408–417 (1996).

[BPR1] S. Basu, R. Pollack and M.-F. Roy, *A new algorithm to find a point in every cell defined by a family of polynomials.* In: Quantifier Elimination and Cylindrical Algebraic Decomposition, B. Caviness and J. Johnson eds., Springer-Verlag, to appear.

[BPR2] S. Basu, R. Pollack and M.-F. Roy, *Computing a set of points meeting every cell defined by a family of polynomials on a variety.* In: Algorithmic foundations of robotics, K. Y. Goldberg, D. Halperin, J.-C. Latombe, R. H. Wilson eds., 537–555. A. K. Peters (1995).

[BPR3] S. Basu, R. Pollack and M.-F. Roy, *On the number of cells defined by family of polynomials on a variety.* Mathematika, 43, 120–126 (1996).

[BPR4] S. Basu, R. Pollack and M.-F. Roy, *On the combinatorial and algebraic complexity of quantifier elimination.* Proc. 35th IEEE Symp. on Foundations of Computer Science, 632–641 (1994).

[BPR5] S. Basu, R. Pollack and M.-F. Roy, *On the combinatorial and algebraic complexity of quantifier elimination.* Journal of the ACM, 43, 1002–1045 (1996).

[BPR6] S. Basu, R. Pollack and M.-F. Roy, *Computing roadmaps of semi-algebraic sets.* Proc. 28th ACM Annual Symp. on Theory of Comp., 168–173 (1996).

[BPR7] S. Basu, R. Pollack and M.-F. Roy *On computing a set of points meeting every cell of a family of polynomials on a variety.* Journal of Complexity, 13, 28–37 (1997).

[BPR8] S. Basu, R. Pollack and M.-F. Roy, *Computing roadmaps of semi-algebraic sets on a variety.* Foundations of Computational Mathematics, Rio de Janeiro, 1997, F. Cucker and M. Shub, eds., 1–15. Springer-Verlag (1997).

[Ben] M. Ben-Or, *Lower bounds for algebraic computation trees.* Proc. 15th ACM Annual Symp. on Theory of Comp., 80–86 (1982).

[Ber] D. Bernstein, *The number of roots of a system of equations.* Funct. Anal. and Appl., 9, 1–4 (1975).

[BCR] J. Bochnak, M. Coste and M.-F. Roy, Géométrie Algébrique Réelle. Springer-Verlag (1987). Real Algebraic Geometry (second edition, in English). Springer-Verlag, to appear.

[Ca1] J. Canny, The Complexity of Robot Motion Planning. MIT Press (1987).

[Ca2] J. Canny, *Some algebraic and geometric computations in PSPACE.* Proc. 20th ACM Annual Symp. on Theory of Comp., 460–467 (1988).

[Ca3] J. Canny, *Some practical tools for algebraic geometry.* Technical report in Spring school on robot motion planning at Rodez, France, PROMOTION ESPRIT, 39–52 (1993).

[Ca4] J. Canny, *Computing road maps in general semi-algebraic sets*, The Computer Journal, 36, 504–514 (1993).

[CaR] J. Canny and A. Rege, *A practical algorithm for geometry theorem proving.* ACM Symp. on Computational Geometry, June 1995, Vancouver.

[CGV] J. Canny, D. Grigor'ev and N. Vorobjov *Finding connected components of a semi-algebraic set in subexponential time.* Appl. Algebra in Engineering, Communication and Computing, 2, 217–238 (1992).

[Chi] A. Chistov, *Polynomial time computation of the dimension of algebraic varieties in zero characteristic.* J. Symbolic Comput., 22 (1), 1–25 (1996).

[Col] G. E. Collins, *Quantifier elimination for real closed fields by cylindrical algebraic decomposition.* Lect. Notes in Comp. Sci., 33, 515–532, Springer-Verlag (1975).

[C] M. Coste, *Effective semi-algebraic geometry.* Lect. Notes in Comp. Sci., 391, 1–27. Springer-Verlag (1989).

[DLo] J. A. De Loera, *Triangulations of polytopes and computational algebra.* Ph.D. Dissertation, Cornell University (1995). (program PUNTOS available via anonymous ftp from ftp://geom.umn.edu/priv/ deloera/).

[Des] R. Descartes, *Géométrie.* 1636. In: A Source Book in Mathematics, 90–93. Harvard University Press (1969).

[GaV] A. Gabrielov and N. Vorobjov, *Complexity of stratification of semi-pfaffian sets.* Discrete and Comput. Geom., 4, 71–91 (1995).

[GoP] J. E. Goodman and R. Pollack, *Upper bound for configuration of polytopes in* $\mathbb{R}^d$. Discrete and Comput. Geom., 1, 219–227 (1986).

[GRR] L. Gonzalez-Vega, F. Rouillier and M.-F. Roy, *Symbolic recipes for polynomial system solving.* In: Some tapas for computer algebra, A. Cohen ed., Springer, to appear.

[GRRT] L. Gonzalez-Vega, F. Rouillier, M.-F. Roy and G. Trujillo, *Symbolic recipes for real solutions.* In: Some tapas for computer algebra, A. Cohen ed. Springer, to appear.

[GPW] J. E. Goodman, R. Pollack and R. Wenger, *Geometric transversal theory.* New Trends in Discrete and Computational Geometry, J. Pach ed., 163–198. Springer-Verlag (1993).

[GRi] L. Gournay and J. J. Risler, *Construction of roadmaps of semi-algebraic sets.* Appl. Algebra in Engineering, Communication and Computing, 4, 239–252 (1993).

[Gri] D. Grigor'ev, *The complexity of deciding Tarski algebra.* J. Symbolic Comput., 5, 65–108 (1988).

[GKV] D. Grigor'ev, M. Karpinsky and N. Vorobjov, *Complexity lower bounds on testing membership to a polyhedron by algebraic decision trees.* Proc. ACM Symp. on Theory of Comp., Montreal, 635–644 (1994).

[GV1] D. Grigor'ev and N. Vorobjov, *Solving systems of polynomial inequalities in subexponential time.* J. Symbolic Comput., 5, 37–64 (1988).

[GV2] D. Grigor'ev and N. Vorobjov, *Counting connected components of a semi-algebraic set in subexponential time.* Comput. Complexit, 2, 133–186 (1992).

[Gud] D. A. Gudkov, *The topology of real algebraic varieties.* Rus. Math. Surv., 29 (4), 1–79 (1974).

[Hard] R. M. Hardt, *Semi-algebraic local triviality in semi-algebraic mappings.* Amer. J. Math., 102, 291–302 (1980).

[Harn] A. Harnack, *Uber di Vieltheiligkeit der ebenen algebraischen Kurven.* Math. Annal., 10, 189–198 (1876).

[HRR] J. Heintz, T. Recio and M.-F. Roy, *Algorithms in real algebraic geometry and applications to computational geometry.* Discrete and Computational Geometry: Papers from the DIMACS Special Year. DIMACS series in Discrete Mathematics and Theoretical Computer Science AMS-ACM, 6, 137–163 (1991).

[HRS1] J. Heintz, M.-F. Roy and P. Solerno, *On the complexity of semi-algebraic sets.* Proc. IFIP 89, San Francisco, 293–298. North-Holland (1989).

[HRS2] J. Heintz, M.-F. Roy and P. Solerno, *Sur la complexité du principe de Tarski-Seidenberg.* Bull. Soc. Math. France, 118, 101–126 (1990).

[HRS3] J. Heintz, M.-F. Roy and P. Solerno, *Single exponential path finding in semi-algebraic sets, Part I: The case of a regular bounded hypersurface.* Proc. AAECC-8 Tokyo, 1990. Lect. Notes in Comp. Sci., 508, 180–186. Springer-Verlag (1991).

[HRS4] J. Heintz, M.-F. Roy and P. Solerno, *Single exponential path finding in semi-algebraic sets II: The general case.* Algebraic geometry and its applications. C. L. Bajaj ed., 467–481. Springer Verlag (1993).

[HRS5] J. Heintz, M.-F. Roy and P. Solerno, *Description of the connected components of a semi-algebraic set in single exponential time.* Journal of Discrete and Comput. Geometry, 11, 121–140 (1994).

[Her] C. Hermite, *Remarques sur le théorème de Sturm.* C. R. Acad. Sci. Paris, 36, 52–54 (1853).

[It1] I. Itenberg, *Contre-exemples à la conjecture de Ragsdale,* C. R. Acad. Sci. Paris, Ser. I, 317, 277–282 (1993).

[It2] I. Itenberg, *Counter-examples to Ragsdale conjecture and T-curves.* Contemporary Mathematics, 182, 55–72 (1995).

[It3] I. Itenberg, *Viro's method and T-curves.* In: Algorithms in Algebraic Geometry and Applications, L. Gonzalez Vega and T. Recio eds., Progress in Math., 43, 177–192. Birkhäuser (1996).

[ItR] I. Itenberg and M.-F. Roy, *Multivariate Descartes' rule.* Beiträge zur Algebra und Geometrie, 37 (2), 337–346 (1996).

[ItSh] I. Itenberg and E. Shustin, *Singular points and limit cycles of planar polynomial vector fields*, preprint. Rennes (1997).

[ItVi] I. Itenberg and O. Viro, *Patchworking algebraic curves disproves the Ragsdale conjecture.* Mathematical Intelligencer, 18 (4), 19–28 (1996).

[Kh1] A. G. Khovansky, *On a class of systems of transcendental equations.* Soviet Math. Dokl., 22 (3), 762–765 (1980).

[Kh2] A. G. Khovansky, *Real analytic varieties with the finiteness property and complex Abelian integrals.* Funct. Anal. and Appl., 18 (2), 119–127 (1984).

[Kh3] A. G. Khovansky, *Fewnomials.* Translation of Mathematical Monographs, AMS (1991).

[KV1] J. J. Koenderink and A. J. Van Doorn, *The singularities of the visual mapping.* Biological cybernetics, 24, 51–59 (1976).

[KV2] J. J. Koenderink and A. J. Van Doorn, *The internal representation of solid shape with respect to vision.* Biological cybernetics, 32, 211–216 (1979).

[Lat] J.-C. Latombe, Robot motion planning. Kluwer Academic Publishers (1990).

[LiW] T. Y. Li and X. Wang, *On multivariate Descartes' rule—a counter-example.* To appear: Beiträge zur Algebra und Geometrie, 39 (1998).

[LR1] T. Lickteig and M.-F. Roy, *Semi-algebraic complexity of quotients and sign determinations of reminders.* Journal of Complexity, 12 (4), 545–571 (1996).

[LR2] T. Lickteig and M.-F. Roy, *Cauchy index computations*, Calcolo, 33, 337–371 (1996).

[Mal] J. Malik, *Interpreting line drawings of curved objects.* Int. Journal of Comp. Vision, 1, 73–103 (1987).

[Mat] Y. Mattiassevitch, *On recursive unsolvability of Hilbert's tenth problem.* In: Proceedings of Fourth International Congress on Logic, Philosophy and Methodology of Science, Bucharest, 1971, 89–110. North Holland (1973).

[Mil] J. Milnor, *On the Betti numbers of real algebraic varieties.* Proc. AMS, 15, 275–280 (1964).

[Mne] N. E. Mnev, *The universality theorems on the classification problem of configuration varieties and convex polytopes varieties.* Lect. Notes in Math., 1346, 527–544. Springer-Verlag (1988).

[MPR] J. L. Montana, L. M. Pardo and T. Recio, *The non scalar model of complexity in computational geometry.* In: Effective Methods in Algebraic Geometry, T. Mora and C. Traverso eds., Progress in Math., 94, 347–362. Birkhäuser (1991).

[MoRan] T. Mostowski and E. Rannou, *Complexity of the computation of the canonical Whitney stratification of an algebraic set in* $\mathbb{C}^n$. AAECC 9. Lect. Notes in Comp. Sci., 539, 281–291 (1991).

[Ole] O. A. Olenick, *Estimates on the Betti numbers of real algebraic hypersurfaces.* Mat. Sb., 28 (70), 635–640 (1951).

[Ore] S. Orevkov, Private communication.

[ORo] J. O'Rourke, This volume.

[PaR] L. M. Pardo and T. Recio, *Rabin's width of a complete proof and the width of a semi-algebraic set.* Proc. EUROCAL 87. Lect. Notes in Comp. Sci., 378, 456–462. Springer-Verlag (1987).

[PRS] P. Pedersen, M.-F. Roy and A. Szpirglas, *Counting real zeros in the multivariate case.* Computational algebraic geometry, Eyssette and Galligo eds., Progress in Math., 109, 203–224. Birkhäuser (1993).

[PSt] P. Pedersen and B. Sturmfels, *Mixed monomial bases.* In: Algorithms in Algebraic Geometry and Applications, L. Gonzalez Vega and T. Recio eds., Progress in Math., 143, 307–316. Birkhäuser (1996).

[Pet] I. Petrowsky, *On the topology of real plane algebraic curves.* Ann. Math., 39 (1), 187–209, (1938).

[PR] R. Pollack and M.-F. Roy, *On the number of cells defined by a set of polynomials.* C. R. Acad. Sci. Paris, Ser. I, 316, 573–577 (1993).

[Rag] V. Ragsdale, *On the arrangement of the real branches of plane algebraic curves.* Amer. J. Math., 28, 377–404 (1906).

[Ran] E. Rannou, *Complexité d'algorithmes de stratifications.* Thèse, Université de Rennes I (1993).

[Re1] J. Renegar, *On the computational complexity and geometry of the first order theory of the reals.* J. of Symbolic Comput., 13 (3), 255-352 (1992).

[Re2] J. Renegar, *Recent progress on the complexity of the decision problem for the reals.* Discrete and Computational Geometry: Papers from the DIMACS Special Year. DIMACS series in Discrete Mathematics and Theoretical Computer Science, AMS-ACM., 6, 287–308 (1991).

[Ri] J.-J. Risler, *Construction d'hypersurfaces réelles [d'après Viro].* Séminaire Bourbaki, 763 (1992).

[Ro1] F. Rouillier, *Algorithmes efficaces pour l'étude des zéros réels des systèmes polynomiaux.* Thèse, Université de Rennes I (1996).

[Ro2] F. Rouillier, *PoSSo-RealSolving (Systems of polynomials).* Proceedings of the PoSSo-workshop, Paris (1995).

[R] M.-F. Roy, *Basic algorithms in real algebraic geometry: from Sturm theorem to the existential theory of reals.* Lectures on Real Geometry in memorial of Mario Raimondo, Expositions in Mathematics, 23, 1–67. de Gruyter (1996).

[RvE] M.-F. Roy and T. van Efelterre, *Aspects graph of bodies of revolution with algorithms of real algebraic geometry.* In: Algorithms in Algebraic Geometry and Applications, L. Gonzalez Vega and T. Recio eds., Progress in Math., 143, 353–364 (1996).

[RV] M.-F. Roy and N. Vorobjov, *Computing the complexification of a semi-algebraic set,* ISSACC 96 Proceedings, 26–39 (1996).

[SS] J. Schwartz and M. Sharir, *On the 'piano movers' problem II. General techniques for computing topological properties of real algebraic manifolds.* Advances in Applied Mathematics, 4, 298–351 (1983).

[Sei] A. Seidenberg, *A new decision method for elementary algebra.* Ann. Math., 60, 365–374 (1954).

[Sh] E. Shustin, *Critical points of real polynomials, subdivisions of Newton polyhedra and topology of real algebraic hypersurfaces.* Amer. Math. Soc. Transl. (2), 173, 203–223 (1996).

[SY] J. M. Steele and A. C. Yao, *Lower bounds for algebraic decision trees.* Journal of Algorithms, 3, 1–8 (1982).

[Str1] V. Strassen, *Die Berechungskomplexität von elementarsymmetrischen Funktionen und Interpolationskoeffizienten.* Numer. Math., 20, 238–251 (1973).

[Str2] V. Strassen, *The complexity of continued fractions.* SIAM J. Comp., 12 (1), 1–27 (1983).

[Stu] C. Sturm, *Mémoire sur la résolution des équations numériques.* Inst. France Sc. Math. Phys., 6 (1835).

[St1] B. Sturmfels, *Viro's theorem for complete intersections.* Annali della Scoula Normale Superiore di Pisa, (4), 21 (3), 377–386 (1994).

[St2] B. Sturmfels, *On the Newton polytope of the resultant.* Journal of Algebraic Combinatorics, 3, 207–236 (1994).

[Syl] J. T. Sylvester, *On a theory of syzygetic relations of two rational integral functions, comprising an application to the theory of Sturm's function.* Trans. Roy. Soc. London, (1853). Reprint in: Sylvester: Collected Math Papers. Chelsea Pub. Comp. NY vol. 1, 429–586 (1983).

[Tar] A. Tarski, A Decision Method for Elementary Algebra and Geometry. University of California Press (1951).

[Tho] R. Thom, *Sur l'homologie des variétés réelles.* In: Differential and combinatorial topology, 255–265. Princeton University Press (1965).

[vD] L. van den Dries, *O-minimal structures on the field of real numbers.* Jber. d. Dt. Math.-Verein, 98, 165–171 (1996).

[Vi1] O. Viro, *Gluing of algebraic hypersurfaces, smoothing of singularities and construction of curves.* Proceedings Leningrad International Topological Conf. (Leningrad, Aug. 1983). Nauka. Leningrad, 149–197 (1983) (Russian).

[Vi2] O. Viro, *Gluing of plane real algebraic curves and constructions of curves of degrees 6 and 7.* Lect. Notes in Math., 1060, 187–200. Springer-Verlag (1984).

[Vi3] O. Viro, *Progress in the topology of real algebraic varieties over the last six years.* Rus. Math. Surv., 41 (3), 55–82 (1986).

[Vi4] O. Viro, *Plane real algebraic curves: constructions with controlled topology.* Algebra i analiz 1, 1–73 (1989) (Russian), English translation in *Leningrad Math. J.*, 1 (5), 1059–1134 (1990).

[Vi5] O. Viro, *Patchworking real algebraic varieties*, preprint. Uppsala University (1994).

[V] N. Vorobjov, *Bounds of real roots of a system of algebraic equations.* Notes of Sci. Seminars of Leningrad Department of Math., Steklov Inst., 137, 7–19 (1984).

[Wal] R. J. Walker, Algebraic Curves. Princeton University Press (1950).

[War] H. E. Warren, *Lower bounds for approximation of nonlinearmanifolds.* Trans. Amer. Math. Soc., 133, 167–178 (1968).

[Wilk] A. Wilkie, *Model completeness results for expansion of the real field by restricted Pfaffian functions and the exponential function.* J. of the Amer. Math. Soc., 9, 1051–1094 (1996).

[Wils] G. Wilson, *Hilbert sixteenth problem.* Topology, 17, 53–73 (1978).

[Yao] A. Yao, *Decision tree complexity and Betti numbers.* Proc. ACM Symp. on Theory of Comp., 615–624 (1994).

INSTITUT DE RECHERCHE MATHÉMATIQUE DE RENNES, CAMPUS DE BEAULIEU, 35042 RENNES CEDEX, FRANCE

E-mail address: Ilia.Itenberg@univ-rennes1.fr

INSTITUT DE RECHERCHE MATHÉMATIQUE DE RENNES, CAMPUS DE BEAULIEU, 35042 RENNES CEDEX, FRANCE

E-mail address: Marie-Francoise.Coste-Roy@univ-rennes1.fr

Contemporary Mathematics
Volume **223**, 1999

Open Problems in the Combinatorics of Visibility and Illumination

Joseph O'Rourke

ABSTRACT. A compendium of thirty-eight open problems is presented: on point and edge visibility, floodlight illumination, and visibility graphs.

The "art gallery theorem," that $\lfloor n/3 \rfloor$ guards suffice and are sometimes necessary to visually cover a polygon of n vertices [**Chv**] [**O**], typifies the interplay between geometry and combinatorics in a wide variety of subsequent results. The open problems in this collection are intellectual kins to these art gallery-like results. The problems represent a personal selection from the frontier of research on the combinatorics of visibility; no attempt is made to be comprehensive. Updates will be maintained on the Web.[1]

In general, a point x can *see*, is *visible* to, or *illuminates* a point y if the line segment xy is not obstructed. (The metaphors of "visibility" and "illumination" are used interchangeably.) Often this means that xy cannot contain any point of a set of obstacles, but sometimes xy is permitted to contain boundary points of the obstacles (line-of-sight grazing contact). Other nuances to the definition of visibility are appropriate for certain problems, and will be detailed below.

Throughout γ will be used to denote the constant of proportionality for quantities of the form $\gamma n + c$ for some constant c, where n is the primary input variable (e.g., the number of vertices). Precise fractional bounds will be supplemented by their approximate decimal equivalents.

1. Point and edge visibility problems

1.1. Art gallery theorems. Most these problems concern simple polygons. Grazing contact is permitted, so a guard (i.e., a point) g sees an interior point x if gx is nowhere exterior, and sees an exterior point y if gy is nowhere interior. A *vertex guard* is a point located at a vertex of the polygon. An *edge guard* is an edge e of the polygon, which sees a point x if there is a point $y \in e$ such that xy is nowhere exterior. Edge guard problems can be viewed as illumination by fluorescent lightbulbs.

1991 *Mathematics Subject Classification.* Primary 52C99.
Key words and phrases. Visibility, illumination, visibility graphs, computational geometry.
Supported by NSF grant CCR-9421670.

[1]URL: `http://cs.smith.edu/~orourke/`.

© 1999 American Mathematical Society

1. Orthogonal Prison Yard. How many vertex guards are needed to cover both the interior and the exterior of a simple orthogonal polygon? An *orthogonal polygon* is one whose edges meet at right angles. The best lower and upper bounds are $\lceil (5n-10)/16 \rceil$ and $\lfloor 5n/12 \rfloor + 2$ [**HK**]. ($\gamma \in [0.31, 0.42]$.)
2. Edge Guard Problem. How many edge guards are needed to cover the interior of a simple polygon? The best lower and upper bounds (for $n > 11$) are $\lfloor n/4 \rfloor$ and $\lfloor 3n/10 \rfloor$ [**S2**]. ($\gamma \in [0.25, 0.30]$.)
3. How many edge guards are needed to cover the interior of a star-shaped polygon? Toussaint showed that $\lfloor n/5 \rfloor$ are necessary [**O**]. ($\gamma \in [0.2, 0.3]$.)
4. Prove or disprove that $\lfloor n/4 \rfloor$ edge guards suffice to cover a star-shaped polygon of n vertices [**O2**].
5. The problem of finding the minimum number of guards needed to cover a given simple polygon is NP-complete [**LL**] [**O**]. Is there a polynomial approximation algorithm with some guarantee of performance? Since set cover can be approximated within a $O(\log n)$ factor, improving upon $O(\log n)\cdot$OPT could be a goal.
6. In a forest of a finite number of disjoint circular trees whose boundaries are perfect mirrors, it is true that some light rays from a lit match always escape to infinity? [**St**]. The trees are disks which reflect light at its incident angle.
7. How many edge guards are needed to cover an n-vertex polyhedral terrain? A *polyhedral terrain* T is a polyhedral surface that meets every vertical line in at most one point. Point x sees y if the segment xy is nowhere strictly below T. The best lower and upper bounds are $\lfloor (4n-4)/13 \rfloor$ [**BSTZ**] and $\lfloor n/3 \rfloor$ [**ERC**]. ($\gamma \in [0.31, 0.33]$.)

1.2. Exterior boundary illumination. These problems ask for the fewest exterior point light sources to illuminate the boundary of a collection of objects. Rectangles are *isothetic* if their sides are parallel to coordinate axes.

8. How many points are need to illuminate n disjoint isothetic rectangles in the plane? Fejes Tóth proved that $4n-7$ points suffice to illuminate $n \geq 4$ disjoint compact convex sets in the plane [**Fej**], but restricted cases often require fewer lights. For rectangles, examples are known that require $n-1$ lights, and $\lfloor (4n+4)/3 \rfloor$ lights suffice [**CRCU**]. ($\gamma \in [1.00, 1.33]$.) Here visibility excludes grazing contact.
9. How many points are need to illuminate n disjoint triangles in the plane? The bounds known here are exactly the same as for rectangles, for different reasons [**CRCU**].
10. How many points are needed to illuminate n disjoint segments in the plane? Examples are known that require $\lceil 4n/9 \rceil$ lights [**Z**], and for $n \geq 11$, it has been proven that $\lceil 2n/3 \rceil - 3$ suffice [**CRCUZ**]. ($\gamma \in [0.44, 0.67]$; it has been conjectured that $\frac{1}{2}$ is the right value.)
11. Hadwiger-Levi Covering. What is the fewest number of exterior points sufficient to illuminate any closed convex body in three dimensions? The conjecture is that 8 (in general, 2^d) points suffice, but this is only proven for 3-polytopes with some affine symmetry [**Bez**]. Illumination here requires the lines of sight to penetrate to the interior.

2. Floodlights

A *floodlight* of angle α (an α-*floodlight*) is a light that projects from a point in a (contiguous) cone of angle at most α. A floodlight placed in a simple polygon with its apex at a vertex is called a *vertex light*; at most one light may be placed at a vertex.

12. Stage Illumination. Given a line segment (the stage) and a set of n floodlights of fixed apexes and apertures, find a sub-exponential algorithm to decide if the floodlights can be rotated to illuminate the stage [**BGL**], or prove intractability. It is only known that the problem is decidable via an $c^{O(n)}$ algorithm [**Roy**].
13. Do five vertex $\pi/5$-floodlights suffice to illuminate every convex polygon of $n > 5$ vertices? It is known that three $\pi/3$-, or four $\pi/4$-floodlights suffice [**OSS**].
14. For what value of α is it the case that any set of k vertex floodlights, whose total angle sum is at least α, suffices to cover any convex polygon of $n \geq k$ vertices? It is known that $\alpha = 2\pi$ does suffice, as such floodlights can cover the entire plane [**BGL**], and that $\alpha = \pi$ does not suffice [**OSS**].
15. Find a subquadratic algorithm for finding an optimal pair of floodlights to illuminate a convex polygon, i.e., floodlights whose angle sum is as small as possible. An $O(n^2)$ algorithm is available [**ECU**].
16. Are the apexes of an optimal set of k floodlights that illuminate a convex polygon always at vertices of the polygon? Open for $k \geq 3$ [**ECU**].
17. How many π-floodlights (apexes at distinct non-exterior points) are sufficient to cover any polygon of n vertices? It is known that $\lceil \frac{2}{5}(n-3) \rceil$ suffice [**CT**], but it has been conjectured that $\lfloor n/3 \rfloor$ suffice [**U**].
18. A similar question may be asked for vertex π-floodlights: how many suffice to cover any polygon of n vertices? Here it is known that $\lfloor 3n/5 \rfloor$ are sometimes needed [**U**], and that $n-2$ suffice; but no $\gamma < 1$ is known to suffice except for special shapes [**O3**]. ($\gamma \in [0.6, 1.0]$.) No angle $\alpha < \pi$ suffices, even with one light per vertex [**ECOUX**].

3. Visibility graphs

3.1. Polygon vertex visibility graphs. Here visibility permits boundary contact: x sees y if the segment xy is nowhere exterior to the polygon. The *polygon vertex visibility graph* of a simple polygon P has a node for each vertex of P, with an arc between two nodes if the corresponding vertices are visible to one another. Note the definition of visibility implies that every edge of P is included as an arc of the visibility graph (and therefore the graph is Hamiltonian). For problems where the graph is the input, usually it is assumed given labeled with vertex indices in the order of a counterclockwise boundary traversal.

19. How many different labeled vertex visibility graphs are there of polygons of n vertices? The number is in $\Omega(n^{(2-\epsilon)n})$ and $O(n^{3n})$ [**HN**].
20. Develop an algorithm to recognize whether a polygon vertex visibility graph is planar. Necessary and sufficient conditions are known [**LC**], but they seem not to lead to a polynomial algorithm.

21. Do general-position polygons realize all visibility graphs? I.e., is there a polygon vertex visibility graph that can only be realized with collinear vertices? There are polygon vertex-edge visibility graphs that do require collinearities [**OS2**].
22. Do Ghosh's four necessary conditions [**Gho**] suffice to characterize polygon vertex visibility graphs?

3.2. Recognition complexity.

23. Is recognition of polygon vertex visibility graphs in NP? It is only known to be in $PSPACE$ [**Eve**].
24. Can any polygon vertex visibility graph be realized by a polygon drawn on an exponential-size grid? If so, visibility graph recognition is in P [**LS**].
25. Is recognition of polygon vertex pseudo-visibility graphs in P? It is only known to be in NP [**OS**]. A *pseudo-visibility* graph is based on a pseudoline arrangement determined by the vertices, where lines of sight are only along the pseudolines.
26. What is the computational complexity of computing the chromatic number of a polygon vertex visibility graph G? Finding a maximum independent set or a minimum vertex cover are both NP-complete problems when restricted to visibility graphs [**LS**]. Of course $\chi(G) \geq 3$ because G can be triangulated.

3.3. Segment visibility graphs. The most-studied version of segment visibility graph is the *endpoint visibility graph* G: for a set S of closed, disjoint line segments, G has a node for each segment endpoint, and an arc between two nodes x and y if they can see each other in the sense that $[x,y] \cap S = \{x,y\}$, or $[x,y]$: the intersection is either just the two endpoints, or the entire closed segment. Note that here visibility is blocked by grazing contact with a segment, but that G contains an arc corresponding to each segment in S.

27. Does the endpoint visibility graph for a set of disjoint non-collinear segments in the plane always contain a Hamiltonian cycle [**Mir**]?
28. Does the endpoint visibility graph for a set of shellable disjoint non-collinear segments in the plane always contain a Hamiltonian cycle [**OR**]? A set of segments is *shellable* when it is possible to order the segments $s_1, s_2, \ldots$ such that s_i lies in the exterior of the convex hull of $\{s_1, s_2, \ldots, s_{i-1}\}$.
29. What is the complexity of recognizing whether a set of disjoint non-collinear segments in the plane admits a simple circuit? A *simple circuit* is a simple (non-crossing) Hamiltonian circuit in the segment endpoint visibility graph that includes every segment. This is known to be an NP-complete problem when the segments are non necessarily disjoint, and therefore may form polygonal chains [**R**].

3.4. Rectangle visibility graphs. There are several varieties of "rectangle visibility graphs," sometimes known as RVG's. G is a *rectangle visibility graph* if it can be realized by closed isothetic rectangles in the plane, with pairwise disjoint interiors, with nodes corresponding to rectangles, and arcs corresponding to visibility, under the following definition of visibility: two rectangles see one another iff there is either a horizontal or a vertical unobstructed "beam" of visibility of finite width connecting them. The "beam"-visibility requirement is often called "ϵ-visibility" in the literature. Variations are obtained by permitting this beam to be a segment, or demanding that the rectangles be entirely disjoint, or that they have no collinear sides.

30. Is every rectangle visibility graph a doubly-linear graph [**HSV**]? A graph is *doubly-linear* if it can be realized in the plane as the union of two plane graphs drawn with straight lines. (All rectangle visibility graphs are the union of two planar graphs, for the horizontal and for the vertical visibility edges.)

31. Is every graph with aboricity two a rectangle visibility graph [**BDHS**]? The *aboricity* of a graph G is the minimum k such that G can be decomposed into k forests.

32. Is every maximum-degree 4 graph a collinear rectangle visibility graph [**BDHS**]? A rectangle representation is called *noncollinear* if no line contains two distinct rectangle sides. It is *collinear* if collinearities are permitted. $K_{4,4}$ is not a noncollinear rectangle visibility graph, although it is a collinear one [**DH**].

3.5. Visibility graphs in $2\frac{1}{2}$D and 3D. Here it is reasonable to extend the ϵ-visibility definition to require a finite-volume beam of visibility for an arc of the graph.

33. What is the largest complete graph realizable by disjoint isothetic rectangles in $2\frac{1}{2}$D with vertical visibility (all rectangles are parallel to the xy-plane)? It is known that K_{22} is realizable and K_{56} is not [**BEH**] [**FHW**].

34. What is the largest complete graph that is a box visibility graph? Such a graph is realized by disjoint isothetic boxes in three-dimensions with lines of sight parallel to a coordinate axis. It is known that K_{42} is realizable and $K_{10^{75}}$ is not [**BJMO**].

35. Is it true that in every collection of isothetic boxes in three dimensions, there is one whose removal reduces the number of edges in the associated box visibility graph? This property holds for rectangles in the plane [**JMO**].

3.6. Point visibility graphs. A *point visibility graph* [**S**] for a simple polygon P is a "continuous" graph with an uncountable number of nodes, one for each point in P, with an arc between nodes x and y iff the segment xy is nowhere exterior.

36. Find an algorithm to decide whether or not two polygons have isomorphic point visibility graphs. An $O(n^2)$ algorithm to detect such isomorphism between spiral polygons is known [**MS**].

37. Is it true that if two polygons have isomorphic point visibility graphs, then there is an isomorphism between the graphs that is also a homeomorphism? This is a conjecture in [**MS**].

38. Can a node corresponding to a reflex vertex be identified in a given polygon point visibility graph [**MS**]?

Acknowledgements. I thank Günter Rote, Thomas Shermer, Jorge Urrutia, and the referees for comments.

References

[BDHS] P. Bose, A. Dean, J. Hutchinson and T. Shermer, *On rectangle visibility graphs.* In S. North, editor, Graph Drawing (Proc. GD '96), volume 1190 of Lecture Notes Comput. Sci., pages 25–44. Springer-Verlag, 1997.

[BEH] P. Bose, H. Everett, M. E. Houle, S. Fekete, A. Lubiw, H. Meijer, K. Romanik, G. Rote, T. Shermer, S. Whitesides and C. Zelle, *A visibility representation for graphs in three dimensions.* Manuscript, 1997.

[Bez] K. Bezdek, *Hadwiger-Levi's covering problem revisited.* In J. Pach, editor, New Trends in Discrete and Computational Geometry, volume 10 of Algorithms and Combinatorics, pages 199–233. Springer-Verlag, 1993.

[BGL] P. Bose, L. Guibas, A. Lubiw, M. Overmars, D. Souvaine and J. Urrutia, *The floodlight problem.* In Proc. 5th Canad. Conf. Comput. Geom., pages 399–404, 1993.

[BJMO] P. Bose, A. Josefczyk, J. Miller and J. O'Rourke, *K_{42} is a box visibility graph.* In Snapshots of Computational and Discrete Geometry, volume 3, pages 88–91. School Comput. Sci., McGill Univ., Montreal, PQ, July 1994. Technical Report SOCS-94.50.

[BSTZ] P. Bose, T. Shermer, G. Toussaint and B. Zhu, *Guarding polyhedral terrains.* Comput. Geom. Theory Appl., 7:173–185, 1997.

[Chv] V. Chvátal, *A combinatorial theorem in plane geometry.* J. Combin. Theory Ser. B, 18:39–41, 1975.

[CRCU] J. Czyzowicz, E. Rivera-Campo and J. Urrutia, *Illuminating rectangles and triangles in the plane.* J. Combin. Theory Ser. B, 57:1–17, 1993.

[CRCUZ] J. Czyzowicz, E. Rivera-Campo, J. Urrutia and J. Zaks, *Illuminating and guarding segments in the plane.* Discr. Math., 137:147–153, 1995.

[CT] G. Csizmadia and G. Tóth, *Note on an art gallery problem.* Comput. Geom. Theory Appl., to appear, 1998.

[DH] A, M. Dean and J. P. Hutchinson, *Rectangle-visibility representations of bipartite graphs.* In R. Tamassia and I. G. Tollis, editors, Graph Drawing (Proc. GD '94), volume 894 of Lecture Notes Comput. Sci., pages 159–166. Springer-Verlag, 1995.

[ECOUX] V. Estivill-Castro, J. O'Rourke, J. Urrutia and D. Xu, *Illumination of polygons with vertex floodlights.* Inform. Process. Lett., 56:9–13, 1995.

[ECU] V. Estivill-Castro and J. Urrutia, *Two-floodlight illumination of convex polygons.* In Proc. 4th Workshop Algorithms Data Struct., volume 955 of Lecture Notes Comput. Sci., pages 62–73. Springer-Verlag, 1995.

[ERC] H. Everett and E. Rivera-Campo, *Edge guarding polyhedral terrain.* Comput. Geom. Theory Appl., 7:201–203, 1997.

[Eve] H. Everett, *Visibility graph recognition.* Report 231/90, Dept. Comput. Sci., Univ. Toronto, Toronto, ON, 1990. Ph.D. Thesis.

[Fej] L. Fejes Tóth, *Illumination of convex discs.* Acta Math. Acad. Sci. Hungar., 29:355–360, 1977.

[FHW] S. P. Fekete, M. E. Houle and S. Whitesides, *New results on a visibility representation of graphs in 3-D.* In F. Brandenburg, editor, Graph Drawing (Proc. GD '95), volume 1027 of Lecture Notes Comput. Sci., pages 234–241. Springer-Verlag, 1996.

[Gho] S. K. Ghosh, *On recognizing and characterizing visibility graphs of simple polygons.* Discrete Comput. Geom., 17:143–162, 1997.

[HK] F. Hoffmann and K. Kriegel, *A graph coloring result and its consequences for some guarding problems.* In Proc. 4th Annu. Internat. Sympos. Algorithms Comput., volume 762 of Lecture Notes Comput Sci., pages 78–87. Springer-Verlag, 1993.

[HN] F. Hurtado and M. Noy, *On the number of visibility graphs of simple polygons.* In Snapshots of Computational and Discrete Geometry, volume 3, pages 100–107. School Comput. Sci., McGill Univ., Montreal, PQ, July 1994. Technical Report SOCS-94.50.

[HSV] J. P. Hutchinson, T. Shermer and A. Vince, *On representations of some thickness-two graphs.* In F. Brandenburg, editor, Graph Drawing '95, volume 1027 of Lecture Notes Comput. Sci., pages 324–332. Springer-Verlag, 1995.

[JMO] A. Josefczyk, J. Miller and J. O'Rourke, *Arkin's conjecture for rectangle and box visibility graphs.* Technical Report 036, Dept. Comput. Sci., Smith College, July 1994.

[LC] S.-Y. Lin and C. Chen, *Planar visibility graphs.* In Proc. 6th Canad. Conf. Comput. Geom., pages 30–35, 1994.

[LL] D. Lee and A. Lin, *Computational complexity of art gallery problems.* IEEE Trans. Inform. Theory, 32:276–282, 1986.

[LS] Y. Lin and S. Skiena, *Complexity aspects of visibility graphs.* Internat. J. Comput. Geom. Appl., 5:289–312, 1995.

[Mir] A. Mirzaian, *Hamiltonian triangulations and circumscribing polygons of disjoint line segments.* Comput. Geom. Theory Appl., 2:15–30, 1992.

[MS] G. MacDonald and T. Shermer, *Isomophism of spiral polygons.* Discrete Comput. Geom., 16:277–304, 1996.

[O] J. O'Rourke, Art Gallery Theorems and Algorithms. Oxford University Press, New York, NY, 1987.

[O2] J. O'Rourke, *Guarding corpulent stars.* In Snapshots of Computational and Discrete Geometry, volume 3, pages 63–71. School Comput. Sci., McGill Univ., Montreal, PQ, July 1994. Technical Report SOCS-94.50. School Comput. Sci., McGill Univ., Montreal, PQ, July 1994. Technical Report SOCS-94.50.

[OR] J. O'Rourke and J. Rippel, *Two segment classes with Hamiltonian visibility graphs.* Comput. Geom. Theory Appl., 4:209–218, 1994.

[O3] J. O'Rourke, *Vertex π-lights for monotone mountains.* Proc. 9th Canad. Conf. Comput. Geom., pages 1–5, 1997.

[OS] J. O'Rourke and I. Streinu, *Vertex-edge pseudo-visibility graphs: Characterization and recognition.* In Proc. 13th Annu. ACM Sympos. Comput. Geom., pages 119–128, 1997.

[OS2] J. O'Rourke and I. Streinu, *The vertex-edge visibility graph of a polygon.* Comput. Geom. Theory Appl., 8:1–17, 1997.

[OSS] J. O'Rourke, T. Shermer, and I. Streinu, *Illuminating convex polygons with vertex floodlights.* In Proc. 7th Canad. Conf. Comput. Geom., pages 151–156, 1995.

[R] D. Rappaport, *Computing simple circuits from a set of line segments is NP-complete.* In Proc. 3rd Annu. ACM Sympos. Comput. Geom., pages 322–330, 1987.

[Roy] M.-F. Roy, Personal communication. At AMS Discrete Comput. Geom. SRC Conference, July, 1996.

[S] T. Shermer, *Recent results in art galleries.* Proc. IEEE, 80:1384–1399, September 1992.

[S2] T. Shermer, *A tight bound on the combinatorial edge guarding problem.* In Snapshots of Computational and Discrete Geometry, volume 3, pages 191–223. School Comput. Sci., McGill Univ., August 1994.

[St] I. Stewart, *Mathematical recreations.* Sci. Amer., 275:100–103, August 1996. Problem due to Janos Pach.

[U] J. Urrutia, *Art gallery and illumination problems.* In J.-R. Sack and J. Urrutia, editors, Handbook on Computational Geometry, North-Holland, to appear, 1998.

[Z] J. Zaks, Personal communication. 1996.

Department of Computer Science, Smith College, Northampton, MA 01063
E-mail address: `orourke@cs.smith.edu`

Contemporary Mathematics
Volume **223**, 1999

Halving Lines and Perfect Cross-matchings

János Pach and József Solymosi

ABSTRACT. It is shown that a set P of $2n$ points in general position in the plane admits a perfect matching with pairwise crossing segments if and only if it has precisely n halving lines. As a consequence, one can give a $O(n \log n)$-time algorithm which decides whether there exists such a matching in P and, if so, finds it.

1. Preliminaries

Let $P = \{p_1, p_2, \ldots, p_{2n}\}$ be a set of $2n$ points in the plane in general position, i.e., no three points are collinear. A line $p_i p_j$ is said to be a *halving line* of P if both open half-planes bounded by $p_i p_j$ contain precisely $n-1$ points. The number of halving lines of P is denoted by $h(P)$.

Taking an arbitrary line through any point of P and turning it around by at most 180 degrees, it always arrives at a position where it becomes a halving line. Thus, we have $h(P) \geq n$, and equality holds, e.g., when P is the vertex set of a convex $2n$-gon.

It is an intriguing open problem to determine the asymptotic behavior of $h(n) = \max_P h(P)$, where the maximum is taken over all $2n$-element sets in general position in the plane. It is known that

$$c_1 n \log n \leq h(n) \leq c_2 n^{4/3}$$

for suitable constants $c_1, c_2 > 0$ (see [**L**], [**EL**], [**D**]). This function plays an important role in the analysis of many algorithms in computational geometry (cf. [**E**]).

We say that two segments *cross* if they have an interior point in common. Let $c(P)$ denote the maximum number of pairwise crossing segments $\overline{p_i p_j}$ whose endpoints belong to P. Obviously, $c(P) \leq n$ holds for every $2n$-element set P. If $c(P) = n$, we say that P has a *perfect cross-matching*. This is the case, for example, when P is the vertex set of a convex $2n$-gon.

1991 *Mathematics Subject Classification.* Primary 52C10, 05C35, 05D15.

Key words and phrases. k-sets, matching, crossing.

Supported by NSF grant CR-94-24398, PSC-CUNY Research Award 663472, and OTKA-T-020914.

Supported by the DIMATIA Center, Prague.

© 1999 American Mathematical Society

Let $c(n) = \min_P c(P)$, where the minimum is taken over all $2n$-element sets in general position in the plane. We have

$$c(n) \geq c_3\sqrt{n},$$

for some positive constant c_3, but there is no sublinear upper bound known for $c(n)$ (see [**A**],[**P**]). In fact, in [**A**] it was shown that every $2n$-element set in general position has an at least $c_3\sqrt{n}$-element subset which not only admits a perfect cross-matching, but also satisfies a much stronger condition. In this strong sense the result is best possible [**V**]. It looked difficult to improve the lower bound on $c(n)$, because we had no good characterization of perfectly cross-matchable sets.

The aim of this note is to give such a good characterization and to design an efficient algorithm which decides whether a set admits a perfect cross-matching.

2. Characterization of perfectly cross-matchable sets

In this section, we would like to point out a simple relation between $c(P)$ and $h(P)$: the first quantity attains its maximum if and only if the second attains its minimum. More precisely, we have the following.

THEOREM 1. *A set of $2n$ points in general position in the plane admits a perfect cross-matching if and only if it has precisely n halving lines.*

PROOF. Suppose first that $P = \{p_1, p_2, \ldots, p_{2n}\}$ has a perfect cross-matching (i.e., n pairwise crossing segments) $\overline{p_{2i-1}p_{2i}}$, $1 \leq i \leq n$. The extension of each of these segments is a halving line, because each of them separates the two endpoints of all other segments $\overline{p_{2i-1}p_{2i}}$. We will show that P has no other halving lines.

Assume, in order to obtain a contradiction, that (say) p_1p_3 is also a halving line. We may suppose without loss of generality that p_1p_2 is horizontal, p_2 is to the right of p_1, and that p_{2i} is below and p_{2i-1} is above p_1p_2, for every $2 \leq i \leq n$. Since each segment $\overline{p_{2i-1}p_{2i}}$ $(3 \leq i \leq n)$ has to cross $\overline{p_1p_2}$, if $\overline{p_{2i-1}p_{2i}}$ has an endpoint to the left of p_1p_3, then its other endpoint must lie to the right of p_1p_3. However, both p_2 and p_4 are on the right-hand side of p_1p_3. This implies that the number of points to the right of p_1p_3 exceeds by at least 2 the number of points to the left of it, contradicting our assumption that p_1p_3 is a halving line.

Suppose next that P has precisely n halving lines. Since there is *at least* one halving line through every point p_k, we obtain that there must be *exactly* one. Thus, we can assume without loss of generality that the complete list of halving lines is $p_{2i-1}p_{2i}$ $(1 \leq i \leq n)$. We will show that the segments $\overline{p_{2i-1}p_{2i}}$ $(1 \leq i \leq n)$ are pairwise crossing.

Assume, for contradiction, that $\overline{p_1p_2}$ and $\overline{p_3p_4}$ have no interior point in common. By renumbering the points if necessary, we may also suppose that p_1p_2 is horizontal, p_2 is to the right of p_1, $\overline{p_3p_4}$ is entirely above the line p_1p_2, and that p_3 is closer to p_1p_2 than p_4 is. Notice that a minor counter-clockwise turn around p_3 will bring the line $\ell = p_3p_4$ into a position, where there are precisely n points on its right-hand side. (Indeed, p_4 will be added to the set of points to the right of p_3p_4.) If we continue to turn ℓ around p_3 in the counter-clockwise direction, we arrive at a position where ℓ becomes parallel to p_1p_2, i.e., it becomes horizontal. At that moment, there are at most $n-2$ points above ℓ (these points form a subset of the set of all points different from p_3 which lie above the halving line p_1p_2). Hence, there is an intermediate position $\ell = p_3p_k$ for some $k \neq 4$, in which the number of points on the right-hand side of ℓ is precisely $n-1$. This means that p_3p_k is a halving line which does not appear in the complete list of halving lines, $p_{2i-1}p_{2i}$ $(1 \leq i \leq n)$. Contradiction. □

Actually, this argument also yields the uniqueness of the perfect cross-matching.

THEOREM 2. *Any set of points in general position in the plane admits at most one perfect cross-matching.*

PROOF. As we have shown, every perfect cross-matching of P consists of exactly those segments between two points of P, whose extensions are *halving lines* of P. □

3. Algorithm

The above characterization of perfectly cross-matchable sets allows us to design an $O(n \log n)$-time algoritm which decides whether a set of $2n$ points satisfies this property and, if so, finds a perfect cross-matching for it.

Let P be a $2n$-element point set in general position in the plane, which is the union of two n-element sets, P_1 and P_2, separated by a straight line (say, by the y-axis). For any non-vertical line ℓ, let $P_i(\ell^+)$ (resp. $P_i(\ell^-)$) denote the set of points in P_i lying above (resp. below) ℓ. A line ℓ not passing through any point of P is called a *ham-sandwich cut* for P, if

$$|P_1(\ell^+)| = |P_2(\ell^-)| = \lfloor n/2 \rfloor.$$

It was shown by Megiddo [**M**] that one can always find such a line ℓ in $O(n)$ steps (see also [**LM**]).

Any matching of P that has a segment to the left of the y-axis, has another one to the right of it, and these two segments cannot cross. Thus, if there exists a perfect cross-matching for P, then all of its segments must cross the y-axis and, similarly, they must also cross the ham-sandwich cut ℓ. Consequently, a perfect cross-matching M for P is the union of a perfect cross-matching M_1 for $P_1(\ell^+) \cup P_2(\ell^-)$ and a perfect cross-matching M_2 for $P_1(\ell^-) \cup P_2(\ell^+)$. Let M_i^+ and M_i^- denote the *upper envelope* and the *lower envelope* (i.e., the pointwise maximum and pointwise minimum) of the lines supporting the segments of M_i, respectively $(i = 1, 2)$. Clearly, M_i^+ and M_i^- are unbounded convex polygonal paths, with at most $\lfloor n/2 \rfloor$ vertices each. (See Figure 1.)

We need the following corollary of Theorem 2.

CLAIM. The set P admits a perfect cross-matching M if and only if the following conditions are satisfied.

(1) $P_1(\ell^+) \cup P_2(\ell^-)$ admits a perfect cross-matching M_1 and $P_1(\ell^-) \cup P_2(\ell^+)$ admits a perfect cross-matching M_2.

(2) The convex hull $\operatorname{conv} P_2(\ell^+)$ is above the polygonal path M_1^+, and $\operatorname{conv} P_1(\ell^-)$ below M_1^-. Similarly, $\operatorname{conv} P_1(\ell^+)$ is above M_2^+, and $\operatorname{conv} P_2(\ell^-)$ below M_2^-.

Then, we have $M = M_1 \cup M_2$.

PROOF. We have seen before that if P admits a perfect cross-matching M, then it satisfies condition (1) and $M = M_1 \cup M_2$ holds. By Theorem 2, M_1 and M_2 are uniquely determined. To see that (2) is necessary, too, assume that (say) $P_2(\ell^+)$ has a point p below M_1^+. Then p lies below the supporting line of at least one segment $\overline{qq'} \in M_1$. Let p' denote the element of $P_1(\ell^-)$ connected to p in M_2. Then $\overline{pp'} \cap \overline{qq'} = \emptyset$, contradicting our assumption that any two segments of M cross.

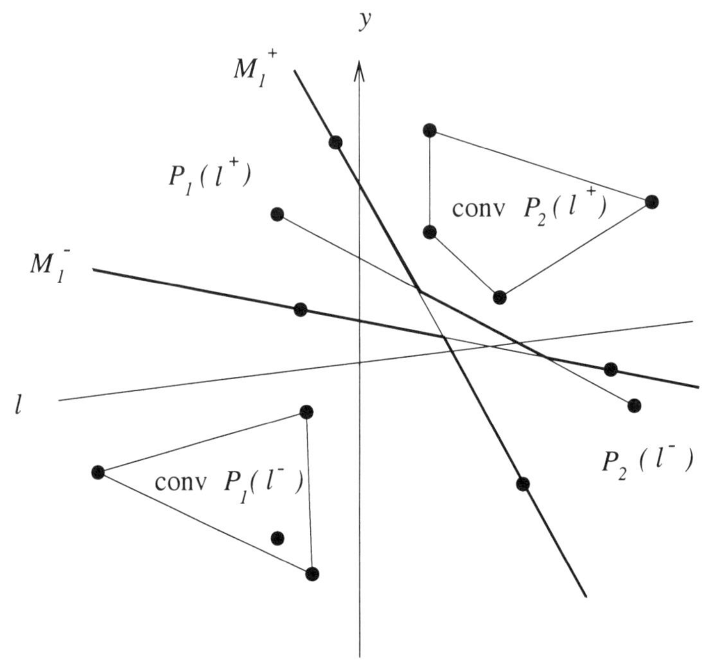

FIGURE 1.

Suppose next that conditions (1) and (2) are satisfied. Then $M = M_1 \cup M_2$ is a perfect cross-matching for P. Indeed, if there were two disjoint segments, $\overline{pp'} \in M_1$ and $\overline{qq'} \in M_2$, such that (say) $\overline{qq'}$ is below (resp. above) the line pp', then conv$P_2(\ell^+)$ would not lie above the polygonal path M_1^+ (resp. conv$P_1(\ell^-)$ would not lie below M_1^-), contradicting condition (2). □

Let M^+ and M^- denote the upper and the lower envelope of all lines supporting the segments of $M = M_1 \cup M_2$, respectively. Clearly, M^+ can be obtained as the upper envelope of M_1^+ and M_2^+, and M^- can be obtained as the lower envelope of M_1^- and M_2^-.

It is well known that one can compute the union and the intersection of two convex polygons of at most n sides in time $O(n)$ ([**PH**], [**S**]). Thus, if we know conv$P_i(\ell^+)$, conv$P_i(\ell^-)$, M_i^+, and M_i^- for $i = 1, 2$, then in linear time we can determine conv$P_i (i = 1, 2)$, M^+ and M^-. If any of the conditions of the Claim is not satisfied, we conclude that P does not admit a perfect cross-matching.

So one can use a divide-and-conquer algorithm to decide whether $P = P_1 \cup P_2$ admits a perfect cross-matching and, if yes, to compute it simultaneously with convP_i $(i = 1, 2)$, M^+ and M^-. At each stage it takes linear time to find a ham-sandwich cut ℓ and to do the merge step.

We obtained the following.

THEOREM 3. *There is an $O(n \log n)$ time, $O(n)$ space algorithm which decides whether a set of $2n$ points in general position in the plane admits a perfect cross-matching and, if so, computes it.*

Clearly, any decision tree that determines the perfect cross-matching of a planar point set of $2n$ points (if it exists) has height $\Omega(n \log n)$. In this sense Theorem 3 is asymptotically tight.

References

[A] B. Aronov, P. Erdős, W. Goddard, D. J. Kleitman, M. Klugerman, J. Pach and L. J. Schulman, *Crossing families*, Combinatorica **14** (1994), 127–134.

[D] T. Dey, *Improved bounds for planar k-sets and related problems*, Discrete and Computational Geometry 19 (1998), 373–382.

[E] H. Edelsbrunner, Algorithms in Combinatorial Geometry, Springer-Verlag, New York, 1987.

[EL] P. Erdős, L. Lovász, A. Simmons and E. Straus, *Dissection graphs of planar point sets*, in: A Survey of Combinatorial Theory (G. Srivastava, ed.), North-Holland, Amsterdam, 1973, 139–149.

[LM] C-Y. Lo, J. Matoušek and W. Steiger, *Algorithms for ham-sandwich cuts*, Discrete and Computational Geometry **11** (1994), 433–452.

[L] L. Lovász, *On the number of halving lines*, Annales Universitatis Scientarium Budapest, Eötvös, Sectio Mathematica **14** (1971), 107–108.

[M] N. Megiddo, *Partitioning with two lines in the plane,* J. Algorithms **6** (1985), 430–433.

[P] J. Pach and P. K. Agarwal, Combinatorial Geometry, J. Wiley & Sons, New York, 1995.

[PH] F. P. Preparata and S. J. Hong, *Convex hulls of finite sets of points in two and three dimensions,* Communications ACM **20** (1977), 87–93.

[S] M. I. Shamos, *Computational Geometry*, Ph.D. Thesis, UMI No. 7819047, Yale University, New Haven, CT, 1978.

[V] P. Valtr, *On mutually avoiding sets,* in: The Mathematics of Paul Erdős, II (R. L. Graham and J. Nešetřil, eds.), Algorithms and Combin. **14**, Springer, Berlin, 1997, 324–332.

MATHEMATICAL INSTITUTE OF THE HUNGARIAN ACADEMY OF SCIENCES, H-1364 BUDAPEST, P.O.B. 127

E-mail address: pach@cims.nyu.edu

MATHEMATICAL INSTITUTE OF THE HUNGARIAN ACADEMY OF SCIENCES, H-1364 BUDAPEST, P.O.B. 127

E-mail address: solymosi@cs.elte.hu

Contemporary Mathematics
Volume **223**, 1999

Three-dimensional Grid Drawings of Graphs

János Pach, Torsten Thiele, and Géza Tóth

ABSTRACT. A three-dimensional *grid drawing* of a graph G is a placement of the vertices at distinct integer points so that the straight-line segments representing the edges of G are pairwise non-crossing. It is shown that for any fixed $r \geq 2$, every r-colorable graph of n vertices has a three-dimensional grid drawing that fits into a box of volume $O(n^2)$. The order of magnitude of this bound cannot be improved.

1. Introduction

In a *grid drawing* of a graph, the vertices are represented by distinct points with integer coordinates and the edges are represented by straight-line segments connecting the corresponding pairs of points. Grid drawings in the plane have a vast literature [**BETT**]. In particular, it is known that every planar graph of n vertices has a two-dimensional grid drawing that fits into a rectangle of area $O(n^2)$, and this bound is asymptotically tight [**FPP**],[**S**].

The possibility of three-dimensional representations of graphs was suggested by software engineers [**MRC**]. The analysis of the volume requirement of such representations was initiated in [**CELR**], where the following statement was proved. Every graph of n vertices has a three-dimensional grid drawing in a rectangular box of volume $O(n^3)$, and this bound cannot be improved. To establish the first half of the statement, it is sufficient to consider representations of *complete graphs*. Cohen et al. used a generalization of a well-known construction of Erdős showing that the vertices of a complete graph K_n can be placed at the points

$$(i, i^2 \bmod p, i^3 \bmod p), \; 1 \leq i \leq n,$$

where p is a prime between n and $2n$. Since no four of these points lie in the same plane, the resulting straight-line drawing of K_n has no crossing edges.

A complete r-partite graph is called *balanced*, if any two of its classes have the same number of points, or their sizes differ by one. Let $K_r(n)$ denote a balanced complete r-partite graph with $n \geq r$ vertices. That is, the vertex set of $K_r(n)$

1991 *Mathematics Subject Classification.* Primary 05C35, 52C10, 68R10.
Key words and phrases. Graph drawing, moment curve, chromatic number, box.
Supported by NSF grant CCR-94-24398 and PSC-CUNY Research Award 667339.
Supported by the Deutsche Forschungsgemeinschaft.
Supported by OTKA-T-020914, OTKA-F-22234 and the Alfred Sloan Foundation.

© 1999 American Mathematical Society

splits into r disjoint classes, $V_1, V_2, \ldots, V_r$, such that $|V_i| = \lfloor n/r \rfloor$ or $\lceil n/r \rceil$, and two vertices are connected by an edge if and only if they belong to different V_i's.

It was pointed out by Cohen et al. [**CELR**] that $K_2(n)$ has a three-dimensional grid drawing within a box of volume $O(n^2)$, and they asked whether this bound is optimal. T. Calamoneri and A. Sterbini [**CS**] proved that any such drawing requires a box whose volume is at least $\Omega(n^{3/2})$. Furthermore, they have shown that $K_3(n)$ and $K_4(n)$ also permit three-dimensional grid drawings of volume $O(n^2)$, and conjectured that the same is true for $K_r(n)$, for any fixed $r > 4$.

The aim of the present note is to answer the question of Cohen et al. and to verify the conjecture of Calamoneri and Sterbini. A graph is called *r-colorable* if its vertices can be colored by r colors so that no two adjacent vertices receive the same color. Equivalently, G is r-colorable if it is a subgraph of a complete r-partite graph.

THEOREM. *For every $r \geq 2$ fixed, any r-colorable graph of n vertices has a three-dimensional grid drawing that fits into a rectangular box of volume $O(n^2)$. The order of magnitude of this bound cannot be improved.*

2. Proof of the theorem

To prove the second assertion, it is enough to establish an $\Omega(n^2)$ lower bound on b, the number of integer points in a box B accommodating a three-dimensional grid drawing of the balanced complete bipartite graph $K_2(n)$ with vertex classes V_1, V_2. Clearly, $K_2(n)$ is r-colorable for any $r \geq 2$. Fix a grid drawing and consider the set of all vectors pointing from a vertex of V_1 to a vertex of V_2. The number of such vectors is $\lceil \frac{n}{2} \rceil \cdot \lfloor \frac{n}{2} \rfloor \geq \frac{n^2-1}{4}$, and no two of them can be identical, otherwise the corresponding four points would induce a parallelogram whose diagonals cross each other. (In fact, no two such vectors can point in the same direction.) On the other hand, the total number of vectors determined by two gridpoints in B is smaller than $8b$. Thus, $8b > \frac{n^2-1}{4}$, as required.

The proof of the upper bound is based on the following.

LEMMA. *For any $r \geq 2$ and for any n divisible by r, the balanced complete r-partite graph $K_r(n)$ has a three-dimensional grid drawing that fits into a rectangular box of size $r \times 4n \times 4rn$.*

PROOF. Let p be the smallest prime with $p \geq 2r - 1$ and set $N := p \cdot \frac{n}{r}$. Note that $p < 4r$ and thus $N < 4n$. For any $0 \leq i \leq r-1$, let

$$V_i := \{(i, t, it) : 0 \leq t < N, t \equiv i^2 \pmod{p}\}.$$

These sets are pairwise disjoint, and each of them has precisely $\frac{N}{p} = \frac{n}{r}$ elements. Connect any two points belonging to different V_i's by a straight-line segment. The resulting drawing of $K_r(n)$ fits into a rectangular box of size $r \times 4n \times 4rn$, as desired.

It remains to show that no two edges of this drawing cross each other. Suppose, for contradiction, that there are two crossing edges, e and e'. We distinguish three different cases, according to the number of distinct classes V_i, the endpoints of e and e' belong to.

Case 1: The endpoints of e and e' are from four distinct classes, $V_{i_1}, V_{i_2}, V_{i_3}$, and V_{i_4}.

Let $(i_\alpha, t_\alpha, i_\alpha t_\alpha), 1 \le \alpha \le 4$, be the corresponding endpoints, so that $t_\alpha \equiv i_\alpha^2 \pmod p$. Then these points lie in a plane, and the determinant

$$D = \begin{vmatrix} 1 & i_1 & t_1 & i_1t_1 \\ 1 & i_2 & t_2 & i_2t_2 \\ 1 & i_3 & t_3 & i_3t_3 \\ 1 & i_4 & t_4 & i_4t_4 \end{vmatrix}$$

vanishes. Therefore, it must also vanish modulo p. However, modulo p this determinant reduces to the Vandermonde determinant. Thus,

$$D \equiv \begin{vmatrix} 1 & i_1 & i_1^2 & i_1^3 \\ 1 & i_2 & i_2^2 & i_2^3 \\ 1 & i_3 & i_3^2 & i_3^3 \\ 1 & i_4 & i_4^2 & i_4^3 \end{vmatrix} = \prod_{1\le\beta<\alpha\le 4} (i_\alpha - i_\beta),$$

which is non-zero modulo p, a contradiction.

Case 2: The endpoints of e and e' are from three different classes, V_i, V_j, V_k.

Assume without loss of generality that two of these points, $(i, t_1, it_1), (i, t_2, it_2)$, belong to V_i, and the other two are $(j, s, js) \in V_j$ and $(k, u, ku) \in V_k$. These four points cannot be coplanar (hence e and e' cannot cross each other), because the corresponding test determinant

$$\begin{vmatrix} 1 & i & t_1 & it_1 \\ 1 & i & t_2 & it_2 \\ 1 & j & s & js \\ 1 & k & u & ku \end{vmatrix} = (j-i)(k-i)(t_2-t_1)(s-u)$$

is non-zero. To see this, observe that $s - u \equiv j^2 - k^2 \equiv (j-k)(j+k) \not\equiv 0 \pmod p$. This is the point where we use the assumption that $0 \le j, k \le r-1 \le (p-1)/2$.

Case 3: The endpoints of e and e' are from two different classes, V_i and V_j.

Let these points be $(i, t_1, it_1), (i, t_2, it_2) \in V_i$ and $(j, s_1, js_1), (j, s_2, js_2) \in V_j$. Now the corresponding test determinant

$$\begin{vmatrix} 1 & i & t_1 & it_1 \\ 1 & i & t_2 & it_2 \\ 1 & j & s_1 & js_1 \\ 1 & j & s_2 & js_2 \end{vmatrix} = (j-i)^2(t_2-t_1)(s_1-s_2)$$

does not vanish, therefore e and e' cannot cross each other. This contradiction completes the proof of the Lemma. □

Now we return to the proof of the upper bound of the theorem. In fact, we can deduce a more precise statement.

COROLLARY. *There exists a $c > 0$ such that for any $r \ge 2$, any r-colorable graph of n vertices has a three-dimensional grid drawing that fits into a rectangular box of volume cr^2n^2.*

PROOF. Fix an r-colorable graph G with n vertices. If each color class contains at most $\lceil \frac{n}{r} \rceil$ vertices, then G is a subgraph of $K_r(n)$ and, by the Lemma, it has a three-dimensional grid drawing fitting into a box of size $r \times 4n \times 4rn$. Otherwise, we split every color class into smaller parts such that all but one of them have *exactly* and the last one *at most* $\lceil \frac{n}{r} \rceil$ points. This defines a decomposition of the vertex set of G into at most $2r-1$ classes, whose sizes do not exceed $\lceil \frac{n}{r} \rceil$, and no two points belonging to the same class are connected by an edge. In other words, G is a subgraph of a balanced complete $(2r-1)$-partite graph K with $(2r-1)\lceil \frac{n}{r} \rceil < 2n+2r$ vertices. Applying the Lemma to K, the Corollary follows. □

3. Remarks and open problems

A. The rectangular box used in the proof of the Theorem has two sides of size $O(n)$. We can use rectangular boxes of different shapes to represent $K_2(n)$.

PROPOSITION. *There is a three dimensional grid drawing of $K_2(n)$ which fits into a rectangular box of size $O(n) \times O(\sqrt{n}) \times O(\sqrt{n})$.*

PROOF. Let V_1 and V_2 be the vertex classes of $K_2(n)$ and let

$$V_1 = \{(i,0,0) \mid 0 \le i \le \lceil n/2 \rceil - 1\},$$
$$V_2 = \{(0,a,b) \mid 0 \le a, b \le k, (a,b) = 1\}$$

where $k \approx \pi\sqrt{n/6}$.

Then

$$|V_1| = \lceil n/2 \rceil,$$
$$|V_2| = \sum_{i=1}^{k} \phi(i) \approx \frac{3k^2}{\pi^2} \approx n/2$$

(see [**HW**]). To see that there is no crossing, we observe that the points of V_2 lie on a horizontal plane and there are no two of these points on a line through $(0,0,0)$ (see Figure 1). □

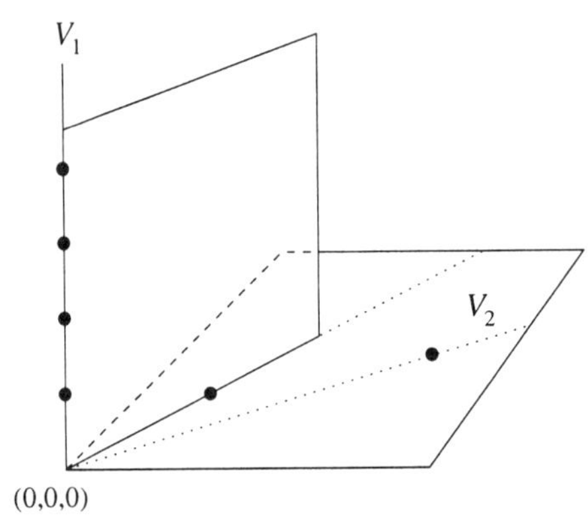

FIGURE 1. Embedding of $K_2(n)$

B. It would be interesting to determine $\boldsymbol{S}_r = \boldsymbol{S}_r(n)$, the set of triples (s_1, s_2, s_3) for which every r-colorable graph of n vertices has a grid drawing that fits into a box of size $s_1 \times s_2 \times s_3$. In particular, what is the smallest $s = s(r)$ with $(s,s,s) \in S_r$? It is not hard to see that $s_i s_j \ge \lfloor n/2 \rfloor$ and $s_i > \frac{r}{5}$ holds for every $(s_1, s_2, s_3) \in \boldsymbol{S}_r$ $(r \ge 2)$.

C. Since any graph of n vertices with fixed maximum degree $r-1$ is r-colorable, it follows from the Theorem that any such graph permits a grid drawing in a box of volume $O(n^2)$. It seems likely that for every fixed r, this bound can be substantially improved. We cannot even decide if every graph with maximum degree 3 has a grid drawing of volume $O(n)$.

References

[BETT] G. Di Battista, P. Eades, R. Tamassia and I. G. Tollis, *Algorithms for drawing graphs: an annotated bibliography,* Computational Geometry: Theory and Applications **4** (1994), 235–282.

[CELR] R. F. Cohen, P. Eades, T. Lin and F. Ruskey, *Three-dimensional graph drawing,* in: Graph Drawing (GD '94, R. Tamassia and I. G. Tollis, eds.), Lecture Notes in Computer Science **894**, Springer-Verlag, Berlin, 1995, 1–11. Also in: Algorithmica **17** (1997), 199–208.

[CS] T. Calamoneri and A. Sterbini, *Drawing* 2-, 3-, and 4-*colorable graphs in $O(n^2)$ volume,* in: Graph Drawing (GD '96, S. North, ed.), Lecture Notes in Computer Science **1190**, Springer-Verlag, Berlin, 1997, 53–62.

[FPP] H. de Fraysseix, J. Pach and R. Pollack, *How to draw a planar graph on a grid,* Combinatorica **10** (1990), 41–51.

[HW] G. H. Hardy and E. M. Wright, An Introduction to the Theory of Numbers, University Press, Oxford, 1954.

[MRC] J. Mackinley, G. Robertson and S. Card, *Cone trees: Animated 3d visualizations of hierarchical information,* in: Proceedings of SIGCHI Conference on Human Factors in Computing, 1991, 189–194.

[S] W. Schnyder, *Embedding planar graphs on the grid,* in: Proceedings of the 1st Annual ACM-SIAM Symposium on Discrete Algorithms, 1990, 138–148.

COURANT INSTITUTE, NYU AND HUNGARIAN ACADEMY OF SCIENCES
E-mail address: pach@cims.nyu.edu

COURANT INSTITUTE, NYU AND FU BERLIN
E-mail address: thiele@math.fu-berlin.de

COURANT INSTITUTE, NYU AND DIMACS
E-mail address: geza@math-inst.hu

Contemporary Mathematics
Volume **223**, 1999

On Polygonal Covers

Michel Pocchiola and Gert Vegter

ABSTRACT. A polygonal cover of a finite collection of pairwise disjoint convex compact sets in the plane is a finite collection of non-overlapping bounded convex polygons such that each polygon covers exactly one convex set. We show that computing a polygonal cover with worst case minimal number of sides reduces to computing a pseudo-triangulation of the collection of convex sets. We obtain a similar reduction for two related problems concerning convex compact sets in the plane : computing a lighting set and computing a family of separating lines. Our main tool is the notion of *screen graph* (introduced in this paper) associated with a pseudo-triangulation.

1. Introduction

A *polygonal cover* of a finite collection of pairwise disjoint convex compact sets in the plane (we refer to a convex compact set as a convex disk) is a finite collection of non-overlapping bounded convex polygons such that each polygon covers exactly one convex disk; see Figure 1.

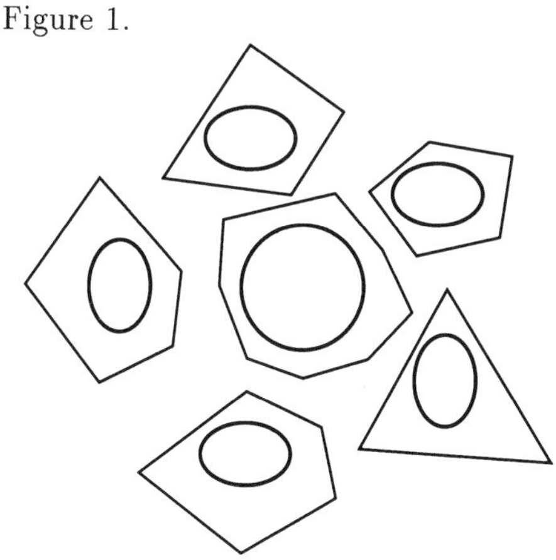

FIGURE 1: A polygonal cover of a collection of six disks.

1991 *Mathematics Subject Classification.* Primary 68Q25, 68U05, 52A10, 52C10, 52B10, 05C90.

Key words and phrases. Pseudotriangle, pseudo-triangulation, screen graph, polygonal cover, translation query, art gallery theorem, packing, covering.

A preliminary version of this work has appeared in the *Proceedings of the 12th Annu. ACM Symposium on Computational Geometry*, Philadelphia, May 1996, pages 291–300.

© 1999 American Mathematical Society

In 1950, L. Fejes Tóth [**F**] has shown the existence of polygonal covers with a total of no more than $6n - 9$ sides for $n \geq 3$ disks, this bound being tight in the worst case [**ERS**]. He used this result in order to show that the upper density of a packing with congruent copies of a convex disk in the plane is bounded by the ratio of the area of the disk to the area of a hexagon of minimum area circumscribed around the disk (see [**PA**] and the reference cited therein). More recently, polygonal covers have been used to show that, in any collection of n pairwise disjoint convex disks, there is a disk separated by a line from at least $n/12$ other disks (see [**HK**] and [**AKP, CER, CRUZ**] for more separation problems) and to design efficient algorithms to answer translation queries for sets of polygons (see [**dEW**]).

The proof of L. Fejes Tóth proceeds by growing the convex disks unboundedly in all directions but the growth (for each direction) is limited by the condition that the convex disks remain pairwise interior disjoint. In this way the convex disks expand into convex polygons that fill the plane except for a finite number of gaps that are also convex polygons. A simple application of the Euler's relation for planar graph leads to the announced upper bound. The purpose of this note is to give a constructive proof of the result of L. Fejes Tóth. Our main result is the following.

THEOREM 1. *A polygonal cover with worst case minimal number of sides for a collection of n pairwise disjoint convex disks is computable in $O(n \log n)$ time under the assumption that each disk has constant computational complexity.*

Our proof is based on properties of the so-called pseudo-triangulations introduced by the authors [**PV, PV2**] to design efficient algorithms to compute visibility graphs. In Section 2 we review the terminology on pseudo-triangulations. In Section 3 we give the proof of Theorem 1: the main ingredient of the proof is the notion of *screen graph* (introduced in Section 3) associated with a pseudo-triangulation. Finally, in Section 4 we elaborate on two related problems concerning pairwise disjoint convex disks in the plane, namely computing a lighting set and computing a family of separating lines.

2. Background material

Consider a finite collection of pairwise disjoint convex disks in the plane $\mathbb{R}^2$. A *bitangent* is a closed line segment whose supporting line is tangent to two disks at its endpoints. A bitangent is said to be *interior* if its supporting line separates the two convex disks to which it is tangent, otherwise the bitangent is said to be *exterior*. It is called *free* if it lies in free space, i.e., the complement of the union of the interior of the disks.

For ease of the exposition we assume that the boundary of each disk is a smooth closed curve such that for each ϕ, $0 \leq \phi \leq 2\pi$, the curve has exactly one point where its (directed) tangent makes an angle of ϕ with the positive x-axis; furthermore we assume that no line is tangent to three disks and that the arrangement of the supporting lines of the bitangents is simple. Note that according to our assumptions two bitangents are disjoint or they cross at an interior point (but their supporting lines are crossing).

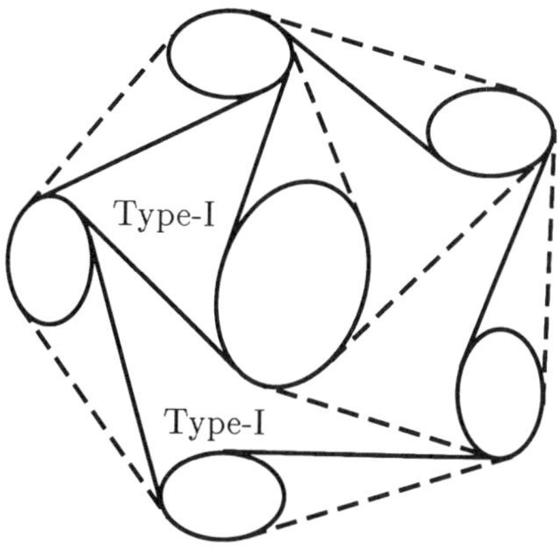

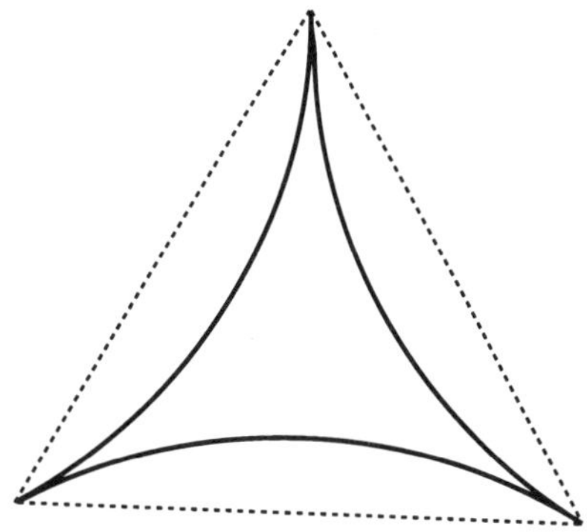

FIGURE 2: A pseudo-triangulation (left)—exterior bitangents are shown dashed and Type-I pseudotriangles are labeled Type-I—and a pseudotriangle (right).

A *pseudo-triangulation* ([**PV, PV2**]) is a maximal subset of non-crossing free bitangents. The boundaries of the convex disks and the bitangents in a pseudo-triangulation induce a regular cell decomposition of the plane, still called a pseudo-triangulation; see the left part of Figure 2. According to [**PV, PV2**], a bounded free face of a pseudo-triangulation is a *pseudotriangle*, that is a simply connected subset of the plane whose boundary consists of three smooth convex curves that are tangent at their endpoints, and which is contained in the triangle formed by the three endpoints of these convex curves; the convex curves are called the *sides* of the pseudotriangle, and their endpoints are called the *cusp points* of the pseudotriangle; see the right part of Figure 2. The number of pseudotriangles is $2n-2$ for n disks, and the number of bitangents is $3n-3$. A pseudo-triangulation of a collection of n disks is computable in $O(n \log n)$ time under the assumption that each disk has constant computational complexity in the sense that the tangent to a disk with a given direction, the common tangents to two disks or the common tangents to a disk and a point are computable in constant time. See [**PV, PV2**] for more details. (If we assume that the convex disks are composed of a total of m arcs, each of constant computational complexity, then a pseudo-triangulation is computable in time $O(m + n \log m)$ time.)

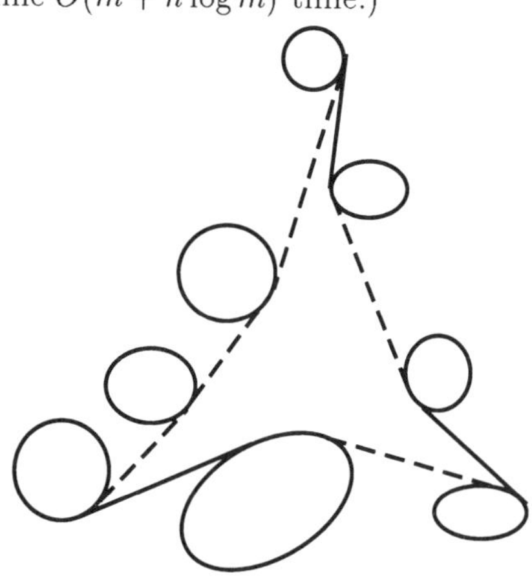

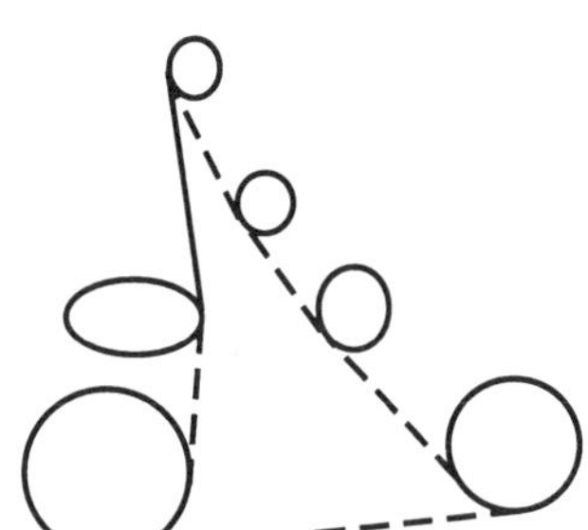

FIGURE 3: A Type-I (left) and a Type-II (right) pseudotriangle.

The sides of the pseudotriangles (of a pseudo-triangulation) are sequences of alternate bitangents and segments of boundaries of the convex disks (*arcs* for short), possibly reduced to single bitangents or to single arcs. A pseudotriangle is said to be of type I if none of its sides is a bitangent, otherwise the pseudotriangle is said to be of type II; see Figure 3. In this latter case one of the sides of the pseudotriangle is a (necessarily exterior) bitangent. Conversely any exterior bitangent is the bitangent side of a unique Type-II pseudotriangle: hence we have a one-to-one correspondence between the set of exterior bitangents and the set of Type-II pseudotriangles. (It follows that the number of exterior bitangents in a pseudo-triangulation of a collection of n disks is at most $2n-2$, and consequently the number of interior bitangents is at least $n-1$.) The boundary of a Type-I pseudotriangle contains exactly three interior bitangents, namely the bitangents at its cusp points. The boundary of a Type-II pseudotriangle contains exactly one interior bitangent, namely the bitangent at the opposite cusp point of the bitangent side. The following lemma gives a lower bound on the number of exterior bitangents in a pseudo-triangulation.

LEMMA 1. *Consider a pseudo-triangulation of a collection of at least three convex disks. We claim that*

1. *The number of exterior bitangents is at least 3 (even 4 if the number of exterior bitangents lying on the convex hull of the convex disks is 2).*
2. *The number of exterior bitangents is at least 4 under the assumption that the supporting line of one of the interior bitangent separates a disk from the other disks and is disjoint from the other interior bitangents. (If the number of disks is at least 4 then, under the same assumption, the number of exterior bitangents is at least 5.)*

PROOF. Consider the graph whose nodes are the convex disks and whose edges are the pairs of convex disks connected by an interior bitangent of some pseudo-triangulation. Clearly this graph is a plane graph without multiple edges; thus its number of edges is at most $3n-6$, where n is the number of disks. Since the total number of bitangents in a pseudo-triangulation is $3n-3$ the first part of claim (1) follows. A similar argument applies to the other claims. □

Collections of n interior-disjoint circles whose contact graphs are triangular planar graphs on n points are examples of configurations of n disks with exactly three free exterior bitangents (according to the 'Koebe-Andreev-Thurston Circle Packing Theorem' [**Z**, page 117] or [**PA**, pages 95–96] any triangular planar graph on n points is realizable as the contact graph of a collection of n interior-disjoint circles).

3. Polygonal covers via screen graphs

The idea is to generate a convex tessellation of the plane that covers and separates the convex disks. To this end we start from the geometric graph G consisting of the union of interior bitangents of a pseudo-triangulation (see Figure 4 part (a)), and we extend its half-edges along their supporting lines until their endpoints hit the current extension $G(t)$ of the geometric graph G. We envision the extension as a continuous process $t \in [0,+\infty) \mapsto G(t) \subset \mathbb{R}^2$ with $G(0)=G$ and $G(t) \subseteq G(t')$ for $t \le t'$, during which the half-edges are extended continuously and concurrently.

If the extension process is not subject to further constraints there is, in general, no reason for the tessellation corresponding to $G(+\infty)$ to cover and separate the convex disks, which is our goal. (For example if we first extend the topmost interior bitangent in the left part of Figure 3, its extension is going to split the lowest disk.)

In order to generate a tessellation that covers and separates the convex disks we are going to proceed in two stages: $G \to SG \to ESG$. The first step produces the *screen graph* SG (see Figure 4 part (*b*)), named because it "screens" the disks from one another (see Lemma 2). This step is subject to further constraints, that turn the extension phase into a deterministic process producing a unique result. The second step does follow freely the continuous plan, and produces what we'll call an *extension* ESG of the screen graph, which is not unique (see Figure 4 part (*c*) for an illustration of this point) but which does cover and separate the disks. Note that according to our assumptions there is at most one region of the resulting tessellation which is a half-plane. The screen graph is defined as follows.

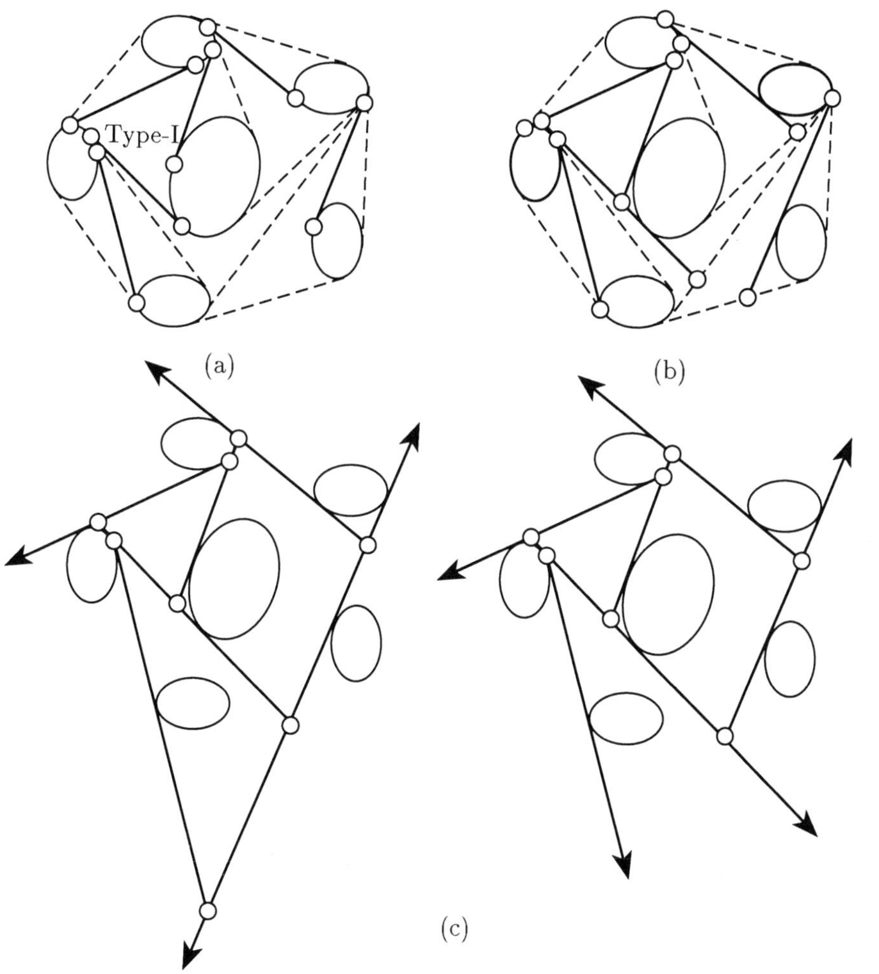

FIGURE 4: (*a*) The interior bitangents of a pseudo-triangulation (with a sole Type-I pseudotriangle), (*b*) its screen graph, and (*c*) two extensions of the screen graph.

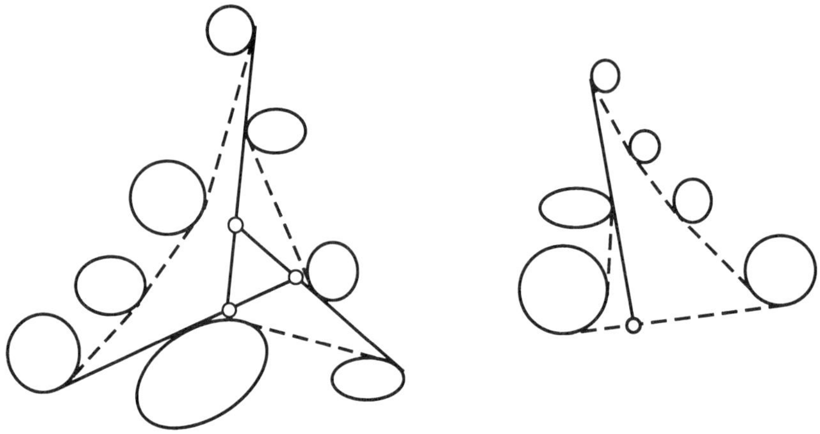

FIGURE 5: The edges of the screen graph inside a Type-I (left) or a Type-II (right) pseudotriangle.

1. The three interior bitangents of each Type-I pseudotriangle are extended up to and including the intersection points of their supporting lines. Note that the extensions of the bitangents lie inside the pseudotriangle because tangent lines to a pseudotriangle cross inside the pseudotriangle. A Type-I pseudotriangle is therefore split into three curved triangles, one per side of the pseudotriangle, plus the straight triangle whose vertices are the intersection points of the supporting lines of the three bitangents; see the left part of Figure 5.
2. The interior bitangent of each Type-II pseudotriangle is extended up to include the intersection point of its supporting line with the bitangent side of the pseudotriangle. A Type-II pseudotriangle is therefore split into two curved triangles, one per non-bitangent side of the pseudotriangle; see the right part of Figure 5.

By construction, any closed line segment joining two non-bitangent sides of a pseudotriangle, other than the interior bitangents of the pseudo-triangulation that defines the screen graph, intersects the screen graph transversally. (The intersection point may be an endpoint of the line segment.) In particular this is the case for the exterior bitangents of the pseudo-triangulation that defines the screen graph. This simple observation has the following key extension.

LEMMA 2. *Any closed line segment joining two distinct convex disks, other than the interior bitangents of the pseudo-triangulation that defines the screen graph, intersects the screen graph transversally. (The intersection point may be an endpoint of the line segment.)*

PROOF. Consider a closed line segment, with endpoints p and q, joining two distinct convex disks and not reduced to an interior bitangent of the pseudo-triangulation (that defines the screen graph). It is sufficient to prove the lemma in the case where the interior of the line segment is disjoint from the obstacles. As noted previously, our claim is true if the line segment reduces to an exterior bitangent of the pseudo-triangulation. So we assume now that the line segment contains neither interior nor exterior bitangent of the pseudo-triangulation.

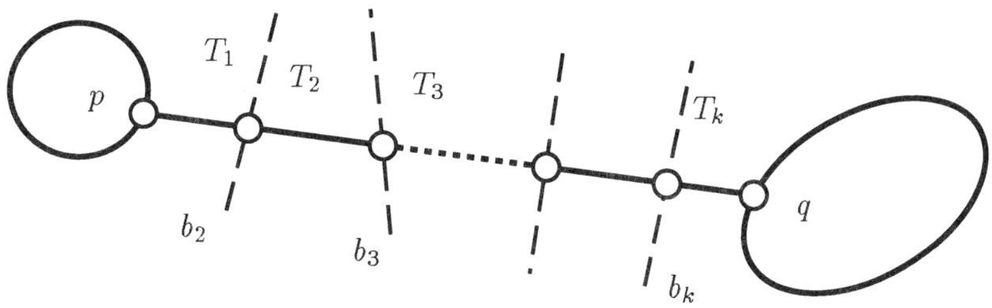

FIGURE 6: Illustration of the proof of Lemma 2. If b_2 and b_3, for example, are bitangent sides of T_1 and T_3, respectively, then T_2 is crossed by joining two of its non-bitangent sides. If b_2 is not a bitangent side of T_1, then T_1 is crossed by joining two of its non-bitangent sides because p does not lie on a bitangent side of T_1.

Let $T_1, b_2, T_2, b_3, \ldots, T_k$, with $k \geq 1$, be the sequence of alternate pseudotriangles and bitangents crossed by the interior of the line segment. See Figure 6 for an illustration. Because a bitangent is a bitangent side of at most one pseudotriangle, because we have k pseudotriangles T_i, but only $k-1$ bitangents b_i, and because the endpoints of the line segment lie on non-bitangent sides of T_1 and T_k, respectively, at most $k-1$ of the $2k$ crossed sides of the crossed pseudotriangles are bitangent sides. This leaves at least one pseudotriangle crossed by joining two of its non-bitangent sides. As noted previously this implies that the line segment intersects the screen graph transversally. This proves our lemma. □

The screen graph is clearly computable in $O(n)$ time (assuming that each pseudotriangle is represented by its three cusp points). Once the screen graph has been constructed one of its possible extensions can be computed in $O(n \log n)$ time using a straightforward adaptation of the Bentley-Ottmann sweep algorithm to compute the intersections of a collection of line segments. From the above lemma we easily deduce the following theorem.

THEOREM 2. *Any extension of the screen graph induces a convex tessellation of the plane that covers and separates the convex disks.*

PROOF OF THEOREM 1. In view of Theorem 2 (and the above discussion on the computational complexity issues) it suffices to show that the polygonal cover induced by an extension of the screen graph has a worst case minimal number of sides, i.e., at most $6n-9$ sides for $n \geq 3$ disks. To count the number of sides we count the number of corners of the polygonal cover. Assume first that no region of the tessellation is a half-plane. It is convenient to put one corner on each infinite edge of the tessellation so that these additional corners span a convex polygon surrounding all the disks. Note that the vertices of the screen graph are of degree one or three. The number of vertices of degree three is three times the number of Type-I pseudotriangles while the number of vertices of degree one is the number of Type-II pseudotriangles. Each vertex of degree three of the screen graph accounts for one corner and each vertex of degree one of the screen graph generates a vertex of its extension that accounts for 2 corners. Therefore the number of sides (or corners) is $2a + 3(2n-2-a) = 6n-6-a$ where a is the number of Type-II pseudotriangles or, equivalently, the number of exterior bitangents. Since, according to Lemma 1, a is at least three the result follows. Assume now that a region (which is necessarily unique) of the tessellation is a half-plane : we put one corner on each infinite edge of the tessellation plus one corner in the half-plane region so that these additional corners span a convex polygon surrounding all the disks. A similar counting argument shows that the number of sides of the polygonal cover is $6n - 6 - a + 1$. Since, according to Lemma 1, claim (2), a is at least four, the result follows.

4. Two related problems

4.1. Separating lines. A family of *separating lines* for a collection of convex disks is a collection of lines such that each pair of disks is separated by at least one of the lines of the family; see Figure 7.

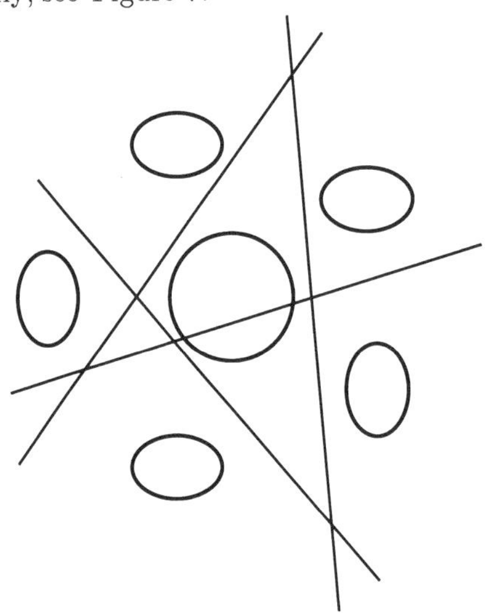

FIGURE 7: A family of separating lines.

It is well known (and easy to prove) that the family of supporting lines of the sides of a polygonal cover of the collection of convex disks is a family of separating lines. From the previous discussion we deduce the following theorem.

THEOREM 3. *The collection of supporting lines of the interior bitangents of a pseudo-triangulation of a collection of n pairwise disjoint convex disks is a family of separating lines for the collection of disks. In particular, such a family can be computed in $O(n \log n)$ time under the assumption that each disk has constant computational complexity.*

4.2. L. Fejes Tóth illumination problem revisited. Partitioning theorems and algorithms are key tools to solve Art gallery problems [**BS, HKK, O, S**]. For example Fisk's proof [**Fi**] of 'Chvátal's Art Gallery Theorem' [**C**], which asserts that $\lfloor n/3 \rfloor$ lighting points are occasionally necessary and sometimes sufficient to illuminate a polygon with n vertices, can be summarized as follow: computing a lighting set of at most $\lfloor n/3 \rfloor$ lighting points for a given polygon with n vertices reduces in $O(n)$ time to computing a triangulation of the polygon. We give now a similar reduction for L. Fejes Tóth's art gallery theorem [**F2**], which asserts that the boundary of a set of $n \geq 4$ pairwise disjoint convex disks can be illuminated by $4n - 7$ points. The proof of L. Fejes Tóth consists of locating the lighting points at suitable vertices of the gaps of a polygonal cover obtained by the growing process of the convex disks described in the introduction (see [**F2**] for the details). Since an extension of a screen graph can be viewed as the result of a growing process applied to the convex disks we can use (and determine efficiently), as Fejes Tóth did, at most $4n - 7$ of its vertices to illuminate the boundary of the convex disks.

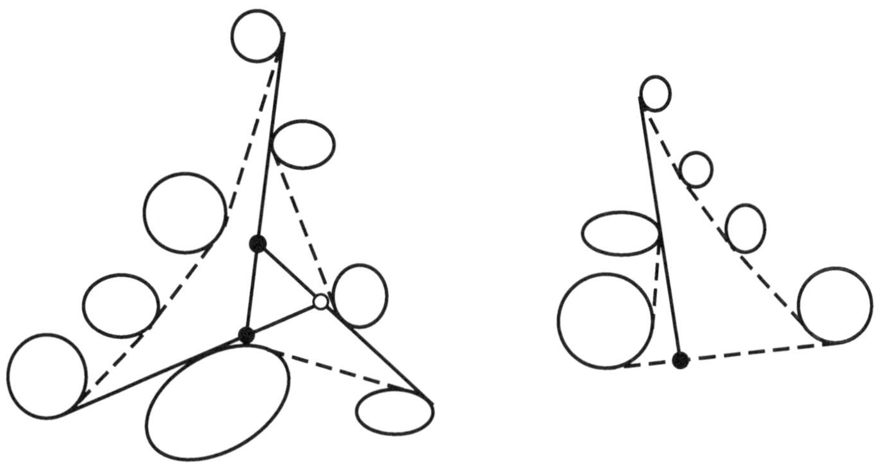

FIGURE 8: The boundary of a Type-I pseudotriangle can be illuminated by two of the three vertices of its screen graph (left). The boundary of a Type-II pseudotriangle can be illuminated by the vertex of its screen graph (right).

However one can obtain a simpler solution using only a subset of the vertices of a screen graph. Our result is the following.

THEOREM 4. *Computing a lighting set with at most* $4n-7$ *lighting points for a collection of* $n \geq 4$ *pairwise disjoint convex disks reduces in linear time to computing (the screen graph of) a pseudo-triangulation of the collection of convex disks. In particular, such a lighting set can be computed in* $O(n \log n)$ *time under the assumption that each disk has constant computational complexity.*

PROOF. Since the boundary points of the convex disks are the boundary points of the pseudotriangles (forget for a moment the points lying on the convex hull of the convex disks) it is sufficient to find a lighting set for the pseudotriangles of a pseudo-triangulation. How many lighting points are necessary for a pseudotriangle? Two in general (computable in $O(1)$ time) but only one if a side of the pseudotriangle reduces to a line segment, as illustrated in Figure 8. Let a be the number of pseudotriangles with a line segment side or, equivalently, the number of exterior bitangents in the pseudo-triangulation. From the above discussion it follows that a lighting set with $4n - 4 - a$ lighting points exists and is computable in $O(n)$ time from the pseudo-triangulation. According to Lemma 1, the number a is at least three, from which the result follows. (We take care to send the lighting points on the convex hull far enough to infinity in suitable directions to guarantee that they also illuminate the boundary of the convex hull; in case a region of the tessellation is a half-plane we have to use the extra lighting point given by the second claim of Lemma 1 to illuminate the whole boundary of the convex hull). □

L. Fejes Tóth also provides collections of $n \geq 4$ convex disks which cannot be illuminated by less than $4n - 7$ points but leaves open a practical characterization of all cases in which this number of lighting points is necessary. From our discussion we see that a necessary condition is that the number of free exterior bitangents is three in which case they all lie on the boundary of the convex hull of the disks—here we use Lemma 1. (Slight perturbations of) collections of n interior-disjoint circles whose contact graphs are triangular planar graphs on n points are examples of configurations of n disks with exactly three free exterior bitangents (according to the Koebe-Andreev-Thurston Circle Packing Theorem' [**Z**, page 117] or [**PA**, pages 95–96] any triangular planar graphs on n points is realizable as the contact graph of a collection of n interior-disjoint circles). However this condition is not sufficient since configurations of circles require only $2n-2$ lighting points (see [**F2**]).

Acknowledgments. We thank Joseph O'Rourke whose comments and suggestions have greatly improved the readability of the paper.

References

[AKP] N. Alon, M. Katchalski and W. R. Pulleyblank, *Cutting disjoint disks by straight lines.* Discrete Comput. Geom., 4:239–243, 1989.

[BS] I. Bjorling-Sachs and D. L. Souvaine, *An efficient algorithm for guard placement in polygons with holes.* Discrete Comput. Geom., 13:77–109, 1995.

[C] V. Chvátal, *A combinatorial theorem in plane geometry.* J. Combin. Theory Ser. B, 18:39–41, 1975.

[CER] J. Czyzowicz, H. Everett and J.-M. Robert, *Separating translates in the plane: Combinatorial bounds and an algorithm.* Report DIM-94-2, Université du Québec à Chicoutimi, Chicoutimi, 1994.

[CRUZ] J. Czyzowicz, E. Rivera-Campo, J. Urrutia and J. Zaks, *Separating convex sets in the plane.* Discrete Comput. Geom., 7:189–195, 1992.

[dEW] M. de Berg, H. Everett and H. Wagener, *Translation queries for sets of polygons.* Internat. J. Comput. Geom. Appl., 5:221–242, 1995.

[ERS] H. Edelsbrunner, A. D. Robison and X. Shen, *Covering convex sets with non-overlapping polygons.* Discrete Math., 81:153–164, 1990.

[F] L. Fejes Tóth, *Some packing and covering theorems.* Acta Scientiarum Mathematicarum (Szeged), 12/A:62–67, 1950.

[F2] L. Fejes Tóth, *Illumination of convex discs.* Acta Math. Acad. Sci. Hungar., 29(3–4):355–360, 1977.

[Fi] S. Fisk, *A short proof of Chvátal's watchman theorem.* J. Combin. Theory Ser. B, 24:374, 1978.

[HKK] F. Hoffmann, M. Kaufmann and K. Kriegel, *The art gallery theorem for polygons with holes.* In Proc. 32nd Annu. IEEE Sympos. Found. Comput. Sci., pages 39–48, 1991.

[HK] R. Hope and M. Katchalski, *Separating plane convex sets.* Math. Scand., 66:44–46, 1990.

[O] J. O'Rourke, Art Gallery Theorems and Algorithms. Oxford University Press, New York, NY, 1987.

[PA] J. Pach and P. K. Agarwal, Combinatorial Geometry. John Wiley & Sons, New York, NY, 1995.

[PV] M. Pocchiola and G. Vegter, *Topologically sweeping visibility complexes via pseudo-triangulations.* Discrete Comput. Geom., 16:419–453, December 1996.

[PV2] M. Pocchiola and G. Vegter, *The visibility complex.* Internat. J. Comput. Geom. Appl., 6(3):279–308, 1996.

[S] T. C. Shermer, *Recent results in art galleries.* Proc. IEEE, 80(9):1384–1399, September 1992.

[Z] G. M. Ziegler, Lectures on Polytopes, volume 152 of Graduate Texts in Mathematics. Springer-Verlag, 1994.

DÉPARTEMENT DE MATHÉMATIQUES ET INFORMATIQUE, ECOLE NORMALE SUPÉRIEURE, URA 1327 DU CNRS, 45 RUE D'ULM 75230 PARIS CEDEX 05, FRANCE
E-mail address: pocchiola@dmi.ens.fr

DEPT. OF MATH. AND COMP. SC., UNIVERSITY OF GRONINGEN, P.O. BOX 800, 9700 AV GRONINGEN, THE NETHERLANDS
E-mail address: gert@cs.rug.nl

Contemporary Mathematics
Volume **223**, 1999

The Universality Theorems for Oriented Matroids and Polytopes

Jürgen Richter-Gebert

ABSTRACT. Universality Theorems are exciting achievements in the theories of polytopes and oriented matroids. This article surveys the main developments in that context. We explain the basic constructions that lead to Universality Theorems. In particular, we show that one can use the Universality Theorem for rank 3 oriented matroids to obtain a Universality Theorem for 6-dimensional polytopes.

1. Universality theorems

Oriented matroids and polytopes are most fundamental objects in combinatorial geometry. A major breakthrough in both fields was the development of so called *Universality Theorems.* Intuitively speaking, a universality theorem states that realization spaces of oriented matroids (resp. polytopes) can be very complicated objects. For any reasonable complexity measure (like algorithmic complexity, topological complexity, algebraic complexity, etc.) questions related to realization spaces are as difficult as the corresponding problem for general systems of polynomial inequalities.

It is the purpose of this article to sketch the main developments and achievements in this field over the last decade, to clarify the main concepts and to give an idea of the proof techniques that were applied. In principle, this article could be used as a "quick reference guide" for the constructions (not for the proofs) that lead to the Universality Theorems for oriented matroids and polytopes.

1.1. Oriented matroids. Oriented matroids are combinatorial models of vector configurations in the real linear space $\mathbb{R}^n$. An oriented matroid contains all relevant data about the relative positions of vectors in a configuration. There are two kinds of oriented matroids: *realizable* oriented matroids (that come from concrete vector configurations), and *non-realizable* oriented matroids (which can be considered as combinatorial or topological generalizations of point configurations).

1991 *Mathematics Subject Classification.* Primary 52B40; Secondary 14P10, 51A25, 52B30.
Key words and phrases. Oriented matroids, polytopes, realization spaces, universality, semi-algebraic sets.
Supported by the DFG Gerhard-Hess-Forschungsförderungspreis awarded to G.M. Ziegler.

© 1999 American Mathematical Society

Here we will be concerned with the realizable case only. We use "+" and "−" as shorthand for "+1" and "−1."

The oriented matroid of a rank d vector configuration $\boldsymbol{V} = (\boldsymbol{v}_1, \ldots, \boldsymbol{v}_n) \in \mathbb{R}^{d \cdot n}$ is the pair $\mathcal{M}_{\boldsymbol{V}} := (\{1, \ldots, n\}, \mathcal{L}_{\boldsymbol{V}})$, with

$$\mathcal{L}_{\boldsymbol{V}} = \left\{ (\text{sign}\langle \boldsymbol{v}_1, \boldsymbol{y} \rangle, \ldots, \text{sign}\langle \boldsymbol{v}_n, \boldsymbol{y} \rangle) \mid \boldsymbol{y} \in \mathbb{R}^d \right\} \subseteq \{-, 0, +\}^n.$$

The elements of $\mathcal{L}_{\boldsymbol{V}}$ are the *covectors* of $\boldsymbol{V}$. Each single covector is a sign vector that encodes how an oriented linear hyperplane

$$H(\boldsymbol{y}) := \{\boldsymbol{x} \in \mathbb{R}^d \mid \langle \boldsymbol{x}, \boldsymbol{y} \rangle = 0\}$$

partitions the vectors in $\boldsymbol{V}$ (with the exception that $H(0, \ldots, 0)$ is the entire space, and not a hyperplane). The vectors *on* $H(\boldsymbol{y})$ are marked "0," those vectors on the *positive* side of $H(\boldsymbol{y})$ are marked "+", and those vectors on the *negative* side of $H(\boldsymbol{y})$ are marked "−." The set $\mathcal{L}_{\boldsymbol{V}}$ gives the complete collection of all such partitions. In general, an oriented matroid is a pair $\mathcal{M} := (\{1, \ldots, n\}, \mathcal{L})$, with $\mathcal{L} \subseteq \{-, 0, +\}^n$ that satisfies a system of certain axioms, which model the combinatorial behavior of hyperplane partitions (see [**BLSWZ, BS**]). The partial order on $\mathcal{L}$ that is induced by the order relations "$- \prec 0$" and "$+ \prec 0$" becomes a lattice $(\mathcal{L}, \prec)$, where we add an artificial maximal element.

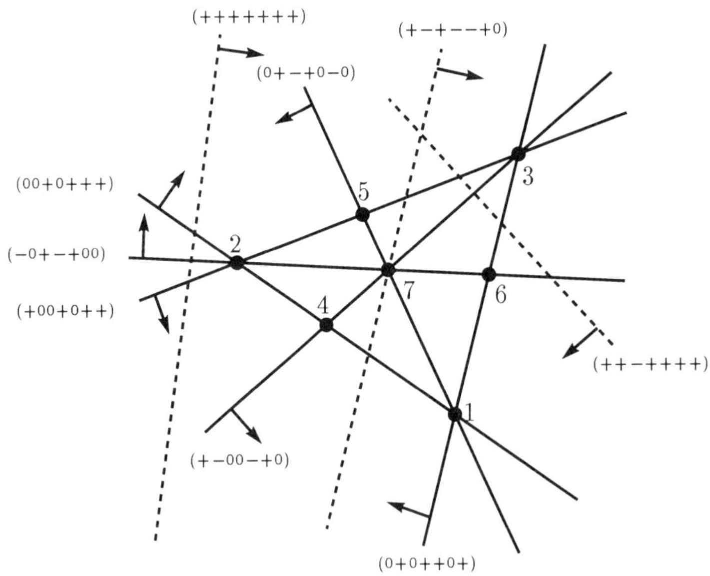

FIGURE 1: Affine picture of an oriented matroid.

We will often use vector configurations in order to represent *affine point configurations*. For planar affine configurations this is done by the usual embedding into the $(z = 1)$-plane. A point (x, y) in the affine plane is then associated with a vector $(x, y, 1)$. The covectors of the affine point configuration are obtained as the partitions by oriented *affine* hyperplanes. Planar point configurations (with not all points on a line) correspond to rank 3 vector configurations. Figure 1 shows the affine picture of a planar configuration and some of the covectors.

An oriented matroid can—in the realizable case—be considered as the *combinatorial type* of the corresponding point configuration (in the same way as a face lattice is the combinatorial type of a polytope). In particular, the oriented matroid $\mathcal{M}_{\boldsymbol{V}}$ completely describes which subsets of $\boldsymbol{V}$ form a linear basis. In Figure 1 for instance $(1,2,3)$ forms a basis. A *realization* of $\mathcal{M}$ is a vector configuration $\boldsymbol{V}$ with $\mathcal{M}_{\boldsymbol{V}} = \mathcal{M}$. The *realization space* of $\mathcal{M}$ is the set of all its realizations considered as a topological space with the natural Hausdorff topology. However, we usually factor out linear transformations, which trivially operate on all realization spaces. Let $b = (1,\dots,d)$ be a basis of $\mathcal{M}$. The realization space of $\mathcal{M}$ w.r.t. b is the set

$$\mathcal{R}(\mathcal{M},b) := \Big\{V \in \mathbb{R}^{n\cdot d} \mid \mathcal{M} = \mathcal{M}_{\boldsymbol{V}} \text{ and } \boldsymbol{v}_i = \boldsymbol{e}_i \text{ for } i = 1,\dots,d\Big\}.$$

Here $\boldsymbol{e}_1,\dots,\boldsymbol{e}_d$ are the d unit vectors of $\mathbb{R}^d$. In other words, the realization space of an oriented matroid $\mathcal{M}$ is the set of all point configurations $\boldsymbol{V}$ with $\mathcal{M}_{\boldsymbol{V}} = \mathcal{M}$, modulo linear transformations. In fact, up to a rational change of coordinates, it does not matter which particular basis is chosen. Therefore it is possible to speak of *the* realization space of an oriented matroid.

In our example of Figure 1 the realization space with respect to the fixed basis $(1,2,3)$ is the set of all locations of points $4,\dots,7$ that produce the same set of covectors. If the position of point 7 is given, then the positions of $4,\dots,6$ are determined by collinearity conditions. Thus the realization space is described by all admissible choices of point 7. We get the right oriented matroid if 7 is chosen in the interior of the triangle $(1,2,3)$. Thus from a topological point of view the realization space is just an open disk.

1.2. Ringel's isotopy conjecture. One of the problems that initiated major research activities in the area of oriented matroids was the *Isotopy Conjecture* of G. Ringel that was posed in 1956 [**Ri**]. His "conjecture" (which originally was just a question) was stated in terms of planar line arrangements (which are — via polarity—equivalent to rank 3 vector configurations). A collection of lines $\boldsymbol{L}$ partitions the plane into a number of cells. The combinatorial type $\mathcal{L}(\boldsymbol{L})$ of a planar (and labeled) line arrangement $\boldsymbol{L}$ is the combinatorial structure of the cell complex that is associated with $\boldsymbol{L}$. This combinatorial type can be derived from (and is up to reorientation equivalent to) the oriented matroid of the point configuration that corresponds to $\boldsymbol{L}$ via polarity: The face lattice of this cell complex is isomorphic to the lattice $(\mathcal{L}_{\boldsymbol{V}}, \prec)$ of the corresponding vector configuration $\boldsymbol{V}$. In terms of line arrangements Ringel's Conjecture may be stated as follows.

CONJECTURE 1.1. Any two line arrangements of identical combinatorial type $\mathcal{L}$ can be continuously deformed into each other, such that every intermediate configuration is again a line arrangement of the same combinatorial type $\mathcal{L}$.
This conjecture can be easily rephrased in terms of rank 3 oriented matroids:

CONJECTURE 1.2. The realization spaces of realizable rank 3 oriented matroids are path connected.

In other words, this conjecture asks the following: Given two rank 3 vector configurations $\boldsymbol{V} \in \mathbb{R}^{3n}$ and $\boldsymbol{W} \in \mathbb{R}^{3n}$ with the same oriented matroid $\mathcal{M} = \mathcal{M}_{\boldsymbol{V}} = \mathcal{M}_{\boldsymbol{W}}$. Is there a continuous function $f : [0,1] \to \mathbb{R}^{3n}$ such that $f(0) = \boldsymbol{V}$, $f(1) = \boldsymbol{W}$ and $\mathcal{M}_{f(t)} = \mathcal{M}$ for all $t \in [0,1]$. It is also easy to derive an affine version that asks for the isotopy of planar point configurations.

The isotopy conjecture was shown to fail heavily: In 1986 N. E. Mnëv proved that realization spaces of rank 3 oriented matroids may essentially have the homotopy type of any finite simplicial complex [**M**]. In particular, they may have arbitrarily many connected components. In the same series of articles (a collection of papers from the Rohlin Seminar in St. Petersburg) where Mnëv's general result was published, K. Suvorov [**Su**] presented a *small* example consisting of only 14 points that counters the Isotopy Conjecture. We here give another such example (found by the author [**R2**]). It consists also of 14 points but compared to Suvorov's example it has the additional properties of being symmetric and constructible. A picture of this configuration is drawn in Figure 2 (on the right).

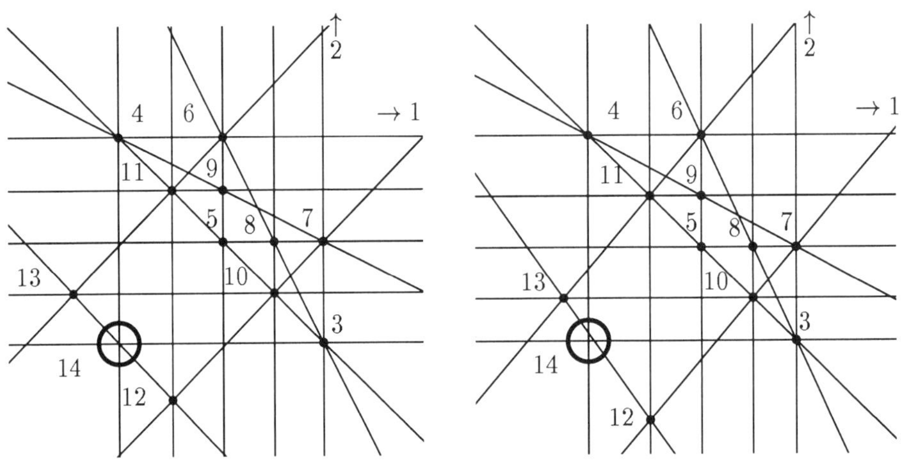

FIGURE 2: A point configuration that violates the isotopy property.

Two of the points (namely 1 and 2) are chosen at infinity on the x-axis and on the y-axis. Then two points 3 and 4 are chosen such that they are not collinear with 1 or 2. The four points $1 \ldots 4$ form a projective basis for the configuration. Point 5 is chosen on the segment that joins 3 and 4. This point is located close to the middle of this segment—but not at the middle! The points $6 \ldots 14$ (in this order) are determined uniquely as the intersections of lines through previously constructed points. In particular, point 14 is the intersection of the lines $(1,3)$ and $(2,4)$. Call the resulting oriented matroid $\mathcal{M}_{\mathrm{R}}$. After the choice of the projective basis $1, \ldots, 4$ the only free parameter in the construction is the position of point 5 on the line $3 \vee 4$. If this point is chosen exactly in the middle of 3 and 4, then the three points $12, 13, 14$ are collinear (Figure 2 on the left). At every position close to the middle these three points are oriented counterclockwise. Thus the realization space of $\mathcal{M}_{\mathrm{R}}$ is a disconnected set: an open interval with one point deleted.

The oriented matroid $\mathcal{M}_{\mathrm{R}}$ also represents a smallest known example of an oriented matroid that has a combinatorial automorphism that cannot be realized geometrically.

1.3. Mnëv's universality theorem. The results of Mnëv concerning realization spaces of oriented matroids are by far deeper than the construction of sporadic examples with complicated realization spaces. One could say that with respect to any "reasonable complexity measure" related to the realizability problem rank 3 oriented matroids behave as complicated as possible. We just mention a few of these measures.

- The realizability problem is as complicated as solving arbitrary systems of polynomial equations and inequalities.
- Realization spaces of oriented matroids may have the homotopy type of an arbitrary finite simplicial complex.
- All algebraic numbers are needed to realize all (realizable) oriented matroids.

The crucial point in the proof of such theorems is to find a technique that translates the problem of *solving a system of polynomial equations and inequalities* into the realizability problem of a certain oriented matroid.

The proofs of the following three versions of the Universality Theorem for oriented matroids need increasing technical effort. However, the underlying construction that gives the desired results is essentially the same for all three versions. We will later on explain the main ingredients of this construction.

THEOREM 1. *(Universality for oriented matroids)*

(i) *There is a polynomial algorithm that takes as input a system Ω of polynomial equations and strict inequalities with integer coefficients and produces an oriented matroid $\mathcal{M}(\Omega)$ such that the realizability problem for $\mathcal{M}(\Omega)$ is equivalent to the solvability problem of Ω.*

(ii) *For every basic primary semialgebraic set V defined over $\mathbb{Z}$ there is an oriented matroid $\mathcal{M}$ whose realization space is homotopy equivalent to V.*

(iii) *For every basic primary semialgebraic set V defined over $\mathbb{Z}$ there is an oriented matroid $\mathcal{M}$ whose realization space is stably equivalent to V.*

We will present a sketch of a proof of the simplest version (i) later in Section 2. The transition from version (i) to version (ii) can be made by the observation that every realization of $\mathcal{M}(\Omega)$ corresponds to a solution of Ω, and conversely, every solution of Ω corresponds to a contractible set of realizations of $\mathcal{M}(\Omega)$. Continuity arguments prove that the the realization space of $\mathcal{M}(\Omega)$ form the total space of a trivial fibration of the solution space of Ω.

The technically hardest version is to prove that the solution space of Ω and the realization space of $\mathcal{M}(\Omega)$ are indeed stably equivalent. Stable equivalence is a very strong and restrictive concept of homotopy equivalence that makes also statements about the algebraic nature of the equivalence relation. We will clarify this in the next section.

1.4. What is "stable equivalence?" Stable equivalence is an equivalence relation on semialgebraic sets. The idea behind the concept of stable equivalence is that semialgebraic sets that only differ by a "trivial fibration" and a rational change of coordinates should be considered as stably equivalent, while semialgebraic sets that differ in certain "characteristic properties" should not turn out to be stably equivalent. In particular, stable equivalence should preserve the homotopy type, and respect the algebraic complexity and the singularity structure.

The concept of *stable equivalence* has been used by different authors. However, the precise definitions they used (see [**G, M, S**]) vary substantially in their technical content. We here use a version that is stronger than previously used variants.

To be more specific we start with the notion of a basic semialgebraic set. Let $\Omega = (\{f_i\}_{0<i\leq r}, \{g_i\}_{0<i\leq s}, \{h_i\}_{0<i\leq t})$ be a finite collection of polynomials

$$f_1,\dots,f_r,g_1,\dots,g_s,h_1,\dots,h_t \in \mathbb{Z}[x_1,\dots,x_n]$$

with integer coefficients. The *basic semialgebraic set* $V(\Omega) \in \mathbb{R}^n$ is the set

$$V = V(\Omega) := \left\{ \boldsymbol{x} \in \mathbb{R}^n \mid \begin{array}{l} f_i(\boldsymbol{x}) = 0 \text{ for } i = 1,\dots,r \\ g_i(\boldsymbol{x}) < 0 \text{ for } i = 1,\dots,s \\ h_i(\boldsymbol{x}) \leq 0 \text{ for } i = 1,\dots,t \end{array} \right\}$$

defined as the solution of a finite number of polynomial equations and polynomial inequalities. A basic semialgebraic set is called *primary*, if the defining equations contain no non-strict inequalities (i.e., $t = 0$ in the above notion).

Let $V \subseteq \mathbb{R}^n$ and $W \subseteq \mathbb{R}^{n+d}$ be basic semialgebraic sets with $\pi(W) = V$, where π is the canonical projection $\pi\colon \mathbb{R}^{n+d} \to \mathbb{R}^n$ that deletes the last d coordinates. V is a *stable projection* of W if W has the form

$$W = \left\{(\boldsymbol{v},\boldsymbol{v}') \in \mathbb{R}^{n+d} \mid \boldsymbol{v} \in V \text{ and } \phi_i^{\boldsymbol{v}}(\boldsymbol{v}') > 0;\ \psi_j^{\boldsymbol{v}}(\boldsymbol{v}') = 0 \text{ for all } i \in X;\ j \in Y\right\}.$$

Here X and Y denote finite (possibly empty) index sets. For $i \in X$ and $j \in Y$ the functions $\phi_i^{\boldsymbol{v}}$ and $\psi_j^{\boldsymbol{v}}$ are affine functionals whose parameters depend polynomially on V. Thus we have

$$\phi_i^{\boldsymbol{v}}(\boldsymbol{v}') = \langle(\phi_i^1(\boldsymbol{v}),\dots,\phi_i^d(\boldsymbol{v}))^T, \boldsymbol{v}'\rangle + \phi_i^{d+1}(\boldsymbol{v})$$

$$\psi_j^{\boldsymbol{v}}(\boldsymbol{v}') = \langle(\psi_j^1(\boldsymbol{v}),\dots,\psi_j^d(\boldsymbol{v}))^T, \boldsymbol{v}'\rangle + \psi_j^{d+1}(\boldsymbol{v})$$

with polynomial functions $\phi_i^1(\boldsymbol{v}),\dots,\phi_i^{d+1}(\boldsymbol{v})$ and $\psi_j^1(\boldsymbol{v}),\dots,\psi_j^{d+1}(\boldsymbol{v})$.

If V is a stable projection of W, then all the fibers $\pi^{-1}(\boldsymbol{v})$ are (non-empty) relative interiors of *polyhedra* (i.e., sets that are obtained by intersecting a finite number of open halfspaces and hyperplanes). In particular, if the sets X and Y are empty we get $W = V \times \mathbb{R}^d$. If the functionals ϕ_i and ψ_i are constant and V is the interior of a convex polytope then W is itself the interior of a polyhedral set, that projects onto V.

Two basic semialgebraic sets V and W are *rationally equivalent* if there exists a homeomorphism $f\colon V \to W$ such that both functions f and f^{-1} are rational functions (with rational coefficients). We may consider a rational equivalence as a kind of "reparametrization" of the set.

DEFINITION 1.3. Two basic semialgebraic sets V and W are *stably equivalent* if they lie in the equivalence class generated by stable projections and rational equivalence. We then write $V \approx W$.

The basic properties of stable equivalence are summarized in the following theorem.

THEOREM 2. *Let $V \subseteq \mathbb{R}^n$ and $W \subseteq \mathbb{R}^m$ be a pair of stably equivalent semialgebraic sets and let A be a subfield of the algebraic numbers of characteristic zero. We have*

(i) *V and W are homotopy equivalent.*

(ii) *V and W have similar singularity structure.*

(iii) *$V \cap A^n = \emptyset \equiv W \cap A^m = \emptyset$.*

The situation in Figure 3 serves as a motivating example for the definition of stable projections. We consider two different configuration spaces:

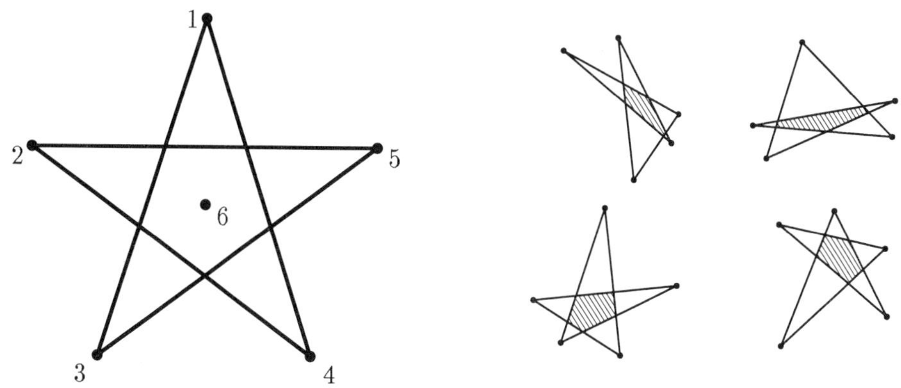

FIGURE 3: The paradigm for stable projections.

$$R_1 = \Big\{(\boldsymbol{p}_1, \dots, \boldsymbol{p}_5) \in \mathbb{R}^{2\cdot 5} \quad \Big| \quad (\boldsymbol{p}_1, \dots, \boldsymbol{p}_5) \text{ in this order forms a convex pentagon}\Big\},$$

$$R_2 = \Big\{(\boldsymbol{p}_1, \dots, \boldsymbol{p}_5, \boldsymbol{p}_6) \in \mathbb{R}^{2\cdot 6} \quad \Big| \quad (\boldsymbol{p}_1, \dots, \boldsymbol{p}_5) \in R_1 \text{ and } \boldsymbol{p}_6 \text{ is in the central cell of the pentagram}\Big\}.$$

The space R_1 is a stable projection of R_2. No matter how we realize the exterior pentagon, the region in which $\boldsymbol{p}_6$ can be properly inserted never vanishes. Moreover this region is bounded by affine hyperplanes whose parameters depend polynomially on the parameters of the exterior pentagon (Figure 3 right).

2. Universality for oriented matroids

Here we present a very short sketch of a proof of Mnëv's Universality Theorem (as stated in Theorem 1.2 version(i)). We will only concentrate on the main ideas and neglect technical difficulties. For this we follow a beautiful treatment of this issue that was presented by P. Shor [**S**].

What is the aim of our proof? We start with a polynomial inequality system Ω:

$$f_1(\boldsymbol{x}) = 0, \dots, f_r(\boldsymbol{x}) = 0, f_{r+1}(\boldsymbol{x}) > 0, \dots, f_s(\boldsymbol{x}) > 0.$$

Here $f_1, \dots, f_s$ are polynomials in m variables $\boldsymbol{x} = (x_1, \dots, x_m)$ taken from the ring $\mathbb{Z}[x_1, \dots, x_m]$. Each of these polynomials can be generated from the variables $x_1, \dots, x_m$ and the unit 1 by a finite sequence of elementary additions and multiplications. For instance

$$4x_1x_2 + 2x_1^2 = ((((1+1)+(1+1)) \cdot x_1) \cdot x_2) + (((1+1) \cdot x_1) \cdot x_1).$$

We consider the minimal number of such elementary operations that are needed to represent f as the coding length of f. The sum of the coding lengths of $f_1, \ldots, f_s$ is the coding length of Ω.

Now we want to construct an oriented matroid $\mathcal{M}(\Omega)$ that is realizable if and only if Ω has a solution. (To obtain also the topological version of the Universality Theorem it must be possible to show that the solution space of Ω and the realization space of $\mathcal{M}(\Omega)$ are stably equivalent.) The construction mainly consists of two parts:

- In a first step the system Ω is transferred into an equivalent inequality system Ω' in m' variables. Ω' should have the following properties:
  - In the whole solution space of Ω' the variables $x_1, \ldots, x_{m'}$ together with all intermediate results that are needed during the computation of the polynomials of Ω' have a *fixed linear order.*
  - The solution space of Ω is a trivial fibration of the solution space of Ω'. (Thus in particular these two spaces are homotopy equivalent.)
- In a second step, the computation of the polynomials of Ω' is decomposed into elementary additions and multiplications. These elementary operations are then geometrically modeled by the classical von Staudt Constructions.

Every realization of the resulting point configuration corresponds to a solution of Ω' and thus to a solution of Ω. The "linear order property" of Ω' is used to ensure that the combinatorial type of the corresponding point configuration can be kept fixed over the whole solution space of Ω'. It can not be overemphasized that the main difficulty in the proof of the Universality Theorem for oriented matroids does *not* lie in the application of the von Staudt Constructions (as falsely mentioned by several authors). The problem is really to find an algebraic technique that allows one to get complete control on the ordering of the values of the variables (and the intermediate results) for every possible solution. There are four approaches known to solve this problem:

- Mnëv's approach as presented in [**M**] is based on *perturbation techniques* that are applied to the situation around the origin in a homogenized version of Ω. This approach has the disadvantage that it is only applicable to the case where Ω contains only inequalities. The general case was treated by Mnëv's PhD thesis. There he used substitution techniques similar to the ones introduced by Shor (see below).
- In Mnëv's treatment of the general case and in Shor's version of the proof *substitution techniques* were used. The idea is to first decompose Ω into elementary operations. Then one replaces every variable x_i by a variable $y_i = x_i + a$. If the new indeterminant a is chosen large enough, then all variables y_i are known to be larger than 1. The validity of the original equations and inequalities has to be preserved by adding a certain set of new elementary equations and new variables. In a third step each of the new variables y_i is replaced by a variable $z_i = y_i + b^i$. If b is large enough then all variables z_i are linearly ordered. Again by adding new equations and additional variables the validity of the original equations is guaranteed. One has to be very careful to ensure that a well-ordering holds also among the newly added variables.

- H. Günzel's proof of the *Universal Partition Theorem* (a generalization of the Universality Theorem that was claimed by Mnëv [**M2**]) bypasses the linear order problem [**G**]. He uses (on a geometric level) *individual projective scales* for all intermediate results in the computation of the original system Ω. Thus in a certain sense the necessary geometric dissection of points is done already geometrically. No algebraic reduction is needed. However, by this Günzel cannot completely control the topology of the resulting realization space. The realization space that is generated by this approach is essentially $N \times M$, where N is the solution space of Ω and M is a non-controllable smooth manifold.
- An approach that merges the ideas of Shor and Günzel was given by the author in [**R**]. The individual projective scales are all taken on one line. Separation is obtained geometrically. By this a proof of the Universality Theorem is given that can be used in a straight-forward way to also prove the Universal Partition Theorem in its sharpest form.

2.1. Shor's normal form. We will here use the final result of Shor's substitution procedure. He reduces the system Ω to a very simple standard form of a particular kind: *all variables are linearly ordered and only elementary additions and multiplications occur as equations.* The price that has to be paid for this reduction is that one has to introduce many new variables. A sketch of a non-constructive version of the following result is also implicit in Mnëv's original PhD thesis.

DEFINITION 2.1 A *Shor normal form* is a triple $\mathcal{S} = (n, \mathbf{A}, \mathbf{M})$ where $n \in \mathbb{N}$ and $\mathbf{A}, \mathbf{M} \in \{1, \dots, n\}^3$ such that for $(i, j, k) \in \mathbf{A} \cup \mathbf{M}$ we have $i \leq j < k$.

To every Shor normal form $\mathcal{S}$ we associate a semialgebraic set $V(\mathcal{S}) \in \mathbb{R}^n$ as the solution of the inequalities

$$1 < x_1 < x_2 < \dots < x_n$$

and the equations

$$x_i + x_j = x_k \quad \text{for } (i, j, k) \in \mathbf{A} \text{ and}$$

$$x_i \cdot x_j = x_k \quad \text{for } (i, j, k) \in \mathbf{M}.$$

THEOREM 3. (Shor [**S**].) *Let $W \subseteq \mathbb{R}^m$ be a primary basic semialgebraic set defined over $\mathbb{Z}$. Then there exists a Shor normal form $\mathcal{S}(W) = \mathcal{S} = (n, \mathbf{A}, \mathbf{M})$ such that the semialgebraic set $V(\mathcal{S}) \subseteq \mathbb{R}^n$ is stably equivalent to W. Furthermore,*

(i) *$\mathcal{S}$ can be computed from the defining relations of W in a time that is polynomially bounded in the coding length of the polynomials defining W,*

(ii) *there exists a polynomial function f such that $f(V(\mathcal{S}(W))) = W$.*

2.2. Von Staudt constructions. The use of von Staudt Constructions is a classical method to build a link between algebra and geometry. The idea behind these constructions is very simple. Consider the drawings in Figure 4. (Lines that seem to be parallel should *really* be parallel in these pictures)

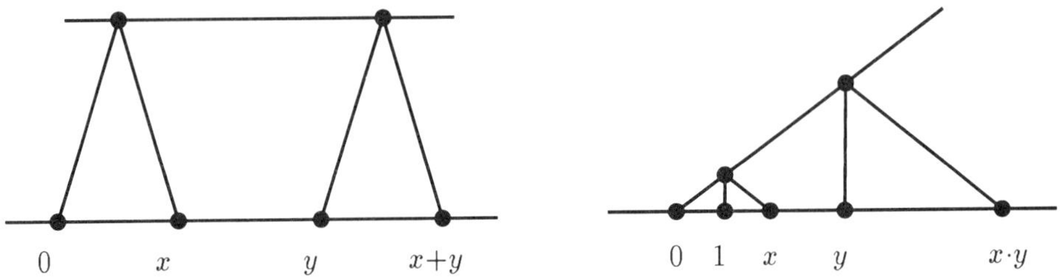

FIGURE 4: The classical von Staudt constructions.

The first picture shows how *addition* can be modeled geometrically. If an additive unit 0 is fixed for any two values x and y on the base line one can easily construct the position of the sum $x + y$. The second picture shows the corresponding construction for *multiplication*. We have to fix an additive unit 0 and a multiplicative unit 1. If x and y are given the construction shows how to obtain the position of $x \cdot y$.

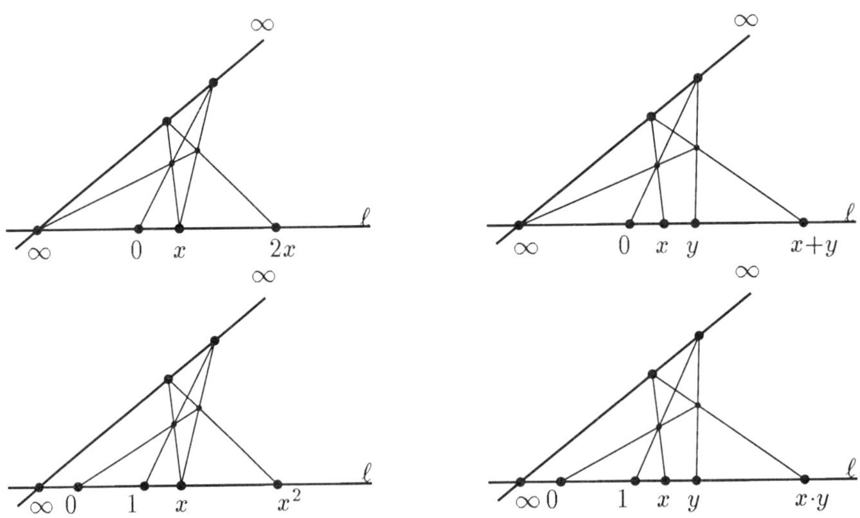

FIGURE 5: Von Staudt constructions in a projective setting.

A projective version of these "geometric calculations" is given in Figure 5. The line at infinity ∞ is here taken as a finite line. The parallelity of two lines a, b translates to the property that a, b, and ∞ have a point in common. The points 0, 1 and ∞ form a basis of a *projective scale* on the base line ℓ of the construction. The numerical value $\nu(x)$ of a variable x can be calculated as the cross ratio

$$\nu(x) = \frac{|\overrightarrow{x0}||\overrightarrow{1\infty}|}{|\overrightarrow{x\infty}||\overrightarrow{10}|}$$

Here $|\overrightarrow{ab}|$ is the directed length from a to b. All together we obtain four combinatorially different geometric situations for the cases $x+x$, $x+y$, $x \cdot x$, and $x \cdot y$. Each of them is given by adding four new points to the points that lie on the base line. Two of the new points (say p and q) lie on the line ∞. The two other points lie on lines that connect p or q to points on the base line ℓ. If p and q are chosen to be "close together" then all four new points are "close together."

2.3. Combining the pieces. We finally have to combine the concepts of Shor normal forms and the von Staudt constructions to get the construction that proves the Universality Theorem for oriented matroids. Again, the idea is very simple:

- We start with the defining system of polynomial equations and inequalities Ω.
- Via Theorem 2.2 we translate this into a corresponding Shor normal form $S(\Omega)$.
- For every element of the linear order $0 < 1 < x_1 < x_2 < \ldots < x_n < \infty$ of $S(\Omega)$ we take a point on the base line ℓ. We furthermore take a line ∞ through the point ∞ on ℓ
- For every elementary arithmetic operation of $S(\Omega)$ (i.e., for $x_i + x_j = x_k$ or for $x_i \cdot x_j = x_k$) we add successively the corresponding von Staudt Construction from Figure 5. The four points that are added for a particular elementary operation have to be "farther out" and "much closer together" than all the previously added points (by this the combinatorial type is stabilized).

This construction can be carried out on a purely combinatorial level. To every Shor normal form $S(\Omega)$ the whole procedure determines a *unique* oriented matroid $\mathcal{M}(\Omega) = \mathcal{M}(S(\Omega))$. This oriented matroid $\mathcal{M}(S(\Omega))$ is realizable if and only if the system of equations and inequalities in $S(\Omega)$ is solvable. After fixing the points 0, 1, and ∞ on ℓ every solution of the system $S(\Omega)$ corresponds to locations of the points on ℓ in one possible realization of $\mathcal{M}(S(\Omega))$. Conversely, every realization of $\mathcal{M}(S(\Omega))$ corresponds to a solution of $S(\Omega)$. This proves that the problem of solving Ω is polynomial time translatable into the problem of finding a realization of $\mathcal{M}(\Omega)$. A more detailed and technical analysis proves that the resulting realization space is indeed stably equivalent to the solution space of S (and thus to the solution space of Ω). In particular if Ω has no solutions at all then $\mathcal{M}(\Omega)$ is not realizable.

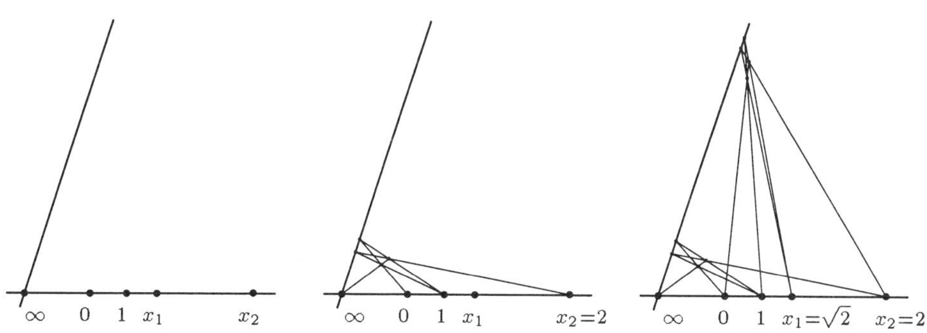

FIGURE 6: A point configuration for $x_1 = \sqrt{2}$.

Figure 6 illustrates the whole construction for a very simple Shor normal form:

$$0 < 1 < x_1 < x_2, \quad x_2 = 1 + 1, \quad x_2 = x_1 \cdot x_1.$$

This system has the unique (non-rational!) solution $x_1 = \sqrt{2}$, $x_2 = 2$. In a first step the linear order is set up. Then a construction for $x_2 = 1 + 1$ is added. Finally, a construction for $x_2 = x_1 \cdot x_1$ is added. Here the newly added four points are chosen "far out" along ∞ and "close together."

3. Universality for polytopes

After the Universality Theorem for oriented matroids was established in 1986, one of the big open problems was how to transfer this result also to other structures in combinatorial geometry. Already in his original paper [**M**] Mnëv gave an application to the theory of polytopes. He proved that for every basic primary semialgebraic set V there is a dimension d and a d-polytope with just $d + 4$ vertices whose realization space is stably equivalent to V. Recently, the author of the present paper proved a similar theorem for 4-dimensional polytopes. It is the purpose of this section to sketch the basic techniques that are needed for these constructions. We will also demonstrate how it is possible to prove a Universality Theorem for 6-dimensional polytopes using of Mnëv's construction for oriented matroids.

3.1. Realization spaces of polytopes. A d-dimensional *polytope* $\boldsymbol{P}$ (d-polytope for short) is the convex hull of a spanning set of points in $\mathbb{R}^d$. A *face* of $\boldsymbol{P}$ is the intersection of $\boldsymbol{P}$ with an affine hyperplane that does not meet the interior of $\boldsymbol{P}$. Faces of dimension $d - 1$ are the *facets* of $\boldsymbol{P}$. One-dimensional faces are the *edges* of $\boldsymbol{P}$. Zero-dimensional faces are the *vertices* of $\boldsymbol{P}$. A polytope is completely determined by the position of its vertices. Thus we may identify $\boldsymbol{P}$ with the point configuration given by its vertex set $(\boldsymbol{p}_1, \dots, \boldsymbol{p}_n)$. The set of all faces of $\boldsymbol{P}$ ordered by inclusion is the *face lattice* of $\boldsymbol{P}$ (here we have to add the empty set and $\boldsymbol{P}$ itself in order to actually obtain a lattice). The face lattice of $\boldsymbol{P}$ plays the role of the combinatorial type of $\boldsymbol{P}$. A *realization* of $\boldsymbol{P}$ is a polytope $\boldsymbol{Q}$ with the same (labeled) face lattice as $\boldsymbol{P}$.

As in the case of oriented matroids it is natural to ask for the *realization space* of a given combinatorial type (i.e., for the set of all geometric polytopes of this combinatorial type). Again we factor out trivial components of the realization space by fixing a suitable basis. The set $b = (0, 1, \dots, d)$ is an *affine basis* of $\boldsymbol{P}$ if the vertex set $(0, \dots, d)$ is affinely independent in every realization of $\boldsymbol{P}$. (For instance, we get a basis of a 3-dimensional polytope by taking a vertex 0 and three vertices 1, 2, 3 that are adjacent to 0 along edges.) For a d-polytope $\boldsymbol{P}$ with n vertices and a basis $(0, \dots, d)$ we formally define the realization space as:

$$\Big\{\boldsymbol{Q} = (\boldsymbol{e}_0, \boldsymbol{e}_1, \dots, \boldsymbol{e}_d, \boldsymbol{q}_{d+1}, \dots, \boldsymbol{q}_{n-1}) \mid \boldsymbol{Q} \text{ is a realization of } \boldsymbol{P}\Big\}.$$

Here $\boldsymbol{e}_0$ is the origin $(0, \dots, 0)$ and $\boldsymbol{e}_1, \dots, \boldsymbol{e}_d$ are the d unit vectors of $\mathbb{R}^d$.

It is a classical result (but none the less a very remarkable fact) that realization spaces of 3-dimensional polytopes are trivial topological sets. This follows by careful inspection into the original proof of Steinitz's Theorem [**St, SR**]. Steinitz's Theorem states that the edge graphs of 3-dimensional polytopes are exactly the simple, planar and 3-connected graphs (see Figure 7). The classical proof of Steinitz constructs a 3-polytope with prescribed edge graph G by starting with a tetrahedron and iteratively adding new vertices and facets. The parameters of each newly added element can always be chosen in an open and contractible region. This implies the results on realization spaces.

A similarly nice result cannot be expected for polytopes in general dimension. We already mentioned that Mnëv proved a Universality Theorem for d-polytopes with $d+4$ vertices. However, also for small dimensions several sporadic examples were known that contrast with the nice behavior in the 3-dimensional case. We just mention the famous example of Bokowski, Ewald and Kleinschmidt: a simplicial 4-polytope with only 10 vertices whose realization space is disconnected [**BEK**].

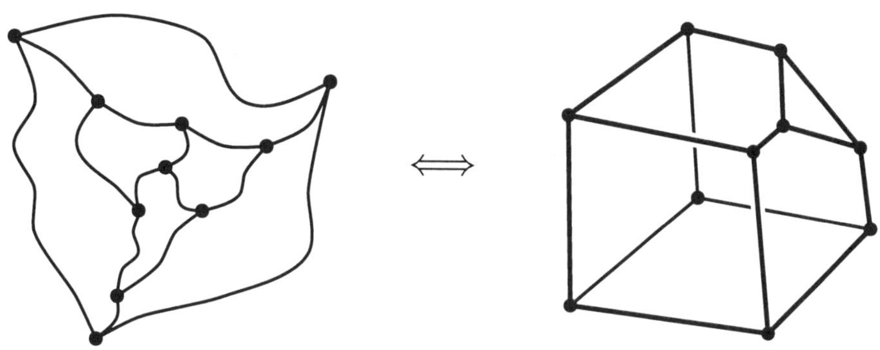

FIGURE 7: The essence of Steinitz's Theorem.

We now will sketch the general constructions that explain all these interesting effects and finally lead to a Universality Theorem already for 4-dimensional polytopes (and thus for polytopes in all dimensions greater than 3).

3.2. Two fundamental constructions. The construction of all known sporadic examples with interesting realization properties, as well as the Universality Theorem for d-polytopes with $d+4$ vertices, as well as the Universality Theorem for 4-polytopes can be traced back to the application of two innocent looking basic constructions: *Lawrence extensions* and *connected sums*. We here will briefly sketch these two constructions.

Lawrence extensions are a tool that transfers incidence information of point configurations to the face lattices of polytopes (compare [**BM**]). A basic Lawrence extension can be described as follows.

- Start with a d-polytope $\boldsymbol{P} = \mathbf{conv}(\boldsymbol{p}_1, \dots, \boldsymbol{p}_n)$—embedded in an affine hyperplane H in $\mathbb{R}^{d+1}$—and with a point $\boldsymbol{q}$ on H outside $\boldsymbol{P}$.
- Take a line ℓ that is transversal to H and meets $\boldsymbol{q}$. On ℓ take new points $\overline{\boldsymbol{q}}$ and $\underline{\boldsymbol{q}}$.
- The convex hull of $\overline{\boldsymbol{q}}$, $\underline{\boldsymbol{q}}$, and $\boldsymbol{P}$ is the Lawrence extension $\boldsymbol{\Lambda}(\boldsymbol{P}, \boldsymbol{q})$.

Every Lawrence extension increases the dimension by one and the number of vertices by two (compare Figure 8). The main properties of Lawrence extensions can be easily described in terms of *shadow boundaries*. The shadow boundary $\mathbf{shadow}(\boldsymbol{P}, \boldsymbol{q})$ of a polytope $\boldsymbol{P}$ as it is seen from a point $\boldsymbol{q}$ (outside of $\boldsymbol{P}$) is the set of all faces of $\boldsymbol{P}$ that contain $\boldsymbol{q}$ in all their supporting hyperplanes. Intuitively speaking, these are all faces of $\boldsymbol{P}$ that are entirely projected to the boundary of the shadow of $\boldsymbol{P}$ by a lamp located at $\boldsymbol{q}$. The corresponding shadow boundary of the polytope in Figure 8 consists of the two edges supported by l_1 and l_2, and the vertices that bound these edges. The Lawrence extension $\boldsymbol{\Lambda}(\boldsymbol{P}, \boldsymbol{q})$ has the following properties:

- $\boldsymbol{\Lambda}(\boldsymbol{P}, \boldsymbol{q})$ is a polytope.
- $\boldsymbol{P}$ is a facet of $\boldsymbol{\Lambda}(\boldsymbol{P}, \boldsymbol{q})$.
- $(\overline{\boldsymbol{q}}, \underline{\boldsymbol{q}})$ is an edge of $\boldsymbol{\Lambda}(\boldsymbol{P}, \boldsymbol{q})$.
- For every realization $\boldsymbol{\Lambda}' = \mathbf{conv}(\overline{\boldsymbol{q}}', \underline{\boldsymbol{q}}', \boldsymbol{P}')$ of $\boldsymbol{\Lambda}(\boldsymbol{P}, \boldsymbol{q})$ there exists a point $\boldsymbol{q}'$ in the supporting hyperplane of $\boldsymbol{P}'$ such that $\mathbf{shadow}(\boldsymbol{P}', \boldsymbol{q}')$ and $\mathbf{shadow}(\boldsymbol{P}, \boldsymbol{q})$ are combinatorially isomorphic.

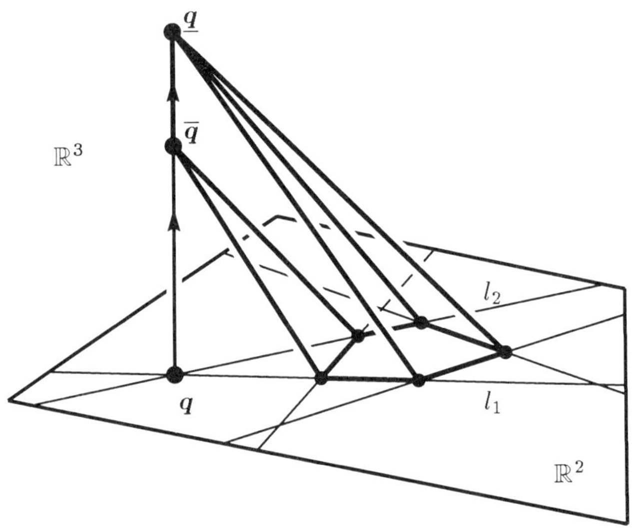

FIGURE 8: A Lawrence extension of a pentagon.

Lawrence extensions can be nicely used to generate polytopes with *non-prescribable faces*. We exemplify this by the well known Kleinschmidt polytope [**K**]. We start with an octahedron $\mathcal{O}$. However, we realize $\mathcal{O}$ in a way in which the path that is darkened in Figure 9 (left) forms the shadow boundary that is seen from a point $\boldsymbol{q}$ at (or close to) infinity. A projection (seen from $\boldsymbol{q}$) of such a realization is shown in Figure 9 (right). In particular, such a realization *cannot* be the regular octahedron: the four points 1, 2, 3, and 4 are never coplanar—their oriented volume is always positive. Now we consider the Lawrence extension $\boldsymbol{P}_{\mathrm{K}} = \boldsymbol{\Lambda}(\mathcal{O}, \boldsymbol{q})$ (the Kleinschmidt polytope). This polytope is a 4-polytope with 8 vertices that contains an octahedral facet. However, no matter how we realize $\boldsymbol{P}_{\mathrm{K}}$ the octahedral facet cannot be regular (since we can reconstruct a point for which the shadow boundary is the darkened path). In particular, again 1, 2, 3, and 4 are never coplanar.

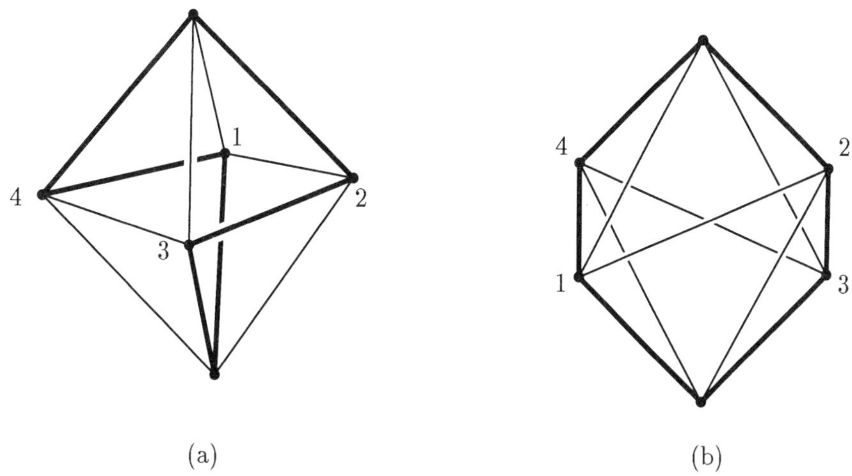

FIGURE 9: Construction of Kleinschmidt's polytope.

The second fundamental construction is the *connected sum* operation. This is an operation for gluing two polytopes of the same dimension without increasing the dimension. Such an operation is necessary if we want to stay within the realm of a fixed dimension. For this we simply start with two polytopes $\boldsymbol{P}$ and $\boldsymbol{Q}$ that contain facets $F_{\boldsymbol{P}}$ and $F_{\boldsymbol{Q}}$ which are projectively equivalent. In particular $F_{\boldsymbol{P}}$ and $F_{\boldsymbol{Q}}$ both have identical combinatorial type F. We apply a projective transformation τ to $\boldsymbol{Q}$ such that $\boldsymbol{P} \cup \tau(\boldsymbol{Q})$ is a polytope that identifies $\boldsymbol{P}$ and $\tau(\boldsymbol{Q})$ along $F_{\boldsymbol{P}}$ and $F_{\boldsymbol{Q}}$. The resulting polytope is the connected sum $\boldsymbol{R} = \boldsymbol{P} \#_F \boldsymbol{Q}$ (compare Figure 10). Observe that the complete operation makes sense also on a purely combinatorial level. The facets of the connected sum $\boldsymbol{P} \#_F \boldsymbol{Q}$ are exactly all facets of $\boldsymbol{P}$ and of $\boldsymbol{Q}$ except $F_{\boldsymbol{P}}$ and $F_{\boldsymbol{Q}}$.

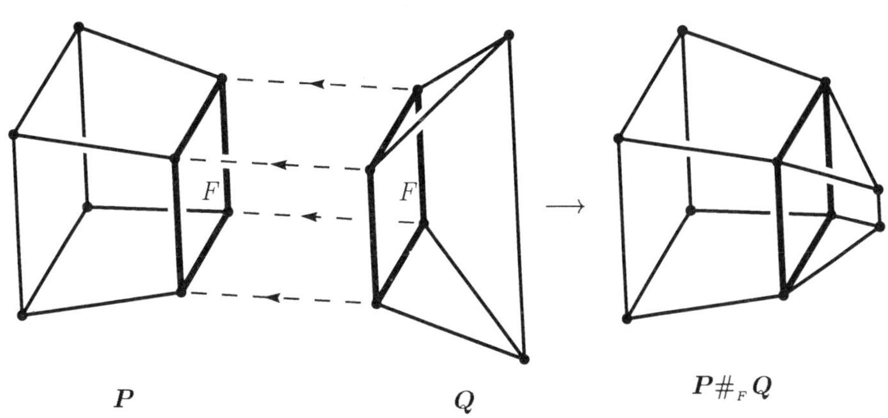

FIGURE 10: Connected sum of a cube and a triangular prism.

It is *not* always possible to take an arbitrary realization of $\boldsymbol{R}$ and cut it by a hyperplane into two pieces that are combinatorially isomorphic to $\boldsymbol{P}$ and $\boldsymbol{Q}$. For instance, in a realization of the connected sum of Figure 10 the four points of the quadrangle along which the gluing is performed must not necessarily stay coplanar. This is only the case if the facet along which the gluing is performed is *necessarily flat*. A d-polytope is necessarily flat if every embedding of its $(d-1)$-skeleton in $\mathbb{R}^{d+1}$ has affine dimension d. For instance triangles are the only necessarily flat 2-polytopes. Prisms and pyramids are necessarily flat in any dimension. Now, the crucial point for realization spaces is the following: Let $\boldsymbol{P}$ and $\boldsymbol{Q}$ be polytopes where the shape of the facets $F_{\boldsymbol{P}} \approx F$ and $F_{\boldsymbol{Q}} \approx F$ cannot be arbitrary (as the octahedron in $\boldsymbol{P}_{\mathrm{K}}$) and where these facets are necessarily flat. Every realization of the sum $\boldsymbol{R}$ can always be dissected into two pieces combinatorially isomorphic to $\boldsymbol{P}$ and $\boldsymbol{Q}$. Thus $\boldsymbol{R}$ must at the same time satisfy both requirements on the shape of F (those coming from $\boldsymbol{P}$, and those coming from $\boldsymbol{Q}$). The realization space of $\boldsymbol{R}$ may be more complicated than those of $\boldsymbol{P}$ and $\boldsymbol{Q}$. In other words, connected sums may be used to superpose non-trivial obstructions in a realization space. For instance the realization spaces of $\boldsymbol{P}$ and $\boldsymbol{Q}$ may be connected, while the realization space of $\boldsymbol{R}$ is not. In all approaches to a universality theorem for polytopes in fixed dimension which are known so far, this trick is used to create complicated realization spaces out of elementary obstructions. (For instance connected sums are used to superpose the elementary arithmetic operations from Shor's normal form.)

As an example for the power of connected sums we now use them to explain the construction of the famous Bokowski, Ewald, Kleinschmidt polytope $\boldsymbol{P}_{\mathrm{BEK}}$ with disconnected realization space. For this we simply glue the polytope $\boldsymbol{P}_{\mathrm{K}}$ (with the non-prescribable octahedral facet $\mathcal{O}$) and its mirror image $-\boldsymbol{P}_{\mathrm{K}}$. The polytope $\boldsymbol{P}_{\mathrm{BEK}}$ is the connected sum of these two polytopes along the facet $\mathcal{O}$. If the facet $\mathcal{O}$ was necessarily flat (which it isn't) then $\boldsymbol{P}_{\mathrm{BEK}}$ would not be realizable at all. This can be seen as follows: if the points of $\mathcal{O}$ in a realization $\boldsymbol{P}'$ of $\boldsymbol{P}_{\mathrm{BEK}}$ lay in an affine 3-space then we could use this 3-space to cut $\boldsymbol{P}'$ into two pieces which are realizations of $\boldsymbol{P}_{\mathrm{K}}$ and $-\boldsymbol{P}_{\mathrm{K}}$, respectively. However, this would force the orientation of $(1,2,3,4)$ to be positive and negative at the same time, which is impossible.

In order to prove the non-connectedness of the realization space of $\boldsymbol{P}_{\mathrm{BEK}}$ take any realization $\boldsymbol{P}'$ of $\boldsymbol{P}_{\mathrm{BEK}}$ (it is realizable, and this is a non-trivial fact!). In this realization $(1,2,3,4)$ has a certain orientation. The mirror image $-\boldsymbol{P}'$ of $\boldsymbol{P}$ is a realization of $\boldsymbol{P}_{\mathrm{BEK}}$ as well. In $-\boldsymbol{P}'$ the points $(1,2,3,4)$ have opposite orientation as in $-\boldsymbol{P}'$. Thus $\boldsymbol{P}'$ cannot be continuously deformed into $-\boldsymbol{P}'$ since during the deformation the points $(1,2,3,4)$ had to be coplanar at some point, which is impossible.

In a sense the polytope $\boldsymbol{P}_{\mathrm{BEK}}$ is very similar to the oriented matroid $\mathcal{M}_{\mathrm{R}}$ of Section 2. Both objects have a combinatorial symmetry which cannot be metrically realized. This is the deep reason for the non-connectedness of their realization spaces.

3.3. Polytopes with $d+4$ vertices. Very briefly we sketch the construction of Mnëv for a Universality Theorem for d-polytopes with $d+4$ vertices. The construction is based on the fact that, via Lawrence extensions, the structure (and the realization properties) of point configurations can be completely transferred to the world of polytopes of sufficiently high dimension. The idea is to take an arbitrary set of polynomial equations and inequalities Ω. One starts with the corresponding rank 3 oriented matroid $\mathcal{M}(\Omega)$, and uses Lawrence extensions to translate $\mathcal{M}(\Omega)$ into a polytope with similar realization properties. For this we have to learn what it means to do a succession of Lawrence extensions. We iteratively define

$$\begin{aligned}\boldsymbol{\Lambda}(\boldsymbol{P},\emptyset) &= \boldsymbol{P},\\ \boldsymbol{\Lambda}(\boldsymbol{P},\{\boldsymbol{q}_1,\dots,\boldsymbol{q}_n\}) &= \boldsymbol{\Lambda}(\boldsymbol{\Lambda}(\boldsymbol{P},\{\boldsymbol{q}_1,\dots,\boldsymbol{q}_{n-1}\}),\boldsymbol{q}_n).\end{aligned}$$

During the iteration step the points $\{\boldsymbol{q}_1,\dots,\boldsymbol{q}_{n-1}\}$ are canonically embedded into the next higher dimension.

Let us assume that $\mathcal{M}(\Omega)$ has n points and that it has a realization $\boldsymbol{Q} = (\boldsymbol{q}_1,\dots,\boldsymbol{q}_n) \in \mathbb{R}^{3n}$. Naively, one could transfer $\mathcal{M}(\Omega)$ into a polytope by the following procedure. We assume that (in the affine picture) the points $\boldsymbol{q}_{n-2},\boldsymbol{q}_{n-1},\boldsymbol{q}_n$ form a triangle $\boldsymbol{P}_\Delta$ whose interior is not crossed by any other hyperplanes spanned by points in $\boldsymbol{Q}$ (such triangles do always exist and are called *mutations*). Then we construct the polytope $\boldsymbol{\Lambda}(\boldsymbol{P}_\Delta,\{\boldsymbol{q}_1,\dots,\boldsymbol{q}_{n-3}\})$. This polytope has the same realization space as $\mathcal{M}(\Omega)$. However, the polytope that is generated in this way has $2n-3$ vertices and dimension $n-1$. Thus the number of points is approximately twice the dimension.

We can obtain a better result if we first dualize the oriented matroid $\mathcal{M}(\Omega)$. (for an introduction to oriented matroid duality see [**BLSWZ**].) The dual oriented matroid $\mathcal{M}^*(\Omega)$ has the same realization properties as $\mathcal{M}(\Omega)$. It has also n vertices but it has rank $n-3$. Now we apply the same procedure. We assume that $\boldsymbol{Q}^* = (\boldsymbol{q}_1^*,\dots,\boldsymbol{q}_n^*) \in \mathbb{R}^{(n-3)n}$ is a realization of $\mathcal{M}^*(\Omega)$. We furthermore assume that (in the affine picture) the points $\boldsymbol{q}_4^*,\boldsymbol{q}_5^*,\dots,\boldsymbol{q}_n^*$ form a simplex $\boldsymbol{P}_\Delta$ whose interior is not crossed by any other hyperplanes spanned by points in $\boldsymbol{Q}$ (again, these *mutations* do always exist). The simplex $\boldsymbol{P}_\Delta$ has dimension $n-4$ and $n-3$ vertices.

Then we construct the polytope $\boldsymbol{\Lambda}(\boldsymbol{P}_\Delta,\{\boldsymbol{q}_1,\boldsymbol{q}_2,\boldsymbol{q}_3\})$. This polytope has still (up to stable equivalence) the same realization space as $\mathcal{M}(\Omega)$. This polytope has dimension $n-1$ and $n+3$ vertices. After a careful topological analysis we obtain:

THEOREM 4. (UNIVERSALITY FOR d-POLYTOPES WITH $d+4$ VERTICES) *For every basic primary semialgebraic set V defined over $\mathbb{Z}$ there is a dimension d and a d-polytope with $d+4$ vertices, whose realization space is stably equivalent to V.*

3.4. 4-polytopes. The geometric constructions that are needed for a proof of the Universality Theorem for 4-polytopes are much more elaborate. So far no way is known to make directly use of Mnëv's result. Since we later on give a complete description of the constructions for the proof in dimension 6, we will only briefly sketch the 4-dimensional case. A more detailed sketch is given in the research report [**RZ**]. A complete proof (and many ramifications) are given in [**R3**]. Again the idea to prove a Universality Theorem is to make use of Shor normal forms. To encode a Shor normal form we need ...

- ... a way to encode values of variables into realization parameters of a polytope.
- ... a way to enforce the chain of inequalities $0 < 1 < x_1 < x_2 < \dots < x_n$.
- ... polytopes that enforce geometric relations for *addition* and *multiplication.*
- ... a way to superpose all the different obstructions.

This is how these different problems are handled in the proof given in [**R3**].

- The values $x_1, x_2, \ldots, x_n$ are encoded as the line slopes of a $2(n+3)$-gon. Each variable corresponds to a pair of parallel opposite sides. Three extra slopes are added to encode the projective basis 0, 1, ∞.
- The chain of inequalities $0 < 1 < x_1 < x_2 < \ldots < x_n$ is automatically enforced by the fact that the slopes of the edges of a convex polygon must be well ordered.
- Von Staudt constructions have no direct analogue in terms of 4-polytopes. Thus we have to construct 4-polytopes that encode *addition* and *multiplication* "from scratch." This is done as follows: Lawrence extensions are used to generate 4-polytopes (so called *basic building blocks*) that have small but useful realization properties. One of them has a hexagon whose shape is not arbitrarily prescribable. Other basic building blocks have the ability to produce copies of an n-gon and to transfer information. Via connected sums these basic building blocks are composed in order to get polytopes $\boldsymbol{P}^{x+y}$ and $\boldsymbol{P}^{x\cdot y}$ that serve as analogues to the von Staudt Constructions: In every realization of $\boldsymbol{P}^{x+y}$ (of $\boldsymbol{P}^{x\cdot y}$) the edge slopes of a 2-face are such that they encode additive (resp. multiplicative) relations.
- Again using connected sums these polytopes for addition and for multiplication are combined to a large polytope $\boldsymbol{P}(S)$ that encodes the Shor normal form S.

It is a remarkable fact that the whole construction of $\boldsymbol{P}(S)$ can be expressed exclusively in terms of connected sums and Lawrence extensions. The realization space of $\boldsymbol{P}(S)$ is stably equivalent to the solution space of S. For every realization of $\boldsymbol{P}(S)$ the slopes of the decisive $2(n+3)$-gon are (up to a projective transformation) a solution of S. Conversely, to every solution of S there exists a (contractible) set of realizations of $\boldsymbol{P}(S)$. Again, after a careful topological analysis we obtain:

THEOREM 5. (UNIVERSALITY FOR 4-POLYTOPES) *For every basic primary semialgebraic set V defined over $\mathbb{Z}$ there is a dimension* 4*-polytope, whose realization space is stably equivalent to V.*

4. From oriented matroids to 6-polytopes

For a long time it was believed that there must be a way to derive the Universality Theorem for polytopes in fixed dimension as a consequence of the Universality Theorem for oriented matroids. It is still an open problem whether this is possible for the case of 4-polytopes. We here present a proof for the Universality Theorem in dimension 6 (this is obviously a weaker result than the 4-dimensional case) that follows these lines. Again a central role is played by Lawrence extensions and connected sums. Zonotopes are used as the link between oriented matroids and polytopes. We follow the treatment in [**R3**].

4.1. Zonotopes. In the first section we have learned how an oriented matroid $\mathcal{M}_{\boldsymbol{V}}$ is associated to a vector configuration $\boldsymbol{V}$. We now associate to $\boldsymbol{V}$ a second object, the *zonotope* $\boldsymbol{Z}_{\boldsymbol{V}}$. A zonotope is a special type of polytope that is generated as the Minkowski sum of finitely many line segments. Equivalently, one can characterize zonotopes by the property that all 2-faces are centrally symmetric. The last characterization implies that for each edge on a zonotope there is a complete "belt" of edges of equal lengths and directions. These are all translates of a corresponding generating line segment in the Minkowski sum representation. The zonotope of a vector configuration $\boldsymbol{V} \in \mathbb{R}^{d \cdot n}$ is defined by

$$\boldsymbol{Z_V} := \sum_{i=1}^{n} [-\boldsymbol{v}_i, +\boldsymbol{v}_i].$$

Here $[-\boldsymbol{v}_i, +\boldsymbol{v}_i]$ is the line segment between $-\boldsymbol{v}_i$ and $+\boldsymbol{v}_i$ in $\mathbb{R}^d$, and the sum is interpreted as the Minkowski sum. The set of edges of $\boldsymbol{Z_V}$ that are parallel to the generating vector $\boldsymbol{v}_i$ is called the i-th belt.

To avoid unnecessary technicalities caused by degenerate situations, from now on we assume that our vector configurations contain no *loops* (i.e., vectors $\boldsymbol{v} = (0, \dots, 0)$) and contain no pairs of *parallel elements* (i.e., vectors $\boldsymbol{v}_i = \lambda \boldsymbol{v}_j$ with $i \neq j$). There is a close relationship between the oriented matroid $\mathcal{M}_{\boldsymbol{V}}$ and the face lattice of $\boldsymbol{Z_V}$. For a sign-vector $\sigma \in \{-, 0, +\}^n$ we define

$$(\boldsymbol{Z_V})_\sigma := \sum_{\sigma_i = +} \boldsymbol{v}_i - \sum_{\sigma_i = -} \boldsymbol{v}_i + \sum_{\sigma_i = 0} [-\boldsymbol{v}_i, +\boldsymbol{v}_i].$$

In particular, $(\boldsymbol{Z_V})_{(0,\dots,0)} = \boldsymbol{Z_V}$. The following theorem is the well known standard connection between zonotopes and oriented matroids. A Proof can be found for instance in [**R3**].

THEOREM 6. *The faces of $\boldsymbol{Z_V}$ are exactly the sets $(\boldsymbol{Z_V})_\sigma$ with $\sigma \in \mathcal{L}_{\boldsymbol{V}}$.*

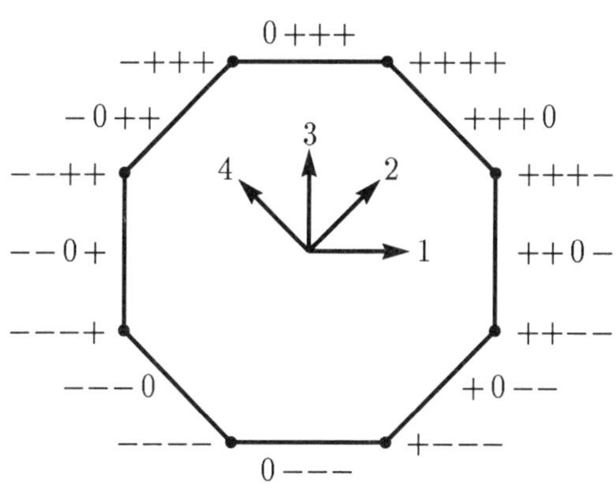

FIGURE 11: A vector configuration and its zonotope.

The last theorem shows that it is natural to label the faces of $\boldsymbol{Z_V}$ by the covectors of $\mathcal{M}_{\boldsymbol{V}}$. The face lattice of $\boldsymbol{Z_V}$ is therefore isomorphic to the lattice $(\mathcal{L}, \prec)$. A face $(\boldsymbol{Z_V})_{\sigma_1}$ is contained in $(\boldsymbol{Z_V})_{\sigma_2}$ if and only if $\sigma_1 \prec \sigma_2$. The vertices of $\boldsymbol{Z_V}$ correspond to the sign vectors $\sigma \in \mathcal{L}_{\boldsymbol{V}} \cap \{-,+\}^n$ (the atoms of $(\mathcal{L}, \prec)$). The facets of $\boldsymbol{Z_V}$ correspond to the cocircuits in $\mathcal{L}_{\boldsymbol{V}}$ (the coatoms of $(\mathcal{L}, \prec)$). Up to translation any zonotope is of the form $\boldsymbol{Z_V}$. Thus also for a general zonotope it is natural to label the faces by the covectors of $\boldsymbol{V}$. Figure 11 demonstrates the connection between the vector configuration, its covectors and the corresponding zonotope. Here the vector configuration consists of 4 non-parallel vectors in $\mathbb{R}^2$. The corresponding zonotope $\boldsymbol{Z}$ turns out to be an 8-gon. The faces of $\boldsymbol{Z}$ are labeled by the covectors. The entire zonotope gets a label (0000).

Closely related to zonotopes is the concept of *planets*. Planets arise from zonotopes by those parallel displacements of the facets that do not alter the combinatorial type. Thus the face lattices of planets are identical to the face lattices of zonotopes. In a natural way we label the faces of a planet that comes from $\boldsymbol{Z_V}$ by the covectors $\mathcal{L} = \mathcal{L}_{\boldsymbol{V}}$. Planets as well as zonotopes have belts of parallel edges. However, in comparison to zonotopes, it is not required that the edges in a belt have equal lengths.

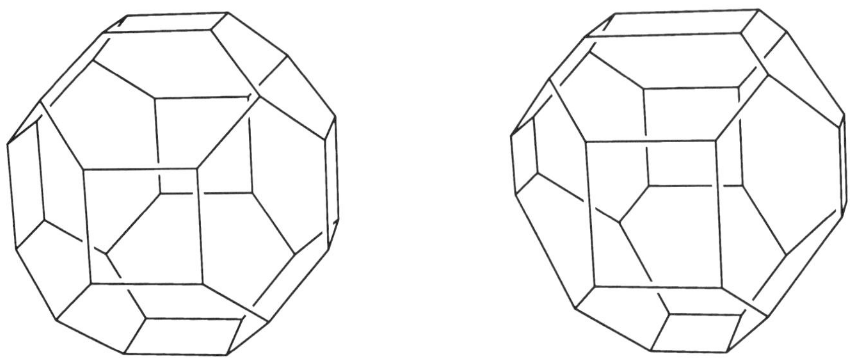

FIGURE 12: The permutahedron and a corresponding planet.

Figure 12 shows a zonotope (the permutahedron) and a planet whose face lattice is isomorphic to the face lattice of the permutahedron. The corresponding vector configuration that generates this zonotope is the one given in Figure 1. Properties of the oriented matroid have their direct counterpart in the face lattice of the corresponding zonotope. For instance the fact that three points are collinear translates to the fact that there exists a hexagonal facet with the corresponding edges.

A three dimensional zonotope considered as a *polytope* has a trivial realization space (as a consequence of Steinitz's Theorem). The polytope setting does not require that the edge belts of the zonotope stay parallel. However, if we restrict ourselves to the category of *planets* and consider all realizations of $\boldsymbol{Z_V}$ that are indeed planets, then this realization space is stably equivalent to the realization space of $\mathcal{M}_{\boldsymbol{V}}$. More formally we define the *planet realization space* as follows. Let $\boldsymbol{Z}$ be a zonotope and B be a (polytope) basis of $\boldsymbol{Z}$. The *planet realization space* $\mathcal{P}(\boldsymbol{Z}, B) \subseteq \mathcal{R}(\boldsymbol{Z}, B)$ is the space of all planets in $\mathcal{R}(\boldsymbol{Z}, B)$.

$$\mathcal{P}(\boldsymbol{Z}, B) = \Big\{\boldsymbol{P} \in \mathcal{R}(\boldsymbol{Z}, B) \mid \boldsymbol{P} \text{ is a planet}\Big\}.$$

From every planet realization in $\mathcal{P}(\boldsymbol{Z_V}, B)$ we can easily derive a realization of the corresponding oriented matroid $\mathcal{M}_{\boldsymbol{V}}$. Let $\boldsymbol{P}$ be a planet that is combinatorially equivalent to $\boldsymbol{Z_V}$, with $\boldsymbol{V} = (\boldsymbol{v}_1, \ldots, \boldsymbol{v}_n)$. Recall that the faces of $\boldsymbol{P}$ are labeled by sign vectors of length n. The labels of vertices have no zero-entry at all, the labels of edges have exactly one zero entry. The labels σ^+, σ^- of the endpoints of an edge σ that is in the i-th belt of $\boldsymbol{P}$ differ in exactly one entry (namely the entry at position i). Assume $\sigma_i^+ = +$ and $\sigma_i^- = -$. To the edge σ we associate the difference vector of its endpoints $\boldsymbol{w}_\sigma = \boldsymbol{p}_{\sigma^+} - \boldsymbol{p}_{\sigma^+}$. If two edges σ and τ belong to the same belt, then the corresponding vectors $\boldsymbol{w}_\sigma$ and $\boldsymbol{w}_\tau$ differ only by a positive scalar multiple. Thus up to a positive factor all edges that belong to one belt (i.e., all edges for which, say, the i-th entry is zero) produce identical vectors. If we take one such vector $\boldsymbol{w}_i$ from each belt i, then we obtain a realization $\boldsymbol{W} = (\boldsymbol{w}_1, \dots, \boldsymbol{w}_n)$ of $\mathcal{M}_{\boldsymbol{V}}$.

A more careful analysis of the situation proves the following fact. A proof can be found in [**R3**].

THEOREM 7. *Let $\boldsymbol{V} \in \mathbb{R}^{3n}$ be a vector configuration without loops or parallel elements. Let b be a basis of $\boldsymbol{V}$ and B be a (polytope) basis of $\boldsymbol{Z}_{\boldsymbol{V}}$. Then*

$$\mathcal{R}(\mathcal{M}_{\boldsymbol{V}}, b) \approx \mathcal{P}(\boldsymbol{Z}_{\boldsymbol{V}}, B).$$

4.2. The construction. Having established the last theorem, the construction that transfers universality results from oriented matroids to polytopes is not too complicated. The problem is that in general there are realizations of $\boldsymbol{Z}_{\boldsymbol{V}}$ as *polytopes* that are not *planets*. This implies that the corresponding spaces $\mathcal{R}(Z, B)$ and $\mathcal{P}(Z, B)$ are not at all stably equivalent. For instance, an oriented matroid $\mathcal{M}_{\boldsymbol{V}}$ that comes from our construction in Section 2 has rank 3. The corresponding zonotope $\boldsymbol{Z}_{\boldsymbol{V}}$ has dimension 3. Thus the space $\mathcal{P}(\boldsymbol{Z}_{\boldsymbol{V}}, B)$ may be arbitrary complicated, while the space $\mathcal{R}(\boldsymbol{Z}_{\boldsymbol{V}}, B)$ is trivial (since it is the realization space of a 3-polytope).

We will construct a 6-polytope $P(\mathcal{M}_{\boldsymbol{V}})$ that contains $\boldsymbol{Z}_{\boldsymbol{V}}$ as 3-face F. The structure of $P(\mathcal{M}_{\boldsymbol{V}})$ will force that in every realization of $P(\mathcal{M}_{\boldsymbol{V}})$ the face F is (up to projective equivalence) indeed a planet. The construction again goes via Lawrence extensions and connected sums.

The idea is the following. For $i = 1, \dots, n$ let $\boldsymbol{q}_i$ be the point at infinity in which the edges of the i-th belt meet. The Lawrence extension $\boldsymbol{\Lambda}(\boldsymbol{Z}_{\boldsymbol{V}}, \boldsymbol{q}_i)$ is a 4-polytope, which contains $\boldsymbol{Z}_{\boldsymbol{V}}$ as a 3-face. In every realization of $\boldsymbol{\Lambda}(\boldsymbol{Z}_{\boldsymbol{V}}, \boldsymbol{q}_i)$ the supporting lines of the i-th belt of this 3-face meet in a point (Figure 13 right).

If (i, j, k) is a non-basis of $\mathcal{M}_{\boldsymbol{V}}$, then the points $\boldsymbol{q}_i, \boldsymbol{q}_j, \boldsymbol{q}_k$ are collinear. Thus the iterated Lawrence extension $\boldsymbol{P}_{(i,j,k)} = \boldsymbol{\Lambda}(\boldsymbol{Z}_{\boldsymbol{V}}, \{\boldsymbol{q}_i, \boldsymbol{q}_j, \boldsymbol{q}_k\})$ is a 6-polytope that has the following properties (compare Figure 13 left):

- $\boldsymbol{Z}_{\boldsymbol{V}}$ is a 3-face of $\boldsymbol{P}_{(i,j,k)}$.
- In every realization of $\boldsymbol{P}_{(i,j,k)}$ the supporting lines of the belts i, j and k meet in three points $\boldsymbol{q}'_i$, $\boldsymbol{q}'_j$, and $\boldsymbol{q}'_k$.
- $\boldsymbol{q}'_i$, $\boldsymbol{q}'_j$, and $\boldsymbol{q}'_k$ are collinear.

Let $b_1, \dots, b_m$ be the non-bases of $\mathcal{M}_{\boldsymbol{V}} = \mathcal{M}(\Omega)$, where $\boldsymbol{V}$ is now a configuration that comes from the constructions of Section 2. By performing connected sums that identify the polytopes $\boldsymbol{P}_{b_1}, \dots, \boldsymbol{P}_{b_m}$ along the common 3-face $\boldsymbol{Z}_{\boldsymbol{V}}$ (this is technically a bit difficult) we obtain our desired 6-polytope $\boldsymbol{P}(\Omega)$. This polytope $\boldsymbol{P}(\Omega)$ has the following properties.

- $\boldsymbol{Z}_{\boldsymbol{V}}$ is a 3-face of $\boldsymbol{P}(\Omega)$.
- In every realization of $\boldsymbol{P}(\Omega)$ and for every $i \in \{1, \dots, n\}$ the supporting lines of edges of the i-th belt meet in a point $\boldsymbol{q}'_i$.
- For every non-basis (i, j, k) if $\mathcal{M}(\Omega)$ the points $\boldsymbol{q}'_i$, $\boldsymbol{q}'_j$, and $\boldsymbol{q}'_k$ are collinear.

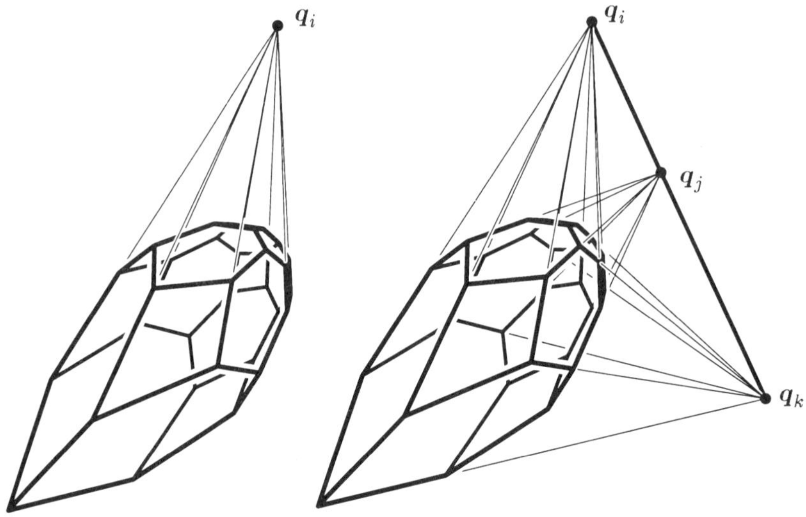

FIGURE 13: Lawrence extensions of a zonotope.

Furthermore we need the following decisive property of our construction of $\mathcal{M}(\Omega)$: Let $\boldsymbol{q}_1, \ldots, \boldsymbol{q}_n$ be n points in $\mathbb{R}^d$; $d > 2$ such that $\boldsymbol{q}_i, \boldsymbol{q}_j, \boldsymbol{q}_k$ are collinear whenever (i, j, k) is a non-basis of $\mathcal{M}(\Omega)$, then all points $\boldsymbol{q}_1, \ldots, \boldsymbol{q}_n$ are coplanar.

Combining this fact with the properties of $\boldsymbol{P}(\Omega)$ shows that for an arbitrary realization $\boldsymbol{P}'$ of $\boldsymbol{P}(\Omega)$ the points $\boldsymbol{q}'_1, \ldots, \boldsymbol{q}'_n$ lie in a plane H (see Figure 14). If we apply a projective transformation that maps H to the plane at infinity, then the 3-face that corresponds to $\boldsymbol{Z_V}$ becomes indeed a planet. This planet can be used to derive a realization of $\mathcal{M}(\Omega)$. (Indeed already $\boldsymbol{q}'_1, \ldots, \boldsymbol{q}'_n$ form a realization of $\mathcal{M}(\Omega)$.)

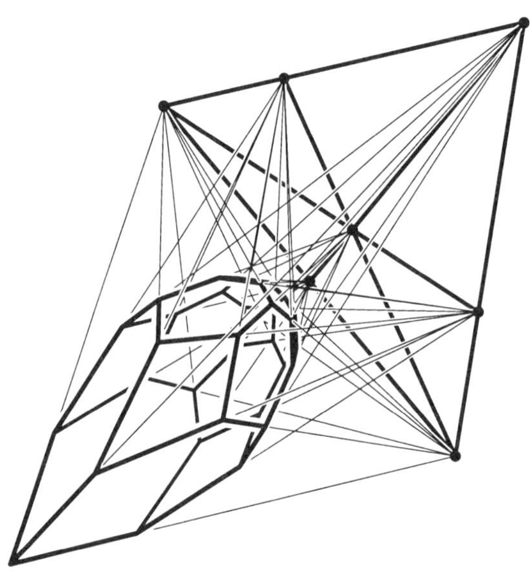

FIGURE 14: The entire construction applied to the permutahedron.

In conclusion the construction has the following crucial two properties: From every realization of $\boldsymbol{P}(\Omega)$ we can derive a solution of Ω. Conversely, to every solution of Ω (after fixing an affine basis of $\boldsymbol{P}(\Omega)$) there exist a contractible set of realizations of $\boldsymbol{P}(\Omega)$ that corresponds to the solution. Again a more detailed analysis shows:

THEOREM 8. (UNIVERSALITY FOR 6-POLYTOPES) *For every basic primary semialgebraic set V defined over $\mathbb{Z}$ there is a dimension 6-polytope, whose realization space is stably equivalent to V.*

References

[BLSWZ] A. Björner, M. Las Vergnas, B. Sturmfels, N. White and G. M. Ziegler, *Oriented Matroids*, Encyclopedia of Mathematics, Vol. 46, Cambridge University Press 1993.

[BM] L. J. Billera and B. S. Munson, *Polarity and inner products in oriented matroids*, European J. Combinatorics, **5** (1984), 293–308.

[BS] J. Bokowski and B. Sturmfels, *Computational Synthetic Geometry*, Lecture Notes in Mathematics **1355**, Springer-Verlag, Berlin Heidelberg 1989.

[BEK] J. Bokowski, G. Ewald and P. Kleinschmidt, *On combinatorial and affine automorphisms of polytopes*, Israel J. Math. bf 47 (1984), 123–130.

[G] H. Günzel, *The universal partition theorem for oriented matroids*, Preprint TH-Aachen (1994), 29p.; Discrete Comput. Geometry, to appear.

[K] P. Kleinschmidt, *On facets with non-arbitrary shapes*, Pacific J. Math., **65** (1976), 511–515.

[M] N. E. Mnëv, *The universality theorems on the classification problem of configuration varieties and convex polytopes varieties*, in: Viro, O.Ya. (ed.): Topology and Geometry—Rohlin Seminar, Lecture Notes in Mathematics **1346**, Springer, Heidelberg 1988, 527–544.

[M2] N. E. Mnëv, *The universality theorems on the oriented matroid stratification of the space of real matrices.* in: Applied Geometry and Discrete Mathematics–The Victor Klee Festschrift, (P. Gritzmann, B. Sturmfels, eds.), DIMACS Series in Discrete Mathematics and Theoretical Computer Science, Amer. Math. Soc., Providence, RI, **6** (1991), 237–243.

[R] J. Richter-Gebert, *Mnëv's universality theorem revisited*, Proceedings of the Séminaire Lotharingien de Combinatorie 1995, 211–225.

[R2] J. Richter-Gebert, *Two interesting oriented matroids*, Documenta Mathematica, Journal der DMV **1** (1996), 137–148.

[R3] J. Richter-Gebert, *Realization Spaces of Polytopes*, Lecture Notes in Mathematics **1643**, Springer, Heidelberg 1966.

[RZ] J. Richter-Gebert and G. M. Ziegler, *Realization spaces of 4-polytopes are universal*, Bulletin of the AMS (research report), **32** (1995), 403–412.

[Ri] G. Ringel, *Teilung der Ebene durch Geraden oder topologische Geraden*, Math. Zeitschr **64** (1956), 79–102.

[S] P. Shor, *Geometry and Discrete Mathematics—The Victor Klee Festschrift*, (P. Gritzmann, B. Sturmfels, eds.), DIMACS Series in Discrete Mathematics and Theoretical Computer Science, Amer. Math. Soc., Providence, RI, **4** (1991), 531–554.

[SR] E. Steinitz and H. Rademacher, *Vorlesungen über die Theorie der Polyeder*, Springer Verlag, Berlin 1934; reprint, Springer Verlag 1976.

[St] E. Steinitz, *Polyeder und Raumeinteilungen*, in: Encyclopädie der Mathematischen Wissenschaften. Band 3 (Geometrie) Teil 3AB12, 1–139, 1922.

[Su] P. Y. Suvorov, *Isotopic but not rigidly isotopic plane systems of straight lines*, in: Viro, O.Ya. (ed.): Topology and Geometry—Rohlin Seminar, Lecture Notes in Mathematics **1346**, Springer, Heidelberg 1988, 545–556.

[Z] G. M. Ziegler, *Lectures on Polytopes*, Graduate Texts in Mathematics **152**, Springer-Verlag New York 1995. Corrections, updates, and more: `http://www.math.tu-berlin.de/~ziegler`

TECHNICAL UNIVERSITY BERLIN, FB MATHEMATIK, SEKR. 6-1, STRASSE DES 17. JUNI 135, D-10623 BERLIN, GERMANY

E-mail address: richter@math.tu-berlin.de

Contemporary Mathematics
Volume **223**, 1999

Periodic and Aperiodic Tilings of E^n

Marjorie Senechal

Dedicated to Branko Grünbaum

ABSTRACT. The theory of tilings, both periodic and nonperiodic, has been extended in many ways over the last decade. We discuss some of the most important recent developments.

1. Introduction

The remarkable growth and progress in discrete and computational geometry over the past decade was catalyzed not only by a major American Mathematical Society conference in Santa Cruz and the birth of the journal *Discrete & Computational Geometry*, but also by the publication of a beautiful and comprehensive treatise, *Tilings and Patterns* [**GS**]. This landmark book, many years in the making, was instrumental in transforming tiling theory—and the theory of patterns—into an important branch of discrete geometry. Today it is widely recognized that tilings and patterns inform and are informed by almost every field of mathematics. They also find application in almost every branch of science. By clarifying, codifying, and unifying a wildly scattered and uneven literature, *Tilings and Patterns* has made an important contribution to many scientific fields, as well as to mathematics.

Of equal importance, Grünbaum and Shephard gave the first comprehensive and detailed presentation of the new field of aperiodic tilings, including the tilings of Robinson, Ammann, and Penrose, and also the extensive unpublished work of Conway on the latter. It is still a valuable reference for these tilings and for many of the issues involved in understanding them.

Tilings and Patterns deals exclusively with tilings and patterns in the Euclidean plane. This may seem strange in an age when mathematical fashion demands that theories be set in spaces as abstract as their architectures will bear. In fact, many of the basic concepts and definitions, and some fundamental theorems, of *Tilings*

1991 *Mathematics Subject Classification.* Primary 52C.

Key words and phrases. Periodic, nonperiodic, matching rules, local rules.

I would like to thank the referee for many helpful comments.

Figures 1, 4, 5, and 9 reprinted with permission from *Handbook of Discrete and Computational Geometry*, Jacob E. Goodman and Joseph O'Rourke, eds., CRC Press, 1997. Copyright CRC Press, Boca Raton, Florida, ©1997.

© 1999 American Mathematical Society

and Patterns can easily be extended to any dimension. However, the restriction to the Euclidean plane was wise at that time, for at least three reasons. First, two dimensional tilings and patterns are easily visualized. (Although visual thinking is one of the key tools for understanding tilings and patterns, visualizing geometrical objects in three—not to mention n—dimensions presented serious challenges for most of us in that primitive computer graphics era.) Second, relatively little was known about tilings and patterns in dimensions greater than two when Grünbaum and Shephard began writing the book. And finally, E^2 is rich enough to keep us all busy for a very long time to come. Tilings of and patterns in the plane present unlimited challenges. It has been known since 1966 [**Be**] that "the tiling problem" for the plane—that is, "the question of whether a given finite set of polygonal tile shapes will or not tile the plane"—is computationally undecidable, and this, says Penrose [**P**], implies "that there is an unlimited variety of ever-more sophisticated ways of covering the Euclidean plane with a finite number of distinct shapes." The plane, then, is a well-equipped laboratory in which to study the mathematics of long-range order.

A key problem in the study of long-range order in tilings and patterns is to understand the relation between local properties, such as configurations of bounded radius, and global ones, such as periodicity, aperiodicity, and so forth. This relation plays a key role in the characterization and classification of tilings.

2. Definitions

Instead of following *Tilings and Patterns* and restricting ourselves to E^2, we will define our terms more generally; we can gain additional insight into tilings and patterns in the plane by considering their counterparts in E^n.

DEFINITION 1. A tiling $\mathcal{T}$ of E^n is a countable family of closed sets, called tiles, whose interiors are pairwise disjoint and whose union is E^n.

We assume that each tile is homeomorphic to a closed n-ball, and that the intersection of any pair of distinct tiles is homeomorphic to a closed k-ball for some $k \in \{0, 1, \dots, n-1\}$. These intersections are the *k-faces* of $\mathcal{T}$. The $(n-1)$-faces of $\mathcal{T}$ are its *facets*, the 1-faces its *edges*, and the 0-faces are the *vertices of* $\mathcal{T}$.

It is necessary to define the faces of a tiling $\mathcal{T}$, as distinct from the faces of its tiles, because there may not be any natural way to define the faces of a tile independently of a tiling in which it appears (think, for example, of Escher's lizards). Even when the tile is a polytope this distinction is important, because its faces need not coincide with faces of a tiling (This is the case in a plane tiling by squares in which one row of squares is shifted relative to the others). When the tiles are polytopes and and the $(n-1)$-faces of the tiles are the facets of $\mathcal{T}$, the tiling is said to be *facet-to-facet*. Facet-to-facet implies face-to-face for all faces of lower dimension [**GR**].

DEFINITION 2. The protoset of a tiling $\mathcal{T}$ is a family $\mathcal{P}$ of shapes, called prototiles, such that every tile in $\mathcal{T}$ is an isometric copy of an element of $\mathcal{P}$.

If $\mathcal{P}$ is the protoset of the tiling $\mathcal{T}$, we say that $\mathcal{P}$ *admits* $\mathcal{T}$. Many protosets admit more than one tiling, some uncountably many. We will always assume that $\mathcal{P}$ is finite. This implies that our tilings are *normal* in the sense of Grünbaum and Shephard : there are real numbers $u > 0$ and $U > 0$ such that every tile contains a ball of radius u in its interior and lies in the interior of some ball of radius U.

Normal tilings are *locally finite*: every ball of finite radius meets only finitely many tiles.

$\mathcal{T}$ is said to be *monohedral* if it has a single prototile. Monohedral tilings would appear to be the simplest of all tilings, but even they present difficult challenges. For example,

- Hilbert, as part of his 18th problem, asked whether monohedral tilings are necessarily *isohedral* ($\mathcal{T}$ is isohedral if its symmetry group acts transitively on its tiles). The first counterexample was found in 1915 [**Re**]; a counterexample due to Penrose [**G**] is shown in Figure 1.

- Keller conjectured that, for every n, in every tiling of E^n by congruent cubes there is at least one pair of cubes that share a whole facet. The conjecture is true for $n \leq 6$ [**Pe**] and false for $n \geq 10$ [**LS**]; the cases $n = 7, 8, 9$ are still open.

- Voronoï [**V**] conjectured that every convex polytope that tiles E^n by translation is affine equivalent to the Voronoï cell (defined below) of an n-dimensional lattice. Delaunay proved that the conjecture is true for $n \leq 4$ [**De**]; some special cases have also been settled in the affirmative but in its general form the conjecture is still open.

- The "einstein" problem is that of finding a single prototile(s) that admit only nonperiodic tilings of E^n. There is no such tile when $n = 1$ (see [**Se**]. Very recently [**P**], Penrose has designed an einstein—a marked hexagon—in the plane; it raises important questions about the concept of matching rules (see §4). There is a family of einsteins in three-dimensional space (see [**S, D**], but these tiles admit tilings invariant under symmetries with a translation component (a so-called screw rotation); one would like to have a tile whose tilings do not admit symmetries of infinite order. The einstein problem is still open in higher dimensions.

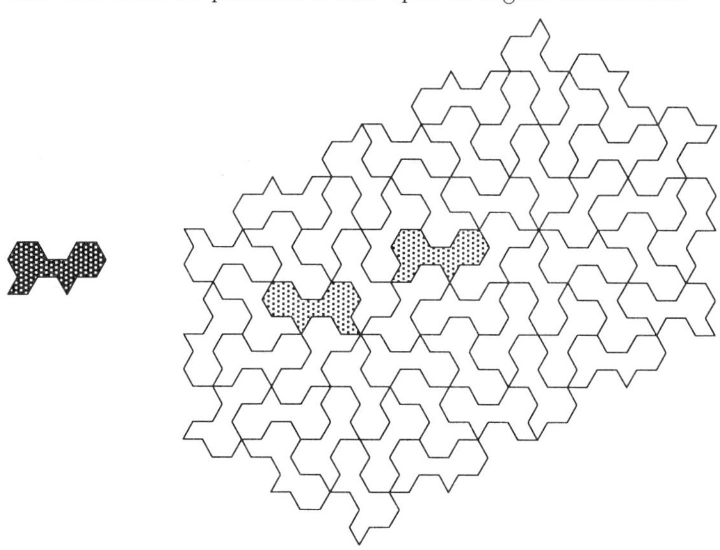

FIGURE 1: A monohedral tiling that is not isohedral.

We now define various local properties of tilings. We will be interested in various bounded configurations of tiles. The *(first) corona* $C(f_k)$ of a k-face $\mathbf{f_k}$ is the set of tiles in $\mathcal{T}$ that have nonempty intersection with $\mathbf{f_k}$; the *second corona* of

$\mathbf{f_k}$ is the set of tiles that meet $C(f_k)$, and so forth. The first corona of a vertex ($k = 0$) is often called a *vertex star* (Figure 2, left). The first corona of a tile is shown in Figure 2, center. The *facet corona* of a tile $T \in \mathcal{T}$ is the set of tiles in $\mathcal{T}$ that share facets with T (Figure 2, right). It is also sometimes useful to consider *centered coronas* of tiles: the centered k^{th} coronas of a tile T in $\mathcal{T}$ is the set of all tiles T' for which there exists a path of no more than $k+1$ tiles beginning with T and ending with T'.

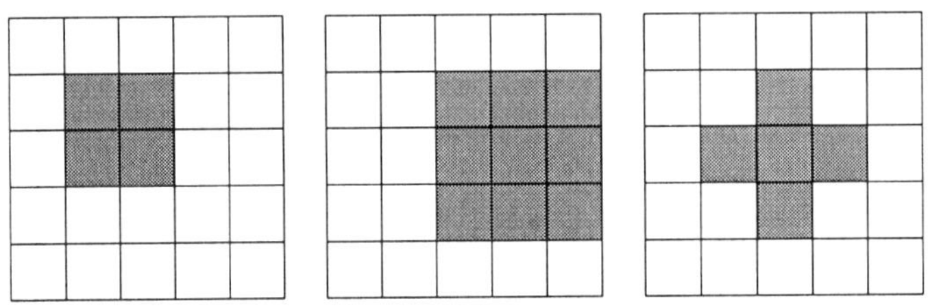

FIGURE 2: Some local configurations in the tiling of the plane by squares. left: a vertex star; center: the first corona of a tile; right: a facet corona.

DEFINITION 3. The kth corona atlas of a tiling $\mathcal{T}$ is the set of isometry classes of kth coronas of $\mathcal{T}$.

DEFINITION 4. Two tilings are said to belong to the same local isomorphism class if, up to similarity, they have the same k^{th} corona atlases, for all $k > 0$.

3. Delaunay sets

Tilings and Patterns is best known for its exposition of tilings; the theory of patterns has received much less attention. But in fact the two are inextricable: every tiling gives rise to several patterns, for example, the pattern whose motifs are the vertices of the tiles, and many patterns—especially point patterns—can be associated to tilings in a natural way. Grünbaum and Shephard built periodicity into their definition of pattern (though not into their definition of tiling). We need to consider more general patterns. For point patterns, the right degree of generality seems to be that of a Delaunay set.

Recall that a point set $X \in E^n$ is *uniformly discrete* if there is a real number $r_c > 0$ such that each open ball of radius r_c contains at most one point of X, and is *relatively dense* in E^n if there is a real number $R_c > 0$ such that each closed ball of radius R_c contains at least one point of Λ.

DEFINITION 5. A Delaunay set $\Lambda \in E^n$ is a point set that is uniformly discrete and relatively dense (in E^n).

Intuitively, the points of a Delaunay set Λ cannot be arbitrarily close together nor can they be arbitrarily far apart.

The notion of relative density can be extended to bounded subsets of Delaunay sets or configurations of tiles. We will say that a bounded subset C of points in a Delaunay set Λ (or tiles in a tiling $\mathcal{T}$) is relatively dense in Λ (or $\mathcal{T}$) if there is some $m > 0$ such that every ball of radius m contains an isometric copy of C.

The counterpart, for Delaunay sets, of a kth corona atlas of a tiling is a ρ-star.

DEFINITION 6. Let Λ be a Delaunay set and $x \in \Lambda$. The ρ-star of x is the set

$$ST_\Lambda(x,\rho) := \Lambda \cap B(x,\rho),$$

with the point x distinguished in some way.

DEFINITION 7. The ρ-atlas of a Delaunay set Λ is the set of isometry classes of ρ-stars of Λ.

We can associate many Delaunay sets to any normal tiling $\mathcal{T}$. For example, normality implies that the set of centers of incircles of the tiles of $\mathcal{T}$ is Delaunay, and in many cases the set of vertices of $\mathcal{T}$ is also a Delaunay set. Conversely, we can associate tilings to a Delaunay set Λ in many different ways. The best known, and one of the most useful, of these tilings is the Voronoï tiling $\mathcal{V}_\Lambda$ induced by Λ. The tiles of this tiling are the Voronoï cells of the points of Λ.

DEFINITION 8. For $x \in \Lambda$, the Voronoï cell of x is the closed set

$$V(x) = \{y \in E^n;\ |x-y| \leq |y-z|,\ \forall\, z \in \Lambda\}.$$

$V(x)$ is an intersection of half-spaces, so it is convex. Since Λ is relatively dense in E^n, $V(x)$ is bounded and thus is a polytope. The collection of Voronoï cells, for all $x \in \Lambda$, is a facet-to-facet tiling of E^n.

It is an easy exercise to show that $V(x)$ is completely determined by the points of Λ that lie in the $2R_c$-star of x:

$$V(x) = \{y \in E^n;\ |x-y| \leq |x-z|,\ \forall\, z \in ST_\Lambda(x, 2R_c)\}.$$

It follows from the construction of $V(x)$ that x and the points of Λ whose Voronoï cells share a vertex v with $V(x)$ lie on a sphere with center v.

DEFINITION 9. The convex hull of the points of Λ whose Voronoï cells share a vertex v with $V(x)$ is the *Delaunay cell* $D(v)$ of v.

Like $V(x)$, $D(v)$ is a polytope; however, unlike $V(x)$, which always contains x, $D(v)$ need not contain v (not even when Λ is a lattice!)

The collection of Delaunay cells of the vertices of a Voronoï tiling $\mathcal{V}$ constitute a second facet-to-facet tiling, the Delaunay tiling $\mathcal{V}^*_\Lambda$ induced by Λ. As our use of $*$ suggests, the Voronoï and Delaunay tilings are combinatorially dual. That is, there exists an inclusion-reversing bijection ψ between the k-faces of $\mathcal{V}_\Lambda$ and the $(n-k)$-faces of $\mathcal{V}^*_\Lambda$. (This duality is in fact stronger: for each $f_k \in \mathcal{V}_\Lambda$, f_k and $\psi(f_k)$ are orthogonal.) It is easy to show that $\mathcal{V}^{**}_\Lambda = \mathcal{V}_\Lambda$.

Finally, we will need the notion of the symmetry group of a tiling or Delaunay set. The classical definition of this group is the set $S(\mathcal{T})$, or $S(\Lambda)$, of isometries that map $\mathcal{T}$ or Λ onto itself. By the symmetry group of a tile T of $\mathcal{T}$ we mean the induced group, the subgroup of $S(\mathcal{T})$ that maps T to itself.

4. Classification of tilings

The classification of tilings is a subtle problem. Shall we classify them by the number of protiles? By the shapes of the prototiles? By the edge skeleton of the tiling? By symmetry? Traditionally, the first step has been to distinguish between periodic and nonperiodic tilings.

4.1. Periodic tilings. Bieberbach's theorem [**Bi, Bi2**] implies that if $S(\mathcal{T})$ is infinite, then $\mathcal{T}$ is periodic. More precisely, $S(\mathcal{T})$ is a crystallographic group, which means that it contains a maximal abelian subgroup isomorphic to Z^n. (The definition of "crystallographic group" can be formulated in several equivalent ways, including the following: a crystallographic group is one that has a point orbit that is a Delaunay set [**DLS**].) The crystallographic groups have been enumerated for $n = 1, 2, 3, 4$ [**BBNWZ**].

A tiling whose symmetry group contains no translations (except 0) will be called nonperiodic, or aperiodic. (We will not need to consider tilings whose symmetry groups are "subperiodic", that is, which contain a subgroup isomorphic to Z^j, for some $j \in \{1, \ldots, n-1\}$.)

As we noted above, to say that $\mathcal{T}$ is isohedral means that its tiles belong to a single orbit of $S(\mathcal{T})$; the classical term "regular" is also used. More generally,

DEFINITION 10. A tiling $\mathcal{T}$ is k-regular if its tiles or points fall into a finite number k of orbits under the action of its symmetry group.

A normal periodic tiling is obviously k-regular; conversely, k-regularity implies periodicity ([**DLS, VW**]).

Symmetry is only one of several tools for the classification of tilings. The edge graph of a tiling is at least as important as its symmetry, in terms of the information that it encodes. Note that two tilings of the plane with the same symmetry group may have different edge graphs, and conversely, a single edge graph can always be realized as a tiling in various ways, even as tilings with different symmetry groups. There are eleven possible edge graphs for isohedral tilings of the plane. Known as *Laves nets*, they are dual to the graphs of uniform tilings by polygons first derived by Kepler (see [**GS**]).

One of the chief contributions of *Tilings and Patterns* was to show how global properties (symmetry groups, edge graphs) and local properties (facet coronas) "interact" to provide a satisfactory classification system of isohedral tilings of the plane.

To see why facet coronas as well as groups and graphs are needed, consider Figure 3, from *Tilings and Patterns*. Here we see three different isohedral tilings, all with the same symmetry group (in IUCr notation, pgg) and the same combinatorial edge skeleton (all three are homeomorphic to an edge-to-edge tiling of the plane by squares). Not only that, even the tiles have the same symmetry in all three cases (the trivial group). Yet these tilings are different. One way to understand this difference is to note that one cannot construct a map between the tiles of one of these tilings and the tiles of another that preserves the symmetry relations between the pairs of tiles. Since the tilings are isohedral, each tile in any one of them can be mapped to each of its adjacents by some symmetry of that tiling, and since the tile symmetry is trivial, the mapping is unique. Now notice that in

(a) each tile is mapped to two of its four neighbors by translation, and onto the other two by glide reflection. In

(b) however, each tile is related to all four neighbors by glide reflection, while in

(c) glide reflection is needed for two adjacents, and rotation for the two others. Thus a classification by global features is not enough: we also need to take local features into account.

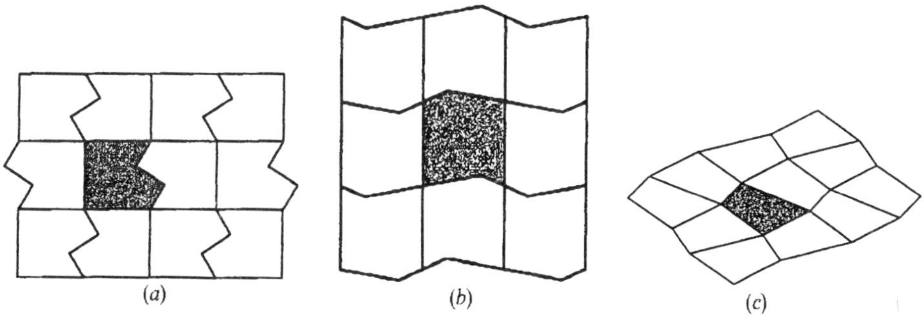

FIGURE 3: Three different isohedral tilings with the same symmetry group, the same skeleton, and the same tile symmetry group.

Grünbaum and Shephard introduced *incidence symbols* to describe facet coronas (see Figure 4). An incidence symbol consists of two strings of signs: the first is a "tile symbol" that encodes the number of edges of the tile and the way that the edges are related by symmetry, and the second is an "adjacency symbol" that encodes the way in which each tile is related to its adjacents.

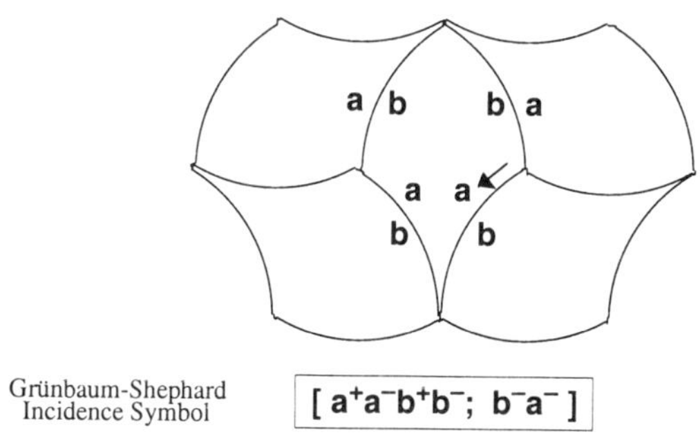

FIGURE 4: The Grünbaum and Shephard incidence symbol.

To construct the tile symbol for an isohedral tiling of the plane $\mathcal{T}$, we begin with any edge of a single tile T. First we label this edge and all edges equivalent to it under the rotational subgroup of the stabilizer of the tile with a symbol, say the superscripted letter a^+; edges equivalent to those edges under reflection receive the same letter with the opposite superscript, say a^-. If an edge is equivalent to itself under reflection, no superscript is needed. (Thus, for example, the edges of a square would be labeled a, a, a, a, and those of a rhombus, a^+, a^-, a^+, a^-.) If any edges of T remain unmarked, we choose one of them and assign it the label b^+ (or b), and continue as above. The tile symbol is the list of edge symbols, in cyclic order. Next we construct the adjacency symbol by listing, in the same order, the labels of the edges of the adjacent tiles, as prescribed by the symmetries that map T onto them. In the three examples of Figure 3, there are no tile symmetries, so in each case the tile symbol is (a^+, b^+, c^+, d^+). The incidence symbols for these tilings are:

(a) $[a^+b^+c^+d^+; c^-b^+a^-d^+]$;

(b) $[a^+b^+c^+d^+; c^-d^-a^-b^-]$;

(c) $[a^+b^+c^+d^+; b^-a^-c^+d^+]$. In this way the three tilings are clearly distinguished.

In the Grünbaum and Shephard classification scheme, two isohedral tilings are said to be equivalent if and only if they have

(a) the same symmetry group,

(b) the same combinatorial edge skeleton, and

(c) the same incidence symbol.

Notice that the tiles of the two tilings may differ in shape, but not in symmetry. Isogonal (vertex-transitive) and isotoxal (edge-transitive) tilings are classified in a similar way. (A tiling is isogonal if its symmetry group G acts transitively on its vertices; it is isotoxal if G acts transitively on its edges.)

"In addition to providing us with a *classification* of the isohedral, isogonal, and isotoxal tilings it [the method of incidence symbols] also enables us to *enumerate* the different types that exist," explained Grünbaum and Shephard. In fact, they gave the first complete enumerations of these classes of tilings of the plane.

To enumerate the isohedral tilings, for example, we choose a skeleton type, a crystallographic group compatible with that skeleton, and a tile group (a finite symmetry group for the prototile that is compatible with both the skeleton and the crystallographic group). We then "mark" the tile, that is, we decorate it with a motif that has the symmetry the tile group. The list of marked isohedral tilings is found by testing all possible incidence symbols for compatibility with edge skeletons and symmetries. The final step in the classification is to produce representatives of equivalence classes of *unmarked* tilings. This is done by transferring the information conveyed by the marks, namely tile symmetry, to the tile shape, by suitably deforming the tile edges. However—and this is where earlier authors had stumbled—it turns out that twelve of the marked tilings have no unmarked representatives (try designing a tiling with incidence symbol $[a^+b^+c^+d^+; a^-b^-c^-d^-]$). Thus we must distinguish between marked and umarked tilings.

THEOREM 1. *(Grünbaum and Shephard). There are precisely:*
- *93 types of marked, and 81 types of unmarked, isohedral tilings;*
- *93 types of marked, and 91 types of unmarked, isogonal tilings;*
- *30 types of marked and 26 types of unmarked normal isotoxal tilings*

of the Euclidean plane.

Despite these successes, time has not been kind to incidence symbols. One obvious limitation is the difficulty in extending the concept to higher dimensional tilings. The computer-friendly dimension-independent "Delaney-Dress symbols," illustrated in Figure 5, are more widely used today. (This Delaney is Matthew Delaney, not Boris Nikolaevitch Delaunay.) Huson, Dress, and others have used them to enumerate 2-isohedral and 3-isohedral tilings of the plane, as well as several large families of tilings in E^3 ([**H, DHM, DH**]). The ongoing work of this group is the first comprehensive approach to the enumeration of three-dimensional tilings.

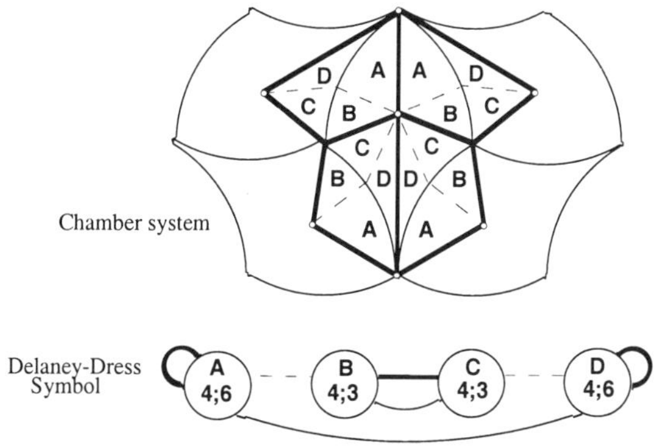

FIGURE 5: Delaney-Dress symbols.

A second limitation is that incidence symbols do not fully characterize isohedral tilings (while Delaney-Dress symbols do). The same incidence symbol may be compatible with two different edge networks (for example, $[a^+b^+c^+d^+; b^+a^+d^+c^+]$ can be realized in the $[4^4]$ and [3.4.6.4] networks—see IH31 and IH55, in Table 6.2.1 of *Tilings and Patterns.*) It is not even obvious that a tiling whose tiles admit an incidence symbol is necessarily periodic. That this is in fact the case is a consequence of the famous "Local Theorem" of Delaunay and his colleagues ([**DDSG, DS**]), which states:

THEOREM 2. *A monohedral tiling $\mathcal{T}$ of n-dimensional space (Euclidean, hyperbolic, or spherical) is isohedral if and only if there exists an integer $k > 0$ such that:*
(a) All centered k^{th} coronas in $\mathcal{T}$ are pairwise congruent;
(b) The symmetry groups of the centered $(k-1)^{st}$ and k^{th} coronas are identical.

It can be shown that when $n = 2$, the congruence of the first centered coronas already suffices to ensure isohedrality [**SD**].

We briefly mention a quite different classification scheme for tilings, one that has proved useful in the theory of nonperiodic tilings and provides insight into periodic tilings as well. To bring some order into the aperiodic zoo, [**BSJ**] introduced the idea of families of *mutually locally derivable* (MLD) tilings. Two tilings are said to belong to the same MLD class if the tiles of either can be partitioned (if necessary) and the pieces regrouped *in situ* to produce the other. (Escher was the all-time master of this technique.) As a mundane example, the edge to edge tilings of the plane by regular hexagons and by equilateral triangles are mutually locally derivable: six triangles sharing a vertex can be grouped into a regular hexagon, and conversely, the hexagons can be decomposed into six equilateral triangles. The nonperiodic Penrose tilings by kites and darts and by rhombs are a more interesting example.

Although incidence symbols do not seem to be as useful a tool for classification as once was hoped, they still have much to teach us. As we will see in the §4, the distinction between marked and unmarked tiles is crucial for understanding the relation between local and global properties of nonperiodic tilings.

4.2. Nonperiodic tilings. In contrast to periodic tilings, the classification of nonperiodic tilings is an intractable problem: as Penrose has noted, it follows from Berger's results that "the different kinds of tiling arrangements that are enforced by subtly constructed prototile shapes must, in a clear sense, defy classification" [**P**].

Nevertheless, some useful distinctions can be made. Just as we distinguish between periodic and nonperiodic tilings, we can also distinguish, among nonperiodic tilings, between those that are *repetitive* and those that are not.

DEFINITION 11. A tiling $\mathcal{T}$ is repetitive if for all positive integers k, the kth coronas of $\mathcal{T}$ are relatively dense in $\mathcal{T}$.

Nonrepetitive tilings include spiral tilings, which have a finite number of "centers," and periodic tilings with "faults." Repetitive tilings include not only the periodic ones (every periodic tiling is repetitive), but also Penrose tilings and most of the other nonperiodic tilings that have been studied over the past two decades.

Two large families of repetitive nonperiodic tilings have been studied intensively. They are generally known as *hierarchical tilings* and *projected tilings. Tilings and Patterns* contains a great deal of information about hierarchical tilings (though this term is not used); projected tilings are a later development.

4.3. Hierarchical tilings. At a Discrete Geometry conference at Oberwolfach in May 1997, Ludwig Danzer posed the following problem: "Give a useful and precise definition of a hierarchical tiling in d dimensions, with $d \geq 2$." This is problem is still open.

Hierarchical tilings are ubiquitous in nature as well as in the tiling literature. Most of them, especially the "natural" ones, have not been studied by mathematicians yet. This suggests that we should not define "hierarchical" too narrowly, lest we rule out structures that may turn out to be interesting. One wants to capture the following idea. There are prototile sets that have the *composition property*: the tiles (let us call them level-0 tiles) can be amalgamated into larger tiles, called level-1 tiles, whose level-1 tiles can be amalgamated into level-2 tiles, and so on *ad infinitum*. For example, 2^n unit hybercubes in E^n can be amalgamated into a hypercube of edgelength two, and so on. We include the case when the original tiles must first be decomposed into smaller shapes. For example, the level-1 tiles of the Penrose tilings by thick and thin rhombs are composed, respectively, of $1 + 2(1/2)$ thick and $2(1/2)$ thin tiles, and $2(1/2)$ thick and $2(1/2)$ thin tiles (Figure 6). A tiling whose tiles can be amalgamated in such a way that the tiles at all levels form tilings (no gaps and no overlaps) is said to be hierarchical.

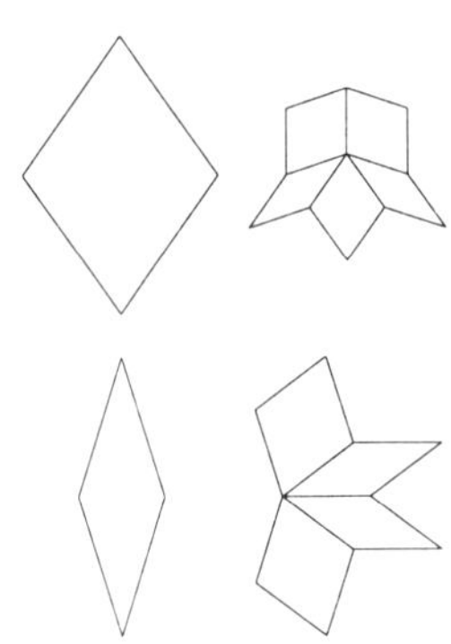

FIGURE 6: Left: thick and thin Penrose rhombs; right: thick and thin rhombs can be inscribed in these cluster (with top and bottom vertices matching those of the cluster), indicating the relation of level-1 tiles to (decomposed into triangles) level-0 tiles.

Some hierarchical tilings, like facet-to-facet tilings of E^n by congruent hypercubes, are periodic; others are nonperiodic. It is reasonable to require that the number of prototiles at each level be finite, but it is in general unnecessarily restrictive to require that the protosets at each level be similar (for interesting examples in which the prototiles at different levels are not similar, see [**GS**]).

If the prototiles at all levels are similar, that is, if the prototiles at level $j+1$ are those of level j inflated by some real number $\lambda > 1$, and if the tilings at all levels are in the same local isomorphism class, the tiling is said to be *self-similar* with expansion factor λ. If $\mathcal{T}$ is self-similar, then there is a simple test for nonperiodicity ([**GS, So**]):

THEOREM 3. *A self-similar tiling $\mathcal{T}$ is nonperiodic if and only if it has the uniqueness property: each level-j tile belongs to a uniquely determined level-$(j+1)$ tile, and so forth.*

In fact, the uniqueness property alone implies nonperidicity; self-similarity is not necessary.

The uniqueness property for self-similar tilings prescribes a decomposition of level-1 tiles into level-0 tiles. We assume that our tilings have the *strong uniqueness property*: decomposition depends only on the prototiles and not on the location of the tiles in the tiling. Then we have a well-defined list of allowed embeddings of level-0 tiles in level-1 tiles. This, in turn, suggests two ways of generating self-similar tilings, beginning with any single tile T.

(i) In the first method, called *inflation* by Grünbaum and Shepherd, we subdivide T, inflate the small tiles to the sizes of the original prototiles, subdivide and inflate again, and so on, *ad infinitum*.

(ii) In the second method, dubbed "up-down generation" by de Bruijn [**deB2**], we embed T in a level-1 tile, then embed that tile in a level-2 tile, and so on. (At each step of this second process we decompose the new (larger) tile into level-0 tiles; uniqueness implies that we can do this unambiguously.)

The second method can be represented graphically by a finite-state automaton whose nodes represent the prototiles and whose directed edges represent the allowed embeddings. Each infinite path corresponds to a particular sequence of embeddings of the starting tile at successively higher levels and (except for a finite number of special cases) represents a tiling of the infinite plane. Two tilings are isometric if and only if their sequences are cofinal. Thus, by a diagonalization argument, we reach the remarkable conclusion that there is an uncountable infinity of tilings with the same hierarchical structure.

The protoset of a self-similar tiling with the strong uniqueness property can be associated to a linear map. Let $\mathcal{P} = \{P_1, \dots, P_j\}$. Then

$$\lambda P_i = m_{i1}P_1 + \cdots + m_{ij}P_j, \quad i = 1, \dots, j$$

where m_{ik} is the number of copies of P_k in the level-1 protile P_i and λ is the inflation factor. The matrix M of this map contains a great deal of information about the tiling. We assume that M is primitive, that is, for some positive integer k all entries of M^k are positive. This ensures that no matter which prototile we start with, copies of every prototile will eventually appear in the tiling; it also ensures that the tiling is repetitive. The expansion factor λ is the leading eigenvalue of M (its existence is ensured by the primitivity of M); the components of the corresponding eigenvector are the relative freqencies with which the prototiles $P_1, \dots, P_j$ appear in the tiling.

The matrix also gives us important information about the diffraction spectrum of the tiling, which is of great importance in quasicrystal theory (see [**SBGC, Se**]) but beyond the scope of this chapter.

It is natural to ask whether every primitive integer matrix can be associated to a hierarchical tiling. This is a difficult question because the matrix gives no information about the shapes of the tiles, and it is highly nontrivial to decide whether a given prototile set has the composition property. However, there are some partial results in this direction [**B**]. A linear map is said to be *expansive* if all of its eigenvalues are greater than one.

THEOREM 4. *(Bandt). Every linear expansive map* $Z^2 \longrightarrow Z^2$ *defines a family of "reptiles" (prototiles of monohedral self-similar tilings).*

Bandt's proof is constructive, and the tiles of the tilings he constructs have fractal boundaries. These tilings are periodic.

More recently, Kenyon has shown that for any algebraic number γ, $|\gamma| > 1$, there is an "inflationary similarity-tiling" of the plane by polygons with expansion factor γ [**K**].

4.4. Projected tilings. The manuscript of *Tilings and Patterns* was essentially completed before de Bruijn introduced *projection* as a way of deriving the Penrose tilings in 1981 [**deB**]. Over the past decade, the "projection method" has been studied intensively (and generalized extensively), both as a tool for studying tilings and as a model for the structure of quasicrystals and other nonperiodic crystals [**Se**].

The projection method, in its simplest "canonical" form, constructs tilings by dualizing sections of periodic tilings. We begin with a lattice L in E^n and its associated Voronoï tiling $\mathcal{V}_L$. First we intersect $\mathcal{V}_L$ with a translate of a lower-dimensional subspace. Let $\mathcal{E}$ be a d-dimensional subspace, where $0 < d < n$ and let $\vec{\gamma} \in E^{n-d}$, the orthogonal complement of $\mathcal{E}$. Let $\mathcal{F}$ be the set of faces of $\mathcal{V}_L$ that have nonempty intersection with $\mathcal{E} + \vec{\gamma}$. To construct the projected tiling, we associate to $\mathcal{F}$ the corresponding set $\mathcal{F}^*$ of faces of the Delaunay tiling $\mathcal{V}_L^*$. If we choose $\vec{\gamma}$ so that $\mathcal{E} + \vec{\gamma}$ does not intersect any face of $\mathcal{V}_L$ of dimension less than $n-d$, then $\mathcal{F}^*$ will not contain any faces of dimension greater than k. The following theorem [**Sc**] guarantees [that the projection of $\mathcal{F}^*$ onto $\mathcal{E}$ is a tiling.

THEOREM 5. *(Schlottmann). Let* $\mathcal{V}_L$ *be the Voronoï tiling of a lattice in* E^n, *let* Π *be the projector of* E^n *onto a d-dimensional subspace* $\mathcal{E}$, *and let* $\mathcal{F}$ *and* $\mathcal{F}^*$ *be as above. Then* $\Pi(\mathcal{F}^*)$ *is a tiling of* $\mathcal{E}$, *dual to the section* $\mathcal{V}_L \cap \mathcal{E}$.

Figure 7 illustrates the simplest possible case of the projection method: here L is the square lattice in E^2 and $\mathcal{E}$ is a line. This simple example illustrates the general situation quite well. De Bruijn showed that we obtain the Penrose rhomb tilings by projection when when L is the integer lattice in E^5 and $\mathcal{E}$ is one of two two-dimensional planes orthogonal to $(1,1,1,1,1)$. The software QuasiTiler (available from the Geometry Center at the University of Minnesota,

http//:www.geom.umn.edu)

allows you to construct tilings of the plane projected from the integer lattices in any dimension up to 13 onto two-flats of any orientation.

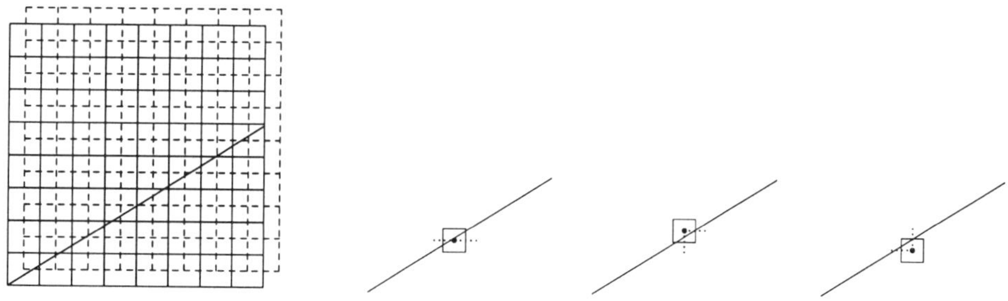

FIGURE 7: (a) L is the square lattice in E^2; its Voronoï cells are squares (solid lines) and its Delaunay cells (dashed lines) are also squares. The one-dimensional subspace $\mathcal{E}$ selects a subset of the vertices, edges, and tiles of the Voronoï tiling (i.e., those faces that it cuts). (b) To each zero-, one-, or two-dimensional face of the Voronoï tiling there corresponds a two-, one-, or zero-dimensional face of the Delaunay tiling. The latter set of faces, projected onto $\mathcal{E}$, constitute the tiling.

If $\mathcal{E}$ is irrational—that is, if it contains at most one point of L—then the projected tiling is nonperiodic. A projected tiling is always repetitive. By suitably varying $\vec{\gamma}$ we obtain a countable infinity of tilings in the same local isomorphism class.

The vertices of a projected tiling are a Delaunay set. In fact, they are a subset of an orbit of a finitely generated Z-module (the projection of the entire lattice is such an orbit). Such a vertex set is called an *almost-lattice* [**M**] because its difference set is contained in the union of a finite number of translates of the set itself ($\Lambda - \Lambda \subset \Lambda - F$), analogous to the closure property of lattices ($L - L = L$).

Loosely speaking, the analogue for projected tilings of the matrix M for hierarchical tilings is the "window" or "acceptance domain" of the projection, in the sense that many properties of the tiling are encoded in it. We denote the orthogonal projector into E^{n-d} by $\Pi^\perp$, and the image of $V(0)$ (the Voronoï cell of the lattice origin) under $\Pi^\perp$ by K.

THEOREM 6. *In the canonical construction, the following are equivalent:*
(a) the Voronoï cell of $x \in L$ *has nonempty intersection with* $\mathcal{E} + \vec{\gamma}$*;*
(b) $x \in L$ *lies in the cylinder* $(K + \vec{\gamma}) \oplus \mathcal{E}$*;*
(c) $\Pi^\perp(x) \in (K + \vec{\gamma})$.

By inspecting the window K, we can determine the relative frequencies not only of the tiles but also of the elements of the corona atlases, for any positive value of k.

The projection method has been generalized in various ways, for example by allowing K to be any compact subset of E^{n-d}, or using multiple windows. Another variant, known as *multigrid dualization*, produces a still larger family of tilings. A multigrid, as defined by de Bruijn [**deB**], is a superposition of a finite number of infinite families of equispaced parallel hyperplanes. Each family is characterized, up to $\pm$, by its interplanar normal vector; the collection of such vectors is the multigrid star. (For example, any section of the regular tiling of E^n by hypercubes is a multigrid; its star is the projection, onto the subspace that defines the section, of the orthonormal basis of E^n.) We dualize this section in the usual way, choosing for the edges of the dual tiling the vectors of the multigrid star.

More generally, we can define a multigrid to be an arrangment of a finite collection of infinite families of "parallel" pseudoplanes (here "parallel" means that no two members of a family intersect); the pseudoplanes need not be stretchable. The family of tilings dual to such arrangement is clearly larger than those produced by projection; it is only beginning to be studied ([**CS**, **Se2**]).

The Penrose tilings (and many other projected tilings) are both hierarchical and projected, but in general the two methods do not produce the same families of tilings. In order for a projected tiling to be hierarchical, the subspace $\mathcal{E}$ must be spanned by algebraic vectors (because the expansion factor λ is an algebraic number). There is no such requirement for projected tilings in general. On the other hand, not every hierarchical tiling is projected. One of the most interesting examples of a hierarchical tiling that is not projected is the self-similar monohedral "pinwheel" tiling shown in Figure 8 [**R**]; it is not projected (from any finite dimensional space) because its tiles appear in infinitely many different orientations.

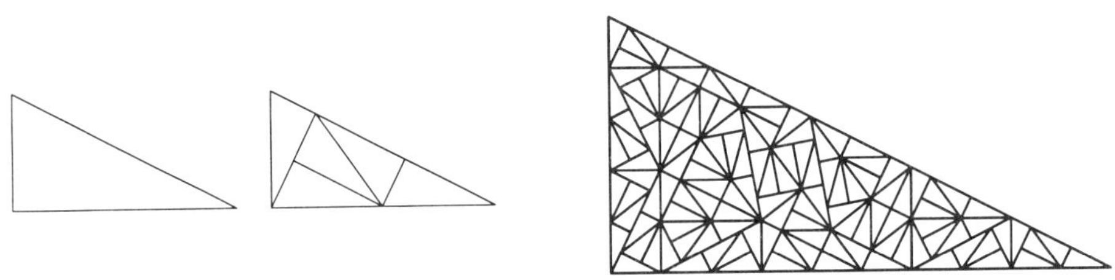

FIGURE 8: (a) The decomposition of the pinwheel prototile; (b) a portion of the tiling.

Returning to classification, we will see in the next section that the problem of classifying nonperiodic tilings is much more complicated than the isohedral case, even if we restrict ourselves to special families such as self-similar or projected tilings. In the first place, the objects that we should classify are not the tilings themselves but local isomorphism classes of tilings.

We will denote the local isomorphism class to which a tiling $\mathcal{T}$ belongs by $\overline{\mathcal{T}}$. Obviously, the kth corona atlas of $\mathcal{T}$ is also the kth corona atlas of $\overline{\mathcal{T}}$ and conversely.

This suggests that tilings should be classified by local properties, but it is not clear that this is always possible. Some local isomorphism classes, such as the class of Penrose tilings, are completely characterized by their facet coronas, while others, such as the Amman tilings (see below) are not characterized by k^{th} coronas for any finite k. Even in cases where the family is characterized by the first facet corona (and thus by an incidence symbol), it may turn out that the symbol does not characterize the MLD class of this family. For example, the Penrose tilings by rhombs and by kites and darts, which are mutually locally derivable (Figure 9), have different incidence symbols.

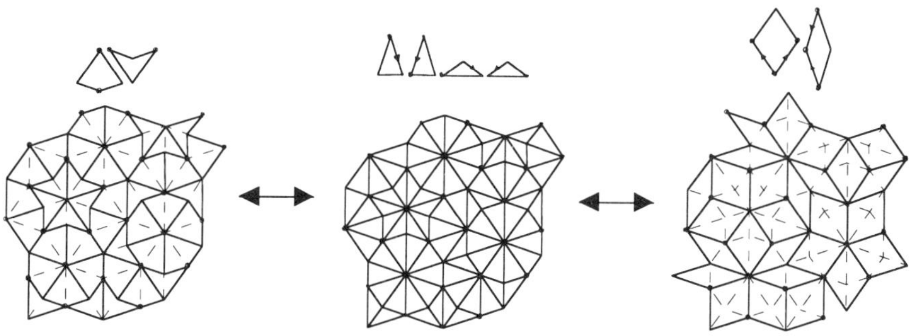

FIGURE 9: The Penrose kite and dart tilings and rhomb tilings are mutually locally derivable.

5. Matching rules and local rules

The Penrose tiles and tilings are paradigmatic: they are hierarchical, they can be obtained by projection, and consequently they have all the properties enjoyed by tilings constructed by either of these methods. They are also paradigmatic in another respect: the local isomorphism class of Penrose tilings can be reconstructed from the individual tiles by juxtaposing them in accordance with Penrose's "matching rules." The Penrose rules can be expressed by placing double and single arrows on the tile edges, as shown in Figure 10. When the rhombs are juxaposed, both the type (single or double) and direction of the arrows must agree.

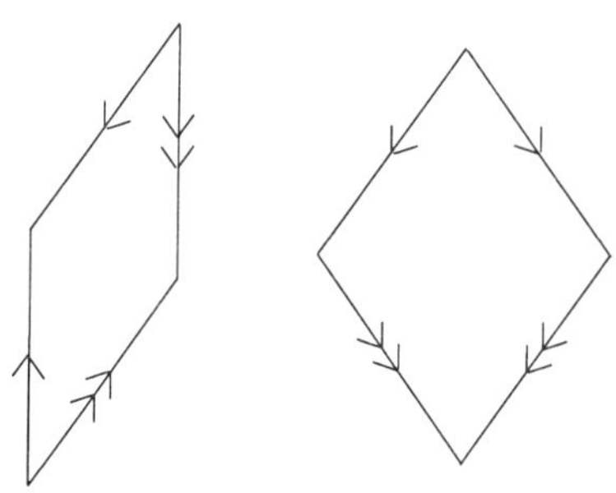

FIGURE 10: The Penrose matching rules can be expressed by double and single arrowed edges.

Matching rules are very attractive: they turn the "tiling game" into a sophisticated non-deterministic jigsaw puzzle. The first examples of what Grünbaum and Shephard called "aperiodic protosets"—protosets that admit only nonperiodic tilings—were presented in terms of matching rules. But today we can easily construct unlimited numbers of nonperiodic tilings, by the methods described above, and it is by no means obvious that matching rules exist for these tilings. What are matching rules, and where do they come from?

In fact, "matching rules" are at least as difficult to define as hierarchical tilings. The idea is a set of instructions for juxtaposing individual tiles, expressed by marking the edges of the tiles, or coloring the vertices, or both. Tiles "equipped" with matching rules are *marked tiles*. Thus matching rules for a protoset $\mathcal{P}$ prescribe the set of coronas admitted by $\mathcal{P}$. We will denote the marked prototiles by $\overline{\mathcal{P}}$.

The most important case is when the rules are *perfect*.

DEFINITION 12. A set of matching rules for a protoset $\mathcal{P}$ is said to be perfect if the marked protoset admits a single local isomorphism class of tilings.

Which unmarked tilings admit matching rules? Goodman-Strauss has found an algorithm that produces perfect matching rules for hierarchical tilings (subject to mild restrictions) [**Go**]. However, in most cases each prototile P_i must be marked in several different ways, thus increasing the number of prototiles. Le Tu has shown that perfect matching rules exist for large families of projected tilings when $\mathcal{E}$ is spanned by algebraic vectors [**LeTu**]; again, the number of prototiles is usually increased by the markings.

A protoset with perfect matching rules is either periodic or admits uncountably many distinct tilings. More precisely, Dolbilin has shown [**Do**] that

THEOREM 7. *If a finite protoset admits tilings of exactly one local isomorphism class, and if that family $\mathcal{S}$ of tilings is closed (in a certain technical sense), then either $|\mathcal{S}| = 1$ and that unique tiling is periodic, or else $|\mathcal{S}| = 2^{\aleph_0}$, where $|\mathcal{S}|$ counts the number of trnaslation-equivalence classes of tilings in $\mathcal{S}$.*

If the set of admissible k^{th} coronas (for some fixed k) for an unmarked protoset $\mathcal{P}$ is finite, it is said to constitute a set of *local rules* for the tilings it admits. We emphasize that local rules do not involve markings of any sort. The kth coronas (for all k) of any tiling constructed by inflation or projection are completely determined by the construction, and there are finitely many of them (for each k), so all tilings obtained by either of these methods obey local rules.

DEFINITION 13. A local isomorphism class $\overline{\mathcal{T}}$ of tilings is said to obey perfect local rules if there exists a positive integer k such that any tiling all of whose k^{th} coronas belong to the set of kth coronas of $\overline{\mathcal{T}}$ is an element of $\overline{\mathcal{T}}$.

If $\overline{\mathcal{P}}$ is a set of perfect matching rules for a protoset $\mathcal{P}$, does it determine perfect local rules for the tilings it admits? That is, if we construct a tiling $\mathcal{T}$ with the marked prototiles $\overline{\mathcal{P}}$ and then erase the marks, is $\overline{\mathcal{T}}$ completely characterized by some k^{th} corona atlas alone? A much-studied example is the family of Ammann "octagonal" tilings (Figure 11a), discussed extensively in *Tilings and Patterns* but not with this question in mind. Superficially, the Ammann tilings seem to be very similar to the Penrose tilings. Like the Penrose tilings, they are hierarchical and can be obtained by projection (in this case from E^4). There are two Ammann prototiles, both rhombs. The rhomb prototiles can be equipped with matching rules that force a single local isomorphism class (Figure 11(b)).

But the analogy stops here. The Penrose rhomb tilings obey perfect local rules which are in fact equivalent to the matching rules: the edge arrowings of Figure 10 admit precisely seven facet coronas (Figure 12), and these coronas determine the edge arrowings unambiguously. Thus every tiling whose facet coronas belong to this set is a Penrose tiling [**deB3**]. In contrast, the set of tilings admitted by the Ammann corona atlases *for arbitrarily large* k include periodic as well as nonperiodic tilings [**Bu**].

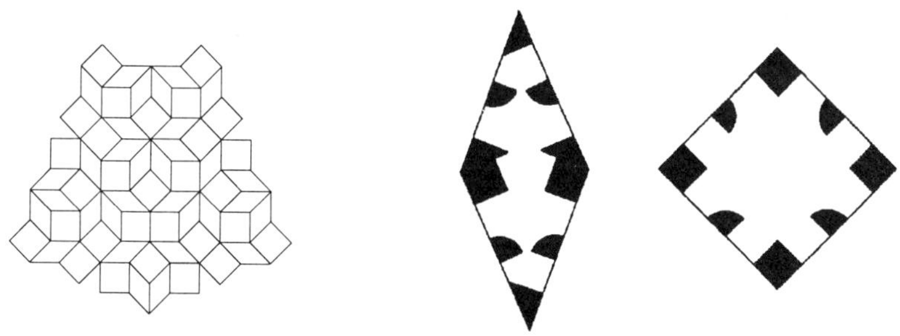

(a) (b)

FIGURE 11: (a) A portion of the Amman tiling; (b) the Ammann matching rules, expressed as markings. Both edges and vertices must be matched.

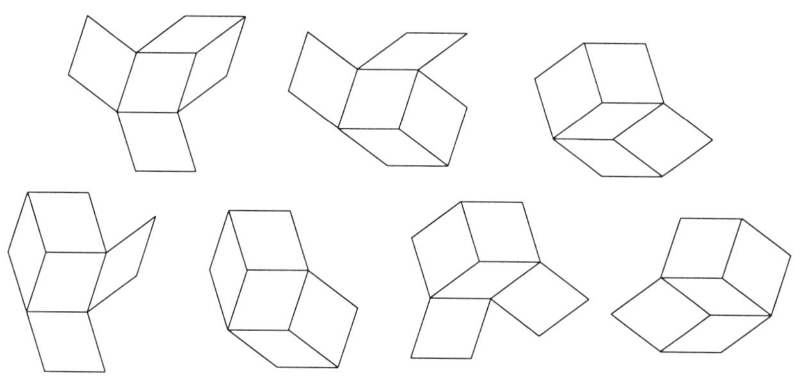

FIGURE 12: The seven facet coronas for the Penrose tiles unambiguously determine the arrowings shown in Figure 10.

The pinwheel tilings are another example of a family of tilings that has perfect matching rules but no perfect local rules. Lagarias [**L**] has recently shown that, for Delaunay sets, the distinction between isometry and translation classes of ρ-stars (see Definition 6) vanishes when the atlas is perfect: a Delaunay set has a perfect atlas under isometries if and only if it has a perfect atlas under translations (perfect ρ-atlases for Delaunay sets are defined in the obvious way.) The vertex set of the pinwheel tiling does not have a perfect atlas under translations, since its tiles appear in infinitely many orientations. It follows that the pinwheel tilings do not obey perfect local rules. Nevertheless, there are matching rules for the pinwheel tiles that enforce the hierarchical structure (and thus the local isomorphism class) [**R**].

Thus the important distinction between marked and unmarked tilings noted by Grünbaum and Shephard a decade ago in their classification of isohedral (and *ipso facto* periodic) tilings reappears in a crucial way in the classification of nonperiodic tilings.

Finally, we mention Penrose's new tiling [**P**], whose single prototile, a marked regular hexagon, obeys what Penrose calls a "second order matching rule;" the details are fairly complicated so we will not discuss them further here.

6. A new direction: the growth of the sizes of atlases

Projected tilings, though they may be nonperiodic, have vertex sets that are almost-lattices, and as such are still very "orderly." Some hierarchical tilings are almost-lattices, but others are not. Indeed, the term "nonperiodic" applies to a broad spectrum of order types, most of which remain to be discovered. Finding ways to characterize these order types is an important problem.

One approach to this problem, still in its infancy, is to study the growth rates of the sizes of the kth corona atlases of tilings, for increasing values of k. This problem is more tractable for Delauany sets, where we can study the function

$$N_\Lambda(\rho) := |ST_\Lambda(x, \rho)|, \quad x \in \Lambda.$$

We know that if $N_\Lambda(\rho)$ is bounded then Λ is periodic [**DLS**]. What can we say about Λ if $N_\Lambda(\rho)$ is linear? If it has polynomial growth? Exponential?

The theory of periodic and nonperiodic tilings and patterns is still developing. *Tilings and Patterns* continues to be a source of inspiration.

References

[BSJ] M. Baake, M. Schlottmann and P. Jarvis, *Quasiperiodic tilings with tenfold symmetry and equivalence with respect to local derivability*, Journal of Physics A, 24 (1992), 4637–54.

[B] C. Bandt, *Self-similar sets 5. Integer matrices and fractal tilings of R^n*. Proc. of AMS, 112 no. 2, (1991), 549–562.

[Be] R. Berger, Memoirs of the American Mathematical Society No. 66, American Mathematical Society (1996), Providence.

[Bi] L. Bieberbach (1910), *Über die Bewegungsgruppen der n-dimensionalen Euklidischen Räume mit einem endlichen Fundamentalbereich.* Göttinger Nachrichten (1910), 75–84.

[Bi2] L. Bieberbach, *Über die Bewegungsgruppen der Euklidischen Räume.* Mathematische Annallen (1912), 72, 400–12.

[BBNWZ] H. Brown, R. Bülow, J. Neubüser, H. Wondratschek and H. Zassenhaus, Crystallographic Groups of Four Dimensional Space (1978), Wiley, New York.

[CS] H. Crapo and M. Senechal, *Tilings with related zonotopes.* Math. Comp. Modelling, 26 (1995), 8–10, 59–73.

[deB] N. G. de Bruijn (1981), *Algebraic theory of Penrose's non-periodic tilings of the plane.* Proceedings of the Koninklijke Nederlandse Akademie van Wetenschappen Series A, 84 (*Indagationes Mathematicae*, 43) (1981), 38–66.

[deB2] N. G. de Bruijn (1990), *Updown generation of Penrose tilings.* Indagationes Mathematicae, New Series, (1990), 1 (2) 201–19.

[deB3] N. G. de Bruijn, *Remarks on Penrose tilings*, to appear in The Mathematics of Paul Erdös, Springer-Verlag, 1994.

[Bu] S. E. Burkov, *Absence of weak local rules for the planar quasicrystalline tiling with 8-fold symmetry.* Communications in Mathematical Physics (1988), 119, 667–75.

[D] L. Danzer, "*A family of 3D-space fillers not permitting any periodic or quasiperiodic tiling.*" Proceedings of APERIODIC 94 (1995), 11–17.

[De] B. N. Delone (Delaunay), *Sur la partition régulière de l'espace à 4 dimensions.* Izv. akad. Nauk SSSR Otedl. Fiz.-Mat. Nauk, 7 (1929), 79–110, 147–164.

[DDSG] B. N. Delone (Delaunay), N. P. Dolbilin, M. I. Shtogrin and R. V. Galiulin (1976), *A local criterion for regularity of a system of points.* Reports of the Acadamy of Sciences of the U.S.S.R. (in Russian), 227. (English translation: Soviet Math. Dokl., 17 (1976), No. 2, 319–22.

[Do] N. Dolbilin, *The countability of a tiling family and the periodicity of a tiling.* Discrete and Computational Geometry 13, (1995), 405–414.

[DHM] A. W. M. Dress, D. H. Huson and E. Molnár, *The classification of face-transitive 3-D tilings.* Acta Crystallographica A49 (1993), 806–817.

[DH] O. Delgado and D. Huson, *Tiling space by platonic solids I*, preprint.

[DLS] N. P. Dolbilin, J. C. Lagarias and Marjorie Senechal, *Multiregular point systems.* Discrete and Computational Geometry, to appear.

[DS] N. Dolbilin and D. Schattschneider, *The local theorem for tilings*, to appear in the Proceedings of the Fields Institute for Research in Mathematical Sciences.

[G] Martin Gardner, *"Mathematical Games."* Scientific American, 233 (1975), 112–115.

[Go] C. Goodman-Strauss, *Matching rules and substitution tilings*, to appear in the Annals of Mathematics.

[GR] P. M. Gruber and S. S. Ryshkov, *"Facet-to-facet implies face-to-face."* European Journal of Combinatorics, 10 (1989), 83–84.

[GS] B. Grünbaum and G. C. Shephard, Tilings and Patterns. W. H. Freeman, New York, 1987.

[H] D. H. Huson, *The generation and classification of Tile-k-atransitive tilings of the Euclidean plane, the sphere, and the hyperbolic plane.* Geometriae Dedicata 47 (1993), 269–296.

[K] R. Kenyon, *Inflationary tilings with a similarity structure.* Comment. Math. Helf, 69, no. 2, (1994), 169–198.

[LS] J. C. Lagarias and P. Shor, *Keller's cube-tiling conjecture is false in high dimensions.* Bulletin Amer. Math. Soc. 27 (1992), 279–283.

[L] J. C. Lagarias, *Geometric models for quasicrystals. II. Local rules under isometries*, preprint.

[LeTu] Le Tu Quoc Thang, *Local rules for pentagonal quasicrystals.* Discrete and Computational Geometry, 14 (1995), 31–70.

[LeTu2] Le Tu Quoc Thang, *Local rules for quasiperiodic tilings*, in The Mathematics of Aperiodic Order, R. Moody, ed, Kluwer Academic Publishers (NATO Series) (1997), 331–366.

[M] R. Moody, *Meyer sets and their duals*, in The Mathematics of Aperiodic Order, R. Moody, ed., Kluwer Academic Publishers (NATO Series), (1997), 403–441.

[P] R. Penrose, *"Remarks on tiling: details of a* $(1+\epsilon+\epsilon^2)$*-aperiodic set,"* in The Mathematics of Aperiodic Order, R. Moody, ed., Kluwer Academic Publishers (NATO Series) (1997), 467–497.

[Pe] O. Perron, *Uber lückenlose Ausfüllung des n-dimensionalen Raumen durch kongruente Würfel.* Math. Z. 46 (1940), 1–26.

[R] C. Radin, *The pinwheel tilings of the plane.* Annals of Mathematics 139 (1994), 661-702.

[Re] K. Reinhardt, *Zur Zerlegung der euklidischen Räume in kongruente Polytope.* Sitzungsberichte der Preuss. Akademie der Wissenschaften Berlin, (1928), 150–5.

[SD] D. Schattschneider and N. V. Dolblin, *One corona is enough for the Euclidean plane*, to appear in the Proceedings of the Fields Institute for Research in Mathematical Sciences.

[S] P. Schmitt, *An aperiodic prototile in space*, informal note (1988).

[Sc] M. Schlottmann, *Periodic and quasiperiodic Laguerre tilings.* International Journal of Modern Physics, 7, Nos. 6 and 7, 1351–63.

[Se] M. Senechal, Quasicrystals and Geometry. Cambridge, Cambridge University Press, 1995

[Se2] M. Senechal, *A critique of the projection method*, in The Mathematics of Aperiodic Order, R. Moody, ed., Kluwer Academic Publishers (NATO Series), (1997), 521–548.

[SBGC] D. Shechtman, I. Blech, D. Gratias and J. W. Cahn, (1984), *Metallic phase with long-range orientational order and no translational symmetry.* Physical Review Letters, (1984), 53, 1951–3.

[So] B. Solomyak, *Nonperiodic self-similar tilings have the unique composition property*, preprint.

[SS] D. Schattschneider and M. Senechal, *The search for an aperiodic tile*, in preparation.

[VW] L. Vatnika and G. Williams, *Multiregular tilings*, preprint.

[V] G. Voronoï (1909), *Nouvelles applications des paramètres continus à la théorie des formes quadratiques, II.* Journal für die reine und angewandte Mathematik, (1909), 136, 67–181.

Department of Mathematics, Smith College, Northampton, MA 01063
E-mail address: senechal@minkowski.smith.edu

Contemporary Mathematics
Volume **223**, 1999

The Early Years of Computational Geometry—a Personal Memoir

Michael Ian Shamos

ABSTRACT. A first-person account of the development of geometric algorithms at the Yale University Department of Computer Science and Carnegie Mellon University during 1972-1978 by one of the architects of computational geometry through his interactions with other researchers such as Dan Hoey, Stanley Eisenstat, H. R. Strong, Gideon Yuval and Jon Bentley.

Computational geometry (CG) as a discipline is now a quarter-century old and has given rise to several entire journals, scores of conferences and thousands of publications. Because many sources credit me with having organized the subject and formulated and solved some of its early problems [**S**], it seems appropriate for me to relate the story of how computational geometry was born.

Most practitioners of CG are not aware that all of my work was motivated by a single problem that I tried to solve while a graduate student at Yale in the early 1970s. The problem was difficult because not only did the necessary tools for solving it not exist, but the problem itself had never been examined from the point of view of computational complexity. However, I cannot explain how I came upon the problem, let alone why I became obsessed with building a new discipline of computer science, without offering some biographical material.

My upbringing in New York City was science-oriented. My father was Chairman of the Physics Department at New York University and is now emeritus. He was an experimental physicist with a facility for mathematics and I spent a lot of time tinkering in his cosmic ray laboratory on the roof of NYU's Main Building. A Geiger counter was always on to warn of any spill of radioactive material, which contributed to the excitement. The place was loaded with test equipment for making sensitive electrical measurements and constant pain with trying to make it work turned me eventually to a career in theory.

My mathematical education was in the hands of my grandfather who, as a civil engineer and surveyor, laid out rights of way for the New York Central Railroad. His training was classical and in weekly geometry sessions he taught me the algorithmic

1991 *Mathematics Subject Classification.* Primary 68U05; Secondary 68-03, 01-08.

Key words and phrases. Computational geometry, analysis of algorithms, history, closest-points problem, Voronoi diagram, geometric intersection.

The author thanks an anonymous referee for pointing out embarrassing errors and making important suggestions to improve the presentation.

©1998 Michael Ian Shamos

methods of Euclid. I was fascinated then, as I am now, at the power of ruler and compass constructions and was struck by the close connection in Greek geometry between an algorithm and its corresponding proof of correctness. Sundays with him were a world of reverse curves, drafting instruments and seven-place log tables. My grandfather showed me the workings of his planimeter, which I still possess. This device, now remembered only in museums, could figure the area enclosed by a simple curve just by tracing a stylus over the curve's boundary. I was amazed that a mechanical device could do this, and I still remember the machine's dials turning as the stylus moved over a piece of paper. At the time, I couldn't imagine how this was enough to calculate the area. This was perhaps my first tangible experience with geometry and computation. It was with great satisfaction 20 years later that I found an optimal algorithm to compute the area of a plane polygon,[1] using the computer as a "discrete planimeter."

I entered the Horace Mann-Barnard School (then the Horace Mann School for Boys) in 1958 and was lucky to have Bob Moses as my first formal math teacher. He left the school a few years later to become a civil-rights worker and now runs the Algebra Project in the Boston area, a grass-roots effort to teach inner-city children the wonders of mathematics. He was and is a true educator. During my freshman year, he gathered up a few students, tossed away the textbook, and had us experimenting with matrix algebra while our less fortunate colleagues were struggling with quadratic equations. This early exposure to matrices was important background for me in solving the polygon area problem, which for a century had been formulated as computing the determinant of a matrix of special form.

While at Horace Mann, I was selected to participate in the Columbia University Science Honors Program, which provided high-school students from all over New York City with advanced science courses taught by Columbia faculty members. It was there that I was introduced to computers by two researchers from NASA, Danny Gardner and Paul Schneck. We were allowed access to the IBM 650 at the T. J. Watson Computation Laboratory. This machine had no disk memory and its only input and output were in the form of punched cards, yet it could compute much faster than mechanical calculators. We soon graduated to the IBM 7094, which at least was transistorized and my friend Allan Cooper, who now works for Microsoft, took the bold step one day of walking into the IBM Data Center at 59th Street and Madison Avenue and asking whether IBM would let the two of us use the machine for free. In 1962, IBM was charging $900 an hour for a 7094. The manager was so astounded at the request that he granted it, and for the next year I spent my off-hours there on a variety of programming projects.

NYU also had a 7094 that was being run by one of my father's faculty acquaintances, Jack Schwartz. His name is now familiar to you as the author of many papers in CG, but you may not know that he was a computer science pioneer 35 years ago. Jack was always bubbling with ideas. In 1963, many years before parallel computing was taken seriously, he circulated a memo describing an interconnection of 1024 mainframes to allow them to cooperate in the solution of numerical problems. He gave Allan and me summer jobs at NYU's Courant Institute. Jack had noticed that most programs written in assembly language ran for a few milliseconds before causing the machine to crash. It was nearly impossible to determine

[1] This is easy to do in linear time. The problem is to reduce the number of multiplications required to a minimum.

where the program had gone wrong because of a lack of debugging tools. (The concept of a debugging tool did not even exist then.) He asked us to write a system program, which we named SENTRY, that would monitor assembly language programs to trap illegal operations and address references. This was essentially the first software debugger, the concept of which you may credit to Jack Schwartz.

All of this computing activity managed to get me admitted to Princeton. I went there to study physics, but was sidetracked somewhat by mathematics. The Physics Department did not accept majors until junior year, while Math allowed sophomore concentration, so I majored in Physics and minored in Mathematics. This created something of a course overload, which I solved by managing never to take any courses in the English Department. My mother to this day wonders how I got through Princeton's liberal arts curriculum without taking a literature course.

The quadrangle housing physics and math was alive with geniuses whom one would pass without fanfare on the sidewalk. Alonzo Church, who was technically in the Philosophy Department, was there. (Back then, when professors spoke reverently of Church's Thesis, I thought it was the document he wrote to get a Ph.D.!) C. C. Gillespie, the historian of science, Solomon Lefschetz, a founder of topology, A. S. Wightman, the mathematical physicist, Marvin Goldberger and Sam Treiman of the Goldberger-Treiman relations (the phrase had nothing to do with their personal lives) could all be found within a few yards of one another. If knowledge could be acquired through osmosis, this was the place to be.

As a sophomore, I enrolled in Albert W. Tucker's Topology Seminar. He was a pioneer in the field, and made major contributions to game theory and operations research as well. He is remembered fondly by everyone except me. For the seminar I wrote a paper in which I tried to define a metric on plane closed curves that would correspond to an intuitive notion of closeness of fit. The work was, alas, more intuitive than rigorous, a defect to which I also readily admit in my later research. Tucker would have none of it. I was summoned to his office by the University police. While this sounds frightening, it was the standard method of delivering important messages to students on campus. It never occurred to me that there might be anything amiss and I even dreamed that he might want to collaborate on extending the results. This fantasy was quickly dispelled. He asked me to explain my paper to him, and questioned me sternly on every statement in it. After an hour, he sat back, apparently pacified, and said, "The reason I called you in here was to determine if you were pulling my leg or whether you are just stupid. I am relieved to find it was the latter, so you will receive a passing grade. If you had been trying to fool me, you would have failed the course." While he was caustic, Tucker actually did me a favor by demonstrating conclusively that I had no future in topology, a subject of which I had no inherent grasp.

My seminal undergraduate experience was working as a research assistant to John Wheeler, who was best known as Feynman's Ph.D. advisor and for inventing the term "black hole." Long hours in laboratory courses had turned my taste decidedly toward the theoretical. But how I could I get Wheeler to accept me as a student? His field was general relativity, which was not even taught until graduate school. For this reason, although he had a very active research group that met weekly, he had only one other undergraduate assistant.

I prepared in secret by spending the first semester of junior year writing a review of variational methods in general relativity and made an appointment to show it to Wheeler. He leafed through the paper, but it was impossible to tell whether it had made an impression. As a review, it contained no new results. He then said, "Let's take a walk." I later learned that this meant he was about to pose a problem—Wheeler did most of his work with colleagues while on foot. He continued, "Back in 1945, Feynman and I proved that electrodynamics can be derived without the use of fields. That is, everything can be explained in terms of bodies acting on one another at a distance. I have always wondered whether one could develop a similar theory of gravitation. You can find our paper in the *Reviews of Modern Physics* that year. Read it through and then come see me so I can help you get started."[**WF**] I was dumbfounded. Here Wheeler had been wondering about a question for 20 years, and he expected me to answer it.

That was the critical moment in my research life. If Wheeler thought I could do it, he must be right. It was as if he had held on to the problem for two decades waiting for someone at the right level to work on it. It wasn't difficult enough for a Ph.D. student, and didn't merit a significant amount of Wheeler's time, but it was a good challenge for an advanced undergraduate. It was invigorating to think that a famous scientist felt I could create new theory when I had never done so before.

Since I didn't have any original ideas about action-at-a-distance gravitation at the time, I made a huge effort studying the work of others. Unfortunately (or fortunately), there were no published papers that even mentioned the concept. However, I developed the habit of reading scientific journals obsessively, an addiction fueled by the fact that the Math-Physics library was open 24 hours a day. I even skimmed over papers I couldn't begin to understand, just to absorb their notation and remember their results.

This behavior came in handy one day in the Spring of 1968. At the regular meeting of Wheeler's research group, someone posed the question whether the existence of matter necessarily implied the existence of electromagnetic force. It was of course known that matter implies gravitation—that is what general relativity is all about. But could one have a universe entirely of neutrons, for example, devoid of electromagnetism? I immediately answered that electromagnetism must exist in any universe, since Einstein had proven it. The group eyed me suspiciously. They were on intimate terms with Einstein's work; some of them, like Wheeler, had been on intimate terms with Einstein himself. Nobody recalled such a result. I said, "Oh, yes, it appeared in a paper of his in the *Revistas Mexicanas de Física* in the early 50s. I'll be happy to run over to the library and bring back a copy." Wheeler said they would all be obliged if I would do so. I returned with it about 20 minutes later and the paper was passed around the table. It indeed contained the assertion that electromagnetism must exist. The group was able to spot an error in the paper rather quickly, and they assured me the problem was still unsolved despite Einstein's claim. From this episode I acquired a reputation for having an encyclopedic familiarity with the published literature. Ritualistic reading of publications would stand me in good stead five years later as I tried to assemble hundreds of geometric problems for computational attack.

Wheeler's intuition about action-at-a-distance was correct. There is a formulation of gravity theory that does not require fields and is perfectly consistent with "classical" general relativity. This work formed the subject of my undergraduate thesis [**Sh**]. The problem was rather abstruse, as the very existence of gravitational radiation, the subject of the thesis, had not been demonstrated at the time and it is still not known whether the phenomenon exists. Later, with the help of Morton Tavel at the Vassar College Physics Department, I extended the work to acoustics [**ST**]. Tavel was fascinated because the acoustic result is completely counterintuitive. Neither electrodynamics nor gravitation requires a medium other than vacuous space in which waves can travel. Acoustics requires a medium; otherwise pressure waves cannot exist. It therefore seems impossible to have action-at-a-distance in acoustics. Yet it *is* possible; however, the velocity of the waves is zero unless there is a medium present, so the result is consistent with classical acoustics.

My hobby for a time at Princeton was computing, which then was under the firm control of the engineering school, a foreign domain to us physicists. Because of my earlier work as a system programmer for Jack Schwartz and a somewhat irreverent attitude toward rules and regulations, I engaged in some activity that would now be considered "hacking," and was expelled for the rest of my undergraduate days from the university computing center for a prank that was innocently intended but had the effect of wiping out a month's worth of computer accounting records. The cost to the university to reconstruct the lost data was approximately equal to my year's tuition. The incident did not diminish my enthusiasm for computers, but extinguished any opportunity I might have had to act on it.

Upon graduation in 1968, I went to work immediately for IBM in East Fishkill, New York at their semiconductor manufacturing plant. This was IBM's largest facility in the world, with over 20,000 employees, dwarfing the more famous "Main Plant" in Poughkeepsie where mid-range mainframes were assembled. Fishkill turned sand into logic. Literally. They grew silicon crystals for computer modules right there in the plant. My job was to write quality control software to analyze electronic test results performed in real-time on semiconductor circuits. Familiarity with test equipment gained in my father's laboratory was useful here; none of the other programmers had any idea what an ohm was. The assignment may not sound fascinating, but it was. Our resources were essentially unlimited, and the problem was practical and vast. The factory turned out the logic devices used in a majority of IBM computers (then a majority of the world's computers) and costs, yields and quality were essential concerns. I labored happily at the task until Vietnam intervened and I was obliged in 1970 to go into government service.

What for many turned into tragedy was fortuitous for me and probably led directly to the development of CG. I attempted to enlist in the Army, but for a variety of reasons, including poor eyesight, I wound up as a commissioned officer in the U.S. Public Health Service at the National Institutes of Health (NIH) in Bethesda, Maryland. I was assigned to the National Cancer Institute (NCI), to the division responsible for developing and testing chemotherapy drugs.

The trick in chemotherapy is to kill cancer cells without destroying too many normal cells. This is a sensitive matter of toxicity and involves getting the right dose of the right compounds delivered to the right place in the body for the right amount of time. NCI spent huge sums (in those days their budget was about $1 billion a year) studying chemical structures, trying to determine which portions of a compound acted on cancer cells and which portions that caused healthy cells to die could be eliminated. Richard Feldmann of the NIH Division of Computer Research and Technology (DCRT) had developed an interactive system to display and search chemical substructures. This was revolutionary work. Indexing of chemical compounds had previously been primitive, and there was no effective way to look up, for example, all known compounds that contained a particular configuration of atoms or radicals. In Feldmann's system, all you had to do was draw a structure on the computer screen with a light pen and you were presented with a list of chemicals that incorporated that structure.

I was assigned as the liaison between Feldmann and NCI. One of the problems he had not yet solved was how to draw chemical structures on the computer screen in a presentable way. They were technically accurate in the sense that all of the correct bonds were shown, but they didn't look nice. I was seduced by the problem. If you open any book on organic chemistry, you will find it filled with beautiful drawings of chemical structures. These are prepared by artists. The issue is how to get a computer to draw them that way. Feldmann explained to me that chemical structures are graphs with labeled nodes, so if I wanted to figure out how to do it I had better read up on graph theory.

I had no acquaintance with that subject, although I did know what it was. There were very few texts available, but the most accessible appeared to be Harary [**H**]. I obtained a copy and spent nearly a year working through most of the exercises. The book contains this intriguing statement on p. 106: "One of the most fascinating areas of study in the theory of planar graphs is the interplay between considering a graph as a combinatorial object and as a geometric figure." That was exactly what we were studying at NIH. Harary went on to mention the result of Istvan Fáry that every planar graph can be embedded in the plane so that all of its edges are straight line segments [**F**]. It started me wondering how one would go about actually finding such an embedding.

I attended courses at DCRT that made me realize that I needed to go to graduate school in computer science. My fiancée, Julie Van Allen, learned that Yale had just started a department and was looking for students. I applied and was accepted under the unusual condition that I would have no fellowship and had to pay graduate tuition. I was, however, eligible for Veteran's benefits that would pay living expenses. For my first two years at Yale, I labored under the detriment of being the only graduate student in the Computer Science Department who actually paid tuition to be there.

The department at Yale was small, filled with bright faculty members and extremely research oriented. Where else could one work with Alan Perlis, the creator of Algol, and Ned Irons, inventor of syntax-directed compiling, Martin Schultz, the numerical analyst who wrote the fundamental text on spline theory, and attend lectures by Sam Winograd, a pioneer in arithmetic complexity? Yale's approach to starting new departments is that they are expected to be the equal of any in the University and certainly the equal of any outside the University!

Yale's attitude toward Ph.D. qualifying exams reflected this seriousness of purpose. We were expected to complete them all during the first year and then devote every moment thereafter to research. The Computer Science Ph.D. qualifiers were exams like no other. We were tested in three subjects: Numerical Analysis, Hardware and Programming, and Computational Complexity. Each exam was take-home and lasted six days, to be turned in at noon. Upon handing in one exam, we received the next, and so forth—18 straight days of exams each semester, for a total of 36 days of qualifiers in a nine-month year! And don't imagine that the fact that the exams were take-home was any help; all of the problems given were unsolved and the answers could not be found in any book or paper. Eating and sleeping were our own affair; if we chose to engage in these activities it was at the expense of the exams. In 1975, when I came to Carnegie Mellon, I learned that their qualifying exam lasted 24 hours, which they thought was quite a challenge.

After Yale, I still laugh at the concept. I passed all the qualifiers with distinction during the first year, which enhanced my reputation in the Department, and I was free to do research full-time.

During that first year, three events took place that altered my view of computer science. The first was the publication of Tarjan's linear-time algorithm for determining graph planarity [**T**]. It is difficult to appreciate now how revolutionary this result was back then. All previous work on planarity testing had been based on the mathematically elegant but computationally clumsy characterization due to Kuratowski that a graph is planar if it does not contain a subgraph homeomorphic to the complete graph K_5 or the complete bipartite graph $K_{3,3}$. Direct application of the theorem leads (with some difficulty) to an $O(n^5)$ algorithm. To do it in linear time was mind-boggling, but it wasn't just the speed of Tarjan's method that caught my attention. It was that he did not base his algorithm on any of the classical theorems characterizing planar graphs. His approach was completely constructivist—he just started to lay out the graph so that if it were planar he would produce a plane embedding and if the graph were not planar the algorithm would report failure. This taught me that developing fast algorithms sometimes requires going beyond traditional mathematics.

The second important event was the appearance of Richard Karp's paper, "Reducibility Among Combinatorial Problems" [**K**], in which he showed that a large collection of important and seemingly unrelated problems could be solved in polynomial time if any one of them could. His method of mutual reducibility seemed a powerful and efficient way (from the point of view of the theorist's time, not computational effort) to use algorithms in multiple contexts. My plan for CG was modeled directly after Karp's structure.

The third inspiration was the publication of Volume III of Knuth, "Sorting and Searching" [**Kn**]. Here was a huge collection of potential thesis topics! In those days, it was commonly accepted that if you could solve a level-50 problem in Knuth, you would get a Ph.D. What amazed me about these unsolved problems was how simple some of them appeared to be. I do not, of course, mean simple to solve, but simple to state and apparently quite fundamental. One of them was the "post-office" problem: Given n points in the plane, how quickly can it be determined which of them is closest to a new given point p? I couldn't believe this hadn't been solved, since it seemed so important for geographic information systems, but it hadn't. An intimately related problem is the closest-points problem: given n points in the plane, which two are closest together? It was a fixation with solving this simple problem that ultimately led to the algorithmic development of the Voronoi diagram.

I joined the graph theory seminar, which met weekly to discuss current problems and papers. I was assigned to present Tarjan's algorithm to the group and suggest a research problem related to it. When I thought of planarity, I immediately recalled the straight-line result of Fáry, and decided that it might be time to develop a fast algorithm to produce straight-line embeddings of planar graphs. The seminar advisor, Stan Eisenstat, thought this was a worthy problem, and I began working on it. Tarjan's algorithm did not seem to be useful immediately, so I went back to Fáry's original paper, which gave a constructive method, and tried to implement it.

Fáry's construction requires repeatedly finding a point interior to the kernel of a star-shaped polygon. I had never heard of a kernel before, nor even of a star-shaped polygon, so I set out to find more about them in the literature. The most lucid explanation was in Yaglom and Boltyanskii [**YB**]. A star-shaped polygon is one having an interior point p with the property that the straight line segment joining p with each boundary point lies wholly within the polygon. The kernel of the polygon is the union of all such points p. A star-shaped polygon thus is one having a non-empty kernel.

Essentially the only classical result on star-shaped polygons at the time was Krasnoselsky's Theorem, which states that a polygon P is star-shaped if and only if, for every four vertices of P there is a point such that the segment joining the point to each of the four vertices lies within P. This is a combinatorial characterization, and although it gives an algorithmic test for star-shapedness, it provides no assistance in constructing the kernel. The material in Yaglom and Boltyanskii made it clear that the kernel is just the intersection of the interior half-planes of P. (Each edge e of a simple polygon P determines two half-planes. The interior half-plane is the one that lies to the interior side of e.) I therefore had a simple algorithmic prescription for the kernel. An n-gon has n edges, so finding the kernel of an n-gon is no more difficult than forming the intersection of n half-planes.

In 1973 I thought it would be a simple matter to go to the literature to find an algorithm for intersecting half-planes and the problem would be solved. So I started looking. To say that I searched high and low would be an understatement. Beginning with Collected Algorithms from CACM, I then went through every computer journal in Yale's substantial collection. There was nothing. I couldn't even find a good algorithm for determining whether a two line segments intersect, a seemingly trivial calculation that presents some unexpected pitfalls. At this point the choice was to abandon the problem or develop my own tools for solving it. Giving up was not a genuine option. Instead, I developed the obvious algorithm, which was to start with one half-plane, and intersect it repeatedly with each other half-plane in turn. This required observing that the intersection of k half-planes is a convex (but possibly unbounded) convex figure having at most k edges. The intersection of one of these objects with a half-plane can easily be formed in $O(k)$ time, which results in an $O(n^2)$ algorithm for the kernel of an n-gon. Applying the result to Fáry's construction gave an $O(n^3)$ algorithm for finding a straight-line embedding of a planar graph. Viewed today, this is an extremely crude algorithm, but it was the first time anyone had performed any algorithmic analysis of the problem. I had no insight then into how the procedure might be improved.

In the Spring of 1973, I presented this result to the Yale Computer Science Department (CSD) at a seminar in which each graduate student was required to report his research progress. Alan Perlis, the late spiritual leader of the Department, attended these seminars faithfully. Perlis exuded genius. He had founded and built the CMU Computer Science Department and was persuaded by Yale to assist in the establishment of a department there. His statements on even the most mundane topics were fascinating, and he peppered his lectures with pithy statements that his students still remember today. (One of them is "ALGOL is a blight."—quite a statement coming from one of its creators.) It was speculated among the graduate students that Perlis was the only Yale faculty member who would be able to pass all the Ph.D. qualifying exams if he had to.[2]

[2] I learned later that this sort of postulating is common among graduate students everywhere. When I arrived at CMU, I was immediately told that Alan Newell was the only faculty member who could pass the qualifying exams there.

Perlis, who had a particular interest in geometry problems, listened attentively to my talk. Toward the end, I made the mistake of saying that I thought the $O(n^2)$ algorithm for intersection of half-planes might be optimal. Perlis instantly made a face at me, and it was clear he was skeptical. He was certainly justified, and not only because faster algorithms have subsequently been found. At the time, there was no computational problem having at most n inputs and n outputs for which a polynomial-time algorithm was known and for which anyone had proven a lower bound worse than $\Omega(n \log n)$. After the talk was over, I went to Stan Eisenstat and confessed that Perlis's reaction had shaken me. Eisenstat had an advance proof copy of Aho, Hopcroft and Ullman's *Analysis of Computer Algorithms*, which contained extensive material on divide-and-conquer. Why not try it?—Eisenstat suggested—nothing else had worked. Divide the n half-planes into two sets, each having $n/2$ half-planes, form their intersection separately, and see where that goes, he urged.

The next day I went back to Eisenstat and said that the critical merge step was intersecting two convex $n/2$-gons P and Q, but there appeared to be no published literature on that problem. He thought about it, pronounced it trivial to solve in linear time, and sketched an algorithm on the blackboard. Either P and Q have intersecting boundaries, they are disjoint, or one contains the other. In the first case, Eisenstat gave a procedure that started from an intersection point of the boundaries of P and Q and stitched together the boundary of their intersection, using two pointers to keep track of progress along P and Q. I remember it was very elegant, but neither of us was ever able to prove it correct, and it ignored the question of how to find that crucial initial intersection point. However, Eisenstat convinced me that a linear algorithm was possible, and then it was fairly simple to find one.

In the spring of 1973 I was fortuitously introduced to Dan Hoey, an encounter that certainly influenced the course of the field. Hoey was a Yale undergraduate in Computer Science with a profound understanding of things computational. He was always in the machine room nursing the PDP-10 and keeping the Department's Mergenthaler typesetting machine in working order. He was brilliant and friendly but somewhat abstruse. He obviously had deep ideas, but was a bit shy about explaining them orally and even less willing to write anything down. His taste ran to mathematical puzzles and games having a mathematical component, like backgammon. (Along these lines, Hoey later produced extensive analyses of the problem of Rubik's cube. Together we computed tables of bearing-off strategies in backgammon.) He was interested in geometry problems, and saw funny diagrams of mine coming out of the Versatec plotter. We struck up a friendship and spent the next two years challenging each other with problems, inventing algorithms, and writing a host of computer programs, only some of which related to geometry.

One day Shimon Even, a guest lecturer from Haifa, presented a seminar on algorithms for finding minimum spanning trees in graphs. He went over the algorithms of Kruskal and Prim, and even discussed the problem of the maximum spanning tree, which provoked some debate about why anyone would ever want to find one. By that time, I was viewing the entire world through geometric glasses. In my scheme, every problem involving numbers was a geometry problem. This is correct in a rather trivial way. Every problem involving scalars or n-vectors can be considered as a problem involving points in n dimensions. The only issue is whether it is useful in a particular case to take this view. After Even's lecture, I wanted to look at the spanning tree problem in the Euclidean plane, that is, where geometric points are taken to be vertices of the complete graph and the edge weights correspond to pairwise Euclidean distance.

The traditional approach forces one to consider all $n(n-1)/2$ edges of the graph and leads to an $O(n^2 \log n)$ algorithm. I felt that the Euclidean problem was so highly constrained, the edge weights not being independent but determined only by the $2n$ coordinates of the points (even fewer degrees of freedom if allowance is made for rotation, translation and change of scale). The question was how to use these constraints to give a faster algorithm. I introduced the problem to Hoey and said that there was a lot more about MST's we needed to know. For a mathematician this would have been a summons to engage in some deep thought. For computer scientists, it was an invitation to start programming. I suggested coding up the Kruskal and Prim algorithms to see how they worked for the Euclidean case. We did this on the PDP-10, using a Tektronix display for output, so we could see the MST's actually being constructed an edge at a time.

The results were productive. We quickly saw that the maximum degree of any vertex in a Euclidean MST was six. This fact is easily proven once it is known, but it had not previously been published. (We later learned that the result, along with many others, appeared in an internal Bell Laboratories technical memorandum by Ron Graham.) Although we discovered many interesting MST properties, we made no progress in developing a fast algorithm. With the appropriate data structure, Prim's algorithm leads to a quadratic algorithm. Because Kruskal proved that the MST always contains a shortest edge of its associated graph, solving the MST problem is at least as hard as solving the closest-points problem, a useful fact later on.

In the summer of 1973, Julie and I got married and rented a house in the woods of Hamden, CT, a northern suburb of New Haven, and acquired a large Belgian sheepdog named Nassau. The property was next to a state land preserve and was extremely quiet. During the year that we lived there, I do not recall hearing a sound. This environment was conducive to contemplative research. After Julie went to sleep at night, the dog and I would stay up working until about 4:00 a.m. He was very patient with me, and listened with head cocked as I explained the difficulties presented by some new problem. Our honeymoon year (we're now in our 24th) was blissful, and her encouragement was crucial in keeping me going when it would have been so easy to give up geometry for more traditional problems.

It was around this time that I started calling my ragtag collection of results "Computational Geometry," unaware that the term had been used at least twice before for different but related disciplines. In their 1969 book *Perceptrons*, Marvin Minsky and Seymour Papert used the phrase to refer to analysis of raster images, and Robin Forrest in England used it in the context of surface modelling and *CAD/CAM*. All of these disciplines can properly be called computational geometry because they all involve solving geometric problems by computer. In this memoir, I use CG to mean the subclass of discrete algorithmic problems in [**PS**] and the journal *Discrete & Computational Geometry*.

In 1973, David Dobkin and Richard Lipton joined the faculty. They worked on a host of problem in computational complexity, including some geometric ones. Their chief contribution to my work at the time was a short preprint entitled "On Some Generalizations of Binary Search," in which they showed how to determine in which region of the plane formed by n intersecting lines a particular point p was located [**DL**]. They gave an $O(\log n)$ algorithm, if polynomial-time preprocessing of the lines is allowed. The idea is central to geometric searching, and it spurred CG considerably. Lipton and Dobkin have by now had long distinguished careers in theoretical computer science, and are responsible for many geometric results.

Dobkin was the head of my thesis committee, and my scientific relationship with him has been the subject of some speculation. Our personal interaction was often stormy, but we never had any disagreement over the significance of geometry problems or whether the work deserved a doctorate. I felt I was up against a Catch-22 situation—no matter how many new results I came up with, the committee wanted additional ones. Such is the risk of defining a new field for one's Ph.D. thesis—it's difficult to tell when you're done. From my vantage point now, the fact that I was prodded to produce more was beneficial. It was important for the thesis to have enough material and rigorous proofs to convince those outside Yale that computational geometry was a worthy subject.

I devoted the fall semester of 1973 to tidying up work on convex polygons and convex hulls. The planar convex hull problem was known by folklore in the graduate computer science departments until Ron Graham published an elegant paper on the problem in 1972 [**G**]. His was the first complexity analysis of a problem in computational geometry, and he gave an $O(n \log n)$ algorithm. It became apparent early that finding the convex hull of n points in the plane is equivalent to sorting, and I showed that every sorting algorithm has a natural interpretation as a convex hull algorithm. Because of this equivalence, it is easy to prove that hull-finding requires $\Omega(n \log n)$ in the worst case. However, it is not immediately clear that simply identifying the hull points, as opposed to listing them in sequential order, is as difficult, but Andy Yao later showed that to be the case also [**Y**]. With respect to convex polygons, I showed how to find the diameter and width (the shortest possible distance between two parallel lines enclosing the polygon) in linear time, which through the use of the convex hull leads to $O(n \log n)$ algorithms for the diameter and width of point sets in the plane. These results were completely new. In hindsight they seem rather elementary, but the question of how difficult it is computationally to find the diameter of a set of points had never even been asked, let alone answered.

I formulated a large number of problems that seemed interesting and whose solutions could be used to attack other problems. Given two polygons P and Q, are they congruent? Can P be made to fit inside Q? What is the smallest triangle (in a area) containing P? What is the largest triangle (in perimeter) that lies entirely within Q? How many unit circles are required to cover P? What is the smallest number of disjoint convex polygons whose union is Q? Is P simple? Some of these I had no idea how to solve, but I thought they should be looked at from the viewpoint of algorithmic complexity.

I also decided to do a complete review of the literature. I manually scanned every issue of the *Communications of the ACM*, *Journal of the ACM*, *SIAM Review*, *Journal of Combinatorial Theory*, *Discrete Mathematics*, *American Mathematical Monthly*, and dozens of textbooks, looking for geometry problems to solve by computer.

While I was at Yale, the University awarded John Wheeler, my undergraduate advisor, an honorary degree. He appeared at the same ceremony at which I was to receive a master's degree in computer science, so we had a chance to reminisce. Wheeler had always taken a geometric approach to physics, especially cosmology, and was intrigued by CG. We briefly discussed the possibility of using combinatorial geometry as the basis of quantization of space, but nothing has yet come of the idea.

By June 1974, I had collected or posed about 75 problems, which I described in a typewritten manuscript entitled, "Problems in Computational Geometry"[**Sh4**]. This was meant primarily as a workbook for me, and it set forth a short statement of each problem, gave a brief description of the obvious, or naive, algorithm for the problem, followed by a discussion of the current state of the art. A secondary objective of this collection was to convince the CSD that computational geometry was a field, and not just a disconnected set of problems. What I did not count on was the power of the Xerox machine. I sent out about five copies of the manuscript to colleagues at other universities and research institutions. I estimate that several hundred more were made thereafter by the recipients and those to whom they passed copies. This was a boon to the field, for it caused many people to begin thinking about and working on these problems. Science Citation Index reveals that the workbook has been cited scores of times in the literature even though it was never formally published.

Yale saw that Hoey and I were getting along well and arranged for us to have office space together on the top floor of Dunham Laboratory on Hillhouse Avenue, a stone's throw from the building in which the Department is now located. This was providential. One day in the spring of 1974, I walked into the office to see Hoey drawing some strange diagrams with colored markers on acetate. As usual, he was excited and intense and had a cigarette dangling from his mouth. When I asked what he was doing he said, "Given n points in the plane, each point is surrounded by a convex polygon, not necessarily closed, such that if you're in a polygon than the closest point of the set to you is the one that owns that polygon. I call them proximal polygons." Hoey had independently discovered Voronoi diagrams!

He had marker ink all over his hands and showed me his drawings. The pictures themselves were messy, but their underlying structure was beautiful. They looked somewhat familiar to me, and I told Hoey that they looked like objects from solid-state physics that we called Wigner-Seitz cells. This set me off on a literature search to see what had been written about these structures. Hoey and I spent the next several months investigating their properties. I eventually found that they had been discovered by G. Voronoi and described by him in a 1908 paper [**V**]. They have been rediscovered independently many times since. Hoey and I needed some name for the overall structure. He liked the term "proximal diagram," but I knew that whatever name we adopted would soon be in common use and I thought the name of their inventor should be included, so I lobbied for "Voronoi diagram," and the term has stuck. It had never been used before. Despite my reverence for his work, I have never seen Voronoi's paper, since the particular volume that contained it was missing from Yale's Mathematics Library. Likewise "Delaunay triangulation" was a phrase of my own creation.

The Voronoi diagram (VD) looks complicated, but Hoey realized that it only contains a linear number of pieces. Thus in principle a lot of fast algorithms could be obtained if the VD could be found quickly. For example, the two closest points of a set can be found in linear time once the VD is known.

We had tried everything to cut down the time to find closest points in the plane. It was maddening that the two farthest points (the diameter) could be found in $O(n \log n)$ time, but we could do no better that quadratic for closest points. Why was "close" difficult when "far" was easy? I spent a lot of time worrying about that one. The answer, as I suspected, is that there is no real difference between near and far in a very general sense, as was later shown by Kevin Brown in his CMU Ph.D. thesis [**B**]. Hoey and I tried cutting sets of points in half, in quarters, into a $\sqrt{n}$ by $\sqrt{n}$ array of rectangles, and other bizarre divisions. In the worst case, none yielded anything faster than quadratic. We would have been satisfied with $O(n^{1.9})$! At least that would have provided some hope, instead of the brick wall that we faced.

Likewise, Hoey and I were unable initially to produce any fast algorithms for the Voronoi diagram. Using the algorithm for intersecting half-planes, it is easy but tedious to develop an $O(n^2 \log n)$ algorithm. We were able to shave this down to $O(n^2)$ through careful housekeeping, but then progress ground to a dead halt. The difficulty was largely psychological. Since we weren't sure that any subquadratic algorithm for closest points was possible, there wasn't much point in looking for a fast VD procedure. Nevertheless, I was so taken with the potential of the Voronoi diagram that I continued to study its properties, and found that it could be used to find all closest points, smallest perimeter triangle, convex hull, minimum spanning tree and a triangulation of a point set. I also showed that it could also be used to give the first $O(\log n)$ solution to Knuth's post-office problem.

Since even Yale embraced the idea that solving a level-50 problem in Knuth was worth a Ph.D., I knew I was done. In January 1975, I presented a thesis proposal to the Yale CSD called, simply, "Computational Geometry." By that time, I had obtained so many results that it was apparent that the work was sufficient for a thesis, and the proposal was well received. I felt confident that I would be able to finish the thesis by May and made plans to return to IBM, from which I had been on leave for five years. I had no desire to return to East Fishkill—it's a lovely town, but not a center for algorithm research. I wanted to work at IBM's Research Center at Yorktown Heights, NY, so I requested the opportunity to give a research talk as part of my job interview there.

The talk took place on January 28, 1975, with many acquaintances from the IBM Computer Science and Mathematics Departments in attendance. One of them was Ray Strong, whom I knew from a Yale seminar on Petri nets. During the talk, I described efforts to solve the closest points problem and said that I felt irked that we had not be able to do better than $O(n^2)$, despite having "tried everything." After the talk, Strong invited me back to his office to discuss an idea he came up with during my lecture. When we got there, he drew a set of points on the board and outlined a procedure. First find the median x-coordinate of the n points in linear time by the method of Blum et al. The vertical line M having this coordinate divides the set in half. Now find the closest pair of points in the left set and in the right set and let the smaller of the two distances be δ. Now draw lines at distance δ parallel to and on either side of the line M. Strong said it was sufficient to consider only points that lie in the two vertical slabs formed by these lines.

Well, Hoey and I had seen these slabs many times as we had tried the same construction in applying divide-and-conquer to the problem. So I immediately said to Strong, "But in the worst case all n points will lie in the slabs, and you get no improvement." He said, "That's true, but the points in the slabs now have a special structure. They're sparse. Within a slab, no two points are closer together than δ." My response was, "So what? How can you use that fact to do anything?" There followed one of those moments that scientists live for—a flash of insight that produces eternal clarity. Strong said, "That means that for every point in the left slab you only have to look at points in the right slab that are within distance δ, and there can be at most a *constant* number of them. For every point in the left you don't have to look at every point in the right, so the merge step won't be quadratic." This was the whole key to the closest-point problem.

There was a little more work to do; we had to find a procedure for locating that constant number of points (at most six, it turned out) to be examined for each point in each slab. This we did by sorting points by y-coordinate at each recursive step, which led to an $O(n \log^2 n)$ algorithm. I wasn't happy about the extra factor of $\log n$, but I drove back to New Haven ecstatic in the knowledge that the closest-point problem was subquadratic. That suggested that a fast algorithm for the Voronoi diagram might be possible. I showed Strong's result to Hoey and we huddled to see whether we could squeeze out that extra factor of $\log n$. It turns out that if the set of points is sorted initially by y-coordinate, no further sorting steps are required, and the merge step of the divide-and-conquer can be performed in linear time, which gives an $O(n \log n)$ algorithm.

After this, I made a renewed assault on constructing the Voronoi diagram. With Eisenstat's encouragement, I took another look at divide-and-conquer. Suppose we divide a set of n points into two sets, L and R, splitting at the median x-coordinate. Now form the Voronoi diagrams of L and R separately. I had gotten this far many months earlier; the big stumbling block lay in the merge step: how do you join the two diagrams together quickly? By playing with colored pens on transparency film, I could see that a polygonal boundary P always exists between L and R such that all of the Voronoi diagram of L lying to the right of P can be discarded and all of the Voronoi diagram of R lying to the left of P can be discarded. The final diagram consists of the surviving portions plus P. But how to find P?

The answer is that P starts at the convex hull of the set, where it is formed by the perpendicular bisector of a hull edge joining a point of L and a point of R. It then proceeds monotonically downward. Monotonicity is important; it allows P to be constructed without backtracking. Since the Voronoi diagram contains only a linear number of points and edges, a non-backtracking algorithm must finish in linear time. These were the only insights required and the collection of closest-point problems was completed about a week after my visit to IBM.

This was none too soon. In February of each year the ACM Computer Science Conference was held. Its principal function, rumor had it, was to serve as a job fair. Department chairmen from all over the country attended to watch potential faculty candidates describe their research. It's a very good way to judge both the depth of a person's work and her ability to communicate. In 1975 the conference was held at the Statler Hilton in Washington, DC, to which I ventured in hope of a decent faculty appointment. Yale had a policy of not making offers to its own graduate students, and in any event the transition from having been a student in a department to faculty in the same department is difficult to manage. IBM, on which I had pinned my hopes, surprised me by telling me that they were "reducing their head count," and it would be much more convenient for them if I just resigned instead of coming back. So, after seven years with IBM, five on full-time unpaid leave, I had to look for a job.

I prepared a short talk on the Voronoi diagram to deliver at the conference, complete with the color drawings that are so effective in communicating its properties. When I arrived at the hotel, I got a copy of the conference proceedings to see who else would be speaking at my session. To my surprise, there was another paper on a geometry problem. Two operations researchers were studying a maximin facilities locations problem: given n points in the plane, where should a new point be located, within the convex hull of the others, so it is as far as possible from any of the n original points? An obvious application of this problem has to do with locating competing stores or noxious facilities such as sewage treatment plants. The authors, I learned from their abstract, would present an $O(n^3)$ algorithm the next day in my session.

I had never heard of the problem, but anything related to closest points screamed "Voronoi diagram" to me. The abstract in the conference program was tantalizing because it described only the result, not the method used to obtain it. I felt that this would be a trial by fire for Voronoi diagrams, since it was the first closest-point problem I hadn't known about at the time I found an $O(n \log n)$ algorithm for its construction. If the VD could be used to solve an entirely new problem, it was clearly a powerful tool.

The night before the talk, I stayed up to work on the facilities location problem. It was immediately clear that if the facility could be located at a point interior to the convex hull, then it must be at a vertex of the Voronoi diagram. (If you move away from a Voronoi point, you're getting *closer* to some point, not farther away.) The troublesome part was that if the point lay on the convex hull, it might be anywhere on the boundary. It took another hour to prove that in that case it must lie at an intersection of a convex hull edge and a Voronoi edge. That completed, I had an $O(n \log n)$ algorithm for a problem whose posers were to present a cubic algorithm the next day. While this was exciting, I worried about whether it was fair to the other authors to present it. I did so as politely as possible, and Joseph Traub, the chairman of the Computer Science Department at Carnegie Mellon, who was in the audience, invited me for a job interview.

When I came to CMU for the interview, Traub had arranged for me to see George Fix, the chairman of the Mathematics Department. Fix was a former colleague of Martin Schultz and had worked on triangulation algorithms for the finite element method, so he was receptive to CG. Later that day, after a hasty conference with Fix, Traub offered me a joint appointment in the Computer Science and Mathematics Departments. I was thrilled at this honor, but later learned the bad news—I had to maintain a full teaching load in both departments! I looked at the catalog of the Math Department and noticed that Combinatorics had not been taught during the preceding five years. I immediately proposed to resurrect it. Fix said he would have to put it to a vote of the faculty. It turned out that he had quite a lobbying job. The CMU Math Department was heavily dominated by analysts. When Fix explained to them that combinatorics was only being taught every five years, one of them grumbled, "Why so often?"

I had a few months left at Yale, and returned to a problem that had been nagging me for two years: how can you tell whether a polygon is self-intersecting? It is easy to check by examining each pair of edges to determine whether they intersection but this gives a quadratic algorithm. Could it be done faster? I posed the problem to Hoey, and after a week of intense introspection fueled by coffee and cigarettes and interrupted only by an occasional game of backgammon, Hoey had a beautiful idea. He first generalized the problem to that of determining whether any two of a set S of n line segments intersect, and then observed that if none of the segments is vertical, then with respect to any given value of x (say, $x = b$), all segments of S that intersect the line $x = b$ are totally ordered by the "above" relation. That is, of any pair of segments c and d that intersect $x = b$ at abscissas e and f, respectively, then $c > d$ if $e > f$ and $c < d$ if $e < f$, while $c = d$ if $e = f$. As the value of b changes, two segments can only change places with respect to one another in the total order if they intersect somewhere. Hoey also realized that two segments intersect iff they are adjacent in the total ordering for some value of b. At this point in my thesis research, I was familiar enough with the subject to know that this was a totally original concept. No one had previously defined such a relation.

Hoey suggested sweeping a vertical line through S from left to right, maintaining a list of segments that intersect the line in the order determined by the "above" relation. A segment is added to the list when its left endpoint is encountered and is removed when its right endpoint in encountered. Before a segment is added, it is checked against its neighbors in the ordering to see if it intersects either of them. When a segment is removed, any pair of segments that then become adjacent must be checked for intersection. If each segment is added and eventually deleted from the list without an intersection being found, then S is intersection-free.

It remained to find a data structure in which the total ordering could be maintained, in which n insertions and n deletions could be made, and the successor and predecessor of each segment in the total ordering could be found n times, all in less than quadratic time. This task fell to me, along with proving the algorithm correct. At the time, AVL trees were barely known, and then only through a marginal reference in Knuth. 2-3 trees had just been invented. Either of these data structures would fill the bill, but 2-3 trees were easier to code, so we set about programming. The algorithm worked like a charm and was amazingly fast. It seemed clear that the algorithm worked, but my job of proving it was somewhat tricky because of the need to be careful about degeneracies. I was able to show that it always finds a leftmost endpoint, and wrote up the paper [**SH2**]. This sweep-line algorithm turned out to be a fundamental technique in CG that had no real progenitor. It formalized the tools necessary to solve the hidden-line problem, which had previously been attacked through a hodge-podge of ad hoc methods.

At about the same time, Hoey developed the generalized Voronoi diagram. I had earlier shown him the farthest point VD along with a proof that, unlike the closest-point VD, its graph is outerplanar. We had also experimented with Voronoi diagrams of line segments, polygons, sets of points and line segments, circles and other geometric figures. We also had made such diagrams in the and metrics. Many of these were later rediscovered and published by other researchers. But no one had yet considered the locus of closeness to multiple points. Hoey invented the higher-order VD and gave an argument that the sum of the number of Voronoi edges in all VD's of all orders on a given set of n points in the plane is only $O(n^3)$. (This was a slight surprise, since the total number of subsets of n points is 2^n.) I produced a precise enumeration, which appeared in our Voronoi diagram paper [**SH**]. This is the paper of which I am most proud, since it is so closely modeled after Karp's and achieved the desired result of unifying problems that had not even been mentioned previously in the same breath.

My thesis was not complete when I left Yale in June 1975, but I didn't see the point in remaining a graduate student. I didn't realize that I would get sidetracked by other research and responsibilities, and wouldn't actually turn in a thesis for more than another two years [**Sh2**]. I finally received a Ph.D. from Yale in 1978, exactly three years after leaving. My grandfather lived to be 93 years old, and turned 88 shortly before I got my diploma. He made the trip to New Haven for the graduation ceremony with pride and understood quite well that the degree was for geometric work inspired by our Sunday lessons with his planimeter.

Hoey stayed on at Yale for a time and then became a graduate student at CMU. He had many other interests beside geometry, and now works for the Naval Research Laboratory in Washington. In the meantime, I met two additional collaborators. The first was Jon Bentley, who in 1975 was a graduate student at the University of North Carolina. As his thesis work, he undertook to generalize the sparsity algorithm for finding closest points into higher-dimensional space. His insight was that you could solve an n-point problem in d dimensions by solving two $n/2$ point problems in d dimensions (the divide step) and one n-point problem in d-1 dimensions (the merge step). This gave an $O(n \log^{d-1} n)$ algorithm instead of the rote $O(dn^2)$, and was a major advance [**BS2**]. I told Traub that we needed Bentley at CMU and his appointment was arranged in short order.

Bentley and I worked on a host of problems in geometry, statistics, and combinatorics. Most of them are described in his prodigious published output. Within a few years, he produced over 50 papers. He and Thomas Ottmann generalized the Shamos-Hoey line segment intersection algorithm to enumerate the intersections in $O((k+n)\log n)$ time, a result subsequently improved by others. Bentley and I taught and worked with students all day and then assembled at my house each night at 11:30 to watch reruns of the television series *Kojak* and do research with my dog Nassau, the world's best canine geometer. We were devoted fans of the show, and Bentley, who was constantly finding analogies between search algorithms and detective work, was not above quoting dialogue from *Kojak* in his scientific papers.

Bentley and I worked primarily on blazingly fast algorithms, since we were interested in large practical problems involving huge amounts of data, such as satellite images. For such work even $O(n \log n)$ algorithms are too slow—real-time processing must take no more than linear time. This means in some cases having to accept approximate answers. But not in all. Bentley and I spent a lot of time on average-case analysis. It bothered us that so many such analyses depended on unreasonable assumptions about the statistical distribution of the inputs. We finally found a way to construct the convex hull of a plane set in average linear time regardless of the form of the distribution of the points, provided that the distribution is continuous [**BS**]. From then on we were hooked, and hardly ever looked at another algorithm unless it was $O(n)$.

Jay Kadane of the CMU Statistics Department liked our work on average-case analysis and discrete algorithms for statistical problems and arranged for me to be appointed to a faculty position there also. I thus became the only person at CMU ever to hold appointments in the Computer Science, Mathematics and Statistics Departments simultaneously. (Don't worry, my first experience with a joint appointment left scars. I was promised that I wouldn't have to teach any extra courses, just an annual seminar in computational statistics.) In my view, all statistical problems are geometry problems. Some of my colleagues felt just the reverse was true.

My other collaborator was Gideon Yuval, a visitor to CMU from Israel. Yuval was an intuitive thinker, much like Hoey, and had no taste for formal proofs. He had flashes of great insight, and when he turned his mind to a problem for any length of time one could be sure that interesting ideas would emerge. One day I mentioned to him a stumbling block in the theory of lower bounds. At the time, all non-trivial lower bounds on computation time were based either on fan- in, algebraic independence or information-theoretic arguments. For example, sorting requires $\Omega(n \log n)$ time because every comparison tree containing all $n!$ permutations of n objects has depth at least $n \log n$. There was no problem in existence having n inputs and a single number as output for which a polynomial-time algorithm was known for which any non-linear lower bound had been proven. This bothered me tremendously, since many geometry problems yield single numbers as results, and I couldn't demonstrate any good lower bounds for them.

Yuval asked me which problem I had in mind, and I said the mean distance among n points in the plane. (That is, the sum of the pairwise distances divided by the number of pairs.) This was a maddening problem because the average square of the pairwise distances can be found trivially in linear time. It was also difficult since no techniques were known for dealing with complexity in terms of square roots, except some old results of Hilbert not immediately applicable to this particular problem. Yuval went off to think about it. The next day he said, "Of course, it must take quadratic time in the worst case." This was striking news, since no quadratic lower bound on any polynomial-time problem was then known. (Quadratic, that is, in the number of inputs.) I challenged him to prove it, but proof was not Yuval's strong point. He instead gave an intuitive argument that dazzled me. He said to consider any algorithm that computes the function D that is the sum of the pairwise distances from the $2n$ coordinates of the points. Each distance is a square root. Now consider D as a function of $2n$ complex rather than real variables. The Riemann surface of D must have $\Omega(2^{n(n-1)/2})$ sheets, and therefore at least $n(n-1)/2$ square root operations must be required to produce a Riemann surface of this complexity.

I was familiar enough with Riemann surfaces from undergraduate school, but I had never heard the term used in connection with computational complexity. It seemed clear that Yuval's idea was either very profound or totally ridiculous. After all, what did an ordinary geometry problem have to do with complex variables? I struggled over the next weeks not only to produce a proof, but also to develop a model of computation in which the results would make sense. The "square root" as an elemental operation existed in no known model of computation. The key is having a set of operations in which only the square root operation can cause a bifurcation of the Riemann surface. Though unexpected, the Shamos-Yuval result is as tight as can be. For n points in the plane, exactly $n(n-1)/2$ square roots are necessary and sufficient to compute the mean distance [**SY**]. The result is also very robust in the sense that all kinds of auxiliary operations can be added to the model without reducing the lower bound. In addition to arithmetic operations, one may feel free to add exponentiation, trigonometric functions, hyperbolic functions, Bessel functions and many other special functions. None of these is sufficient to reduce the number of square roots required even by one.

It has always been my practice in looking at research problems to examine the simplest, almost trivial, cases to obtain a full understanding. In geometry, this often meant looking at one-dimensional problems. How difficult is it to compute the mean distance among n points on a line? Here no square roots are involved. If the points are sorted, it is easy. The interval between point k and point $k+1$ is part of the interval connecting all points whose left endpoint is between 1 and k and whose right endpoint is between $k+1$ and n, inclusive. This means that the sum of the distances is just a weighted sum of the intervals between adjacent points. However, since the final result is a single number, how can we prove that sorting is required?

Yuval also provided the key to this one-dimensional case. He observed that in the arithmetic decision tree model, one can always compute partial derivatives of any computed quantity with respect to any of the input variables without any additional comparisons. He then explained that the partial derivatives of the mean distance with respect to the x-coordinates of the points can be used to reconstruct the values of k, that is, the indexes of the points in sorted order. Therefore, any mean distance algorithm can be transformed into a sorting algorithm without performing even one extra comparison. This gives an $\Omega(n \log n)$ lower bound for the mean distance on one dimension. This result is very powerful, for it is a virtual generator of lower bound results. Perhaps because it lies buried in a conference proceedings, it has been utilized only rarely and appears to be unknown to most researchers [**SY**].

During my second three-year appointment at CMU, I began to grow restless. It was difficult to obtain funding for geometry research because so few people were engaged in it that the government agencies assumed it was not important. A shining exception was the Office of Naval Research (ONR), which was a firm supporter from the beginning. Even while I was a graduate student at Yale, ONR allowed me to serve as co-principal investigator with David Dobkin. I was disappointed at the apparently slow uptake of CG by computer scientists, even though I had enthusiastic colleagues in Jon Bentley, Ron Graham, Franco Preparata, D.T. Lee, Leo Guibas, Gottfried Toussaint and Bernard Chazelle, and many others in Europe, such as Herbert Edelsbrunner and Thomas Ottmann.

Being at CMU, I was exposed to a lot of forefront software that was under development there. I noticed that after a project's final report was turned in, nothing else would be done with the resulting software. I thought this was wasteful and in 1979 I became a reluctant entrepreneur. I formed a software company to market the Scribe document production system, which was developed by Brian Reid, then a graduate student at CMU. His system was the forerunner of desktop publishing, and he eventually received the Grace Murray Hopper Award of the ACM for it. Growing the company turned my attention to business matters, and I eventually found it useful to obtain a law degree. These activities were inconsistent with being a full-time faculty member, so in 1981 I shifted to adjunct status and spent the next 17 years combining a legal career with starting software companies. At the beginning of 1998, I returned to CMU to lead a project called the Universal Library, which can be thought of as an electronic "library of libraries" a gateway to all digital Internet collections. A license to practice law, once obtained, is rarely relinquished, so I am still a partner in the Webb Law Firm, the largest intellectual property practice in Western Pennsylvania, where my concentration, logically, is in computer law.

The existence of my textbook with Franco Preparata [**PS**] is due to the intercession of Ron Graham. Every new discipline needs some time to mature before a sensible introductory book can be written, and indeed in the case of computational geometry none had appeared during the decade following 1975, a year in which many results were published. Graham, an generous and unselfish supporter of research in combinatorics and discrete geometry, thought that it was about time for a book. He would have been perfectly qualified to write one, but he felt that the first book should come from the architects of the subject. After 1981 I was in no position to take on the task, so Graham cajoled Preparata into doing it.

I had met Franco at the SIGACT conference in Albuquerque in 1975 when I presented my first geometry paper [**Sh3**]. We subsequently attended numerous conferences together but had never actually worked on a problem together. He was kind enough to invite me to teach at the Winter School of Computational Geometry in Erice, Italy in 1978. (By curious coincidence, this was the last time I saw Yuval—on the ancient Carthaginian island of Motya, off the western coast of Sicily.) When Franco took up the book project, he proposed to start with the text of my Ph.D. thesis, supplementing it with additional material and adding new chapters. Even though the resulting book contains many of my words and results, it was really written by Preparata, and he is now working on a new edition.

I have no regrets about having retired from active work in computational geometry. I did as much as I could, and feel pleased about having laid out the field and constructed its first tools. Many recent results by others are so complex that I am certain I could never have discovered them myself, and the subject is certainly flourishing. Those who are quick to link cause and effect might note that CG did not really take off until I left it!

Like all fishermen, I sigh at the ones that got away. I wish I had been a participant in some results that escaped me completely: Megiddo's linear-time algorithm for linear programming in two and three dimensions, Lee and Preparata's linear-time algorithm for the kernel of a simple polygon, the results of Tarjan, Chazelle et al. on triangulating polygons, and the Lipton-Tarjan algorithm for separating planar graphs. I came so close with Megiddo. He was visiting CMU for a year in the Graduate School of Industrial Administration, where operations research was done. He called me up one day from across campus and asked whether I knew that my $O(n \log n)$ algorithm for the smallest circle enclosing n points in the place was not optimal. I had never heard of him and thought the call was a prank! Now there was a missed opportunity! And it would have been nice to have appreciated the importance of motion planning back then. However, I made my contribution and moved on, and it is a pleasure now to watch the oak trees that have grown from acorns planted in the 1970s.

I marvel that everything in my life preceding the work in computational geometry seemed to have prepared me for it, and am grateful for the confluence of people who contributed their time, ideas and encouragement. They did it because they loved the problems, and worked on them without any thought of having proper credit ascribed to them. Therefore I hope this account clarifies who did what in the early years.

References

[BS] J. L. Bentley and M. I. Shamos, *Divide and conquer for linear expected time.* Info. Proc. Lett. 7:87–91 (1977).

[BS2] J. L. Bentley and M. I. Shamos, *Divide and conquer in multidimensional space.* Proc. Eighth Annual ACM Symposium on Automata and Theory of Computing (May, 1976) 220–230.

[B] K. Q. Brown, *Geometric Transformations for Fast Geometric Algorithms.* Ph.D. Thesis, Department of Computer Science, Carnegie Mellon University (1979).

[DL] D. P. Dobkin and R. J. Lipton, *On some generalizations of binary search.* Yale University Computer Science Department Technical Report (1974).

[F] I. Fáry, *On straight line representations of planar graphs.* Acta Sci. Math. Szeged. 11:229–233 (1948).

[G] R. L. Graham, *An efficient algorithm for determining the convex hull of a finite planar set.* Info. Proc. Lett. 1:132–133 (1972).

[H] F. Harary, Graph Theory. Addison-Wesley, Reading, MA (1971).

[K] R. M. Karp, *Reducibility among combinatorial problems.* In Complexity of Computer Computations. Edited by R. E. Miller and J. W. Thatcher. Plenum Press, New York (1972).

[Kn] D. E. Knuth, The Art of Computer Programming Volume III: Sorting and Searching. Addison-Wesley, Reading, MA (1968).

[PS] F. P. Preparata and M. I. Shamos, Computational Geometry: An Introduction. Springer-Verlag, New York (1985).

[S] R. Sedgewick, Algorithms. Addison-Wesley, Reading, MA (1983).

[Sh] M. I. Shamos, *An Absorber Theory of Gravitational Radiation.* Senior undergraduate thesis, Physics Department, Princeton University (1968).

[Sh2] M. I. Shamos, *Computational Geometry.* Ph.D. thesis, Department of Computer Science, Yale University, 1978. Available from University Microfilms, Ann Arbor, MI.

[Sh3] M. I. Shamos, *Geometric complexity.* Proc. Seventh Annual ACM Symposium on Automata and Theory of Computation (May, 1975) 224–233.

[Sh4] M. I. Shamos, Problems in Computational Geometry. Unpublished book manuscript (1974, revised 1977). Distributed in photocopy.

[SH] M. I. Shamos and D. Hoey, *Closest-point problems.* Proc. Sixteenth IEEE Symposium on Foundations of Computer Science (Oct., 1975) 151–162.

[SH2] M. I. Shamos and D. Hoey, *Geometric intersection problems.* Proc. Seventeenth Annual IEEE Symposium on Foundations of Computer Science (Oct., 1976) 208–215.

[ST] M. I. Shamos and M. A. Tavel, *An absorber theory of acoustic radiation.* J. Acoust. Soc. of America 54:46–49 (1973).

[SY] M. I. Shamos and G. Yuval, *Lower bounds from complex function theory.* Proc. Seventeenth Annual IEEE Symposium on Foundations of Computer Science (Oct., 1976) 268–273.

[T] R. E. Tarjan, *Depth-first search and linear graph algorithms.* SIAM J. Comput. 1:146–160 (1972).

[V] G. Voronoi, *Nouvelles applications des paramères continus à la théorie des formes quadratiques.* J. reine angew. Math. 134, 198–287 (1908).

[WF] J. A. Wheeler and R. P. Feynman, *Interaction with the absorber as the mechanism of radiation.* Rev. Mod. Phys. 17:157 (1945).

[YB] I. M. Yaglom and V. G. Boltyanskii, Convex Figures. Holt, Reinhart and Winston, New York (1961).

[Y] A. C. Yao, *A lower bound to finding convex hulls.* J. ACM 28:780–787 (1981).

School of Computer Science, Carnegie Mellon University, Pittsburgh, PA 15213
E-mail address: shamos@cs.cmu.edu

Contemporary Mathematics
Volume **223**, 1999

Arrangements of Surfaces in Higher Dimensions

Micha Sharir

ABSTRACT. We review recent progress in the study of arrangements of surfaces in higher dimensions. This progress involves new and nearly tight bounds on the complexity of lower envelopes, single cells, zones, and other substructures in such arrangements, and the design of efficient algorithms (near optimal in the worst case) for constructing and manipulating these structures. We then present applications of the new results to motion planning, Voronoi diagrams, visibility, and geometric optimization.

1. Introduction

The combinatorial, algebraic, and topological analysis of arrangements of surfaces in higher dimensions has become one of the most active areas of research in computational geometry during the past 5 years. In this paper we will review the recent progress in the study of combinatorial and algorithmic problems related to such arrangements.

Given a set Γ of n surfaces in $\mathbb{R}^d$, the arrangement $\mathcal{A}(\Gamma)$ that they induce is the decomposition of $\mathbb{R}^d$ into maximal connected regions (cells) of dimensions $0, 1, \ldots, d$, such that each region is contained in the intersection of a fixed subset of Γ and is disjoint from all the other surfaces. For example, the arrangement of a set L of lines in the plane consists of *vertices* (0-dimensional cells) which are the intersection points of the lines, *edges* (1-dimensional cells) which are relatively open intervals along the lines delimited by pairs of consecutive intersection points, and *faces* (2-dimensional cells) which are the connected components of $\mathbb{R}^2 \setminus \bigcup L$.

Arrangements of lines and of hyperplanes have been studied earlier quite extensively. In fact, Edelsbrunner's book from 1987 **[E]** is essentially devoted to the study of such arrangements. However, it has recently been realized that the more general

---

1991 *Mathematics Subject Classification.* Primary 05C99, 51M99, 52B99, 52C10, 52C99, 68Q20, 68Q25, 68R99, 90C27.

Key words and phrases. Computational geometry, combinatorial geometry, arrangements, lower envelopes, Voronoi diagrams, geometric optimization, Davenport-Schinzel sequences, motion planning, algebraic surfaces.

Work on this paper has been supported by NSF Grants CCR-94-24398 and CCR-93-11127 and by grants from the U.S.-Israeli Binational Science Foundation and the G.I.F., the German-Israeli Foundation for Scientific Research and Development.

This is an update of an article **[S3]** that appeared in Volume 955 of *Lecture Notes in Computer Science*; permission granted by Springer-Verlag.

© 1999 American Mathematical Society

arrangements of (curved) surfaces are more significant and have a wider range of applications, because many geometric problems in diverse areas can be reduced to problems involving such arrangements. We will give here four such examples:

Motion planning. Assume that we have a robot system B with d degrees of freedom, i.e., we can represent each placement of B as a point in d-space. Suppose that the workspace of B is cluttered with obstacles, whose shapes and locations are known. For each combination of a geometric feature (vertex, edge, face) of an obstacle and a similar feature of B, define their *contact surface* as the set of all points in d-space that represent a placement of B in which contact is made between these specific features. Let Z be a point corresponding to a given initial *free* placement of B, in which it does not intersect any obstacle. Then the set of all free placements of B that can be reached from Z via a collision-free continuous motion will obviously correspond to the cell containing Z in the arrangement of the contact surfaces. Thus, the robot motion planning problem leads to the problem of computing a single cell in an arrangement of surfaces in higher dimensions. The *combinatorial complexity* of this cell, i.e., the total number of lower-dimensional faces appearing on its boundary serves as a trivial lower bound for the running time of the motion planning problem (assuming the entire cell has to be output). It turns out that in most instances this bound can be almost matched by suitable algorithms.

Generalized Voronoi diagrams. Let S be a set of n 'simply-shaped'[1] pair-wise-disjoint compact convex objects in d-space, and let ρ be some metric. The *Voronoi diagram* $\mathrm{Vor}(S)$ of S under the metric ρ is defined, as usual, as the decomposition of d-space into *Voronoi cells* $V(s)$, for $s \in S$, where

$$V(s) = \{x \in \mathbb{R}^d \mid \rho(x, s) \leq \rho(x, s') \text{ for all } s' \in S\}.$$

For each $s \in S$ define a function $f_s(x) = \rho(x, s)$, for $x \in \mathbb{R}^d$ (where $\rho(x, s) = \min_{y \in s} \rho(x, y)$). Define $F(x) = \min_{s \in S} f_s(x)$, for $x \in \mathbb{R}^d$; we refer to F as the *lower envelope* of the f_s's. If we project the graph of F onto $\mathbb{R}^d$, we obtain a decomposition of $\mathbb{R}^d$ into connected cells, so that over each cell F is attained by a fixed set of functions. This decomposition is nothing but $\mathrm{Vor}(S)$, as follows directly by definition, and has already been noted in [**ES**]. In other words, the study of Voronoi diagrams in d dimensions is equivalent to the study of the lower envelope of a collection of surfaces in $\mathbb{R}^{d+1}$. Planar Voronoi diagrams have been studied intensively (see [**Au, E, LS, PrS**]), but very little is known about generalized Voronoi diagrams in higher dimensions.

Transversals. Let S be a set of 'simply-shaped' compact convex objects in $\mathbb{R}^d$. A hyperplane h is called a *transversal* of S if h intersects every member of S. Let $\mathcal{T}(S)$ denote the space of all hyperplane transversals of S. We wish to study the structure of $\mathcal{T}(S)$. To facilitate this study, we apply the geometric duality described, e.g., in [**E**], which maps hyperplanes to points and vice versa, and which preserves incidences and the above/below relationship between points and hyperplanes. Specifically, this duality maps a point $(a_1, \ldots, a_d)$ to the hyperplane $x_d = a_1x_1 + \cdots + a_{d-1}x_{d-1} + a_d$, and a hyperplane $x_d = b_1x_1 + \cdots + b_{d-1}x_{d-1} + b_d$ to the point $(-b_1, \ldots, -b_{d-1}, b_d)$. Let h be the hyperplane $x_d = a_1x_1 + \cdots +$

[1] Formally, this means that each object in S is defined as a Boolean combination of a constant number of polynomial equalities and inequalities of constant maximum degree; we also refer to such objects as semialgebraic sets of *constant description complexity*.

$a_{d-1}x_{d-1} + a_d$, and suppose that h intersects an object $s \in S$. Translate h up and down until it becomes tangent to s. Denote the resulting upper and lower tangent hyperplanes by

$$x_d = a_1x_1 + \cdots + a_{d-1}x_{d-1} + U_s(a_1, \ldots, a_{d-1})$$

and

$$x_d = a_1x_1 + \cdots + a_{d-1}x_{d-1} + L_s(a_1, \ldots, a_{d-1}),$$

respectively. Then we have

$$L_s(a_1, \ldots, a_{d-1}) \leq a_d \leq U_s(a_1, \ldots, a_{d-1}).$$

Now if h is a tranversal of S, we must have

$$\max_{s \in S} L_s(a_1, \ldots, a_{d-1}) \leq a_d \leq \min_{s \in S} U_s(a_1, \ldots, a_{d-1}).$$

In other words, in dual space, $\mathcal{T}(S)$ is the region 'sandwiched' between a lower envelope of a collection of n surfaces and an upper envelope of another collection of n surfaces.

Geometric optimization. Many problems in geometric optimization can be solved by reducing them to problems involving the analysis of substructures in arrangements. One such example is the problem of computing the *width* of a set S of n points in $\mathbb{R}^3$; this is the smallest distance between two parallel planes enclosing S between them. This problem has been studied in a series of recent papers [**AAS, AS3, CEGS2**], where more details can be found. Following standard techniques, it suffices to solve the decision problem where we are given a parameter $w > 0$ and wish to determine whether the width of S is $\leq w$. For this, we need to test all 'antipodal' pairs of edges of the convex hull C of S (these are pairs of edges that admit parallel supporting planes), and determine whether there exists such a pair whose supporting planes lie at distance $\leq w$. In the worst case the number of such pairs may be $\Theta(n^2)$, and the goal is to avoid having to test explicitly all such pairs. This is achieved by a step that partitions the set of all antipodal pairs of edges into a collection of complete bipartite graphs, so that, for each such graph $E \times E'$, it suffices to determine whether there exist $e \in E$, $e' \in E'$, such that the distance between the lines containing e and e', respectively, is $\leq w$. This problem can be solved by mapping the lines containing the edges in E' into points in 4-space (lines in three dimensions have four degrees of freedom), and by mapping the lines containing the edges of E into trivariate functions, so that the distance between a line ℓ' in the first set and a line ℓ in the second set is $\leq w$ if and only if the point that represents ℓ' lies below the graph of the function that represents ℓ. We have thus landed in a problem where we are given a collection of points and a collection of surfaces in 4-space, and we need to determine whether there exists a point that lies below the upper envelope of the surfaces. Full details can be found in the papers cited above.

As these examples show, many problems in combinatorial and computational geometry can be rephrased in terms of certain substructures (single cells, lower envelopes, regions enclosed between an upper envelope and a lower envelope, etc.) in arrangements of surfaces in higher dimensions. The past decade has seen a significant expansion in the study of arrangements. This work has been described in the recent book [**SA**] and in several other surveys [**GS, H2, HS4, S3**].

The main theme in the study of arrangements is to analyze the combinatorial complexity of such substructures, with the goal of obtaining bounds on this complexity that are significantly smaller than the complexity of the full arrangement. A second theme is the design of efficient algorithms for constructing such substructures.

In this study there are three main relevant parameters: the number n of surfaces, their maximum algebraic degree b, and the dimension d. The approach taken here is 'combinatorial', in which we want to calibrate the dependence of the complexity of the various structures and algorithms on the number n of surfaces, assuming that the maximum degree, as well as any other factor that does not depend on n, is constant. In this way, all issues related to the algebraic complexity of the problem are 'swept under the rug'. These issues should be (and indeed have been) picked up in the complementary 'algebraic' mode of research, where the dependence on the maximum degree b is more relevant; this area is known as computational real algebraic geometry; see [**HRR, SS**] for studies of this kind.

2. Complexity of lower envelopes

During the past five years, significant progress has been made on the problem of bounding the complexity of the lower envelope of a collection of multivariate functions. This problem has been open since 1986, when it was shown in [**HSh**] that the combinatorial complexity of the lower envelope of n univariate continuous functions, each pair of which intersect in at most s points, is at most $\lambda_s(n)$, the maximum length of an (n,s)-*Davenport–Schinzel sequence*. This bound is slightly super-linear in n, for any fixed s (for example, it is $\Theta(n\alpha(n))$ for $s = 3$, where $\alpha(n)$ is the extremely slowly growing inverse of Ackermann's function [**HSh**]; see also [**ASSh, SA**]). Since the complexity of the full arrangement of such a collection of functions can be $\Theta(n^2)$ in the worst case, this result shows that the complexity of the lower envelope is smaller than the overall complexity of the arrangement by nearly a factor of n.

It was then conjectured that a similar phenomenon occurs in higher dimensions. That is, the combinatorial complexity of the lower envelope of a collection $\mathcal{F}$ of n 'well-behaved' d-variate functions should be close to $O(n^d)$ (as opposed to $\Theta(n^{d+1})$, which can be the complexity of the entire arrangement of the function graphs). More precisely, according to a stronger version of this conjecture, this quantity should be at most $O(n^{d-1}\lambda_s(n))$, for some constant s depending on the shape of the functions in $\mathcal{F}$.

This problem was almost completely settled in 1994 [**HS2, S2**]: Let $\mathcal{F}$ be a collection of (possibly partially-defined) d-variate functions, such that all functions in $\mathcal{F}$ are algebraic of constant maximum degree and, in case of partial functions, the domain of definition of each function is a semi-algebraic set defined by a constant number of polynomial equalities and inequalities of constant maximum degree. (We refer to such a region as having *constant description complexity*.) Then, for any $\varepsilon > 0$, the combinatorial complexity of the lower envelope of $\mathcal{F}$ is $O(n^{d+\varepsilon})$, where the constant of proportionality depends on ε, d, and on the maximum degree of the functions and of the polynomials defining their domains.[2] Thus, apart from a

[2] In this paper we will state similar complexity bounds that depend on $\varepsilon > 0$, without saying repeatedly 'for any $\varepsilon > 0$'; the meaning of such a bound is that it holds for any $\varepsilon > 0$, where the constant of proportionality depends on ε (and perhaps also on other problem constants), and usually tends to infinity when ε tends to zero. For algorithmic complexity, the meaning of such a bound is that the algorithm can be fine-tuned, as a function of ε, so that its complexity obeys the stated bound.

small remaining gap, the (weak form of the) above conjecture has been settled in the affirmative.

Here is a sketch of the proof for the special case where $d = 2$ and each triple of function graphs intersect in at most two points. The idea of the proof is to charge each vertex v of the envelope to some 'nearby' vertices of the arrangement of the function graphs, and to use this charging scheme to derive a recurrence for the number of vertices of the envelope. In more detail, let us denote by $\mathcal{F}$ also the collection of function graphs in $\mathbb{R}^{d+1}$, let $E_{\mathcal{F}}$ denote the lower envelope of $\mathcal{F}$, let $M_{\mathcal{F}}$ denote the xy-projection of $E_{\mathcal{F}}$, and let $\varphi(\mathcal{F})$ denote the number of vertices of $M_{\mathcal{F}}$. A vertex of $E_{\mathcal{F}}$ is called *inner* if it lies in the relative interior of three function graphs of $\mathcal{F}$. Besides projections of inner vertices, $M_{\mathcal{F}}$ also contains *boundary* vertices, which lie on the projection of at least one boundary edge of a graph in $\mathcal{F}$. We denote by $\varphi^*(\mathcal{F})$ the number of inner vertices. We also put $\varphi(n) = \max_{|\mathcal{F}|=n} \varphi(\mathcal{F})$ and $\varphi^*(n) = \max_{|\mathcal{F}|=n} \varphi^*(\mathcal{F})$, where the maximum is taken over all collections $\mathcal{F}$ satisfying the above properties (with a fixed bound on the 'description complexity' of the function graphs).

To bound the number of boundary vertices, we fix an arc e bounding the domain of some function in $\mathcal{F}$, construct the 'vertical curtain' $C_e = e \times \mathbb{R}$, and intersect all other function graphs with C_e to obtain a collection $\mathcal{F}_e$ of arcs of constant description complexity, so each pair of them intersect in at most some constant number, q, of points. We observe that each boundary vertex on e corresponds to a vertex of the lower envelope of $\mathcal{A}(\mathcal{F}_e)$. The theory of Davenport-Schinzel sequences [**HSh, SA**] implies that the number of boundary vertices on e is $O(\lambda_{q+2}(n))$. Repeating this analysis for each of the $O(n)$ boundary arcs e, we conclude that the overall number of boundary vertices is $O(n\lambda_{q+2}(n))$.

Next, let v be an inner vertex of $E_{\mathcal{F}}$, lying on the graphs of three functions $f_1, f_2, f_3 \in \mathcal{F}$. Let β_{ij} denote the curve of intersection of the graphs of f_i and f_j, for $1 \le i < j \le 3$. Since the triple of these graphs intersect at most twice, we may assume that no such intersection exists in the halfspace $x < x(v)$ (the other case is fully symmetric). Also, it is easily seen that at least one of the intersection curves, say β_{12}, has the property that as we trace it from v in the decreasing x-direction, it lies above $E_{\mathcal{F}}$ near v, i.e., the graph of f_3 passes below it.

We fix a constant parameter t, depending on the prespecified ε and on the complexity of the individual function graphs in $\mathcal{F}$. We follow β_{12} from v in decreasing x-direction, and mark on it the vertices of $\mathcal{F}(\mathcal{F})$ that lie on it. We stop the tracing as soon as one of the following events occurs.

(i) We reach an x-extreme point of β_{12};

(ii) We reach a point that lies above a boundary edge e of some function graph in $\mathcal{F}$;

(iii) We have encountered t vetices of $\mathcal{A}(\mathcal{F})$.

In case (i) we charge v to the stopping point. Since there are $O(n^2)$ such points and each can be charged at most twice, the number of vertices of this kind is $O(n^2)$. In case (ii) we again charge v to the stopping point w, and observe that w is a vertex of $\mathcal{A}(\mathcal{F}_e)$ at *level* $< t$ (i.e., at most t intersection arcs lie below w). A standard application of the classical Clarkson-Shor probabilistic technique [**CS**] (detailed in [**S**]) implies that the number of such vertices w of $\mathcal{A}(\mathcal{F}_e)$ is $O(t^2\lambda_{q+2}(n/t)) = O(t\lambda_{q+2}(n))$. Hence, summing over all boundary edges e, we conclude that the number of inner vertices v of $E_{\mathcal{F}}$ of this kind is $O(nt\lambda_{q+2}(n))$. Finally, in case

(iii) we charge v to the block of the t vertices of $\mathcal{A}(\mathcal{F})$ that we have encountered. Again, each of these vertices lies at level $< t$, and a similar application of the Clarkson-Shor technique implies that the number of these vetices, in the entire $\mathcal{A}(\mathcal{F})$, is $O(t^3\varphi^*(n/t))$. Since each of these vertices can be charged at most a constant number of times, it easily follows that the number of inner vertices v of the third type is $(1/t) \cdot O(t^3\varphi^*(n/t)) = O(t^2\varphi^*(n/t))$. Putting everything together, we obtain the recurrence

$$\varphi^*(n) = O(t^2\varphi^*(n/t)) + O(tn\lambda_{q+2}(n)),$$

whose solution, assuming t is a sufficiently large constant, is $O(n^{2+\varepsilon})$.

The proof of the general case is more involved, but the appoach is similar to the one sketched above. Full details can be found in [**SA**].

Prior to the derivation of this bound, the above conjectures have been confirmed only in some special cases, including the case in which the graphs of the functions are d-simplices in $\mathbb{R}^{d+1}$, where a tight worst-case bound, $\Theta(n^d\alpha(n))$, was established in [**E2, PS**]. (The case $d = 1$, involving n segments in the plane, where the bound is $\Theta(n\alpha(n))$, had been analyzed earlier, in [**HSh, WS**].) There are also some even more special cases, like the case of hyperplanes, where the maximum complexity of their lower envelope is known to be $\Theta(n^{\lfloor(d+1)/2\rfloor})$, by the so-called *Upper Bound Theorem* for convex polytopes [**McM**]. The case of balls also admits a much better bound, using a standard lifting transformation (see [**E**]).

The new analysis technique has later been extended in various ways, to be described below. These additional results include near-quadratic bounds on the complexity of a single cell in a three-dimensional arrangement of surfaces and on the complexity of the region enclosed between two envelopes of bivariate functions, and efficient algorithms for constructing lower envelopes and single cells. The new results have been applied to obtain improved algorithmic and combinatorial bounds for a variety of problems; we will mention some of these applications later on.

3. Algorithms for lower envelopes

Once the combinatorial complexity of lower envelopes of multivariate functions has been (more or less) resolved, the next task is to derive efficient algorithms for computing such lower envelopes. One of the strongest specifications of such a computation is as follows. We are given a collection $\mathcal{F}$ of d-variate algebraic functions satisfying the above conditions. We want to compute the lower envelope $E_{\mathcal{F}}$ and store it in some data structure, so that, given a query point $p \in \mathbb{R}^d$, we can efficiently compute the value $E_{\mathcal{F}}(p)$, and the function(s) attaining $E_{\mathcal{F}}$ at p. (This is, for example, the version that is needed for solving the three-dimensional width problem described above.) Of course, we need to assume here an appropriate model of computation, where various primitive operations on a constant number of functions can be each performed in constant time. As mentioned above, the exact arithmetic model in computational real algebraic geometry is an appropriate choice.

This task has recently been accomplished for the case of bivariate functions in several papers [**ASS, BD, BD2, dBDS, S2**]. Some of these techniques use randomized algorithms, and their expected running time is $O(n^{2+\varepsilon})$, which is comparable with the maximum complexity of such an envelope. The simplest algorithm is probably the one given in [**ASS**]. It is deterministic and uses divide-and-conquer. That is, it partitions the set of functions into two subsets of roughly equal size, and computes recursively the lower envelopes E_1, E_2 of these two subsets. It then projects these envelopes onto the xy-plane, to obtain two respective planar maps M_1, M_2, and forms the overlay M of M_1 and M_2. Over each face f of M there are only two functions that can attain the final envelope (the function attaining E_1 over f and the function attaining E_2, so we compute the lower envelope of these two functions over f, and repeat this step for all faces of M. It is easy to see that the cost of this step is proportional to the number of faces of M. In general, overlaying two planar maps of complexity N each, can result in a map whose complexity is $\Theta(N^2)$, which may be $\Omega(n^4)$ in our case. Fortunately, it was shown in [**ASS**] that for lower envelopes, the overlay of M_1 and M_2 has complexity $O(n^{2+\varepsilon})$. In other words, the worst-case bound on the complexity of the overlay of (the xy-projections of) two lower envelopes of bivariate functions is asymptotically the same as the bound for the complexity of a single envelope. This implies that the complexity of the above divide-and-conquer algorithm is $O(n^{2+\varepsilon})$.

In higher dimensions, the only result known so far is that lower envelopes of trivariate functions satisfying the above properties can be computed, in the above strong sense, in randomized expected time $O(n^{3+\varepsilon})$ [**AAS**]. For $d > 3$, it is also shown in [**AAS**] that all vertices, edges and 2-faces of the lower envelope of n d-variate functions, as above, can be computed in randomized expected time $O(n^{d+\varepsilon})$. It is still an open problem whether such a lower envelope can be computed within similar time bounds in the above stronger sense. Another, more difficult problem is to devise *output-sensitive* algorithms, whose complexity depends on the actual combinatorial complexity of the envelope. This problem is really hard even for collections of nonintersecting triangles in 3-space. It would also be interesting to develop algorithms for certain special classes of functions, where better bounds are known for the complexity of the envelope, e.g., for envelopes of piecewise-linear functions (see below for more details).

4. Single cells

Lower envelopes are closely related to other substructures in arrangements, notably *single cells* and *zones*. (The lower envelope is a portion of the boundary of the bottommost cell of the arrangement.) In two dimensions, it was shown in [**GSS**] that the complexity of a single face in an arrangement of n arcs, each pair of which intersect in at most s points, is $O(\lambda_{s+2}(n))$, and so is of the same asymptotic order of magnitude as the complexity of the lower envelope of such a collection of arcs. Again, the prevailing conjecture is that the same holds in higher dimensions. That is, the complexity of a single cell in an arrangement of n algebraic surfaces in d-space satisfying the above assumptions is close to $O(n^{d-1})$, or, in a stronger form, this complexity should be $O(n^{d-2}\lambda_s(n))$, for some appropriate constant s. The weaker version of this conjecture for the 3-dimensional case has recently been confirmed in [**HS3**]: Let $\mathcal{A}$ be an arrangement of n two-dimensional surface patches in $\mathbb{R}^3$, all of them algebraic of constant description complexity. It was proved in [**HS3**] that, for any $\varepsilon > 0$, the complexity of a single cell in $\mathcal{A}$ is $O(n^{2+\varepsilon})$, where the constant of proportionality depends on ε and on the maximum degree of the surfaces and of their boundaries. The proof is based on an extension of the argument developed for lower envelopes, but one has to add several nontrivial ingredients, in order to handle the more complex topology of a single cell. The analysis in [**HS3**] seems to extend to higher dimensions, except for some key steps that seem to require the introduction of new algebraic geometry techniques. So far, this extension is still open.

The results of [**HS3**] mentioned above easily imply that, for fairly general robot systems with three degrees of freedom, the complexity of the space of all free placements of the system, reachable from a given initial placement, is $O(n^{2+\varepsilon})$, a significant improvement over the previous, naive bound $O(n^3)$. The corresponding algorithmic problem, of devising an efficient (near-quadratic) algorithm for computing such a cell, has recently been solved in [**SSh**]. We will say more about this result when we discuss vertical decompositions below. Prior to this result, several other near-quadratic algorithms were proposed for some special classes of surfaces [**ASS, ArS2, H, HS**]. For example, the paper [**HS**] gives a near-quadratic algorithm for the single cell problem in the special case of arrangements that arise in the motion planning problem for a (nonconvex) polygonal robot moving (translating and rotating) in a planar polygonal region. However, this algorithm exploits the special structure of the surfaces that arise in this case (namely, that any cross-section of such an arrangement at a fixed orientation of the polygon is polygonal), and does not extend to the general case. The algorithm given in [**ASS**] also provides a near-quadratic solution for the case where all the surfaces are graphs of totally-defined continuous algebraic bivariate functions (so that the cell in question is xy-monotone).

In higher dimensions, we mention the special case of a single cell in an arrangement of n $(d-1)$-simplices in $\mathbb{R}^d$. It was shown in [**ArS2**] that the complexity of such a cell is $O(n^{d-1}\log n)$; a simplified proof was recently given in [**T**]. For $d=3$, this bound is much sharper than the general bound stated above; the best lower bound known for this complexity is $\Omega(n^{d-1}\alpha(n))$, so a small gap between the upper and lower bounds still remains.

5. Zones

Given an arrangement $\mathcal{A}$ of surfaces in $\mathbb{R}^d$, and another surface σ_0, the *zone* of σ_0 is the collection of all cells of the arrangement $\mathcal{A}$ that σ_0 crosses, and the complexity of the zone is the sum of complexities of all these cells. The 'classical' Zone Theorem [**E, ESS**] asserts that the maximum complexity of the zone of a hyperplane in an arrangement of n hyperplanes in $\mathbb{R}^d$ is $\Theta(n^{d-1})$, where the constant of proportionality depends on d. This has been extended in [**APS**] to the zone of an algebraic or convex surface (of any dimension $p<d$) in an arrangement of hyperplanes. The bound on the complexity of such a zone is $O(n^{\lfloor (d+p)/2\rfloor}\log^c n)$, and $\Omega(n^{\lfloor (d+p)/2\rfloor})$ in the worst case, where $c=1$ when $d-p$ is odd and $c=0$ when $d-p$ is even. It is not clear whether the logarithmic factor is really needed, or that it is just an artifact of the proof technique.

The result of [**HS3**] can easily be extended to obtain a bound of $O(n^{2+\varepsilon})$, on the complexity of the zone of an algebraic surface σ_0 (of constant description complexity) in an arrangement of n algebraic surfaces in $\mathbb{R}^3$, as above. Intuitively, the proof proceeds by cutting each of the given surfaces along its intersection curve with σ_0, and by shrinking the surface away from that curve, thus leaving a 'tiny' gap there. These modifications transform the zone of σ into a single cell in the arrangement of the new surfaces, and the result of [**HS3**] can then be applied. (The same technique has been used earlier in [**EGPPSS**], to obtain a near-linear bound on the complexity of the zone of an arc in a 2-dimensional arrangement of arcs.) Once the bound on the complexity of a single cell is extended to higher dimensions, it should lead right away to a similar bound for a zone of a surface. A similar technique implies that the complexity of the zone of an algebraic or convex surface in an arrangement of n $(d-1)$-simplices in $\mathbb{R}^d$ is $O(n^{d-1}\log n)$ [**ArS2, T**].

6. Generalized Voronoi diagrams

One of the interesting applications of the new bounds on the complexity of lower envelopes is to generalized Voronoi diagrams in higher dimensions. Let S be a set of n pairwise-disjoint convex objects in d-space, each of constant description complexity, and let ρ be some metric. The *Voronoi diagram* Vor(S) of S under the metric ρ has been defined in the introduction. As shown there, the diagram is simply the minimization diagram of the lower envelope of n d-variate 'distance functions' induced by the objects of S.

In the classical case, in which ρ is the Euclidean metric and the objects in S are singletons (points), one can replace these distance functions by a collection of n hyperplanes in $\mathbb{R}^{d+1}$, so the maximum possible complexity of their lower envelope (and thus of Vor(S)) is $\Theta(n^{\lceil d/2 \rceil})$ (see e.g., [**E**]). In more general settings, though, this reduction is not possible. Nevertheless, the new bounds on the complexity of lower envelopes imply that the complexity of the diagram is $O(n^{d+\varepsilon})$. While this bound is nontrivial, it is conjectured to be too weak. For example, this bound is near-quadratic for planar Voronoi diagrams, but the complexity of practically any kind of planar Voronoi diagram is known to be only $O(n)$.

In three dimensions, the above-mentioned bound for point sites and Euclidean metric is $\Theta(n^2)$. It has been a long-standing open problem whether a similar quadratic or near-quadratic bound holds in 3-space for more general objects and metrics (here the new results on lower envelopes give only an upper bound of $O(n^{3+\varepsilon})$). Thus the problem stated above calls for improving this bound by roughly another factor of n. It thus appears to be a considerably more difficult problem than that of lower envelopes, and the only hope of making progress here is to exploit the special structure of the distance functions $\rho(x, s)$.

Fortunately, some progress on this problem was made recently. It was shown in [**CKSTW**] that the complexity of the Voronoi diagram is $O(n^2\alpha(n)\log n)$, for the case where the objects of S are lines, and the metric ρ is a convex distance function induced by a convex *polytope* with a constant number of facets. (Note that the L_1 and L_∞ metrics are special cases of such distance functions. Note also that such a distance function is not necessarily a metric, because it will fail to be symmetric if the defining polytope is not centrally symmetric.) The best known lower bound for the complexity of the diagram in this special case is $\Omega(n^2\alpha(n))$. In another recent paper [**BSTY**], it is shown that the maximum complexity of the L_1-Voronoi diagram of a set of n points in $\mathbb{R}^3$ is $\Theta(n^2)$. Finally, it is shown in [**T2**] that the complexity of the three-dimensional Voronoi diagram of point sites under general polyhedral convex distance functions is $O(n^2\log n)$. The most intriguing unsolved problem is to obtain a similar bound for a set S of n lines in space but under the Euclidean metric. The proof technique of [**CKSTW**] breaks down in this case. Another, more tractable open problem is to extend these results to sets of convex polytopes under the same polyhedral convex distance functions.

An interesting special case of these problems involves *dynamic Voronoi diagrams* for moving points in the plane. Let S be a set of n points in the plane, each moving along some line at some fixed velocity. The goal is to bound the number of combinatorial changes of Vor(S) over time. This dynamic Voronoi diagram can easily be transformed into a 3-dimensional Voronoi diagram, by adding the time t as a third coordinate. The points become lines in 3-space, and the metric is a distance function induced by a horizontal disc (that is, the distance from a point $p(x_0, y_0, t_0)$ to a line ℓ is the Euclidean distance from p to the point of intersection of ℓ with the horizontal plane $t = t_0$). Here too the open problem is to derive a near-quadratic bound on the complexity of the diagram. Cubic or near-cubic bounds are known for this problem, even under more general settings [**FL**, **GMR**, **S2**], but subcubic bounds are known only in some very special cases [**C**].

Next, consider the problem of bounding the complexity of generalized Voronoi diagrams in higher dimensions. As mentioned above, when the objects in S are n points in $\mathbb{R}^d$ and the metric is Euclidean, the complexity of Vor(S) is $O(n^{\lceil d/2 \rceil})$. As d increases, this becomes drastically smaller than the naive $O(n^{d+1})$ bound or the improved bound, $O(n^{d+\varepsilon})$, obtained by viewing the Voronoi diagram as a lower envelope in $\mathbb{R}^{d+1}$. The same bound of $O(n^{\lceil d/2 \rceil})$ has recently been obtained in [**BSTY**] for the complexity of the L_∞-diagram of n points in d-space (it was also shown that this bound is tight in the worst case). It is thus tempting to conjecture that the maximum complexity of generalized Voronoi diagrams in higher dimensions is close to this bound. Unfortunately, this was recently shown to be false in [**A**], where a lower bound of $\Omega(n^{d-1})$ is given. The sites used in this construction are convex polytopes, and the distance is either Euclidean or a polyhedral convex distance function. For $d = 3$, this lower bound does not contradict the conjecture made above, that the complexity of generalized Voronoi diagrams should be at most near-quadratic in this case. Also, in higher dimensions, the conjecture mentioned above is still not refuted when the sites are singleton points. Finally, for the general case, the construction of [**A**] still leaves a gap of roughly a factor of n between the known upper and lower bounds.

7. Union of geometric objects

A subproblem related to generalized Voronoi diagrams is as follows. Let S and ρ be as above (say, for the 3-dimensional case). Let K denote the region consisting of all points $x \in \mathbb{R}^3$ whose smallest distance from a site in S is at most r, for some fixed parameter $r > 0$. Then $K = \bigcup_{s \in S} B(s, r)$, where $B(s, r) = \{x \in \mathbb{R}^3 \mid \rho(x, s) \leq r\}$. We thus face the problem of bounding the combinatorial complexity of the union of n objects in 3-space (of some special type). For example, if S is a set of lines and ρ is the Euclidean distance, the objects are n congruent infinite cylinders in 3-space. In general, if the metric ρ is a distance function induced by some convex body P, the resulting objects are the *Minkowski sums* $s \oplus (-rP)$, for $s \in S$, where $A \oplus B = \{x + y \mid x \in A,\ y \in B\}$. Of course, this problem can also be stated in any higher dimension.

Since it has been conjectured that the complexity of the whole Voronoi diagram should be near-quadratic (in 3-space), the same conjecture should apply to the (simpler) structure K (whose boundary can be regarded as a 'cross-section' of the diagram at 'height' r; it does indeed correspond to the cross-section at height r of the lower envelope that represents the diagram). Recently, this conjecture has been confirmed in [**ArS3**], in the special case where both P and the objects of S are convex polyhedra (see also [**HY**] for an earlier study of a special case, with a slightly better bound). Let us discuss this result in more detail. An earlier paper [**ArS**] has studied the case involving the union of k arbitrary convex polyhedra in 3-space, with a total of n faces. It was shown there that the complexity of the union is $O(k^3 + nk\log^2 k)$, and can be $\Omega(k^3 + nk\alpha(k))$ in the worst case. The upper bound was subsequently improved to $O(k^3 + nk\log k)$ [**ASTa**], which still leaves a small gap between the upper and lower bounds. In the subsequent paper [**ArS3**], these bounds were improved in the special case where the polyhedra in question are Minkowski sums of the form $s_i \oplus P$, where the s_i's are k pairwise-disjoint convex polyhedra, P is a convex polyhedron, and the total number of faces of these Minkowski sums is n. The improved bounds are $O(nk\log k)$ and $\Omega(nk\alpha(k))$. They are indeed near-quadratic, as conjectured.

However, the case where P is a ball (namely, the case of the Euclidean distance) is still open. The simplest unsolved instance of this problem is to establish a near-quadratic upper bound for the complexity of the union of n congruent infinite cylinders in 3-space. Recently, some progress has been made on this problem in [**AS4**], where a bound $O(n^{8/3+\varepsilon})$ has been obtained.

In higher dimensions, it was recently shown in [**BSTY**] that the maximum complexity of the union of n axis-parallel hypercubes in d-space is $\Theta(n^{\lceil d/2\rceil})$, and this improves to $\Theta(n^{\lfloor d/2\rfloor})$ when all the hypercubes have the same size.

8. Vertical decomposition

In many algorithmic applications, one needs to decompose a d-dimensional arrangement, or certain portions thereof, into a small number of subcells, each having constant description complexity. In a typical setup where this problem arises, we need to process in a certain manner an arrangement of n surfaces in d-space. We choose a random sample of r of the surfaces, for some sufficiently large constant r, construct the arrangement of these r surfaces, and decompose it into subcells as above. Since no such subcell is crossed by any surface in the random sample, it follows by standard ε-net theory [**Cl, HW, KPW**] that with high probability, none of these subcells is crossed by more than $O(\frac{n}{r}\log r)$ of the n given surfaces. (For this result to hold, it is essential that each of these subcells have constant description complexity.) This allows us to break the problem into recursive subproblems, one for each of these subcells, solve each subproblem separately, and then combine their outputs to obtain a solution for the original problem. The efficiency of this method crucially depends on the number of subcells. The smaller this number is, the faster is the resulting algorithm. (We note that the construction of a 'good' sample of r surfaces can also be performed deterministically, e.g., using the techniques of Matoušek [**M**].)

The only general-purpose known technique for decomposing an arrangement of surfaces into subcells of constant description complexity is the *vertical decomposition* technique. In this method, we erect a vertical 'wall' up and down (in the x_d-direction) from each $(d-2)$-dimensional face of the arrangement, and extend these walls until they hit another surface. This results in a decomposition of the arrangement into subcells so that each subcell has a unique top facet and a unique bottom facet, and each vertical line cuts it in a connected (possibly empty) interval. We next project each resulting subcell on the hyperplane $x_d = 0$, and apply recursively the same technique within each resulting $(d-1)$-dimensional projected cell, and then 'lift' this decomposition back into d-space, by extending each subcell

c in the projection into the vertical cylinder $c \times \mathbb{R}$, and by cutting the original cell by these cylinders. We continue the recursion in this manner until we reach $d = 1$, and thereby obtain the vertical decomposition of the given arrangement. The resulting subcells have the desired properties. Furthermore, if we assume that the originally given surfaces are algebraic of constant maximum degree, then the resulting subcells are semi-algebraic and are defined by a constant number of polynomials of constant maximum degree (although the latter degree can grow quite fast with d). In what follows, we ignore the algebraic complexity of the subcells of the vertical decomposition, and will be mainly interested in bounding their number as a function of n, the number of given surfaces.

It was shown in [**CEGS**] that the number of cells in such a vertical decomposition of the entire arrangement is $O(n^{2d-3}\beta(n))$, where $\beta(n)$ is a slowly growing function of n (related to the inverse Ackermann's function). However, the only known lower bound is the trivial $\Omega(n^d)$, so there is a considerable gap here, for $d > 3$; for $d = 3$ the two bounds nearly coincide. Improving the upper bound appears to be a very hard task. This problem has been open since 1989; it seems difficult enough to preempt, at the present state of knowledge, any specific conjecture on the true maximum complexity of the vertical decomposition of arrangements in $d > 3$ dimensions.

The bound stated above applies to the vertical decomposition of an entire arrangement of surfaces. In many applications, however, one is interested in the vertical decomposition of only a portion of the arrangement, e.g., a single cell, the region lying below the lower envelope of the given surfaces, the zone of some surface, a specific collection of cells of the arrangement, etc. Since, in general, the complexity of such a portion is known (or conjectured) to be smaller than the complexity of the entire arrangement, one would like to conjecture that a similar phenomenon applies to vertical decompositions. Very recently, it was shown in [**SSh**] that the complexity of the vertical decomposition of a single cell in an arrangement of n surface patches in 3-space, as above, is $O(n^{2+\varepsilon})$. As mentioned above, this leads to a near-quadratic algorithm for computing such a single cell, which implies that motion planning for fairly general systems with three degrees of freedom can be performed in near-quadratic time, thus settling a major open problem in the area. A similar near-quadratic bound has been obtained in [**ASS**] for the vertical decomposition of the region enclosed between a lower envelope and an upper envelope of bivariate functions. Another recent result [**AES**] gives a bound on the complexity of the vertical decomposition of the first k levels in an arrangement of surfaces in 3-space, which is only slightly worse than the worst-case complexity of this undecomposed portion of the arrangement. A challenging open problem is to obtain improved bounds for the complexity of the vertical decomposition of the region lying below the lower envelope of n d-variate functions, for $d \geq 3$.

Finally, an interesting special case is that of hyperplanes. For such arrangements, the vertical decomposition is a too cumbersome construct, because there are other easy methods for decomposing each cell into simplices, whose total number is $O(n^d)$. Still, it is probably a useful exercise to understand the complexity of the vertical decomposition of an arrangement of n hyperplanes in d-space. A recent result of [**GHMS**] gives an almost tight bound of $O(n^4 \log n)$ for this problem in 4-space, but nothing significantly better than the general bound is known for $d > 4$. Another interesting special case is that of triangles in 3-space. This has been studied by [**dBGH, T**], where almost tight bounds were obtained for the case of a single cell ($O(n^2 \log^2 n)$), and for the entire arrangement ($O(n^2\alpha(n)\log n + K)$, where K is the complexity of the undecomposed arrangement). The first bound is slightly better than the general bound of **[SSh]** mentioned above. The paper **[T]** also derives sharp complexity bounds for the vertical decomposition of many cells in such an arrangement, including the case of all nonconvex cells.

9. Other applications

We conclude this survey by mentioning some additional applications of the new advances in the study of arrangements. We have already discussed in some detail the motion planning application, and have seen how the new results lead to a near-optimal algorithm for the general motion planning problem with three degrees of freedom. Here we discuss three other kinds of applications: to visibility problems in three dimensions, to geometric optimization, and to transversals.

9.1. Visibility in three dimensions. Let us consider a special case of the so-called *aspect graph* problem, which has recently attracted much attention, especially in the context of three-dimensional scene analysis and object recognition in computer vision. The aspect graph of a scene represents all topologically-different views of the scene. For background and a survey of recent research on aspect graphs, see **[BDy]**. Here we will show how the new complexity bounds for lower envelopes, with some additional machinery, can be used to derive near-tight bounds on the number of views of polyhedral terrains.

Let K be a *polyhedral terrain* in 3-space; that is, K is the graph of a continuous piecewise-linear bivariate function, so it intersects each vertical line in exactly one point. Let n denote the number of edges of K. A line ℓ is said to *lie over* K if every point on ℓ lies on or above K. Let $\mathcal{L}_K$ denote the space of all lines that lie over K. (Since lines in 3-space can be parametrized by four real parameters, we can regard $\mathcal{L}_K$ as a subset of 4-space.) Using an appropriate parametrization, the *lower envelope* of $\mathcal{L}_K$ consists of those lines in $\mathcal{L}_K$ that touch at least one edge of K. Assuming general position of the edges of K, a line in $\mathcal{L}_K$ (or any line, for that matter) can touch at most four edges of K. We estimate the combinatorial complexity of this lower envelope, in terms of the number of its *vertices*, namely those lines in $\mathcal{L}_K$ that touch four distinct edges of K. It was shown in **[HS2]** that the number of vertices of $\mathcal{L}_K$, as defined above, is $O(n^3 \cdot 2^{c\sqrt{\log n}})$, for some absolute positive constant c.

We give here a sketch of the proof. We fix an edge e_0 of K, and bound the number of lines of $\mathcal{L}_K$ that touch e_0 and three other edges of K, with the additional proviso that the three other contact points all lie on one fixed side of the vertical plane passing through e_0. We then multiply this bound by the number n of edges, to obtain a bound on the overall number of vertices of $\mathcal{L}_K$. We first rephrase this problem in terms of the lower envelope of a certain collection of surface patches in 3-space, one patch for each edge of K (other than e_0), and then exploit the results on lower envelopes reviewed above.

The space $\mathcal{L}_{e_0}$ of oriented lines that touch e_0 is 3-dimensional: each such line ℓ can be specified by a triple (t, k, ζ), where t is the point of contact with e_0, and $k = \tan\theta$, $\zeta = -\cot\phi$, where (θ, ϕ) are the spherical coordinates of the direction of ℓ, that is, θ is the orientation of the xy-projection of ℓ, and ϕ is the angle between ℓ and the positive z-axis.

For each edge $e \neq e_0$ of K, let σ_e be the surface patch in $\mathcal{L}_{e_0}$ consisting of all points (t, k, ζ) representing lines that touch e and are oriented from e_0 to e. Note that if $(t, k, \zeta) \in \sigma_e$ then $\zeta' > \zeta$ iff the line (t, k, ζ') passes below e. It thus follows that a line ℓ in $\mathcal{L}_{e_0}$ is a vertex of the lower envelope of $\mathcal{L}_K$ if and only if ℓ is a vertex of the lower envelope of the surfaces σ_e in the $tk\zeta$-space, where the height of a point is its ζ-coordinate. It is easy to show that these surfaces are algebraic of constant description complexity. Actually, it is easily seen that the number s of intersections of any triple of these surfaces is at most 2. The paper [**HS2**] studies the special case of lower envelopes of collections of such algebraic surface patches in 3-space, with the extra assumption that $s = 2$. It is shown there that the complexity of the lower envelope of such a collection is $O(n^2 \cdot 2^{c\sqrt{\log n}})$, for some absolute positive constant c, a bound that is slightly better than the general bound stated above. These arguments immediately complete the proof. (This bound has been independently obtained by Pellegrini [**P**], using a different proof technique.) Recently, de Berg [**dB**] has given a lower bound construction, in which the lower envelope of $\mathcal{L}_K$ has complexity $\Omega(n^3)$, implying that the upper bound stated above is almost tight in the worst case.

We can extend the above result as follows. Let K be a polyhedral terrain, as above. Let $\mathcal{R}_K$ denote the space of all rays in 3-space with the property that each point on such a ray lies on or above K. We define the lower envelope of $\mathcal{R}_K$ and its vertices in complete analogy to the case of $\mathcal{L}_K$. By inspecting the proof sketched above, one easily verifies that it applies equally well to rays instead of lines. Hence we obtain that the number of vertices of $\mathcal{R}_K$, as defined above, is also $O(n^3 \cdot 2^{c\sqrt{\log n}})$.

We can apply this bound to obtain a bound of $O(n^5 \cdot 2^{c'\sqrt{\log n}})$, for some constant $c' > c$, on the number of topologically-different orthographic views (i.e., views from infinity) of a polyhedral terrain K with n edges. We omit here details of this analysis, which can be found in [**HS2**]. The paper [**dBHO**] gives a lower bound construction that produces $\Omega(n^5\alpha(n))$ topologically-different orthographic views of a polyhedral terrain, so the above bound is almost tight in the worst case. It is also instructive to note that, if K is an arbitrary polyhedral set in 3-space with n edges, then the maximum possible number of topologically-different orthographic views of K is $\Theta(n^6)$ [**PD**].

Consider next the extension of the above analysis to bound the number of perspective views of a terrain. As shown recently in [**AS**], the problem can be reduced to the analysis of $O(n^3)$ lower envelopes of appropriate collections of 5-variate functions. This leads to an overall bound of $O(n^{8+\varepsilon})$, for the number of topologically-different perspective views of a polyhedral terrain with n edges. This bound is also known to be almost tight in the worst case, as follows from another lower bound construction given in [**dBHO**]. Again, in contrast, If K is an arbitrary polyhedral set with n edges, the maximum possible number of topologically-different perspective views of K is $\Theta(n^9)$ [**PD**].

9.2. Geometric optimization. In the past few years, many problems in geometric optimization have been attacked by techniques that reduce the problem to a problem involving arrangements of surfaces in higher dimensions. These reduced problems sometimes call for the construction of, and searching in lower envelopes or other substructures in such arrangements; one such example, that of computing the width in three dimensions, has been sketched in the introduction. Hence the area of geometric optimization is a natural extension, and a good application area, of the study of arrangements, as described above. See [**AS2**] for a recent survey on geometric optimization.

One of the basic techniques for geometric optimization is the *parametric searching* technique, originally proposed by Megiddo [**Me**]. It has been used to solve a wide variety of geometric optimization problems, including many of those that involve arrangements. Some specific results of this kind include:

- **Selecting distances in the plane:** Given a set S of n points in $\mathbb{R}^2$ and a parameter $k \leq \binom{n}{2}$, find the k-th largest distance among the points of S [**AASS**]. Here the problem reduces to the construction and searching in 2-dimensional arrangements of congruent disks.
- **The segment center problem:** Given a set S of n points in $\mathbb{R}^2$, and a line segment e, find a placement of e that minimizes the largest distance from the points of S to e [**ESh**]. Using lower envelopes of certain special kinds of bivariate functions, and applying a more careful analysis, the problem can be solved in $O(n^{1+\varepsilon})$ time, improving substantially a previous near-quadratic solution given in [**AEST**].
- **Extremal polygon placement:** Given a convex polygon P and a closed polygonal environment Q, find the largest similar copy of P that is fully contained in Q [**ST**]. This is just an extension of the corresponding motion planning problem, where the size of P is fixed. The running time of the algorithm is almost the same as that of the motion planning algorithm given in [**KS, KST**].
- **Width in three dimensions** (see also the introduction): Compute the width of a set S of n points in $\mathbb{R}^3$; this is the smallest distance between two parallel planes enclosing S between them. This problem has been studied in a series of papers [**AAS, AS3, CEGS2**], and the current best bound is $O(n^{3/2+\varepsilon})$ [**AS3**]. The technique used in attacking this and the two following problems reduce them to problems involving lower envelopes in 4 dimensions, where we need to construct and to search in such an envelope.
- **Biggest stick in a simple polygon:** Compute the longest line segment that can fit inside a given simple polygon with n edges. The current best solution is $O(n^{3/2+\varepsilon})$ [**AS3**] (see also [**AAS, AST**]).
- **Smallest-width annulus:** Compute the annulus of smallest width that encloses a given set of n points in the plane. Again, the current best solution is $O(n^{3/2+\varepsilon})$ [**AS3**] (see also [**AAS, AST**]).
- **Geometric matching:** Consider the problem where we are given two sets S_1, S_2 of n points in the plane, and we wish to compute a minimum-weight matching in the complete bipartite graph $S_1 \times S_2$, where the weight of an edge (p, q) is the Euclidean distance between p and q. One can also consider the analogous nonbipartite version of the problem, which involves just one set S of $2n$ points, and the complete graph on S. The goal is to explore the underlying geometric structure of these graphs, to obtain faster algorithms than those available for general abstract graphs.

  It was shown in [**V**] that both the bipartite and the nonbipartite versions of the problem can be solved in time close to $O(n^{2.5})$. Recently, a fairly sophisticated application of vertical decomposition in 3-dimensional arrangements, given in [**AES**], has improved the running time for the bipartite case to $O(n^{2+\varepsilon})$. This technique does not yet extend to the nonbipartite case, which remains an interesting open problem.

This list is by no means exhaustive.

9.3. Transversals. Let S be a set of n compact convex sets in $\mathbb{R}^d$, each of constant description complexity. The space $\mathcal{T}(S)$ of all hyperplane transversals of S has been defined in the introduction. It was shown there that this space, in a dual setting, is the region in $\mathbb{R}^d$ enclosed between the upper envelope of n surfaces and the lower envelope of n other surfaces, where each surface in the first (resp. second) collection represents the locus of the lower (resp. upper) tangent hyperplanes to an object of S. The results of [**ASS**] imply that the complexity of $\mathcal{T}(S)$ is $O(n^{2+\varepsilon})$ in three dimensions. No similarly sharp bounds are known in higher dimensions. The results of [**ASS**] concerning the complexity of the vertical decomposition of such a 'sandwiched' region imply that $\mathcal{T}(S)$ can be constructed in $O(n^{2+\varepsilon})$ time.

The problem can be generalized by considering lower-dimensional transversals. For example, in three dimensions we can also consider the space of all line transversals of S. By mapping lines in $\mathbb{R}^3$ into points in 4-space, and by using an appropriate parametrization of the lines, the space of all line transversals of S can also be represented as the region in $\mathbb{R}^4$ enclosed between an upper envelope and a lower envelope of two respective collections of surfaces. Since no sharp bounds are known for the complexity of such a region in 4-space, the exact calibration of the complexity of the space of line transversals in 3-space is still an open problem.

References

[AAS] P. K. Agarwal, B. Aronov and M. Sharir, *Computing envelopes in four dimensions with applications*, SIAM J. Comput. 26 (1997), 1714–1732.

[AASS] P. K. Agarwal, B. Aronov, M. Sharir and S. Suri, *Selecting distances in the plane*, Algorithmica 9 (1993), 495–514.

[AES] P. K. Agarwal, A. Efrat and M. Sharir, *Vertical decompositions of shallow levels in arrangements and their applications*, Proc. 11th ACM Symp. on Computational Geometry (1995), 39–50. Also to appear in SIAM J. Comput.

[AEST] P. K. Agarwal, A. Efrat, M. Sharir and S. Toledo, *Computing a segment-center for a planar point set*, J. Algorithms 15 (1993), 314–323.

[ASS] P. K. Agarwal, O. Schwarzkopf and M. Sharir, *The overlay of lower envelopes in 3-space and its applications*, Discrete Comput. Geom. 15 (1996), 1–13.

[AS] P. K. Agarwal and M. Sharir, *On the number of views of polyhedral terrains*, Discrete Comput. Geom. 12 (1994), 177–182.

[AS2] P. K. Agarwal and M. Sharir, *Efficient algorithms for geometric optimization*, to appear in ACM Computing Surveys.

[AS3] P. K. Agarwal and M. Sharir, *Efficient randomized algorithms for some geometric optimization problems*, Discrete Comput. Geom. 16 (1996), 317–337.

[AS4] P. Agarwal and M. Sharir, On the complexity of the union of congruent cylinders in 3-space, manuscript, 1997.

[ASSh] P. Agarwal, M. Sharir and P. Shor, *Sharp upper and lower bounds for the length of general Davenport Schinzel sequences*, J. Combin. Theory, Ser. A 52 (1989), 228–274.

[AST] P. K. Agarwal, M. Sharir and S. Toledo, *New applications of parametric searching in computational geometry*, J. Algorithms 17 (1994), 292–318.

[APS] B. Aronov, M. Pellegrini and M. Sharir, *On the zone of a surface in a hyperplane arrangement*, Discrete Comput. Geom. 9 (1993), 177–186.

[A] B. Aronov, personal communication, 1995.

[ArS] B. Aronov and M. Sharir, *The union of convex polyhedra in three dimensions*, Proc. 34th IEEE Symp. on Foundations of Computer Science (1993), 518–527.

[ArS2] B. Aronov and M. Sharir, *Castles in the air revisited*, Discrete Comput. Geom. 12 (1994), 119–150.

[ArS3] B. Aronov and M. Sharir, *On translational motion planning of a convex polyhedron in 3-space*, SIAM J. Comput. 26 (1997), 1785–1803.

[ASTa] B. Aronov, M. Sharir and B. Tagansky, *The union of convex polyhedra in three dimensions*, SIAM J. Comput. 26 (1997), 1670–1688 (a revised version of [**ArS**]).

[Au] F. Aurenhammer, *Voronoi diagrams—A survey of a fundamental geometric data structure*, ACM Computing Surveys 23 (1991), 346–405.

[BD] J. D. Boissonnat and K. Dobrindt, *andomized construction of the upper envelope of triangles in* $\mathbb{R}^3$, Proc. 4th Canadian Conf. on Computational Geometry (1992), 311–315.

[BD2] J. D. Boissonnat and K. Dobrindt, *On-line randomized construction of the upper envelope of triangles and surface patches in* $\mathbb{R}^3$, Comp. Geom. Theory Appls. 5 (1996), 303–320.

[BSTY] J. D. Boissonnat, M. Sharir, B. Tagansky and M. Yvinec, *Voronoi diagrams in higher dimensions under certain polyhedral convex distance functions*, Proc. 11th ACM Symp. on Computational Geometry (1995), 79–88. Also to appear in Discrete Comput. Geom.

[BDy] K. W. Bowyer and C. R. Dyer, *Aspect graphs: An introduction and survey of recent results*, Int. J. of Imaging Systems and Technology 2 (1990), 315–328.

[CEGS] B. Chazelle, H. Edelsbrunner, L. Guibas and M. Sharir, *A singly exponential stratification scheme for real semi-algebraic varieties and its applications*, Proc. 16th Int. Colloq. on Automata, Languages and Programming (1989), 179–193.

[CEGS2] B. Chazelle, H. Edelsbrunner, L. Guibas and M. Sharir, *Diameter, width, closest line pair, and parametric searching*, Discrete Comput. Geom. 10 (1993), 183–196.

[C] L. P. Chew, *Near-quadratic bounds for the* L_1 *Voronoi diagram of moving points*, Proc. 5th Canadian Conf. on Computational Geometry (1993), 364–369.

[CKSTW] L. P. Chew, K. Kedem, M. Sharir, B. Tagansky and E. Welzl, *Voronoi diagrams of lines in three dimensions under a polyhedral convex distance function*, Proc. 6th ACM-SIAM Symp. on Discrete Algorithms (1995), 197–204. Also to appear in J. Algorithms.

[Cl] K. L. Clarkson, *New applications of random sampling in computational geometry*, Discrete Comput. Geom. 2 (1987), 195–222.

[CS] K. L. Clarkson and P. W. Shor, *Applications of random sampling in computational geometry*, II, Discrete Comput. Geom. 4 (1989), 387–421.

[dB] M. de Berg, personal communication, 1993.

[dBDS] M. de Berg, K. Dobrindt and O. Schwarzkopf, *On lazy randomized incremental construction*, Discrete Comput. Geom. 14 (1995), 261–286.

[dBGH] M. de Berg, L. Guibas and D. Halperin, *Vertical decomposition for triangles in 3-space*, Discrete Comput. Geom. 15 (1996), 35–61.

[dBHO] M. de Berg, D. Halperin, M. Overmars and M. van Kreveld, Sparse arrangements and the number of views of polyhedral scenes, Manuscript, 1991.

[E] H. Edelsbrunner, Algorithms in Combinatorial Geometry, Springer-Verlag, Heidelberg 1987.

[E2] H. Edelsbrunner, *The upper envelope of piecewise linear functions: Tight complexity bounds in higher dimensions*, Discrete Comput. Geom. 4 (1989), 337–343.

[EGPPSS] H. Edelsbrunner, L. Guibas, J. Pach, R. Pollack, R. Seidel and M. Sharir, *Arrangements of curves in the plane: topology, combinatorics, and algorithms*, Theoret. Comput. Sci. 92 (1992), 319–336.

[ES] H. Edelsbrunner and R. Seidel, *Voronoi diagrams and arrangements*, Discrete Comput. Geom. 1 (1986), 25–44.

[ESS] H. Edelsbrunner, R. Seidel and M. Sharir, *On the zone theorem for hyperplane arrangements*, SIAM J. Comput. 22 (1993), 418–429.

[ESh] A. Efrat and M. Sharir, *A near-linear algorithm for the planar segment center problem*, Discrete Comput. Geom. 16 (1996), 239–257.

[FL] J.-J. Fu and R. C. T. Lee, *Voronoi diagrams of moving points in the plane*, Internat. J. Comput. Geom. Appl. 1 (1994), 23–32.

[GHMS] L. Guibas, D. Halperin, J. Matoušek and M. Sharir, *On vertical decomposition of arrangements of hyperplanes in four dimensions*, Discrete Comput. Geom. 14 (1995), 113–122.

[GMR] L. Guibas, J. Mitchell and T. Roos, *Voronoi diagrams of moving points in the plane*, Proc. 17th Internat. Workshop Graph-Theoret. Concepts Computer Science, Lecture Notes in Comp. Sci., vol. 570, Springer-Verlag, pp. 113–125.

[GS] L. Guibas and M. Sharir, *Combinatorics and algorithms of arrangements*, in New Trends in Discrete and Computational Geometry, (J. Pach, Ed.), Springer-Verlag, 1993, 9–36.

[GSS] L. Guibas, M. Sharir and S. Sifrony, *On the general motion planning problem with two degrees of freedom*, Discrete Comput. Geom. 4 (1989), 491–521.

[H] D. Halperin, *On the complexity of a single cell in certain arrangements of surfaces in 3-space*, Discrete Comput. Geom. 11 (1994), 1–33.

[HS] D. Halperin and M. Sharir, *Near-quadratic bounds for the motion planning problem for a polygon in a polygonal environment*, Discrete Comput. Geom. 16 (1996), 121–134.

[HS2] D. Halperin and M. Sharir, *New bounds for lower envelopes in three dimensions with applications to visibility of terrains*, Discrete Comput. Geom. 12 (1994), 313–326.

[HS3] D. Halperin and M. Sharir, *Almost tight upper bounds for the single cell and zone problems in three dimensions*, Discrete Comput. Geom. 14 (1995), 285–410.

[H2] D. Halperin, *Arrangements*, in Handbook of Discrete and Computational Geometry (J. E. Goodman and J. O'Rourke, Editors), CRC Press, Boca Raton, FL, 1997, 389–412.

[HS4] D. Halperin and M. Sharir, *Arrangements and their applications in robotics: Recent developments*, Proc. Workshop on Algorithmic Foundations of Robotics (K. Goldberg et al., Editors), A. K. Peters, Boston, MA, 1995, 495–511.

[HY] D. Halperin and C.-K. Yap, *Complexity of translating a box in polyhedral 3-space*, Proc. 9th Annu. ACM Sympos. Comput. Geom. 1993, 29–37.

[HSh] S. Hart and M. Sharir, *Nonlinearity of Davenport–Schinzel sequences and of generalized path compression schemes*, Combinatorica 6 (1986), 151–177.

[HW] D. Haussler and E. Welzl, *ε-nets and simplex range queries*, Discrete Comput. Geom. 2 (1987), 127–151.

[HRR] J. Heintz, T. Recio and M. F. Roy, *Algorithms in real algebraic geometry and applications to computational geometry*, in Discrete and Computational Geometry: Papers from DIMACS Special Year, (J. Goodman, R. Pollack, and W. Steiger, Eds.), American Mathematical Society, Providence, RI, 137–163.

[KS] K. Kedem and M. Sharir, *An efficient motion planning algorithm for a convex rigid polygonal object in 2-dimensional polygonal space*, Discrete Comput. Geom. 5 (1990), 43–75.

[KST] K. Kedem, M. Sharir and S. Toledo, *On critical orientations in the Kedem-Sharir motion planning algorithm*, Discrete Comput. Geom. 17 (1997), 227–239.

[KPW] J. Komlós, J. Pach and G. Woeginger, *Almost tight bound on epsilon-nets*, Discrete Comput. Geom. 7 (1992), 163–173.

[LS] D. Leven and M. Sharir, *Intersection and proximity problems and Voronoi diagrams*, in Advances in Robotics, Vol. I, (J. Schwartz and C. Yap, Eds.), 1987, 187–228.

[M] J. Matoušek, *Approximations and optimal geometric divide-and-conquer*, J. Comput. Syst. Sci. 50 (1995), 203–208.

[McM] P. McMullen and G. C. Shephard, Convex Polytopes and the Upper Bound Conjecture, Lecture Notes Ser. 3, Cambridge University Press, Cambridge, England, 1971.

[Me] N. Megiddo, *Applying parallel computation algorithms in the design of serial algorithms*, J. ACM 30, 852–865.

[PS] J. Pach and M. Sharir, *The upper envelope of piecewise linear functions and the boundary of a region enclosed by convex plates: Combinatorial analysis*, Discrete Comput. Geom. 4 (1989), 291–309.

[P] M. Pellegrini, *On lines missing polyhedral sets in 3-space*, Proc. 9th ACM Symp. on Computational Geometry (1993), 19–28.

[PD] H. Plantinga and C. Dyer, *Visibility, occlusion, and the aspect graph*, Internat. J. Computer Vision, 5 (1990), 137–160.

[PrS] F. Preparata and M. Shamos, Computational Gemetry: An Introduction, Springer-Verlag, Heidelberg, 1985.

[SS] J. T. Schwartz and M. Sharir, *On the Piano Movers' problem: II. General techniques for computing topological properties of real algebraic manifolds*, Advances in Appl. Math. 4 (1983), 298–351.

[SSh] O. Schwarzkopf and M. Sharir, *Vertical decomposition of a single cell in a 3-dimensional arrangement of surfaces*, Discrete Comput. Geom. 18 (1997), 269–288.

[S] M. Sharir, *On k-sets in arrangements of curves and surfaces*, Discrete Comput. Geom. 6 (1991), 593–613.

[S2] M. Sharir, *Almost tight upper bounds for lower envelopes in higher dimensions*, Discrete Comput. Geom. 12 (1994), 327–345.

[S3] M. Sharir, *Arrangements in higher dimensions: Voronoi diagrams, motion planning, and other applications*, Proc. Workshop on Algorithms and Data Structures, Ottawa, Canada, August, 1995, Lecture Notes in Computer Science, Vol. 955, Springer-Verlag, 109–121.

[SA] M. Sharir and P. K. Agarwal, Davenport-Schinzel Sequences and Their Geometric Applications, Cambridge University Press, New York, 1995.

[ST] M. Sharir and S. Toledo, *Extremal polygon containment problems*, Comput. Geom. Theory Appls. 4 (1994), 99–118.

[T] B. Tagansky, *A new technique for analyzing substructures in arrangements*, Discrete Comput. Geom. 16 (1996), 455–479.

[T2] B. Tagansky, *The Complexity of Substructures in Arrangements of Surfaces*, Ph.D. Dissertation, Tel Aviv University, July 1996.

[V] P. M. Vaidya, *Geometry helps in matching*, SIAM J. Comput. 18 (1989), 1201–1225.

[WS] A. Wiernik and M. Sharir, *Planar realization of nonlinear Davenport–Schinzel sequences by segments*, Discrete Comput. Geom. 3 (1988), 15–47.

School of Mathematical Sciences, Tel Aviv University, Tel Aviv 69978, Israel, and Courant Institute of Mathematical Sciences, New York University, New York, NY 10012, USA

E-mail address: sharir@math.tau.ac.il

Contemporary Mathematics
Volume **223**, 1999

Geometric Discrepancy Theory

Joel Spencer

ABSTRACT. Bounds on discrepancy in a geometric setting are often much tighter than for general set systems. Arithmetic progressions also exhibit this geometric behavior. Methods and bounds are reviewed for several cases emphasizing similarities in the methodologies.

1. Basics

We begin with the notion of discrepancy. Let $\mathcal{A}$ be a family of subsets of a set Ω, whose elements will be called vertices. A *two-coloring* is a map $\chi : \Omega \rightarrow \{-1, +1\}$. For any $A \subseteq \Omega$ we define $\chi(A) = \sum_{a \in A} \chi(a)$. Less formally, we think of $-1, +1$ as Red and Blue so that $|\chi(A)|$ gives how evenly balanced A has been colored by the two colors. The formal definition of discrepancy $disc(\mathcal{A})$ is then given by

$$disc(\mathcal{A}) = \min_{\chi} \max_{A \in \mathcal{A}} |\chi(A)| \tag{1}$$

The study of discrepancy nearly always breaks down into the study of lower bounds and of upper bounds on discrepancy. The following expanded definitions may prove helpful. The *upper* bound $disc(\mathcal{A}) \leq K$ means there exists a two-coloring $\chi : \Omega \rightarrow \{-1, +1\}$ so that $|\chi(A)| \leq K$ for all $A \in \mathcal{A}$. The *lower* bound $disc(\mathcal{A}) \geq K$ means that given any two coloring $\chi : \Omega \rightarrow \{-1 + -1\}$ there is some $A \in \mathcal{A}$ with $|\chi(A)| \geq K$.

In this survey we shall examine the notion of discrepancy in three contexts.

• General Set Systems. Here the object will be to give an upper bound on discrepancy in terms of the sizes of the $A \in \mathcal{A}$, the number of vertices, the degrees of the vertices, etc. In this general context one doesn't hope for lower bounds.

• Number Theorem. Here we consider the specific case of $\Omega = \{1, \dots, n\}$ and the $A \in \mathcal{A}$ being the arithmetic progressions in Ω.

• Geometry. This is central to our current discussion. We take Ω to be a finite subset of Euclidean space R^d. We can take $\mathcal{A}$ to be the intersection of Ω with the halfspaces, the balls, the rectangles, or any of a number of geometric objects. We'll look particularly at Ω a set of vertices in the plane and $\mathcal{A}$ the family of halfspaces.

1991 *Mathematics Subject Classification.* Primary 52C10; Secondary 11B25, 05C65.
Key words and phrases. Discrepancy, probabilistic methods.

© 1999 American Mathematical Society

2. Way off!

The definition of the upper bound with its emphasis on existence leads quite naturally to an Erdős-type existence argument.

THEOREM 1. *If $\mathcal{A}$ consists of m sets on n vertices then*

$$disc(\mathcal{A}) \leq [2n\ln(2m)]^{1/2} \tag{2}$$

PROOF. A uniformly chosen χ has

$$\Pr[|\chi(A)| \geq \alpha] < 2e^{-\alpha^2/2|A|}$$

by standard Large Deviation inequalities, see, e.g., the Appendix of [**AS**]. Setting $\alpha = [2n\ln(2m)]^{1/2}$ this probability is at most $\frac{1}{m}$ for each $A \in \mathcal{A}$, as $|A| \leq n$. Thus with positive probability none of these events occurs and so there exists χ with all $|\chi(A)| \leq \alpha$, as desired. □

It is instructive to apply this bound in two important cases. Consider the geometry example of n vertices in the plane with halfspaces. There are $m = O(n^2)$ halfspaces. Therefore the discrepancy is $O(n^{1/2}\ln^{1/2} m) = O(n^{1/2}\ln^{1/2} n)$. Consider also the Number Theory example of arithmetic progressions on n vertices. There are $m = O(n^2 \ln n)$ progressions (n choices for the first element and, given the difference d, at most n/d choices for the length) and again the discrepancy is $O(n^{1/2}\ln^{1/2} m) = O(n^{1/2}\ln^{1/2} n)$.

These results, written in O-from, are technically correct since they only give an upper bound on discrepancy. But both are far from the true value which, we shall see, is $\Theta(n^{1/4})$ in both cases. In geometric settings the discrepancy is often far lower than basic set theoretic considerations would give.

3. Floating colors

The *degree* of a vertex is the number of sets $A \in \mathcal{A}$ that contain it. Here we give a remarkable and far reaching theorem of J. Beck and T. Fiala [**BF**].

THEOREM 2. *If all vertices have degree at most D then $\mathcal{A}$ has discrepancy at most $2D - 1$.*

PROOF. No assumptions are made about the total number of vertices, the total number of sets, or the sizes of the individual sets. An Erdős-type existence argument (at least, in straightforward style) could not give this result - a very large set randomly colored would be highly unlikely to have discrepancy less than $2D$. The proof uses nothing more than linear algebra!

To each vertex v we associate a variable x_v which will change in time. Initially all x_v are set to zero. At all times all x_v lie in the closed interval $[-1, +1]$. If $x_v = \pm 1$ then v is called fixed. Once v is fixed, the x_v never changes and becomes the final color of v. If $-1 < x_v < +1$ then v is called floating. A set $A \in \mathcal{A}$ is called dangerous if it has more than D floating points; otherwise it is called safe. Analogous to our previous notation we define $\chi(A) = \sum_{v \in A} x_v$ which shall change in time. We require at all times that

$$\chi(A) = 0 \text{ for all dangerous sets } A \in \mathcal{A}$$

Call this Condition BF. Condition BF holds initially since all $x_v = 0$ so all $\chi(A) = 0$. Now suppose Condition BF holds at a certain time. Each floating vertex is in at most D dangerous sets (indeed each vertex is in at most D sets) and each dangerous set contains more than D floating vertices so the number of dangerous sets is strictly less than the number of floating vertices. Think of the x_v, v floating, as variables and Condition BF as a system of linear equations on these variables. This system is underdetermined, with more variables than equations. Thus we can move along a line of solutions for condition BF in Euclidean space. We do so until one of the x_v reaches ± 1 and stop. This gives a new set of values for the x_v with at least one of the previously floating v now fixed so that some of the previously dangerous sets may now be considered safe. We continue this procedure until all of the v are fixed. (Near the end we may well have no dangerous A. At such a stage we can simply set those $x_v = \pm 1$ arbitrarily.)

We claim these final x_v give a two-coloring for which all $|\chi(A)| < 2D$. Consider any $A \in \mathcal{A}$. At the first time it becomes safe it has $\chi(A) = 0$ and at most D floating vertices. Each floating vertex v can have x_v change by less than two from that value to its final value, e.g., from $x_v = -0.999$ to $x_v = +1$. As such $\chi(A)$ changes by less than $2D$ and so the final $|\chi(A)|$ is less than $2D$. Being an integer, it is at most $2D - 1$. □

The above proof is most algorithmic. Given the x_v at any time find, by standard Linear Algebra packages, a nontrivial (i.e., not all zero) solution to the underdetermined homogeneous system

$$\sum_{v \in A \cap FL} y_v = 0,$$

where FL is the set of floating vertices and A ranges over the dangerous sets. Then let λ be that real of maximal absolute value so that all $|x_v + \lambda y_v| \leq 1$, v floating. Reset $x_v \leftarrow x_v + \lambda y_v$.

We illustrate the power of the Beck-Fiala theorem with another theorem of Beck [**B**]. Let Ω be a set of n vertices in the plane and let $\mathcal{A}$ consist of those rectangles with axes parallel to the X and Y-axes. We show here that the discrepancy is $O(\ln^4 n)$. (Beck actually did somewhat better with the exponent.) We fix an n vertex set Ω. For ease of exposition we make the simplifying assumptions that no two vertices of Ω have the same X or Y coordinate, that n is precisely a power of two, and that all vertices are in the first quadrant.

A critical idea, which appears often in Geometric Discrepancy, is a definition of canonical sets. These must be such that the family of canonical sets has small discrepancy and all sets $A \in \mathcal{A}$ can be simply expressed in terms of the canonical sets. In our case consider the $n = 2^t$ different X-coordinates, ordered $x_1 < x_2 < \ldots < x_n$. We split this set into two halves, four fourths, eight eighths, $\ldots$, $n/2$ pairs, down to n singletons. More formally for $0 \leq s \leq t$ and $1 \leq i \leq 2^s$ set

$$X_{s,i} = \{x_u : (i-1)n2^{-s} < u \leq in2^{-s}\}$$

(this includes the entire set as $X_{0,1}$) and analogously (but with their own ordering) define the $Y_{s,i}$. The canonical rectangles are then all products $X_{s,i} \times Y_{s',i'}$ of such sets. Each vertex $v = (x_v, y_v)$ lies in precisely $t+1$ sets $X_{s,i}$ and $t+1$ sets $Y_{s',i'}$ and so lies in $(t+1)^2$ canonical rectangles. Set $D = (t+1)^2 \sim \ln^2 n$. From the Beck-Fiala theorem there is a two coloring χ so that all canonical rectangles C have $|\chi(C)| \leq 2D - 1 = O(\ln^2 n)$.

Now consider any aligned rectangle cornered at the origin, say $[0, x] \times [0, y]$. Say $x_l \leq x < x_{l+1}$ and write l in base two as $l = 2^{t_1} + \ldots + 2^{t_r}$ with $t_1 > t_2 > \ldots > t_r$.

We partition $\{x_1, \ldots, x_l\}$ into r "canonical" sets, the first 2^{t_1} values, the next 2^{t_2} values, ... , the last 2^{t_r} values, all of which are of the form $X_{s,i}$. Similarly on the Y-axis we have $y_{l'} \leq y < y_{l'+1}$ and we partition $\{y_1, \ldots, y_{l'}\}$ into r' canonical sets. Now the rectangle $[0, x] \times [0, y]$ is partitioned into the rr' canonical products. But $rr' = O(\ln^2 n)$. Every rectangle R is the disjoint union of $O(\ln^2 n)$ canonical rectangles C so the two coloring χ given by the Beck-Fiala theorem has

$$|\chi(A)| \leq \sum |\chi(C)| = O(\ln^4 n)$$

Finally, consider the vertices A in any aligned rectangle $[x', x] \times [y', y]$. Let B, C, D, E be the vertices in rectangles $[0, x] \times [0, y]$, $[0, x] \times [0, y')$, $[0, x') \times [0, y]$ and $[0, x') \times [0, y')$ respectively. Then

$$\chi(A) = \chi(B) - \chi(C) - \chi(D) + \chi(E)$$

so that $|\chi(A)| = O(\ln^4 n)$.

4. Two somewhat similar lower bounds

4.1. Arithmetic progressions. In 1964 K.F. Roth [**R**] gave a lower bound $\Omega(n^{1/4})$ for the discrepancy of the arithmetic progressions on $\{1, \ldots, n\}$. This is a real gem!

Fix an arbitrary (not random) coloring χ of $\{1, \ldots, n\}$. Set $Q = n^{1/2}$ and $L = Q/2$, ignoring roundoff for convenience. For $1 \leq j \leq n$ and $1 \leq d \leq Q$ define

$$\Delta_{j,d} = \chi(j + d) + \chi(j + 2d) + \ldots + \chi(j + Ld)$$

the discrepancy of the arithmetic progression $j + d, j + 2d, \ldots, j + Ld$. We "wrap around" at n so that if $j + id > n$ we mean by $\chi(j + id)$ the value $\chi(j + id - n)$. As $Ld < n$ we can only wrap around once ($j + id < 2n$ always) so this set is the union of at most two actual arithmetic progressions, so if $\Delta_{j,d} \geq \Delta$ there will be an actual arithmetic progression with discrepancy at least $\frac{1}{2}\Delta$. As constants do not concern us here we concentrate on the Δ. Roth actually shows that the average value of $\Delta_{j,d}^2$ (over the above ranges of j, d) is $\Omega(n^{1/2})$. We may think of this then as a kind of variance bound.

For convenience we define $e(x) = e^{2\pi i x}$.

Using Fourier analysis Roth writes

$$\sum_{d=1}^{Q} \sum_{j=1}^{n} \Delta_{j,d}^2 = \sum_{d=1}^{Q} \int_0^1 |F_d(t)|^2 dt \tag{3}$$

with the Fourier transform

$$F_d(t) = \sum_{j=1}^{n} \Delta_{j,d} e(jt) \tag{4}$$

Set

$$A(t) = \sum_{j=1}^{n} \chi(j) e(jt)$$

and

$$\Psi_d(t) = \sum_{s=1}^{L} e(-st)$$

so that (reflecting Δ being the convolution of χ and $\{d, 2d, \ldots, Ld\}$ in the transform)

$$F_d(t) = A(t)\Psi_d(t)$$

so that

$$\sum_{d=1}^{Q}\sum_{j=1}^{n}\Delta_{j,d}^2 = \int_0^1 |A(t)|^2 \sum_{d=1}^{Q} |\Psi_d(t)|^2 dt \tag{5}$$

Since A is itself a Fourier transform

$$\int_0^1 |A(t)|^2 dt = \sum_{j=1}^{n} \chi^2(j) = n \tag{6}$$

To deal with the Ψ_d Roth uses a little number theory. For any real number $t \in [0, 1]$ there are integers a, q with $1 \leq q \leq Q$ so that

$$|qt - a| \leq \frac{1}{(Q+1)}$$

Why? Consider the numbers $0, t, 2t, \ldots, Qt$ modulo one - they split the unit "circle" into $Q + 1$ parts so some two of them it, jt are within $\frac{1}{Q+1}$ so, setting $q = |j - i|$, qt is within $\frac{1}{Q+1}$ of zero which means, not considering modulo one, that qt is within $\frac{1}{Q+1}$ of an integer a. This is a classic result, used (dividing both sides by q) for approximating an arbitrary t by rational numbers $\frac{a}{q}$.

For any $t \in [0, 1]$ let q, a be as above. Note the function e has period one and let $\epsilon = qt - a$ so that

$$\Psi_q(t) = e(\epsilon) + e(2\epsilon) + \ldots + e(L\epsilon)$$

Formally this is a geometric series for which we can find the precise sum. Less formally, though still formalizable, the values $e(j\epsilon)$ are all lying on the unit circle (in the Complex plane) but with $|L\epsilon| \leq \frac{L}{Q} = \frac{1}{2}$ they all lie in the same half plane and so they don't "cancel" but rather $|\Psi_q(t)| \geq cL = \Omega(n^{1/2})$. Thus

$$\sum_{d=1}^{Q} |\Psi_d(t)|^2 \geq |\Psi_q(t)|^2 = \Omega(n)$$

bounding the sum by the one particularly large term. Inserting this bound

$$\sum_{d=1}^{Q}\sum_{j=1}^{n}\Delta_{j,d}^2 \geq \int_0^1 cn|A(t)|^2 dt \geq cn^2 \tag{7}$$

so that the average value of $\Delta_{j,d}^2$ is at least $c'n^{1/2}$ and therefore there exist some j, d with $|\Delta_{j,d}|$ least $\Omega(n^{1/4})$.

4.2. Half spaces. We consider here a specific set of n points in the plane - a square grid. For convenience we assume n an odd perfect square and set

$$\Omega = \{(in^{-1/2}, jn^{-1/2}) : |i|, |j| < \frac{1}{2}n^{1/2}\}$$

so that $\Omega \subset [-\frac{1}{2}, +\frac{1}{2}] \times [-\frac{1}{2}, +\frac{1}{2}]$. Let $\mathcal{A}$ denote the Half Spaces. Here we give our version of J. Pach's [**PA**] version of Chazelle, Matoušek and Sharir's [**CMS**] version of Alexander's [**A**] theorem that the discrepancy is at least $cn^{1/4}$. We have designed our argument to maximize the similarities with Roth's result. Still, the results are not really equivalent and their similarities and dissimilarities remain intriguing.

Fix a two coloring χ, our notation being that $\chi(P)$ is the color of the point $P \in \Omega$. Note χ is arbitrary, not random. Set $w = \epsilon n^{-1/2}$ with ϵ a small absolute constant, $\epsilon = 10^{-3}$ would do. Now we want to pass a *random sliver* of width w through our vertices. What do we mean by this? Since constants do not concern us here is one approach.

Our vertices lie well inside the unit circle of radius one. Define a random line l at direction θ chosen uniformly $\theta \in [0, \pi)$ and at directional distance b from the origin, $b \in [-1, +1]$ chosen uniformly. Algebraically the line has equation

$$x \sin\theta + y\cos\theta = b \tag{8}$$

Make this random line l the center of a random sliver l^w by placing two parallel lines $w/2$ on each side as the boundaries. Let l^+ be the left side of the sliver, l^- the right. (Formally, letting $f(x,y) = x\sin\theta + y\cos\theta$, we give l^+ equation $f(x,y) \in [b - \frac{w}{2}, b)$, l^- equation $f(x,y) \in [b, b + \frac{w}{2})$ and l^w equation $f(x,y) \in [b - \frac{w}{2}, b + \frac{w}{2})$. The use of halfopen slivers avoids technical overlap problems.) $\Delta^+(l)$ denote the summation of $\chi(P)$ over all $P \in l^+$ and $\Delta^-(l)$ denote the same summation over $P \in l^-$ and set $\Delta(l) = \Delta^+(l) - \Delta^-(l)$. We shall show

$$E[\Delta^2(l)] > cn^{1/2} \tag{9}$$

with the expectation over the random choice of l.

Assume (9). Then there is a specific l with

$$|\Delta(l)| = |\Delta(l^+) - \Delta(l^-)| \geq c'n^{1/4}$$

and so some sliver l' (l^+ or l^-, maybe both) has discrepancy $\Omega(n^{1/4})$. But consider the two halfspaces H, H' whose boundaries are the two sides of l', both in the same direction so that, say $H \subseteq H'$. Then $|\chi(H) - \chi(H')| = |\chi(l')|$ so at least one of them has discrepancy $\Omega(n^{1/4})$, completing the argument.

Showing (9) may be regarded as a variance calculation. For any vertex $P \in \Omega$ set $Y_P = \chi(P)$ if $P \in l^+$; $Y_P = -\chi(P)$ if $P \in l^-$; and $Y_P = 0$ if P does not lie in the random sliver at all. Then

$$\Delta(l) = \sum_{P \in \Omega} Y_P$$

so we square and set

$$E[\Delta^2(l)] = VAR + COV$$

with the "variance terms"

$$VAR = \sum_P E[Y_P^2]$$

and the "covariance terms"

$$COV = \sum_{P \neq Q} E[Y_P Y_Q] \tag{10}$$

For the variance terms we note Y_P^2 is simply the indicator for P lying in the random sliver. Consider any particular P, fix θ, and consider the random sliver at angle θ.

Of the uniformly chosen $b \in [-1, +1]$ a subinterval of length precisely w of them have sliver containing P. Therefore $E[Y_P^2] = 2/w$ and

$$VAR = \sum_P \frac{2}{w} = \frac{2n}{w} > c\epsilon n^{1/2}$$

Now we want to show that the covariance terms $E[Y_P Y_Q]$ are small, that $|COV|$ is small compared to VAR. The idea is that conditional on P lying in, say, l^+ vertex Q is roughly equally likely to lie in l^+ as l^- and these terms will roughly cancel. Note, however, that when P, Q are extremely close together P lying in l^+ makes it very likely that Q also lies in l^+. We let d denote the distance between P and Q and bound $|E[Y_P Y_Q]|$ in terms of d. It will be convenient to parametrize d by

$$w = d\alpha$$

Note that $d \geq n^{-1/2}$ and our deliberate choice of w assures that $\alpha \leq \epsilon$. We note that $\chi(P), \chi(Q)$ affect this expectation by multiplicative ± 1 factors so for simplicity we may assume $\chi(P) = \chi(Q) = +1$.

Let I^w be the indicator for P, Q lying in a random sliver l^w and I^+, I^- the same for P, Q lying in l^+, l^- respectively. Then

$$Y_P Y_Q = I^w - 2(I^+ + I^-)$$

Let $I^{w/2}$ be the indicator for P, Q lying in a random sliver of width $w/2$, analogous to I^w. Then $E[I^+] = E[I^-] = E[I^{w/2}]$ so

$$E[Y_P Y_Q] = E[I^w] - 4E[I^{w/2}] \tag{11}$$

Let θ denote the angle random line l makes with PQ so that we may think of $\theta \in [0, \pi)$ chosen uniformly. (Alternatively, by symmetry we may assume PQ is a horozontal line.) Let $E_\theta[I^w]$ be the expectation for this given θ, and similarly for $E_\theta[I^{w/2}]$. Recall the random line sweeps through length two. Now P, Q are $d \sin\theta$ apart as the random line l sweeps through. If $\sin\theta > \alpha$ then the sliver never contains them both so the expectation is zero. If $\sin\theta \leq \alpha$ then $E_\theta[I^w] = \frac{1}{2}[w - d\sin\theta]$ Our range of critical θ is up to $\sin^{-1}(\alpha)$ and

$$E[I^w] = \frac{1}{\pi}\int_0^{sin^{-1}(\alpha)} \frac{1}{2}[w - d\sin\theta]d\theta \tag{12}$$

If we approximate $\sin\theta$ by θ in this region near zero we *would* get

$$E[I^w] \sim \frac{1}{2\pi}\int_0^\alpha [w - d\theta]d\theta = \frac{w\alpha}{4\pi}$$

Under this approximation $4E[I^{w/2}] \sim E[I^w]$ (note $\alpha = w/d$ halves) so $E[Y_P Y_Q] \sim 0$. Precise evaluation of (12) gives

$$E[I^w] = \frac{1}{2\pi}\left(w\sin^{-1}(\alpha) + d[\sqrt{1-\alpha^2} - 1]\right) \tag{13}$$

Taylor Series (in α) gives that this is $\frac{w\alpha}{4\pi} + cw\alpha^3$ plus smaller order terms. It is the second term that dominates $E[Y_P Y_Q]$ and gives

$$E[Y_P Y_Q] = O(w\alpha^3) = O(w^4 d^{-3}) \tag{14}$$

Now COV, given by (10), is the sum of terms $E[Y_P Y_Q]$. Recall we have set $w = \epsilon n^{-1/2}$ with ϵ small. Consider the contribution to (10) from a given P.

We bound the contribution by letting Q range over *all* points of the plane with $Q - P = (i,j)n^{-1/2}$ with i,j integral. These contribute

$$(\epsilon n^{-1/2})^4 n^{3/2} \sum (i^2+j^2)^{-3/2}$$

to COV, where the summation is over integers $(i,j) \neq (0,0)$. However, the inner summation converges to a constant. Roughly speaking, the contribution to COV can be estimated up to constant by the contribution from adjacent pairs P, Q. The contribution from a given P is now $O(\epsilon^4 n^{-1/2})$. As there are n vertices P, $COV = O(\epsilon^4 n^{1/2})$.

By selecting ϵ small enough we can make COV less than half the size of VAR so that $E[\Delta^2] = \Omega(\epsilon^4 n^{1/2})$ which is want we wanted to prove.

5. Entropy

Given a family $\mathcal{A}$ of subsets of Ω a *partial coloring* is a map $\chi : \Omega \to \{-1, 0, +1\}$. Those $x \in \Omega$ with $\chi(x) \neq 0$ are called colored, those with $\chi(x) = 0$ are called uncolored. We speak, roughly, of a *substantial* partial coloring as one for which a positive proportion of the $x \in \Omega$ are colored. We define $\chi(A) = \sum_{a \in A} \chi(a)$ as before. We now give a general criterion that assures a substantial partial coloring for which all $\chi(A)$ are appropriately small.

The entropy of a discrete random variable X is defined by

$$ENT[X] = \sum_{\alpha} -p_\alpha \log_2 p_\alpha$$

where α ranges over the possible values of X, $p_\alpha = \Pr[X = \alpha]$, and we interpret $0 \log_2 0$ as 0. We also write

$$H(p) = -p\log_2 p - (1-p)\log_2(1-p)$$

for $p \in [0,1]$, the entropy of the indicator random variable taking value 1 with probability p. We shall use three basic properties of this well studied function.

- Subadditivity. $ENT[(X_1, \ldots, X_s)] \leq ENT[X_1] + \ldots + ENT[X_s]$
- Bulge. If $ENT[X] \leq K$ then some $p_\alpha \geq 2^{-K}$
- Optimal Uniformity. If X takes on s values then $ENT[X] \leq \log_2 s$ with equality when X is uniform.

Define the $2b$-roundoff function $R_b(z)$ as the nearest integer to $\frac{z}{2b}$. For definiteness we'll take the smaller integer in case of ties. Note $R_b(z) = 0$ for $z \in (-b, +b]$, $R_b(z) = -3$ for $z \in (-7b, -5b]$ and in general the value $R_b(z)$ determines an interval of length $2b$ for z.

Let S_m, as standard, denote the sum of m independent identical variables, each uniformly ± 1. For our purposes note S_m is the distribution of $\chi(A)$ where A is an m-set and $\chi : A \to \{-1, +1\}$ is uniform. Now define

$$ENT[m,b] = ENT[R_b(S_m)] \tag{15}$$

THEOREM 3. *Let $A_1, \ldots, A_s \subseteq \Omega$ with $|\Omega| = n$, $|A_i| = m_i$ If*

$$\sum_{i=1}^{s} ENT[m_i, b_i] < \epsilon n \tag{16}$$

and

$$H(\gamma) < 1 - \epsilon \text{ with } \gamma < \frac{1}{2}$$

then there exists a partial χ *on* Ω *with*

$$|\chi(A_i)| \le b_i, 1 \le i \le s$$

and at least $2\gamma n$ *of the* $x \in \Omega$ *colored.*

In applications, as constants do not concern us, we can take $\epsilon = \gamma = 0.1$ *to meet the second criteria so that if* $\sum ENT[m_i, b_i] < n/10$ *then there is a substantial partial coloring with all* $|\chi(A_i)| \le b_i$.

PROOF. Let $\chi : \Omega \to \{-1, +1\}$ be uniform and set $X_i = R_{b_i}(\chi(A_i))$ and $X = (X_1, \dots, X_s)$. Then $ENT[X_i] = ENT[m_i, b_i]$ and by subadditivity $ENT[X] \le \sum ENT[m_i, b_i] \le \epsilon n$. By bulge there is a value of X that has probability at least $2^{-\epsilon n}$ of appearing. The probability space consists of 2^n equally likely colorings so there is a family C of at least $2^{(1-\epsilon)n}$ colorings which all have the same X_i. Suppose $\chi_1, \chi_2 \in C$ and define

$$\chi(x) = \frac{\chi_1(x) - \chi_2(x)}{2}$$

Then χ is a partial coloring with possible values $+1, 0, -1$. For each A_i

$$\chi(A_i) = \frac{\chi_1(A_i) - \chi_2(A_i)}{2}$$

But since χ_1, χ_2 have the same X_i value $\chi_1(A_i), \chi_2(A_i)$ lie on a common interval of length $2b_i$ so that $|\chi(A_i)| \le b_i$.

How many points are colored? We may consider $\chi_1 : \Omega \to \{-1, +1\}$ as a vertex of $\{-1, +1\}^n$ and consider that space as the Hamming cube with the Hamming metric ρ. Then the number of i with $\chi(i) \ne 0$ is the number of i with $\chi_1(i) \ne \chi_2(i)$ which is precisely the Hamming distance $\rho(\chi_1, \chi_2)$. We select χ_1, χ_2 at maximal distance. Here we can use a beautiful theorem of Kleitman [**K**] that any subset C of the Hamming cube of size at least $\sum_{i=0}^{t} \binom{n}{t}$ (with $t < \frac{n}{2}$) has diameter at least $2t$, the extreme case being the ball around a point of radius t. Our condition on ϵ, γ assures (with some technical calculation) that $2^{(1-\epsilon)n} \ge \sum_{i \le n\gamma} \binom{n}{t}$ so the diameter is at least $2n\gamma$ and so we can find χ with at least $2n\gamma$ points colored. □

A limitation of this result is that it gives only a partial coloring and generally one wants a full coloring $\chi : \Omega \to \{-1, +1\}$. This can *sometimes* be fixed by iterating the procedure, finding new χ on the uncolored vertices until all vertices have been colored. A positive aspect of the result is that condition (16) gives a kind of cost function for the b_i—one can adjust the b_i for the various A_i up and down to try to meet whatever conditions one has.

To apply our Theorem we need a good bounds on $ENT[m, b]$. For m large S_m is roughly $m^{1/2}N$ where N is standard normal. We use the parametrization $b = \lambda m^{1/2}$. We can bound

$$\Pr[R_b(S_m) \ne 0] \le \Pr[|S_m| > \lambda m^{1/2}] < 2e^{-\lambda^2/2}$$

While $R_b(S_m)$ takes on an arbitrarily high number of values the contribution to its entropy (for λ large) is dominated by the $R_b(S_m) = \pm 1$ terms and $ENT[R_b(S_m)] = O(\lambda^2 e^{-\lambda^2/2})$. The key points will be that this is bounded by a constant for any fixed λ and that it rapidly goes to zero as λ becomes large. With $\lambda < 1$ set

$s = \lambda^{-1}$ which we assume, for convenience, to be integral. We split the information ($=$ entropy) in $R_b(S_m)$ into the information in $R_{bs}(S_m)$ and the value of $R_b(S_m)$ modulo s. The first has constant entropy and the second, as it takes on s values, has entropy at most $\log_2 s$ by Optimal Uniformity. By Subadditivity the total entropy is $O(1) + \log_2 s = O(\log \lambda^{-1})$ as $\lambda \to 0$.

We summarize with some suboptimal results that hold for *all* m, b: $ENT[m, b] \le g(bm^{-1/2})$ where $g : R \to R$ is a function with growth $g(\lambda) = O(e^{-\lambda^2/3})$ as $\lambda \to \infty$ and $g(\lambda) = O(\log(\lambda^{-1}))$ as $\lambda \to 0$.

5.1. Six standard deviations suffice. I suppose every mathematician has a particular result that he or she is most proud of. This is mine.

THEOREM 4. [**S**] *Let $\mathcal{A}$ consist of n sets $A_1, \ldots, A_n$ on an n-set Ω. Then*

$$disc(\mathcal{A}) \le cn^{1/2}$$

where c is an absolute constant.

We take λ a constant with $g(\lambda) < 0.2$ above. As all $|A_i| \le n$ we may take all $b_i = \lambda n^{1/2}$ so that (16) is satisfied and there is a substantial partial coloring with all $|\chi(A_i)| \le \lambda n^{1/2}$. Now we iterate the procedure giving substantial partial colorings of still uncolored vertices. This part, which we omit here, is not trivial since there will be $O(\ln n)$ iterations and standard methods would increase the discrepancy by a $\ln n$ factor. One actually shows that at each iteration the bound on the discrepancy goes down in such a way that the sum of all the bounds remains $O(n^{1/2})$.

5.2. Arithmetic progressions. Let $\mathcal{A}$ be our Number Theory example, the arithmetic progressions on $\Omega = \{1, \ldots, n\}$.

THEOREM 5. [**MS**] *$disc(\mathcal{A}) = O(n^{1/4})$ Combined with Roth's* [**R**] *result (more than thirty years earlier!) this resolves, up to constants, $disc(\mathcal{A})$.*

We decompose $\mathcal{A}$ into canonical sets somewhat analogously to Beck's aligned rectangle decomposition. Indeed, the decomposition was given first by Beck. For each $1 \le d \le n$ and each $1 \le i \le d$ we split the progression $\{i, i+d, i+2d, \ldots\}$ into intervals of sizes $1, 2, 4, 8, \ldots$. Formally for each $0 \le t$ and each $u \ge 0$ we let $\{i + sd : u2^t \le s < (u+1)2^t\}$ be a canonical set whenever it lies entirely in Ω. Note canonical set sizes are only powers of two. Any arithmetic progression A is a subinterval of some $\{i, i+d, i+2d, \ldots\}$ and so can be written $B - D$ where B, D are subintervals beginning at i. Then B (and also D) can be split into canonical intervals of different sizes. For example, the first $13 = 2^3 + 2^2 + 2^0$ elements are the first 2^3 followed by the next 2^2 followed by the next 2^0 elements. Thus a coloring χ for which $|\chi(C)| \le f(|C|)$ for all canonical C has $|\chi(A)| \le |\chi(B)| + |\chi(D)| \le 2\sum f(2^t)$ for all progressions A.

Some calculation gives that there are $O(n^2 l^{-2})$ canonical sets of size l. We shall use the parametrization $l = xn^{1/2}$ so there are $O(x^{-2}n)$ canonical sets of size $xn^{1/2}$. (If we take n a power of two then x ranges over positive and negative powers of two.) We'd like a substantial partial coloring with the $xn^{1/2}$-sets having discrepancy at most some b_x where we parametrize $b_x = s_x n^{1/4}$. Condition (16) now becomes

$$\sum_x x^{-2} n ENT[xn^{1/2}, s_x n^{1/4}] < 0.1n$$

or, more conveniently,

$$\sum_x x^{-2} g(s_x x^{-1/2}) < 0.1$$

Now we have considerable room. We choose $s_1 = \lambda$ large so that $g(\lambda)$ is small. For $x > 1$ we can take $s_x = x^{-0.1}$ as then $\sum_{x \geq 1} x^{-2} \log(x^{0.6})$ would converge. For $x \leq 1$ we can take $s_x = x^{0.1}$ as then $\sum_{x \leq 1} x^{-2} g(x^{-0.4})$ is convergent since $g \to 0$ so rapidly. With care we can assure that the entire sum is at most 0.1. Now for any arithmetic progression A

$$|\chi(A)| \leq 2 \sum_x b_x \leq 2n^{1/4} \sum_x s_x = O(n^{1/4})$$

by our choice of s_x.

Here again finding a full coloring χ with discrepancy $O(n^{1/4})$ via iteration is a tricky matter—one needs on the i-th iteration to give a substantial partial coloring with discrepancy at most $c_i n^{1/2}$ where $\sum c_i$ itself converges. Again, we omit this section.

5.3. Half spaces redux. Here we consider upper bounds for the discrepancy of halfspaces. Let $\Omega \subset R^d$ be an n-set and let $\mathcal{A}$ be the collection of halfspaces on Ω.

THEOREM 6. [**M**] *disc*$(\mathcal{A}) = O(n^{\frac{1}{2} - \frac{1}{2d}})$

This matches the lower bound for discrepancy. We shall restrict to the plane, $d = 2$, where the lower bound was given previously. Here the decomposition is a bit tricky. Call a subfamily $\mathcal{B} \subseteq \mathcal{A}$ r-separated if $|B \Delta B'| \geq r$ for all $B, B' \in \mathcal{B}$. A geometric theorem of Chazelle and Welzl [**CW**] gives, for the plane, that $|\mathcal{B}| \leq c(n/r)^2$. Suppose $n = 2^s$ for convenience. For $0 \leq i \leq s$ let $\mathcal{B}_i$ be a maximal 2^i-separated subfamily of $\mathcal{A}$. We take $\mathcal{B}_s = \{\emptyset\}$. Now let $E \in \mathcal{A}$ be arbitrary. There exists $E_0 \in \mathcal{B}_0$ with $|E \Delta E_0| \leq 1$ as otherwise we could add E to $\mathcal{B}_0$. (If $E \in \mathcal{B}_0$ take $E_0 = E$.) Inductively we find $E_i \in \mathcal{B}_i$ with $|E_i \Delta E_{i-1}| \leq 2^i$, ending with $E_s = \emptyset$. Now

$$E = E \Delta E_s = (E \Delta E_0) \Delta (E_0 \Delta E_1) \Delta \ldots \Delta (E_{s-1} \Delta E_s)$$

The canonical sets (which are not themselves usually in $\mathcal{A}$) are all the sets of the form $E_{i-1} \Delta E_i$, $0 \leq i \leq s$ given by this procedure. Each $l = 2^i$ gives $c(n/l)^2$ canonical sets, each of size at most l.

The situation is identical to that of arithmetic progressions! We can give a substantial partial coloring so that, with $l = xn^{1/2}$, the canonical sets corresponding to a particular x value have discrepancy at most $s_x n^{1/2}$. Again any $E \in \mathcal{A}$ is expressed in terms of canonical sets corresponding to different x and so all $|\chi(E)|$ are bounded by $2n^{1/2} \sum s_x$ which we can make $O(n^{1/2})$.

In this case the iteration is particularly simple. Having colored some of the points after i iterations we are left with a set of n_i uncolored points. Applying the above argument we can give a substantial partial coloring of these n_i points so that all halfspaces have discrepancy at most $cn_i^{1/2}$, c the constant in the O-expression above. But because the colorings are substantial $n_i < n(0.9)^i$. The total discrepancy is at most $c \sum_i n_i^{1/2} \leq cn^{1/2} \sum_i (0.9)^{i/2}$. The last sum is convergent so the total discrepany is $O(n^{1/2})$.

5.4. A final conjecture. I close with a favorite open problem. Let

$$v_1, \ldots, v_n \in R^n$$

with all $|v_i|_2 \leq 1$. The *Komlós conjecture* is that there exists a map $\chi : \{1, \ldots, n\} \to \{-1, +1\}$ so that

$$|\chi(1)v_1 + \ldots + \chi(n)v_n|_\infty \leq K$$

where K is an absolute constant, not dependent on n.

While this problem remains open the entropy method comes close. Write $v_j = (a_{1j}, \ldots, a_{nj})$, so that the v_j are the column vectors of an $n \times n$ matrix $A = [a_{ij}]$. Set $\sigma_i = [\sum_{j=1}^n a_{ij}^2]^{1/2}$ so that $\sum_i \sigma_i^2 = \sum_i \sum_j a_{ij}^2 = \sum_j |v_j|_2^2 \leq n$.

Let $\chi : \{1, \ldots, n\} \to \{-1, +1\}$ be uniformly chosen. Define $L_i = \sum_{j=1}^n \chi(i)a_{ij}$ so that

$$\chi(1)v_1 + \ldots + \chi(n)v_n = (L_1, \ldots, L_n)$$

Note the L_i are random variables with mean zero and variance σ_i^2. Let K be a large constant and R_K the $2K$-roundoff function previously defined. Some technical work gives that L_i can be treated basically like a normal distribution with mean zero and variance σ_i^2 and in particular that $ENT[R_K(L_i)] \leq g(K\sigma_i^{-1})$ where g is the function defined at the start of the section. Then the random variable $L = (R_k(L_1), \ldots, R_K(L_n))$ has, by Subadditivity,

$$ENT[L] \leq \sum_i ENT[R_K(L_i)] \leq \sum_i g(K\sigma_i^{-1})$$

The average value of σ_i^2 is at most one. If all $\sigma_i = 1$ then $ENT[L] \leq ng(K) < 0.1n$ by taking K large. We don't know that all of the σ_i are the same. Fortunately, a little calculation gives that $g(Kx^{-1}) < 0.1x^2$ (taking K a large constant) for all positive x – indeed, when $x \to \infty$ or $x \to 0$ the asymptotics leave plenty of room. Since we do have $\sum \sigma_i^2 \leq n$

$$\sum_i g(K\sigma_i^{-1}) \leq \sum_i 0.1\sigma_i^2 \leq 0.1$$

so $ENT[L] \leq 0.1n$.

Our methods now give (as we showed originally in **[S]** by more cumbersome means) a substantial partial coloring χ with

$$|\chi(1)v_1 + \ldots + \chi(n)v_n|_\infty \leq K$$

Unfortunately, unlike in our previous examples, all attempts to extend the partial coloring to a full coloring by iteration have (thus far!) led to an additional factor of $\ln n$. The Komlós Conjecture remains a most intriguing problem.

References

[A] R. Alexander, *Geometric methods in the study of irregularities of distribution.* Combinatorica 10:115–136 (1990).

[AS] N. Alon and J. Spencer, The Probabilistic Method, John Wiley (1992).

[B] J. Beck, *Balanced two-colourings of finite sets in the square I.* Combinatorica 1:327–335 (1981).

[BF] J. Beck and T. Fiala, *Integer making theorems.* Discrete Applied Math 3:1–8 (1981).

[CMS] B. Chazelle, J. Matoušek and M. Sharir, *An elementary approach to lower bounds in geometric discrepancy.* Discrete & Computational Geometry 13: 363–381 (1995).

[CW] B. Chazelle and E. Welzl, *Quasi optimal range searching in spaces of finite VC-dimension.* Discrete & Computational Geometry 4:467–489 (1989).

[K] D. Kleitman, *On a combinatorial problem of Erdős*. J. Combinatorial Theory 1:209–214 (1966).
[M] J. Matoušek, *Tight upper bounds for the discrepancy of halfspaces*. Discrete & Computational Geometry 13:593–601 (1994).
[MS] J. Matoušek and J. Spencer, *Discrepancy in arithmetic progressions*. J. American Math. Soc. 9:195–204 (1996).
[PA] J. Pach and P. Agarwal, Combinatorial Geometry, John Wiley (1995).
[R] K. F. Roth, *Remark concerning integer sequences*. Acta Arithmetica 9:257–260 (1964).
[S] J. Spencer, *Six standard deviations suffice*. Trans. Amer. Math. Soc. 289:679–706 (1985).

COURANT INSTITUTE, NEW YORK, NY 10012
E-mail address: spencer@cs.nyu.edu

Contemporary Mathematics
Volume **223**, 1999

Proof of Reay's Conjecture on Certain Positive-dimensional Intersections

Helge Tverberg

ABSTRACT. In 1979 Reay conjectured that any set of $2(k-1)d+2$ points in real d-space can be split into k sets such that their convex hulls intersect in a set of positive dimension. We prove this conjecture and study the sets with one point less and no such splitting.

1. Introduction

In 1921, Radon [**R3**] proved his classical result on any set S of $d+2$ points in $\mathbf{R}^d$: It can be partitioned into two disjoint subsets S_1 and S_2 so that $\operatorname{conv} S_1 \cap \operatorname{conv} S_2 \neq \emptyset$. Here $\operatorname{conv} S_i$ means as usual, the convex hull of S_i, i.e., the minimal convex set containing S_i.

This result, now called Radon's theorem, is a central convexity result and we recommend the reader to consult Eckhoff's eminent survey [**E2**] for the many ensuing developments. We'll focus on a development which started in 1952, with Rado's result [**R2**], on a set S of $d+2+2^d(k-2)$ points, where k is an integer ≥ 2. He proved that in this case S can be partitioned into $S_1 \cup \ldots \cup S_k$ so that

$$\operatorname{conv} S_1 \cap \ldots \cap \operatorname{conv} S_k \neq \emptyset. \tag{1}$$

Birch [**B**] conjectured that (1) might be achieved even if S has only $(k-1)(d+1)+1$ points, and he proved this for $d=2$. It is easy to see that fewer points won't do in general. The present author proved Birch's conjecture in 1966, [**T**], (see also the recent simple proof by Onn and Sarkaria in [**S**]).

Variants of the last result soon started to appear. One way to go would be to require a higher-dimensional intersection, not just a non-empty one, in (1). In particular Reay, [**Re**], conjectured in 1979 that if $|S| = 2(k-1)d+2$, then one can make the intersection at least 1-dimensional. A partition which achieves this will henceforth be called a *good* partition.

Reay showed that his conjecture, if true, is best possible, by producing a set S, of $2(k-1)d+1$ points, which admits no good partition. A natural way to produce a set S with that many points is to choose $k-1$ points on each coordinate half-axis, and then add the origin. This works here: For if there is a good partition of S,

1991 *Mathematics Subject Classification.* Primary 52A37; Secondary 52A10, 52A20.
Key words and phrases. Finite point sets, convex hulls, positive-dimensional intersections.

© 1999 American Mathematical Society

then some point $(a_1, \ldots, a_d)$ in (1) must be different from the origin and thus, say, $a_d > 0$. But one of the parts, S_1, say, has no points on the positive x_d-axis, and so $\operatorname{conv} S_1$ is contained in the half-space $x_d \leq 0$. Thus $\operatorname{conv} S_1$ does not contain $(a_1, \ldots, a_d)$, as was assumed.

2. Proof of Reay's conjecture

We first remark that the conjecture was proved by Reay himself for $d = 2$ and by Eckhoff [**E**] for $k = 2$. Just as Reay's example above is suggested by the number $2(k-1)d+1$, it seems fair to say that our proof is suggested by the number $2(k-1)d+2$, which equals $2[(k-1)d+1]$. The number in square brackets is just the number of points needed to get a non-empty intersection in (1), in $\mathbf{R}^{d-1}$, for a k-partition. And the most natural way to produce such a set from the given set S, is to split S evenly by a hyperplane H, match up the parts and then take the intersection with H of each segment defined by a matching pair. Here are the details.

Assume that S admits no good k-partition. This means that for each k-partition of S the intersection in (1) is empty or a single point. These single points form a finite set I. We assume the coordinate system to be chosen such that the points of S have different x_d-coordinates. Let p_i be the point with x_d-coordinate a_i, where

$$a_1 < a_2 < \ldots < a_{2N},$$

with $N = (k-1)d + 1$. We now choose a hyperplane H, given by $x_d = c$, with $a_N < c < a_{N+1}$. Here c is chosen such that $H \cap I = \emptyset$. Let q_i be the point $H \cap \{p_i, p_{i+N}\}$. The set $\{q_1, \ldots, q_N\}$ can have fewer than N points, as there may be coincidences. Therefore we appeal to a seemingly stronger version of the 1966 result, with possibly multiple points, which follows from the original version by an easy limit argument. It says that there is a partition of $\{1, 2, \ldots, N\}$ into parts $J_1, \ldots, J_k$ such that, with $Q_j = \{q_i \mid i \in J_j\}$, $\cap \operatorname{conv} Q_j \neq \emptyset$. Let q be any point in that intersection. Then q is in H, as all the q_i's are in H. If we define the sets S_j by

$$S_j = \{p_i \mid i \in J_j\} \cup \{p_{i+N} \mid i \in J_j\},$$

it is clear that $\operatorname{conv} Q_j \subseteq \operatorname{conv} S_j$, for $j = 1, \ldots, k$. Thus q also belongs to the set I, contradicting the fact that $H \cap I = \emptyset$. This finishes the proof. We remark that the same idea of connecting neighboring dimensions by use of halving hyperplanes was used by Rado in the proof of his result mentioned above, and accounts for the factor 2^d occurring there.

3. A conjecture on the extremal case

At the end of section 1 we described Reay's example of a set of $2(k-1)d+1$ points without a good partition. Such a set, or an affine image of it, we'll call a *Reay set*. We venture the conjecture that if S has $2(k-1)d+1$ points, but admits no good partition, then S is a Reay set. This conjecture is supported by the fact that it holds for $k = 2$, for $d = 2$ (the first non-obvious cases), and when S is close enough to a given Reay set. The case $k = 2$ actually follows from Eckhoff's work in [**E**], but here a direct proof will be given.

3.1. The case $k = 2$. Let S, with $|S| = 2d+1$ and without any good partition, be given. Assume first that some vertex, p, of $\operatorname{conv} S$, is not collinear with any pair of the remaining points of S. Then we let H and H' be parallel supporting hyperplanes of $\operatorname{conv} S$, with $H \cap S = p$. Let S' be the central projection from p of $S \setminus \{p\}$ into H'. S' will consist of $2d$ distinct points, because of the assumption on p. Let H'' be a hyperplane of H', with d points of S' in each of the open half-spaces it defines, and extend H'' to a hyperplane H''' of $\mathbf{R}^d$, through p. Clearly H'' can be chosen such that $H''' \cap I = \emptyset$, where I is the finite set defined in the proof of Reay's conjecture. Proceeding as in that proof a contradiction is now obtained.

It can thus be assumed that each vertex p is collinear with two further points p', p'' from S, with p' between p and p''. If q is another vertex, $q' = p'$. For if $q' \neq p'$, and $\{p', q'\} \cap \{p, p'', q, q''\} = \emptyset$, then these two sets may be extended to give a good partition of S. If $\{p', q'\} \cap \{p, p'', q, q''\} \neq \emptyset$ then, say, $p' = q''$. But then $\operatorname{conv}\{q', q''\}$ is contained in $\operatorname{conv}\{q, p, p''\}$. This gives a good partition of S unless $\{q', q''\} \cap \{q, p, p''\} \neq \emptyset$, which happens only if $q' = p''$. In this case $\operatorname{conv}\{q', q''\} \subseteq \operatorname{conv}\{p, q\}$, and $\{q', q''\} \cap \{p, q\} = \emptyset$, so again a good partition exists.

We have thus seen that there is a point p'_0 such that for each vertex p the ray from p towards p'_0 meets S in a further point p'', after p'_0. Since S can be assumed d- dimensional one can choose d vertices $p_1, \dots, p_d$ so that the vectors $p'_0 p_i$ are linearly independent. Then the set $\{p'_0, p_1, \dots, p_d, p'_1, \dots, p'_d\}$ is a Reay set and constitutes all of S.

3.2. The case when S is close to a Reay set. Let S' be a given Reay set, with points $p'_1, \dots, p'_{2N-1}$. Then there is an $\epsilon > 0$ such that if $S = \{p_1, \dots, p_{2N-1}\}$, with $|p_i - p'_i| < \epsilon$ for $i = 1, \dots, 2N-1$, and S has no good partition, then S is a Reay set. In order to see this, we first note that, by the result just proved, the restriction $k \geq 3$ can be imposed. We can assume that S' consist of the origin and $k-1$ points on each coordinate half-axis. Let the numbering of S' be such that p'_i is the point on the positive x_i-axis furthest away from the origin, and p'_{i+d} the similar point on the negative x_i-axis.

Consider now the set $S^* = S - \{p_1, \dots, p_{2d}\}$. If ϵ is small enough S^* will be in the interior of $\operatorname{conv}\{p_1, \dots, p_{2d}\}$, since a similar statement is true for S'. Now S^* admits no good $(k-1)$-partition, as this would give a good k-partition of S, using $\{p_1, \dots, p_{2d}\}$ as the kth part. Thus S^* is a Reay set (since we can reason inductively here). Let p_{2d+1} be the point of S which is close to the x_d-axis and has the second-largest value of x_d, and consider the set $S^{**} = S^* \setminus \{p_{2d+1}\} \cup \{p_d\}$. Assume first that S^{**} has a good $(k-1)$-partition. As $k \geq 3$ one of the parts does not use the point p_d. This part is then contained in $\operatorname{conv}\{p_{2d+1}, p_1, \dots, p_{d-1}, p_{d+1}, \dots, p_{2d}\}$, by the corresponding property of S', and again one would get a good partition of S. S^{**} thus also forms a Reay set. A comparison of S^* and S^{**} shows that S has $2k-1$ points on the line through p_d and p_{2d+1}, with the middle one being close to the origin. By symmetry it follows that S is a Reay set.

3.3. The case $d = 2$. In this case we use Birch's method from [**B**] to find a necessary and sufficient condition for a set W, with $3k-1$ points, to admit no good k- partition. It is then shown that S, with $4k-3$ points, is a Reay set iff each of its $(3k-1)$- subsets satisfies this necessary and sufficient condition.

The condition in question is as follows: There must be two lines L_i, defining open half planes H_i^+, H_i^-, so that for $i = 1, 2$,

$$|L_i \cap W| \geq k+1,$$
$$(2) \qquad |H_i^+ \cap W| \leq k-1,$$
$$|H_i^- \cap W| \leq k-1.$$

It is clear that the condition is sufficient. For if a pair of points is in the intersection (1), it cannot be in H_i^+ or H_i^-. Thus it is in L_i, $i = 1, 2$, and hence in $L_1 \cap L_2$.

For the necessity proof, we introduce $R(W)$, the *Rado set* corresponding to W. $R(W)$ is, by definition, the intersection of those closed half-planes which contain at least $2k$ points from W. The name is after Rado who introduced it, (in more general form) in [**R**]. $R(W)$ is clearly compact and convex. It is also non-empty, for if not, then by Helly's theorem three of the defining half-planes would have an empty intersection, so that their complements would cover W. But this is impossible, as $3(k-1) < 3k-1 = |W|$.

If p is any point in $R(W)$, then any half-plane containing p will contain at least k points from W. This property of $R(W)$ is characteristic, so it gives a second definition. Therefore $B(W)$, the *Birch set* of W, i.e., those points which can occur in the intersection (1) for a k-partition of W, is contained in $R(W)$. In [**B**] Birch used Rado's concept to prove that the Birch set is nonempty (for $|W| \geq 3k-2$), hence the name.

$R(W)$ can also be defined as the intersection of those convex polygons, possibly degenerate, which have all vertices in W and contain exactly $2k$ points from W. For a point in $\mathbf{R}^2$ is easily seen to be outside some such polygon iff it is outside one of the defining half-planes for $R(W)$. This shows that unless W is contained in a line, $R(W)$ is the intersection of closed half-planes with at least two points of W on the boundary.

Since $B(W) \subseteq R(W)$ it is clear that if $R(W)$ is a single point, then W has no good partition. In every other case, there is one. This is clear if W is contained in a line, so assume $\dim(W) = 2$. Choose a point x in the boundary of $R(W)$, but not in W itself. We'll see that $x \in B(W)$. There are two points u, v in W so that x is in the open segment uv. Let the remaining $3(k-1)$ points of W be $w_i = (r_i, \theta_i)$ where

$$0 \leq \theta_1 \leq \theta_2 \leq \dots \leq \theta_{3k-3} < 2\pi,$$

where x is taken as the origin for polar coordinates. Put

$$S_i = \{w_i, w_{i+k-1}, w_{i+2k-2}\}$$

for $i = 1, \dots, k-1$, and $S_k = \{u, v\}$. Then $x \in \operatorname{conv} S_k$. If, say, $x \notin \operatorname{conv} S_1$, one could assume

$$0 < \theta_1 \leq \theta_k \leq \theta_{2k-1} < \pi.$$

But then the open upper half-plane would contain $w_1, \dots, w_{2k-1}$. After a slight turning, if necessary, it would also contain u or v, so that some closed half-plane contained in it would contain at least $2k$ points from W, but not x, which is a contradiction.

We now check that if $R(W)$ is a single point, then the condition (2) is satisfied. We have seen that $R(W)$ is the intersection of finitely many closed half-planes.

The intersection of the corresponding open half-planes must be empty, as otherwise $R(W)$ would be two-dimensional. Thus, by Helly's theorem, there must be some two or three of these open half-planes with empty intersection. If there are three, already two of these must have empty intersection. For if not, the complements of the corresponding closed half-planes would cover W (except for maybe one point, the intersection of the closed half-planes), but this is impossible, as $3(k-1) < |W|-1$.

Among the closed half-planes intersecting in $R(W)$ there is hence a pair with non-intersecting interiors, i.e., those interiors are H_1^+ and H_1^- for some line L_1. $R(W)$ thus equals the intersection of the closed half-lines $H_j \cap L_1$, where the H_j are the remaining ones of the closed half-planes defining $R(W)$. As $R(W)$ is a single point, Helly's theorem in L_1 shows that two of the corresponding open half-lines have an empty intersection. As each of the open half-planes defining them has a non-empty intersection with H_1^+ and H_1^-, it follows from the preceding paragraph that they are H_2^+ and H_2^- for some line L_2. This shows that the condition (2) is satisfied.

It remains to be shown that each $(3k-1)$-subset of S satisfies (2) if and only if S is a Reay set. The "if" part is obvious, so let now the $(3k-1)$-subsets of S satisfy (2). Then there are at least two *heavy* lines, i.e., lines meeting S in at least $k+1$ points. If four exist they cannot be concurrent in a point of S, as $1+4k > |S|$. Thus by the principle of inclusion and exclusion, one would have $|S| \geq 4(k+1) - 6 + 0$, a contradiction. If there are exactly three heavy lines, intersecting S in S^1, S^2 and S^3, we remove $|S^i| - k$ points from S^i, $i = 1, 2$, using the same point twice if $S^1 \cap S^2 \neq \emptyset$. The set thus obtained does not satisfy (2), and so

$$4k - 3 - (|S^1| - k + |S^2| - k - |S^1 \cap S^2|) \leq 3k - 2.$$

From this inequality, together with

$$|S^1| + |S^2| + |S^3| - |S^2 \cap S^3| - |S^1 \cap S^3| - |S^1 \cap S^2| + |S^1 \cap S^2 \cap S^3| \leq 4k - 3$$

we get the contradiction

$$\begin{aligned} |S^3| &\leq k + |S^2 \cap S^3| + |S^1 \cap S^3| - |S^1 \cap S^2 \cap S^3| - 2 \\ &\leq k - |S^1 \cap S^2 \cap S^3|. \end{aligned}$$

Finally, if there are just two heavy lines attempts to make one of them light show that

$$4k - 3 - (|S^i| - k) \leq 3k - 2, \text{ for } i = 1, 2,$$

i.e., $|S^i| \geq 2k - 1$, $i = 1, 2$. This shows that $|S^1| = |S^2| = 2k - 1$, $|S^1 \cap S^2| = 1$, $S^1 \cup S^2 = S$. Furthermore the points of S^i must be evenly distributed to both sides of $S^1 \cap S^2$, i.e., S is a Reay set. For if one of the open half-lines of, say, S_1, defined by $S_1 \cap S_2$ contains fewer than $k-1$ points of S, we could remove $k-2$ points from S, including those mentioned, and obtain a $(3k-1)$-set in S, not satisfying (2).

Note added in proof (February 1998): The conjecture stated at the beginning of Section 3 has very recently been proved by Roudneff [**Ro**]. He also proves, for $d+4$, that with $(k-1)(d+1)+1+i$ points in general position one gets a splitting giving at least an i-dimensional intersection, for $i < 5$. See [**E2**, pp. 428–9], for more information for general d.

References

[B] B. J. Birch, *On 3N points in a plane*, Proc. Cambridge Philos. Soc. **55** (1959), 289–293.

[E] J. Eckhoff, *On a class of convex polytopes*, Israel J. Math. **23** (1976), 332–336.

[E2] J. Eckhoff, *Helly, Radon and Carathéodory type theorems*, Handbook of Convex Geometry (P. M. Gruber and J. M. Wills, eds.), vol. A, North-Holland, Amsterdam, 1993, 389–448.

[R] R. Rado, *A theorem on general measure*, J. London Math. Soc. **26** (1946), 291–300.

[R2] R. Rado, *Theorems on the intersection of convex sets of points*, J. London Math. Soc. **27** (1952), 320–328.

[R3] J. Radon, *Mengen konvexer Körper, die einen gemeinsamen Punkt enthalten*, Math. Ann. **83** (1921), 113–115.

[Re] J. Reay, *Several generalizations of Tverberg's theorem*, Israel J. Math. **34** (1979), 238–244.

[Ro] J.-P. Roudneff, Personal communication, December 1997.

[S] K. S. Sarkaria, *Tverberg's theorem via number fields*, Israel J. Math. **79** (1992), 317–320.

[T] H. Tverberg, *A generalization of Radon's theorem*, J. London Math. Soc. **41** (1966), 123–128.

DEPARTMENT OF MATHEMATICS, UNIVERSITY OF BERGEN, N-5007 BERGEN, NORWAY
E-mail address: tverberg@mi.uib.no

Contemporary Mathematics
Volume **223**, 1999

Progress in Geometric Transversal Theory

Rephael Wenger

ABSTRACT. Let $\mathcal{A}$ be a family of convex sets in $\mathbb{R}^d$. A line transversal to $\mathcal{A}$ is a line which intersects every member of $\mathcal{A}$. More generally, a k-transversal to $\mathcal{A}$ is an affine subspace of dimension k which intersects every member of $\mathcal{A}$. This paper discusses recent progress in geometric transversal theory, including necessary and sufficient conditions, Helly-type theorems, piercing or Gallai numbers, topological and combinatorial structure, and induced orderings.

1. Introduction

A k-transversal to a family of convex sets in $\mathbb{R}^d$ is an affine subspace of dimension k, called a *k-flat*, (such as a point, line, plane or hyperplane) which intersects every member of the family. The study of k-transversals dates back to papers by Vincensini in 1935 [**V**] and Santaló in 1940 [**S**] on line transversals in the plane but really developed only in the late 1950's and early 1960's with contributions by Danzer, Debrunner, Grünbaum, Hadwiger, Klee and Valentine [**DGK**]. There was incremental progress since then, most notably Eckhoff's thesis in 1969 [**E**] and work by Katchalski and Lewis in the early eighties [**Ka, KL, KL2, L**]. In the last ten years there has been an explosion of results, including necessary and sufficient conditions for hyperplane transversals, Helly-type theorems for line transversals to translates, piercing or Gallai numbers for hyperplane transversals, bounds on the complexity of the space of transversals and the number of orders induced by transversals, and algorithms for finding transversals. This paper will cover some of the major recent results in geometric transversal theory and describe some of the open problems suggested by these results.

2. The space of k-transversals

Let $\mathcal{A}$ be a family of convex sets in $\mathbb{R}^d$. The set of k-transversals to $\mathcal{A}$ forms a topological space, denoted $\mathcal{T}_k^d(\mathcal{A})$, lying in the "affine Grassmannian" of all k-flats in $\mathbb{R}^d$. This space can be easily visualized for line transversals in the plane by mapping each non-vertical line transversal $\{(x, y) : y = mx + b\}$ to the point (m, b) in dual space. (See Figure 1.)

1991 *Mathematics Subject Classification.* Primary 52A35.
Key words and phrases. k-transversal, Helly-type theorem, piercing or Gallai number, geometric permutation, upper and lower envelope, convexity.
Supported by NSA grant MDA904-97-1-0018.

© 1999 American Mathematical Society

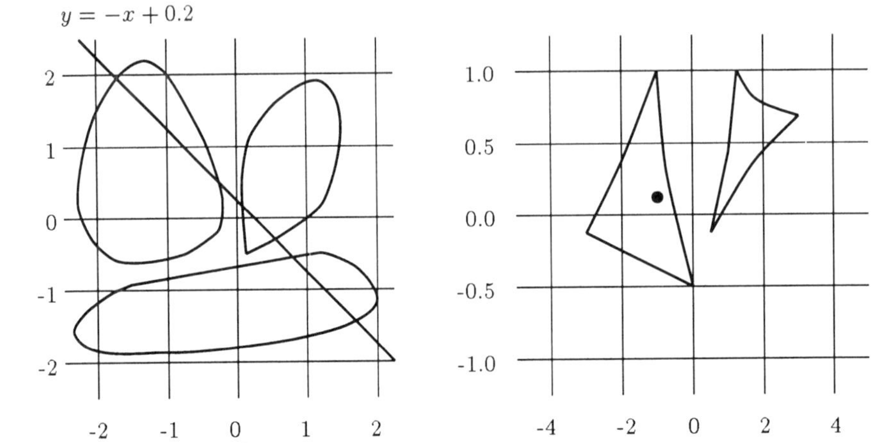

FIGURE 1. A family $\mathcal{A}$ of convex sets and its transversal space $\mathcal{T}_1^2(\mathcal{A})$.

As the example in Figure 1 shows, the space of transversals is not necessarily connected, even for line transversals in the plane. The space of line transversals to a family of n convex sets in the plane can have as many as $2n - 2$ connected components [**ES, KLZ**].

Let $\mathcal{A}$ be a family of pairwise disjoint convex sets in $\mathbb{R}^d$. A directed line transversal intersects the elements of $\mathcal{A}$ in a fixed order. An undirected line induces a pair of linear orderings on $\mathcal{A}$ corresponding to its two possible orientations. Katchalski [**Ka2**] studied pairs of linear orderings induced by line transversals and called them *geometric permutations.*

If two line transversals induce different geometric permutations on $\mathcal{A}$, then they must lie in different connected components in the space of line transversals to $\mathcal{A}$. Thus each connected component is associated with a single geometric permutation of $\mathcal{A}$. In the plane, the converse is also true. Two lines which induce the same geometric permutation on a finite family $\mathcal{A}$ of pairwise disjoint compact convex sets in $\mathbb{R}^2$ must lie in the same connected component in $\mathcal{T}_1^2(\mathcal{A})$ [**W3**]. In $\mathbb{R}^3$, this is no longer true. Two lines which induce the same geometric permutation on a family $\mathcal{A}$ in $\mathbb{R}^3$ may lie in different connected components in $\mathcal{T}_1^3(\mathcal{A})$.

The generalization of geometric permutations to k-transversals requires a generalization of linear orderings to higher dimensions. In the early eighties, Goodman and Pollack [**GP**] introduced and explored the notion of the *order type* of a set of points which is the family of orientations (-1, 0 or $+1$) of its $(d+1)$-tuples. For example, the order type of a set of points in the plane is the family of left or right turns (or no turn) made by every three points of the set. The order type of a set of points in $\mathbb{R}^d$ describes the relative positions of the points in $\mathbb{R}^d$ in the same way that a linear ordering of points in $\mathbb{R}$ captures their relative positions in $\mathbb{R}$.

Order types are special cases of *acyclic oriented matroids.* A rank $d+1$ acyclic oriented matroid on a set $\mathcal{A}$ is a family of orientations of the $(d+1)$-tuples of $\mathcal{A}$ with the property that the orientations of the $(d+1)$-tuples of every subset of $\mathcal{A}$ of size $d+3$ match the orientations of the $(d+1)$-tuples of some set of $d+3$ points in $\mathbb{R}^d$ [**BLWSZ**, p.140]. In other words, an acyclic oriented matroid behaves locally (i.e., for subsets of size $d+3$) like an order type of points in $\mathbb{R}^d$. If all the $(d+1)$-tuples have non-zero orientation, this property suffices to define an oriented matroid; otherwise, additional conditions relating the zero and non-zero orientations are needed. (A rank $d+1$ oriented matroid which is not acyclic "locally" matches the orientation of vectors in $\mathbb{R}^{d+1}$.) Not every acyclic oriented matroid can be represented as the order type of some set of points. Those that can are called *realizable*. (See [**BLWSZ**] for an introduction to oriented matroid theory.)

Let τ be an oriented k-transversal of a family $\mathcal{A}$ of convex sets. For each $a \in \mathcal{A}$, choose a point $p_a \in a \cap \tau$. Define a rank $k+1$ realizable acyclic oriented matroid on $\mathcal{A}$ by setting the orientation of every $(k+1)$-tuple of $\mathcal{A}$ to equal the orientation in τ of the corresponding $(k+1)$-tuple of points of $\{p_a\}$. Goodman and Pollack noted that if no $k+1$ members of $\mathcal{A}$ have a $(k-1)$-transversal, then none of the orientations are zero and so this oriented matroid is independent of the choices of the points p_a. Thus if no $k+1$ members of $\mathcal{A}$ have a $(k-1)$-transversal, then τ induces an oriented matroid on $\mathcal{A}$.

A family $\mathcal{A}$ of convex sets is *k-separated* if no $k+2$ members of $\mathcal{A}$ have a k-transversal. Note that a family of convex sets is 0-separated if and only if the sets are pairwise disjoint. An oriented k-transversal induces a unique rank $k+1$ oriented matroid on a $(k-1)$-separated family of convex sets.

Just as a line induces a pair of linear orderings on a family of pairwise disjoint sets, a k-transversals induces a pair of oriented matroids on a $(k-1)$-separated family corresponding to its two orientations. Unfortunately, the term geometric permutation is misleading when applied to pairs of oriented matroids which do not resemble permutations in any way.

Two k-transversals which induce different pairs of oriented matroids on $\mathcal{A}$ must lie in different connected components of $\mathcal{T}_k^d(\mathcal{A})$. The converse is not true in general but does hold for hyperplane transversals [**W3**]. Just as line transversals in $\mathbb{R}^2$ which induce the same geometric permutation lie in the same connected component, two $(d-1)$-transversals which induce the same pair of oriented matroids on a finite $(d-2)$-separated family $\mathcal{A}$ of compact convex sets in $\mathbb{R}^d$ lie in the same connected component of $\mathcal{T}_{d-1}^d(\mathcal{A})$.

3. Necessary and sufficient conditions

A 0-transversal to a family $\mathcal{A}$ of convex sets is simply a point common to all the sets. The well-known Helly's Theorem can be reformulated as necessary and sufficient conditions for the existence of point transversals.

THEOREM 1. Helly's Theorem [**He**] *Let $\mathcal{A}$ be a finite family of at least $d+1$ convex sets in $\mathbb{R}^d$. Family $\mathcal{A}$ has a point transversal (a point in common) if and only if every $d+1$ members of $\mathcal{A}$ have a point transversal (a point in common.)*

Helly's Theorem motivated many other theorems of a similar form: A family of objects has property $\mathcal{P}$ if and only if every m members of the family have property $\mathcal{P}$. These theorems are sometimes called *Helly-type* theorems.

Vincensini's 1935 paper [**V**] gave a Helly-type theorem for the existence of line transversals in the plane, a theorem which unfortunately was false. In 1940 Santaló [**S**] showed that for any m there is a family $\mathcal{A}$ of convex sets in $\mathbb{R}^2$ such that every m members of $\mathcal{A}$ have a line transversal but $\mathcal{A}$ has no line transversal. Necessary and sufficient conditions were given by Santaló and others for special families such as parallelotopes with edges parallel to the coordinate axes [**S**], but it was not until 1957 that Hadwiger gave the first general conditions for the existence of line transversals in the plane. Hadwiger's Transversal Theorem is:

THEOREM 2 (Hadwiger's Transversal Theorem [**H**]). *Let $\mathcal{A}$ be a finite family of pairwise disjoint convex sets in $\mathbb{R}^2$. Family $\mathcal{A}$ has a line transversal if and only if there exists a linear ordering of $\mathcal{A}$ such that every three members of $\mathcal{A}$ are intersected by a directed line in the given order.*

As was noted in Section 2, the space of line transversals to a family $\mathcal{A}$ of convex sets in the plane can have many different connected components. To some extent, this explains why there is no Helly-type theorem for line transversals in the plane. The transversal space is simply too complicated. However, every connected component in the space of line transversals of $\mathcal{A}$ is associated with a unique pair of orderings of $\mathcal{A}$. By prespecifying a linear ordering, Theorem 2 focuses on the existence of a single connected component. Interestingly enough, Theorem 2 does not guarantee that some directed line transversal intersects $\mathcal{A}$ in the given linear ordering. Every four members of $\mathcal{A}$ must be intersected by a directed line in the given order to ensure that result.

Let $\mathcal{A}$ be a $(d-1)$-separated family of compact convex sets in $\mathbb{R}^d$. Every connected component in the space of hyperplane transversals of $\mathcal{A}$ is associated with a unique pair of oriented matroids on $\mathcal{A}$. By prespecifying an oriented matroid, one can focus on the existence of a single connected component of $\mathcal{T}^d_{d-1}(\mathcal{A})$. Using order types (realizable acyclic oriented matroids) Goodman and Pollack in 1988 generalized Hadwiger's Transversal Theorem to hyperplane transversals. Their proof only uses local properties of order types and so applies equally well to acyclic oriented matroids as is stated here.

THEOREM 3 ([**GP2**]). *Let $\mathcal{A}$ be a finite $(d-2)$-separated family of convex sets in $\mathbb{R}^d$. Family $\mathcal{A}$ has a hyperplane transversal if and only if there exists a rank $d+1$ acyclic oriented matroid of $\mathcal{A}$ such that every $d+1$ members of $\mathcal{A}$ are intersected by an oriented hyperplane consistently with that oriented matroid.*

An oriented hyperplane h intersects a family $\mathcal{B} \subseteq \mathcal{A}$ consistently with an oriented matroid $\mathcal{M}$ on $\mathcal{A}$ if the oriented matroid $\mathcal{M}'$ induced by h on $\mathcal{B}$ is a submatroid of $\mathcal{M}$, i.e., the orientation of every $(d+1)$-tuple in $\mathcal{M}'$ equals the orientation of the corresponding $(d+1)$-tuple in $\mathcal{M}$. Theorem 3 does not guarantee that $\mathcal{A}$ has an oriented transversal which induces the given oriented matroid on $\mathcal{A}$.

The condition that $\mathcal{A}$ is $(d-2)$-separated generalizes the pairwise disjointness condition in Hadwiger's original theorem. In 1989, Wenger [**W2**] removed this condition for line transverals in $\mathbb{R}^2$ and subsequently Pollack and Wenger [**PW**] removed the condition for hyperplane transversals in $\mathbb{R}^d$. Their theorem also generalizes conditions by Katchalski [**Ka**] for the existence of hyperplane transversals based on the existence of line transversals to every three sets and applies to families of connected sets, not just convex ones.

THEOREM 4 ([**PW**]). *Let $\mathcal{A}$ be a finite family of connected sets in $\mathbb{R}^d$. Family $\mathcal{A}$ has a hyperplane transversal if and only if for some k, $0 \le k < d$, there exists a rank $k+1$ realizable acyclic oriented matroid of $\mathcal{A}$ such that every $k+2$ members of $\mathcal{A}$ are intersected by an oriented k-flat consistently with that oriented matroid.*

Since $\mathcal{A}$ may not be $(k-1)$-separated, an oriented k-flat τ which intersects $\mathcal{B} \subseteq \mathcal{A}$ may not induce a unique oriented matroid on $\mathcal{B}$. Nevertheless, we say that a k-flat τ intersects $\mathcal{B} \subseteq \mathcal{A}$ *consistently* with an oriented matroid $\mathcal{M}$ on $\mathcal{A}$ if for every $b \in \mathcal{B}$ there is a point $p_b \in b \cap \tau$ such that the orientations of $(k+1)$-tuples in $\{p_b\}$ match the orientations of corresponding $(k+1)$-tuple of $\mathcal{M}$.

The proof by Hadwiger of Theorem 2 and the generalization by Goodman and Pollack entails shrinking the convex sets in $\mathcal{A}$ until some $d+1$ sets have only a single hyperplane transversal h consistent with the given oriented matroid. Those $d+1$ sets pin the hyperplane h and one can show that h intersects every other convex set in $\mathcal{A}$.

The proof by Pollack and Wenger of Theorem 4 is quite different. Each rank $k+1$ realizable acyclic oriented matroid corresponds to the family of orientations of some set P of points in $\mathbb{R}^k$. Pollack and Wenger construct an antipodal mapping from the set of unit vectors in $\mathbb{R}^d$ to $\mathbb{R}^k$ based on the family of convex sets $\mathcal{A}$ and the point set P. By the Borsuk-Ulam theorem, one of these vectors maps to zero. The proof concludes by showing that this vector is the normal of a hyperplane transversal of $\mathcal{A}$.

The proof by Pollack and Wenger involves explicitly representing the oriented matroid by a set of points $P \subseteq \mathbb{R}^k$. Thus the oriented matroids in Theorem 4 must be realizable. Anderson and Wenger removed this realizability condition to give the theorem in its most general form:

THEOREM 5 ([**AW**]). *Let $\mathcal{A}$ be a finite family of connected sets in $\mathbb{R}^d$. Family $\mathcal{A}$ has a hyperplane transversal if and only if for some k, $0 \leq k < d$, there exists a rank $k+1$ acyclic oriented matroid of $\mathcal{A}$ such that every $k+2$ members of $\mathcal{A}$ are intersected by an oriented k-flat consistently with that oriented matroid.*

A rank $k+1$ oriented matroid can be represented by an arrangement of oriented pseudospheres (topological spheres whose intersections are topological spheres of the proper dimension) in $\mathbb{S}^k$, the k-sphere [**BLWSZ**, Chapter 5]. Based on the family of convex sets $\mathcal{A}$, Anderson and Wenger construct a lower semi-continuous antipodal mapping from the set of unit vectors $\mathbb{S}^{d-1} \subseteq \mathbb{R}^d$ to faces in this pseudosphere representation in $\mathbb{S}^k$. They prove that some unit vector must map to the null face, $\emptyset$, and that this vector is the normal of a hyperplane transversal of $\mathcal{A}$.

Hadwiger's Transversal Theorem and its generalizations give conditions for the existence of hyperplane transversals ($(d-1)$-transversals) in $\mathbb{R}^d$. What about k-transversals for values of k other than $d-1$? Two lines may induce the same pair of orderings on a family $\mathcal{A}$ in $\mathbb{R}^3$ but lie in different connected components in $\mathcal{T}_1^3(\mathcal{A})$. Thus specifying a given ordering may not isolate a single connected component in $\mathcal{T}_1^3(\mathcal{A})$. Not suprisingly, Hadwiger's Transversal Theorem does not generalize to line transversals in $\mathbb{R}^3$. Aronov, Goodman, Pollack and Wenger showed that for any number m there exists a family $\mathcal{A}$ of convex sets in $\mathbb{R}^3$ and a linear ordering on $\mathcal{A}$ such that every m convex sets are met by a line consistently with that linear ordering but $\mathcal{A}$ has no line transversal [**GPW**].

PROBLEM 1. *Give necessary and sufficient conditions for the existence of line transversals to a family of convex sets in $\mathbb{R}^3$ (or more generally k-transversals to a family of convex sets in $\mathbb{R}^d$ where $0 < k < d-1$.)*

4. Helly-type transversal theorems

Santaló [**S**] showed that there is no Helly-type theorem for line transversals to families of convex sets in the plane. Hadwiger and Debrunner [**HDK**] proved that there were no such theorems even if the convex sets were pairwise disjoint and Lewis **[L]** constructed counter-examples for families of pairwise disjoint line segments.

However, for special types of families there are Helly-type theorems for line transversals. In a 1957 paper, Danzer **[D]** proved that if $\mathcal{A}$ was a family of pairwise disjoint congruent circles and every five members of $\mathcal{A}$ have a line transversal then $\mathcal{A}$ has a line transversal. A year later Grünbaum **[G]** gave a similar theorem for pairwise disjoint congruent squares and conjectured that the theorem was true for any family of pairwise disjoint translates of a convex set.

Starting around 1980, Katchalski worked on Grünbaum's conjecture, publishing a number of papers on line transversals **[Ka, KL, KL2]**. His efforts resulted in a 1986 paper **[Ka2]** in which he proved a weak form of Grünbaum's conjecture showing that if every 128 members of a family of pairwise disjoint translates had a line transversal then the family had a line transversal. Finally, in 1988, thirty years after Grünbaum published his conjecture, Tverberg presented a proof.

THEOREM 6 (**[T]**). *Let $\mathcal{A}$ be a finite family of pairwise disjoint translates of a convex set in $\mathbb{R}^2$. Family $\mathcal{A}$ has a line transversal if and only if every five members of $\mathcal{A}$ have a line transversal.*

Tverberg first proved the conjecture for families of six pairwise disjoint translates. These six translates can be characterized by six points in the plane representing the six directions of translation. A careful analysis of the different orders in which lines intersect five of the six translates and of different configurations of six points in the plane proves the conjecture for families of size six. Larger families of more than six convex sets can be transformed by shrinking into families where some subset of five translates has a unique line transversal. Since Grünbaum's conjecture is established for six translates, this unique line transversal must meet every other translate and thus must be a line transversal for the entire family.

Katchalski's proof of a weak form of Theorem 6 is quite different. A family of n pairwise disjoint convex sets in $\mathbb{R}^2$ can have up to $2n-2$ different geometric permutations (Theorem 20.) However, Katchalski, Lewis and Liu **[KLL]** proved that a family $\mathcal{A}$ of pairwise disjoint translates can have at most eight geometric permutations, independent of the size of the family. They later reduced the constant eight to three **[KLL2, KLL3]**. In addition, any two of these geometric permutations must agree on all but five of their elements. These conditions on geometric permutations can be used to construct an ordering of $\mathcal{A}$ from the line transversals of subsets of $\mathcal{A}$ of size 128 such that every three members of $\mathcal{A}$ are intersected by a directed line in the given order. Applying Hadwiger's Transversal Theorem, Theorem 2, gives the desired result.

As discussed in Section 2, each geometric permutation of $\mathcal{A}$ corresponds to a single connected component in $\mathcal{T}_1^2(\mathcal{A})$. Thus, Katchalski, Lewis and Liu implicitly proved that $\mathcal{T}_1^2(\mathcal{A})$ has at most three connected components if $\mathcal{A}$ is a family of pairwise disjoint translates. This partly explains why there is a Helly-type theorem for line transversals to pairwise disjoint translates.

A convex set is *ρ-stubby*, $\rho \geq 1$, if it is contained in a ball of radius ρ and contains a ball of radius one. Katchalski's proof of his Helly-type theorem generalizes to finite families $\mathcal{A}$ of pairwise disjoint ρ-stubby sets. For each ρ there is a constant c_ρ such that a finite family $\mathcal{A}$ of pairwise disjoint ρ-stubby convex sets has a line transversal if and only if every c_ρ members of $\mathcal{A}$ have a line transversal. Any family of pairwise disjoint translates can be mapped by affine transformation to a 2-stubby family [**J**]. Thus the Helly-type theorem for line transversals of ρ-stubby sets leads directly to a Helly-type theorem for line transversals of translates.

Recently, Robert [**R**] replaced the pairwise disjointness condition in Theorem 6 by a condition that the intersection of any j translates is empty. For every j there is a constant c_j such that if $\mathcal{A}$ is a finite family of convex translates and the intersection of every j translates in $\mathcal{A}$ is empty, then $\mathcal{A}$ has a line transversal if and only if every c_j translates have a line transversal.

Under certain conditions, Helly-type theorems can be used to derive linear expected time algorithms for related algorithmic problems. In [**SW**], Sharir and Welzl described a technique called Generalized Linear Programming which they applied to many problems related to convex programming. In particular, they gave an algorithm for finding a point in the intersection of n half-spaces in $\mathbb{R}^d$ and proved that their algorithm ran in $O(d^3 2^d n)$ expected time. This bound is linear in the number of half-spaces, n, although exponential in d. (Matoušek, Sharir and Welzl [**MSW**] subsequently proved a subexponential $O(nde^{4\sqrt{d\ln(n+1)}})$ bound on the expected running time of this algorithm.) Amenta [**Am**] showed that Generalized Linear Programming could be applied to solve in linear expected time many other algorithmic problems related to Helly-type theorems including the problem of finding a line transversal for translates. (See also [**Dy, Dy2, Dy3, EW, Me, Me2**] for deterministic algorithms which solve specific Helly-type problems in linear time.)

Katchalski conjectured that Theorem 6 generalizes to families $\mathcal{A}$ of translates in higher dimensions. If this were true, the number of connected components of $\mathcal{T}_k^d(\mathcal{A})$ and the number of oriented matroids on $\mathcal{A}$ induced by oriented k-transversals is probably bounded by a fixed constant depending only on k and d.

PROBLEM 2. *Do there exist numbers m, m' and m'' such that for any family $\mathcal{A}$ of pairwise disjoint translates of a convex set in $\mathbb{R}^3$:*

- *$\mathcal{A}$ has a line transversal if and only if every m members of $\mathcal{A}$ have a line transversal;*
- *the number of geometric permutations of $\mathcal{A}$ is at most m';*
- *the number of connected components of $\mathcal{T}_1^3(\mathcal{A})$ is at most m''?*

(More generally, do there exist similar bounds for k-transversals to $(k-1)$-separated families of translates in $\mathbb{R}^d$?)

5. Piercing numbers

In 1957, Hadwiger and Debrunner gave the following variation of Helly's Theorem:

THEOREM 7 ([**HD**]). *For every $p \geq q \geq d+1$ where $p(d-1) < (q-1)d$: If $\mathcal{A}$ is a finite family of at least p convex sets in $\mathbb{R}^d$ and out of every p members of $\mathcal{A}$ some q have a point in common, then some set of $p-q+1$ points intersects every member of $\mathcal{A}$.*

The value of $p-q+1$ is tight and cannot be reduced. The smallest number of points required to intersect every member of $\mathcal{A}$ is called the *piercing number* or sometimes the *Gallai number* of $\mathcal{A}$ after a question of T. Gallai on the smallest number of needles required to pierce all members of any family of pairwise intersecting circular disks in $\mathbb{R}^2$.

Without the condition $p(d-1) < (q-1)d$ it is not necessarily true (and sometimes false) that some set of $p-q+1$ points intersects every member of $\mathcal{A}$ but Hadwiger and Debrunner conjectured that for every p and q one could replace $p-q+1$ by some constant c for which the theorem would hold. In the following ten years, this was shown for families of axes parallel parallelotopes [**HDK**] and for families of homothets [**W**] but it was not until 1992 that Alon and Kleitman presented a proof for the general case.

THEOREM 8 ([**AKl**]). *For every $p \geq q \geq d+1$, there exists a positive integer $c_{p,q,d}$ such that: If $\mathcal{A}$ is a finite family of at least p convex sets in $\mathbb{R}^d$ and out of every p members of $\mathcal{A}$ some q have a point in common, then some set of $c_{p,q,d}$ points intersects every member of $\mathcal{A}$.*

Let $\mathcal{A}$ be a family of $n \geq p$ convex sets in $\mathbb{R}^d$ such that out of every p members of $\mathcal{A}$ some q have a point in common. Alon and Kleitman use Katchalski and Liu's fractional version of Helly's Theorem [**KLi**] to show that some point intersects at least βn members of $\mathcal{A}$ where β is a fixed constant depending solely on p, q and d. (See also [**AK, E3, K**] for sharp quantitative proofs.) They then apply duality in linear programming to construct a family Y of points such that every member of $\mathcal{A}$ intersects at least $\beta|Y|$ elements of Y. Finally, using weak ϵ-nets for convex hulls they find a set X whose size is a fixed constant depending solely on β and d such that the convex hull of any subset of Y of size $\beta|Y|$ contains some member of X. Since every convex set in $\mathcal{A}$ contains at least $\beta|Y|$ elements of Y, every convex set in $\mathcal{A}$ intersects X. (See [**KA**] for a new, purely combinatorial proof by the same authors.)

Although Helly's theorem does not generalize to hyperplane transversals, Alon and Kalai showed that Theorem 8 does.

THEOREM 9 ([**AK2**]). *For every $p \geq q \geq d+1$, there exists a positive integer $c_{p,q,d}$ such that: If $\mathcal{A}$ is a finite family of at least p convex sets in $\mathbb{R}^d$ and out of every p members of $\mathcal{A}$ some q have a hyperplane transversal, then some set of $c_{p,q,d}$ hyperplanes intersects every member of $\mathcal{A}$.*

The proof of Theorem 9 follows the proof of Theorem 8. Let $\mathcal{A}$ be a family of $n \geq p$ convex sets in $\mathbb{R}^d$ such that out of every p members of $\mathcal{A}$ some q have a hyperplane transversal. Alon and Kalai show that some hyperplane intersects at least βn members of $\mathcal{A}$ and for some family Y of hyperplanes every member of $\mathcal{A}$ intersects at least $\beta|Y|$ elements of Y. They construct a β-cutting for Y, i.e., a set of d-dimensional simplices partitioning $\mathbb{R}^d$ whose size is a fixed constant depending solely on β and d such that the interior of every simplex intersects fewer than $\beta|Y|$ members of Y. (See [**M**].) Every member of $\mathcal{A}$ intersects at least $\beta|Y|$ members of Y so every member of $\mathcal{A}$ intersects at least one simplex facet. If X is the set of hyperplanes containing the simplex facets, then every member of $\mathcal{A}$ intersects X.

Eckhoff gave almost exact minimal values for $c_{p,p,2}$.

THEOREM 10 ([**E2**]). *Let $\mathcal{A}$ be a finite family of convex sets in $\mathbb{R}^2$. If every four members of $\mathcal{A}$ have a line transversal, then there is a set of two lines which intersects every member of $\mathcal{A}$.*

THEOREM 11 ([**E4**]). *Let $\mathcal{A}$ be a finite family of convex sets in $\mathbb{R}^2$. If every three members of $\mathcal{A}$ have a line transversal, then there is a set of four lines which intersects every member of $\mathcal{A}$.*

Eckhoff conjectured, although he could not prove, that the number four in the conclusion of Theorem 11 could be reduced to three.

Alon and Kalai conjectured that Theorem 9 could be generalized to k-transversals in $\mathbb{R}^d$. A first step would be to show that the existence of a k-transversal to every r members of a family of $\mathcal{A}$ implies the existence of a k-tranversal to $r+1$ members of $\mathcal{A}$. They posed the following problems.

PROBLEM 3.

- *For $r \geq 5$, do there exist integers n_r such that: If $\mathcal{A}$ is a finite family of at least n_r convex sets in $\mathbb{R}^3$ and every r members of $\mathcal{A}$ have a line transversal, then some $r+1$ members of $\mathcal{A}$ have a line transversal? (More generally, if $\mathcal{A}$ is in $\mathbb{R}^d$ and every r members of $\mathcal{A}$ have a k-transveral, does some set of $r+1$ members of $\mathcal{A}$ have a k-transversal?)*
- *For $p \geq q \geq 5$, do there exist integers $c_{p,q}$ such that: If $\mathcal{A}$ is a finite family of at least p convex sets in $\mathbb{R}^3$ and out of every p members of $\mathcal{A}$ some q have a line transversal, then some set of $c_{p,q}$ lines intersects every member of $\mathcal{A}$? (More generally, if $\mathcal{A}$ is in $\mathbb{R}^d$ and out of every p members of $\mathcal{A}$ some q have a k-transveral, does some set of $c_{p,q,k,d}$ k-flats intersects every member of $\mathcal{A}$?)*

6. Combinatorial complexity

Let $\mathcal{A}$ be a family of convex sets in $\mathbb{R}^d$. Much of the early work on transversals was motivated by Helly's theorem and directed toward giving conditions for the existence of a k-transversal of $\mathcal{A}$. More recently, researchers have studied the structure of $\mathcal{T}_k^d(\mathcal{A})$, the space of all k-transversals of $\mathcal{A}$. Computer scientists, in particular, have been interested in explicitly constructing representations of this set.

The boundary of $\mathcal{T}_k^d(\mathcal{A})$ consists of k-transversals which are tangent to one or more members of $\mathcal{A}$. A *face* of this boundary is a maximally connected region of k-flats which are tangent to the same members of $\mathcal{A}$. The *combinatorial complexity* of $\mathcal{T}_k^d(\mathcal{A})$ is the number of such faces. A major problem has been to bound this combinatorial complexity for various families of convex sets, particularly convex sets bounded by algebraic surfaces, and thus bound the size of an explicit representation of $\mathcal{T}_k^d(\mathcal{A})$.

Most progress has been made in bounding the complexity of $\mathcal{T}_{d-1}^d(\mathcal{A})$, the space of hyperplane transversals to $\mathcal{A}$. Let $\mathcal{D}(h)$ be the mapping which takes each "non-vertical" hyperplane

$$h = \{(x_1, \dots, x_d) : x_d = \sigma_1 x_1 + \boldsymbol{\sigma_2 x_2} + \cdots + \sigma_{d-1} x_{d-1} + \sigma_d\}.$$

to the point $(\sigma_1, \dots, \sigma_d)$ in dual space. The set of non-vertical hyperplane transversals to $\mathcal{A}$ is represented by a set of points in this dual space. (See Figure 1.)

To construct this set in dual space, consider a single compact convex set $a \in \mathcal{A}$. The non-vertical hyperplanes tangent to set a map to two unbounded surfaces in the dual space. The hyperplanes intersecting set a map to the points between these two surfaces. (See Figure 2.)

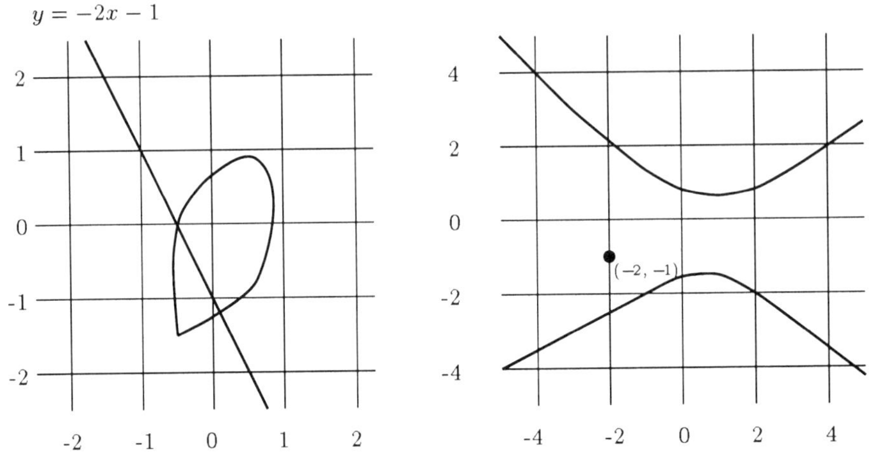

FIGURE 2. A convex set and its transversals.

More precisely, define

$$\begin{aligned} \phi_a^+(\sigma_1, \ldots, \sigma_{d-1}) &= \max\{\sigma_d : \mathcal{D}^{-1}(\sigma_1, \ldots, \sigma_{d-1}, \sigma_d) \cap a \neq \emptyset\} \text{ and} \\ \phi_a^-(\sigma_1, \ldots, \sigma_{d-1}) &= \min\{\sigma_d : \mathcal{D}^{-1}(\sigma_1, \ldots, \sigma_{d-1}, \sigma_d) \cap a \neq \emptyset\}. \end{aligned}$$

For fixed $\{\sigma_1, \ldots, \sigma_{d-1}\}$ the set

$$\{\mathcal{D}^{-1}(\sigma_1, \ldots, \sigma_{d-1}, \sigma_d) : \phi_a^+(\sigma_1, \ldots, \sigma_{d-1}) \leq \sigma_d \leq \phi_a^-(\sigma_1, \ldots, \sigma_{d-1})\}$$

is a set of parallel hyperplanes which intersect a.

For each $a \in \mathcal{A}$, the graphs of functions ϕ_a^+ and ϕ_a^- define two surfaces in $\mathbb{R}^d$. The "corridor" of points between those surfaces represents the set of hyperplanes intersecting a. (If a is not closed, the boundaries of the corridors may not represent hyperplanes intersecting a. If a is not bounded, the functions ϕ_a^+ and ϕ_a^- may not be defined at all points.) The intersection of all these corridors represents the set of non-vertical hyperplane transversals to $\mathcal{A}$. Since each corridor is bounded by a "top" and "bottom" surface, the points corresponding to hyperplane transversals are the points below all the "top" surfaces and above all the "bottom" surfaces. Formally, the space of non-vertical hyperplane transversals to a family $\mathcal{A}$ of compact convex sets is

$$\{\mathcal{D}^{-1}(\sigma_1, \ldots, \sigma_{d-1}, \sigma_d) : \min_{a \in \mathcal{A}} \phi_a^+(\sigma_1, \ldots, \sigma_{d-1}) \geq \sigma_d \geq \max_{a \in \mathcal{A}} \phi_a^-(\sigma_1, \ldots, \sigma_{d-1})\}.$$

The function $\min_{a \in \mathcal{A}} \phi_a^+(\sigma_1, \ldots, \sigma_{d-1})$ is called the *lower envelope* of the functions $\{\phi_a^+\}$. Similarly, $\max_{a \in \mathcal{A}} \phi_a^-(\sigma_1, \ldots, \sigma_{d-1})$ is the *upper envelope* of the functions $\{\phi_a^-\}$. A representation of the space of non-vertical hyperplane transversals of $\mathcal{A}$ can be constructed by computing the lower and upper envelopes of $\{\phi_a^+\}$ and $\{\phi_a^-\}$, respectively, and then intersecting these envelopes.

Upper and lower envelopes have been studied extensively over the past fifteen years, particularly by Micha Sharir and his students and colleagues who made great progress in bounding their combinatorial complexity. This work led to asymptotically tight bounds in 1989 by Pach and Sharir on the combinatorial complexity of upper and lower envelopes of piecewise linear functions. Edelsbrunner, Guibas and Sharir applied these bounds to the combinatorial complexity of the space of hyperplane transversals to a family of convex polytopes:

THEOREM 12 ([**PS, EGS**]). *Let $\mathcal{A}$ be a family of convex polytopes in $\mathbb{R}^d$ with a total of n_f faces. The combinatorial complexity of the space of hyperplane transversals to $\mathcal{A}$ is $O(n_f^{d-1}\alpha(n))$.*

$\alpha(n)$ is the slow growing inverse of the Ackermann function.

Let $\lambda_s(n)$ be the maximum length of an (n, s) Davenport-Schinzel sequence. (See [**SA**] for the definition and discussion of Davenport-Schinzel sequences and Ackermann's function and their relationship to upper and lower envelopes.) $\lambda_s(n)$ is almost but not quite linear in n and is $O(n\alpha(n)^{O(\alpha(n)^{(s-2)/2})})$. Atallah and Bajaj bounded the complexity of line transversals in the plane in terms of $\lambda_s(n)$.

THEOREM 13 ([**AB**]). *Let $\mathcal{A}$ be a family of n compact connected sets in $\mathbb{R}^2$ such that any two members of $\mathcal{A}$ have at most s common supporting lines. The combinatorial complexity of the space of line transversals to $\mathcal{A}$ is $O(\lambda_s(n))$.*

In $\mathbb{R}^3$, Agarwal, Schwarzkopf and Sharir bounded the complexity of the space of plane transversals to convex sets bounded by algebraic surfaces with degree less than or equal to some fixed constant.

THEOREM 14 ([**ASS**]). *Let $\mathcal{A}$ be a family of convex sets in $\mathbb{R}^3$ bounded by algebraic surfaces of bounded degree. The combinatorial complexity of the space of plane transversals to $\mathcal{A}$ is $O(n^{2+\epsilon})$, for any $\epsilon > 0$.*

Of course, the hidden constant in the O-notation depends upon ϵ and the maximum degree of the algebraic surfaces.

For hyperplane transversals to families of balls, the known bounds on the combinatorial complexity is drastically lower.

THEOREM 15 ([**HIIRY**]). *Let $\mathcal{A}$ be a family of n $(d-1)$-balls in $\mathbb{R}^d$. The combinatorial complexity of the space of hyperplane transversals to $\mathcal{A}$ is $O(n^{\lceil d/2\rceil})$.*

It is not known if this bound is asymptotically tight.

The proof by Houle et al. is based on the fact that a convex polytope in $\mathbb{R}^{d+1}$ which is the intersection of n half-spaces has $O(n^{\lceil d/2\rceil})$ faces. Represent each hyperplane

$$h = \{(x_1, \dots, x_d) : \sigma_1 x_1 + \sigma_2 x_2 + \cdots + \sigma_d x_d + \sigma_{d+1} = 0\}$$

uniquely by the $(d+1)$-tuple $(\sigma_1, \dots, \sigma_{d+1})$ where $\sigma_1^2 + \cdots + \sigma_d^2 = 1$. (Note this representation is different from the previous one.) The hyperplanes which intersect a ball centered at $(c_1, c_2, \dots, c_d)$ with radius r are:

$$\{(\sigma_1, \dots, \sigma_{d+1}) : r \geq (\sigma_1 c_1 + \cdots + \sigma_d c_d + \sigma_{d+1}) \geq -r \text{ and } \sigma_1^2 + \cdots + \sigma_d^2 = 1.\}$$

Thus each ball defines two linear inequalities or, equivalently, two half-spaces. The hyperplane transversals for n balls is the intersection of $2n$ half-spaces, two for each ball, and the quadric surface $\sigma_1^2 + \cdots + \sigma_d^2 = 1$. The intersection of the $2n$ half-spaces has $O(n^{\lceil d/2\rceil})$ faces and the intersection with the quadric surface has the same complexity.

Finally, some bounds are known for the space of line transversals in $\mathbb{R}^3$. A series of papers led to nearly tight bounds by Agarwal on the complexity of line transversals to convex polytopes in $\mathbb{R}^3$:

THEOREM 16 ([**A**]). *Let $\mathcal{A}$ be a family of convex polytopes in $\mathbb{R}^3$ with a total of n_f faces. The combinatorial complexity of the space of line transversals to $\mathcal{A}$ is $O(n_f^3 \log n)$.*

A very recent result by Agarwal, Aronov and Sharir bounds the complexity of the space of line transversals to balls in $\mathbb{R}^3$.

THEOREM 17 ([**AAS**]). *Let $\mathcal{A}$ be a family of n balls in $\mathbb{R}^3$. The combinatorial complexity of the space of line transversals to $\mathcal{A}$ is $O(n^{3+\epsilon})$, for any $\epsilon > 0$.*

In higher dimensions, there are no published tight or nearly tight bounds on the combinatorial complexity of the space of transversals.

PROBLEM 4.

- *What is the combinatorial complexity of the space of k-transversals to a family of convex polytopes in $\mathbb{R}^d$ for $d \geq 4$ and $k < d - 1$?*
- *What is the combinatorial complexity of the space of line transversals to a family of convex sets bounded by algebraic surfaces of bounded degree in $\mathbb{R}^3$ or the space of k-transversals in $\mathbb{R}^d$ for $d \geq 4$? (This is not even known for hyperplane transversals when $d \geq 4$.)*

Most of the above results bound the complexity of the space of transversals in terms of the complexity of the underlying objects, either the total number of polytope faces or the maximum degree of the algebraic curves bounding the convex sets. This is because even two convex sets in the plane could have an arbitrary number of common tangent lines, each of which is a face in the space of line transversals. Thus there is no finite bound which is only in terms of n on the complexity of the space of k-transversals to an arbitrary family of n convex sets. However, if $\mathcal{A}$ is a family of pairwise disjoint compact convex sets in the plane, then any two sets in $\mathcal{A}$ have exactly four common tangent lines. By Theorem 13, the combinatorial complexity of the space of line transversals to $\mathcal{A}$ is $O(\lambda_4(n))$. In fact, a closer analysis gives the combinatorial complexity as $O(n)$.

Two sets are pairwise disjoint if they have no point transversal. The proper generalization for m sets in higher dimensions is that they have no $(m-2)$-transversal. In this case, in place of four tangent lines to two convex sets, Cappell et al. showed that the space of tangent hyperplanes to $m \leq d$ compact strictly convex sets in $\mathbb{R}^d$ form 2^{m-1} copies of $\mathbb{S}^{d-m}$, the $(d-m)$-sphere.

THEOREM 18 ([**CGPPSW**]). *Let $\mathcal{A}$ be an $(m-2)$-separated family of $m \leq d$ compact strictly convex sets in $\mathbb{R}^d$. The space of hyperplanes simultaneously tangent to each member of $\mathcal{A}$ is homeomorphic to the union of 2^{m-1} copies of $\mathbb{S}^{d-m}$.*

The condition of strict convexity can be dropped when m equals d. It is not known if this condition is required for m less than d.

The proof of Theorem 18 involves shrinking the m convex sets in $\mathcal{A}$ to m points. The set of hyperplanes through $m \leq d$ points in general position is homeomorphic to $\mathbb{S}^{d-m}$. After some topological analysis, it follows that the set of hyperplane tangents which separate $\mathcal{A}$ in a particular manner is also homeomorphic to $\mathbb{S}^{d-m}$. Since there are 2^{m-1} ways in which hyperplanes separate $\mathcal{A}$ (including not separating $\mathcal{A}$ at all) the space of hyperplane transversals is homeomorphic to 2^{m-1} copies of $\mathbb{S}^{d-m}$. Lewis, Von Hohenbalken and Klee [**LVK**] gave a simple, elegant proof for the case m equal to d using Brouwer's fixed point theorem.

Theorem 18 implies that the hyperplane tangents to separated families behave essentially like topological spheres. By analyzing the complexity of the boundary of the union of a set of topological spheres, Cappell et al. gave the following bound:

THEOREM 19 ([**CGPPSW**]). *Let $\mathcal{A}$ be a $(d-2)$-separated family of n compact strictly convex sets in $\mathbb{R}^d$. The combinatorial complexity of the space of hyperplane transversals to $\mathcal{A}$ is $O(n^{d-1})$.*

The only known lower bounds are for line transversals in the plane in which case the bound is tight.

No bounds are known on the complexity of the space of k-transversals to a $(k-1)$-separated family of convex sets in $\mathbb{R}^d$. Anderson noted that the space of k-flats which are simultaneously tangent to a $(k-1)$-separated family of $k+1$ balls in $\mathbb{R}^d$ is $(S^{d-k-1})^{k+1}$, the product of $k+1$ spheres of dimension $d-k-1$ [**An**].

PROBLEM 5. *Let $\mathcal{A}$ be a $(k-1)$-separated family of compact convex sets in $\mathbb{R}^d$.*

- *What is the topological structure of the space of k-flats simultaneously tangent to $m \leq k+1$ members of $\mathcal{A}$?*
- *What is the combinatorial complexity of the space of k-transversals to $\mathcal{A}$?*

7. Counting geometric permutations

Let $\mathcal{A}$ be a $(k-1)$-separated family of convex sets in $\mathbb{R}^d$. Two k-transversals which induce different pairs of oriented matroids on $\mathcal{A}$ must lie in different connected components of $\mathcal{A}$. Thus a lower bound on the number of pairs of oriented matroids induced by k-transversals of $\mathcal{A}$ is also a lower bound on the number of connected components of $\mathcal{A}$. When k equals $d-1$, each connected component is associated with a unique pair of oriented matroids on $\mathcal{A}$ induced by a hyperplane transversal of $\mathcal{A}$. The number of connected components equals the number of such pairs. Given a $(k-1)$-separated family of convex sets in $\mathbb{R}^d$, how many different pairs of oriented matroids on $\mathcal{A}$ are induced by k-transversals of $\mathcal{A}$?

In 1985, Katchalski, Lewis and Zaks [**KLZ**] constructed families of $n \geq 4$ pairwise disjoint convex sets in $\mathbb{R}^2$ which have $2n-2$ geometric permutations. Five years later, Edelsbrunner and Sharir showed that $2n-2$ was the maximum achievable.

THEOREM 20 ([**ES**]). *Let $\mathcal{A}$ be a family of $n \geq 4$ pairwise disjoint compact convex sets in $\mathbb{R}^2$. The maximum number of pairs of linear orderings induced by line transversals of $\mathcal{A}$ is $2n-2$.*

Represent the directions in the plane as points on the unit circle. Edelsbrunner and Sharir noted that each directed line transversal can can be translated to the right until it is tangent to at least one convex set $a \in \mathcal{A}$. Label the direction of that line transversal "a". This labelling divides the transversal directions into labelled arcs which form a cyclic sequence around the unit circle. This cyclic sequence has the property that $a \ldots b \ldots a \ldots b \ldots$ is not a subcycle for any $a, b \in \mathcal{A}$. These sequences are $(n,2)$ Davenport-Schinzel cycles and their maximum length is $2n-2$. (See [**SA**, p.7].) Since each geometric permutation corresponds to one or more arcs, $2n-2$ is also a bound on the number of geometric permutations.

Any bound on the combinatorial complexity of the space of k-transversals to $\mathcal{A}$ is also a bound on the number of oriented matroids induced by k-transversals of $\mathcal{A}$. In particular, Theorem 18 implies the following:

THEOREM 21 ([**CGPPSW**]). *Let $\mathcal{A}$ be a $(d-2)$-separated family of n convex sets in $\mathbb{R}^d$. The maximum number of oriented matroids induced by oriented hyperplane transversals of $\mathcal{A}$ is $O(n^{d-1})$.*

As with Theorem 18, it is not known if this bound is tight.

Finally, Goodman, Pollack and Wenger bounded the number of oriented matroids induced by k-transversals in $\mathbb{R}^d$ for general k and d.

THEOREM 22 ([**GPW3**]). *Let $\mathcal{A}$ be a $(k-1)$-separated family of n compact convex sets in $\mathbb{R}^d$ where $k \leq d-2$. The maximum number of oriented matroids induced by oriented k-transversals of $\mathcal{A}$ is $O(n^{k(k+1)(d-k)})$.*

The $O(n^{k(k+1)(d-k)})$ term actually bounds the maximum size of a set of incompatible oriented matroids induced by k-transversals to subsets of $\mathcal{A}$ and is probably not tight. Two oriented matroids of two possibly different subsets $\mathcal{A}', \mathcal{A}'' \subseteq \mathcal{A}$ are *incompatible* if the two oriented matroids restricted to their common elements, $\mathcal{A}' \cap \mathcal{A}''$, are different. Note also that when k equals $d-1$, the $O(n^{k(k+1)(d-k)})$ term evaluates to $O(n^{(d-1)d})$ which does not match the $O(n^{d-1})$ term in Theorem 21.

The order in which a directed line l intersects two sets separated by a hyperplane h depends upon the order in which l intersects the two half-spaces defined by h. For each pair of convex sets $a, a' \in \mathcal{A}$, choose a hyperplane strictly separating a from a''. A careful analysis of the number of different ways a line can intersect these $\binom{n}{2}$ hyperplanes gives an $O(n^{2(d-1)})$ bound on the maximum number of oriented matroids induced by line transversals of $\mathcal{A}$.

Similarly, a family $\mathcal{A}$ of compact convex sets is $(k-1)$-separated if and only if every j sets in $\mathcal{A}$ can be strictly separated by a hyperplane from every other $k+1-j$ sets. The oriented matroid (consisting of a single orientation) induced by an oriented k-transversal on $k+1$ sets is determined by the manner in which the k-transversal intersects these separating hyperplanes, i.e., the $2^k - 1$ hyperplanes which separate every j sets from every other $k+1-j$ sets for every $1 \leq j \leq k$. In fact, the oriented matroid depends only on the relationship between the k-transversals and the normal vectors to these hyperplanes. For each pair of subsets $\mathcal{A}', \mathcal{A}'' \subseteq \mathcal{A}$ of size j and $k+1-j$, respectively, choose a hyperplane strictly separating $\mathcal{A}'$ and $\mathcal{A}''$. A careful analysis of the number of different ways a k-flat can intersect these $(2^k-1)\binom{n}{k+1}$ hyperplanes gives the $O(n^{k(k+1)(d-k)})$ bound in Theorem 22.

The only lower bounds known are by Katchalski, Lewis and Liu for line transversals in $\mathbb{R}^d$.

THEOREM 23 ([**KLL3**]). *There exist families $\mathcal{A}$ of n pairwise disjoint convex sets in $\mathbb{R}^d$ whose directed line transversals induce $\Omega(n^{d-1})$ linear orderings of $\mathcal{A}$.*

Even for line transversals in $\mathbb{R}^3$, the asymptotic upper and lower bounds, $O(n^4)$ and $\Omega(n^2)$, do not match.

PROBLEM 6. *Give asymptotically tight upper and lower bounds on the maximum number of linear orderings induced by line transversals of n pairwise disjoint convex sets in $\mathbb{R}^3$ (or more generally give tight bounds on the maximum number of oriented matroids induced by k-transversals of $(k-1)$-separated convex sets in $\mathbb{R}^d$.)*

8. Convexity on the affine Grassmannian

In a 1995 paper entitled "Foundations of a theory of convexity on affine Grassmannian manifolds" [**GP3**], Goodman and Pollack suggest yet another way of exploring geometric transversals. Instead of asking for conditions for the existence of transversals or studying the structure of the space of transversals, they ask under what conditions is a set $\mathcal{F}$ of k-flats the space of transversals to some family of convex sets? They answer that $\mathcal{F}$ is a space of transversals to some family $\mathcal{A}$ of convex sets if and only if every k-flat "surrounded" by $\mathcal{F}$ is in $\mathcal{F}$. A k-flat f is *surrounded* by $\mathcal{F}$ if $f \in \mathcal{F}$ or there is some j-flat g containing f such that every $(j-1)$-flat in g which contains f strictly separates two k-flats of $\mathcal{F}$ lying in g; i.e., f is "trapped" by the elements of $\mathcal{F}$ lying in g. The family $\mathcal{A}$ of convex sets need not be finite and the convex sets in $\mathcal{A}$ may not be compact.

Given a family $\mathcal{F}$ of k-flats in $\mathbb{R}^d$, let $\mathcal{C}(\mathcal{F})$ be the set of all k-flats surrounded by $\mathcal{F}$ or, alternatively, the smallest space of k-transversals containing $\mathcal{F}$. $\mathcal{C}(\mathcal{F})$ shares many properties with the convex hull of a set of points. For one thing, if $\mathcal{F}$ is a family of 0-flats or points, then $\mathcal{C}(\mathcal{F})$ is simply the convex hull of the point set $\mathcal{F}$. As with convex hulls of point sets, the operator $\mathcal{C}$ is monotone (increasing) and idempotent, i.e.:

- $\mathcal{F} \subseteq \mathcal{C}(\mathcal{F})$;
- $\mathcal{F}_1 \subseteq \mathcal{F}_2 \Rightarrow \mathcal{C}(\mathcal{F}_1) \subseteq \mathcal{C}(\mathcal{F}_2)$; and
- $\mathcal{C}(\mathcal{C}(\mathcal{F})) = \mathcal{C}(\mathcal{F})$.

It also has an "anti-exchange" property:

- If $f, f' \notin \mathcal{C}(\mathcal{F})$ and $f \in \mathcal{C}(\mathcal{F} \cup \{f'\})$ while $f' \in \mathcal{C}(\mathcal{F} \cup \{f\}$, then $f = f'$.

Finally, it is invariant under nonsingular affine transformations, so

- $\mathcal{C}(\sigma(\mathcal{F})) = \sigma(\mathcal{C}(\mathcal{F}))$ for nonsingular affine transformations σ.

$\mathcal{C}$ also satisfies a property similar to the Krein-Milman property for point sets. The *extreme* k-flats of $\mathcal{F}$, denoted $\operatorname{ext}(\mathcal{F})$, are the k-flats $f \in \mathcal{F}$ which are not surrounded by $\mathcal{F} \setminus \{f\}$, i.e., $f \notin \mathcal{C}(\mathcal{F} \setminus \{f\})$.

- If $\mathcal{F} = \mathcal{C}(\mathcal{F})$ and $\mathcal{F}$ is compact, then $\mathcal{F} = \mathcal{C}(\operatorname{ext} \mathcal{F})$.

Because of these many similarities with point convexity, Goodman and Pollack called $\mathcal{C}$ the convex hull operator for the affine Grassmannian. A set of k-flats $\mathcal{F}$ is convex if $\mathcal{F} = \mathcal{C}(\mathcal{F})$, or, equivalently, if $\mathcal{F}$ is the space of k-transversals to some family of convex point sets. However, there is one major difference between convex point sets and convex sets of k-flats. Convex point sets are always connected while a space of k-transversals may have many connected components. Goodman and Pollack showed that this difference is inevitable since there is no way of defining convex sets of k-flats which have the properties listed above and are always connected.

As we noted, $\mathcal{C}(\mathcal{F})$ is not necessarily connected. However, if $\mathcal{F}$ is connected, is $\mathcal{C}(\mathcal{F})$ connected? Equivalently, if $\mathcal{F}$ is convex (i.e., $\mathcal{F} = \mathcal{C}(\mathcal{F})$) is every connected component of $\mathcal{F}$ convex? Goodman, Pollack and Wenger showed this was false by constructing a connected set $\mathcal{F}$ of lines parallel to the (x,y)-plane in $\mathbb{R}^3$ which surround the z-axis and no other line. $\mathcal{C}(\mathcal{F})$ equals $\mathcal{F} \cup \{z\text{-axis}\}$ which is clearly not connected. Alternatively, the connected component $\mathcal{F}$ of $\mathcal{C}(\mathcal{F})$ is not convex, since $\mathcal{F} \neq \mathcal{C}(\mathcal{F})$.

The set $\mathcal{F}$ given by Goodman et al. was not closed. What if $\mathcal{F}$ is both closed and connected? What if it is compact and connected? If $\mathcal{F}$ is compact and convex, is every connected component of $\mathcal{F}$ convex? Goodman, Pollack and Wenger showed that the last statement was true for lines in $\mathbb{R}^3$ under stronger conditions.

THEOREM 24 ([**GPW2**]). *Let $\mathcal{A}$ be a finite family of pairwise disjoint compact convex point sets in $\mathbb{R}^3$. If $\mathcal{L}$ is a connected component of $\mathcal{T}_1^3(\mathcal{A})$, then $\mathcal{L} = \mathcal{C}(\mathcal{L})$.*

Moreover, $\mathcal{L}$ is itself the space of line transversals to some finite family of pairwise disjoint compact convex point sets.

Goodman et al. first proved that every two lines in $\mathcal{C}(\mathcal{L}) \subseteq \mathcal{T}_1^3(\mathcal{A})$ induce the same geometric permutation on $\mathcal{A}$. They then showed that if two line transversals which induce the same geometric permutation of $\mathcal{A}$ intersect, then the lines lie in the same connected component of $\mathcal{T}_1^3(\mathcal{A})$. Finally, they proved that if l is some line in $\mathcal{C}(\mathcal{L})$, then l is connected to a parallel line in $\mathcal{C}(\mathcal{L})$ which intersects a line in $\mathcal{L}$. Thus $\mathcal{C}(\mathcal{L})$ is connected and hence equals $\mathcal{L}$.

PROBLEM 7.

- *Is it true that if $\mathcal{A}$ is a finite 1-separated family of compact convex sets in $\mathbb{R}^4$ and $\mathcal{F}$ is a connected component of $\mathcal{T}_2^4(\mathcal{A})$, then $\mathcal{F} = \mathcal{C}(\mathcal{F})$? (More generally, if $\mathcal{A}$ is $(k-1)$-separated and $\mathcal{F}$ is a connected component of $\mathcal{T}_k^d(\mathcal{A})$, then is $\mathcal{F} = \mathcal{C}(\mathcal{F})$?)*
- *Is it true that if $\mathcal{F}$ is a compact, connected family of lines in $\mathbb{R}^3$, then $\mathcal{C}(\mathcal{F})$ is connected, (or, more generally, that $\mathcal{C}(\mathcal{F})$ is connected for any compact, connected family of k-flats?)*

9. Surveys

The classical 1963 survey "Helly's Theorem and its relatives" by Danzer, Grünbaum and Klee [**DGK**] contains much of the early work on geometric transversal theory. Two recent survey papers, "Helly, Radon and Carathéodory type theorems" by Eckhoff [**E5**] and "Geometric transversal theory" by Goodman, Pollack and Wenger [**GPW**] describe the progress made in the eighties. The *Handbook of Discrete and Computational Geometry* contains a chapter "Helly-Type Theorems and Geometric Transversals" by Wenger [**W4**]. Much of the material on the combinatorial complexity of the space of transversals can be found in Sharir and Agarwal's book *Davenport-Schinzel Sequences and Their Geometric Applications* [**SA**], including definitions and analyses of Davenport-Schinzel sequences and upper and lower envelopes.

References

[A] P. K. Agarwal, *On stabbing lines for polyhedra in* 3d. Comput. Geom. Theory Appl. 4 (1994), 177–189.

[AAS] P. K. Agarwal, B. Aronov and M. Sharir, *Line traversals of balls and smallest enclosing cylinders in three dimensions.* In Proc. 8th ACM-SIAM Sympos. Discrete Algorithms (1997), pp. 483–492.

[ASS] P. K. Agarwal, O. Schwarzkopf and M. Sharir, *The overlay of lower envelopes and its applications.* Discrete Comput. Geom. 15 (1996), 1–13.

[AK] N. Alon and G. Kalai, *A simple proof of the upper bound theorem.* European J. Combin. 6 (1985), 211–214.

[AK2] N. Alon and G. Kalai, *Bounding the piercing number.* Discrete Comput. Geom. 13 (1995), 245–256.

[AKl] N. Alon and D. Kleitman, *Piercing convex sets and the Hadwiger–Debrunner (p,q)-problem.* Adv. Math. 96 (1992), 103–112.

[Am] N. Amenta, *Helly-type theorems and generalized linear programming.* Discrete Comput. Geom. 12 (1994), 241–261.

[An] L. Anderson, Personal communication, 1996.

[AW] L. Anderson and R. Wenger, *Oriented matroids and hyperplane transversals.* Adv. Math. 119 (1996), 117–125.

[AB] M. J. Atallah and C. Bajaj, *Efficient algorithms for common transversals.* Inform. Process. Lett. 25 (1987), 87–91.

[BLWSZ] A. Björner, M. Las Vergnas, N. White, B. Sturmfels and G. Ziegler, Oriented Matroids. Cambridge University Press, Cambridge, 1993.

[CGPPSW] S. E. Cappell, J. E. Goodman, J. Pach, R. Pollack, M. Sharir and R. Wenger, *Common tangents and common transversals.* Adv. Math. 106 (1994), 198–215.

[D] L. Danzer, *Über ein Problem aus der kombinatorischen Geometrie.* Arch. Math. 8 (1957), 347–351.

[DGK] L. Danzer, B. Grünbaum and V. Klee, *Helly's theorem and its relatives.* In Convexity, vol. 7 of Proc. Symp. Pure Math., American Mathematical Society, Providence, 1963, pp. 101–180.

[Dy] M. E. Dyer, *Linear time algorithms for two- and three-variable linear programs.* SIAM J. Comput. 13 (1984), 31–45.

[Dy2] M. E. Dyer, *On a multidimensional search technique and its application to the Euclidean one-centre problem.* SIAM J. Comput. 15 (1986), 725–738.

[Dy3] M. E. Dyer, *A class of convex programs with applications to computational geometry.* In Proc. 8th Annu. ACM Sympos. Comput. Geom. (1992), pp. 9–15.

[E] J. Eckhoff, *Transversalenprobleme vom Gallai'schen Typ*, Ph.D. dissertation, Georg-August-Universität, Göttingen, 1969.

[E2] J. Eckhoff, *Transversalenprobleme in der Ebene.* Arch. Math. 24 (1973), 195–202.

[E3] J. Eckhoff, *An upper bound theorem for families of convex sets.* Geom. Dedicata 19 (1985), 217–227.

[E4] J. Eckhoff, *A Gallai-type transversal problem in the plane.* Discrete Comput. Geom. 9 (1993), 203–214.

[E5] J. Eckhoff, *Helly, Radon and Carathéodory type theorems.* In Handbook of Convex Geometry. North-Holland, 1993, pp. 389–448.

[EGS] H. Edelsbrunner, L. Guibas and M. Sharir, *The upper envelope of piecewise linear functions: algorithms and applications.* Discrete Comput. Geom. 4 (1989), 311–336.

[ES] H. Edelsbrunner and M. Sharir, *The maximum number of ways to stab n convex nonintersecting sets in the plane is $2n-2$.* Discrete Comput. Geom. 5 (1990), 35–42.

[EW] P. Egyed and R. Wenger, *Ordered stabbing of pairwise disjoint convex sets in linear time.* Discrete Appl. Math. 31 (1991), 133–140.

[GP] J. E. Goodman and R. Pollack, *Multidimensional sorting.* SIAM J. Comput. 12 (1983), 484–507.

[GP2] J. E. Goodman and R. Pollack, *Hadwiger's transversal theorem in higher dimensions.* J. Amer. Math. Soc. 1 (1988), 301–309.

[GP3] J. E. Goodman and R. Pollack, *Foundations of a theory of convexity on affine Grassmann manifolds.* Mathematika 42 (1995), 305–328.

[GPW] J. E. Goodman, R. Pollack and R. Wenger, *Geometric transversal theory.* In New Trends in Discrete and Computational Geometry, J. Pach, Ed., vol. 10 of Algorithms and Combinatorics. Springer-Verlag, Heidelberg, Germany, 1993, pp. 163–198.

[GPW2] J. E. Goodman, R. Pollack and R. Wenger, *On the connected components of the space of line transversals to a family of convex sets.* Discrete Comput. Geom. 13 (1995), 469–476.

[GPW3] J. E. Goodman, R. Pollack and R. Wenger, *Bounding the number of geometric permutations induced by k-transversals.* J. Combin. Theory Ser. A 75 (1996), 187–197.

[G] B. Grünbaum, *On common transversals.* Arch. Math. 9 (1958), 465–469.

[H] H. Hadwiger, *Über Eibereiche mit gemeinsamer Treffgeraden.* Portugal. Math. 16 (1957), 23–29.

[HD] H. Hadwiger and H. Debrunner, *Über eine Variante zum Hellyschen Satz.* Arch. Math. 8 (1957), 309–313.

[HDK] H. Hadwiger, H. Debrunner and V. Klee, Combinatorial Geometry in the Plane. Holt, Rinehart & Winston, New York, 1964.

[He] E. Helly, *Über Mengen konvexer Körper mit gemeinschaftlichen Punkten.* Jber. Deutsch. Math.-Verein 32 (1923), 175–176.

[HIIRY] M. E. Houle, H. Imai, K. Imai, J.-M. Robert and P. Yamamoto, *Orthogonal weighted linear L_1 and L_∞-approximation and applications.* Discrete and Applied Mathematics 43 (1993), 217–232.

[J] F. John, *Extremum problems with inequalities as subsidiary conditions.* In Studies and Essays Presented to R. Courant on his 60th Birthday. Interscience, New York, 1948, pp. 187–204.

[K] G. Kalai, *Intersection patterns of convex sets.* Israel J. Math. 48 (1984), 161–174.

[Ka] M. Katchalski, *Thin sets and common transversals.* J. Geometry 14 (1980), 103–107.

[Ka2] M. Katchalski, *A conjecture of Grünbaum on common transversals.* Math. Scand. 59 (1986), 192–198.

[KL] M. Katchalski and T. Lewis, *Cutting families of convex sets.* Proc. Amer. Math. Soc. 79 (1980), 457–461.

[KL2] M. Katchalski and T. Lewis, *Cutting rectangles in the plane.* Discrete Math. 42 (1982), 67–71.

[KLL] M. Katchalski, T. Lewis and A. Liu, *Geometric permutations and common transversals.* Discrete Comput. Geom. 1 (1986), 371–377.

[KLL2] M. Katchalski, T. Lewis and A. Liu, *Geometric permutations of disjoint translates of convex sets.* Discrete Math. 65 (1987), 249–259.

[KLL3] M. Katchalski, T. Lewis and A. Liu, *The different ways of stabbing disjoint convex sets.* Discrete Comput. Geom. 7 (1992), 197–206.

[KLZ] M. Katchalski, T. Lewis and J. Zaks, *Geometric permutations for convex sets.* Discrete Math. 54 (1985), 271–284.

[KLi] M. Katchalski and A. Liu, *A problem of geometry in R^n.* Proc. Amer. Math. Soc. 75 (1979), 284–288.

[KA] D. Kleitman and N. Alon, *A purely combinatorial proof of the Hadwiger Debrunner (p,q) conjecture.* Elect. J. Combin. 4 (1997).

[L] T. Lewis, *Two counterexamples concerning transversals for convex subsets of theplane.* Geom. Dedicata 9 (1980), 461–465.

[LVK] T. Lewis, B. von Hohenbalken and V. Klee, *Common supports as fixed points.* Geometriae Dedicata 60 (1996), 277–281.

[M] J. Matoušek, *Cutting hyperplane arrangements.* Discrete Comput. Geom. 6 (1991), 385–406.

[MSW] J. Matoušek, M. Sharir and E. Welzl, *A subexponential bound for linear programming.* Algorithmica 16 (1996), 498–516.

[Me] N. Megiddo, *Linear programming in linear time when the dimension is fixed.* J. ACM 31 (1984), 114–127.

[Me2] N. Megiddo, *On the ball spanned by balls.* Discrete Comput. Geom. 4 (1989), 605–610.

[PS] J. Pach and M. Sharir, *The upper envelope of piecewise linear functions and the boundary of a region enclosed by convex plates: combinatorial analysis.* Discrete Comput. Geom. 4 (1989), 291–309.

[PW] R. Pollack and R. Wenger, *Necessary and sufficient conditions for hyperplane transversals.* Combinatorica 10 (1990), 307–311.

[R] J.-M. Robert, *Geometric orderings of intersecting translates and their applications.* Comput. Geom. Theory Appl. 7 (1997), 59–72.

[S] L. Santaló, *Un teorema sobre conjuntos de paralelepipedos de aristas paralelas.* Publ. Inst. Mat. Univ. Nac. Litoral 2 (1940), 49–60.

[SA] M. Sharir and P. K. Agarwal, Davenport-Schinzel Sequences and Their Geometric Applications. Cambridge University Press, New York, 1995.

[SW] M. Sharir and E. Welzl, *A combinatorial bound for linear programming and related problems.* In Proc. 9th Sympos. Theoret. Aspects Comput. Sci. (1992), vol. 577 of Lecture Notes Comput. Sci., Springer-Verlag, pp. 569–579.

[T] H. Tverberg, *Proof of Grünbaum's conjecture on common transversals for translates.* Discrete Comput. Geom. 4 (1989), 191–203.

[V] P. Vincensini, *Figures convexes et variétés linéaires de l'espace euclidien à n dimensions.* Bull. Sci. Math. 59 (1935), 163–174.

[W] G. Wegner, *Über eine kombinatorisch-geometrische Frage von Hadwiger und Debrunner.* Israel J. Math. 3 (1965), 187–198.

[W2] R. Wenger, *A generalization of Hadwiger's transversal theorem to intersecting sets.* Discrete Comput. Geom. 5 (1990), 383–388.

[W3] R. Wenger, Geometric permutations and connected components. Tech. Rep. TR-90-50, DIMACS, 1990.

[W4] R. Wenger, *Helly-type theorems and geometric transversals.* In Handbook of Discrete and Computational Geometry. CRC Press LLC, 1997.

The Ohio State University, Columbus, OH 43210, U.S.A.
E-mail address: wenger@cis.ohio-state.edu

Contemporary Mathematics
Volume **223**, 1999

Recent Progress on Polytopes

Günter M. Ziegler

ABSTRACT. This is a survey of recent progress in five active areas of research connected to convex polytopes—together with seven "challenge problems."

This is a discussion of five very active and important areas of research on the (combinatorial) theory of (convex) polytopes, with reports about recent progress, and a selection of seven "challenge" problems that I hope to see solved soon:

Universality Theorems: for polytopes of constant dimension: see Richter-Gebert's work!
Challenge: Can all 3-dimensional polytopes be realized with coordinates of polynomial size?
Challenge: Provide a Universality Theorem for simplicial 4-dimensional polytopes.

Triangulations: and subdivisions of polytopes.
Challenge: Decide whether all triangulations on a fixed point set in general position can be connected by bistellar flips.

0/1-polytopes: and their combinatorial structure.
Challenge: The "0/1 Upper Bound Problem": Is the maximal number of facets of 0/1-polytopes bounded by an exponential function in the dimension?

Neighborly polytopes: Explicit constructions and extremal properties.
Challenge: Is every polytope a quotient of a neighborly polytope?

Monotone paths: and the simplex algorithm for linear programming.
Challenge: The "Monotone Upper Bound Problem": What is the maximal number of vertices of a monotone path on a d-dimensional polytope with n facets?
Challenge: Is there a polynomial upper bound for the running time of the RANDOM-EDGE simplex algorithm?

Disclaimer. This discussion is (solely) concerned with the **combinatorial** theory of **convex** polytopes—**recent** progress, and it is a **personal selection** of topics,

---

1991 *Mathematics Subject Classification.* Primary 52-02.
Key words and phrases. Convex polytopes, universality, triangulations and bistellar flips, 0/1-polytopes, neighborly polytopes, monotone paths.
Supported by a DFG Gerhard-Hess-Forschungsförderungspreis and by the German Israeli Foundation (G.I.F.) grant I-0309-146.06/93.

© 1999 American Mathematical Society

problems and directions that I consider to be interesting. It is meant to be very informal, and cannot provide more than a sketch that hopefully makes you ask for more details and look at some of the references. Background material is in **[Gr1]** and in **[Z]**. Also watch for:

- The new edition of the "classic," Grünbaum **[Gr1]**,
- the *updates and more* offered for **[Z]** on the Web,
- Ewald's new (1996) book **[Ewal]**, and
- Richter-Gebert's very recent (1997) book **[Ri]**.

Notation. In the following our notation will be that $P \subseteq \mathbb{R}^d$ is a convex polytope of dimension d (a d-*polytope*) with m vertices and n facets.

1. Universality

The *realization space* $\mathcal{R}(P)$ of P is the space of all matrices that coordinatize a polytope that is combinatorially equivalent to P, modulo affine equivalence. This space can be given as a space of all matrices that satisfy a certain set of "determinantal" equations and strict inequalities, as in our example.

EXAMPLE 1.1 (Pentagon) We will construct $\mathcal{R}(C_2(5))$, where $C_2(5) \subseteq \mathbb{R}^2$ is a convex pentagon with vertices labeled $1, 2, 3, 4, 5$ in counter-clockwise order. The affine equivalence allows one to fix the first three vertices, for example as $\binom{x_1}{y_1} = \binom{0}{1}$, $\binom{x_2}{y_2} = \binom{0}{0}$, and $\binom{x_3}{y_3} = \binom{1}{0}$. With this we can write down the realization space of a pentagon as

$$\mathcal{R}(C_2(5)) \;=\;$$
$$\left\{ \begin{pmatrix} 0 & 0 & 1 & x_4 & x_5 \\ 1 & 0 & 0 & y_4 & y_5 \end{pmatrix} : \begin{array}{ll} x_4 > 0, & x_5 > 0, \\ y_4 > 0, & x_4 y_5 - x_5 y_4 + x_5 - x_4 > 0, \\ x_4 + y_4 > 0, & x_4 y_5 - x_5 y_4 + y_4 - y_5 > 0 \end{array} \right\} \subseteq \mathbb{R}^4.$$

Equivalently, one could fix the affine basis by explicitly fixing x_1, x_2, x_3 and y_1, y_2, y_3 and then using the conditions

$$\det \begin{pmatrix} 1 & 1 & 1 \\ x_i & x_j & x_k \\ y_i & y_j & y_k \end{pmatrix} > 0 \text{ for all } i < j < k.$$

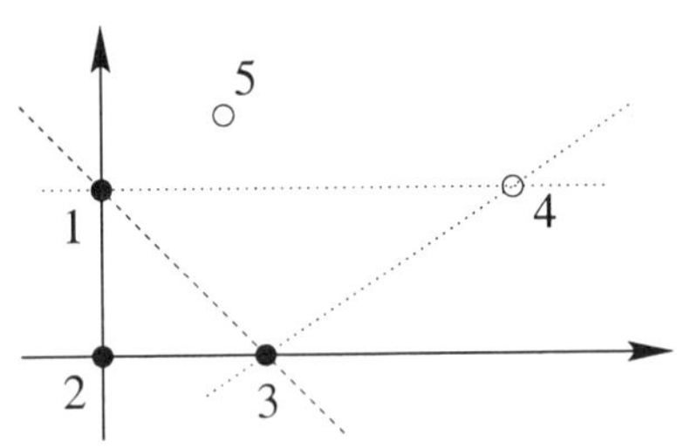

In our figure, assume that the points $1, 2, 3$ have been fixed. The coordinate axes together with the dashed line bound an open polyhedron, which is the set of possible positions for the point 4. After that point has been fixed, the y-axis together with the two dotted lines bound the possible positions for 5, which again is an open polyhedron. From this one can see that $\mathcal{R}(C_2(5))$ is a 4-dimensional open semialgebraic set that has the topological type of an open 4-ball. Also, we can inductively construct rational coordinates (with "small" denominators and numerators) for the vertices of a pentagon. Similarly, for all $m \geq 3$ we get that $\mathcal{R}(C_2(m))$ is a topological $2(m-3)$-ball with "small rational points."

Steinitz's classical theorem of 1922 [**SR**], [**Ri**] states that the realization spaces of 3-polytopes are as "nice" as in the 2-dimensional case: they have the topological type of $(e-6)$-dimensional open balls (where $e(P)$ denotes the number of edges).

On the algebraic side, one can derive (not from Steinitz' proof!) that there is a singly-exponential bound for the vertex coordinates of 3-polytopes [**OS**]. For example, all combinatorial types of rational 3-polytopes that have a triangle face can be realized with their m vertices placed in $\{0, 1, 2, \ldots, 43^m\}^3$, as shown by Richter-Gebert [**Ri**]. The lower bound for the coordinate size needed is embarassingly low: of order $\Omega(m^{3/2})$, which is what one needs for a pyramid over an $(m-1)$-gon. This leads us to our first challenge problem.

CHALLENGE 1. Can all 3-polytopes be realized with integral coordinates whose size is bounded by a polynomial in the number of vertices?

The situation in dimensions $d \geq 4$ is radically different. From Mnëv's Universality Theorem (1988) one can derive that for polytopes with $d+4$ vertices the realization spaces are "universal." However, this does not answer the question for the situation in any fixed dimension.

I will not give a definition of "universal" here—this is usually expanded into "can be stably equivalent to an arbitrary primary semialgebraic set defined over $\mathbb{Z}$." Thus "universal" implies "(nearly) arbitrarily bad," in at least three respects:

- in terms of topology: all homotopy types of finite simplicial complexes do occur,
- in terms of singularities: all types of singularities of primary semialgebraic sets (defined over $\mathbb{Z}$) come up, and
- in terms of arithmetic properties: rational coordinates do not always exist, and no finite-dimensional extension of $\mathbb{Q}$ suffices to find coordinates for all combinatorial types.

I want to emphasize here that the "right" (in many respects) definition of *stably equivalent* and hence of *universal* is very important and original work by Richter-Gebert [**Ri**], [**RZ**]; all the previously used notions are either too strong to yield correct theorems, or they are too weak to derive all the three types of consequences, or they are unnatural or too difficult to handle.

In 1994, Richter-Gebert completed a result that was long sought-after: a Universality Theorem for polytopes of some fixed dimension. (See also [**Gun**].)

THEOREM 1.1 (Richter-Gebert [**Ri**]). *The realization spaces of 4-dimensional polytopes are universal.*

CHALLENGE 2. Is there a Universality Theorem for the realization spaces of *simplicial* 4-polytopes?

Richter-Gebert's method cannot be applied here, since it depends on incidence theorems in the 2-skeleton, where simplicial polytopes don't have any. One can, however, expect that the realization spaces of simplicial 4-polytopes are not nice in general: on the one hand, the realization spaces of simplicial d-polytopes with $d+4$ vertices are complicated (in the topological sense) [**BS**, Sect. 6.2], furthermore (exactly) one "non-trivial" example of a simplicial 4-polytope is known, the BEK polytope [**BEK**], [**BS**, Sect. 6.2], with a disconnected realization space.

Richter-Gebert's work started with the analysis of some other very "small" and special 4-polytope. The crucial building block for his constructions turned out to be the 4-dimensional polytope X^* with 8 facets and 12 vertices whose polar is given by this affine Gale diagram:

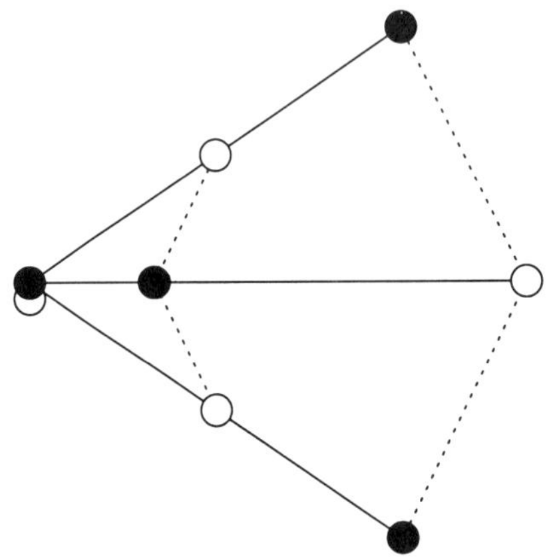

The polytope X^* has a hexagon 2-face whose vertices satisfy a projective condition: two "opposite" edges and the corresponding diagonal intersect in one point (or are parallel), thus answering (in the negative) a problem posed in [**Z**, Prob. 5.11*]. See Richter-Gebert [**Ri**, p. 91], for a primal picture (Schlegel diagram) of this polytope. However, our little picture here may illustrate the power of the theory of Gale diagrams (most of it due to Perles [**Gr1**], see [**Z**, Lect. 6]), it allows one to analyze this interesting 4-dimensional polytope P in terms of a really simple 2-dimensional picture!

2. Triangulations and bistellar flips

Fix a finite set $V \subseteq \mathbb{R}^d$ and let $P := \text{conv}(V)$ be its convex hull. For most of this section we will assume that V is in general position (i.e., no $d+1$ of the points are contained in a hyperplane), in order to make the following definitions as simple as possible. A *triangulation* of V is a triangulation of P whose vertex set is contained in V. (A nice new survey on triangulations of polytopes is Lee [**Lee2**].)

A *(bistellar) flip* in a triangulation is a certain type of local change: In the general position case this operation replaces a subcomplex that represents the d-dimensional "picture of" a front side of a $(d+1)$-dimensional simplex by its back side.

A triangulation is *regular* if it is a "correct" picture of the front side of a simplicial $(d+1)$-polytope. This is a global property that may be destroyed (or created) by a flip.

For example, our next figure represents a planar set V of 11 points, whose convex hull is a 7-gon. The straight lines indicate a triangulation that uses only 10 of the 11 points. One dotted line (that would replace the straight line it crosses) represents one possible flip that one could visualize as replacing the two front faces of a tetrahedron by its two back faces. Both triangulations of the figure (before and after the flip) are regular (Exercise!).

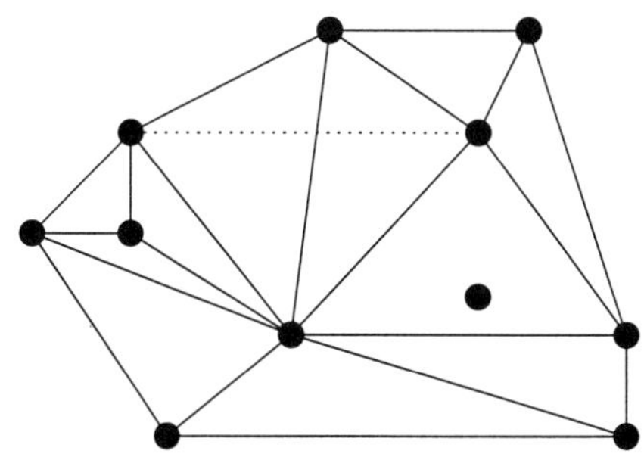

Triangulations are of tremendous importance for topics that range from the construction of splines (PDE, CAD, ...) to algebraic geometry (the resolutions of singularities of toric varieties). In view of this there is a surprising variety of very basic, very interesting and very open problems connected to the set of triangulations of a finite point set V.

To describe some of these, we consider the graph $G(V)$ whose vertex set is the (finite) set of all triangulations of V, and whose edges correspond to all possible bistellar flips. For example, if V is the vertex set of a pentagon, then the graph $G(V)$ that one obtains is a five-cycle, as one can see from our figure.

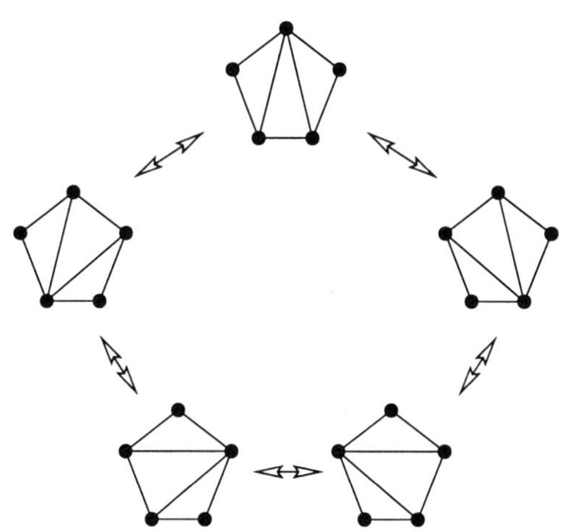

The induced subgraph of $G(V)$ that contains the *regular* triangulations of V and their bistellar flips will be denoted by $G_r(V)$.

THEOREM 2.1 (Gel'fand-Kapranov-Zelevinsky [**GKZ**]). *For every set* $V \subseteq \mathbb{R}^d$ *of* $m > d$ *points the graph of regular triangulations* $G_r(V)$ *is the graph of an* $(m - d - 1)$*- dimensional polytope, the "secondary polytope"* $\Sigma(V)$.

Thus (by "Balinski's Theorem") the graph $G_r(V)$ is $(m - d - 1)$-vertex-connected; in particular, it is connected and its minimum degree is at least $m - d - 1$. The following configuration (found in 1996, see also [**dL**]) indicates that the same is not true if we consider the complete graph $G(V)$.

EXAMPLE 2.2 (de Loera, Santos & Urrutia [**dLSU**]) For $d = 3$ and $m = 13$, let V consist of the 12 midpoints of edges and of the center of a regular cube. Then P is a polytope (truncated cube) whose boundary may be triangulated by adding 6 additional, independent edges, as in our figure:

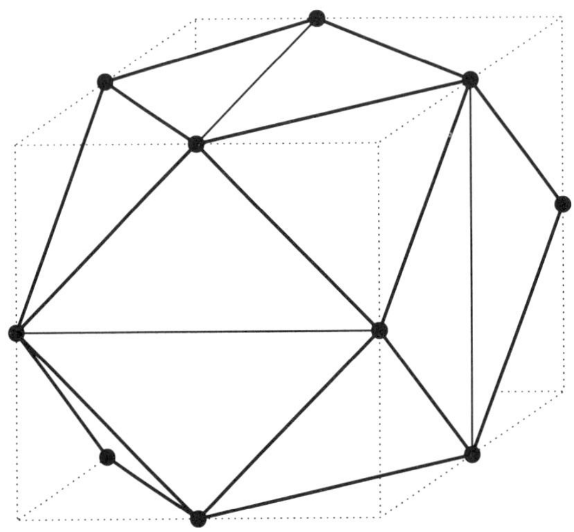

(The same figure can be found in Coxeter [**Cox**, p. 152]!) From this we obtain a triangulation of P for which every tetrahedron uses the center point of the original cube. (This is a triangulation with 13 vertices, $30 + 12 = 42$ edges, $20 + 30 = 50$ triangles and 20 tetrahedra.) It is easy to check that only $6 = m - d - 4$ flips are possible. (Thus, in particular, the triangulation cannot be regular.) The point configuration can be perturbed to general position, maintaining this property. It can also be lifted to a $(d' = 4)$-configuration of $m' = 14$ points in convex position in $\mathbb{R}^4$, where we still have only $6 = m' - d' - 4$ flips.

This was an example with "few" flips: but is there always at least one flip possible?

CHALLENGE 3. Given m points in $\mathbb{R}^d$ (in general position), is the graph of all triangulations connected by bistellar flips?

This is known to be true for $d \leq 2$. It was proved for $m - d \leq 3$ by Lee [**Lee**], who showed that then all triangulations are regular. For all cyclic polytopes $C_d(m)$ it is true by very recent work of Rambau [**Ra**]; except for that, the problem is open. I expect a negative answer.

In the case of cyclic polytopes one can use the "higher Stasheff-Tamari posets" introduced by Edelman & Reiner [**ER**], natural partial orders on the set of all triangulations of a cyclic polytope $C_d(m)$, by putting them into correspondence with triangulated hypersurfaces in $C_{d+1}(m)$:

$$\begin{array}{c}\text{hypersurfaces in } C_{d+1}(m)\\ \updownarrow \quad \updownarrow \quad \updownarrow \quad \updownarrow\\ \text{triangulations of } C_d(m)\end{array}$$

Rambau's proof proceeds by establishing a sequence of simple properties of cyclic polytopes and their triangulations: interestingly enough, many of them are easily seen to be *false* in general. For example, a key step is to establish that for the

$(d+1)$-simplices of $C_{d+1}(m)$, the "on top of" relation generates a partial order. (This is false for a perturbation of the "capped prism" of Lee [**Lee2**, Sect. 3.3].) But the fact that proofs don't generalize doesn't show that the general conjecture is false ...

3. Extremal properties of 0/1-polytopes

A 0/1-*polytope* is a polytope of the form $P = \text{conv}(V)$, where V is a set of 0/1-vectors, $V \subseteq \{0,1\}^d$. Every $v \in V$ is a vertex of $P = \text{conv}(V)$, and thus P is what one calls a *subpolytope* of the usual unit cube, $\text{conv}(\{0,1\}^d)$.

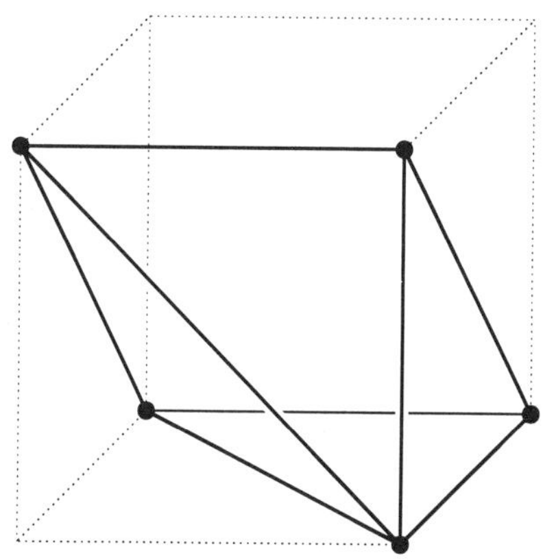

As an example, our figure shows a square pyramid that arises as a 0/1-polytope in the unit 3-cube. However, in the study of 0/1-polytopes one should not rely too much on low-dimensional geometric intuition: things only become interesting (and complicated) in high dimensions.

About "general" 0/1-polytopes not much is known. However, there are many details known about the "special" 0/1-polytopes that are associated with problems of combinatorial optimization such as the Traveling Salesman Problem (see Grötschel & Padberg [**GP**]) and Max-Cut Problems (see Deza & Laurent [**DL**]). However, the division between "general" and "special" 0/1-polytopes is somewhat artificial, since a recent result of Billera & Sarangarajan [**BiSa**] shows that *every* 0/1-polytope is (affinely equivalent to) a face of some asymmetric TSP-polytope.

The study of general 0/1-polytopes gets some of its motivation from core problems of combinatorial optimization such as TSP and Max-Cut: for example, one would like to know how "typical" the special 0/1-polytopes are. For that, it is interesting to describe the properties of "random 0/1-polytopes" as well as extremal properties, such as the following innocent(-looking) little "upper bound problem":

CHALLENGE 4. How fast does $f(d)$, the maximal number of facets of a d-dimensional 0/1-polytope grow with d? In particular, is there a constant C such that $f(d) < C^d$ for all d?

For small d, one can enumerate all possibilities, and this yields $f(d) = 2^d$ for $d \leq 4$ and $f(5) \geq 40$. The best known general bounds (as of February 1998),

$$(3.26)^d < f(d) \leq O((d-2)!).$$

Here the lower bound is derived from direct sums of "centered" 0/1-polytopes, see [**KRSZ**]. The constant 3.26 is derived from a specific example of a 12-dimensional

0/1-polytope with 1489211 facets—see [**K2**] for this and other current records in small dimensions. The upper bound, by T. Fleiner, V. Kaibel and G. Rote [**FKR**] uses estimates of the surface area.

How can one obtain 0/1-polytopes with "many" facets? It may well be that "random" 0/1-polytopes—with the right number of vertices—have many facets, but that may be hard to verify.

Perhaps one has more luck with very explicit constructions, such as cut polytopes? Here $\mathrm{CUT}(k) \subseteq \mathbb{R}^{\binom{k}{2}}$ is the convex hull of the incidence vectors of all cuts of a complete undirected graph K_k. After all, for small k the cut polytopes $\mathrm{CUT}(k)$ have "many" facets: CUT(7) has dimension $d = \binom{7}{2} = 21$ and $(1.743)^d$ facets, while CUT(8) has dimension $d = \binom{8}{2} = 28$ and at least $(1.985)^d$ facets, and CUT(9), of dimension $d = \binom{9}{2} = 36$, has more than $(2.309)^d$ facets; see Christof's library of small polytopes SMAPO [**Chr**].

4. Constructing neighborly polytopes

Neighborly polytopes (4-polytopes such that any two vertices are adjacent; 6-polytopes such that every set of three vertices determines a triangle 2-face, etc.) are important. They have a prominent role in the theory of convex polytopes due to their extremal properties in terms of their numbers of faces, as given by the Upper Bound Theorem of McMullen [**Z**, Thm. 8.23], which got a notable strengthening recently in Novik [**Nov**].

Neighborly polytopes exist; in fact, the cyclic polytopes $C_d(n)$—our prime examples—are easy to write down in explicit coordinates (using the moment curve) and to analyze (using Vandermonde determinants). Also, we know that there are really many neighborly polytopes (Shemer [**Sh**]) and in some models of random polytopes it seems that even random polytopes are often neighborly. Gale wrote "the likelihood of getting a neighborly polytope increases rapidly with the dimension of the space" a long time ago [**Gale**, p. 262], but despite several attempts this has not been proven in any form.

In fact, it is not that easy to write down many neighborly polytopes. However, in recent work Kortenkamp [**K**] has provided a new method to construct neighborly polytopes with $m = d + 4$ vertices—they can be constructed and analyzed in terms of their 2-dimensional affine Gale diagrams. For this, take any finite configuration of m "black" points in the plane (no three on a line). From it we get a "balanced" configuration of $m + 1$ black points and $m - 1$ white points:

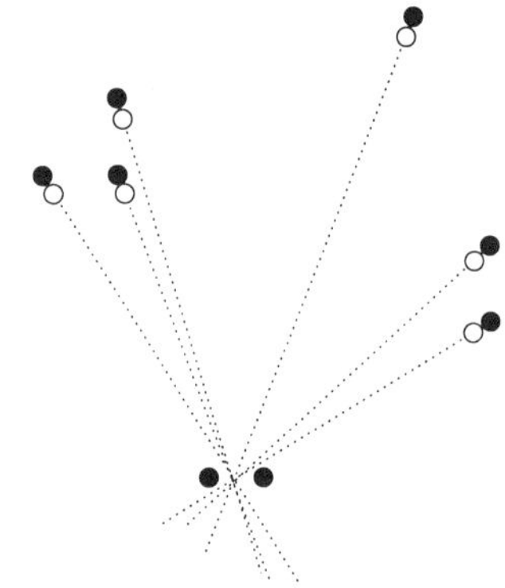

Here a configuration of points in general position is "balanced" if for every hyperplane spanned by some of them one has

surplus of black points on one side = surplus of white points on the other side.

This condition also characterizes the Gale diagrams of neighborly polytopes, and thus (with the usual gymnastics and formalism of Gale diagrams) our little figure proves the following theorem.

THEOREM 4.1 (Kortenkamp [**K**]). *Every simplicial d-polytope with $m \leq d+4$ vertices is equivalent to a quotient (iterated vertex figure) of a neighborly $(2d+4)$-polytope with $2d+8$ vertices.*

To extend this result to arbitrary simplicial polytopes one "only" needs a construction that embeds every point configuration in general position in $\mathbb{R}^d$ into a "balanced" configuration.

CHALLENGE 5. (Perles [**Gr1**], [**Stu**]). Is every simplicial d-polytope (with vertices in general position) a quotient (iterated vertex figure) of a neighborly polytope?

Note, however, that some simple-sounding embedding problems are notoriously difficult, such as the problem of embedding any matroid of rank 3 into a finite projective plane (this is believed to be possible; see Wanner & Ziegler [**WZ**]), or the problem of embedding every line arrangement in the plane into a simplicial line arrangement (this is believed to be impossible in general; see Grünbaum[**Gr2**, p. 9].

5. Monotone paths on polytopes

A *linear program* is the task to find, with respect to a linear height function $c^t x$, a highest vertex of the set $P \subseteq \mathbb{R}^d$ of solutions of a set of n linear inequalities. We deal here with a very geometric setting of linear programming and the simplex algorithm (as in [**Z**, Sect. 3.2]). That is, with the usual reductions we may assume without loss of generality:

- P is a bounded d-dimensional polytope,
- P has n facets, all inequalities are facet-defining,
- P is simple ("primal non-degeneracy"),
- P has no horizontal edges ("dual non-degeneracy"); in particular the minimal and the maximal ("optimal") vertex with respect to $c^t x$ are unique, and
- the minimal vertex of P with respect to $c^t x$ is known ("Phase I is done").

Our version of the simplex algorithm starts at the minimal vertex, and a *pivot rule* (which has only "local" information) chooses a path consisting of edges that improve the objective function $c^t x$ (a *monotone path*) until the optimal vertex is reached. At the core of linear programming theory we find the following two questions:

- Is there always a short (in terms of number of edges) path to the optimal vertex?

  (A very strong version of the Hirsch conjecture [**Z**, p. 87] would need such a path that has at most $n - d$ edges), and

- Can a simplex algorithm find one?
  (Is there a pivot rule for which the number of steps is bounded by a polynomial function of n and d? This would provide a strongly polynomial algorithm for linear programming!)

Unfortunately, virtually every *deterministic* pivot rule has been shown to be exponential in the worst case on certain "deformed product programs": see Amenta & Ziegler [**AZ**]. However, it seems that this has been proved for *none* of the natural randomized pivot rules. The simplest one (to state) is the RANDOM-EDGE rule: at every vertex choose, with equal probability, one of the increasing edges that leave the vertex.

CHALLENGE 6. Is the expected number of steps of the RANDOM-EDGE simplex algorithm bounded by a polynomial function of n and d?

Looking for especially "bad" linear programs, one is first led to the classical examples of linear programming theory, starting with the Klee-Minty cubes: Here RANDOM-EDGE is polynomial, with quadratic upper bound (which is easy to see) and with a nearly-quadratic lower bound (which takes some effort to show [**GHZ**]).

But what are the "worst" linear programs for the simplex algorithm? This question leads one, for each d and n, to consider the following hierarchy of geometric extremal problems:

$f_1(d,n)$: the maximal number of vertices of a 2-dim. projection	$\leq$	$f_2(d,n)$: the maximal number of vertices on a monotone path	$\leq$	$f_3(d,n)$: the maximal number of vertices

This is a very natural hierarchy: $f_1(d,n)$ is the largest number of steps for the simplex algorithm with the Gass-Saaty/Borgwardt shadow vertex rule, $f_2(d,n)$ is the largest number of steps for the simplex algorithm with the most stupid choice of pivots, and $f_3(d,n)$ is a geometric upper bound that is known, by the Upper Bound Theorem. (There is a very similar hierarchy for 0/1-polytopes; see Section 3 and [**KRSZ**].)

It is not at all clear whether the hierarchy collapses, that is, whether $f_1(d,n) = f_2(d,n) = f_3(d,n)$ holds for all n and d. This is true for $d \leq 3$. Also the three functions grow similarly fast—like polynomials of degree $\lfloor \frac{d}{2} \rfloor$ for constant d, and like exponential functions for constant $\frac{n}{d}$. However, for $d = 4$ we have indications for a gap: here we know that polars of cyclic polytopes achieve the maximal numbers of $f_3(d,n)$ vertices, they do not achieve the maximal number of vertices in a 2-dimensional shadow [**AZ**].

CHALLENGE 7. Determine $f_2(d,n)$: what is the maximal number of vertices of a monotone path on a d-dimensional polytope with n facets?

In this area of research we have a wonderful mix of optimization (LP: simplex algorithm), geometry (polytopes) and combinatorics (graphs, enumeration) that poses enough challenges for the future ...

References

[AZ] N. Amenta and G. M. Ziegler, *Deformed products and maximal shadows*, this volume, pp. 57–90.

[BiSa] L. J. Billera and A. Sarangarajan, *All* 0-1 *polytopes are traveling salesman polytopes*, Combinatorica **16** (1996), 175–188.

[BEK] J. G. Bokowski, G. Ewald and P. Kleinschmidt, *On combinatorial and affine automorphisms of polytopes*, Israel J. Math. **47** (1984), 123–130.

[BS] J. Bokowski and B. Sturmfels, *Computational Synthetic Geometry*, Lecture Notes in Mathematics, **1355** (1989), Springer, Heidelberg.

[Chr] T. Christof, *SMAPO—Library of linear descriptions of small problem instances of polytopes in combinatorial optimization*, http://www.iwr.uni-heidelberg.de/iwr/comopt/soft/SMAPO/SMAPO.html

[Cox] H. S. M. Coxeter, *Regular Polytopes*, Second edition, Macmillan 1963; corrected reprint, Dover 1973.

[dL] J. A. De Loera, *Nonregular triangulations of products of simplices*, Discrete Comput. Geometry **15** (1996), 253–264.

[dLSU] J. A. De Loera, F. Santos and J. Urrutia, *The number of geometric bistellar neighbors of a triangulation*. Discrete Comput. Geom., to appear.

[DL] M. Deza and M. Laurent, *Geometry of Cuts and Metrics, Algorithms and Combinatorics*, Vol. 15, Springer-Verlag, Berlin Heidelberg 1997.

[ER] P. H. Edelman and V. Reiner, *The higher Stasheff-Tamari posets*, Mathematika **43** (1996), 127–154.

[Ewal] G. Ewald, *Combinatorial Convexity and Algebraic Geometry*, Graduate Texts in Mathematics **168**, Springer-Verlag, New York 1996.

[FKR] T. Fleiner, V. Kaibel and G. Rote, *Upper bounds for the number of facets of a* 0-1-*polytope*, in preparation; personal communication 1997.

[GHZ] B. Gärtner, M. Henk and G. M. Ziegler, *Randomized simplex algorithms on Klee-Minty cubes*, Preprint, TU Berlin, May 1996, 21 pages.

[Gale] D. Gale, *Neighboring vertices on a convex polyhedron*, in: "Linear Inequalities and Related Systems" (H. W. Kuhn and A. W. Tucker, eds.), Annals of Math. Studies **38**, Princeton University Press, Princeton 1956, pp. 255–263.

[GKZ] I. M. Gel'fand, M. M. Kapranov and A. V. Zelevinsky, *Discriminants of polynomials in several variables and triangulations of Newton polyhedra*, Leningrad Math. Journal **2** (1991), 449–505.

[GP] M. Grötschel and M. Padberg, *Polyhedral theory/Polyhedra computations*, in E. L. Lawler, J. K. Lenstra, A. H. G. Rinnoy Kan., D. B. Shmoys (eds.), "The Traveling Salesman Problem," Wiley 1988, 251–360.

[Gr1] B. Grünbaum, *Convex Polytopes*, Interscience, London 1967; revised edition (V. Klee and P. Kleinschmidt, eds.), Graduate Texts in Math., Springer-Verlag, in prep.

[Gr2] B. Grünbaum, *Arrangements of hyperplanes*, in Proc. Second Louisiana Conference on Combinatorics and Graph Theory, Baton Rouge 1971.

[Gun] H. Günzel, *On the universal partition theorem for* 4-polytopes, Preprint no. 64, RWTH Aachen, November 1995, 26 pages.

[K] U. Kortenkamp, *Every simplicial d-polytope with at most d+4 vertices is a quotient of a neighborly polytope*, Discrete Comput. Geom. **18** (1997), 455–462.

[K2] U. H. Kortenkamp, *Small* 0/1-*polytopes with many facets*, http://www.math.tu-berlin.de/~hund/01-Olympics.html

[KRSZ] U. Kortenkamp, J. Richter-Gebert, A. Sarangarajan and G. M. Ziegler, *Extremal properties of* 0/1-*polytopes*, Discrete & Comput. Geometry **17** (1996), in print.

[Lee] C. W. Lee, *Regular trinagulations of convex polytopes*, in Applied Geometry and Discrete Mathematics—The Victor Klee Festschrift (P. Gritzmann, B. Sturmfels, eds.) DIMACS Series in Discrete Mathematics and Theoretical Computer Science, Amer. Math. Soc. **4** (1991), 443–456.

[Lee2] C. W. Lee, *Subdivisions and triangulations of polytopes*, CRC Handbook of Discrete and Computational Geometry (J. E. Goodman, J. O'Rourke, eds.), CRC Press, Boca Raton (Florida) 1997, pp. 271–290.

[Nov] I. Novik, *Upper bound theorems for simplicial manifolds*, Preprint 1996, 33 pages.

[OS] S. Onn and B. Sturmfels, *A quantitative Steinitz' theorem*, Beiträge zur Algebra und Geometrie/Contributions to Algebra and Geometry **35** (1994), 125-129.

[Ra] J. Rambau, *Triangulations of cyclic polytopes and higher Bruhat orders*, Mathematika, to appear.

[Ri] J. Richter-Gebert, *Realization Spaces of Polytopes*, Lecture Notes in Mathematics **1643**, Springer-Verlag, Heidelberg 1996.

[RZ] J. Richter-Gebert and G. M. Ziegler, *Realization spaces of 4-polytopes are universal*, Bulletin of the American Mathematical Society **32** (1995), 403–412.

[Sh] I. Shemer, *Neighborly polytopes*, Israel J. Math. **43** (1982), 291–314.

[SR] E. Steinitz and H. Rademacher, *Vorlesungen über die Theorie der Polyeder*, Springer-Verlag, Berlin 1934; reprint, Springer-Verlag 1976.

[Stu] B. Sturmfels, *Some applications of affine Gale diagrams to polytopes with few vertices*, SIAM J. Discrete Math. **1** (1988), 121–133.

[VS] A. M. Vershik and P. V. Sporyshev, *Asymptotic behavior of the number of faces of random polyhedra and the neighborliness problem*, Selecta Math. Sov. **11** (1992), 181–201.

[WZ] Th. Wanner and G. M. Ziegler, *Supersolvable and modularly complemented matroid extensions*, European J. Combinatorics **12** (1991), 341–360.

[Z] G. M. Ziegler, *Lectures on Polytopes*, Graduate Texts in Mathematics **152**, Springer-Verlag, New York 1995; Revised printing 1997.
Updates, corrections, and more, `http://www.math.tu-berlin.de/~ziegler`

Dept. Mathematics 6-1, Technische Universität Berlin, 10623 Berlin, Germany
E-mail address: `ziegler@math.tu-berlin.de`

Contemporary Mathematics
Volume **223**, 1999

APPENDIX

Application Challenges to Computational Geometry

CG IMPACT TASK FORCE REPORT

ABSTRACT. With rapid advances in computer hardware and visualization systems, geometric computing is creeping into virtually every corner of science and engineering, from design and manufacturing to astrophysics to molecular biology to fluid dynamics. This report assesses the opportunities and challenges this presents for the field of computational geometry in the years ahead. Can CG meet the algorithmic needs of practitioners? Should it look to applied areas for new sources of problems? Can CG live up to its potential and become a key player in the vast and diverse world of geometric computing? These are some of the questions addressed in this document. It was prepared by a group of computer scientists, engineers, and mathematicians with extensive experience in geometric computing. This report is intended as a wake-up call rather than an agenda setter. It is hoped it will engage a community-wide discussion on the future of computational geometry.

1. Preamble

The fraction of computing falling under the loosely defined rubric of "geometric computation" has been on the rise and is likely to become dominant in the next decade. Computer graphics, manufacturing, scientific visualization, computer vision, astrophysics, molecular biology, and fluid mechanics are just a few in a crowd of avid users of geometric computing. Where does computational geometry fit into all this?

Twenty-odd years ago, the nascent field of computational geometry set sail on a mission to build general tools—analytical and computational—to satisfy the

The Computational Geometry Impact Task Force: Nina Amenta (Xerox PARC), Tetsuo Asano (Osaka Electro-Comm. U.), Gill Barequet (Tel Aviv U.), Marshall Bern (Xerox PARC), Jean-Daniel Boissonnat (INRIA), John Canny (U.C. Berkeley), Bernard Chazelle (**Chair**, Princeton U.), Ken Clarkson (AT&T Bell Laboratories), David Dobkin (Princeton U.), Bruce Donald (Cornell U.), Scot Drysdale (Dartmouth U.), Herbert Edelsbrunner (U. Illinois at Urbana-Champaign), David Eppstein (U.C. Irvine), A. Robin Forrest (U. East Anglia), Steve Fortune (AT&T Bell Laboratories), Ken Goldberg (U.C. Berkeley), Michael Goodrich (Johns Hopkins U.), Leonidas J. Guibas (Stanford U.), Pat Hanrahan (Stanford U.), Chris M. Hoffmann (Purdue U.), Dan Huttenlocher (Cornell U.), Hiroshi Imai (U. Tokyo), David Kirkpatrick (UBC), D.T. Lee (Northwestern U.), Kurt Mehlhorn (Max Planck Inst.), Victor Milenkovic (U. Miami), Joe Mitchell (SUNY at Stony Brook), Mark Overmars (U. Utrecht), Richard Pollack (Courant Institute, NYU), Raimund Seidel (U. Saarbrücken), Micha Sharir (Tel Aviv U. and NYU), Jack Snoeyink (UBC), Godfried Toussaint (McGill U.), Seth Teller (MIT), Herb Voelcker (Cornell), Emo Welzl (ETH Zürich), and Chee Yap (Courant Institute, NYU).

© 1999 American Mathematical Society

algorithmic needs of geometric computing [**Ed, GO, Mehl, Mul, OR, PS, SU**]. The intention was to create a body of knowledge to which computer programmers could turn for help when wrestling with geometric problems. The vision was that of a two-way pipeline: applications areas feeding CG with important practical problems, and CG in turn providing answers in the way of algorithmic tools and mathematically sound analyses.

What is the state of the "pipeline" today? Before answering this question, one may want to step back a little. Any new field needs to build itself a home and a home needs foundations. On that score one may strike a celebratory note. Indeed, computational geometry has met with considerable success in resolving the asymptotic complexity of basic geometric problems (e.g., convex hulls, Voronoi diagrams, triangulations, low-dimensional optimization) [**Chaz**]. Most of the classical problems posed in the first decade of the field have been solved. Many of them entailed building entirely new sub-areas in the field of algorithm design. For example, the twin use of randomization and derandomization has been the main force behind some truly remarkable success stories (e.g., optimal algorithms for convex hull and Voronoi diagrams, linear-time solutions of LP-type problems). Multidimensional searching, triangulation, geometric sampling, the theory of Davenport-Schinzel sequences, and zone theorems are some of the crowning intellectual achievements in the field of CG.

Computational geometry has been remarkably dynamic, productive and creative. In fact, in the area of algorithms and data structures, CG is arguably where most of the action has been in the last decade. Building theoretical foundations was undoubtedly the proper course to chart at the outset: CG can look back to its accomplishments with legitimate pride.

In the midst of its success, however, the field is standing at a crossroads. There are two options: CG can use its successes as justification for keeping the pursuit of theoretical investigations as the centerpiece of its agenda. Or it can move towards building an effective pipeline with geometric computing. In the first case, CG might simply fall out of the economic loop altogether: it might even shrink to the status of a recreational activity. In the second case, CG might grow to become as indispensable to geometric computing as, say, civil engineering is to bridge-building. The choice is for the CG community to make.

Why such alarm? The sobering reality is that, notwithstanding a few exceptions, the pursuit of practical solutions that specifically address the users' needs has lagged behind. At the same time, the computational geometry community has been largely unsuccessful in reaching out and communicating its more practical discoveries to potential users. Technology transfer has been slow and sporadic. To put it bluntly, the much-vaunted pipeline looks more like a pipe dream.

It need not be so. An effective pipeline is probably a feasible project: one certainly worth undertaking given the stakes. Among the problems to overcome first are the "obvious" ones: CG'ers must start thinking not only in terms of asymptotic complexity but also in terms of code robustness, precision, CPU times, standards, benchmarks, distribution, etc. Perhaps less obvious, though just as important, are communication problems of a more cultural nature: For example, why is it that in the graphics and computer vision literatures, rotating an object is considered (rightly so) a linear algebra problem, but computing a visibility graph or matching a geometric pattern is rarely considered a computational geometry problem? There are many computational-geometric questions in vision, graphics and robotics that are not recognized within those fields to belong in the province of CG'ers.[1] For this CG deserves most of the blame. It has failed to address the problems that people in practice want to see solved. In the absence of a pipeline, CG can never hope to be the place practitioners turn to for help. The sobering fact is that, realistically, the pipeline is mostly for CG'ers to build.

Does geometric computing need CG? It is too easy and tempting for CG'ers to convince themselves that it is the case. We try not to succumb to this temptation here. We simply state our belief that computational geometry is the *only* arena dedicated solely to algorithmic pursuits in geometric computing: discrete geometers want to understand countable geometric structures; combinatorial geometers want to count or approximate or enumerate them; practitioners want to produce working code for solving specific geometric problems. Only computational geometers make the algorithmic understanding of geometric problems their central preoccupation. We believe such pursuits can be the key to breaking current and future computational bottlenecks in many areas of engineering. This document gives abundant evidence to support this view.

If so, then why has CG broken so few "computational bottlenecks" that practitioners care about? Perhaps CG has been reluctant to cast itself in the demeaning role of "algorithm caterer" at the service of practitioners. This narrow view of CG is both false and dangerous. CG's link to the real world is essential to its intellectual vitality. The current feebleness of that link has been harmful to the field as a whole. Not only has it squelched the impact of CG in the engineering community, but research opportunities have been missed, scientific challenges have been overlooked. The intellectual landscape of CG has much to gain from close interaction with the real-world practice of geometric computing.

It is not our intent to pit theory and practice against each other, and decree the relative worthiness of each. We also want to make it absolutely clear that we are firm believers in the crucial importance of theoretical work. If anything we believe that the existence of an effective pipeline would further motivate all sorts of new theoretical investigations which the current insularity of the field has hindered. Our view is that a mature field of computational geometry should be home to a diversity of interests, some theoretical and some practical, most of them feeding on and contributing to the others.

It is particularly important to create a platform linking theoretical research at one end to production-quality, usable software at the other end. Such a platform exists in combinatorial optimization, where some of the most impressive theoretical advances (such as interior-point methods) have had tremendous impact in practice. CG is nowhere near this happy state of affairs.

To get there will require a concerted effort. The prevailing winds are favorable, however. A combination of factors, from the intellectual maturing of CG to more prosaic forces such as the job market and funding agencies, are causing a subtle but profound reorientation of the field. Instead of letting these factors alone dictate

[1]Defined loosely as the community gathered around such conferences as ACM SOCG, CCCG, and journals such as D&CG, IJCG&A, CGT&A, or more broadly, the geometric components of Algorithmica, J. Algorithms, SIAM J. Computing, JACM, etc. We recognize that many people not attached to that community do bona-fide computational-geometric work nevertheless. The focus of this report, however, is on the community that makes the design and analysis of geometric algorithms its primary occupation.

this reorientation, this document attempts to make the CG community itself the main agent of reform.

This report includes a set of specific recommendations (§**2**), followed by a sample of problem areas where computational-geometric techniques can have a major impact:

- §**3:** Computer Graphics and Imaging
- §**4:** Shape Reconstruction
- §**5:** Computer Vision
- §**6:** Geographical Information Systems
- §**7:** Mesh Generation
- §**8:** Robotics
- §**9:** Manufacturing and Product Design
- §**10:** Robustness
- §**11:** Molecular Biology
- §**12:** Astrophysics

This list of problem areas is not comprehensive by any means. It is a compendium of short case-studies, which, it is hoped, will stimulate research and in a modest way help to redirect computational geometry towards more practical ends. To illustrate what might be achieved we also include a few concrete success stories, where cutting-edge research in computational geometry has had a pivotal role in the production of geometric software.

In Section §13 we list some important resources for computational geometers (journals, conferences, Web sites, mailing lists, software).

We wish to close this section on a note of optimism. Computational geometers are blessed with a field whose scope dwarfs most other areas of computer science. Cryptography, graph algorithms, optimization, computational biology, etc., are all critically important, but they are relatively well-focused areas. By contrast, computational geometry spreads its wings over the entire computing spectrum. This breadth is an asset. But it is also a challenge. Indeed, the field is so vast that it is fragmented and has trouble recognizing itself under one roof. Computational geometry is diverse and so is its community. Within this diversity, however, we must strive to build an identity and unity of purpose. This is the main objective of our effort.

This is not meant to imply that all worthy computational-geometric research is to be done under the aegis of the CG community. Researchers in graphics, modeling, manufacturing, biology, etc, have their own traditions of fine geometric work: there is no need for this to change. What must be remedied is the lack of communication between practitioners and theoreticians of geometric computing.

It is imperative to make CG research more responsive to the needs of users. But, again let us restate our belief that not all computational geometers need to work in directly applicable research. Long-term structural explorations are also needed. Our point is not to emphasize practice to the detriment of theory. On the contrary, our chief goal is to broaden—not simply to shift—the band which CG occupies in the spectrum of computer science: Instead of a collection of small houses scattered across town and oblivious to one another, we envision CG as a large edifice with a theoretical wing and a practical wing and many aisles (the "pipeline") connecting them. Our purpose is not to bring down one of the wings, but to consolidate the edifice.

2. Recommendations

It will take more than a change of heart to reform computational geometry. Structural changes are necessary. Happily some of them are already underway. We identify four broad categories: (1) Production and distribution of usable (and useful) geometric codes; (2) Joint forums between CG and applied areas; (3) Experimental research; (4) Reform of the reward structure in CG.

REMARK: Educational matters were left out from our discussion. Teaching and research in CG at the undergraduate and graduate levels are critical issues that must be addressed. But the great differences among national education systems precluded a discussion that would have been of much meaning to the international CG community at large.

1. PRODUCTION AND DISSEMINATION OF GEOMETRIC CODE: Journals are the time-honored repository of scientific knowledge. But what about software? Journals are woefully inadequate. An on-line library of geometric code available through the Internet would be a useful starting point. Many questions need to be ironed out. Should the library conform to rigid formats and should contributions be refereed (ie, pass a number of standardized tests) before being archived? Or should it follow the freewheeling, open-door policy of most archives on the Web? What programming languages should be used? What documentation and maintenance service should be expected? Should data type conventions be enforced?

   Applied areas such as graphics and geometric modeling have produced vast amounts of geometric code over the years, some of it publicly available. Sometimes, differences in programming languages and standards might make integration difficult, but this need not be always the case. At the very least, how to tap into such resources effectively should be addressed. Our sense is that many of these issues cannot be settled once and for all before trying out various alternatives on a smaller scale. There are models to learn from in optimization, computer algebra, and numerical analysis. But it is unclear that any of them can be replicated verbatim to fit the needs of CG. For example, the uniquely complex data typing in CG places formidable hurdles in the way to correctness, robustness and portability. Only preliminary efforts have been attempted in CG so far, and to plan *the* perfect system in a vacuum is perhaps utopian. People should feel encouraged to try out their ideas and time will tell what works and what does not.

2. INTERDISCIPLINARY FORUMS: CG'ers should attend conferences in applied areas and practitioners should be invited to address the main CG conferences. This is already happening. Clearly, this is not enough. To provide fresh grist for CG's mill, means must be found so that CG'ers are given opportunities to hear about computational-geometric bottlenecks in applications areas. For example, one could organize joint conferences between CG and specialized topics. Special issues of non-CG journals could be devoted to such meetings. At the same time, the potential users of CG'ers' output

must be kept informed of new developments. A series of "CG gems" books patterned after the computer graphics gems might be useful. Or at the least the graphics gems books, which already include significant amounts of geometry, should be given greater attention (and perhaps receive more contributions) by CG'ers. The current CG newsletters and mailing lists might spawn special wide-distribution issues (announcing newly released software or recent developments of interest to practitioners).

3. EXPERIMENTATION: To feed the CG pipeline with usable results, experimental work must be integrated within mainstream CG research. Actually, it is one of the great assets of CG that it lends itself so naturally to experimental work. Regrettably, that asset has been grossly neglected. The standard programming cycle (design, code, debug, test and benchmark) is painfully slow, so the number of iterations tends to be small. The field of optimization has shown that practical algorithmic innovations tend to require repeated iterations through the programming cycle. Because this does not happen often enough in computational geometry, algorithmic innovations tend to be mostly of a theoretical nature. The unusual slowness of the implementation cycle is due to the complexity of geometric data types and the lack of adequate software tools (code libraries, visual debuggers, dataset catalogs, etc.).

   Quality experimental research (as practiced in, say, biology or physics) must satisfy highly demanding criteria. Any geometric algorithm that is claimed to be the "method of choice" should be not only implemented and tested, but benchmarked against its competitors. Claims of strengths should be backed up by credible evidence and weaknesses should be identified. None of this is possible as long as every piece of code must be written from scratch, as long as every test input must be produced by hand, and as long as every debugging, visualizing, and measuring tool must be hand-crafted. There is a need for a large-scale effort in building software tools for computational-geometric experimentation. Standards should be set by which to judge the quality of experimental work.

   To allow for benchmarking, representative data sets should be collected and archived. Input data should enable effective robustness and efficiency testing. Raw data should be included as well as highly structured or data-typed data.

   One might ask: experimental computational geometry in applied areas has been alive and well for years. Why do we need to create anew something which already exists? The answer is that in many cases experimental work has been so closely tied to applications that more general pursuits might have been overlooked. An analogy would be that although the oil, food, and pharmaceutical industries each pursue their own brand of experimental chemical engineering tailored to their needs, few chemical engineers would deny the value of unfettered experimentation. The same is true in computational geometry. Unfettered geometric experimentation will complement (not supplant) ongoing experimental work in applied areas.

4. REWARD STRUCTURE. Conferences, workshops and journals should be receptive (as some already are) to experimental work and software building.

One possible suggestion is to run an annual workshop devoted solely to experimentation in CG, featuring geometric results as well as software tools. Researchers in relevant applied fields with a tradition of experimentation (like drug design) could be invited to share their experience with CG. For an experimental culture to take hold, it is essential that quality experimental work should be rewarded through the standard channels (journals, conferences, hiring, promotion, grant awards, etc.) For this to happen, experimental research must be judged and evaluated according to recognized standards. Of course, this is a bit of a chicken-and-egg question, and a certain amount of bootstrapping might be necessary. It probably means that quality standards should be made flexible enough at first to reflect the relative immaturity of CG experimentation. Within a few years, however, firm standards should be put into place.

Together with an experimental culture, a software systems culture must be encouraged to grow within CG. Building novel geometric software that transforms the practice of geometric computing should be considered on par with proving a theorem that changes the mathematical landscape of the field. The reward structure must be adjusted, and CG must learn how to judge non-theoretical research. This is not to say that one should systematically reward any attempt at writing code. On the contrary, yardsticks for distinguishing bad from good code should be introduced, and the standards should be just as rigorous as they are for evaluating mathematical work. Promotion and hiring decisions should reflect these cultural changes. Again, none of these changes can be decreed. A new way of thinking must first take hold.

Finally, CG needs not only to open up to experimental and software-building work but also to rethink its approach towards theoretical research. Several fundamental theoretical questions remain open in geometric optimization, combinatorial geometry, geometric primitives, geometric searching, etc. **[Chaz]**. But the list of open problems whose centrality is so compelling as to draw a consensus—even within the CG community—is not nearly as long as one might think. We believe that many of the more interesting theoretical questions have not been formulated yet. These questions will surface as firmer bridges between CG and applied areas are created. For example, fundamental questions in computational topology might arise from geometric work in biology or astrophysics. Many of the classical computational-geometric concepts, such as convex hulls and Voronoi diagrams, arose from exposure to the natural sciences. Because CG is not a "foundational" science but a part of applied mathematics, it must draw its main inspiration from the physical world that it tries to model.

While CG must look outside for new frontiers to conquer, it must also become more critical of its current theoretical research. Problems whose only merit is to be open should probably be left as such. There are so many more open problems than what CG can ever hope to solve that it should focus on problems of identifiable importance, be they theoretical or practical.

3. Computer graphics and imaging

Rendering and image synthesis are two areas replete with geometric problems, some of which create significant computational bottlenecks. We discuss some of the problems encountered in rendering and we briefly report on a success story in image synthesis, where computational geometry has had a direct successful impact.

Radiosity. Graphics can be divided into two main sub-areas: *Modeling* concerns itself with the description of surfaces and their properties, light, and the media in which these are embedded, while *simulation* applies these computational models to predict behavior. One of the most important types of simulation is *rendering*, in which a description of scene and light sources is processed to generate synthetic imagery corresponding to one or a sequence of specified viewpoints. Rendering affords a rich source of computational-geometric problems. Typically, rendering a scene involves a *visibility* operation, in which the surfaces and/or surface fragments visible to the synthetic eyepoint are identified, followed by a *shading* operation, in which color values, derived from radiometric quantities, are computed at sample points on each visible surface. Computer graphics therefore combines the notion of combinatorial structures with some notion of each element's weight or importance [**DT**].

Rendering is one of the most computationally intensive tasks in computer graphics. For example, some state-of-the-art rendering techniques use radiosity methods to compute global illumination distributions in the scene [**CW**]. Such methods recursively divide the surfaces in the environment into smaller patches, then reduce the rendering equation to a linear system based on the pairwise interaction between these patches. At the inner loop of these methods is the computation of *form factors*. Informally, a form factor is defined for a pair of surface patches as the average solid angle spanned by one patch and visible from points of the other through the obstacles in the scene. Computing visibility in this manner is a well-studied problem in computational geometry, but one for which efficient practical solutions have yet to be found. Fast, practical methods for computing form factors could have a significant impact in rendering. Efficient approximation methods should also be sought [**Pell**]. Current methods use quadtrees or k-D trees to divide each patch into a hierarchy of patches at different resolutions. Interactions between surfaces occur at different resolutions determined by the ratio of their sizes to the distance separating them, and by some estimate of the energy flowing between them.

For the sake of concreteness, here are some specific problems. Note that in any real graphics application, a successful algorithm should be output-sensitive and incremental (as, for example, fly-throughs in complex building environments [**ARB, FST, FS, T**]).

Given a scene of n convex polygons (or balls) in general position:

1. Determine the scene visible from a synthetic eyepoint. If z-buffer hardware is available, it might be advantageous to identify a superset of the surfaces visible from the eyepoint.
2. Given a frame rendering cost model of the form $c_1 V_k + c_2 A_k$, where c_1, c_2 are technology-dependent constants, V_k is the number of vertices of the kth polygon, and A_k is its screen area, identify a superset of polygons visible from an eyepoint so as to minimize the total frame rendering cost. (This is

a fairly good model of the cost to render a polygon using standard computer graphics hardware, eg, an SGI [**FS**].)

3. Given a surface fragment with non-zero area, what (portions of) other surfaces are visible from this fragment? (This is the "area-to-area" visibility problem, which is much harder than the standard "point" visibility problem.)
4. Identify all pairs of intervisible polygons, or a tight superset. For each intervisible pair A, B, identify the "blockers" for the pair, ie, the set of polygons that block some stabline from A to B. (These problems arise in global illumination computations, which attempt to find a radiation distribution at equilibrium among the scene surfaces.)

Note that some of these problems call for hybrid solutions that mix object-space criteria (such as minimizing the combinatorial complexity of the rendered objects) image-space considerations, (such as expending more computation processing objects that occupy large screen areas), or other weighting factors (such as ignoring energy sources that have no discernible effect on the computed solution). It would be highly desirable to have algorithms that are parametrized by "knobs" that could be adjusted manually (or automatically) to reflect the particular technology in use (like the constants c_1, c_2 above).

Existing methods are very slow because they seek to produce exact solutions. Corresponding lower bounds indicate that dramatic speed-ups computing exact solutions are unlikely. Limited error tolerance is often acceptable in practice, however, and efficient approximation methods should be sought. In practice, current techniques are often based on quadtrees, octrees, or BSP trees, all elementary data structures that have been around for decades. There has been considerable progress in the theory of space partitioning recently. Can some of the new ideas be used or extended to provide simple, fast algorithms for visibility computation?

A success story in image synthesis. Thomson Training and Simulation (a subsidiary of Thomson) has developed a new generation of machines for image synthesis. The approach followed departs drastically from its predecessors. Instead of relying on massive, dedicated hardware, the new machines are mostly software-driven and use standard processors (for example, the i860). This change allowed the flexibility to provide new functionalities, such as inserting moving objects or supporting different types of animation. Expectedly, this return to software dependency entailed redesigning and reimplementing a new generation of data structures and algorithms, many of which fell squarely within computational geometry.

A new "polygon engine" was built to allow the user to view collections of thousands of nonconvex polygons with holes at interactive rates. In addition, a fully dynamic algorithm for constructing Delaunay triangulations was developed, which runs in real-time on the new machine. Many of the underlying ideas were borrowed from the recent work on randomized algorithms in computational geometry [**Clark, Mul**]. In particular, the algorithm uses a hierarchical representation, the so-called Delaunay tree [**BT**], to construct a terrain in real-time and automatically adapt the database to the specifications of the graphics device as well as to the distance of the observer from the terrain.

4. Shape reconstruction

In many application domains, it is necessary to reconstruct 3D models of objects from 2D cross-sections: medical imaging, microscopy, geology, and aerospace manufacturing are all heavy users of shape reconstruction [**MK, S**]. Two main approaches have been developed. The first is inspired by image processing techniques and treats the data as a 3D image obtained by piling up the images of the cross-sections. This approach is volume- or voxel-based; the interpolation is usually done in the vicinity of each voxel, or based on a pillar of voxels of all slices with the same (x, y) coordinates [**LC**].

The second approach is more boundary oriented; it attempts to construct a polyhedral model of the object that interpolates the boundaries of the cross-sections. It considers slice entities (contours) and, by considering each pair of successive slices in turn, it then concatenates all the layer interpolations. Hybrid methods have also been tried: for example, one might fit a surface to an unstructured cloud of points in 3D [**BF**].

The first approach has been dominant in the past decade, despite the severe limitations caused by its storage requirements. At UCLA Medical Center, a human skull was reconstructed in this fashion: it involved 800,000 triangles, which was small enough to allow visualization on a computer screen but too large to be manufactured. Efficient geometric solutions began to emerge on the practical scene only in the late eighties.

With few exceptions, all such methods make use of recent data structuring techniques developed by computational geometers. For example, a Voronoi-diagram based method was developed at INRIA (France) [**Bo, BB**]. It involves computing the Delaunay triangulation of pairs of successive slices, then projecting each pair element against the other, thus tetrahedralizing the space between consecutive slices.

The code was commercialized by two companies. One (NOESIS) produces general-purpose software for medical imaging and industrial vision; the other (CRIL) maintains a large library for processing 3D medical images (into which the reconstruction algorithm was included). A non-commercial version can be accessed via `ftp` and has been used at over 300 sites. The program has been used for research on neurons: the cover page of the Journal of Comparative Neurology (April 1994) shows a result obtained with the commercial version of the software. Experimental results on reconstruction of craniofacial cartilage are reported in the Anatomical Record (1991). Other applications of the code are mentioned below:

- Clinical use in radiation therapy planning software (DKFZ German center of cancer research).
- Reconstruction of a shark kidney (Max Planck Institut für Molekulare Biologie).
- Finite-element method mesh for a dog heart (Purdue University).
- Heart movement model (Yale University).
- Brain from MRI (University of Colorado).
- Body surface potential mapping (Dalhousie University).
- Volume calculation of surfaces undergoing wear (Utah Supercomputing Institute).
- Bone reconstruction (Center for Medical Robotics & Computer Assisted Surgery, Carnegie Mellon University).
- Rapid prototyping from medical data (Higher Technical Institute, Cyprus).

- Mesh generation for finite-element method (University of Waikato, New Zealand).
- Tessellations of brain surfaces (Washington University).
- Reconstruction of MRI scans of arteries (University of Waterloo).

An alternative approach was recently pursued in Israel [**BS**] (see this paper for an extensive review of the relevant literature). This technique is based on the idea of *geometric hashing*, which was developed for object recognition problems in computer vision. This software package has been tested intensively on a variety of complex solid reconstruction problems, in medical imaging and in the reconstruction of terrains from topographic elevation contour data. It was shown to be robust, applicable in fairly complex situations, and to produce 'intuitively correct' solutions.

Here are some specific computational geometry problems related to reconstruction from parallel cross sections:

1. Given two parallel slices, when is it guaranteed that there exists a non-self-intersecting polyhedral surface that interpolates between them? How does adding Steiner points (on the edges of the polygons) help?
2. Define a geometric measure that characterizes the "goodness" of the interpolation.
3. Extend any of the existing geometric (contour-based) methods to consider more than two successive slices at each step.

Of course, shape reconstruction does not necessarily proceed from parallel cross sections. Many problems related to geometric probing, object recognition, and pattern matching use various different models of shape, and each presents its own set of problems.

Finally, it is worth reiterating that medical applications motivate a wealth of interesting geometric reconstruction and analysis problems. For example, sometimes an organ needs to be reconstructed so that its volume can be estimated [**CCLB**]. Rapid prototyping of medical models is an area of growth with an abundance of challenging computational-geometric problems [**KSAVG**]. Advances on these problems might have a significant impact in fields such as computer-aided surgery.

5. Computer vision

Most problems in computer vision are inherently computational-geometric. Why they have not been tackled by computational geometers is a mystery. Perhaps the answer lies in the fact that the appropriate mathematical formulations of these problems are elusive. In computer vision, to find the proper level of abstraction in formalizing questions is often an integral part of the research effort. At the same time, it is one of those rare fields of computing (like speech recognition) where the computer always takes a back seat to the human brain: put simply, humans are better at vision than computers. A direct benefit is that humans are good bench-markers and can effectively judge the quality of a program from its output. Two areas which have made considerable use of geometric algorithms are model-based recognition (or pattern matching) and the recovery of three-dimensional structure from two-dimensional images (in particular, stereopsis and structure from motion).

Model-based recognition. In model-based recognition problems, a model of an object undergoes some geometric transformation that maps the model into a sensor coordinate system (say, an image plane or a cylindrical coordinate system from a 3D scanner). The development of efficient algorithms for identifying such transformations is central to many model-based recognition systems. In reconstruction problems such as stereopsis and structure from motion, the geometry of multiple projections of a scene provides constraints that enable efficient algorithms.

There are several approaches to the problem which explicitly rely on results from computational geometry. For instance, in certain approaches to recognition, an object is represented implicitly by a set of two-dimensional projections. When the objects are polyhedra it is useful to be able to bound the number of combinatorially distinct projections (those with different faces, edges or vertices visible). The set of combinatorially distinct views is referred to as the *aspect graph*. Combinatorially precise descriptions of the aspect graph and algorithms for computing it were developed using techniques from computational geometry [**GCS**].

There are a number of other approaches to model-based recognition which employ non-trivial geometric algorithms, and which often draw explicitly on results from computational geometry. The affine hashing method [**LW**] uses a redundant representation of a set of points in order to locate that point set under an affine transformation, in the presence of extraneous data points. The underlying idea is to use each triple of points in the model as an affine basis, and rewrite the other model points in terms of each basis. In order to recognize an object, triples of image points are selected, and for each triple all remaining image points are expressed in terms of the basis. When a correct basis is found, this will result in affine invariant coordinates that are the same (up to sensing error) as one of the encodings of the model. In practice this algorithm has better running time than the hypothesize-and-test approach, which consists of considering each pair of model and image bases explicitly, as is done in the alignment method [**HU**]. However, the sensitivity of the affine hashing approach to sensor errors is difficult to characterize.

Several researchers have developed recognition methods that explicitly account for sensor errors. These methods make considerable use of results on arrangements from computational geometry. Most of these methods represent each point or line segment in an "image" set as a polygonal region (say, the Minkowski sum of the image set with a box). The matching problem is then to search for transformations bringing each point or line segment of the model into such a polygonal region in the image. These problems can be structured as sweeping arrangements, using algorithms from computational geometry [**B, Cass**].

A different approach to model-based matching problems involves the development of cost functions for measuring the difference between two sets of points and line segments under various transformations. Cost functions based on the Hausdorff distance have been investigated in both the computational geometry [**AST, ABB, HKS**] and computer vision [**HKR, Ru**] literatures. The applied methods developed in the vision community are provably good approximation schemes for solving the combinatorial problems that were originally investigated in the computational geometry community.

Image representation. Quantized images are commonly represented as sets of pixels encoding color/brightness information in matrix form. An alternative model is based on contour lines: A contour representation allows for easy retrieval of the full image in bitmap form [**AK**]. It has been used primarily for data compression of an image. The idea is to encode, for each level l, the boundaries of connected regions of pixels at levels greater than or equal to l. It is easy to reconstruct an original image from those boundaries. There exist output-sensitive algorithms for computing contour representations.

One problem is how to store such representations in a compact manner. In practice one seldom needs the entire contour representation. Typical use is in the form of a query asking for the contours matching a given gray level. Current data structuring techniques should be called upon to provide efficient solutions. So far, they have not. Attempts to remedy this are underway.

Here is a typical problem encountered with this kind of representation. Suppose that we wish to erase wrinkles around the eyes of a person in a digitized picture given in contour representation. Because wrinkles might intersect contour lines, these might become disconnected after removal of the wrinkles. To reconnect them is not so easy. Dynamic programming might be a natural approach for this problem. In fact, it has been tried successfully in several experiments. Dynamic programming tends to be costly, however, and faster heuristics should be investigated.

Another problem ripe for a computational-geometric attack is *resolution enhancement*. Suppose we have an image scanner with 6-bit resolution for each of the three colors (RGB), and we wish to increase the resolution to, say, 8 bits. The naive approach, which consists of adding the gray levels of four pictures of the same object, has immediate flaws. Of the numerous solutions proposed in the computer vision literature, most suffer the drawback of causing blurring. As it turns out, the problem has a natural formulation in terms of weighted Voronoi diagrams, a well-studied construction in computational geometry.

Since such a Voronoi diagram is hard to compute especially in the presence of high degeneracy, a different approach might be preferable. An important observation here is that the objects are not continuous but discrete. This is the main difference with interpolation of contour lines in geographic information systems. Indeed, suppose that we want to improve the resolution on an intensity level k by a factor of M. This means partitioning the pixels at intensity level k into M different finer levels. Let S denote the set of all pixels with intensity level k, and B_k (resp. B_{k+1}) denote the set of all contour edges between pixels with intensity levels lower (resp. higher) than k. For each pixel in S we may compute the Euclidean distance to the nearest boundary edges in B_k and B_{k+1}. Using the ratio between these distances we can thus classify those pixels into M finer levels. Without getting into details, it is apparent that several variants of this heuristic can be designed. Efficient implementations and classifications of these heuristics would be very useful.

Image segmentation. A central problem, called *segmentation*, is to distinguish objects from background [**ACKT**]. For intensity images (ie, those represented by point-wise intensity levels) four popular approaches are: threshold techniques, edge-based methods, region-based techniques, and connectivity-preserving relaxation methods.

Threshold techniques, which make decisions based on local pixel information, are effective when the intensity levels of the objects fall squarely outside the range of levels in the background. Because spatial information is ignored, however, blurred region boundaries can create havoc.

Edge-based methods center around contour detection: their weakness in connecting together broken contour lines make them, too, prone to failure in the presence of blurring.

A region-based method usually proceeds as follows: the image is partitioned into connected regions by grouping neighboring pixels of similar intensity levels. Adjacent regions are then merged under some criterion involving perhaps homogeneity or sharpness of region boundaries. Overstringent criteria create fragmentation; lenient ones overlook blurred boundaries and overmerge. Hybrid techniques using a mix of the methods above are also popular.

A connectivity-preserving relaxation-based segmentation method, usually referred to as the *active contour model*, was proposed recently. The main idea is to start with some initial boundary shape represented in the form of spline curves, and iteratively modify it by applying various shrink/expansion operations according to some energy function. Although the energy-minimizing model is not new, coupling it with the maintenance of an "elastic" contour model gives it an interesting new twist. As usual with such methods, getting trapped into a local minimum is a risk against which one must guard; this is no easy task.

In contrast to the heuristic nature of these approaches, computational geometry suggests a more algorithmic tack. One would first formalize a mathematical criterion for the "goodness" of a given segmentation. This would allow us to formulate the segmentation problem as an optimization problem under certain geometric constraints.

The objective function that one would seek to optimize is the *interclass variance* that is used in discriminant analysis. Experimentation seems to suggest that this approach might be very promising. Formally, the criterion is described as follows. Imagine that an image is to be partitioned into two connected sets S_0 and S_1. The average intensity levels of the sets are denoted by μ_0 and μ_1, respectively. Then, the objective is to minimize the potential function

$$V(S_0, S_1) = |S_0||S_1|(\mu_0 - \mu_1)^2.$$

Computational geometry provides several tools (for instance, hand probing, Monge matrix searching) that lead to efficient solutions for optimal segmentation, as long as one is only interested in separation of an object defined by monotone chains. Without such assumptions the problem is NP-hard, and efficient heuristics remain to be found.

Discriminant analysis has also been applied successfully to the problem of transforming an intensity image into a binary black&white picture. This line of research relates directly to what has long been an active line of research in computational geometry, point clustering: Partition a set of points in $\mathbf{R}^d$ into k clusters so that some inter-cluster criterion is minimized.

6. Geographical information systems

Geographic Information Systems (GIS) [**LT**] are increasingly being cited as a motivating application for computational geometry research. According to Unit 1 of the NGCGIA Core Curriculum [**GKP**], "a GIS can be seen as a system of hardware, software and procedures designed to support the capture, management, manipulation, analysis, modeling and display of spatially-referenced data for solving complex planning and management problems."

The key phrase is "spatially-referenced data." A database system that combines data sets of different types from different sources about different features/objects is likely to find that converting data sets into a common form dominates the processing efforts, when it is possible at all. Data that can be assigned a spatial location, however, can be combined by simply displaying data sets simultaneously on a map or computer screen. Add a mechanism to select which data sets to display, and you have a rudimentary GIS.

Many systems have been developed in natural resource management applications—either in connection with remote sensing or from computer-aided design tools—and some have been highly successful in making up-to-date information available for management decisions in a manner that was not possible using a large inventory of paper maps. GIS use in government and marketing applications continues to expand.

Of course, a rudimentary GIS leads to the complaint [**Iles**]: "A GIS is basically a tool that mines data and displays it. It doesn't clean it up, or maintain it, and seldom even looks to see if it's reasonable." Computational geometers see their potential contribution here: in providing the algorithms and data structures to support analysis of geometric data. A forthcoming special issue of *Algorithmica*, edited by J. Snoeyink, will focus on the interface between computational geometry and GIS.

6.1. Snapshots of the parallel history of GIS and CG. While computational geometers might like to see GIS as an instance of the practical implementation of computational geometry, this is not historically accurate. The GIS and CG communities have, for the most part, developed their understanding of geometric computation independently. In some key instances ("topological structure" for storing planar subdivisions, and "TINs"—triangulated terrains, interpolation properties of the Voronoi and Delaunay), development in GIS has preceded that in CG.

Computer cartography emerged in the 1960s. Many basic concepts (map layers, topological structure, TINs) can be traced back to work done in the The Land Inventory branch of the Canadian government or the Harvard Lab for Computer Graphics and Spatial Analysis. ESRI's ARC/INFO, which is the most popular commercial GIS system, got its start with Harvard Lab technology.

Interaction between CG and GIS has increased in recent years. CG papers at the biennial International Symposium on Spatial Data Handling have been generally well received. ESRI, Caliper Corp., and other companies have been hiring from the computational geometry community and taking in research results (including, for example, nearest-neighbor properties of Delaunay triangulations, and the use of Voronoi diagrams of line segments to assist in "buffering" (offsetting) operations). A common complaint has been that publication delays and the lack of textbooks in CG hampers the adoption of research results by other fields such as GIS.

6.2. Data model and data structure debates. Which data model is best for the spatial information stored in a GIS? This question leads to several recurring debates that consider various parameterizations of the spaces of possible data models and data structures. We list three examples: the long-running raster/vector debate is limited by GIS implementation technology, the tiled/seamless debate is concerned with the practical and conceptual views of a large data set, and the object/layer/field debate appeals to human cognition of space.

Raster/vector. Whether data should be stored and processed in raster or vector form has historically been a source of much debate. The debate is perhaps dying down slightly because most systems, at least from a user's perspective, have some provision for handling both. It is still true that not all operations are supported in both—typical systems compute buffer zones around vector features by converting to the raster domain and back.

Raster has many advantages including simplicity of concepts and algorithms, and hardware support. Remotely-sensed data, displayed output, and many digital elevation maps (DEMs) are in raster form. It is good for data that should be sampled evenly. Rasters are huge, do not scale well, and introduce aliasing effects.

Vector data is natural for linear features such as roads and boundaries, and mandatory for accuracy in instances of land survey. The speed of vector computation is most often criticized; geometers should rejoice in this as an opportunity to provide significant improvements. We should remember, however, that these improvements will need to be realizable for "typical data" and will need to provide some feature that raster processing does not provide. Paper descriptions of worst-case optimal algorithms do not constitute significant improvement in a community that wants implementations of solutions.

Tiled/seamless. In vector databases as well as raster databases, the quantity of data becomes overwhelming. Practical considerations dictate that the data must be partitioned and considered in smaller chunks. People who converted to GIS and computer cartography systems from paper maps were accustomed to dealing with map sheets, quadrants, and tiles. Other users, however, demand a seamless structure—maybe even one that allows a wide range of levels of detail. Debates rage about what data structure is best for the partitioning: uniform grid, quadtrees, R-trees, binary space partitions, BANG files.

Object/layer/field. In the last few years some members of the GIS community have argued that the raster/vector debate obscures more essential issues about how people think about and manipulate space. Suppose that you want to describe the agricultural activity in a certain region. There are several spatially-referenced variables that may be of interest: owner, crop grown, soil type, elevation, rainfall, etc.

One might speak of farms in the region as geometric objects that have owner attributes. Fields could be objects with a *crop grown* attribute, and could inherit an owner from the farms that contain them. This is an example where an object-oriented approach to the data is natural: There is clear separation of objects, their representation can be encapsulated, and they form a hierarchy in which representations and attributes can be inherited.

When one adds *soil type*, however, one may find that a field can contain several soil types or that a soil type may encompass several fields. Subdividing fields into objects that can be given a single soil type attribute may or may not be a good idea—it depends on the importance of the soil type variable. It seems more natural to start a new layer. This in turn creates problems where the same object may have different representations in different layers. A stream, for example, may be represented as a change of ownership and a change in soil type.

It might be natural to think of soil type, elevation, and rainfall as functions defined over the region—cropland as a "vector field." Defining data structures to support continuous data models is difficult because data is usually obtained by discrete sampling. Even so, the amount of data can be unwieldy and data can be expensive to collect. The traditional soil map quantizes the domain into subjective categories (clay, clay with gravel, sandy, very sandy, etc.) and creates a raster or partitions the region into cells that are assigned one of these soil types to obtain an object-oriented, vector-based map. An implementation of a vector field idea might represent the proportion of each possible component (clay, gravel, sand, etc.) as elements of a vector. One then needs algorithms to quantize this vector space to present understandable output.

Arguments in this debate usually conclude by recommending raster or vector implementation so that the object/layer/field debate, too, becomes limited by current technology.

6.3. Geometric problems in GIS. Two things should be kept in mind when looking at geometric problems that arise in GIS. First, even though computational geometers are likely to see themselves as solvers of geometric problems, the concepts of space, shape, and geometric computation that develop can be a more important contribution than the solution of any particular problem. Second, the "geometric" is only one aspect of a GIS. In a survey for cartographers, Goodchild [**Good2**] lists four ways in which to view GIS; geometers may want to keep these views in mind when weighing the importance of problems motivated by GIS:

1. Automated Mapping: facilitating the production of standard maps,
2. Map Analysis: providing measurement and overlay tools that are cheaper than traditional methods,
3. Inventory: giving geographic access capabilities to existing governmental and corporate databases,
4. Spatial analysis and spatial decision support: enabling new uses for old data by giving users query and analysis tools.

Data structure tuning. To a GIS salesman, one of the most important features of their product is how fast it can sift through the data and put the layers that the user has selected onto the screen. Because users can select data based on spatial location, attributes, and level of detail, multidimensional access structures (B-trees, k-D trees, quadtrees, R-trees) are important and tuning these is a difficult task: van Oosterom's book on reactive data structures [**vanO**] is an example of what can be done.

Existing legacy data limits the applicability of new research. One cannot expect people to adopt a data structure if it involves expensive conversion. An emerging opportunity, however, is the partitioning of GIS tasks between server and clients. For various reasons (data management, resource needs, etc) it is natural to put spatial data on a large server and have clients request the regions and attributes for the problem that they are interested in. With this architecture, what information should be sent by the server to help clients build geometric structures for local analysis? If the input is line segments, then sending a trapezoidation helps a number of algorithms, but maintaining a trapezoidation on the server may increase the data structure size and involve data conversion. Sorting by first coordinate may be a less-costly improvement.

Generalization. Since almost all GIS systems allow a user to zoom in on data, there is no intrinsic scale associated with the computer representation of a map. Nevertheless, there is an associated level of detail which determines a range of scales at which the data can be displayed without being too crowded or too sparse. *Generalization* is the process whose goal is to extend this range [**BM**]. Several geometric problems fall under the heading of generalization:

LINE SIMPLIFICATION: Approximate a polygonal line by one with less data (but hopefully the same information). This is has been the subject of much research in GIS, image processing, and computational geometry. Interesting problems remain for simplifying several lines (as in a contour map) or for demanding topological properties (e.g., forbidding self-intersection) from the result.

CLUSTERING: Separate geometric objects can be aggregated and represented symbolically as a polygon or as a point.

DISTORTION: Moving or changing the representation of geometric objects to enhance readability of the output. A railway that parallels a river may need to be displayed at lower resolution. A driver would prefer that a winding mountain road have some bends exaggerated on a map rather than being simplified to a line segment.

The more advanced generalization tasks are difficult to define a precise geometric optimization problems. Automating these tasks is often seen as an application for artificial intelligence techniques and "knowledge-based" heuristics. Computational geometry can offer not only efficient implementation of lower-level primitives, but also structures such as Voronoi diagrams and constrained triangulations that provide a more continuous model of space.

Digitizing. Many maps are constructed by a labor-intensive digitization process. One of the exciting applications of computational geometry in GIS is an idea of Chris Gold [**Gold2**]. Rather than force someone to carefully trace what may be a fuzzy boundary anyway (eg, the edge of a forest stand), use the digitizer to spray a bunch of points near the boundary, then use the Voronoi edges that are from bisectors of points in different regions to be the boundary. This speeds up digitization by a factor of 4, avoids problems with dangling or intersecting edges, and still gives reasonable accuracy because the Voronoi edges average between the points.

Building topology. What the GIS community calls "building topology" is to establish the ordering information for a planar subdivision—determining the edges (arcs) around faces (polygons) and vertices (nodes). This task is complicated by the fact that the input data may be huge, imprecise, and occasionally erroneous; it is usually seen as an expensive batch computation process. Gold [**Gold**] has further advocating Voronoi diagrams as way of maintaining topology as a map is constructed. Because of the increase in data and computational effort, this will work best with smaller maps. What is the best way to combine these ideas with tiling ideas to operate on large amounts of data?

Polygon overlay and update. The spatial data in GIS's from ARC/INFO [**ESRI**] down to the Digital Chart of the World [**FGB**] is organized into layers, each of which is a subdivision of the plane with explicitly stored neighbor relations. Custom maps are produced by selecting layers of interest and overlaying them to produce new maps. This "overlay" process is the primary tool for map analysis.

As a theoretical geometric problem, computing the regions defined by a set of segments or pixels is perhaps not so difficult. But as an engineering problem, tuning algorithms is a challenge. When one considers the errors and different levels of generalization in the input, the task becomes more interesting. If the same feature, say, a stream bed, appears in more than one map, an overlay will likely have many spurious polygons in the vicinity of the stream bed. It would then be desirable to remove one copy from the overlay. Even better would be to average the two maps in some principled manner to improve overall accuracy.

A special form of the general polygon matching is to match two maps and try to determine what has changed from one to the other. This is important for observing historical change, checking map completeness, rectifying common boundaries, or using a higher resolution map to improve a lower resolution map. Again, generalization and error make this an intriguing problem.

Label placement. Label placement is a source of many geometric optimization problems. Even when labels are to be placed in fixed positions relative to point features, these are typically NP-hard. Since maps do need labels, heuristics and efficient algorithms for identifying constraints and possible positions are needed.

Spatial accuracy. Current GIS data structures do not support error bounds. This is one of the major weaknesses of analysis carried out in a GIS—one does not know how to asses the quality of the result. One of the major challenges in GIS is a computationally efficient theory of spatial error. There has been much discussion on the topic. In 1966, Perkal [**Perk**], for example, proposed fattening boundaries by convolving them with a disk of radius ϵ; these are usually called buffer zones [**X**], and queries with buffer zones are an important type of overlay. See the survey [**Good3**]. Computation in floating point is one source of error. GIS data sets are riddled with degeneracies, so robust computation can be important.

Spatial analysis. Thus far, GIS's have offered only rudimentary spatial analysis capabilities [**Good**], such as polygon overlay and area computation. More interesting queries require more interesting data structures: e.g., Okabe et al. [**OBS**] list 35 GIS queries that can be answered using different types of Voronoi diagram.

Terrain analysis. Research on representation and computation of terrains has been an active area.

- TIN: Triangulated surfaces were suggested in GIS in 1978, and dubbed TINs or Triangulated Irregular Networks [**PFLM**]. Delaunay triangulations are often suggested (and used) as good TINs. Other (data-dependent) triangulations may sometimes be more appropriate.
- SIMPLIFICATION: Kumler's monograph [**Kum**] compares elevation accuracy of TINs to DEMs and finds that, byte for byte, Digital Elevation Models are better. The simplification methods he uses, however, are probably not the best possible. (Also, error measure that reflect curvature or symmetric volume difference should be considered.) Simplifications with some sort of hierarchical structure are especially useful. See Heckbert and Garland's survey [**Heck**] for many citations to geometry and graphics papers on surface simplification methods.
- DRAINAGE: For many natural resource management tasks, the watershed—or region that drains to the same point—is the natural management unit.

Much of the work on computing drainage networks is raster based; some geometers are looking into preprocessing TINs for drainage queries.

- VIEWSHEDS: Some forestry operations are now required to preserve the visual impact of the forest. Thus, "viewshed" algorithms become necessary. Location problems for microwave and cell phone relays also involve line-of-sight constraints imposed by the terrain.
- FLYBY: Fast graphics engines give the capability to fly-by data models to get a better feeling for the actual terrain (eg, systems such as PCI's EASY/PACE).

Adding dimensions 3 and 4. Adding the spatial and temporal dimensions to the flat maps in a GIS is a subject of current research [**Ra**]. None of the standard GIS's support functions of three variables. In such systems, geological cross sections or floors of buildings must be treated as flat, horizontal data sets which offer no geological or architectural interconnections.

Time creates interesting geometric problems. In an agricultural region, the land will be partitioned into different fields in different years, and land ownership or use changes. Some of the dynamic data structures in computational geometry may have a role to play in allowing efficient storage, query, update, and historical summary of changing land uses.

6.4. Resources in GIS. The most relevant journals are the *International Journal of Geographic Information Systems* (IJGIS) and *Cartography and Geographic Information Systems* (CGIS). Other journals include *American Cartographer, CVGIP: Graphical Models and Image Processing, Cartographica, Geographical Analysis, Journal of Remote Sensing.* Duane Marble maintains a ftp-able bibliography on GIS (in refer and mac, but not in bibtex format) on

`bastet.sbs.ohio-state.edu`

Of course, the Web is a good source of information and data. The NCGIA activities occur at the University of California, Santa Barbara,

`http://www.ncgia.ucsb.edu/,`

in Buffalo, NY `http://www.geog.buffalo.edu/`, and in Maine

`http://ncgia.umesve.maine.edu/.`

Maine has some technical reports on line, including a 1992 update of the research agenda, 92–7.ps. BC environment has the NCGIA Core Curriculum on its Web `http://www.env.gov.bc.ca/gis.`

Xerox Parc has a map program

`http://pubweb.parc.xerox.com/map.`

NAISMap is a GIS on the Web

`http://www-nais.ccm.emr.ca/naismap/naismap.html.`

GRASS (Geographic Resources Analysis Support System) is a public domain GIS developed by the US Army Corps of Engineers

`http://www.cecer.army.mil/grass/GRASS.main.html.`

GRASSlinks is a Web interface

`http://www.regis.berkeley.edu/grasslinks/index.html.`

ORES has an extensive list of data sites and other information. See for example,

`http://www.csn.net/~bthoen/ores/gis/index.html`

For GIS FAQ, see

`http://www.census.gov/geo/gis/faq-index.html`

Finally, for ESRI (ARC/Info), see: `http://www.esri.com/`.

7. Mesh generation

A mesh is a discretization of a geometric domain, e.g., the air around a wing, into small simple shapes called elements [**Ba, BE, HM, HL, Lö, SP**]. A structured mesh is usually a warped grid of boxes, while an unstructured mesh is typically a triangulation. Some advantages of structured meshes that hold generally over most applications, are simplicity, availability of code, and suitability for multigrid and finite difference methods. On the other hand, unstructured meshes conform to the domain more easily and allow element sizes to vary more dramatically. Structured meshes are currently more popular, but unstructured are catching up, especially in the more academically-inclined community.

Some mesh generation goals vary with the application. For example, long skinny elements, aligned with flow, can be quite useful in computational fluid dynamics. Moving features, such as shock fronts and vortices, require changing meshes.

Compared to GIS, mesh generation is still something of a cottage industry. Research and development groups in finite-element methods tend to write their mesh generators. There are a few public-domain codes (e.g., PLTMG, written by Randy Bank, and GEOMPACK, written by Barry Joe), and some commercially available code (e.g., ICEM CFD), but all large manufacturing companies use their own codes. There is no publicly available package that is adequate for generating 3D meshes for computational fluid dynamics. Aerospace engineering experts even go as far as admitting the lack of adequate code for generating 2D meshes for viscous flows. There lies an exciting window of opportunity for computational geometry.

Below are a list of open questions in mesh generation. Some are outstanding theoretical problems; others represent serious computational bottlenecks in practice.

- HEURISTICS FOR POINT-SET TRIANGULATIONS: Theoretical analysis of existing heuristics remains to be done. In practice, the "diagonal-flipping" algorithm works quite well for 2D Delaunay triangulations, but there are no theorems of the form: most initial triangulations are only $O(n)$ flips away from Delaunay. Rivara refinement (split a simplex by adding the midpoint of its longest edge, then recurse on neighbors) works well in 3D, but there is no theorem analogous to Rivara's theorem in the plane, showing that angles stay bounded away from zero.
- SURFACE REPRESENTATION: Investigate appropriate metrics to measure how well a mesh conforms with the surface boundary, and design heuristics for constructing good meshes. When approximating highly curved surfaces, it is important to ensure that edges of the surface triangulation are aligned with the directions of principal curvatures so as to avoid "scalloping effects." One possibility is to make the surface triangulation consist mostly of right-angle triangles, with the two right sides aligned with the directions of principal curvature. It seems natural to require that lengths of the right sides should be inversely proportional to the surface curvature. A theoretical understanding of this question, and more generally, of the problem of fitting a mesh on a complex geometry, would be very useful in computational fluid dynamics.

- VOLUME MESH GENERATION: As expected, small volume elements must be used in regions where rapid variations of the computed solution are expected. But the precise requirements vary from problem to problem. For example, for computations of the scattering of an electromagnetic wave, it is necessary to use a uniformly spaced mesh everywhere and ensure that the cell width is small compared with the wavelength of the incident radiation. For computations of fluid flow, particularly Navier-Stokes calculations at high Reynolds number, there will be extreme variations in cell size and/or aspect ratio.

  The Delaunay triangulation is one of the most popular means of generating 3D meshes of tetrahedra. An effective approach is to insert a mesh node into an existing triangulation, and repeat this procedure until a sufficiently fine mesh has been achieved. Since the quality of the final volume triangulation depends on the positions at which the points are placed, it is natural to investigate point placement strategies. The two-dimensional case is relatively well understood. For example, placing each new mesh point at the circumcenter of a triangle with maximum circumradius leads to a triangulation whose minimum angle is at least 30° and such that the ratio of maximum to minimum edge length is at most 2. Nothing similar is known in three dimensions. It is a major open problem whether point placement techniques can be used effectively to generate high quality 3D triangulations.

  When dealing with viscous flow computations at high Reynolds number, extremely fine meshes are required close to the boundary surface. It is customary to generate meshes so that the point spacing normal to the boundary surface is very close while allowing a much larger spacing tangential to the surface. This may lead to poorly-shaped elements near boundary junctures. New point placement strategies are needed to ensure high-quality meshes in such extreme situations.
- HEXAHEDRAL MESHES: There are several geometric benefits to be gained from using meshes of hexahedra (ie, affinely transformed cubes) such as regularity, angle distribution, anisotropy. Some commercially available systems support such meshes. Unfortunately, the computational geometry of hexahedral mesh generation is so little understood that existing systems suffer major drawbacks. Is there a good algorithm for partitioning a solid into hexahedra that meet face-to-face? The outside-in, or advancing-front, approach starts with an initial boundary mesh and fills out the volume with hexahedra. Unfortunately, many starting configurations are unfillable. Conversely, the inside-out approach fills volume and produces boundary meshes as a side effect. This has the disadvantage of leaving us no choice in the boundary meshing. Intermediate strategies are yet to be investigated.
- ADVANCING FRONT: Some tetrahedral mesh generators use the advancing-front approach as well. These generators typically place Steiner points in successive layers and then use the Delaunay triangulation as the mesh. The same difficulties arise: by meshing the boundary first this method commits to a level of detail too early in the process, and non-tetrahedralizable pockets can form when fronts collide.
- MESH GENERATION AND CAD SYSTEMS: Uniting mesh generation with CAD systems is much needed. The idea is to produce grids directly from

the CAD geometry and at the same time retain independence of grid characteristics. Not only would this solve a compatibility issue, but it should also make the mesh generation task easier. A mesh generator has to "parse" the domain, which is hard to accomplish unless the grid generator is intimately tied to the systems used to represent the geometry. Some systems (e.g., ICEM CFD) have taken steps in that direction but much remains to be done.

- MESH PARTITIONING: A mesh must often be broken up into pieces to allow for efficient parallelization or vectorization. In the latter case, for example, the number of elements in each group must be matched to the target architecture. Should processor architecture influence not just the partitioning of the mesh but also its generation?
- MULTIGRID FOR UNSTRUCTURED MESHES: Hierarchical triangulations, as in Kirkpatrick's point location algorithm, can be used to combine unstructured meshes and multigrid methods. Randomized analysis of the incremental algorithm shows that such hierarchies of small expected size form automatically for Delaunay triangulations. The hierarchy represents the history of the insertion process. The randomized result also hold in three and higher dimensions. Can similar results be obtained for triangulations other than Delaunay or the related regular triangulations?
- ROBUSTNESS: Robustness appears to be a serious issue in mesh generation. In general, the current state of robustness research in computational geometry leaves practitioners unsatisfied. In particular, many are wary of automatic degeneracy handling and prefer explicit treatment of it in their codes. (See Section 10.)
- AUTOMATIC REMESHING: A unified meshing system might be envisioned along the following lines: first, the domain is meshed, based on the geometry alone; then the differential equations are solved. Finally, remeshing occurs, based on an automatic error analysis. The last two steps can be repeated many times. The advantage of such an approach is that the final mesh is produced from an actual approximation of the solution and not an educated guess.

We close this section with a brief mention of alternative approaches to mesh generation. Meshes are used primarily to integrate partial differential equations over complicated domains. The base of numerical solution technologies is surprisingly narrow: finite-element methods, boundary-element methods, finite-difference methods, and monte-carlo methods. Are there alternatives to the methods listed above? Recent research suggests that some might lie in finding more clever representations of the underlying geometry which exploit the physics of the problem (in some cases, medial axes and Voronoi methods may come into play) and then mapping the physics onto the new representations of problem geometry. This results into a reformulation of the problem that is both geometric and numeric, as opposed to the purely numeric approaches currently in favor.

8. Robotics

Computational geometry has already had a noticeable impact in motion planning problems. Years ago the standard approach to motion planning followed one of two methodologies. The STRIPS planning formalism favored in the AI community was based on pre- and post-conditions in the propositional calculus. The lack of geometric grounding made it difficult to instantiate motion, grasping, or assembly actions and obtain guarantees of performance or correctness. In practice such approaches often showed limitations.

Techniques from control engineering were also attempted. The weakness there lay in their tendency to be local and thus to become trapped in local minima. Nevertheless, these methods have been popular in practice because of their simplicity, speed, and integration with the control layer. There has also been a flurry of interest in algorithms for non-holonomic motion planning, where motion constraints are not integrable. A textbook covering the theory of non-holonomic planning has recently appeared [**MLS**]. Path planning problems have been widely studied by computational geometers, with some spectacular theoretical successes. General motion planning problems were first shown to be solvable in polynomial time [**SS**], then with a practical exponent (equal to the number of degrees of freedom) [**C**], and then with a tightened exponent in many cases with the aid of Davenport-Schinzel sequences [**Sh**]. For the case of industrial robots, the degrees of freedom range from four to six. Multi-robot systems and robot hands are much higher. AGVs (autonomous guided vehicles) have 3 degrees of freedom with non-holonomic constraints. These concrete problems are too large for the methods above to solve at this time. But even though asymptotic methods are rarely used in practical systems, many effective planners have been built using concepts from CG. Concepts such as generalized Voronoi diagrams or skeletons, critical orientations, line and plane sweep, and random sampling have been used in heuristic planners that are fast and robust. Latombe's text [**L**] provides a comprehensive overview.

A basic representational problem for robotics and CG more generally is that the world contains many curved objects. Curved 3D configuration spaces arise even when objects are polygonal and motion is restricted to the plane. In graphics, given today's rendering technology, it is reasonable to approximate curved surfaces at varying resolution by planar facets. However, polygonal approximations can lead to errors in mechanical analysis, for example when computing the stability of parts on a table under gravity. To what extent should CG deal with algebraic surface representations? The majority of CG work so far has abstracted away from the algebraic structure, and left the algebraic computations to a black box. There is certainly nothing intrinsic in the problems to make this be so. Indeed, "computational geometry" was not that long ago a field which studied algebraic computations on curves and surfaces. That field has since come to be called "geometric modeling," and it has maintained a close ties with practitioners with a shared interest in geometric CAD systems. If CG is to maintain its claim to the geometric computing turf, it should broaden its definition to include at least the original one.

In addition to path planning, robotics suggests a rich variety of geometric problems. A workshop was initiated in 1994 to encourage interaction between researchers from robotics and computational geometry [**GLHW**]; a second workshop is currently being planned.

Industrial robots. To cope with shrinking product life cycles, computational methods are needed to reduce the time required to configure assembly lines. One idea is to formally model a small "vocabulary" of parameterized modular components (e.g., robot arms, conveyor belts, flexible part feeders, modular fixtures, light beams, 2D vision systems) and develop CG- based algorithms to efficiently compile them into an assembly system for a given CAD product model [**CG**]. Here are a few relevant problem areas:

1. CONVEX DECOMPOSITION: For motion planning, collision checking is currently most efficient for convex polyhedra, so we need better methods for convex decomposition of polyhedra. Algorithms should produce a near-optimal number of convex components. As for other research areas, fast, robust code is needed. Convex decomposition has many applications and is currently a bottleneck. See [**CP**] for a survey of current methods and [**CDST**] for a practical approach to the problem.
2. GRASPING AND FIXTURING: The goal here is to place contacts—fingers or fixture elements—so as to constrain object motion. This is one of the central problems in robotics. A number of basic questions have been answered, such as the minimum number of contacts sufficient[2] to hold any nonrotational part in form closure (4 in 2D, 7 for polyhedra) [**MNP, KMY**] but efficient algorithms are still needed. In modular fixturing, where fixture elements are constrained to a regular lattice, recent results suggest a number of open questions about the existence of solutions for classes of fixtures and parts [**ZG, WC, ORSW**].

   When there is uncertainty in part pose or applied forces, minimizing the number of grasp points can be posed as a convex set covering problem. Recently, CG researchers have described efficient and probably practical algorithms for near-optimal grasps. This goes beyond the previous works which either do not consider optimality, or do not allow search over a large space of possible finger placements. If the CG methods pan out in practice, this would be a strong vindication of CG in robotics.
3. APPROXIMATION ALGORITHMS FOR GRASP CIRCUIT PLANNING: Consider picking up parts as they arrive on a conveyor belt: the gripper may have multiple suction cups that must be "loaded" during each pass: Given a robot gripper with k suction cups and a collection of n parts, rapidly approximate an optimal path for loading the suction cups and depositing the parts in a pallet. Although this problem can be reduced to TSP, recent approximation results were successfully adapted to this and variations where the points to be visited are moving with known velocity [**CMR**]. Approximation algorithms have rarely been applied to problems in industrial robotics but hold great promise.
4. GEOMETRIC PROBING: CAD models are assumed by many algorithms but are often unavailable. Geometric probing can be used to generate such models. Probing hardware includes touch probes, light beams, scanline and raster cameras. Depending on the application and sensor, the probing strategy may compute convex hull, line hull or ray hull. See [**Sk**] for a review. More work is needed on online probing strategies that include models of probe and control uncertainty. The problem of probing to minimize error turns out to be dual to grasping, so the algorithms mentioned there are directly relevant.

[2] The number necessary was demonstrated in 1900.

5. Part Orienting: Algorithmic approaches to orienting ("feeding") parts for industrial assembly are needed. To be practical and cost effective, feeders must be able to rapidly and repeatedly orient parts at subsecond rates. Current feeders are designed by human trial and error. Complete algorithms are needed to take part geometry as input and generate feeding strategies or feeder hardware specifications as output. A complete algorithm for feeding 2D parts with algebraic boundaries is described in [**RG**]. More work is needed to design feeders for 3d parts and to provide accurate mechanical simulation of collisions and statistical behavior of parts arriving in bulk.
6. Assembly Sequence Planning: The problem of disassembling a collection of parts has a strong geometric component. For local motions, a disassembly sequence can be found in polynomial time. The worst case running time is $O(n^4)$, but it is fast in practice [**WKLL**]. More recently, good worst-case bounds have been given for an important special case: where the set of possible disassembly directions is fixed a priori (true for many robotic workcells). The latter methods use lower envelopes and Davenport-Schinzel sequences to efficient find collisions along the disassembly direction. The main limitation so far is the inability to deal with complex motion during disassembly. e.g. a translation and twist motion would not be possible. Extensions of this type, plus the addition of other constraints on the sequence, will make this a rich area for future research.

Autonomous robots. Autonomous robots have a far more difficult time than industrial robots. They must deal with time-varying environments containing unmodeled entities (humans) and with partial, inaccurate or outdated knowledge of their environment. While some researchers have nevertheless tackled the model-building problem for mobile robots (which leaves the robot with a classical motion planning problem on the map), others have sought simpler routes to navigation. These algorithms are typically online algorithms with limited state. In [**Lum**], robots navigating in the plane used local decision rules to provably reach the goal. These methods used performance measures derived from path length and were early examples of competitive ratios in path planning, [**PY**].

Geometric models of uncertainty have been used, based on the existence of landmarks, which are uniquely recognizable features in the environment. In [**LL**], a nearsighted robot with a bad compass wandered between "islands" of certainty near landmarks. In [**Hu**], a robot with better vision moved along corridors in space formed by aligning landmarks.

The argument is made below that the chief goals of autonomous robotics require a further understanding of geometric complexity. A central theme in this line of work is to determine what information is required to solve a task and how to direct a robot's actions to acquire that information. Another key question is to assess the capabilities of a robot in a given environment or class of environments.

These questions can be difficult. Structured environments, such as those found around industrial robots, contribute towards simplifying the robot's task because a great amount of information is implicitly encoded into both the environment and the robot's control program. One goal is to quantify the information encoded under the assumption that, say, the mechanics are quasi-static or the environment is not dynamic. Conversely, one might ask how much information must the control system or planner compute? Successful manipulation strategies often exploit properties of the (external) physical world (e.g., compliance) to reduce uncertainty and hence gain information. Often, such strategies exploit mechanical computation, in which the mechanics of the task circumscribes the possible outcomes of an action by dint of physical laws. Executing such strategies may require little or no computation; in contrast, planning or simulating these strategies may be computationally expensive. Since during execution one might witness very little "computation" in the conventional sense, traditional techniques from computer science have been difficult to apply in obtaining meaningful upper and lower bounds on the true task complexity. It is hoped that a theory can be developed to measure the sensitivity of plans to particular assumptions about the world, and to minimize those assumptions wherever possible.

Because geometry is the natural language for characterizing sensors, tasks, and the complexity of robotics operations, such an undertaking is inherently computational-geometric. More and more, robotics researchers are abandoning off-line models where a priori knowledge of the whole environment is assumed. For example, in the stereotypical off-line model, one might assume that the robot, on booting, reads a geometric model of the world from a disk and proceeds to plan. A preferred alternative now is to consider on-line paradigms where the robot investigates the world and incrementally builds data structures to the external environment. As time evolves, the task effectively forces the agent to move, sense, and update its model of the world. Certainly, from an on-line viewpoint, off-line questions such as "what is the complexity of plan construction for a known environment, given an *a priori* world model?" might often appear secondary, if not downright artificial.

While it is profitable to explore on-line paradigms for autonomous agents and sensorimotor systems, the framework remains to be extended in certain crucial directions. In particular, sensing has never been carefully considered or modeled in the on-line paradigm. The chief difficulty is the absence of tools for measuring the complexity of a robot. Just as complexity classes are characterized by the amount of resources available, one would hope to have relations between the computational power of a robot and its hardware. How much power is gained by adding an extra sensor?

Such questions have been overlooked. As a result, robot builders are able to make unverifiable claims about robot performance and resource consumption. What is missing from autonomous systems is both an accounting method to assess the capabilities of a robot and tools for arguing correctness and completeness for on-line autonomous robot algorithms. The inability to compare systems leads to redundant work and inhibits progress. Computational geometry has established a clear progress metric, while emphasizing completeness and correctness for control algorithms. But the systems for which it has explanatory power are far from real embedded systems. Yet such systems are probably best analyzed and characterized by using geometric methods. This is a direction in which computational geometry could have a significant impact.

Here is a concrete example of an on-line navigation. Some robots are purely reactive and plan their moves without retaining knowledge of their past interactions with the environment. Others build a map, ie, a geometric model of the world it encounters. The former are too weak, the latter are too slow. What are good intermediate strategies? Navigating in a graph with bounded storage is a well-studied problem in computer science. Interesting work in on-line navigation with uncertainty has already been done. To pursue this further and experiment with physical robots would be extremely useful.

The "minimalist" rationale of complexity theory (ie, using the minimum resources to achieve a given goal) has parallels in robotics. Resources are more diverse: they are measured by the number of sensors, sensor probes, actuators, agents, as well as by the amount of computation time and space and communication. In robotics, minimizing the use of resources has become an increasing preoccupation. It has been shown that walking/running machines could be built without static stability, that dextrous manipulation could be performed without sensing. Biped, kneed walker have been built without sensors, computers, or actuators.

Computational geometry is uniquely poised to demonstrate and prove minimalist solutions to robotics problems. Key questions include the following:

- What are minimal configurations (resource-wise) for on-line navigation, parts-feeding, grasping, singulation, recognition, or manipulation.
- Are there sensorless geometric filters[3] for classes of objects? What on-line filters uses a minimum amount of sensing? For example, with a distributed actuator system such as a MEMS array device, one can create a programmable vector-field for parts manipulation: First one centers a part using a radial, inward-squeezing pattern, which centers the part at the origin. Next, one makes the array push in the left direction within a narrow band around the x- axis. The rest of the array pushes in the opposite. This strategy filters small parts to the left and big parts to the right.

Micro-electromechanical systems. A wide variety of micromechanical structures (devices typically in the μm range) have been built recently by using processing techniques known from VLSI industry. Various microsensors and microactuators have been shown to perform successfully. For example, a single-chip air-bag sensor is commercially available; video projections using an integrated, monolithic mirror array have been demonstrated recently. More difficult is the fabrication of devices that can interact and actively change their environment. Problems arise from (i) unknown material properties and the lack of adequate models for mechanisms at very small scales, (ii) the limited range of motion and force that can be generated with microactuators, (iii) the lack of sufficient sensor information with regard to manipulation tasks, and (iv) design limitations and geometric tolerances due to the fabrication process.

The design, analysis, and control of micro-electromechanical systems (MEMS) are inherently geometric in nature. While much is known about building MEMS, little is understood about how to use them to manipulate objects. More precisely, manipulation of solid objects subject to hard or intermittent contact is entirely open. The state of the art is somewhat similar to that of robotics before geometric planning algorithms. Development of control strategies for manipulation by MEMS is a key bottleneck. MEMS actuators are tiny; hence one is forced to consider parallel manipulation strategies by teams or arrays consisting of a large aggregate of micro-actuators. It is believed that based on work on sensorless and near- sensorless manipulation, one can develop geometric theories of manipulation and control

[3] A geometric filter is a device which filters parts, ie, sends them to different spatial destinations, based on some geometric characteristic: for example, circular parts to the left, square parts to the right.

for microactuator arrays. Can one develop sensorless (open-loop) manipulation algorithms for arrays of MEM actuators? Can these strategies be time-invariant, or is a clock necessary? Can MEMS implement geometric filters? Can MEMS arrays orient or pose parts uniquely, without sensing? Can MEMS arrays implement assembly algorithms?

9. Manufacturing and product design

Computational geometry arises at all levels of manufacturing, from design and simulation, to process planning, product inspection and testing. A number of recent resources have emphasized the prominence of geometry in manufacturing [**Hoff2, ITM, VRC, VR, WR, Y4**]. A forthcoming special issue of *Algorithmica* (J. Mitchell, guest editor) focuses on applications of computational geometry to manufacturing.

Solid modeling. A first issue to tackle is solid modeling, by which is meant the computer representation and manipulation of 3D shapes. About 20 years ago, the "geometric coverage" was limited to solids bounded by surfaces from natural quadrics and planes, i.e., built with cones, spheres, cylinders, even with tori. For many years this was considered adequate for mechanical design, notwithstanding the persistent pressure from the aerospace and ship building industry to allow for more complex doubly-curved surfaces. Today, it is common to allow as surface elements subsets of tensor product nonuniform rational B-splines (NURBS), with some systems going further to consider triangular patches and more general types of parametric surfaces.

In the research arena there was an effort at the University of Bath a few years ago that resulted in a modeler with general algebraic surface elements. Algebraic surfaces have many mathematical and computational advantages compared to the parametric elements used now, and are emerging as an active, forefront research area. The commercial project generally cited as being the most advanced is Boeing's design of the 777 aircraft, which was done entirely using geometric and solid modeling (plus engineering analysis, costing analysis, and other nongeometric tasks). The Boeing 777 was modeled in CATIA, a commercial system from Dassault in France that is marketed in the US by IBM. Interestingly, CATIA is based on 20-year-old technology, which is an indication of the lag between research and production in manufacturing.

There are many technical issues that arise from the type of operations one would like to perform on these surfaces, be it as part of a larger conceptual operation, or in its own right. They are addressed by a community that variously hails from applied mathematics and approximation theory to—on the more experimental side—mechanical engineering. A wide variety of classical computational geometry problems arise in the process; their effective solutions would have considerable impact. This has already happened in some cases.

An issue somewhat less visible to academics, but important in practice, is the simple question: How does one express a complex shape? Or more precisely, how can a human define unambiguously a complex shape to an automaton? This is "the user interface problem" in CAD. Current practice has resulted in user interfaces that require, as an integral part, geometric constraint solvers. The issue is partly technical and partly conceptual.

In the technical category one finds such crisply stated questions as, "Given four spheres, how many lines are tangent to all of them?" (Estimates are known, but the exact answer is not.)

In the conceptual category one will occasionally witness discussions among mathematicians on the "fairness" or the "eye-pleasing quality" of a curve in space. While such subjective discussions might tread on slippery grounds, they are nevertheless essential to the design of commercial products where aesthetics matters (as in the automobile industry). Although computational geometry is an unlikely participant in such discussions, it can learn from this that not all design criteria are quantifiable by optimization functions, and that flexibility and experimentation are vital components of design tools.

There are deep and pressing problems that CG'ers, as computer scientists, may wish to ponder. Specifically, there are two "input-and-interaction" modalities for CAD systems: (1) programming interfaces, and (2) "pick-and-click" graphics user interfaces. Media such as the PADL-2 system for solids bring much of the usual power to the job: naming, parameter passing, conditionals, etc. Language interfaces are nicely suited to defining parametric families of parts, but lack the visual feedback that most engineers consider essential. Graphics user interfaces provide plenty of feedback, but almost no abstractive power (names, parameters, etc.); they deal in instances rather than generics. An important challenge is to devise a user interface with the visualization power of modern GUIs and some of the abstractive power of a programming system.

In its maturing stage, computational geometry has focused its attention mostly on linear objects. The motivation was sound: Why deal with curved shapes if we do not even understand polyhedral objects? In manufacturing, such a limitation is simply unacceptable. The time has come to bridge this gap.

At present, the frontiers in solid modeling for manufacturing include how to define surfaces subject to constraints that require variational approaches and differential equations. Another major push is an attempt to integrate shape design with other aspects of engineering design, such as analysis and costing [**Z**].

One particular challenge for CG, worthy serious attention, is *anisotropy*. Loosely speaking, anisotropy refers to the situation in which material properties vary over the interior of an object. (Wood is a simple example of an anisotropic material.) Some new methods of rapid prototyping processes, of which stereolithography is the best known (although not necessarily the best), are cap able of producing parts having anisotropic material properties. The currently popular schemes for representing solids—boundary representations and constructive solid geometry (CSG)—appear to be incapable of supporting anisotropy; thus, new means must be found, since customizable anisotropy is one of the most attractive properties of layered manufacturing processes.

Also, it is still the case that boundary-representation modelers lack robustness: Finite-precision arithmetic can lead to incorrect results (see Section 10). This is an important area of research, together with the design of better visualization tools to facilitate—and ultimately to automate—geometric reasoning.

Solid modeling is the archetypical example of an area of computer science in which object representations and algorithms are inseparable. Boundary representations and triangulated solid representations can be used for describing the same objects, but the algorithms that deal with them are fundamentally different. Representations are often dictated by particular needs. (For instance, boundary representations are natural for graphics display; triangulated boundary representations are required by fast display hardware.) Thus, manufacturing applications must be identified first. Next, the appropriate representations must be chosen. Only then can CG begin to think about efficient algorithms to operate on them. This is a different mode of operation from traditional CG: in classical problems such as polygonal triangulation or 3D convex hulls or 2D Voronoi diagrams, standard representations are more or less equivalent (via linear-time reductions) so it makes sense to think mostly of an object as a geometric entity rather than as its particular representation. Things are quite different in solid modeling.

Intersections of parametric surfaces. Spline representations are used extensively in computer graphics and in modeling "sculptured surfaces" found in automobile bodies, airplane airframes, and ship hulls. Several classical methods have been developed; for instance, Coons Patches, Bézier Surfaces, B-splines. The object has been to allow designers (particularly ones using CAD workstations) to produce the desired surfaces with minimal effort and maximum of control of the surface shape. Algorithms for rendering curves and surfaces on graphics screens are well understood. However, computing intersections of lines, curves, or planes with these surfaces can be rather complex and inefficient. Intersecting two such surfaces is usually very messy. Usually the parametric mapping cannot be analytically inverted, so even the "inverse point solution" of finding the parametric coordinates for a point in 3-space that lies on the surface requires an iterative procedure, and intersections must be approximated by using numerical methods.

Repeated computations of intersections with parametric surfaces tend to be very slow and error-prone. If one wishes to intersect many rays with a parametric surface, or to find if a tool envelope penetrates a parametric surface, it may be desirable to approximate the parametric surface by a polygonal surface. The operations can then be done on this approximated surface. The question naturally arises, "What is a good approximation for this particular application?" Much work has been done in the area of approximating surfaces by triangles, particularly with respect to finite element methods. However, for machining applications the question is not finding the fat triangles that are good for finite-element methods or triangles that render nicely on a graphics screen, but how to get a polygonal surface with a small number of triangles that is at all points within a fixed tolerance of the given surface. A number of heuristics have been used, but it is still an area of active research.

Intelligent CAD/CAM: manufacturing processes. The manufacturing industry has at its disposal a wide variety of processes for constructing objects, including gravity casting, injection molding, layered manufacturing, material removal via conventional (or chemical or electrical) machining, deformation (forging, rolling, extrusion, bending), composition (as in composite materials, sintered ceramics, and the like), spray deposition, etc.

Each manufacturing process imposes certain restrictions on the types of objects that can be built and the manner in which they are built. For example, a sphere is difficult to build using layered manufacturing (e.g., stereolithography) but not using injection molding. Also, a cube can be manufactured using stereolithography without the use of supporting pillars only when placed on one of its facets.

In all of these manufacturing contexts, CAD/CAM systems of growing sophistication are presently being introduced. Geometric computation has become ubiquitous in the manufacturing industry, as more and more real-world objects begin their design life as geometric objects modeled within a computer.

Given an object to be manufactured, two fundamental questions arise, having to do with the *feasibility* of a process and its *optimization*:

- Can it be manufactured using a particular process?
- For a given process, what is the best way to manufacture the object?

The latter question gives rise to many different problems depending on the meaning of "best." The geometry of the object, coupled with the restrictions imposed by the particular manufacturing process under consideration, play a vital role in determining the answer to these questions. Their importance is quite evident. For example, when designing an object to be built by a certain type of manufacturing process, an engineer must always keep in mind the process that is to be used to manufacture the object. This limits the creativity of the engineer, as the question of design feasibility is ever-present. In fact, the engineer is rarely sure whether the object can be built in the first place, since no formal methods exist to determine the feasibility of an object for most manufacturing processes.

At present, many aspects of the manufacturing processes and their ensuing geometric problems are tackled by relying on heuristics in trial-and-error fashion, which necessitates a great deal of human intervention. To quote from an injection molding book [**Pri**].

> "During injection, the mold is tilted into a favorable position that will eliminate surface defects such as bubbles and insure a complete fill. Mold orientation during fill is a cut-and-try process to find the most favorable position."

It is an open problem to decide whether a given object can be built by a particular process: This is a question into which computational geometry should be able to make inroads. Practical solutions especially are highly sought. This form of "automatic design verification" would liberate the designer by informing him or her of the full range of possible processes that can build an object. To make these ideas more concrete, we consider two examples: gravity casting and stereolithography.

- GRAVITY CASTING: In the case of gravity casting, the main motivation for focusing on the geometry of the object to be molded is to remove the "cut-and-try" phase from the process of finding a favorable mold position[4]

  Given a mold, establishing whether there exists an orientation that allows the filling of the mold using only one pin gate[5] as well as determining an orientation that allows the most complete fill are two major problems in the field of injection molding.

  These problems are particularly difficult when one takes into account the fluid dynamics and physics of the molding process; typically, this is handled by costly simulations. To date, only heuristics have been proposed to

[4] A mold refers to the whole assembly of parts that make up a cavity into which liquid (e.g., molten metal) is poured to give the shape of the required component when the liquid hardens.

[5] A pin gate is the point at which the liquid is injected into the mold.

solve the two problems (feasibility and optimization) above. Efficient solutions from computational geometry might provide a handle on the geometric component of the problem.

- STEREOLITHOGRAPHY: Stereolithography, a form of "layered manufacturing," is one of a growing set of processes used in rapid prototyping; for recent surveys of this and related technologies, see [**SBE, YG**].

  The components of the stereolithography manufacturing process (e.g., in the system patented by 3D Systems of Sylmar, CA) consist of a vat of liquid photocurable monomer, a computer-controlled table on a stand that can be moved up and down in the vat and a laser above the vat that can shine on the surface of the liquid and can move in a horizontal plane. The system works as follows:

  In the first stage the table is positioned right below the surface of the plastic and the laser is controlled to move about so that the light shines on the plastic surface and draws the bottom-most cross-section of the object being built. Upon contact the laser light solidifies the plastic, and the first cross-section of the object is formed and rests on the table. In the second stage, the table is lowered by a small amount to allow liquid to cover the hardened layer and the laser then draws the next cross-section of the object. The light from the laser penetrates the liquid just deep enough so that the cross-section is welded to the lower cross-section produced in the previous stage. This process is repeated until the entire object is formed.

  The direction given by a normal to the table pointing to the laser is called the direction of formation for the object. There are objects that can be formed only if the direction of formation is chosen correctly. In practice, supporting structures have to be built as part of the model and be removed in a postprocessing step. Naturally, there are also shapes (say, a sphere) that cannot be formed using stereolithography regardless of the direction of formation chosen.

  Computational geometry has recently made valuable contributions by designing algorithms for determining whether a given polyhedron can be built using various geometric models of the stereolithography process. However, more work is needed to make these models conform to manufacturing reality.

It is unlikely that computational geometry alone will ever supplant the complicated fluid-dynamics simulations performed today by engineers. But to reject or ignore the physics of these problems is also to ignore their very essence. Computational geometers should make the effort of addressing some of these physical problems because they hold rich opportunities for CG. Meanwhile, the hope is that CG can help to reduce the number of costly simulations required. The ultimate usefulness of such algorithms must be determined by collaborative efforts between computational geometers and manufacturing engineers.

Verification of numerical control machine tool programs. A manufacturing process of particular prominence, with a long history, is milling. A Numerical Control (NC) milling machine cuts a part out of a block of metal or other material (the "stock") by moving a cutting tool through space. For 3-ax is machining, a rotating cutting tool, with a vertical axis of rotation, does the cutting. An "NC program" specifies the movement. Typically, the program consists of a "cutter location file," which is a long list of (x, y, z) coordinates for successive tool locations. The tool is moved to the first spatial location, then moves along a straight line (or, say, a circular arc) to the next, then along a straight line from the second location to the third, and so on, until the list of locations is completed. (Actually, the file also contains other information, such as tool sizes, and instructions for when tools must be changed.) For 5-axis machines, two additional coordinates giving the angular orientation of the tool axis are also specified and the tool linearly interpolates both spatial and angular coordinates.

Typically these NC programs are created by tool programmers. The programmers begin with a geometric representation of the part expressed in Constructive Solid Geometry (CSG), or some sort of boundary representation using splines, and generate series of tool movements that are supposed to mill the part from the stock.

As with computer programs, NC programs are error-prone. Bugs can produce cuts that are too deep or leave too much material. Thus, an important practical problem is: "Given a surface of known equation and a file of NC tool movements, does the shape that the tool cuts match the mathematical shape to within a given tolerance?" It has been claimed [**FCDZ**] that "current methods of verifying NC part programs result in one of the highest non-recurring cost factors in producing NC machined parts within the aerospace industry." It is also a significant cost within the aircraft and automotive industries, among others.

The early approach to finding bugs was to use "proofing runs" on wood or foam; this is still a popular technique. Later systems would move a graphical display of a cutter over a display of the part. The user could visually check for errors as the tool moved on the graphics display. Only gross errors could be detected in this way, however. In the 70's, work began on programs that actually detected errors in NC programs or verified that none existed. A good overview of this problem, including a precise mathematical formulation, history, and summary of its current state is found in [**MV**].

Some of these methods use "complete" representations of the current state of the partially cut workpiece. Researchers investigated the feasibility of using CSG systems or solid modelers for the simulation of NC programs [**HV**]. Others developed systems based on octree [**NAB**] or boundary representation [**Sp**] of the workpiece. For each method, every tool movement would update the workpiece, and at the end the workpiece was compared to the desired piece. Unfortunately, all of these approaches were slow. Each tool movement subtracts a fairly complex shape from the current workpiece representation and requires significant updating. For example, for a simple 3-axis movement with a ball-end cutter the swept volume is bounded by parts of two spheres, three cylinders, and two planes; the number of movements is typically on the order of 10,000.

To avoid this problem, a number of people took advantage of the fact that it is not necessary to represent the entire state of the workpiece to verify the correctness of an NC program. The desired shape is known in advance, and this information can be used to simplify the process. Several researchers used an image-space approach, where an extended Z-buffer is provided for each pixel in a given graphical view of the part, and each tool movement updates the pixel information for each pixel that it passes over [**AEF, Van, WW**]. CGTech's Vericut simulator uses such an approach. These methods are faster than the complete methods but they are view-dependent. Because of this, errors might not be visible from a certain view direction and would therefore be missed. (Consider a nearly flat surface viewed edge-on. A large surface would be mapped into a thin line represented by only a few pixels, and errors that occurred where there were no pixels would be missed.)

Another approach is to grow vectors normal to the surface of the part [**Ch**]. These vectors are "mowed down" by the passage of the cutting tool, and their lengths are checked at the end of the run. Vectors that stick out above the surface indicate excess material and vectors cut below the surface point to gouges. One might use pixels to select these vectors (one vector for each pixel) [**OG**]. Or one might use properties of the tool size and the surface curvature to pick fewer vectors while still guaranteeing that they will find all errors greater than a user-set tolerance [**DJSH**].

A recent modification of the extended Z-buffer, called ray representations (ray-reps), has been proposed [**MR**]. A ray-rep is the intersection of a grid of parallel rays in some direction with the part. (Note that the intersection of a ray and a part consists of zero, one, or several line segments.) Given such a representation, all of the operations of CSG can be computed quickly by using a specialized parallel computer for processing ray-reps, called a raycasting engine. All of the approaches using modified Z-buffers or vectors introduce error. Even if the intersection with the tool path movement is calculated exactly (more on this below), in effect the part surface is being replaced by a set of discrete points. What can we say about the correctness of the surface at points where there are no vectors? All of these approaches use enough vectors to be fairly accurate in practice, but with a few exceptions none of them attempt to analyze errors introduced by the approach to check whether they fall within tolerance.

The ray-rep representation seems an ideal candidate for such analysis. Even with the special-purpose parallel computer one wishes not to use more rays that are needed. But how to analyze the errors introduced remains an open question, as is the determination of the optimal number of rays to use. It is also interesting to look at other possible applications of this representation and to analyze the relationship between the spacing of the rays and the error introduced.

A second area that needs more study is the one of intersecting rays with surfaces. Intersecting rays with part surfaces is reasonably well understood. However, when the surfaces are represented parametrically using B-splines or similar methods, computing intersections generally requires a slow interactive technique. One must also compute the intersection of rays with the envelopes swept out by tool movements. This is not too difficult for 3-axis movements of ball-end cutters. But the same cannot be said of the intersection of a ray with a 3-axis toroidal-end cutter movement. Most existing codes use a polygonal approximation to the envelope or a large number of discrete tool locations. Unfortunately, no error analysis is available.

For 5-axis cutter movement, an efficient method has been given for computing the intersection within a given error bound by using 3-axis tool movement as an approximation and relying on tree-based localization technique to quickly decide which of the 3-axis movements is the one intersected first [**Q**]. See also [**LJ**]. What remains open is to find a method for generating a polygonal approximation for the 5-axis envelope with a provable error bound that generates a reasonable number of triangles. Finding such a provably good approximation technique is an interesting and important open problem.

Generation of numerical control machine tool programs. An ultimate goal, of course, is to automatically produce NC programs that are error-free and perform the machining quickly. The generation of error-free NC programs directly from computer-based geometric models would significantly shorten the product development cycle. While some limited attempts at automatic NC program generation have been reported in the literature [**ACd**], a practical system has yet to be achieved. Producing such a system would be a very valuable step, and is quite likely to use a mix of computational geometry and classical optimization.

The special case of pocket machining has received recent attention and has had some success both in the algorithmic understanding of the problem, and in the automatic generation of NC programs. Held's thesis [**Held**] brought to the forefront some of the issues of interest to computational geometers. Offshoots from Held's thesis, including software for offsetting polygonal boundaries, have made it into commercial products, such as *LARK* (available from MTA SZTAKI, a subdivision of the Hungarian Academy of Sciences, and the spin-off company, CADMUS). Provably good approximation algorithms for specific (simplified) classes of problems have been devised by CG'ers for minimizing the total tool motion [**AFM**] and the number of retractions necessary in "zigzag" pocket machining [**AHS**]. But many more problems must be addressed, particularly those that deal with realistic models of machining.

So far, the discussion has been limited to "nominal spatial effects." It has been assumed that the tool moves exactly as specified and removes all material within its swept volume. In real systems, other considerations for verification or generation must be taken into account. One concern is variations from nominal geometry in terms of allowed tolerances and positioning uncertainties. Other considerations are dynamic effects, such as dealing with deformation of the part and tool due to pressure in cutting, tool breakage if the pressure is too great, and tool chatter and wear.

Layout: optimal use of material. Various layout and cutting problems are of immediate relevance to manufacturing, as they involve the optimal use of valuable raw materials. For example, in the parts nesting problem, we must place a given set of polygonal shapes (markers, templates, or patterns) on a sheet of material (sheet metal, cloth, cardboard, etc.), without allowing overlaps, in order to minimize the waste when the shapes are cut out from the material. Motivated by applications in the apparel industry, CG'ers have been investigating these layout problems, with some recent successes [**ML, DM, Mi, Mi4**].

In an application at Boeing, where very expensive sheet metal stock can be wasted in inefficient layouts, a software package ("2NA: 2-Stage Nesting Algorithm") is in current use on the factory floor, built on principles of optimization and basic computational geometry primitives. However, much more work is necessary before automated nesting algorithms will be able to replace human experts in many aspects of the material usage problem.

10. Robustness

Geometric computation must preserve numerical accuracy and topological consistency: on digital computers this is often akin to squaring the circle. Geometric algorithms are usually described assuming that input data is in general position and that exact real arithmetic provides reliable geometric primitives. Often an implementer substitutes floating-point arithmetic for real arithmetic and uses real-world data, which might be degenerate by accident or design. Hence, the correctness proof of the mathematical algorithm does not extend to the program, and the program can fail on seemingly appropriate input data. This is the well-known problem of "nonrobustness" in geometric computing [**Hoff**].

The failure rate of a floating-point implementation can be reduced with various ad-hoc methods, such as tolerancing (taking a nearly-zero expression value as exactly zero has the effect of treating a nearly-degenerate configuration as exactly degenerate) or input perturbation. With such methods a usable, though not completely reliable, implementation can often be obtained. This hardly qualifies as a rigorous approach, however: the effort involved is unpredictable, the required expertise is hard to codify, and there is no guarantee that the implementation will succeed on the next problem instance.

In the past few years some research has addressed robustness issues. Proposed approaches include error analysis, sophisticated tolerancing, symbolic reasoning, or software that provides exact arithmetic on integers, rationals, or even algebraic numbers. There is yet no agreement on the best approach. Indeed, all proposed methods have shortcomings, typically limited applicability or inefficiency. A sophisticated programmer implementing a standard algorithm may well find some of these methods useful and relevant; the situation is more problematic for an unsophisticated programmer with a variant problem.

When addressing these issues, it is important to distinguish between degeneracies and numerical precision.

10.1. Degeneracies. Degeneracies arise from the special position of two geometric objects. For example, two segments in general position either do not intersect or intersect at a point interior to both segments. Two intersecting segments in special position may overlap, may share a common endpoint with or without being collinear, may have one segment endpoint interior to the other segment, etc. Real-world data is likely to be degenerate. For example segment endpoints may be explicitly chosen from a coarse grid, to facilitate interactive design.

The effect of degeneracy is to vastly increase the number of special cases. While a sorting algorithm must deal only with the possibility of two keys being equal, a typical geometric algorithm faces the possibility of dozens or hundreds of different special cases [**EM, EC, Hoff, Y**]. The presence of numerical data, added to the inherent complexity of geometric data types, makes geometric algorithms much harder to implement correctly than combinatorial (say, graph-theoretical) ones. They are also much harder than just purely numerical algorithms (such as those addressed by numerical analysis) many of which consist of large chunks of straight-line code. Since the overall utility of an implementation may depend upon the correct treatment of special cases,the handling of special cases can permeate the implementation. This raises the obvious reliability concern that all cases have been considered and correctly handled.

Symbolic perturbation schemes allow degeneracies to be resolved automatically. Conceptually, each geometric coordinate c_i is replaced with a symbolically perturbed coordinate $c_i + f_i(\epsilon)$, where $\epsilon > 0$ is unknown but very small and the perturbation function f_i is simple, say a polynomial. Substitution of the symbolically perturbed coordinates in a predicate expression results in a polynomial in

ϵ with coefficients determined by the original geometric coordinates. The sign of the expression is given by the sign of the first nonzero coefficient, with coefficients taken in order of increasing powers of ϵ. For many classes of predicates, the f_i can be chosen to resolve all degeneracies.

While symbolic perturbation is certainly a useful tool in the implementation of geometric algorithms, existing schemes are not as applicable as might be desired. First, symbolic perturbation requires exact arithmetic, since the correctness of the perturbation depends upon exact evaluation of arithmetic expressions. Second, symbolic perturbation has been worked out in detail for only a small class of predicates. For example, constructed objects are often disallowed, since the perturbation function for a constructed object depends upon how the object was constructed and is much more complicated than the perturbation function for a primitive object. Perhaps most fundamentally, in a degenerate situation an algorithm implemented using symbolic perturbation does not solve the problem instance, but an arbitrarily-chosen nearby problem instance. This might be inappropriate in some applications. For example, a highly degenerate polytope may see its combinatorial complexity blow up exponentially as a result of a small perturbation.

Although much work remains to make symbolic perturbation generally useful, no other approach to the issue of degeneracy has as much promise.

10.2. Numerical precision. A geometric predicate is determined by the sign of an arithmetic expression. Sign-evaluation is exact in the conceptual model of the real numbers. However, a computer implementation cannot use real numbers. There are two scientifically plausible substitutions for exact real arithmetic. One is to use floating-point arithmetic, and somehow deal with the resulting rounding error. The other is to substitute exact arithmetic on a subset of the reals, such as the integers or the rationals. There is now a growing community of researchers who have interest in "real arithmetic computation," and some of these issues are relevant to the exact computation paradigm in CG. One such conference, "Real Numbers and Computers," will be held for the second time in April 1996.

Floating-point arithmetic. Floating-point arithmetic is widely used because it has many practical advantages. It provides a familiar approximation to the real numbers, with useful properties like automatic scaling. It is widely available on different computers and is well supported by programming languages. Current workstations have highly optimized native floating-point arithmetic, sometimes faster than native integer arithmetic. Floating-point arithmetic is sufficiently widespread in scientific computing that programmers rarely consider other options.

However, some geometric predicates cannot be resolved using floating-point arithmetic. If an instance of a predicate is nearly degenerate, then the value of the corresponding expression can be very small, less than the rounding error in the floating-point evaluation of the expression. Hence the sign of the computed expression may well be erroneous. Usually it is possible to argue that the computed expression value is the true value for slightly perturbed coordinate data. Since coordinate data may well be imprecise originally, the erroneous sign may appear to be innocuous. The difficulty arises with multiple predicate evaluations; there is no guarantee that any single global perturbation produces all the computed predicate values. Indeed, the computed predicate values may be geometrically inconsistent. Catastrophic implementation failure can easily result.

There are two broad categories of methods to deal consistently with floating-point rounding error. One category formalizes the notion of tolerances. A typical strategy might associate inner and outer tolerances with an object, say a point. If the inner tolerances of two points intersect, they are deemed coincident and merged; if the outer tolerances are disjoint, they are deemed separate; if neither case holds, the situation is ambiguous. As long as no ambiguity results during a computation, the result is correct. The hope is that ambiguities arise infrequently; an obvious drawback of this strategy is that is not clear what to do if an ambiguity does arise. Another drawback is that the generalization to complex geometric objects is not straightforward.

Since arithmetic operations have non-constant costs, it makes sense to judge the performance of an algorithm in terms of the precision K needed in the output of the algorithm. This gives rise to the notion of "precision-sensitive algorithms" where the running time is a function of K as well as a function of the input size. Since K can be exponential in the input size, exploiting this new parameter can be quite significant. Notice that precision-sensitivity is the bit-complexity analogue of the very fruitful idea of "output-sensitivity" invented by CG. The paper [**CSY**] first applied this concept to the NP-hard problem of Euclidean shortest paths.

A second floating-point approach is error analysis. This approach is modeled on the error analysis of numerical methods, particularly linear algebra. The goal is to show that a suitably implemented algorithm provides an answer that is in some precise sense near the mathematically correct answer. Error analysis of geometric algorithms requires consideration of both combinatorial and numeric structure. Often it is easy to argue that an algorithm produces combinatorially valid output, at least with suitably relaxed requirements. It has turned out to be much more difficult to argue that the numeric error associated with combinatorial structure is small. Full error analysis has been carried only for a few simple algorithms.

Exact computation. Exact geometric computation [**YD**] requires that every predicate evaluation be correct. This can be achieved either by computing every numeric value exactly, or by some symbolic or implicit numeric representation that allows predicate values to be computed exactly. Exact computation is theoretically possible whenever all the numeric values are algebraic, which is the case for most current problems in computational geometry.

With exact geometric computation, it is no longer reasonable to assume that each arithmetic operation takes constant time, as is the case with floating-point arithmetic. Rather, the cost of an arithmetic operation depends upon its context. Simple geometric predicates can be expressed as the sign-evaluation of an integer polynomial. The required arithmetic bit-length can be estimated from the degree of the polynomial and the bit-length of coordinates. For many predicates involving linear objects (such as orientation predicates) the degree is small, and the required bit-length is relatively minimal. However, even in simple cases software arithmetic is required, say over the integers or rationals, with a resulting increase in performance cost.

The cost of arithmetic also increases because of geometric constructions. A geometric constructor produces a new geometric object from old objects. Typically the coordinates of the new object can be expressed as polynomials in the coordinates of the old objects, and hence bit-length can be estimated from polynomial degree. For example, the coordinates of the intersection point of three planes have bit-length about three times the plane coefficient bit-length; the coefficients of a plane through three such points have bit-length about nine times the original plane coefficient bit-lengths. Hence an algorithm that uses geometric objects constructed to arbitrary depth can require arithmetic with prodigious bit-length, even if the algorithmic predicates are relatively simple.

Complex primitives on linear objects, or simple primitives on curved objects, apparently require that arithmetic be augmented with square roots, and, more generally, arbitrary polynomial roots. Such arithmetic operations can be implemented symbolically using general algebraic techniques such as resultants, Sturm sequences, and root separation bounds. Whether such techniques can be implemented with adequate efficiency is a question of considerable interest.

Because of the potential expense of exact computation, it is often appropriate to avoid it if possible. Any significant application may well have a core requiring exact computation and outer layers that are less demanding. For example, the boolean operations of a solid modeler might require exact computation to guarantee topological consistency, while the computer-graphics rendering of a resulting solid could tolerate some imprecision. Other applications, for example, a geometric theorem-prover, might require exact computation throughout.

To summarize. For algorithms with modest arithmetic bit-length requirements, exact arithmetic is appealing. Exact arithmetic appears not to be used widely, perhaps because of the performance cost of the required software arithmetic. However, computer hardware continues to increase in speed, which may mean that the performance cost becomes less significant. Furthermore, there is evidence that software exact arithmetic can be better designed for computational geometry applications, decreasing its effective cost.

Exact-arithmetic implementation of geometric algorithms would be much more attractive with the development of software arithmetic packages appropriate for computational geometry[**FVW, N**]. There are many issues to be explored: for example, the use of adaptive-precision arithmetic, the granularity of evaluation, algorithms for primitive evaluation, required arithmetic operations. For example, beyond speeding up basic arithmetic operations, more effective optimization techniques could be used at the expression level. Geometric computations invariably involves larger units of numerical computation, such as expressions like determinants or Euclidean lengths. This opens up an exciting new area of software construction in which many designs are possible.

The utility of exact arithmetic would be increased with the development of exact-arithmetic rounding algorithms. Geometric algorithms often produce objects that have both combinatorial and numeric data. For example, the vertices, edges, and faces of a polyhedron in three dimensions have both numeric coordinates and combinatorial incidence structure. The result of an exact computation may specify the numeric coordinates exactly, with coordinates of large bit-length. However, arithmetic on large coordinates is expensive, and an application may be satisfied with a short bit-length approximation to the numeric coordinates, as long as the combinatorial and numeric data are consistent. Rounding the coordinates of a complicated object such as a polyhedron is not straightforward, since its combinatorial structure may be invalidated by small perturbations of its faces or vertices. However, many applications are insensitive to changes in the combinatorial structure. If the structure is permitted to change, there are methods to round polygons or other planar objects made up of line segments to the integer grid [**Hob, GM**] or any nonuniform grid [**Mi2, Mi3**]. In general, rounding algorithms, particularly for curved or higher dimensional structures, are as yet inadequately developed.

There are important applications, such as operations on algebraic curves and surfaces, where bit-length estimates appear to rule out the use of exact arithmetic. It is possible that such bit-length estimates are too pessimistic, either in theory because the underlying algebraic machinery is not developed enough to give sharp estimates, or in practice because instances requiring long bit-length are infrequent. Many predicates that are well-understood in the linear domain (incircle, orientation) become much more complex in the curved domain, and their best evaluation using exact arithmetic is not well understood.

A case-study: dimensional tolerancing and metrology. We close this section by discussing one application area that seems ideally suited to the techniques of computational geometry and the requirements of exact computation [**Y2, Y3**]. Dimensional tolerancing theory and practice form a key cornerstone of modern precision engineering, allowing designers to specify tolerances in a specific way that also, in principle, allows machinists to verify them. It has been observed that the precision requirement on manufacturing is essentially increasing linearly with time. Increased precision in manufacturing has strong positive economic implications, from more efficient use of material to greater reliability in parts. But such benefits are partly wasted if the metrological theory and practice do not keep up with increasing manufacturing standards. This gap is partially responsible for the so-called "CMM crisis" identified by Walker in 1988. Briefly, the crisis concerns the inability of current Coordinate Measuring Machine (CMM) software to reliably compute mathematically-precise outputs, even assuming the input data are reliable. In fact, the reliability of the input data in existing systems are orders of magnitude better than the reliability of the computed values. The goal, then, is to completely remove the unreliability introduced by software. The computational geometry attack on this problem is two-pronged: first, we base algorithms on mathematically correct methods (this is sometimes called "CG methods" in the literature). Second, we implement these algorithms using exact computation in a way that characterizes the input imprecision.

Beyond the basic algorithms in computational metrology, there are some larger system issues. One is the "reference software" project that several national standards agencies (say, in Germany, UK, USA) are attempting to construct. The idea is that CMM vendor software will continue to be proprietary (black boxes) and for contractual purposes, standards will be guaranteed for such software. The National Institute for Standards and Technology has instituted an Algorithms Testing and Evaluation Program for certifying vendor software. The heart of this program is the so-called "reference software." It is clear that the exact computation technology is ideally suited for this work. Note that algorithms in the reference software need not achieve "real-time speeds" to be useful. Another example concerns packages for tolerancing analysis that allow a designer to check the effects of tolerance specifications. Such packages are commercially available but again they lack reliability. An exact-computation version of such a system would be a major advancement.

11. Molecular biology

With the human genome project underway and the promise of a revolutionary approach to understanding diseases, molecular biology is emerging as one of the most critical scientific disciplines today. It is an area likely to be heavily funded for many years, and its connections to computer science are widely recognized. Besides applied and numerical mathematics, the three major areas of computer science most likely to have an impact on research and development in molecular biology are information retrieval, string and pattern matching, and computational geometry. Several basic problems in molecular biology have a strong geometric flavor. These include problems dealing with the spatial structure of proteins and other macromolecules. Three such problems are briefly described below.

Spatial structure of a protein. The difficulty and reliability in determining the geometry or spatial structure of a molecule depends greatly on the type of available data. X-ray crystallography is currently the primary source of structural information, followed next by magnetic resonance imaging (MRI). To derive molecular structure from MRI data, one must resolve atomic positions given a set of approximate pairwise distances. The traditional tools for this problem come from distance geometry [**CH**].

Lacking X-ray and MRI data, one can only resort to structural analysis of unknown proteins by looking for patterns in the amino acid residue sequence that match those of known proteins. For evolutionary reasons, the many proteins occurring in nature share a limited number of common internal structures and folds. Recognizing such patterns and threading the unknown protein onto it greatly simplifies structure determination when X-ray or MRI data are available. Modeling based on such analysis may also be valuable.

The most difficult version of the problem is also the one with the largest potential benefit. It is the ab initio determination of structure from the amino acid residue sequence [**Ri2**]. While one can in principle use Newtonian mechanics to simulate the natural folding of the molecule, the sheer scale of the calculation is daunting. A modest-size protein folding simulation with current algorithms would require in the neighborhood of 10^{19} floating-point operations. A variety of heuristic methods are used to find the minimum energy configuration, including simulated annealing, Monte Carlo, and search with reduced degrees of freedom. As yet, none of these methods have come close to a general solution. Practical methods fail in most cases because the target function (say, the sum of energy potentials) has a large number of local maxima at any level of detail.

Ligand-protein docking. Consider an active site on a usually large receptor molecule, and a ligand molecule, which could be small or large. The general question is how snugly the ligand fits into the active site. This recognition mechanism is ubiquitous in biology and is basic to communication, immune response, and control of metabolism. The quality of fit has a geometric and a chemical component. The geometric component measures how well the surface shapes complement each other as a hand in glove. The chemical component measures how well the secondary forces between ligand and receptor atoms hold the two together. In the simplest version of the problem, both ligand and receptor are assumed to be rigid bodies, and the objective is to find a best alignment of the surfaces describing their boundaries. Very often the rigidity assumption is not realistic and flexible models need to be considered. Abstractly, the docking problem is related to the computer vision problem of matching a 3-dimensional geometric model to range data. The widely used schemes for representing the molecular surface and receptor sites in docking algorithms may not be the best choices. Other descriptions, such as alpha-shapes [**EM2**], and other matching criteria, such as Hausdorff distance rather than least squares, may be more appropriate.

A problem related to docking is the recognition of motifs in proteins, which are substructures of the proteins with similar geometry. Matching motifs may indicate genetic links or active sites with similar properties.

Drug design. Given a specific receptor, the problem is to find or design a binding small ligand molecule. If the receptor site geometry is known, the problem is to find a molecule that satisfies some geometric constraints and is also a good chemical match. After finding good candidates according to those criteria, a docking step with energy minimization can be used to predict binding strength.

If the site geometry is not known, as is often the case, the designer must base the design on other ligand molecules that bind well to the site. If those other molecules are rigid, the problem becomes one of identifying the substructures or active groups that contribute to the fit. By joining the groups with alternative molecular scaffolding, one can build molecules with good or better affinity for the site. The joining process requires solutions of kinematics and distance geometry problems. If the receptor site geometry is unknown and the ligands are flexible, then the designer must first posit the configurations of the ligands in their bound state by assuming that their active groups are in similar places. New drugs are enumerated which can be folded into a similar appearance.

The role of computational geometry. One of the greatest challenges for computational geometry is to contribute to the understanding of nature in terms of geometry and algorithms through the connection offered by molecular biology. The most difficult of the specific problems mentioned above is the simulation of the folding process that would allow the determination of protein structure from the amino acid residue sequence, and a solution may be out of reach at the moment. The other problems seem tractable, and solutions already exist for easy cases and also for some of the hard ones. The only certainty is that a considerable amount of research is still called for.

Computational geometry can help in several ways. The first is to participate in the design of geometric models. The sphere model for atoms is the fundamental idea connecting particle physics with geometry. There are still open questions as to what extent this simplification is sufficiently accurate and how physical questions can be approached through studying large conglomerates of spheres. Based on various extensions of the spherical model [**LR, Ri**], the biology community has developed its own geometric software [**Con**]. The geometry is enhanced by graphics and numerical software visualizing and utilizing the geometric information. New software will have to compete with the available packages, which are already widely used, and it will need to follow data and calling standards so components can be plugged together to solve large problems.

12. Astrophysics

Large ongoing projects, such as the Sloan Digital Sky Survey (SDSS)[**KSNLP, BRSCL, RBSCL**], aim at creating a comprehensive digital map of the northern sky. The project SDSS will result in a complex archive of about 20 terabytes containing exquisitely calibrated digital images of the accessible half of the northern sky, in 5 wavelength bands, from the ultraviolet to the infrared. A processing pipeline will identify and derive parameters (fluxes, sizes, colors) of over 100 million objects, classify them by type (star, galaxy, quasar) and obtain high resolution spectral information for the brightest one million galaxies and 100,000 quasars in the survey area. All this information will be placed into the archive for subsequent access and analysis.

The huge amount of information involved makes storage and access a formidable computational challenge. The data is organized as points in $\mathbf{R}^d$. Typically the dimension d is fairly low, e.g., 8: right ascension, declination, five colors, redshift.

Astrophysical database. The information should be stored in a database that can support geometric queries of the kind:

- RANGE QUERY [**Chaz, Mat**]: Given a d-dimensional rectangular box defined by intervals $\{(a_i, b_i) : 1 \leq i \leq d\}$, where $a_i, b_i \in R_i$ (R_i is the range of the attribute i), find all the objects which lie within the specified box. For example, one may be asked to find all objects within a given right ascension and declination.
- LINEAR COMBINATION QUERY: Given a d-dimensional slab of the form $l \leq \sum a_{i_j} x_{i_j} \leq u$, where $i_j \in J$ (with J the index set defined on all attributes), find all the objects that lie in the specified range. This query involves searching whether linear combinations of given attributes have a value within the range given by l and u. For example, we may have a query of the form: find all objects over the whole sky with a given color, ie, objects which satisfy $u - g \leq 1$, where u and g are color bands.
- CONVEX POLYHEDRON QUERY: Given a d-dimensional convex polyhedron, specified as intersections of halfspaces (linear combinations of attributes), find all the objects which lie inside the polyhedron.
- NEAREST NEIGHBOR QUERY: Given a query object q, find the closest (in the Euclidean sense) object in the database. This is a difficult problem to solve efficiently (see [**AMNSW**] for a good history of the problem). In practice, one can implement a simpler version of the problem called *approximate nearest neighbor* [**AMNSW**] which takes a parameter ε, and finds an object in the database whose distance to q is at most $(1 + \varepsilon)$ times the distance to the actual nearest neighbor.
- k-NEAREST NEIGHBORS QUERY [**AMNSW**]: Given a query object q, find k-nearest objects (approximate) to q in the database.
- PROXIMITY QUERY: Given two regions in d-space and a parameter δ, find all pairs of objects, one in each region, that lie within a distance δ of each other. For example, one may have a query of the form: find all blue galaxies within 2 arcmin of quasars.

One would also be able to compute topological information (genus, fractal dimension) of cell complexes formed by subsets of the stored objects. Traditionally, the relational format has been the favored model for astronomical databases. Such models tend to be slow and unable to accommodate sophisticated geometric query retrieval. In that regard, space partitioning schemes, such as k-D trees, appear much more promising. The appropriate choice of splitting hyperplanes is also a matter of debate. Statistic-based strategies, such as splitting normally to the direction with maximum variance (covariance matrix method), are potential candidates, but many other choices look promising, too. Actually, the current advances in space partitioning raises the hope for yet more efficient algorithms.

For the nearest neighbor problem, Voronoi diagram-based methods are doomed because of the exponential blow in the dimension. Approximation methods have considered [**AMNSW**]. A full taxonomy should be developed so that the best algorithm for any given dimension can be identified. When the dimension is very large, projection methods might work well: pick a random k-flat for small k, and project the points normally to it. Solve the problem in the projection flat. The hope is that a random flat will more or less preserve distance relationships. Current knowledge does not allow us to assess how good such methods are. More sophisticated strategies should also be sought. The size of the answer to the queries could be huge, and so one would like to develop faster (yet accurate) estimation strategies to help the user determine what he is up against. This information can be very useful in practice. To perform accurate estimations, statistical analysis of the distribution of the data in each dimension is required for performing good partitioning. The resulting geometric data structures from the partitioning typically satisfy some nice properties which can be used for performing the estimation efficiently.

The massive amount of data necessitates the use of hierarchical memory organization. External memory models have been proposed which let researchers design I/O efficient algorithms. Recently several I/O efficient algorithms for specifically geometric problems have been proposed which make use of novel techniques [**A, CGR**]. Chiang [**Chi**] gives experimental evidence of the importance of redesigning geometric algorithms to minimize page faults and to optimize the number of I/O accesses.

Computational topology. An important feature of a galaxy is its topology. Does the galaxy distribution consist of isolated high-density clumps (meatball topology), does it have isolated voids surrounded by walls (bubble topology), or does it resemble a sponge, with interlocked high- and low- density regions?

The standard first step is to turn galaxies, which are represented by finite set of points, into smooth surfaces in 3-space. To do so choose a weighting function (a Gaussian) convolving with each point. This induces a continuous density function throughout the sky. Isodensity contours form the surfaces whose topology we wish to compute. The Gauss-Bonnet formula allows us to express the genus of the surface in terms of the integral of the Gaussian curvature. The integral is computed by sampling the surface by a fine enough grid. These techniques work fairly well and have been used to determine the topology of many galaxies from satellite observations [**MFW**].

Alternatively, the points can be connected to form a simplicial complex. Instead of a weighting function we choose a spherical ball around each point and consider the boundary of the union of balls. To compute the components of this surface and the genus of each component, we convert the union of balls into a complex as follows. First, clip the balls to within their Voronoi cells. The resulting cells overlap along their boundaries. The alpha complex represents pairwise, triplewise, and quadruplewise overlap by edges, triangles, and tetrahedra. The complex is homotopy equivalent to the union of balls [**Ed2**], and there is a fast linear space algorithm that computes the three betti numbers. The number of surfaces, their genuses, and how they are nested is readily derived from this information.

If other parameters are taken into account, the dimension of the ambient space shoots up and to reveal the full topology is more difficult. Of all the topological invariants, the homology groups are the most popular, mostly because of their computational tractability. The homology groups describe the cycle structure of a topological space. It would be extremely useful to be able to compute homology groups efficiently for low-dimensional geometric complexes. Outside astrophysics, there are many important applications in pattern recognition and classification in biology and chemistry as well as in robotics and scene analysis. Comparing homologies is a good (if not foolproof) test to rule out topological equivalence: two spaces with different homologies cannot be homeomorphic; of course, the converse is not true.

There are also applications to the classification of time series and the analysis of dynamical systems. Spectral analysis has long been the standard tool for classifying time series. In the absence of sharp harmonics, the signal is dismissed as noise. This might be a mistake. Analysis for nonlinear dynamics can be done by computing topological invariants of the attractor set for the system being measured. To do so one can map the time series into $\mathbf{R}^d$ ($d \approx 10$) by scanning a "comb" of length d across the series and thinking of each d-tuple of entries at the teeth as a point in $\mathbf{R}^d$. By grouping all $(k+1)$-tuples of points at a distance less than ε from one another, one can define k-simplices. Finally, by introducing the obvious boundary operators, this yields a complex whose (Čech) homology is called the ε-homology of the set. Because Čech homology commutes with inverse limits, this is an appropriate method for computing the homology of the attractor. Moreover the use of ε-homology (where simplices are defined only in terms of pairwise distances) should have computational advantages. This is similar to the idea of alpha complex alluded to in the earlier paragraph.

Computing homology groups, or their ranks referred to as betti numbers, can be done by algebraic methods, but it is often inefficient. Exploiting the specific geometry of the manifolds can lead to much faster algorithms. There is a compelling analogy with graph algorithms. Many problems on graphs can be phrased algebraically and solved by inverting linear systems. That is not the way it is done, however. Graph algorithms seek to exploit the inner combinatorial structure of graphs to avoid generic algebraic treatments. Similarly, computational geometry can be used to bypass linear algebra in the computation of homologies. An example is the algorithm for simplicial complexes embedded in $\mathbf{R}^3$ described in [**DE**]. In particular, the use of Mayer-Vietoris sequences should be useful in designing efficient divide-and-conquer schemes for computing the homology of higher-dimensional complexes by computational geometric means.

N-body problem. One of the most central computational problems in physics is the simulation of a particle system with gravitational or Coulombic forces. The first task is to compute the potential field of the system. The potential at each point is the sum of the various potentials determined by the particles. The potential at x contributed by a particle at p is often of the form $q_p/||x-p||_2$, where q_p depends on the strength of the charge (or mass) of the particle at p.

Assuming infinite arithmetic, it is easy to compute the potential in $O(n^2)$ time, where n is the number of particles. A method, called the fast multipole method, based on multipole expansion was developed in [**Gre**] and refined by introducing computational-geometric ideas in distance geometry [**CK**].

To apply the fast multipole method, one must first construct a tree decomposition of the set of particles, along with a set of pairs of nodes in the tree. This step can be shown to require $\Theta(n \log n)$ time in the algebraic model. Given such a decomposition, the fast multipole method requires only linear time. However, the constants involved in the fast multipole method are so large that for realistic numbers of particles, this turns out to be the most time-consuming part of the computation.

The large constant in the fast multipole method arises both from geometry and from the need to manipulate series expansions, the size of which grows with the desired output precision. Despite the large constant, the method has proven useful for simulations involving millions of particles, for which the direct approach is not even feasible. Reduction of this constant would have significant practical implications for particle simulation, both by speeding up existing applications, and extending the useful range of the fast multipole method to smaller sets of particles.

There is a great deal of flexibility in the geometrical decomposition needed for the fast multipole method, and some gains could result from investing more effort in this part of the algorithm. Currently, the tree computed is a fairly standard box decomposition that guarantees a small number of pairs of nodes for which the fast multipole method must be applied. However, for any given point set, there may be another tree that results in a somewhat smaller number of pairs, in which case it may be worthwhile to expend the effort needed to find such a tree.

The efficient implementation of the fast multipole method poses non-trivial technical challenges. For example, in current implementations, one typically constructs a tree in which the leaves are subsets of particles rather than single particles, and computes the potential between these subsets directly. The size of these subsets has a strong effect on the efficiency of an implementation, and must be chosen carefully in order for the method to be of any benefit.

Efficient parallel implementations would be particularly useful, because a particle simulation can require many potential computations, each of which carries the simulation through one small time step. The more steps that can be performed, the more can be learned about the dynamics of large particle systems, so any speed increases will have a direct impact on computational physics.

13. Resources for computational geometers

13.1. Journals. Three journals are devoted to the subject of computational geometry: *Discrete & Computational Geometry*, *International J. Computational Geometry and Applications*, and *Computational Geometry: Theory and Applications*.

Computational geometry papers also often appear in algorithm journals such as *Algorithmica* and *J. Algorithms*, and occasionally are published in general theoretical computer science and discrete mathematics journals including *Discrete Mathematics*, *Information Processing Letters*, *J. ACM*, *J. Combinatorial Theory, Ser. A*, *J. Computer and System Sciences*, *SIAM Journal on Computing*, and *Theoretical Computer Science*.

In addition, geometric papers with a strong applications content may be published in specialized journals aimed at the particular application of the paper. *ACM*

Trans. on Graphics publishes papers in computer graphics. *Computer Aided Geometric Design* is largely concerned with topics related to splines and surface approximation, but has also published several papers on triangulation. *Annals of Mathematics and Artificial Intelligence* published in 1991 a special issue on algorithmic robot motion planning; the same subject is also covered by *J. Intelligent and Robotic Systems*. Some areas of geometry (especially robotics) have close connections with computational algebra and can be found in *J. Symbolic Computation*.

13.2. Conferences. There are several annual conferences and workshops devoted entirely to computational geometry. The oldest of these is the highly selective ACM Symposium on Computational Geometry. Others include the Canadian Conference on Computational Geometry, the ARO- MSI Workshops on Computational Geometry, and the European Workshop on Computational Geometry.

There is usually a large amount of geometry at the major algorithms conferences, including the ACM-SIAM Symposium on Discrete Algorithms (SODA), the Int. Symp. Algorithms and Computation (ISAAC), the European Symp. on Algorithms (ESA), the Workshop on Algorithms and Data Structures (WADS), and the Scandinavian Workshop on Algorithm Theory (SWAT). The three major annual theory conferences, ie, the ACM Symposium on Theory of Computing (STOC), the IEEE Symposium on Foundations of Computer Science (FOCS), and the International Colloquium on Automata, Languages and Programming (ICALP) also usually contain some amount of computational geometry.

The 2nd Federated Computing Conference to be held in Philadelphia, May 1996, is to include not only the ACM Symp. Comp. Geom., but also a new ACM Workshop on Applied Computational Geometry. DIMACS, Schloss Dagstuhl, INRIA, and Tel Aviv University regularly organize workshops in computational geometry. DREI, an NSF-sponsored program for high school teachers is having a special year in computational geometry, which includes a research component (3 workshops), and the Joint Summer Research Conference entitled 'Discrete and Computational Geometry: Ten Years Later,' taking place in July, 1996 at Mt. Holyoke College under the sponsorship of AMS-IMS-SIAM, will have a large computational geometry component.

Pointers to calls for papers, programs, contents, and proceedings for many of these conferences as well as for conferences relating to various applications of computational geometry can be found online at

`http://www.ics.uci.edu/~eppstein/junkyard/jconf.html`

13.3. Web sites. Directly related to the theme of this task force, David Eppstein's web page "Geometry in Action"

`http://www.ics.uci.edu/~eppstein/geom.html`

is devoted to applications and potential applications of computational geometry, and includes pointers to over 100 individual projects and applications in areas such as astronomy, geographic information systems, CAD/CAM, data mining, graph drawing, graphics, medical imaging, metrology, molecular modeling, robotics, signal processing, textile layout, typography, video games, vision, VLSI, and windowing systems. This page also includes pointers to other web sites including the ones discussed here. Eppstein maintains some open problems at

`http://www.ics.uci.edu/~eppstein/junkyard/open.html`

A large bibliography of computational geometry papers, maintained by Bill Jones and Otfried Schwarzkopf, is available at `http://www.cs.ruu.nl/people/otfried/html/geombib.html`

Schwarzkopf also maintains some other web resources including a dictionary of French computational geometry terminology:

`http://www.cs.ruu.nl/people/otfried/html/francais.html`

Other lists of computational geometry resources are maintained at Carleton Univ.

`http://www.scs.carleton.ca/~csgs/resources/cg.html`

and by Jeff Erickson at Berkeley

`http://www.cs.berkeley.edu/~jeffe/compgeom.html`

Various items concerning conferences, seminars, software, bibliography and a list of researchers with web pages can be found in

`http://wwwdim.uqac.uquebec.ca/~jmrobert/CG.html`

which is maintained by Jean-Marc Robert.

13.4. Mailing lists. Several computational geometry mailing lists are distributed at AT&T Bell Laboratories: *compgeom-announce* for announcements about professional activities, *compgeom-discuss* for discussion or questions, and *compgeom-tribune*, a newsletter sent out aperiodically by Hervé Brönnimann. Mail to

`compgeom-xxxx@research.att.com`

will be forwarded to subscribers to those lists. (Mail sent to compgeom-tribune will be forwarded to its editor.) For more information, send mail to

`compgeom-request@research.att.com`

with the message body "send readme" or see an HTML version of the readme file provided at

`http://pine.fernuni-hagen.de/GI/compgeom/compgeom.html`

The compgeom-request server is also capable of some other services including email searches of the computational geometry bibliography.

The Geometry Forum at Swarthmore College maintains several related newsgroups: *geometry.announcements* for announcements of programs and resources for geo meters; *geometry.college* for topics relevant to the study and teaching of college-level geometry; *geometry.forum* for news and discussion of the Forum news service; *geometry.institutes* for proceedings of geometry conferences and institutes; *geometry.pre-college* for K-12 math discussions and online projects; *geometry.puzzles*, for your spare time; *geometry.research*, for advanced geometry investigations and research topics; and *geometry.software.dynamic*, for discussions of work using the Geometer's Sketchpad and/or Cabri Geometre. The contents are also available via mailing lists; to subscribe, send email to

`majordomo@forum.swarthmore.edu`

with message body "subscribe geometry-xxxx." For more information, see

`http://forum.swarthmore.edu/~sarah/topics/about.newsgroups.html`

Messages from all these groups are archived and can be searched at

`http://forum.swarthmore.edu/~sarah/HTMLthreads/index.html`

13.5. Software. Many people have implemented the basic geometric algorithms such as convex hulls and Voronoi diagrams; a few more complicated geometric algorithms are also implemented and available.

Nina Amenta has collected a large number of free computational geometry programs at

`http://www.geom.umn.edu:80/software/cglist/`

The programs are arranged by subject and include convex hulls, Voronoi diagrams, Delaunay triangulation, low dimensional linear programming, triangulation, point location, shape reconstruction, collision detection, constraint solving, and visualization.

Seth Teller has a collection of C code and SGI executables for linear programming, voronoi diagrams, and manipulation of NURBS surfaces at

`http://graphics.lcs.mit.edu/~seth/geomlib/geomlib.html`

Joe O'Rourke has made C code corresponding to the material in his textbook, "Computational Geometry in C," available at

`ftp://grendel.csc.smith.edu/pub/compgeom/`

The code itself is in file "sharfile," and a description of the other files in this directory is in "README."

Various people have provided computational geometry packages for the *Mathematica* symbolic computation package; these routines are available from MathSource at

`http://www.wri.com/mathsource/`

A keyword search on MathSource for "geometry" turned up roughly 40 packages (not all related to computational geometry); of particular interest are "Computational Geometry" by E. C. Martin, "CirclePack" by O. Goodman, and "Vertex Enumeration," by K. Fukuda and I. Mizukoshi.

LEDA, a Library of Efficient Data Types and Algorithms, is a project at the Max Planck Inst. für Informatik, run by Kurt Mehlhorn, that provides a large collection of data types and algorithms in a form that can be used by non-experts. Along with basic non-geometric data structures such as binary search trees, LEDA includes implementations of several geometric algorithms including convex hulls and line segment intersection. For more information see

`http://www.mpi-sb.mpg.de/LEDA/leda.html`

14. Conclusions

Computational geometry enjoys two unique assets: (1) its diversity and potential to affect most forms of computing; (2) its mature algorithmic foundations. The challenge is now to make this potential come true, and build an effective pipeline connecting theory to practice. We believe that this will revitalize the field and open new vistas for geometric research, both of a practical and a theoretical nature. The solutions of the most exciting open problems might be unknown. But the most exciting open problems themselves might be unknown, too. We are confident that creating the pipeline will unveil many of these problems.

15. Postscript

Following the public release of this document, several interesting developments have occurred. As we had hoped, a lively discussion of the issues raised in the report did take place. The reader is encouraged to read an archived compilation of these exchanges at the URL's,

`http://www.cs.duke.edu/~jeffe/compgeom/taskforce.html`

`http://www.inf.fu-berlin.de/~gaertner/discuss.html`

Also, in June 14–15, 1996, the *ACM Workshop on Strategic Directions in Computing Research*, held a special session on computational geometry, chaired by Roberto Tamassia. The report of the Working Group, which is to appear in ACM Computing Surveys 28(4), December 1996, is also accessible on the Web at the URL,

`http://www.cs.brown.edu/people/rt/sdcr/report/report.html`

16. Acknowledgments (by Bernard Chazelle)

In June of 1994, I assembled a task force to reflect on the needs to reform computational geometry and initiate a community-wide debate. Its members were: Nina Amenta (Xerox PARC), Tetsuo Asano (Osaka Electro-Comm. U.), Gill Barequet (Tel Aviv U.), Marshall Bern (Xerox PARC), Jean-Daniel Boissonnat (INRIA), John Canny (U.C. Berkeley), Bernard Chazelle (**Chair**, Princeton U.), Ken Clarkson (AT&T Bell Laboratories), David Dobkin (Princeton U.), Bruce Donald (Cornell U.), Scot Drysdale (Dartmouth U.), Herbert Edelsbrunner (U. Illinois at Urbana-Champaign), David Eppstein (U.C. Irvine), A. Robin Forrest (U. East Anglia), Steve Fortune (AT&T Bell Laboratories), Ken Goldberg (U.C. Berkeley), Michael Goodrich (Johns Hopkins U.), Leonidas J. Guibas (Stanford U.), Pat Hanrahan (Stanford U.), Chris M. Hoffmann (Purdue U.), Dan Huttenlocher (Cornell U.), Hiroshi Imai (U. Tokyo), David Kirkpatrick (UBC), D.T. Lee (Northwestern U.), Kurt Mehlhorn (Max Planck Inst.), Victor Milenkovic (U. Miami), Joe Mitchell (SUNY at Stony Brook), Mark Overmars (U. Utrecht), Richard Pollack (Courant Institute, NYU), Raimund Seidel (U. Saarbrücken), Micha Sharir (Tel Aviv U. and NYU), Jack Snoeyink (UBC), Godfried Toussaint (McGill U.), Seth Teller (MIT), Herb Voelcker (Cornell), Emo Welzl (ETH Zürich), and Chee Yap (Courant Institute, NYU).

Although the Task Force never met physically as a whole, a considerable amount of exchanges and discussions ensued. People outside the Task Force also contributed copiously to its deliberations. The present document summarizes some of these reflections. It evolved from bits and pieces submitted by individual members. The final assembly and editing of the document (i.e., the final typos) are mine.

I thank the Task Force members for their enthusiastic response and the tremendous amount of work they put into it. I also wish to acknowledge the helpful comments of Alok Agarwal, Timothy Baker, Paul Callahan, Martin Held, Robert Lupton, Dinesh Manocha, Prabhakar Raghavan, Kumar Ramaiyer, Subhash Suri, Roberto Tamassia, and Jovan Zagajac. Thanks to all.

References

[AST] P. K. Agarwal, M. Sharir and S. Toledo, *Applications of parametric searching in geometric optimization*, Proc. 3rd ACM-SIAM Symp. Disc. Alg. (1992), 72–82.

[ARB] J. M. Airey, J. H. Rohlf and F. P. Brooks, *Towards image realism with interactive update rates in complex virtual building environments*, ACM Siggraph, Special Issue on 1990 Symp. Interactive 3D Graphics 24 (1990), 41–50.

[ABB] H. Alt, B. Behrends and J. Blömer, *Measuring the resemblance of polygonal shapes*, Proc. 7th Ann. ACM Symp. Comp. Geom. (1991), 186–193.

[A] L. Arge, *The buffer tree: a new technique for optimal I/O-algorithms*, in Workshop on Data Structures and Algorithms, 1995.

[AFM] E. M. Arkin, S. Fekete and J. S. B. Mitchell, *Optimal tours for lawnmowing and milling*, SUNY Stony Brook, 1995. An earlier paper appears in the 5th Canad. Conf. on Comp. Geom., 1993, 461–466.

[AHS] E. M. Arkin, M. Held and C. L. Smith, *Optimization problems related to zigzag pocket machining*, Proc. 6th ACM-SIAM Symp. Disc. Alg. (1996), 419–428.

[ACd] G. T. Armstrong, G. C. Carey and A. dePennington, Numerical Code Generation From a Geometric Modeling System, Solid Modeling by Computers: From Theory to Applications, ed. M. S. Pickett and J. W. Boyse, Plenum Press, New York, 1984.

[AMNSW] S. Arya, D. M. Mount, N. S. Netanyahu, R. Silverman and A. Wu, *An optimal algorithm for approximate nearest neighbor*, Sympos. Disc. Alg., 1994.

[AK] T. Asano and S. Kimura, *Contour representation of an image with applications*, Proc. SPIE's International Symposium on Vision Geometry IV, San Diego (July 1995), 14–22.

[ACKT] T. Asano, D. Z. Chen, N. Katoh and T. Tokuyama, *Polynomial-time solutions to image segmentation*, Proc. of the 7th Ann. SIAM-ACM Conference on Discrete Algorithms (Jan. 1996), 104–113.

[AEF] P. Atherton, C. Earl and C. Fred, *A graphical simulation system for dynamic five-axis NC verification*, Autofact Show of the Society of Manufacturing Engineers, Detroit, Nov. 1987, 2-1 - 2-12.

[B] H. S. Baird, Mode-Based Image Matching Using Location, MIT Press, 1985.

[Ba] T. J. Baker, *Developments and trends in three-dimensional mesh generation*, Appl. Numerical Mathematics 5 (1989) 275–304.

[BS] G. Barequet and M. Sharir, *Piecewise-linear interpolation between polygonal slices*, CVGIP: Image Understanding, to appear.

[BE] M. Bern, and D. Eppstein, *Mesh generation and optimal triangulation*, In "Computing in Euclidean Geometry" (2nd Edition). Eds. D.-Z. Du and F. K. Hwang, World Scientific Press (1995), 47–123.

[Bo] J.-D. Boissonnat, *Shape reconstruction from planar cross sections*, Computer Vision, Graphics and Image Processing 44 (1988), 1–29.

[BF] J.-D. Boissonnat and O. D. Faugeras, *Triangulation of 3-D objects*, Proc. 7th Int. Joint Conf. on Artificial Intelligence (1981), 658–660.

[BB] J.-D. Boissonnat and B. Beiger, *Three-dimensional reconstruction of complex shapes based on the Delaunay triangulation*, eds., Acharya, R. S. and Goldgof, D. B. In "Biomedical Image Processing and Biomedical Visualization," 1905 (1993), SPIE, 964–975.

[BT] J.-D. Boissonnat and M. Teillaud, *On the randomized construction of the Delaunay tree*, Theoret. Comput. Sci. 112 (1993), 339–354.

[BFHCP] W. Bouma, I. Fudos, C. M. Hoffmann, J. Cai and R. Paige, *A geometric constraint solver*, Computer Aided Design 27 (1995), 487–501.

[BG] R. Brost and K. Goldberg, *A complete algorithm for synthesizing modular fixtures for polygonal parts.* IEEE Transactions on Robotics and Automation, 1996, to appear.

[BRSCL] R. Brunner, K. Ramaiyer, A. Szalay, A. Connolly and R. Lupton, *An object-oriented approach to astronomical databases*, Proc. 4th Ann. Conf. Astronomical Data Analysis Software and Systems, Baltimore, 1994.

[BM] B. Buttenfield and R. McMaster, eds., Map Generalization: Making Rules for Knowledge Representation, John Wiley & Sons, 1991.

[CGR] P. Callahan, M. Goodrich and K. Ramaiyer, *Topology B-trees and their applications*, Workshop on Data Structures and Algorithms, 1995.

[CK] P. Callahan and S. R. Kosaraju, *A decomposition of multi-dimensional point-sets with applications to k-nearest-neighbors and n-body potential fields,* Proc. 24th Ann. ACM Sympos. Theory Comput. (1992), 546–556.

[C] J. Canny, The Complexity of Robot Motion Planning, M.I.T. Press, Cambridge, 1988.

[CG] J. Canny and K. Goldberg, *A RISC approach to sensing and manipulation*, Journal of Robotic Systems, 12(6):351–362, June 1995.

[Cass] T. A. Cass, *Feature matching for object localization in the presence of uncertainty*, Proc. IEEE Conf. Computer Vision and Pattern Recognition (1990), 360–364.

[CMR] P. Chalasani, R. Motwani and A. Rao, Approximating tsp variants for robot grasp and delivery, in preparation, July 1994.

[Ch] I. T. Chappel, *A new approach to automatic tool path generation for numerically controlled milling machines*, Proc. 4th Int. Conf. on Manufacturing Engineering, Brisbane, Australia, May 1988, 29–32.

[Chaz] B. Chazelle, *Computational geometry: a retrospective*, In "Computing in Euclidean Geometry" (2nd Edition), Eds. D.-Z. Du and F. Hwang, World Scientific Press (1995), 22–46.

[CDST] B. Chazelle, D. P. Dobkin, N. Shouraboura and A. Tal, *Strategies for polyhedral surface decomposition: an experimental study*, Computational Geometry: Theory and Applications, 1995, to appear.

[CP] B. Chazelle and L. Palios, *Decomposition algorithms in geometry*, Algebraic Geometry and its Applications, C. Bajaj, Ed., Chap. 27, Springer-Verlag (1994), 419–447.

[Chi] Y. J. Chiang, *Experiments on the practical I/O efficiency of geometric algorithms: distribution sweep vs. plane sweep*, Workshop on Data Structures and Algorithms, 1995.

[CSY] J. Choi, J. Sellen and C. Yap, *Precision-sensitive Euclidean shortest path in 3-space*, Proc. 11th Ann. ACM Symp. Comput. Geom. (1995), 350–359.

[Clark] K. L. Clarkson, *Randomized geometric algorithms,* In "Computing in Euclidean Geometry," Eds. D.-Z. Du and F. Hwang, World Scientific Press (1992), 117–162.

[CW] M. F. Cohen and J. R. Wallace, Radiosity and Realistic Image Synthesis, Academic Press Professional, San Diego, CA (1993).

[Con] M. L. Connolly, *Analytical molecular surface calculation*, J. Appl. Cryst. 16 (1983), 548–558.

[CCLB] L. T. Cook, P. N. Cook, K. R. Lee, S. Batnitzky, B. Y. S. Wong, S. L. Fritz, J. Ophir, S. J. Dwyer III, L. R. Bigongiari and A. W. Tempelton, *An algorithm for volume estimation based on polyhedral approximation*, IEEE Trans. Biomedical Engineering 27 (1980), 493–500.

[CH] G. M. Crippen and T. F. Havel, Distance Geometry and Molecular Conformation, Wiley, New York, 1988.

[DM] K. Daniels and V. J. Milenkovic, *Multiple translational containment. Part I: An approximation algorithm*, Algorithmica, special issue on CG in manufacturing, to appear.

[DE] C. J. A. Delfinado and H. Edelsbrunner, *An incremental algorithm for betti numbers of simplicial complexes*, Proc. 9th Ann. Symp. Comput. Geom., 1993, 232–239.

[DJSH] R. L. Drysdale, R. B. Jerard, B. Schaudt and K. Hauck, *Discrete simulation of NC machining*, Algorithmica 4 (1989), 33–60.

[DT] D. Dobkin and S. Teller, *Computer graphics*, in J. E. Goodman and J. O'Rourke, eds, Handbook of Discrete and Computational Geometry, CRC Press, Boca Raton, 1997, 779–795.

[Ed] H. Edelsbrunner, Algorithms in Combinatorial Geometry, Springer-Verlag, 1987.

[Ed2] H. Edelsbrunner, *The union of balls and its dual shape*, Disc. Comput. Geom. 13 (1995), 415–440.

[EM] H. Edelsbrunner and E. P. Mücke, *Simulation of simplicity: a technique to cope with degenerate cases in geometric algorithms*, ACM Trans. Graphics, 9 (1990), 66–104.

[EM2] H. Edelsbrunner and E. P. Mücke, *Three-dimensional alpha shapes*, ACM Trans. Graphics 13 (1994), 43–72.

[EC] I. Emiris and J. Canny, *A general approach to removing degeneracies*, SIAM J. Comput. 24 (1995), 650–664.

[ESRI] ESRI White Paper Series. Environmental Systems Research Institute, Inc. ARC/INFO: GIS Today and Tomorrow, Mar. 1992.

[FGB] Federal Geomatics Bulletin, 4(1), 1992. GIS Division, Energy, Mines and Resources. Ottawa, Canada.

[FVW] S. Fortune and C. Van Wyk, *Efficient exact arithmetic for computational geometry*, Proc. 9th ACM Symp. Comp. Geometry (1993), 163–172.

[FCDZ] R. Fridshal, K. P. Cheng, D. Duncan and W. Zucker, *Numerical control part program verification system*, Proc. Conf. CAD/CAM Technology in Mechanical Engineering, MIT (March 1982), MIT Press, 236–254.

[FST] T. Funkhouser, C. Séquin and S. Teller, *Management of Large Amounts of Data in Interactive Building Walkthroughs*, Proc. 1992 Workshop on Interactive 3D Graphics, 11–20.

[FS] T. Funkhouser, C. Séquin, *Adaptive display algorithm for interactive frame rates during visualization of complex virtual environments*, Computer Graphics 27, Proc. Siggraph (1993), 247–254.

[GCS] Z. Gigus, J. F. Canny and R. Seidel, *Efficiently computing and representing aspect graphs of polyhedral objects*, IEEE Trans. Pat. Anal. Mach. Intel. 13 (1991), 542–551.

[Gold] C. M. Gold, *Problems with handling spatial data—the Voronoi approach*, CISM Journal 45 (1991), 65–80.

[Gold2] C. M. Gold, *An object-based dynamic spatial model, and its application in the development of a user-friendly digitizing system*, Proc. 5th Intl. Symp. Spatial Data Handling, IGU Commission on GIS (1992), 495–504.

[GLHW] K. Goldberg, J.-C. Latombe, D. Halperin, R. Wilson, editors. *The Algorithmic Foundations of Robotics: First Workshop*. A. K. Peters, Boston, MA, 1995.

[Good] M. F. Goodchild, *A spatial analytical perspective on geographical information systems*, Int. J. GIS 1 (1987), 327–334.

[Good2] M. F. Goodchild, *Geographic information systems and cartography*, Cartography 19 (1990), 1–13.

[Good3] M. F. Goodchild, *Issues of quality and uncertainty*, J. C. Muller, editor, Advances in Cartography, Elsevier Applied Science, London (1991), 113–139.

[GKP] M. F. Goodchild, K. K. Kemp and T. K. Poiker, *NCGIA Core Curriculum*, National Center for Geographic Information Analysis, University of California, Santa Barbara CA 93106-4060, USA, 1989. On WWW at `http://www.env.gov.bc.ca/~tclover/giscourse/ToC.html`.

[GO] J. E. Goodman and J. O'Rourke, (eds), Handbook of Discrete and Computational Geometry, CRC Press, Boca Raton, 1997.

[Gre] L. F. Greengard, The Rapid Evolution of Potential Fields in Particle Systems, The MIT Press, 1988.

[GM] L. J. Guibas and D. H. Marimont, *Rounding arrangements dynamically*, Proc. 11th Ann. Symp. Comput. Geom., 1995 (190–199).

[HM] B. Hamann and R. J. Moorhead, *A survey of grid generation methodologies, the National Grid Project, and scientific visualization efforts at the NSF Engineering Research Center for Computational Field Simulation*, to appear in: G. M. Nielson, et al., eds., IEEE Computer Society Press, Los Alamitos, California.

[Heck] P. S. Heckbert and M. Garland, *Fast polygonal approximation of terrains and height fields*, Technical Report CMU-CS-95-181, Carnegie Mellon University, 1995.

[Held] M. Held, *On the computational geometry of pocket machining*, Vol. 500 of Lecture Notes in Computer Science. Springer-Verlag, June 1991.

[Hob] J. Hobby, *Practical segment intersection with finite precision arithmetic*, Manuscript, AT&T Bell Labs, October 1993.

[Hoff] C. M. Hoffmann, *The problems of accuracy and robustness in geometric computation*, IEEE Computer 22 (1989), 31–42.

[Hoff2] C. M. Hoffmann, Geometric and Solid Modeling, Morgan Kaufmann, 1989.

[Hoff3] C. M. Hoffmann, *Geometric approaches to mesh generation*, in "Modeling, Mesh Generation, and Adaptive Numerical Methods for Partial Differential Equations," eds. I. Babuska, J. Flaherty, W. Henshaw, J. Hopcroft, J. Oliger, T. Tezduyar, Springer Verlag, IMA Volumes in Mathematics and its Applications 75 (1995).

[HoffV] C. M. Hoffmann and P. J. Vermeer, Geometric Constraint Solving in R^2 and R^3, Computing in Euclidean Geometry, eds. D. Z. Du and F. Hwang, World Scientific Publishing, 2nd ed., 1994.

[HL] K. Ho-Le, *Finite element mesh generation methods: a review and classification*, Computer-Aided Design 20 (1988), 27–38.

[HV] W. A. Hunt and H. B. Voelcker, *An exploratory study of Automatic verification of programs for numerically controlled machine tools*, Production Automation Project Tech Memo No. 34, University of Rochester, Jan. 1982.

[Hu] S. Hutchinson, *Exploiting visual constraints in robot motion planning*, IEEE Conference on Robotics and Automation (1991), 1722–1727.

[HKS] D. P. Huttenlocher, K. Kedem and M. Sharir, *The upper envelope of Voronoi surfaces and its applications*, Disc. Comp. Geom. 9 (1993), 267–291.

[HKR] D. P. Huttenlocher, G. A. Klanderman and W. J. Rucklidge, *Comparing images using the Hausdorff distance*, IEEE Trans. Pat. Anal. Mach. Intel. 15 (1993), 850–863.

[HU] D. P. Huttenlocher and S. Ullman, *Recognizing solid objects by alignment with an image*, Int. J. Computer Vision, 5 (1990), 195–212.

[Iles] K. Iles, *Data considerations—Some principles for accuracy and reliability of data in a GIS system*, GIS'89 Symposium (1989), Vancouver, Canada, 103.

[ITM] Information Technology for Manufacturing: A Research Agenda, National Academy Press, 1995.

[KSNLP] S. M. Kent, C. Stoughton, H. Newberg, J. Loveday, D. Petravick, V. Gurbani, E. Berman and G. Sergey, *Sloan digital sky survey*, Proc. 3rd Ann. Conf. Astronomical Data Analysis Software and systems, Victoria, British Colombia, 1993.

[KMY] D. Kirkpatrick, B. Mishra and C. K. Yap, *Quantitative Steinitz's theorems with applications to multifingered grasping*, Disc. Comput. Geom. 7 (1992).

[KSAVG] H. M. Klein, W. Schneider, G. Alzen, E. D. Voy and R. W. Günter, *Pediatric craniofacial surgery: Comparison of milling and stereolithography for 3D model manufacturing*, Pediatric Radiology 22 (1992), 458–460.

[Kum] M. P. Kumler, *Intensive comparison of triangulated irregular networks (TINS) and digital elevation models (DEMS)*, Cartographica 31 (1994), monograph 45.

[LW] Y. Lamdan and H. J. Wolfson, *Geometric hashing: A general and efficient model-based recognition scheme*, International Conference on Computer Vision (1988), 238–249.

[L] J.-C. Latombe, Robot Motion Planning, Kluwer Academic Publishers, Boston, 1991.

[LT] R. Laurini and D. Thompson, Fundamentals of spatial information systems. Academic Press, London, 1992.

[LL] A. Lazanas and J.-C. Latombe, *Landmark-based robot navigation*, American Association for Artificial Intelligence Conference (1992), 816–822.

[LR] B. Lee and F. M. Richards, *The interpretation of protein structures: estimation of static accessibility*, J. Mol. Biol. 55 (1971), 379-400.

[LJ] S. X. Li and R. B. Jerard, *5-Axis Machining of Sculptured Surfaces with a Flat-End Cutter*, CAD, 26 (1994), 165–178.

[Lö] R. Löhner, *Finite elements in CFD: what lies ahead*, Int. J. Numer. Meth. Eng. 24 (1987), 1741–1756.

[LC] W. E. Lorensen and H. E. Cline, *Marching cubes: A high resolution 3D surface construction algorithm*, Computer Graphics 21 (1987), 163–169.

[Lum] V. Lumelsky, *Algorithmic and complexity issues of robot motion in an uncertain environment*, Journal of Complexity 3 (1987), 146–182.

[MNP] X. Markenscoff, L. Ni and C. H. Papadimitriou, *The geometry of grasping*, IJRR 9 (1990).

[Mat] J. Matoušek, *Geometric range searching*, ACM Comput. Surveys 26 (1995), 421–461.

[Mehl] K. Mehlhorn, Data Structures and Algorithms 3: Multidimensional Searching and Computational Geometry, Springer-Verlag, Heidelberg, Germany, 1984.

[MR] J. P. Menon and D. M. Robinson, *Advanced verification via massively parallel raycasting*, Manufacturing Review 6 (1993), 141–154.

[MV] J. P. Menon and H. B. Voelcker, *Towards a comprehensive formulation of NC verification as a mathematical and computational problem*, J. Design and Manufacturing 3 (1993), 263–277.

[Mi] V. J. Milenkovic, *Multiple translational containment. Part II: Exact algorithms*, Algorithmica, special issue on CG in manufacturing, to appear.

[Mi2] V. J. Milenkovic, *Robust polygon modeling*, Special issue of Computer-Aided Design on Uncertainties in Geometric Computations, 25 (1993), 546–566.

[ML] V. J. Milenkovic and Z. Li, *A compaction algorithm for nonconvex polygons and its application*, European J. Operations Research, 84 (1995), 539–560,

[Mi3] V. J. Milenkovic, *Practical methods for set operations on polygons using exact arithmetic*, Proc. 7th Canad. Conf. Comput. Geom., 1995 (55–60).

[Mi4] V. J. Milenkovic, *Translational polygon containment and minimal enclosure using linear programming based restriction*, Proc. 28th Ann. ACM Symp. Theory Comput., to appear, 1996.

[MFW] B. Moore, C. S. Frenk, D. H. Weinberg, W. Saunders, A. Lawrence, R. S. Ellis, N. Kaiser, G. Efstathiou and M. Rowan-Robinson, *The topology of the QDOT IRAS redshift survey*, Monthly Notices Royal Astronomical Society, 256 (1992), 477–499.

[MK] H. Müller and A. Klingert, *Surface interpolation from cross sections*, in: Focus on Scientific Visualization (H. Hagen, H. Müller, and G. M. Nielson, eds.), Springer Verlag, Berlin, 1993, 139–189.

[Mul] K. Mulmuley, Computational Geometry: An Introduction Through Randomized Algorithms, Prentice-Hall, 1994.

[MLS] R. Murray, Z. Li and S. Sastry, A Mathematical Introduction to Robotics Manipulation. CRC Press, 1994.

[N] S. Näher, *The LEDA user manual*, Version 3.1 (Jan. 16, 1995). Available by anonymous ftp from `ftp.mpi-sb.mpg.de` in directory `/pub/LEDA`.

[NAB] I. Navazo, A. D. Ayala and P. Brunet, *A geometric modeler based on the exact octree representation of polyhedra*, Computer Graphics Forum, 5, 91–104.

[OBS] A. Okabe, B. Boots and K. Sugihara, *Nearest neighbourhood operations with generalized Voronoi diagrams: A review*. Int. J. GIS 8 (1994), 43–71.

[OG] J. H. Oliver and E. D. Goodman, *Color graphic verification of NC milling programs for sculptured surfaces*, 10th Ann. Automotive Computer Graphics Conference and Exposition, Engineering Society of Detroit, Dec. 1985.

[OR] J. O'Rourke, Computational Geometry in C, Cambridge Univ. Press, 1994.

[ORSW] M. Overmars, A. Rao, O. Schwarzkopf and C. Wentink, *Immobilizing polygons against a wall*, ACM Symp. Comput. Geom. (1995), Vancouver, BC.

[PY] C. H. Papadimitriou and M. Yannakakis, *Shortest paths without a map*, Theoretical Computer Science 84 (1991), 127–150.

[Pell] M. Pellegrini, *Monte Carlo approximation of form factors with error bounded a priori*, Proc. 11th Ann. ACM Symp. Comput. Geom. (1995), 287–296.

[Perk] J. Perkal, *On the length of empirical curves*, Discussion Paper 10, Michigan Inter-University Community of Mathematical Geographers, University of Michigan, Ann Arbor, 1966.

[PFLM] T. K. Peucker, R. J. Fowler, J. J. Little and D. M. Mark, *The triangulated irregular network*, Amer. Soc. Photogrammetry Proc. Digital Terrain Models Symposium (1978), 516–532.

[PS] F. P. Preparata and M. I. Shamos, Computational Geometry: an Introduction, Springer-Verlag, New York, 1988.

[Pri] W. I. Pribble, *Molds for reaction injection, structural foam and expandable styrene molding*, in: Plastics mold engineering handbook, J. H. DuBois and W. I. Pribble (Eds.), Van Nostrand Reinhold Company Inc., New York, 1987.

[Q] J. Quinn, *Accurate Verification of Five-Axis Numerically Controlled Machining*, Ph.D. Thesis, Dartmouth College, May 1993.

[RBSCL] K. Ramaiyer, R. Brunner, A. Szalay, A. Connolly and R. Lupton, *Prototype astronomical database*, Proc. 2nd ACM Workshop on Advances in Geographic Information Systems, Gaithersburg, Maryland, 1994.

[RG] A. Rao and K. Goldberg, *Manipulating algebraic parts in the plane*, IEEE Transactions on Robotics and Automation 11 (1995).

[Ra] J. Raper, ed., Three Dimensional Applications in Geographic Information Systems, Taylor & Francis, 1989.

[Ri] F. M. Richards, *Areas, volumes, packing, and protein structures*, Ann. Rev. Biophys. Bioeng. 6 (1977), 151–176.

[Ri2] F. M. Richards, *The protein folding problem*, Scientific American 264 (1991), 54–63.

[Ru] W. J. Rucklidge, *Locating objects using the Hausdorff distance*, Proc. 5th Int. Conf. on Computer Vision (1995), 457–464.

[SU] J.-R. Sack and J. Urrutia, (eds), Handbook of Computational Geometry, Elsevier Science Publishers, in preparation.

[SP] N. Sapidis and R. Perucchio, *Advanced techniques for automatic finite element meshing from solid models*, Computer-Aided Design 21 (1989) 248–253.

[S] L. L. Schumaker, *Reconstructing 3D objects from cross-sections*, in: Computation of Curves and Surfaces (W. Dahmen, M. Gasca, and C. A. Micchelli, eds.), Kluwer Academic Publishers, 1989, 275–309.

[SS] J. T. Schwartz and M. Sharir, *On the piano movers' problem: 2. general techniques for computing topological properties of real algebraic manifolds*, Advances in Applied Mathematics 4 (1983), 298–351.

[Sh] M. Sharir, *Almost tight upper bounds for lower envelopes in higher dimensions*, Disc. Comput. Geom. 12 (1994), 327–345.

[Sk] S. S. Skiena, *Problems in geometric probing*, Algorithmica 4 (1989), 599–605.

[Sp] R. B. Speen, *A dynamic approach to modeling for numerically controlled verification*, Masters Thesis, Purdue University, 1985.

[SBE] P. Stucki, J. Bresenham, R. Earnshaw, Guest Editors, *Rapid prototyping technology*, Special issue of IEEE Comp. Graphics and Applications 15 (1995), 17–55.

[T] S. Teller, *Visibility Computations in Densely Occluded Polyhedral Environments*, CS Dept., UC Berkeley (1992), TR UCB/CSD 92/708 or http://cs-tr.cs.berkeley.edu:80/TR/UCB:CSD-92-708.

[Van] T. Van Hook, *Real-time shaded NC milling display*, ACM SIGGRAPH, 20, 4, (1986), 15–20.

[vanO] P. van Oosterom, Reactive Data Structures For Geographic Information Systems, Oxford University Press, 1993.

[VRC] H. B. Voelcker, A. A. G. Requicha and R. W. Conway, *Computer applications in manufacturing*, in "Annual Review of Computer Science," Ed. J. F. Traub, Vol. 3. Palo Alto, CA: Annual Reviews Inc. (1988), 349–387.

[VR] H. B. Voelcker and A. A. G. Requicha, *Research in solid modeling at the University of Rochester: 1972-1987*, in "Fundamental Developments of Computer-Aided Geometric Modeling," Ed. L. Piegl, Academic Press Ltd., London, England (1993), 203–254.

[WC] A. Wallack and J. Canny, *Planning for modular and hybrid fixtures*, IEEE International Conference on Robotics and Automation, May 1994.

[WW] W. P. Wang and K. K. Wang, *Real-time verification of multiaxis NC programs with raster graphics*, IEEE Proc. 1986 Int. Conf. Robotics and Automation, San Francisco (April 1986), 166–171.

[WKLL] R. H. Wilson, L. Kavraki, T. Lozano-Perez and J.-C. Latombe, *Two-handed assembly sequencing*, Intl. J. of Robotics Research, 14 (1995), 335–350.

[WR] M. J. Wozny and W. C. Regli, Guest Editors. *Computer science in manufacturing*, Special issue of Comm. ACM 39 (1996), 33–85.

[X] W.-N. Xiang, *A GIS method for riparian water quality buffer generation*, Int. J. GIS 7 (1993), 57–70.

[YG] X. Yan and P. Gu, *A review of rapid prototyping technologies and systems*, Computer-Aided Design 28 (1996), 307–318.

[Y] C. Yap, *Symbolic treatment of geometric degeneracies*, J. Symb. Comput. 10 (1990), 349–370.

[Y2] C. Yap, *Towards exact geometric computation*, Fifth Canadian Conf. on Computational Geometry (1993), 405–419. To appear, Comp. Geom. Theory and Appl.

[Y3] C. Yap, *Exact computational geometry and tolerancing metrology*, In "Snapshots of Computational and Discrete Geometry, Vol 3," Eds. by David Avis and Jit Bose, McGill School of Comp. Sci, Tech. Rep. No.SOCS-94.50, 1994. (A special issue dedicated to Godfried Toussaint)

[Y4] C. Yap, *Report on NSF Workshop on Manufacturing and Computational Geometry*, IEEE Computational Science and Engineering, 2 (1995), 82–84. Full report on web at URL `http://cs.nyu.edu/cs/faculty/yap/index.html`.

[YD] C. Yap and T. Dubé, *The exact computation paradigm*, In "Computing in Euclidean Geometry" (2nd Edition). Eds. D.-Z. Du and F. K. Hwang, World Scientific Press (1995), 452–492.

[Z] J. Zagajac, *A fast method for estimating discrete field values in early engineering design*, to appear in IEEE CG&A.

[ZG] Y. Zhuang and K. Goldberg, *On the existence of solutions in modular fixturing*, International Journal of Robotics Research, to appear.

Selected Titles in This Series

(Continued from the front of this publication)

197 **J. E. Bonin, J. G. Oxley, and B. Servatius, Editors,** Matroid theory, 1996

196 **David Bao, Shiing-shen Chern, and Zhongmin Shen, Editors,** Finsler geometry, 1996

195 **Warren Dicks and Enric Ventura,** The group fixed by a family of injective endomorphisms of a free group, 1996

194 **Seok-Jin Kang, Myung-Hwan Kim, and Insok Lee, Editors,** Lie algebras and their representations, 1996

193 **Chongying Dong and Geoffrey Mason, Editors,** Moonshine, the Monster, and related topics, 1996

192 **Tomek Bartoszyński and Marion Scheepers, Editors,** Set theory, 1995

191 **Tuong Ton-That, Kenneth I. Gross, Donald St. P. Richards, and Paul J. Sally, Jr., Editors,** Representation theory and harmonic analysis, 1995

190 **Mourad E. H. Ismail, M. Zuhair Nashed, Ahmed I. Zayed, and Ahmed F. Ghaleb, Editors,** Mathematical analysis, wavelets, and signal processing, 1995

189 **S. A. M. Marcantognini, G. A. Mendoza, M. D. Morán, A. Octavio, and W. O. Urbina, Editors,** Harmonic analysis and operator theory, 1995

188 **Alejandro Adem, R. James Milgram, and Douglas C. Ravenel, Editors,** Homotopy theory and its applications, 1995

187 **G. W. Brumfiel and H. M. Hilden,** $SL(2)$ representations of finitely presented groups, 1995

186 **Shreeram S. Abhyankar, Walter Feit, Michael D. Fried, Yasutaka Ihara, and Helmut Voelklein, Editors,** Recent developments in the inverse Galois problem, 1995

185 **Raúl E. Curto, Ronald G. Douglas, Joel D. Pincus, and Norberto Salinas, Editors,** Multivariable operator theory, 1995

184 **L. A. Bokut′, A. I. Kostrikin, and S. S. Kutateladze, Editors,** Second International Conference on Algebra, 1995

183 **William C. Connett, Marc-Olivier Gebuhrer, and Alan L. Schwartz, Editors,** Applications of hypergroups and related measure algebras, 1995

182 **Selman Akbulut, Editor,** Real algebraic geometry and topology, 1995

181 **Mila Cenkl and Haynes Miller, Editors,** The Čech Centennial, 1995

180 **David E. Keyes and Jinchao Xu, Editors,** Domain decomposition methods in scientific and engineering computing, 1994

179 **Yoshiaki Maeda, Hideki Omoro, and Alan Weinstein, Editors,** Symplectic geometry and quantization, 1994

178 **Hélène Barcelo and Gil Kalai, Editors,** Jerusalem Combinatorics '93, 1994

177 **Simon Gindikin, Roe Goodman, Frederick P. Greenleaf, and Paul J. Sally, Jr., Editors,** Representation theory and analysis on homogeneous spaces, 1994

176 **David Ballard,** Foundational aspects of "non"standard mathematics, 1994

175 **Paul J. Sally, Jr., Moshe Flato, James Lepowsky, Nicolai Reshetikhin, and Gregg J. Zuckerman, Editors,** Mathematical aspects of conformal and topological field theories and quantum groups, 1994

174 **Nancy Childress and John W. Jones, Editors,** Arithmetic geometry, 1994

173 **Robert Brooks, Carolyn Gordon, and Peter Perry, Editors,** Geometry of the spectrum, 1994

172 **Peter E. Kloeden and Kenneth J. Palmer, Editors,** Chaotic numerics, 1994

171 **Rüdiger Göbel, Paul Hill, and Wolfgang Liebert, Editors,** Abelian group theory and related topics, 1994

170 **John K. Beem and Krishan L. Duggal, Editors,** Differential geometry and mathematical physics, 1994

(See the AMS catalog for earlier titles)